PRINCIPLES OF
ELECTRICAL ENGINEERING
SERIES

Applied Electronics

A FIRST COURSE IN
ELECTRONICS, ELECTRON TUBES,
AND ASSOCIATED CIRCUITS

SECOND EDITION

by

Truman S. Gray

THE M.I.T. PRESS

MASSACHUSETTS INSTITUTE OF TECHNOLOGY
CAMBRIDGE, MASSACHUSETTS

EIGHTH PRINTING, JANUARY, 1963

NINTH PRINTING, MARCH, 1965

Library of Congress Catalog Card Number: 54–6261

APPLIED ELECTRONICS

Foreword

The staff of the Department of Electrical Engineering at the Massachusetts Institute of Technology has for some years been engaged in an extensive program of revising as a unit its entire presentation of the basic technological principles of electrical engineering. This new edition of *Applied Electronics* covers a part of that presentation.

The decision to undertake so comprehensive a plan rather than to add here and patch there came from the belief that the Department's large staff, with its varied interests in teaching and related research, could effect a new synthesis of educational material in the field of electrical engineering and evolve a set of textbooks with a breadth of view not easily approached by an author working individually.

Such a comprehensive revision, it was felt, should be free from the duplications, repetitions, and unbalances so often present in an unintegrated program. It should possess a unity and breadth arising from the organization of a subject as a whole. It should appeal to the student of ordinary preparation and also provide a depth and rigor challenging to the exceptional student and acceptable to the advanced scholar. It should comprise a basic course adequate for all students of electrical engineering regardless of their ultimate specialty. Restricted to material which is of fundamental importance to all branches of electrical engineering, the course should naturally lead into any one branch.

This book and the reorganized program of teaching out of which it has grown are thus products of a major research project to improve educational methods. The rapid development of electronics brought about by the impetus of the recent wars has made desirable revision of the original book to include new and improved devices, techniques, and methods of presentation. During these developments it has become clear that revision of this treatment and extension of it to new areas such as are included in this book should become more and more the responsibility of individual authorities who could relate their work to the over-all structure.

KARL T. COMPTON

Preface

During the years since the first edition of this book was published, electronics has truly come of age. We now rely on it for our comfort, our convenience, and even our lives in diverse fields such as energy conversion, communication, and control. We look to it with justified expectancy for new useful developments of benefit to mankind. The importance of electronics in science and engineering and, correspondingly, in technological education, has thus become even more clearly established than ever before. To facilitate such education, this book aims to lay a foundation for effective engineering application of the basic phenomena of electronics.

The extent of the use of electronics in the different branches of electrical engineering—power, communications, measurement, control, and others—precludes a complete treatment of the subject in a single volume. Hence, this book is not exhaustive; details of application are expected to follow in courses designed for specialization by students in the different branches. This book is for a first basic course. Rigor of thought and analysis, rather than extensiveness of scope, is its intended feature.

New devices, new principles, and new methods of analysis have extended the possibilities for application of electronics. The basic pattern of the field, and hence of this book, remains, however, essentially unchanged from that of the original edition. On the premise that proper application of electronic apparatus requires a working knowledge of the physical phenomena involved in the apparatus, the first part of the book is a discussion of those phenomena. The second part is an explanation of the way the phenomena combine to govern the characteristics, ratings, and limitations of electronic devices, and the third is a consideration of applications common to the several branches of electrical engineering. Finally, the fourth part is a treatment of semiconductor devices, primarily the transistor, in a manner parallel to the previous treatment of vacuum tubes. This arrangement makes practicable use of the book as a textbook in a number of different ways. In its entirety, it is intended to be suitable for a two-semester course. Assigning the early chapters and certain of the later chapters as reference material for reading only, with resultant emphasis upon the chapters that treat the circuit applications of electron tubes and semiconductor devices, makes possible use of the book for a one-semester course. To provide for additional study by particularly apt or advanced students, more material than is usually covered in a first course is presented; and to aid independent study outside the classroom, graphical data on typical electron tubes and answers to representative problems stated at the ends of the chapters are included in appendices.

Most of the functional methods by which electronics is employed in engineering are included. To make the book adequate as a point of departure into independent study and analysis of specialized applications of electronics, emphasis is placed on care in reasoning, with the thought that ease of understanding is synonymous with clarity of conception. Attempt is made to point out all links in the chain of reasoning in order to avoid those gaps that are so easily spanned intuitively by experienced engineers, but are so disturbing to the careful but inexperienced student. In addition to exact logic, this effort involves not advanced mathematics, but rather scrupulous attention both to aids to clearness of thought and to apparently minor details that are elementary but essential. One important aid is precise definitions of symbols and interpretation of them in terms of physical quantities. Among the elementary details requiring attention are the algebraic signs associated with the distinction between actual and reference directions of quantities, and avoidance of the common error of mixing complex numbers and time functions in the same equation. The three categories of mathematical quantities—scalars, complex numbers, and vectors—are distinguished by distinctive type, in accordance with the ASA American Standard Letter Symbols for Electrical Quantities. Since some of the rules for mathematical manipulation of quantities in each of these categories differ from the rules for quantities in the other two categories, such a distinction is essential for clarity. Symbols for the various component currents and voltages in electron-tube circuits are consistent with the recently revised standard for those quantities, and rationalized meter-kilogram-second units for physical quantities are used throughout the book, in accordance with almost universal present-day practice.

During preparation of this revision, it has been a pleasure to recall the contributions of colleagues who shared in supplying preliminary drafts of sections of the original edition. Many of them are now at other educational institutions or with industrial organizations; some, however, are still my close associates. The fact that many of the ideas and concepts in those early drafts continue to be regarded as fundamental and are hence retained in this revised book attests to the soundness of their judgment. I have been greatly aided by discussions with and suggestions from my present colleagues. In particular, I wish especially to thank Professor A. B. Van Rennes and Professor E. F. Buckley for their many constructive suggestions throughout the book, and their able, generous, and untiring aid in reading all the manuscript and the proof. I am also indebted to Professor S. J. Mason and Professor R. E. Scott for their suggestions regarding circuit analysis, and to Professor R. B. Adler for his advice regarding the chapter on semiconductor devices. Dean F. G. Fassett, Jr., has been ever helpful with counsel on presentation and style, and Dean H. L. Hazen and Professor G. S. Brown have provided continual inspiration by their encouragement and support of this work. To all these individuals, and to my wife Isabel for her constant encouragement, assistance, and forbearance, I extend my thanks, with the hope that their helpfulness will be reflected in increased usefulness of the book to students.

TRUMAN S. GRAY

October 27, 1953

Contents

CHAPTER X

AMPLIFIERS WITH OPERATION EXTENDING BEYOND THE LINEAR RANGE OF THE TUBE CHARACTERISTIC CURVES; CLASS AB, CLASS B, and CLASS C AMPLIFIERS 609

CHAPTER XI

VACUUM-TUBE OSCILLATORS 657

CHAPTER XII

MODULATION AND DEMODULATION, OR DETECTION 689

Table of Symbols

In this book a **boldface roman-type** or **script letter** is used to represent a space vector, and an ordinary *italic* or *script letter* to represent its magnitude, for example: $\mathbf{B}$, $\mathscr{E}$, B, $\mathscr{E}$. Similarly, a ***boldface italic letter*** is used to represent a complex number, and an *italic letter* its magnitude, for example: $\boldsymbol{E}$, E. Ordinary italic or script letters are used to represent the ordinary real scalar quantities. For voltage, current, and charge, capital letters generally represent fixed quantities, and lower-case letters represent variable quantities. For transistors, however, an exception is made, as is explained in Art. 4, Ch. XIII. In general, each letter stands for a quantity of a particular kind, and subscripts are used to distinguish several quantities of the same kind from one another. For example, i is used for instantaneous current, and i_b specifies the instantaneous plate current in an electron tube.

The notation used in this book conforms to that standardized by the Institute of Radio Engineers[1] for use with electron tubes and their circuits. In order to make this conformity possible no distinction is made between e and v, or E and V. Any one is used to represent a voltage whether it be that of a source or not.

In the table that follows are listed the more important symbols used in this book. Many of the special symbols obtained through adding subscripts to the letters listed are omitted from this list, but are defined in the text where used. The standardized symbols used to designate voltage and current components encountered in electron-tube circuits are omitted from the main list and appear instead in a table at the end of the list. This table is repeated in Art. 20, Ch. VIII.

Abbreviations used in this book are, in general, those approved by the American Standards Association.[2]

---

[1] *Standards on Abbreviations, Graphical Symbols, Letter Symbols, and Mathematical Signs, 1948* (New York: The Institute of Radio Engineers, 1948), 1–9.

[2] *American Standard Abbreviations for Scientific and Engineering Terms — ASA No. Z10.1* (New York: American Society of Mechanical Engineers, 1941).

TABLE OF SYMBOLS

ENGLISH LETTER SYMBOLS

Symbol			Description	Defined or First Used
Complex	*Scalar*	*Vector*		*Page*

Complex	Scalar	Vector	Description	Page
			Symbol *Description* *Defined or First Used*	
	e_{c1}		Instantaneous control-grid voltage of electron tube	203
	e_{c2}		Instantaneous screen-grid voltage of electron tube	202
	e_{crit}		Critical grid voltage of a thyratron	366
	e_{d0}		Instantaneous rectified voltage	304
	e_{gna}		Sum component in grid-to-ground voltage	505
	e_{gnd}		Difference component in grid-to-ground voltage	505
	e_h		Instantaneous carrier voltage	738
	e_i		Instantaneous input voltage	339
	e_k		Instantaneous cathode-to-ground voltage	429
	e_m		Instantaneous modulating voltage	739
	e_o		Instantaneous output voltage	431
	e_o		Local oscillator voltage	758
	e_s		Instantaneous value of alternating source voltage	280
	e_s		Input signal voltage	430
	e_t		Total radiation emissivity	82
	e_0		Instantaneous control voltage of a triode	667
	e_1		Instantaneous source voltage for anode 1 of polyphase rectifier	304
F		F	Force	4
F			Noise figure	816
	f		A function	58
	f		Frequency	111
	f		Fractional quantity	143
	f_0		Geometric mean frequency	519
	f_1		Lower half-power frequency	515
	f_2		Upper half-power frequency	516
G			Conductance	422
$G(f)$			Frequency spectrum of a wave	746
G_g			Input conductance of vacuum tube	422
$G(\omega)$			Angular-frequency spectrum of a wave	702
	g_m		Mutual conductance, or control-grid-to-plate transconductance of vacuum tube	194
	h		Planck constant	5
I			Constant current	80

Complex	Scalar	Vector	Description	Defined or First Used — Page
I	I		Effective value of alternating current . . .	290
			(See also table at end of this list for standardized symbols for currents encountered in electron-tube circuits.)	
	I_a		Effective value of rectifier anode current . .	309
	I_b		Instantaneous transistor base current . .	797
	I_c		Instantaneous transistor collector current .	789
	I_{dc}		Average rectifier load current	288
	I_e		Instantaneous transistor emitter current .	789
	I_f		Filament current of electron tube	171
	I_L		Effective value of current in inductor . . .	664
	I_n		Rms shot noise current	496
I_o			Output current	573
	I_s		Saturation thermionically emitted current .	80
	i		Angle of incidence.	27
	i		Instantaneous current	30
		i	Vector indicating magnitude and direction of current in a stream.	32
	i_b		Instantaneous plate current of electron tube .	119
	i_c		Instantaneous grid current of electron tube. .	190
	i_c		Instantaneous capacitor current.	317
	i_{c2}		Instantaneous screen-grid current of electron tube	204
	i_d		Instantaneous plate current of composite tube	468
	i_i		Instantaneous input current	338
	J		Current density	77
	J_s		Saturation current density	77
	$J_n(\delta)$		Bessel function of order n and argument δ. .	763
	j		$\sqrt{-1}$	46
	K		Stefan-Boltzmann constant	82
	K		Constant of proportionality	130
	k		Boltzmann constant	68
	k		Constant of proportionality	692
	k		An integer	633
	L		Inductance.	46
	l	1	Length	8
	M		Rate of evaporation of a tungsten filament .	88

Complex	Scalar	Vector	Description	Page
	M		Atomic weight	263
	M		Mutual inductance	558
	M'		Rate of evaporation of a unit tungsten filament	88
	m		Mass	4
	m		An integer	307
	m		Modulation factor	698
	m_e		Rest mass of an electron	3
	N		Number	45
	N		Number of turns	295
	$N(W)$		Distribution function for kinetic energies	68
	$N_x(W_x)$		Distribution function for x-associated kinetic energies	72
	n		Number per unit volume, area, length, or time	31
	n		An integer	307
	P		Average power	82
	P		Output power	310
	P_{ac}		Alternating-current power to load	446
	P_B		Power radiated by plate of electron tube	172
	P_b		Quiescent power input to plate of electron tube	445
	P_{bb}		Power from plate power supply	445
	P_{bs}		Power input to plate of electron tube	616
	P_{cs}		Power input to grid of electron tube	637
	P_{dc}		Direct-current power to load	290
	P_g		Power supplied by source of grid-signal voltage	641
	P_h		Power contained in carrier wave	710
	P_{in}		Input power	290
	P_L		Power to load	445
	P_m		Power output of modulating amplifier	710
	P_p		Plate power dissipation	290
	P_2		Rating of transformer secondary windings	310
	p		Pressure	142
	p		Number of phases	306
	Q		Constant electric charge	6
	Q		Quiescent operating point	196
	Q		Ratio of reactance to resistance for an inductor	340
	Q_e		Magnitude of charge of electron	3

Complex	Scalar	Vector	Description	Page
			Symbol / **Description** / **Defined or First Used**	
	r_k		Radius of cathode of electron tube	50
	r_m		Mutual incremental resistance	801
	r_p		Radius of plate of electron tube	50
	r_p		Dynamic, or incremental, or variational, plate resistance of vacuum tube.	194
	r_{11}		Incremental self-resistance	799
	r_{12}		Incremental transfer resistance	799
	r_{21}		Incremental transfer resistance	799
	r_{22}		Incremental self-resistance	799
	s		Area of a surface	128
	s_p		Area of plate of electron tube	132
	T		Absolute temperature.	68
	T		Period of sinusoidal wave	447
	t		Time	8
u			Complex variable representing velocity in a plane.	46
	u		A fraction	143
	u		Variation in rectified voltage	355
	V		Voltage across tungsten filament	88
	V'		Voltage across unit tungsten filament . . .	88
	V_b		Instantaneous base-to-emitter voltage drop .	797
	V_{bb}		Base-bias supply voltage	796
	V_c		Instantaneous collector-to-base voltage drop .	793
	V_{cc}		Collector-bias supply voltage . . .	793
	V_e		Instantaneous emitter-to-base voltage drop .	793
	V_{ee}		Emitter-bias supply voltage	789
	V_{en}		Instantaneous emitter-to-ground voltage drop	795
	v		Speed	4
		$\mathbf{v}$	Velocity	32
	v_{nc}		Collector noise voltage	814
	v_{ne}		Emitter noise voltage.	814
	v_s		Signal source voltage	789
	W		Energy	63
	W		Power input to a tungsten filament. . . .	88
	W'		Power input to a unit tungsten filament . .	88
	W_a		Potential-energy barrier at surface of a metal .	72
	W_i		Energy level at top of Fermi band	68

Symbol		Description	Defined or First Used
Scalar			Page
η		Efficiency	290
η_p		Plate efficiency of a vacuum tube	448
θ Theta		An angle	36
θ		Impedance angle	409
θ_A		Angle of complex voltage amplification	413
λ Lambda		Wavelength	5
λ		Mean free path	141
μ Mu		Amplification factor of vacuum tube	189
ν Nu		Volume	128
ν		Number of collisions per centimeter	142
π Pi		Ratio of circumference to diameter of circle (3.14159...)	6
ρ Rho		Charge density	128
ϕ Phi		An angle	30
ϕ		Phase angle of current	410
ϕ		Voltage equivalent of work function	73
ϕ_c		Contact potential difference	75
$\phi(t)$		Instantaneous phase angle of modulated wave	691
ψ Psi		Phase angle of voltage	55
ω Omega		Angular frequency	55
ω_c		Angular frequency of carrier wave	692
ω_m		Angular frequency of modulating wave	692
$\omega(t)$		Instantaneous angular frequency of modulated wave	693
ω_0		Resonant angular frequency	548

OTHER SYMBOLS

$\approx$	Approximately equal to	327
$\equiv$	Defined as	193
$\gg$	Large compared with	341
$\ll$	Small compared with	341
Σ	Sum of	636
$\lvert A \rvert$	Magnitude of A	307
Δ	Increment of	29
$\cdot$	Dot product	10
$\times$	Cross product	32
grad	Gradient of	8

SYMBOLS FOR VACUUM-TUBE CIRCUITS

Component	Name and Assigned Positive Reference Direction* of Quantity			
	Voltage rise from cathode to grid	Voltage rise from cathode to plate	Current through the external circuit toward the grid	Current through the external circuit toward the plate
Instantaneous total value	e_c	e_b	i_c	i_b
Quiescent value; steady value when varying component of grid voltage is zero	E_c	E_b	I_c	I_b
Average value of the total quantity	E_{cs}	E_{bs}	I_{cs}	I_{bs}
Instantaneous maximum of the total quantity	E_{cm}	E_{bm}	I_{cm}	I_{bm}
Instantaneous value of the varying component	e_g	e_p	i_g	i_p
Complex effective value of the varying component	E_g	E_p	I_g	I_p
Effective value of the varying component	E_g	E_p	I_g	I_p
Amplitude of the varying component	E_{gm}	E_{pm}	I_{gm}	I_{pm}
Average value of the varying component	E_{g0}	E_{p0}	I_{g0}	I_{p0}
Instantaneous value of the harmonic components	$e_{g1}, e_{g2}, \cdots$	$e_{p1}, e_{p2}, \cdots$	$i_{g1}, i_{g2}, \cdots$	$i_{p1}, i_{p2}, \cdots$
Effective value of the harmonic components	$E_{g1}, E_{g2}, \cdots$	$E_{p1}, E_{p2}, \cdots$	$I_{g1}, I_{g2}, \cdots$	$I_{p1}, I_{p2}, \cdots$
Amplitude of the harmonic components	$E_{g1m}, E_{g2m}, \cdots$	$E_{p1m}, E_{p2m}, \cdots$	$I_{g1m}, I_{g2m}, \cdots$	$I_{p1m}, I_{p2m}, \cdots$

* Amplitudes and ordinary effective values, not complex, in the table have only magnitude. They do not have sense, or direction.

CHAPTER I

Electron Ballistics

Electronics includes in a broad sense all electrical phenomena, for all electric conduction involves electrons. The common interpretation of the term at present, however, is expressed by a standard definition[1] of electronics, which is "that field of science and engineering which deals with electron devices and their utilization." Here an electron device[1] is "a device in which conduction by electrons takes place through a vacuum, gas, or semi-conductor." Electronics has become increasingly important because of its growing application to the problems of the electrical industry. During the early years of the industry, electronic conduction—except for the arc lamp—usually took the form of annoying and somewhat puzzling accidents, such as puncture of insulation, flashover of insulators, and corona leakage current. Recently, however, despite the fact that electronic conduction still has many puzzling aspects, scientists and engineers have found an increasing number of ways in which it can be harnessed, guided, and controlled for useful purposes. Electronics consequently is now as important to the engineer concerned with rolling of steel rails or the propulsion of battleships as to the engineer concerned with the communication of intelligence.

The occurrence of electronic conduction is widespread, and its nature diverse. Sometimes it is unconfined, as in lightning or some arcs; at other times confined, as in the electron tube or the neon sign. Sometimes it is visible, as in the arc light; at other times invisible, as in the vacuum tube. Sometimes the conduction is undesirable and uncontrolled, as in the example of lightning striking a transmission line, or in corona formation on the line. At other times the conduction is intentional, and may be controlled by minute electrical forces, as in the electron tube.

The field of application of electronic phenomena already covers a very wide range of power. The asymmetric nonlinear property common to many types of electronic conduction finds application not only in the radio detector tube, where the power handled is extremely small, but also in the railway mercury-arc rectifier that handles the power to move trains over mountains. The property of certain types of electronic conduction that makes possible the control of a large flow of energy by the expenditure of a relatively small amount of power

[1] "Standards on Electron Tubes: Definitions of Terms, 1950," *I.R.E. Proc.*, *38* (1950), 433.

finds use over ranges of current varying all the way from that involved
in the electrometer vacuum tube capable of measuring currents of
10^{-15} ampere to the enormous bursts of current amounting to thousands of amperes required for modern electric spot-welders and
nuclear particle accelerators.

Electronic conduction through a vacuum or gas takes place by
virtue of the fact that under certain conditions charged particles,
known as electrons and ions, are liberated from electrodes and produced in the gas in the conducting path; and that in the presence of an
impressed electric field these charged particles experience a force
that causes them to move and constitute an electric current. Thus
electronic conduction in a vacuum or gas embraces the following
important physical processes:

(a) the liberation of charged particles from electrodes,
(b) the motion of the particles through the space between the
electrodes,
(c) the production of charged particles in the space between the
electrodes, and
(d) the control of the flow of the particles by the electric field caused
by electrodes interposed in the space, or by the magnetic field
produced by an external means.

Practical circuit elements that embody possible combinations of the
foregoing processes are almost always nonlinear, and effective utilization of such elements in circuits requires an analysis suited to their
nonlinearity.

In the application of electronic devices the engineer must have a
knowledge of their characteristics and limitations. As in most electrical
equipment, the electrical aspects of the design of these devices are
often not the limiting ones; chemical, thermal, mechanical, and
physical phenomena often govern their rating. A thorough understanding of the physical principles underlying the behavior of a device
is therefore necessary in order that intelligent application be made of
it. Accordingly, the first part of this book is devoted to a discussion of
the physical aspects of electronic conduction, the second part is a
description of the electrical characteristics of typical electron tubes,
and the third part is a treatment of the fundamental methods of
circuit analysis and the basic engineering considerations important
in the application of electronic devices.

1. CHARGE AND MASS OF ELEMENTARY PARTICLES

Over a period of years a number of elementary particles of importance in electronic conduction have been identified, and the charge

and mass of each have been measured. A few of the more frequently encountered particles, all of which are constituents of the atom, together with their charges and masses, are listed in Table I.

TABLE I*

Name	Charge	Mass
Electron	$-Q_e$	m_e
Positron	$+Q_e$	m_e
Neutron	0	$1,838m_e$
Proton	$+Q_e$	$1,837m_e$

* The symbols e and m are generally used in the literature for the charge and mass of the electron; however, in this book the symbols shown are used to be consistent with those introduced in *Electric Circuits*.

Note that Q_e is the symbol for the *magnitude* of the charge of an electron—it is a positive number and does *not* include the negative sign associated with the negative charge. The negative sign is indicated separately in all the following analytical work where Q_e appears.

In Table I:

$$Q_e = (1.60203 \pm 0.00034) \times 10^{-19} \text{ coulomb}$$

Electronic constants in mks units[2]   ▶[1]

$$m_e = (9.1066 \pm 0.0032) \times 10^{-31} \text{ kilogram}$$

▶[2]

Of the elementary particles, the electron is basic in the field of electronics, and the charge and mass of the others are expressed in terms of its charge and mass. The neutron and the proton are particles which have the next higher quantity of mass ordinarily observed. Mesons, which are short-lived charged particles found in cosmic-ray and other nuclear studies, have values of mass intermediate between those of the electron and the neutron. Neutrinos, which are postulated to satisfy the requirements of nuclear theory, have neither charge nor mass. Neither mesons nor neutrinos have engineering significance at present. The ratio of charge to mass for the electron appears in many of the theoretical expressions for the motion of charged particles in electric and magnetic fields; hence there are numerous ways of measuring it experimentally. Precise measurements[2] give for the ratio the value

$$Q_e/m_e = (1.7592 \pm 0.0005) \times 10^{11} \text{ coulombs per kilogram}$$

▶[3]

[2] These values are taken from R. T. Birge, "A New Table of Values of the General Physical Constants," *Rev. Mod. Phys.*, 13 (October, 1941), Table *a*, p. 234, and Table *c*, pp. 236–237, with permission.

Use of nuclear resonance as a measuring tool has resulted in a further improvement[3] by a factor of about three in the precision of measurement of this ratio.

All charged particles of importance in engineering have essentially multiples of the charge of the electron or proton. Particles having the mass of a molecule and the charge of an electron or proton are known as *positive or negative ions*, depending on the sign of their charge. Occasionally particles are encountered which have the mass of the molecule and small multiples of the electron's charge. These are called multiple-charged ions. Ions, which are discussed in Ch. III, generally result from collision processes in gases.

The value for the mass m_e given above is for the electron moving with speeds small compared with the speed of light. This value of mass is ordinarily called the *rest mass*, although no experimental measurements of mass have yet been made on an electron at rest. Experiment shows that the apparent mass of the electron increases with its speed. The theory of relativity,[4] which is based on the hypothetical law that "it is of necessity impossible to determine absolute motion of bodies by any experiment whatsoever," predicts that the speed of light is an asymptotic value unattainable by any material body. In other words, the mass of an electron approaches infinity as its speed approaches the speed of light.

The dependence of the mass of any particle on its speed is given by the expression

$$m = \frac{m_0}{\sqrt{1 - (v/c)^2}}, \qquad [4]$$

where

     m **is the mass of the particle in motion,**

     m_0 **is the mass of the particle at rest,**

     v **is the speed of the particle,**

     c **is the speed of light*—$(2.99776 \pm 0.00004) \times 10^8$ meters per second.**

In general, force is given by the time rate of change of momentum; that is,

$$F = \frac{d(mv)}{dt}. \qquad [5]$$

---

[3] H. A. Thomas, R. L. Driscoll, and J. A. Hipple, "Determination of e/m from Recent Experiments in Nuclear Resonance," *Phys. Rev.*, 75 (1949), 922.

[4] *The Encyclopædia Britannica* (14th ed.; New York: Encyclopædia Britannica, Inc., 1938), 89–99.

* See footnote 2 on page 3.

The right-hand side of Eq. 5 reduces to the simple product of mass and acceleration only when the mass is constant. From Eq. 4 it follows that the mass is not increased by so much as 1 per cent until the speed of the particle reaches about 15 per cent of the speed of light. It is evident from subsequent considerations that this speed is not generally reached except in devices with impressed voltages that exceed 6,000 volts. Hence assumption that the mass is constant at the rest value is reasonable in computing the force on charged particles in devices having impressed voltages lower than this value.

Because of the electric and magnetic fields surrounding a moving electron, the mass exhibited in its inertia may be entirely electromagnetic.[5] On this assumption, the radius of the equivalent charged sphere, which has no mass in the ordinary sense, may be calculated to be about 2×10^{-15} meter. This is to be compared with the radius of a molecule, which ranges around 10^{-10} meter.[6]

It is found experimentally that a beam of moving electrons may be diffracted by a metallic crystal in a manner similar to the diffraction of light-waves by a grating.[7] This wave-like behavior of electrons shows that the particle concept is not complete. The wavelength experimentally found to be associated with a moving electron is

$$\lambda = \frac{h}{m_e v}, \qquad [6]$$

where

h is the Planck radiation constant*—$(6.624 \pm 0.002) \times 10^{-34}$ joule second,

m_e is the mass of the electron,

v is the speed of the electron.

One of the valuable features of electronic devices is the rapidity with which they act; it is possible to start, stop, or vary a current with them in as short a time as a small fraction of a microsecond. This rapidity of action results from the extreme agility of the electron— a property associated with the fact that the electron has the large ratio of charge to mass stated in Eq. 3. Although both quantities are small, their ratio is very large—much larger than that of any other charged body dealt with in engineering. The size of this ratio can

[5] H. A. Lorentz, *The Theory of Electrons* (Leipzig: B. G. Teubner, 1909).

[6] R. A. Millikan, *Electrons (+ and —), Protons, Photons, Neutrons, and Cosmic Rays* (2nd ed.; Chicago: The University of Chicago Press, 1947), 184, 188.

[7] C. Davisson and L. H. Germer, "Diffraction of Electrons by a Crystal of Nickel," *Phys. Rev., 30* (1927), 705–740; C. Davisson, "Electron Waves," *J.F.I., 208* (1929), 595–604.

* See footnote 2 on page 3.

perhaps be grasped from a computation of the force of repulsion between one kilogram of electrons located at each of the poles of the earth. Their separation is about 7,900 miles, or 1.27×10^7 meters. Their charge is given by Eq. 3, and, by Coulomb's law, the force of repulsion between them is

$$= \frac{Q_1 Q_2}{4\pi\varepsilon_v d^2} = \frac{(1.76 \times 10^{11})^2}{4\pi \times 8.85 \times 10^{-12} \times (1.27 \times 10^7)^2} = 1.73 \times 10^{18} \text{ newtons} \quad [7]$$

$$= 1.95 \times 10^{14} \text{ tons.} \quad [8]$$

Clearly, the electron's charge-to-mass ratio is enormous to produce such a large force at such a great distance. Consequently, the electric force on an electron in an electrostatic field can overcome the inertia of the electron and produce high velocities in a very short time.

With knowledge of the charge and mass of the particles involved in electronics, it is possible to proceed with the analysis of the motion of particles in electrostatic and magnetostatic fields given in the following articles of this chapter. The source of the charged particles is reserved for consideration in subsequent chapters. At this point it is sufficient to know that electrons are given off by a heated metallic surface, and that electrons and ions are produced in a gas when the process of ionization occurs.

2. The elements of the operation of electron tubes

An *electron tube*[8] is "an electron device in which conduction by electrons takes place through a vacuum or gaseous medium within a gas-tight envelope." Ordinarily it consists of two or more metallic electrodes enclosed in an evacuated glass or metal chamber. The electrodes are insulated from one another. If the chamber is evacuated until the remaining gas molecules have no effect—chemically or electrically—on the operation of the tube, it is called a *vacuum tube*. Other tubes contain gas introduced after the evacuation process has been carried out. These are called *gas tubes* when the amount of gas is sufficient to have an appreciable effect on their electrical characteristics. One of the electrodes, called the *cathode*, serves as a source of electrons by virtue of one or more of the several electron-emission processes discussed in Ch. II. Another electrode, called the *anode* (or *plate*), is usually maintained electrically positive with respect to the cathode. The resulting electric field in the tube exerts a force on the electrons and causes them to move toward the anode, thereby setting up an

[8] "Standards on Electron Tubes: Definitions of Terms, 1950," *I.R.E. Proc., 38* (1950), 433.

electric current in the interelectrode space. A simple circuit involving such a tube is shown in Fig. 1. In the external circuit, the electrons flow from the anode through the voltage source to the cathode; by convention, the electric current is in the opposite direction.

When a device of this general character has only two electrodes, one of which serves as a source of electrons, it is usually termed a *diode*. By appropriate control of the voltages of other electrodes which may be inserted in the chamber, the electric field between the cathode and the anode can be modified. The flow of the electrons is thereby

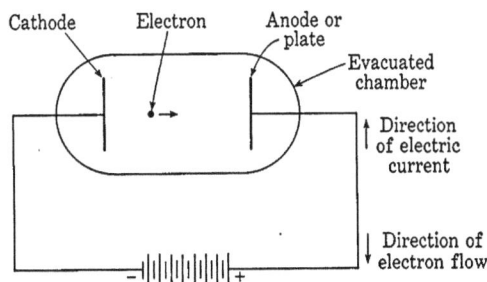

Fig. 1.   A simple electron tube.

changed, and the current in the external circuit can be controlled. These *control electrodes* are often called *grids* because of the form they had in early tubes. Tubes with one control electrode are called *triodes*; those with two control electrodes, *tetrodes*; and so on, in accordance with the total number of active electrodes.

In some electron tubes of the vacuum type, the number of charged particles traversing the interelectrode region is so small that the electric field established in this region by the charge on the particles is negligible in comparison with the field established by the charges on the electrodes. This condition is expressed by the statement that the space charge of the charged particles in motion is negligible. In devices for which this condition is true, the electrostatic force on any single particle may be considered to result wholly from the field that exists in the absence of all the particles. For example, the motion of the single electrons in the vacuum phototube or in the electron beam in a cathode-ray tube may often be computed with sufficient accuracy on the assumption that space-charge effects are negligible.

The following articles of this chapter deal with the behavior of particles in only those devices in which space-charge effects are negligible. The behavior of particles in devices in which space-charge effects are of appreciable importance, and some of the fundamental properties of these devices, are discussed in Ch. III. The paths of the charged particles in electrostatic and magnetostatic fields discussed in this chapter are similar in many respects to the trajectories of projectiles in the gravitational field of the earth—hence the chapter is titled *electron ballistics*.

3. MOTION OF CHARGED PARTICLES IN ELECTROSTATIC FIELDS IN VACUUM

Because the charged particles of interest in electron tubes are so small in comparison with the dimensions of the tubes in which they move, the forces that act upon them may be calculated as though the particles were concentrated at points. Thus the force exerted on such a particle by an electrostatic field is given by

$$\mathbf{F} = Q\mathcal{E}, \tag{9}$$

where

$\mathbf{F}$ is the force acting upon the charged particle,

Q is the charge carried by the particle,

$\mathcal{E}$ is the electric field intensity at the location of the particle.

The quantities $\mathbf{F}$ and $\mathcal{E}$ in this relation are vectors.* Equation 9 specifies (a) that the magnitude of the force is the product of the magnitude of the field intensity and the charge and (b) that the direction of the force is that of the field if the charge is positive and is opposite to the direction of the field if the charge is negative.

If E is the potential at each point in the tube, taken with respect to any arbitrary zero of potential, the gradient[9] of E, written **grad** E, is a vector oriented in the direction in which E increases most rapidly and whose magnitude is the rate of change of E with distance in this direction. Since

$$\mathcal{E} = -\ \mathbf{grad}\ E, \tag{10}$$

Eq. 9 may be written in the form

$$\mathbf{F} = -\ Q\ \mathbf{grad}\ E. \tag{11}$$

If the particle is free to move, it is accelerated according to the equation

$$\mathbf{F} = m\mathbf{a}, \tag{12}$$

or

$$\mathbf{F} = m\ \frac{d^2\mathbf{l}}{dt^2}, \tag{13}$$

* Quantities that are vectors in space are printed in **boldface script or roman (upright)** type.

[9] N. H. Frank, *Introduction to Electricity and Optics* (2nd ed.; New York: McGraw-Hill Book Company, Inc., 1950), 1–14.

where

F is the total force acting on the particle,

m is the mass of the particle,

a is the acceleration of the particle,

l is the displacement of the particle from an arbitrary origin,

dl is the differential displacement of the particle, and is along the path the particle traverses,

t is the time measured from an arbitrary reference instant.

The quantities F, a, and l are vectors, and Eqs. 12 and 13 relate both magnitudes and directions, just as do Eqs. 9, 10, and 11.

The only forces experienced by a particle moving in an evacuated tube are those caused by the fields of force, such as electric, magnetic or gravitational fields, that may be present. In this article it is supposed that no field other than an electrostatic one is present; therefore, Eqs. 9, 11, and 13 may be combined as

$$Q\mathcal{E} = m \frac{d^2\mathbf{l}}{dt^2},$$

[14]

or

$$\frac{d^2\mathbf{l}}{dt^2} = \frac{Q}{m} \mathcal{E},$$

[15]

and

$$\frac{d^2\mathbf{l}}{dt^2} = -\frac{Q}{m} \text{ grad } E.$$

[16]

Equations 15 and 16 do not involve any co-ordinate system. They may be expressed, however, in terms of any desired co-ordinate system. If, for example, a set of rectangular co-ordinate axes is chosen, the equations may be used to express the relations among the components of the vectors along these axes. Thus, if $\mathcal{E}_x$, $\mathcal{E}_y$, and $\mathcal{E}_z$ are the components of $\mathcal{E}$ along the x, y, and z axes, respectively, Eq. 15 becomes

$$\frac{d^2x}{dt^2} = \frac{Q}{m} \mathcal{E}_x,$$

▶[17]

$$\frac{d^2y}{dt^2} = \frac{Q}{m} \mathcal{E}_y,$$

▶[18]

$$\frac{d^2z}{dt^2} = \frac{Q}{m} \mathcal{E}_z.$$

▶[19]

Since the components of **grad** E along the co-ordinate axes are the rates of change of E with distance along these axes,

$$x \text{ component of } \mathbf{grad}\ E = \frac{\partial E}{\partial x}, \tag{20}$$

$$y \text{ component of } \mathbf{grad}\ E = \frac{\partial E}{\partial y}, \tag{21}$$

$$z \text{ component of } \mathbf{grad}\ E = \frac{\partial E}{\partial z}, \tag{22}$$

and

$$\frac{d^2x}{dt^2} = -\frac{Q}{m}\frac{\partial E}{\partial x}, \tag{23}$$

$$\frac{d^2y}{dt^2} = -\frac{Q}{m}\frac{\partial E}{\partial y}, \tag{24}$$

$$\frac{d^2z}{dt^2} = -\frac{Q}{m}\frac{\partial E}{\partial z}. \tag{25}$$

If the initial velocity and position of a charged particle and the potential distribution in the tube are known, it is possible to determine completely the motion of charged particles in electrostatic fields, provided the differential equations just derived can be solved. However, unless the field is uniform, at least one of the field components varies with the co-ordinates, and the equations are nonlinear. In addition, if the speed of the particle is a large fraction of the speed of light, the mass becomes a function of the speed of the particle, and the equations are again nonlinear. The solution of the nonlinear equations may often require the use of graphical, numerical, or mechanical methods.

An alternative and powerful attack on the problem of the motion of a charged particle in electrostatic fields, which yields much information about the motion, is the use of the principle of conservation of energy to derive a relation between the potential and the speed of a charged particle at any point. Let the particle travel from the point P_1 to the point P_2. The differential displacement of the particle along its path is $d\mathbf{l}$. Since the kinetic energy acquired by the particle equals the work done on the particle by the field,[10]

$$\tfrac{1}{2}mv_2{}^2 - \tfrac{1}{2}mv_1{}^2 = \int_{P_1}^{P_2} \mathbf{F}\cdot d\mathbf{l}, \tag{26}$$

[10] For an explanation of the dot-product notation used in the integral of Eq. 26, see a textbook such as N. H. Frank, *Introduction to Electricity and Optics* (2nd ed.; New York: McGraw-Hill Book Company, Inc., 1950), 107.

where

v_2 is the speed of the particle at P_2,

v_1 is the speed of the particle at P_1.

In Eq. 26 and the equations derived from it, the assumption is made that the speed of the particle never exceeds a small fraction of the speed of light. If the speed of the particle is large enough, the mass becomes a function of the speed, in accordance with Eq. 4, and the kinetic energy is no longer given by $\frac{1}{2}mv^2$. The value of **F** from Eq. 9 may be substituted to give the result

$$\tfrac{1}{2}mv_2{}^2 - \tfrac{1}{2}mv_1{}^2 = Q \int_{P_1}^{P_2} \boldsymbol{\mathcal{E}} \cdot d\mathbf{l} = -[QE_2 - QE_1], \qquad [27]$$

where

E_2 is the potential at P_2,

E_1 is the potential at P_1.

Equation 27 may be written

$$\tfrac{1}{2}mv_1{}^2 + QE_1 = \tfrac{1}{2}mv_2{}^2 + QE_2, \qquad [28]$$

which states that the sum of the kinetic energy and the potential energy of the particle does not change during the motion. Equation 28 could have been written directly, since it is a statement of the principle of conservation of energy for a charged particle in an electrostatic field.

An alternative form of Eq. 28 is

$$v_2 = \sqrt{v_1{}^2 - 2\frac{Q}{m}(E_2 - E_1)}, \qquad \blacktriangleright[29]$$

and this form may be used to find the speed of a particle at any point on its path if the speed at any one point is known. In particular, if the point P_1 is taken as the point at which the particle starts from rest, and if the potential of this point is chosen as the reference for potential, then v_1 and E_1 are zero. Since the point P_2 may be any point on the path of the particle, the subscripts may be dropped from the symbols relating to it to give the very useful relation

$$v = \sqrt{-2\frac{Q}{m}E} \qquad \blacktriangleright[30]$$

for a particle that starts from rest at a point where the potential is zero.

From Eq. 30 it appears that the speed and the kinetic energy of a particle moving in an electrostatic field depend only upon the total potential through which the particle moves, and not upon the manner in which the potential varies along the entire path. It should be noted, however, that the direction of the particle velocity and the time required for the particle to move a given distance do depend upon the distribution of the electric field. These quantities cannot be determined without the use of information in addition to that contained in Eq. 30.

In the remainder of this article the differential equations and the relation between potential and velocity are used to determine the motion of charged particles in certain configurations of electrostatic fields which have plane symmetry and are of particular interest in electron tubes. It is fortunate that approximate plane or cylindrical symmetry exists in many practical electronic devices, because the symmetry makes determination of the electronic motion in them relatively easy.

3a. *Uniform Field; Zero Initial Velocity.* In Fig. 2a, k and p are the cathode and plate, respectively, of a simple electron tube. These electrodes are assumed to lie in parallel planes separated by a distance d, which is very small relative to the dimensions of the electrodes, so that they may be treated as infinite parallel planes. If a constant voltage is applied across these electrodes, the potential gradient and the field between the plates are constant in time and uniform in space, and are directed perpendicularly to the plates. The potential of the plate with respect to the cathode* is called e_b. Alternatively, e_b is the voltage rise from the cathode to the plate, or the voltage drop from the plate to the cathode. If e_b is positive, the vector $\mathcal{E}$ is directed from the plate to the cathode, and the potential gradient from the cathode to the plate; physically, the force on a positively charged particle between the electrodes is in a direction to move it toward the cathode. Because of its negative charge, however, an electron tends to move from cathode to anode.

The rectangular co-ordinate axes in Fig. 2a are drawn so that the origin is located in the plane of the cathode, and the x axis is perpendicular to the electrode surfaces. The potential distribution in the tube can therefore be described by the graph in Fig. 2b. Suppose that a charged particle is set free at the origin of co-ordinates in the surface of the cathode with zero initial velocity, and the equations of its motion are to be found. Under these conditions, $\mathcal{E}_x$ equals $-(e_b/d)$; and

---

* In general, in this book, constant voltages and currents are denoted by capital letters and variable voltages and currents by lower-case letters. The lower-case e_b is used here in preparation for a future use in which e_b becomes a variable.

since the field is along the x axis only, $\mathcal{E}_y$ and $\mathcal{E}_z$ are both zero. Thus either Eqs. 17, 18, and 19 or Eqs. 23, 24, and 25 become

$$\frac{d^2x}{dt^2} = -\frac{Q}{m}\frac{e_b}{d},$$

[31]

$$\frac{d^2y}{dt^2} = 0,$$

[32]

$$\frac{d^2z}{dt^2} = 0.$$

[33]

In addition, x, y, z, dx/dt, dy/dt, and dz/dt are all zero when t is zero. Since the particle starts at rest and is not accelerated in the y or z

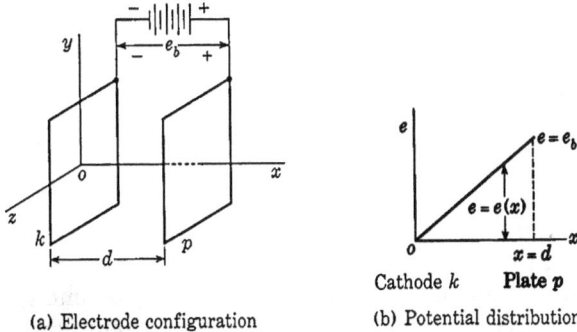

(a) Electrode configuration

(b) Potential distribution

Fig. 2.   Potential distribution between infinite, parallel-plane electrodes.

direction, its motion is confined to the direction along the x axis. By integration of Eq. 31 and use of the initial conditions to evaluate the constants of integration, the equations that describe the motion of the particle are found to be

$$v = -\frac{Q}{m}\frac{e_b}{d}t,$$

[34]

$$x = -\frac{1}{2}\frac{Q}{m}\frac{e_b}{d}t^2,$$

[35]

where v is the speed of the particle evaluated at any point on its path. In this integration it is assumed that the speed of the particle is never more than a small fraction of the speed of light, so that the mass of the particle may be considered a constant.

Equations 34 and 35 may be solved simultaneously to eliminate t and give the relation

$$v = \sqrt{-2 \frac{Q}{m} e},$$ [36]

where

$$e = \frac{e_b}{d} x$$ [37]

and is the potential at any point in the tube with respect to the cathode. It should be noticed that Eq. 36 agrees with Eq. 30 and might, in fact, have been obtained directly from that equation. Furthermore, the derivation of Eq. 30 is more general than that of Eq. 36 because it does not involve the assumption that the electric field is uniform. The speed with which the particle strikes the plate is

$$v_p = \sqrt{-2 \frac{Q}{m} e_b},$$ [38]

and the kinetic energy of the particle as it reaches the plate is

$$\tfrac{1}{2} m v_p^2 = -Q e_b,$$ [39]

which is the decrease in the potential energy of the particle that occurs as the particle moves across the tube. Thus, the increase of kinetic energy of the particle equals the decrease of potential energy and, in accordance with the principle of conservation of energy, the total energy of the particle is constant.

The time of transit from the cathode to the plate is obtained through solving Eq. 35 for t when x equals d, and is

$$t_{kp} = \frac{2d}{\sqrt{-2 \dfrac{Q}{m} e_b}} = \frac{2d}{v_p}.$$ ▶[40]

Alternatively, dividing the total distance traveled by the average speed will give this time of transit. Since the acceleration in the uniform field is constant and the particle starts from rest, the average speed is just half the final speed, and thus the same value for the transit time is obtained.

If e_b and Q are positive, negative values are obtained for v and x in Eqs. 34 and 35, and an imaginary value is obtained for v in Eq. 36. These apparently absurd results are explained by the physical fact that a positively charged particle will not leave a negative electrode to approach a positive electrode. If, however, the charged particle is an electron or other negatively charged particle, Q is itself negative; and if e_b is positive, v is real, x is positive, and the equations describe correctly the movement of the electron toward the plate.

3b. *Uniform Field; Initial Velocity in the Direction of the Field.* In some electronic devices the charged particles cannot be considered to start from rest at the cathode. Consideration of the initial velocity at the cathode is required, for example, in an analysis of the motion of an electron in a vacuum diode, as discussed in Ch. II, or in a multi-electrode device in which an electron set free at one electrode acquires a velocity by moving through a potential difference between one pair of electrodes and then enters the electric field between two other electrodes.

If a particle starts from the origin at the cathode in Fig. 2 with an initial velocity that is in the direction of the field and has a magnitude v_k, then, x, y, z, dy/dt, and dz/dt are zero when t is zero, but dx/dt equals v_k when t is zero. The equations of motion obtained by integration of Eqs. 23, 24, and 25 are then

$$\frac{d^2x}{dt^2} = -\frac{Q}{m}\frac{e_b}{d}, \tag{41}$$

$$\frac{dx}{dt} = -\frac{Q}{m}\frac{e_b}{d}t + v_k, \tag{42}$$

$$x = -\frac{1}{2}\frac{Q}{m}\frac{e_b}{d}t^2 + v_k t, \tag{43}$$

$$y = 0, \tag{44}$$

$$z = 0. \tag{45}$$

The speed with which the particle strikes the plate, v_p, may be determined through application of the principle of conservation of energy. Accordingly, the kinetic energy of the particle as it strikes the plate equals the sum of the initial kinetic energy and that acquired from the field. Thus,

$$\tfrac{1}{2}mv_p^2 = -Qe_b + \tfrac{1}{2}mv_k^2, \tag{46}$$

whence

$$v_p = \sqrt{-\frac{2Qe_b}{m} + v_k^2}. \tag{47}$$

The time of transit from cathode to plate, t_{kp}, is the ratio of the distance to the average speed. Since the acceleration is uniform, the average speed is the average of the initial and final speeds. Hence,

$$t_{kp} = \frac{2d}{v_k + v_p}. \tag{48}$$

3c. *Uniform Field; Any Initial Velocity.* If in the tube of Fig. 2 the initial velocity of the particle is considered to have an arbitrary direction, and if for further generality the origin of co-ordinates is supposed to have any location, the general equations for a charged particle moving in a uniform field aligned with the x axis are obtained. Under these generalized boundary conditions Eqs. 17, 18, and 19 yield by integration

$$\frac{dx}{dt} = \frac{Q}{m}\mathcal{E}t + C_1 , \qquad [49]$$

$$\frac{dy}{dt} = C_2 , \qquad [50]$$

$$\frac{dz}{dt} = C_3 , \qquad [51]$$

and

$$x = \frac{1}{2}\frac{Q}{m}\mathcal{E}t^2 + C_1 t + C_4 , \qquad [52]$$

$$y = C_2 t + C_5 , \qquad [53]$$

$$z = C_3 t + C_6 . \qquad [54]$$

The constants of integration, C_1 through C_6, have values that depend upon two sets of boundary conditions, of which each set is expressible in terms of the three co-ordinates; for example, the boundary conditions are often the initial velocity and initial position of the particle.

Note that the motion of a charged particle in a uniform electrostatic field is strictly analogous to the motion of a material particle in a uniform gravitational field, since the acceleration imparted to the particle by either field is constant. The path of a charged particle, like the trajectory of a projectile, is in general a parabola. If the initial velocity is in the direction of the field, or is zero, the parabola degenerates into a straight line. This fact is seen at once if Eqs. 52, 53, and 54 are recognized as the parametric equations of a parabola. The equations of the parabola may be placed in a more commonly encountered form if the co-ordinate axes are so chosen that the x–y plane is in the direction determined by the field and the initial velocity, and the particle starts from a point in this plane. It is easy to show that the motion is then entirely in the x–y plane. The constants C_3 and C_6 become zero, and the motion is described by the displacements x and y. If the value of t given by Eq. 53 is substituted into Eq. 52, there results an expression of the form

$$x = C_7 y^2 + C_8 y + C_9 , \qquad [55]$$

where C_7, C_8, and C_9 are constants. This form is recognized as the equation of a parabola.

4. UNITS FOR NUMERICAL COMPUTATIONS; THE ELECTRON VOLT

In the numerical solution of any practical problem involving the motion of charged particles in electrostatic and magnetostatic fields, the question of the units to be used always arises. Since the motion of the particle is essentially a problem in mechanics, it is desirable to use a system of units in which measurement of the motion, or the forces that give rise to the motion, is convenient, as well as one in which the electrical units are of convenient size. The international meter-kilogram-second (mks) system provides such units, and all the derived equations in this book hold when numerical values substituted into the equations for the quantities involved are expressed in the mks rationalized system or any other self-consistent rationalized system of units. However, with minor alterations* the equations also hold for any self-consistent unrationalized system, such as the cgs absolute electromagnetic (aem) system or the cgs absolute electrostatic (aes) system. These systems have been, and still are, used frequently in the literature of electronics, and familiarity with the commonly encountered conversion factors for converting from one system to another is desirable. A table of conversion factors is given in Appendix B.[11]

For illustration, consider the expression

$$F = \mathcal{E}Q = ma. \qquad [56]$$

In the three previously mentioned systems of units this equation may be written in the following three ways:

For the mks system,   [57]

$$[F]_{newtons} = [\mathcal{E}]_{volts\ per\ meter}\,[Q]_{coulombs} = [m]_{kilograms}\,[a]_{meters\ per\ sec\ per\ sec}.$$

For the aes system,   [58]

$$[F]_{dynes} = [\mathcal{E}]_{statvolts\ per\ cm}\,[Q]_{statcoulombs} = [m]_{grams}\,[a]_{cm\ per\ sec\ per\ sec}.$$

---

* See Appendix B for an explanation of the distinction between rationalized and unrationalized units. Substitution of $\varepsilon_v/(4\pi)$ for ε_v and $D/(4\pi)$ for D are the only changes necessary to convert the particular equations included in this book into the form for which unrationalized units are applicable. Only equations that involve ε_v and D explicitly need be changed. All others are suitable for either rationalized or unrationalized units.

[11] See also E. E. Staff, M.I.T., *Electric Circuits* (Cambridge, Massachusetts: The Technology Press of M.I.T.; New York: John Wiley & Sons, Inc., 1940), 754–756.

For the aem system, [59]

$$[F]_{dynes} = [\mathcal{E}]_{abvolts\ per\ cm} [Q]_{abcoulombs} = [m]_{grams} [a]_{cm\ per\ sec\ per\ sec}.$$

In Eqs. 57, 58, and 59, the units appended to the brackets around the terms are the three sets of units in which the quantities within the brackets may be expressed.

Frequently the quantities involved in a problem are known in a mixed system of units, and often the result is desired in another system of units. For example, electrical measuring instruments usually indicate potential differences in volts and currents in amperes, which are in the mks system. However, in electronic work some quantities, such as force and length, are often expressed in the cgs system, where the unit of force is the dyne and the unit of length is the centimeter. When the data are known in a hybrid system, conversion factors must be used before a useful numerical result can be obtained.

To illustrate the use of conversion factors, assume that the field intensity $\mathcal{E}$ is given in volts per centimeter, and the charge Q is given in coulombs. It may be desired to know the force F in dynes. Any convenient form of Eq. 56 involving units in which quantities are known or desired may be chosen as a starting point; for instance, Eq. 58 may be chosen as one containing the dyne. Thus

$$[F]_{dynes} = [\mathcal{E}]_{statvolts\ per\ cm} [Q]_{statcoulombs}. \qquad [60]$$

Any change in the units used for a quantity within a bracket, such as F, $\mathcal{E}$, or Q, can be made, provided the proper conversion factor is also included so that the units of the whole quantity within the bracket are not changed. For example,

$$[F]_{dynes} = \left[\mathcal{E}_{volts\ per\ cm} \times \frac{1}{300}\right]_{statvolts\ per\ cm} [Q_{coulombs} \times 3 \times 10^9]_{statcoulombs}. \quad [61]$$

The names of the units outside the brackets can be dropped when new units are chosen for the quantities within the brackets, and the conversion factors can then be combined, giving the desired equation,

$$[F]_{dynes} = [\mathcal{E}]_{volts\ per\ cm} [Q]_{coulombs} \times 10^7. \qquad [62]$$

While at first sight this method may seem lengthy, it is sound and is often useful, especially when a number of quantities are involved and several of them must be converted to different units.

Publications in the field of electronics frequently use these mixed systems of units; hence, conversion factors of 10^7, 10^8, and 10^9 are often encountered in the equations. Consistent systems of units, however, are used for derived equations in this volume, except as specifically noted.

When the value for the ratio of the charge to the mass of the electron given in Eq. 3 is substituted in Eq. 30, the speed of an initially stationary electron is given in terms of the potential difference through which it moves, as

$$v = 5.94 \times 10^5 \sqrt{E} \text{ meters per second,} \qquad \blacktriangleright[63]$$

where E is in volts. This equation holds only for speeds small compared with that of light. The energy of the electron is then

$$\tfrac{1}{2}m_e v^2 = EQ_e \qquad [64]$$

$$= 1.60 \times 10^{-19} \times E \text{ joule.} \qquad [65]$$

Since for reasonable values of voltage the energy of the electron is extremely small, and since its energy is directly proportional to the potential difference through which it moves, it is convenient and customary to adopt as a unit of *energy* the *electron volt*, abbreviated to *ev*, which is the kinetic energy that an initially stationary electron acquires by moving through a potential difference of one volt. Thus

$$1 \text{ ev} = 1.60 \times 10^{-19} \text{ joule.} \qquad \blacktriangleright[66]$$

The electron volt serves as a convenient unit of energy for calculations involving the charge of an electron just as the joule, which might be called the coulomb volt of energy, serves for calculations involving coulombs of charge, or involving amperes.

5. DEFLECTION OF THE ELECTRON BEAM IN A CATHODE-RAY TUBE

One application of the analysis of the behavior of charged particles in an electrostatic field is in some types of electron-beam tubes. As a class such tubes include devices for many different purposes, as summarized in Fig. 3. They find extensive application, especially in television,[12] radar,[13] and in experimental studies of time-varying phenomena.[14] The oscilloscope tube in particular is indispensable for the

[12] V. K. Zworykin and G. A. Morton, *Television* (New York: John Wiley & Sons, Inc., 1940); D. G. Fink, *Television Engineering* (2nd ed.; New York: McGraw-Hill Book Company, Inc., 1952); Scott Helt, *Practical Television Engineering* (New York: Murray Hill Books, Inc., 1950).

[13] L. N. Ridenour, Editor, *Radar System Engineering*, Massachusetts Institute of Technology Radiation Laboratory Series, Vol. 1 (New York: McGraw-Hill Book Company, Inc., 1947); D. G. Fink, *Radar Engineering* (New York: McGraw-Hill Book Company, Inc., 1947).

[14] J. H. Ruiter, Jr., *Modern Oscilloscopes and Their Uses* (New York: Murray Hill Books, Inc., 1949); J. F. Ryder and S. D. Uslan, *Encyclopedia on Cathode-Ray Oscilloscopes and Their Uses* (New York: John F. Ryder Publisher, Inc., 1950).

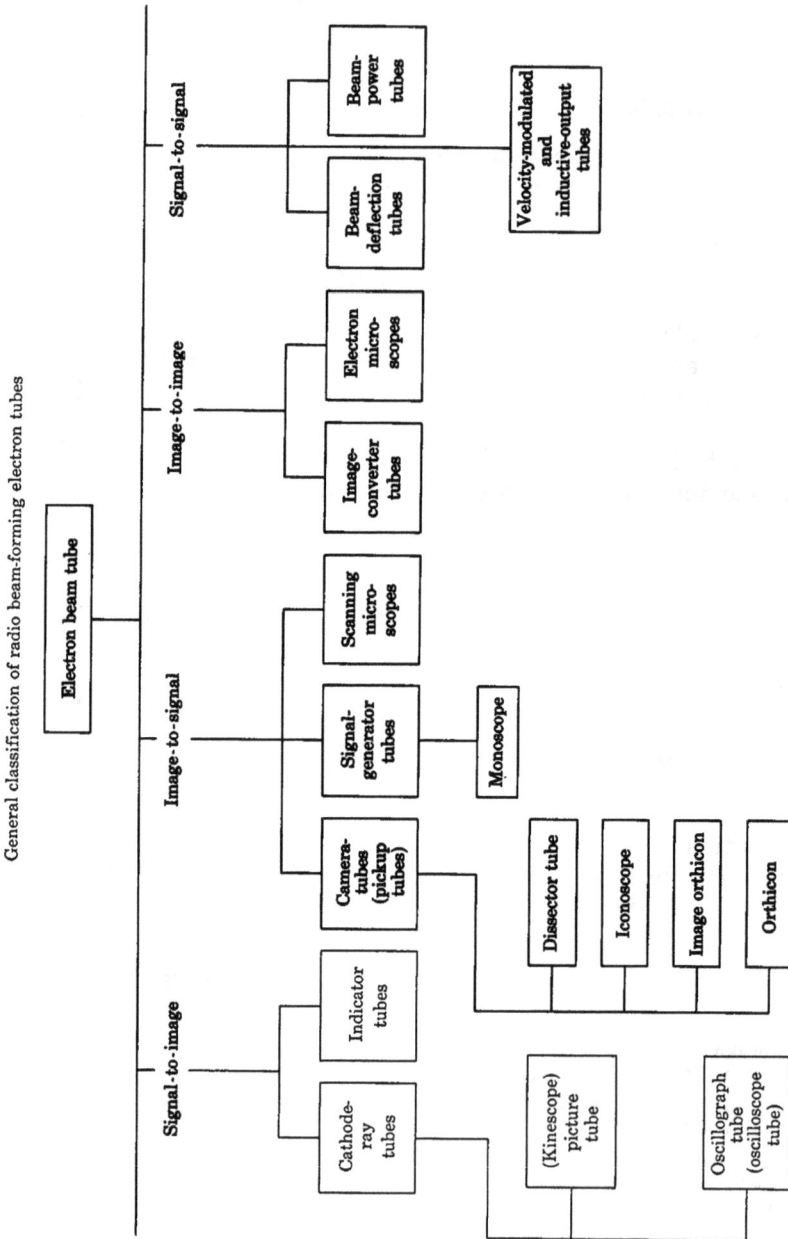

Fig. 3. General classification of electron-beam tubes. (This chart is taken from "Standards on Electron Tubes: Definitions of Terms, 1950," *I.R.E. Proc.*, *38* (1950), *428*, with permission.)

study of cyclic and repetitive transient phenomena in the audible- and low radio-frequency range. High-speed oscilloscopes are available for the study of nonrepetitive transient phenomena having durations as short as a few millimicroseconds.

A photograph of a typical cathode-ray oscilloscope tube and its internal structure is shown in Fig. 4. The essential elements in this

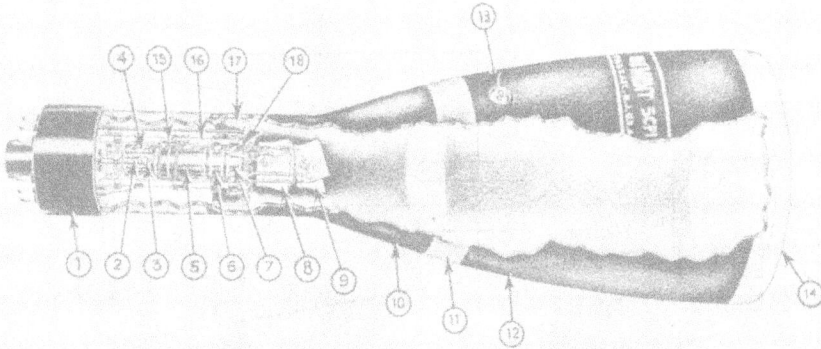

Fig. 4. Typical cathode-ray tube with electrostatic focusing and deflection.
(*Courtesy Allen B. DuMont Laboratories, Inc.*)

1. Base
2. Heater
3. Cathode
4. Control grid (G)
5. Pre-accelerating electrode (connected internally to A_2)
6. Focusing electrode (A_1)
7. Accelerating electrode (A_2)
8. Deflection plate pair ($D_3 D_4$)
9. Deflection plate pair ($D_1 D_2$)
10. Conductive coating (connected internally to A_2)
11. Intensifier gap
12. Intensifier electrode (A_3)
13. A_3 terminal
14. Fluorescent screen
15. Getter
16. Ceramic gun supports
17. Mount support spider
18. Deflection plate structure support

tube are as illustrated in Fig. 5. They comprise: (a) a source of electrons, usually a heated cathode; (b) an arrangement of electrodes termed an *electron gun*, which serves to attract the electrons from the cathode, to focus them into a fine pencil or beam of rays, and to project them from the cathode down the major axis of the tube (hence the name *cathode ray*); (c) an arrangement of electrodes called deflecting plates, or of coils, as is discussed in Art. 7b, located beyond the gun and used to deflect the electron beam; and (d) a target or screen placed in a plane substantially perpendicular to the axis of the gun and coated with a phosphor such as willemite, calcium tungstate, or zinc silicate, which becomes luminescent when struck by the electrons. The whole assembly is enclosed in a glass or metal container having a glass window, and the container is evacuated to a pressure of about 10^{-9} atmosphere.

Electrons emitted by the hot cathode k are accelerated toward the final anode p of the electron gun under the influence of the field established by the anode-to-cathode voltage E_b. In the very simple electron gun of Fig. 5 this anode is shown as a disc with a hole in its center. Although some of the electrons strike an electrode in the gun and return to the cathode through the source of E_b, many of them emerge from the gun as a fine pencil of rays. In the absence of a voltage e_d

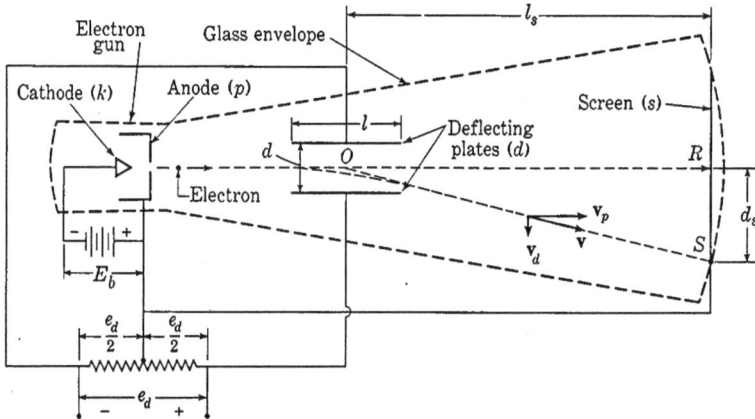

Fig. 5.   Electrostatic deflection in a cathode-ray tube.

between the deflecting plates d, the region beyond the gun is essentially field free by virtue of the shielding effect of the metal container or of a conducting film on the inside of the glass container. The electrons then pass down the axis of the tube through the field-free space and strike the screen s at the point R with substantially the velocity they had on leaving the gun. (Secondary-emission phenomena invalidate this statement when E_b is below about 100 volts or above about 5,000 volts, the exact values depending on the screen material.[15] See Art. 12, Ch. II, for further discussion of this limitation.)

Since the initial velocity of the electrons as they leave the cathode corresponds at the most to one or two electron volts (see Art. 5, Ch. V), the velocity v_p of the electrons that emerge from the gun is practically that corresponding to the change in potential energy E_bQ_e. Thus, from Eq. 30,

$$v_p = \sqrt{2\frac{Q_e}{m_e}E_b} \, . \tag{67}$$

[15] W. B. Nottingham, "Electrical and Luminescent Properties of Phosphors under Electron Bombardment," *J. App. Phys.*, 10 (1939), 73–82.

When a voltage is applied between the deflecting plates, the electrons acquire a velocity component v_d perpendicular to the axis of the tube as they pass through the field in the region between the plates. Instead of striking the screen at the point R, they now strike it at some other point, say S. On the assumption that the fringing of the field at the edges of the deflecting plates can be neglected and that the plates are parallel to the axis of the tube, the field between the plates is uniform and perpendicular to the axis of the tube, and a relation can be found for the deflection d_s of the spot in terms of the anode-to-cathode voltage E_b, the deflecting voltage e_d, and the dimensions of the tube and electrodes.

The axial velocity component v_p of the electron is unchanged by the deflecting field, because the field acts in a direction perpendicular to the direction of that component. The time required for the electron to pass through the deflecting plates is therefore

$$t_d = l/v_p, \qquad [68]$$

where l is the length of the deflecting plates. During this time the electron experiences a constant sidewise acceleration given by $(Q_e/m_e)(e_d/d)$, where d is the separation of the deflecting plates. Since this acceleration is constant, the electron acquires a component of velocity v_d perpendicular to the axis of the tube given by

$$v_d = \frac{Q_e}{m_e} \frac{e_d}{v_p} \frac{l}{d}. \qquad [69]$$

Because of the uniform acceleration, the electron describes a parabolic path, and it emerges with the speed

$$v = \sqrt{v_p{}^2 + v_d{}^2}. \qquad [70]$$

After the electron leaves the region between the plates, its path is a straight line, since it is assumed then to be in a field-free space. If the straight-line path is projected backward, it can be shown to pass through the point O at the center of the plates. Then, by similar triangles,

$$\frac{d_s}{l_s} = \frac{v_d}{v_p}, \qquad [71]$$

where l_s is the distance from the center of the plates to the screen. Substitution of Eq. 69 in Eq. 71 gives

$$d_s = l_s \frac{Q_e}{m_e} \frac{e_d}{v_p{}^2} \frac{l}{d}, \qquad [72]$$

and elimination of v_p by means of Eq. 67 gives

$$d_s = \frac{1}{2} l_s \frac{l}{d} \frac{e_d}{E_b} .$$ [73]

The sensitivity of the tube with respect to the deflection voltage is therefore

Electrostatic deflection sensitivity $= \dfrac{d_s}{e_d} = \dfrac{1}{2} \dfrac{l}{d} \dfrac{l_s}{E_b} .$ ▶[74]

Since only the ratios of lengths and voltages are involved in Eq. 74, the equation holds for any units of length or voltage, provided corresponding quantities are measured in the same units. The sensitivity is evidently decreased as the accelerating voltage E_b is increased.

Often another set of deflecting plates is provided, these plates being so located along the axis of the tube that they deflect the spot as a function of a second deflecting voltage in a direction perpendicular to the deflection caused by the first set of plates. The path of the spot on the screen is then a function of the two deflecting voltages. If an unknown transient voltage is impressed on one set of plates, and a voltage of known waveform is impressed on the other, the path of the spot is a curve giving the unknown voltage as a function of the known. For example, one voltage may be made directly proportional to time, whereupon the path of the spot delineates the waveform of the second voltage as a function of time, and the tube may serve as an oscilloscope.

For the tube to be useful, the path of the spot must be visible or of such a nature that it can be photographed. Hence the anode-to-cathode voltage E_b must be large in order to transmit sufficient energy to the phosphor to make its luminescence visible or sufficiently bright to be photographed. On the other hand, the deflection sensitivity of the beam decreases as E_b is increased, and a compromise between sensitivity and luminosity must therefore be made. As a rule, the observation or recording of transient phenomena of short duration requires a large anode-to-cathode voltage and a consequent sacrifice of sensitivity.

In the foregoing analysis, it is assumed that the deflecting voltage e_d is constant; yet the real utility of the cathode-ray tube is in the study of phenomena involving a time variation of e_d. If the speed of the electrons is so great that the deflecting field does not change appreciably while each electron moves through it, the deflecting field is essentially an electrostatic field for each electron in the beam. However, although the electron is a very agile particle and the component of velocity v_p is large, the time variation of e_d is sometimes so rapid that, during transit through the region between the deflecting plates, the electron experiences a time variation of the deflecting field that is not negligible. In such circumstances the position of the spot on the screen is not a direct indication of the instantaneous phenomena, because the component of velocity v_d acquired by the electron is not the result of a single value of the deflecting voltage.

A guide to the conditions for which the electron velocity is important may be obtained from a numerical example. Suppose the accelerating voltage in the electron gun, E_b, is 1,000 volts and the length l of the deflecting plates is 0.02 meter. The speed of the electrons as determined from Eqs. 3 and 67 is

$$v_p = \sqrt{2 \times 1.76 \times 10^{11} \times 10^3} = 1.87 \times 10^7 \text{ meters per second,} \qquad [75]$$

and the time required for the electron to travel the distance l through the deflecting plates is

$$t_d = \frac{0.02}{1.87 \times 10^7} = 1.07 \times 10^{-9} \text{ second.} \qquad [76]$$

Although this time may at first appear to be small, it is not small compared with the duration of 10^{-7} or 10^{-8} second observed for many transient electrical phenomena, or the period of a radio-frequency wave used for deflection. Hence it follows that the anode-to-cathode voltage E_b must be large not only to insure a bright spot on the screen but also to insure accuracy in the display of phenomena that occur in such a short time, for a large velocity v_p and a correspondingly short time of transit through the deflecting plates are then required.[16]

The beam in a cathode-ray tube may be deflected by a magnetic field instead of an electric field. An analysis of magnetic deflection is given in Art. 7b.

6. ELECTRON OPTICS

A second application of the analysis of the behavior of charged particles in electrostatic fields is that of *electron optics*. The requirement of a cathode-ray oscilloscope tube as well as certain other electronic devices, such as electron microscopes, is that the surface concentration of electron current leaving a given plane in the device be reproduced on some other plane with a surface magnification greater or smaller than unity. Usually the magnification desired is less than unity in oscilloscope and television tubes but is greater than unity in electron

[16] F. M. Gager, "Cathode-Ray Electron Ballistics," *Communications*, *18* (1938), 10; H. E. Hollmann, "Theoretical and Experimental Investigations of Electron Motions in Alternating Fields with the Aid of Ballistic Models," *I.R.E. Proc.*, *29* (1941), 70-79.

microscopes.[17] In oscilloscope and television tubes, the magnification is desired small in order that the electron concentration on the screen be intense and the luminescent spot be small and bright. A small, bright spot makes possible greater detail in the figure traced out on the screen and hence permits greater resolution of the data in electrical transient studies. In the electron microscope, the magnification is desired large for the same reason that it is desired large in optical microscopes.

Considerable attention has been given this problem during recent years, with the result that a new branch of science called electron optics[18] has appeared and has become of great importance in the design of electron microscopes and cathode-ray tubes. The term electron optics comes from the striking analogy that exists between the behavior of light when it passes through refracting media and the behavior of electrons when they pass through electrostatic or magnetostatic fields.

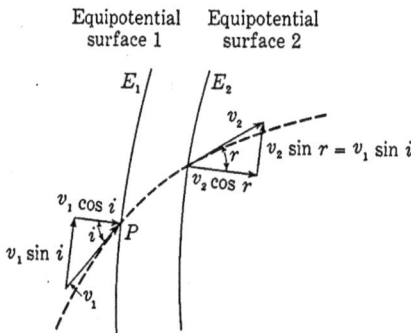

Fig. 6. Electron trajectory illustrating the optical analogy.

As an illustration of this analogy, consider that an infinitesimal region separates the two equipotential surfaces indicated in Fig. 6. Let the potential of surface 1 be E_1 and the potential of surface 2 be E_2; and let the datum of the potentials be the point where an electron that passes through point P on surface 1 had zero velocity. When this electron reaches point P, its speed is

$$v_1 = \sqrt{2 \frac{Q_e}{m_e} E_1} .$$ [77]

If the angle between the path of the electron at P and the normal to

---

[17] R. P. Johnson, "Simple Electron Microscopes," *J. App. Phys.*, *9* (1938), 508–516; V. K. Zworykin, "Electron Optical Systems and Their Applications," *I.E.E.J.*, *79* (1936), 1–10; L. Marton, M. C. Banca, and J. F. Bender, "A New Electron Microscope," *RCA Rev.*, *5* (1940), 232–243.

[18] Several books that include the subject are: E. Brüche and O. Scherzer, *Geometrische Elektronenoptik* (Berlin: Julius Springer, 1934); I. G. Maloff and D. W. Epstein, *Electron Optics in Television* (New York: McGraw-Hill Book Company, Inc., 1938); L. M. Meyers, *Electron Optics* (New York: D. Van Nostrand Company, Inc., 1939); V. K. Zworykin and G. A. Morton, *Television* (New York: John Wiley & Sons, Inc., 1940); V. K. Zworykin, G. A. Morton, E. G. Ramberg, J. Hillier, and A. W. Vance, *Electron Optics and the Electron Microscope* (New York: John Wiley & Sons, Inc., 1945).

the equipotential surface is denoted by i, the velocity component of the electron perpendicular to the equipotential surface is $v_1 \cos i$. This velocity component is in the direction of the electric field at P. Similarly, the velocity component tangential to the equipotential surface and perpendicular to the electric field is $v_1 \sin i$ as indicated in the figure.

When the electron reaches the second equipotential surface its speed is

$$v_2 = \sqrt{2 \frac{Q_e}{m_e} E_2}. \qquad [78]$$

If the path of the electron where it passes through the second equipotential surface makes an angle denoted by r with the normal to the surface, the velocity component in a direction perpendicular to the surface is $v_2 \cos r$ and that tangential to the surface is $v_2 \sin r$. Although in general the equipotential surfaces are curved, they may be considered to be parallel planes in the region traversed by the electron if they are sufficiently close together. Thus, while the electron is passing through the infinitesimal region that separates the two surfaces, the velocity component $v_1 \sin i$ tangential to the surfaces does not change, because it is perpendicular to the direction of the force exerted on the electron by the electric field. Hence,

$$v_1 \sin i = v_2 \sin r, \qquad [79]$$

and

$$\frac{v_1}{v_2} = \frac{\sin r}{\sin i} = \frac{\sqrt{E_1}}{\sqrt{E_2}}. \qquad [80]$$

The angles i and r are analogous to the angles of incidence and refraction in optics, and Eq. 80 is equivalent to Snell's law if it is considered that $\sqrt{E_1}$ and $\sqrt{E_2}$ are analogous to the refractive indices η_1 and η_2 of the first and second media encountered by a light ray.

A complication arises when an attempt is made to formulate an analogy between light optics and electron optics in an actual problem. Instead of the uniform media with well-defined boundaries that are used in optical systems, the electric fields established by the charges on the electrodes in an electron tube present a continuously variable refracting medium for electrons. In most electron-optical systems, the field, even though it is a complicated function of the radius and the position along the axis, possesses approximate cylindrical symmetry. The distribution of such a field is not often readily calculable, but it can often be found by an experimental method involving models,[19] and a point-by-point calculation is then practical for the determination of the electron's path. Typical potential distributions for a space-charge-free aperture lens and a double-cylinder lens are shown in Figs. 7a and 7b.

A second complication sometimes of importance is that the electric field encountered by the particles consists not only of that set up by the charges on the electrodes but also of that resulting from the

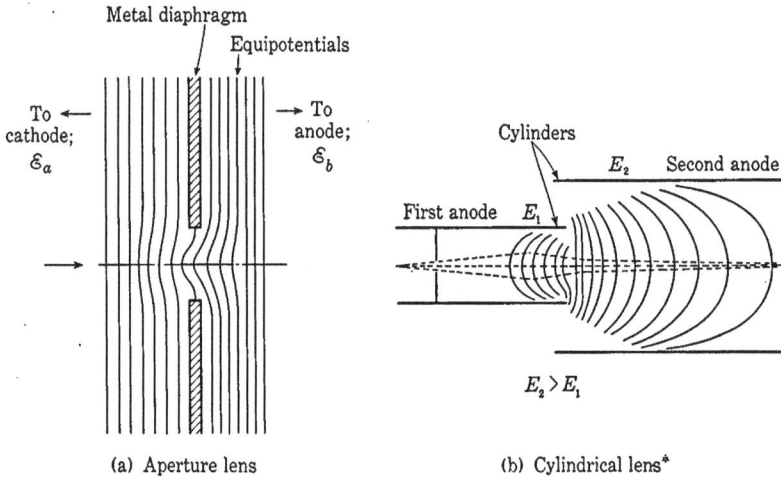

(a) Aperture lens   (b) Cylindrical lens*

Fig. 7.   Electron lenses.*

charges of the particles in the interelectrode space. These modify the total field to a degree depending on the amount and distribution of the charge density in the space. This space-charge effect is frequently of appreciable importance near the cathode of an electron tube but is usually negligible in other regions.

The point-by-point solution for the path of an electron in an electron tube involves, first, a determination of the electric field intensity $\mathcal{E}$ throughout the region and, second, a determination of the motion of

[19] P. H. J. A. Kleynen, "The Motion of an Electron in a Two-Dimensional Electrostatic Field," *Philips Tech. Rev.*, *2* (1937), 338–345; E. D. McArthur, "Experimental Determination of Potential Distribution," *Electronics*, *4* (June, 1932), 192–194; H. Salinger, "Tracing Electron Paths in Electric Fields," *Electronics*, *10* (October, 1937), 50–54; D. Gabor, "Mechanical Tracer for Electron Trajectories," *Nature*, *139* (1937), 373; V. K. Zworykin and J. A. Rajchman, "The Electrostatic Electron Multiplier," *I.R.E. Proc.*, *27* (1939), 558–566.

* This diagram is adapted from D. W. Epstein, "Electron Optical System of Two Cylinders as Applied to Cathode-Ray Tubes," *I.R.E. Proc.*, *24* (1936), Fig. 2, p. 1099, with permission.

an electron in this field. During the time interval Δt_1 in which the electron traverses a portion of its path Δl_1 from, say, point P_1 where l equals l_1 to point P_2 where l equals l_2, the change in its velocity component in the direction of the field is

$$\Delta v_1 = -\frac{Q_e}{m_e} \, \mathcal{E}_1 \, \Delta t_1 , \qquad [81]$$

where $\mathcal{E}_1$ is the electric field intensity encountered between P_1 and P_2, and is assumed to be constant over the interval. If the velocity of the electron at P_1 is resolved into the components v_n perpendicular and v_t parallel to the field, v_n does not change during the time interval Δt_1, and the speed of the electron at P_2 is

$$v_2 = \sqrt{v_n{}^2 + (v_t + \Delta v_1)^2} . \qquad [82]$$

During the interval Δt_1, the distance the electron moves in the direction perpendicular to the electric field is $v_n \, \Delta t_1$, and that in the direction of the field is $v_t \Delta t_1 + \frac{1}{2} \dfrac{-Q_e}{m_e} \mathcal{E}_1 \, \Delta t_1{}^2$. Thus the total distance moved is

$$\Delta l_1 = \sqrt{\left(v_t \Delta t_1 + \frac{1}{2} \frac{-Q_e}{m_e} \mathcal{E}_1 \, \Delta t_1{}^2 \right)^2 + (v_n \, \Delta t_1)^2} . \qquad [83]$$

The position of P_2 relative to P_1 and the velocity at P_2 are thereby computed. During the next succeeding time interval Δt_2, the increment of path Δl_2 traversed by the electron may be computed by a repetition of the process; the velocity v_2 is resolved into components along and perpendicular to the electric field intensity $\mathcal{E}_2$ encountered between P_2 and P_3 and assumed to be constant over that interval. By means of this point-by-point method the paths of rays through the field can be computed through the use of field plots and computation charts, and the refractive properties of the field can be determined. Graphical and machine methods equivalent to the point-by-point method are also used for a determination of the trajectory.[20]

7. MOTION OF CHARGED PARTICLES IN MAGNETOSTATIC FIELDS

A charged particle in motion in a magnetostatic field experiences a force whose direction is perpendicular both to the direction of motion of the particle and to the direction of the field. The magnitude of the

[20] V. K. Zworykin and J. A. Rajchman, "The Electrostatic Electron Multiplier," *I.R.E. Proc.*, *27* (1939), 558–566; J. P. Blewett, G. Kron, F. J. Maginiss, H. A. Peterson, J. R. Whinnery, and H. W. Jamison, "Tracing of Electron Trajectories Using the Differential Analyzer," *I.R.E. Proc.*, *36* (1948), 69–83.

force is proportional to, first, the magnitude of the magnetic flux
density; second, the charge on the particle; and, third, the component of the velocity of the particle that is perpendicular to the
direction of the field. The relationship among the directions of the
quantities for a positively charged particle is shown in Fig. 8, where
F, **v**, and **B** are vectors representing the force, velocity, and magnetic flux density, respectively. The direction of the force on a
positively charged particle is the same as the direction a right-hand
screw having its axis perpendicular to both **v** and **B** would advance
if it were rotated in the same angular direction that would rotate
v into **B** through the smaller of the two angles between those
vectors. Expressed symbolically, the magnitude and sense of the
force is given by

$$F = BQv \sin \phi ,$$ ▶[84]

where

B is the magnitude of the magnetic flux density,

Q is the charge on the particle,

v is the speed of the particle,

ϕ is the smaller of the two angles between the direction of the
magnetic field and the direction of the motion of the
particle.

Since $\sin \phi$ is zero when the charge moves in the direction of the
magnetic field, the force is then zero; but, when the charge moves in a
direction perpendicular to the magnetic field, $\sin \phi$ is unity, and the
force is a maximum given by BQv. Equation 84 indicates that for a
negatively charged particle the sense of the force is negative. In other
words, the direction of the force is opposite to that of the right-hand
screw just explained, and is hence opposite to the direction shown
in Fig. 8.

Since a current is equivalent to a movement of charged particles,
the relationship expressed in Eq. 84 should be equivalent to that
observed experimentally for a conductor carrying a current in a
magnetic field. The equivalence may be shown as follows. When a
conductor carrying a current i is placed in a magnetic field of flux
density B, the conductor experiences a force per unit length given
by the relation

$$F = Bi \sin \phi,$$ [85]

where ϕ is the smaller of the two angles between the direction of the
field and the direction of the current. The direction of this force is
perpendicular to the plane containing the field vector **B** and the

conductor. This relationship is also illustrated by Fig. 8. An electric current in a conductor of small, uniform cross section is equivalent to a stream of electric charges moving along the path of the current, all with the same speed, and is given by

$$i = Qnv, \qquad [86]$$

where

Q is the charge on each particle,

n is the number of charged particles per unit length of the path of i,

v is the speed of the particles,

nv is the number of charged particles that pass a given cross section of the path of i per unit time.

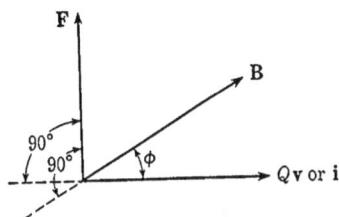

Fig. 8. Direction of the force exerted by a magnetic field on a current or a positively charged moving particle.

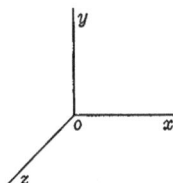

Fig. 9. Rectangular co-ordinate system for Eqs. 92, 93, and 94.

The substitution of i from Eq. 86 in Eq. 85 gives

$$F = BQnv \sin \phi \qquad [87]$$

for the force on a unit length of conductor, or the force on n particles. Since there is no reason why the force on one particle should be greater than that on another, the force on one particle is given by division of Eq. 87 by n; whence,

$$F = BQv \sin \phi \qquad [88]$$

for one particle,[21] an expression identical with Eq. 84.

[21] This derivation of Eq. 88 from the equation for the force on a current-carrying conductor applies only to the force on a single charged particle moving as a part of a uniform stream of charged particles. For a discussion of the force on a single moving charge, see W. R. Smythe, *Static and Dynamic Electricity* (2nd ed.; New York: McGraw-Hill Book Company, Inc., 1950), 565–567 and 574–576.

In the notation of vector analysis,[22] the vector force **F** as given by Eq. 84 is

$$\mathbf{F} = Q(\mathbf{v} \times \mathbf{B}), \qquad [89]$$

and as given by Eq. 85 is*

$$\mathbf{F} = \mathbf{i} \times \mathbf{B}. \qquad [90]$$

As a result of the force, acceleration of the charged particle takes place, and

$$\mathbf{F} = m\mathbf{a} = Q(\mathbf{v} \times \mathbf{B}). \qquad [91]$$

In the rectangular co-ordinate system of Fig. 9, Eq. 91 becomes the set of three differential equations

$$\frac{d^2x}{dt^2} = \frac{Q}{m}\left[B_z \frac{dy}{dt} - B_y \frac{dz}{dt} \right], \qquad \blacktriangleright[92]$$

$$\frac{d^2y}{dt^2} = \frac{Q}{m}\left[B_x \frac{dz}{dt} - B_z \frac{dx}{dt} \right], \qquad \blacktriangleright[93]$$

$$\frac{d^2z}{dt^2} = \frac{Q}{m}\left[B_y \frac{dx}{dt} - B_x \frac{dy}{dt} \right], \qquad \blacktriangleright[94]$$

where B_x, B_y, and B_z are the components of the magnetic flux density along the corresponding three co-ordinate axes. These equations are readily derived from the fact that, when the direction of motion of the charge is perpendicular to the direction of the magnetic field, the force is given by BQv and is perpendicular to these two directions as shown in Fig. 8. The foregoing equations hold in any consistent system of units. For example, in the mks system the force is given in newtons when Q is in coulombs, B is in webers per square meter, v is in meters per second, and x, y, and z are in meters.

An electric field exerts a force on a charged particle whether the particle is at rest or in motion, and the force is always in the direction of the field. On the other hand, according to Eq. 84 or Eq. 89, a magnetic field exerts a force on a charged particle only if the particle is in motion and the direction of the motion is not parallel to the direction of the field. If a force does act on a particle because of a magnetic field,

[22] For an introduction to the cross-product notation, see N. H. Frank, *Introduction to Electricity and Optics* (2nd ed.; New York: McGraw-Hill Book Company, Inc., 1950), 107–108.

* Current is ordinarily defined as a scalar quantity, and current density as a vector. Since the current considered here is supposed to be analogous to a stream of charged particles, the current-density vector associated with the current crossing a given cross section of path is directed along the path at every point on the cross section, and it is therefore possible to define the vector **i** whose magnitude is the current, and whose direction is that of the current path.

the force is perpendicular to the field. Since the force is also perpendicular to the direction of motion of the particle, a magnetic field can do no work on the particle. Thus it is not possible for a magnetic field to change either the kinetic energy or the speed of a charged particle. A magnetic field is capable of changing only the direction of motion of the particle.

In the remainder of this article, the equation for the force on a moving charged particle in a magnetic field, Eq. 84 or Eq. 89, together with the concept that a magnetic field can change the direction but not the magnitude of the velocity of a charged particle, is applied to determine the path of a charged particle in a uniform magnetic field for different initial-velocity conditions.

7a. *Circular Path.* In Fig. 10 a charged particle, assumed to be an electron, is projected into a magnetostatic field having a direction perpendicular to the paper and directed into it. The force exerted by the field is always perpendicular to the direction of motion, and the particle therefore follows a curved path. The force producing the curvature is

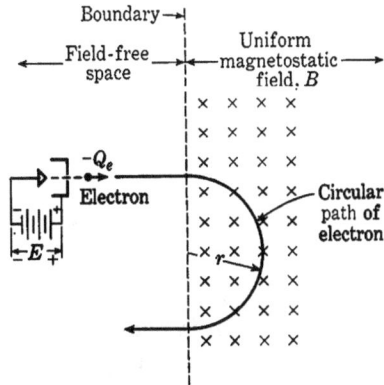

Fig. 10. Circular path of an electron that enters perpendicularly into a magnetostatic field.

$$F = BQv, \qquad \blacktriangleright[95]$$

where v is the speed of the particle at any point. The acceleration caused by this force is, from kinematics, v^2/r, where r is the radius of curvature of the path. By Newton's second law,

$$F = ma = BQv = \frac{mv^2}{r} ; \qquad [96]$$

whence,

$$r = \frac{mv}{BQ} . \qquad [97]$$

In this equation v is a constant speed unchanged by the magnetic field, and Q and m are constants. If, further, the magnetic flux density B is uniform everywhere along the path, the radius is then constant and the path is a circle.

If the particle acquires its speed v from motion through an accelerating electrostatic field of potential difference E, as is indicated in

Fig. 10, the speed is given by Eq. 30, and substitution of this relation in Eq. 97 gives

$$r = \frac{1}{B} \sqrt{-2\frac{m}{Q}E}.$$ [98]

In Eq. 98, Q is negative if the particle carries a negative charge. Thus if the particle is an electron, the value of Q_e/m_e from Eq. 3 may be placed in Eq. 98, and the expression for r becomes

$$r = \frac{1}{B} \sqrt{\frac{2E}{1.76 \times 10^{11}}}$$

$$= 3.37 \times 10^{-6} \frac{\sqrt{E}}{B} \text{ meters},$$ ▶[99]

where E is in volts and B is in webers per square meter.

The relation given in Eq. 98 may be applied for the experimental measurement[23] of the ratio Q/m for electrons or ions. A beam of charged particles produced by a "gun" in the form of a filament and pierced anode, or by some other suitable source, is directed into a ring-shaped tube properly arranged in the magnetic field of Helmholtz coils so that the beam is bent in a circle along the axis of the tube. To determine the path of the beam, the charge that impinges on a collector electrode located in the tube at a fixed distance away from the source of the particles may be measured, sufficient gas may be left in the tube to render the path visible through excitation of the gas, or the beam may be allowed to fall on a fluorescent or photographic plate. The measurements of r, E, and B constitute data for the calculation of Q/m. The magnetic field of the earth may contribute to the bending of the beam, and this possibility must be taken into account in the calculation unless experimental precautions are taken to counteract it.

The principle that the radius depends on the mass-to-charge ratio of the particle as indicated by Eq. 98 has been applied in a device called a *mass spectrometer* for the identification and separation of particles having different values of this ratio. In this apparatus a beam composed of a mixture of ions having different ratios of mass to charge separates along paths of different radii when it passes through a magnetic field. The particle having the desired ratio is then collected on an electrode at the terminus of its path.

Isotopes, which are atoms that have different masses but identical chemical properties and are therefore different forms of the same

---

[23] A. J. Dempster, "Positive-Ray Analysis of Potassium, Calcium, and Zinc," *Phys. Rev.*, **20** (1922), 631–638; F. W. Aston, *Mass Spectra and Isotopes* (2nd ed.; London: Edward Arnold & Co., 1942).

chemical element, were first discovered and identified by use of a mass spectrometer. Separation of natural uranium to obtain the isotope uranium 235 was first accomplished in quantity by application of the same principle.[24]

The mass spectrometer also finds wide industrial use as a means for locating leaks in vacuum systems.[25] The device is connected to the evacuated system so as to sample the gas in it while a small jet of rare gas such as helium is passed over the outside surface. Presence

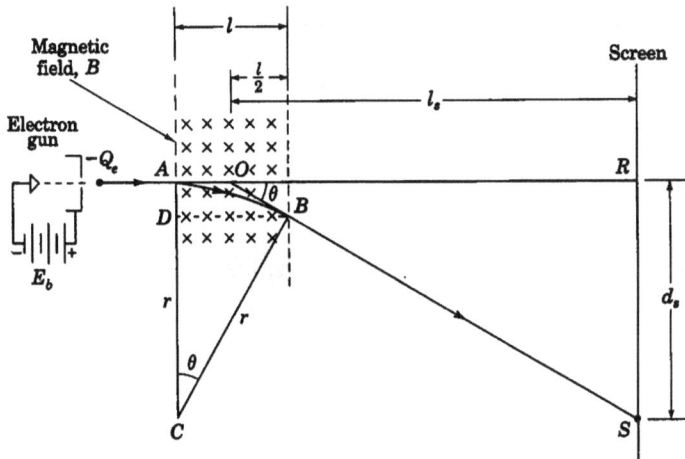

Fig. 11. Magnetic deflection of the beam in a cathode-ray tube.

of that particular gas in the mass-spectrometer indication shows that the leak has been reached by the jet.

7b. Magnetic Deflection in a Cathode-Ray Tube. A magnetic field is frequently used as a means of deflecting the electron beam in a cathode-ray tube. Usually the magnetic field is produced by the instantaneous current in a pair of coils located near the electron gun and oriented so as to direct the field perpendicular to the electron path, as is illustrated in Fig. 11.

If at any instant the field is uniform between the boundaries indicated and remains constant while the electrons pass through it, the beam, which enters the field at A, is deflected to B along a circular path, having a radius r. Thereafter it continues along a straight path

[24] H. D. Smyth, *Atomic Energy for Military Purposes* (Princeton, N.J.: Princeton University Press, 1947), 187–205. J. H. Pomeroy, "Electromagnetic Methods of Separating Isotopes," *The Science and Engineering of Nuclear Power, Vol. II*, Clark Goodman, Editor (Cambridge, Mass.: Addison-Wesley Press, Inc., 1948), Appendix A-2, 301–306.

[25] A. O. Nier, C. M. Stevens, A. Hustrulid, and T. A. Abbot, "Mass Spectrometer for Leak Detection," *J. App. Phys.*, *18* (1947), 30–48.

to produce a spot on the screen at the deflected position S. The circular path AB subtends an angle θ at C. The triangle SOR formed by the screen, the undeflected path, and the deflected path projected backward to O is similar to the triangle BCD because their respective sides are perpendicular. Hence, the angle at O is also θ, and

$$\frac{d_s}{SO} = \frac{l}{r}. \tag{100}$$

For small values of θ, point O lies at the center of l as is indicated on the diagram, and SO is approximately equal to l_s, the distance from the center point to the screen. When the deflection is small, therefore, its value as given by Eq. 100 is

$$d_s = \frac{l l_s}{r}. \tag{101}$$

Substitution for r from Eq. 98 with the values of Q and m for an electron gives

$$\text{Magnetic deflection sensitivity} = \frac{d_s}{B} = l l_s \sqrt{Q_e/(2m_e)} \, \frac{1}{\sqrt{E_b}}, \tag{102}$$

where E_b is the accelerating voltage in the electron gun. Thus the instantaneous deflection is directly proportional to the magnetic flux density B, and, in turn, to the instantaneous current in the coils that produce the field.

Note that, for magnetic deflection, the sensitivity is inversely proportional to $\sqrt{E_b}$. For this reason, magnetic deflection is advantageous in tubes with high accelerating voltages. It is used extensively in television and radar applications.

Magnetic deflection of the beam through large angles requires a more complete analysis than the foregoing because the point O is then not exactly at the center of the field and the distance SO is not closely approximated by l_s. In any practical application, however, the distance l is indefinite because the magnetic field does not have sharp boundaries. Hence, use of a more exact expression than Eq. 102 is seldom justified.

7c. *Cyclotron, Betatron, and Synchrotron.* The angular velocity (about the center of curvature of its path) of a particle moving with a velocity v along a path of radius of curvature r is

$$\frac{d\theta}{dt} = \frac{v}{r}. \tag{103}$$

From Eq. 97, this angular velocity for a moving electron in a magnetic field becomes

$$\frac{d\theta}{dt} = \frac{v}{r} = \frac{BQ}{m}.$$ ▶[104]

The angular velocity is therefore *independent* of the translational velocity. Thus, a fast-moving particle, which travels in a path of large radius of curvature in a given uniform magnetostatic field in accordance with Eq. 97, requires the same time for a complete revolution as

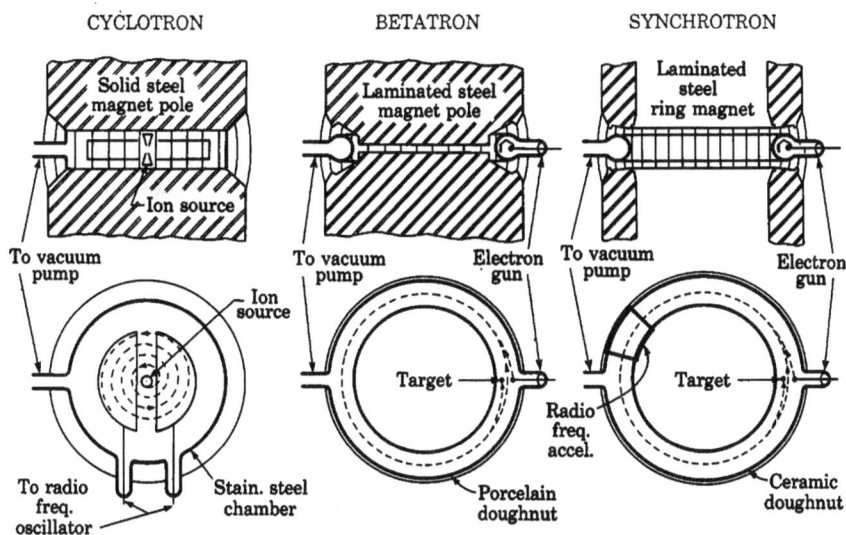

Fig. 12.   Schematic diagrams of the three principal types of cyclic accelerators for elementary particles.*

does a slow-moving particle, which travels in a path of small radius of curvature.

This principle is applied in the *cyclotron*, in which it is used to impart high kinetic energy to charged particles.[26] Figure 12 shows a schematic diagram of the cyclotron. A short cylindrical tank, cut in two along a diameter, forms two metals *dees*—semicylindrical chambers resembling the letter **D** in shape. The dees are insulated from each other in a vacuum chamber located between the poles of a powerful electro-

* This diagram is adapted from I. A. Getting, "Artificial Cosmic Radiation," *The Technology Review, 50* (June, 1948), 436, edited at the Massachusetts Institute of Technology, with permission.

[26] E. O. Lawrence and M. S. Livingston, "A Method for Producing High-Speed Hydrogen Ions, without the Use of High Voltages (Abstract)," *Phys. Rev., 37* (1931), 1707; M. S. Livingston, "The Cyclotron," *J. App. Phys., 15* (1944), 2-19, 128-147; M. S. Livingston, "Particle Accelerators," *Advances in Electronics, Vol. I*, L. Marton, Editor (New York: Academic Press, Inc., 1948), 269-316.

magnet weighing many tons in most cyclotrons, and the direction of the magnetic field is perpendicular to the semicircular faces of the dees. A high-frequency oscillator is connected to the dees, and a source of charged particles such as protons or deuterons is provided near their center as shown on the diagram. When a charged particle leaves the source it is accelerated into the dees in one direction or the other by the electric field set up by the oscillator. Inside the dees, the electric field is small, because it is concentrated mainly across the gap, and the charged particle travels in a circular path at constant speed under the influence of the magnetic field. The frequency of the oscillator and the magnetic field strength are so adjusted that, when the charged particle again reaches the gap, the electric field has reversed. The electric field therefore gives the charged particle a second acceleration across the gap, adding to its speed. The charged particle then travels a path of larger radius through the opposite dee, but reaches the gap after the same time interval. During this interval the electric field has again reversed and is in a direction to accelerate the particle further. This process is repeated many times until finally the charged particle travels in a path near the periphery of the dee. As long as the speed of the particle remains so low that the mass stays essentially constant, a fixed-frequency oscillator can be used because the time of a revolution is constant. Subjected to a hundred or more revolutions with an increase in speed corresponding to several thousand volts twice each revolution, the particles can be given energies corresponding to several million volts, yet the voltage of the source need be only a few thousand volts. The particles are finally led off through a side tube to a target for experimental purposes, such as the nuclear disintegration of elements.

When the speed of the particles such as protons in a cyclotron exceeds that corresponding to energies of 10 or 20 million electron volts, the mass increases appreciably, and the angular velocity decreases as the energy increases. An equilibrium is reached at which the particles cross the gaps at essentially the instants of reversal of the electric field. The particles then receive no further energy from the field and hence travel in a path of constant radius. If the frequency is then decreased, however, the radius and the mass continue to increase in such a manner as to maintain the equilibrium. A cyclotron with such a changing frequency is called a *frequency-modulated cyclotron* or *synchrocyclotron*.[27] It has produced particles having energies of 400 million electron volts.

[27] W. M. Brobeck, E. O. Lawrence, K. R. MacKenzie, E. M. McMillan, R. Serber, D. C. Sewell, K. M. Simpson, and R. L. Thornton, "Initial Performance of the 184-inch Cyclotron of the University of California," *Phys. Rev.*, 71 (1947), 449–450.

Magnetic induction, rather than a radio-frequency electric field, is used for accelerating electrons in a device also illustrated in Fig. 12 called a *betatron*.[28] A changing magnetic flux in it induces a uniform electromotive force tangentially along a circular path for the electrons, and accelerates them to high energies. The electrons are held in an orbit of fixed radius by a second changing magnetic field located at the orbit but perpendicular to it. The flux density of the second field is maintained equal to one-half the average flux density inside the orbit. An evacuated annular chamber in the shape of a hollow dough-nut surrounds the circular path. The magnetic flux for inducing the electromotive force along the path is produced in a laminated steel electromagnet by an alternating or pulsed power source, and the second magnetic field for guiding the electrons is produced in an air gap in the same or a different electromagnet excited by the same source. Bunches of electrons released in the chamber in synchronism with the supply travel many times around the chamber, and gain a few hundred electron volts of energy each time around. Finally the magnetic field is suddenly altered so as to cause the electrons to emerge with energies of more than 100 million electron volts.

The *synchrotron*[29] shown in Fig. 12 is a particle accelerator that utilizes some of the principles of both the cyclotron and the betatron. For accelerating electrons, the machine first acts as a betatron while imparting an initial energy to the electrons. After they reach energies of a few million electron volts, so that their speed is near that of light, additional energy is given them by a radio-frequency electric field existing between electrodes situated along their path in the doughnut-shaped chamber. The frequency of the electric field can be constant because the electrons continue to travel at essentially the speed of light despite further increase in energy. Thus their angular velocity around the center of their orbit is essentially constant. The guiding magnetic field at their orbit must continue to increase in proportion to their mass, however. Increase of the magnetic field inside the orbit is no longer necessary once the betatron action is complete, hence the steel inside the orbit may be allowed to saturate magnetically. The weight and cost of the machine may therefore be reduced below the corresponding values for a betatron. Electron energies of more than 300 million electron volts are produced by electron synchrotrons.

[28] D. W. Kerst, "The Acceleration of Electrons by Magnetic Induction," *Phys. Rev.*, 60 (1941), 47–53; E. E. Charlton and W. F. Westendorf, "A 100-Million Volt Induction Electron Accelerator," *J. App. Phys.*, 16 (1945), 581–593.

[29] V. Veksler, "A New Method of Acceleration of Relativistic Particles," *J. Phys. U.S.S.R.*, 9 (1945), 153–158; E. M. McMillan, "The Synchrotron—A Proposed High Energy Particle Accelerator," *Phys. Rev.*, 68 (1945), 143–144; E. M. McMillan, "The Origin of the Synchrotron," *Phys. Rev.*, 69 (1946), 534.

A synchrotron for accelerating protons must include a changing radio frequency as well as a changing magnetic field at the orbit, because an auxiliary means for accelerating protons to speeds near that of light before injection for synchrotron action is not available. Increase of the frequency by a factor of ten while the guiding magnetic field increases from zero to about 10,000 gausses is expected to produce proton energies of several billion electron volts. Such a *proton synchrotron*[30] is hence also known as a *bevatron* from the initials *bev* for billion electron volts.

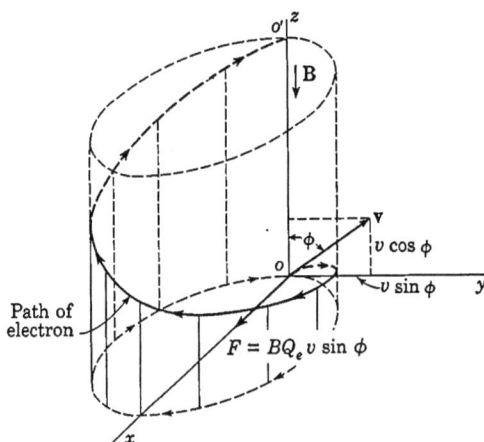

Fig. 13.   Helical path of a moving electron in a uniform magnetostatic field.

7d. *Helical Path.* In each of the foregoing illustrative examples a charged particle moves in a direction perpendicular to that of the magnetic field. If the particle has a constant speed but does not move perpendicularly into the magnetic field, it describes a helical path.[31] This is shown in Fig. 13, where a particle, assumed in the figure to be an electron, starting from the origin has a velocity **v**, which, at the initial instant, lies in the y–z plane and makes an angle ϕ with the z axis. Since Q is negative for this diagram, the angle ϕ on it has the same significance as that between the magnetic flux density **B** and $Q\mathbf{v}$ on Fig. 8. The component of velocity of the particle parallel to the field and along the z axis, $v \cos \phi$, is unaltered by the field. Hence the

[30] M. S. Livingston, J. P. Blewett, G. K. Green, and L. J. Haworth, "Design Study for a Three-Bev Proton Accelerator," *R.S.I.*, *21* (1950), 7–22; E. J. Lofgren, "Berkeley Proton-Synchrotron," *Science*, *11* (March 25, 1950), 295–300.

[31] H. Busch, "Eine neue Methode zur e/m Bestimmung," *Phys. Zeits.*, *23* (1922), 438–441; H. Busch, "Berechnung der Bahn von Kathodenstrahlen im axialsymmetrischen elektromagnetischen Felde," *Ann. d. Phys.*, *81* (1926), 974–993.

particle continues to move in the direction of the z axis with a speed $v \cos \phi$. However, the velocity component $v \sin \phi$ gives rise to a force on the particle, and, although the magnitude of this component remains constant, its direction is altered continuously. By Eq. 97, the projection of the path in the x–y plane is a circle whose radius is

$$r = \frac{mv \sin \phi}{BQ}$$ ▶[105]

The time of one revolution of the projection is

$$t_0 = \frac{2\pi r}{v \sin \phi} = \frac{2\pi m}{BQ}.$$ [106]

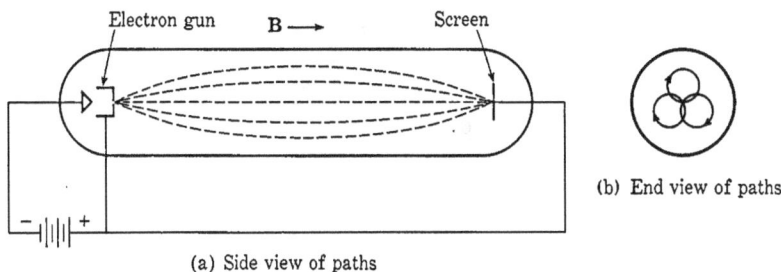

(a) Side view of paths

(b) End view of paths

Fig. 14.   Focusing by a longitudinal magnetic field.

The superposition of the uniform circular motion in the plane normal to **B** and the uniform translational motion parallel to **B** is the condition for a helical path. The vertical distance covered in one revolution is the pitch of the helix, which is given by

$$\text{Pitch} = t_0 v \cos \phi = \frac{2\pi m v \cos \phi}{BQ}.$$ ▶[107]

7e. Magnetic Focusing. The helical paths described above are sometimes utilized for focusing cathode-ray tubes and high-voltage x-ray tubes, and are typical of the trajectories in some magnetic electron lenses.[32] Figure 14 shows the method of focusing a divergent beam of electrons produced by an electron gun. The divergent electrons in the beam are made to follow helical paths by the magnetic field directed along the axis of the tube. The electrons follow paths whose projections on a plane perpendicular to the axis of the tube are circles

[32] V. K. Zworykin and G. A. Morton, *Television* (New York: John Wiley & Sons, Inc., 1940), 117–120; C. J. Davisson and C. J. Calbick, "Electron Lenses," *Phys. Rev.*, *38* (1931), 585; and *42* (1932), 580.

that pass through the axis, as is shown in Fig. 14b. The time of a revolution is independent of the angle of divergence, and, if this angle is small enough for its cosine to be essentially unity, all the electrons have the same translational velocity parallel to the axis of the tube. Thus all the electrons are at the same phase of their revolution at any distance from the source. By adjustment of the magnetic field, which may be produced by a solenoid, all the electrons can be allowed to complete one or any integral number of revolutions during their travel down the tube, and they then focus at the screen.

8. MOTION OF CHARGED PARTICLES IN CONCURRENT ELECTRO-STATIC AND MAGNETOSTATIC FIELDS

When electrostatic and magnetostatic fields simultaneously influence the motion of charged particles, the net force on the particle is the sum of the electrostatic and magnetostatic forces which would appear if each field acted independently. Both forces may be represented as vector quantities; hence, in vector notation, the net force is

$$\mathbf{F} = Q[\mathcal{E} + (\mathbf{v} \times \mathbf{B})] = m\mathbf{a}, \qquad [108]$$

this expression being the sum of Eqs. 9 and 89. On resolution of the vectors into components along the axes of the rectangular co-ordinate system of Fig. 9, Eq. 108 becomes the set of three equations:

$$\frac{d^2x}{dt^2} = \frac{Q}{m}\left[\mathcal{E}_x + B_z\frac{dy}{dt} - B_y\frac{dz}{dt}\right], \qquad \blacktriangleright[109]$$

$$\frac{d^2y}{dt^2} = \frac{Q}{m}\left[\mathcal{E}_y + B_x\frac{dz}{dt} - B_z\frac{dx}{dt}\right], \qquad \blacktriangleright[110]$$

$$\frac{d^2z}{dt^2} = \frac{Q}{m}\left[\mathcal{E}_z + B_y\frac{dx}{dt} - B_x\frac{dy}{dt}\right]. \qquad \blacktriangleright[111]$$

These equations completely describe the motion of a particle in combined electrostatic and magnetostatic fields. Their solution, however, is a simple one only when special field configurations having symmetry are involved.

Since, as is shown in Art. 7, the component of force on a moving charged particle caused by a magnetic field is always perpendicular to the direction of motion, the magnetic field cannot alter the kinetic energy of the particle. If a charged particle moves in a combined electrostatic and magnetostatic field, the changes in its kinetic energy

and speed are the result of the electric field alone, and Eqs. 29 and
30 are applicable just as though the magnetic field were not present.
The paths in concurrent fields are not, in general, circles, since the
speed of a particle is altered as the particle moves through the electric
field. A discussion of several important examples of such motion
follows.

8a. *Motion in Parallel Fields.* Perhaps the simplest example of
motion in combined fields occurs when the fields are parallel. A particle
liberated with zero velocity is then accelerated along the electric
field, but experiences no force from the magnetic field, because the
particle has no velocity component perpendicular to that field. This

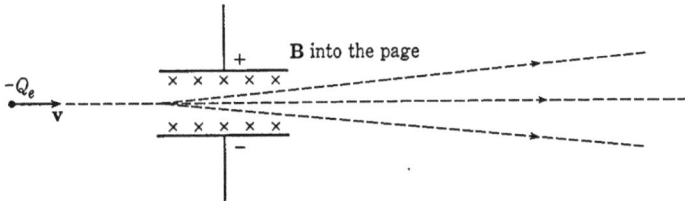

Fig. 15.   Arrangement for measurement of particle velocity.

situation also holds if the particle has an initial velocity parallel to
the direction of the fields, but not if the initial velocity has a component perpendicular to them.

If the particle has an initial velocity component perpendicular to
the parallel fields, the path that results is a "helix" whose pitch
changes with time as the particle is accelerated in the direction of the
electric field. The diameter of the "helix," however, is constant,
because neither of the fields can contribute energy to the component
of velocity perpendicular to them.

8b. *Measurement of Particle Velocity.* When the electric and magnetic fields are perpendicular to each other, several useful forms of
trajectory arise. One of these forms occurs when particles (for example,
the electrons in the beam of a cathode-ray tube) are deflected simultaneously by both fields. In Fig. 15 a stream of particles having a
velocity **v** enters the evacuated region between the parallel deflection
plates in a direction parallel to the plane of the plates. A uniform
electric field intensity $\mathcal{E}$ and a uniform magnetic flux density **B**
are produced in the region between the plates, and are so oriented that
they tend to deflect the beam in opposite directions; that is, the directions of **v**, **B**, and $\mathcal{E}$ are mutually perpendicular. If the magnitudes
of the fields are so adjusted that the forces they exert on the particles
are equal, the particles pass through the region without deflection

from their original path. For this condition,

$$\mathcal{E}Q + Q(\mathbf{v} \times \mathbf{B}) = 0 , \qquad [112]$$

or

$$\mathcal{E}Q = QvB, \qquad [113]$$

since the fields are perpendicular. Hence,

$$v = \mathcal{E}/B. \qquad [114]$$

Since $\mathcal{E}$ and B can be determined, this arrangement of the fields provides a method of measurement of the velocity of the particles, and it was so used in early experiments on the properties of electrons.[33]

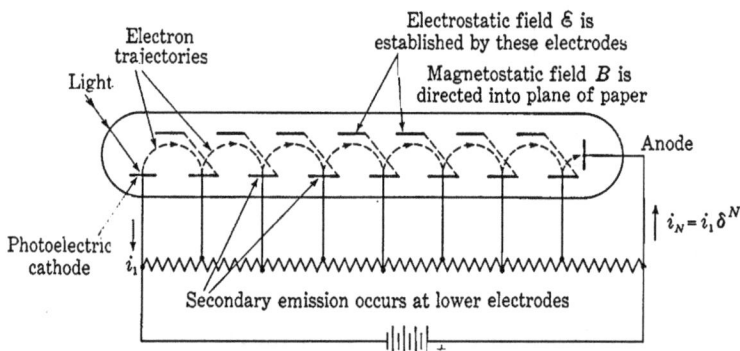

Fig. 16.   A secondary-emission electron-multiplier tube.

Special precautions are necessary to limit the fields effectively to the region between the deflecting plates.

8c. *Secondary-Emission Electron Multiplier.* An important engineering application of a particular form of electron trajectory encountered in crossed electrostatic and electromagnetic fields is found in an early form of a device called an *electron multiplier*.[34] The arrangement of the electrodes and the trajectory of the electrons are shown in Fig. 16. The dotted straight lines show connections between the upper and lower parallel plane electrodes. The direct-voltage source and the voltage divider create an electric field in the tube that tends to accelerate electrons from the cathode on the left to the anode on the right through the set of plates. An electron emitted by the action of light on the photoelectric cathode is drawn toward the first upper plate by the electric field. However, a magnetostatic field set up by an external coil or permanent magnet is directed perpendicularly to the electric field, that is, perpendicularly to and into the page of the

[33] J. J. Thomson, "Cathode Rays," *Phil. Mag.*, 44 (1897), 293–316.

[34] V. K. Zworykin, G. A. Morton, and L. Malter, "The Secondary Emission Multiplier —A New Electronic Device," *I.R.E. Proc.*, 24 (1936), 351–375.

diagram. The path of the electron is therefore bent until it hits the second lower plate. This plate and all the other lower horizontal plates have specially prepared surfaces so that, when one electron strikes them with considerable velocity, several electrons are emitted by the process of secondary emission discussed in Art. 12, Ch. II. These *secondary electrons* start toward the second upper plate, but their paths also are bent by the magnetic field, and they strike the third lower plate, where each causes the emission of several secondary electrons. This process is repeated in several similar stages down the length of the tube, the number of electrons in the path increasing in each stage until finally the stream of electrons is made to strike the anode and is led to the external circuit. If the ratio of the number of secondary electrons that leave a plate to the number of primary ones that strike it is δ, and if the device has N stages, the current at the anode is δ^N times the current that leaves the photoelectric cathode.

The current is thereby multiplied in the tube. By this means very sensitive photoelectric devices can be constructed.

The electric field in the electrode configuration of Fig. 16 is not uniform in magnitude or direction; hence an exact derivation of the trajectory of an electron is difficult. However, the trajectory is somewhat similar to that for an electron released with zero initial velocity at the negative electrode of the parallel-plane configuration of Fig. 17, where a uniform magnetic flux density **B** exists in a direction perpendicular to that of the electric field. By adjustment of the magnetic field strength, the electron path

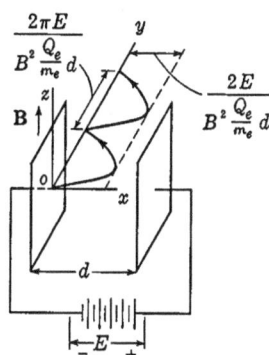

Fig. 17. Cycloidal motion in crossed fields.

can be curved to such an extent that the electron does not reach the positive electrode.

The trajectory of the electron in Fig. 17 is conveniently described by reference to the rectangular co-ordinate axes x, y, and z oriented as shown. Then, if the electrodes are infinite in extent,

$$\mathcal{E}_x = -E/d, \qquad [115]$$

where E is the potential of the right-hand electrode with respect to the left-hand electrode and d is the distance between the electrodes. Also,

$$B_z = B, \qquad [116]$$

and all other components of the crossed electrostatic and magnetostatic fields are zero. Equations 109, 110, and 111 apply in this co-ordinate system, and, upon substitution of the above conditions and

$-(Q_e/m_e)$ for the charge-to-mass ratio of the electron, they become

$$\frac{d^2x}{dt^2} = -\frac{Q_e}{m_e}\left[-\frac{E}{d} + B\frac{dy}{dt}\right],$$ [117]

$$\frac{d^2y}{dt^2} = -\frac{Q_e}{m_e}\left[-B\frac{dx}{dt}\right],$$ [118]

$$\frac{d^2z}{dt^2} = 0.$$ [119]

From Eq. 119 it follows that there is no force on the electron in the z direction; hence the electron moves only in the x–y plane. The position of the electron may therefore be expressed in the notation of complex numbers.* Thus if

$$u = x + jy,$$ [120]

Eqs. 117 and 118 may then be combined to express the motion in terms of this single variable, for multiplication of Eq. 118 by j and addition of it to Eq. 117 give

$$\frac{d^2x}{dt^2} + j\frac{d^2y}{dt^2} = \frac{Q_e}{m_e}\left[\frac{E}{d} - B\frac{dy}{dt} + jB\frac{dx}{dt}\right];$$ [121]

whence,

$$\frac{d^2u}{dt^2} = \frac{Q_e}{m_e}\left[\frac{E}{d} + jB\frac{du}{dt}\right],$$ [122]

or

$$\frac{d^2u}{dt^2} - jB\frac{Q_e}{m_e}\frac{du}{dt} = \frac{Q_e}{m_e}\frac{E}{d}.$$ [123]

The solution† of this differential equation subject to the initial conditions that at

$$t = 0, \quad u = 0, \quad \text{and} \quad v = 0 \quad \text{or} \quad \frac{du}{dt} = 0,$$ [124]

is

$$u = j\frac{E}{Bd}\left[t - j\frac{1}{B(Q_e/m_e)}\left(1 - \epsilon^{jB(Q_e/m_e)t}\right)\right].$$ [125]

---

* **Boldface italic** (slanting) type is used for letters denoting complex numbers.

† This solution may be recognized as similar in form to that for the charge in the problem of an electric circuit containing resistance and inductance in series and subject to a suddenly applied electromotive force E. Thus, if

$$L\frac{d^2q}{dt^2} + R\frac{dq}{dt} = E,$$

the current is

$$i = \frac{dq}{dt} = \frac{E}{R}\left(1 - \epsilon^{-(R/L)t}\right),$$

and by integration,

$$q = \frac{E}{R}\left[t - \frac{L}{R}\left(1 - \epsilon^{-(R/L)t}\right)\right].$$

Solutions for the variables x and y contained in $\boldsymbol{u}$ may be obtained from Eq. 125 through expansion of both sides into their real and imaginary parts. As a first step, the equation may be put in the form

$$\boldsymbol{u} = \frac{E}{B^2(Q_e/m_e)d} \left[jB\frac{Q_e}{m_e}t + \left(1 - \epsilon^{jB(Q_e/m_e)t} \right) \right]. \qquad [126]$$

Expansion of the exponential term into its equivalent trigonometric expressions gives

$$\boldsymbol{u} = x + jy$$
$$= \frac{E}{B^2(Q_e/m_e)d} \left[jB\frac{Q_e}{m_e}t + \left(1 - \cos B\frac{Q_e}{m_e}t - j\sin B\frac{Q_e}{m_e}t \right) \right]. \qquad [127]$$

This equation of complex numbers is equivalent to two equations of real numbers given by equating separately the reals and the imaginaries on the two sides of the equation. Thus

$$x = \frac{E}{B^2(Q_e/m_e)d} \left[1 - \cos B\frac{Q_e}{m_e}t \right], \qquad \blacktriangleright[128]$$

and

$$y = \frac{E}{B^2(Q_e/m_e)d} \left[B\frac{Q_e}{m_e}t - \sin B\frac{Q_e}{m_e}t \right]. \qquad \blacktriangleright[129]$$

These equations for the components of motion parallel to the x and y axes may be recognized as being those for a cycloid. Such a path for the electron is shown in Fig. 17. The x co-ordinate of the electron always lies between zero and

$$\frac{2E}{B^2(Q_e/m_e)d}.$$

The y co-ordinate, however, increases continuously with time. The average velocity along the y axis is E/Bd. The distance along the y axis between the points for which x is zero is

$$\frac{2\pi E}{B^2(Q_e/m_e)d}.$$

If d is greater than the maximum x co-ordinate,

$$\frac{2E}{B^2(Q_e/m_e)d},$$

an electron cannot reach the positive plate. Therefore, for a fixed separation, a magnetic flux density greater than the critical magnetic flux density,

$$B_0 = \frac{1}{d} \sqrt{2Em_e/Q_e},$$ [130]

is sufficient to interrupt the electronic current between the electrodes.

The relation given by Eq. 130 was used to measure[35] the ratio of charge to mass of the particles emitted by the photoelectric effect, which were thus identified as electrons.

Because uniform fields are seldom established with the electrode configuration of limited size employed in practical electronic devices, the expressions derived above are useful only as a guide to the conditions for critical control or production of the desired paths, such as those in the electron multiplier. In the electron multiplier the electric field is nonuniform, and the arrangement of the electrodes produces longitudinal as well as transverse components of varying magnitude with respect to position in the tube. The path of the electrons in that device therefore only approximately resembles a cycloid.

Most present-day electron multipliers do not require an auxiliary magnetic field to produce the desired path of the electrons.[36] The electrodes are shaped so that the electrostatic field alone causes the electrons to impinge upon the electrodes in succession. Figure 28 of Ch. II shows one configuration of electrodes used. Discussion of the operating characteristics of multiplier phototubes is given in Art. 13, Ch. II.

8d. *Magnetron.* Another practical application of the motion of electrons in perpendicular electrostatic and magnetostatic fields occurs in an electron tube called the *magnetron*.[37] Its chief application is as an oscillator to generate high radio-frequency power for such purposes as microwave radar.[38] In its elementary form the magnetron consists of a

[35] J. J. Thomson, "On the Masses of Ions in Gases at Low Pressures," *Phil. Mag.*, 48 (1899), 547–567.

[36] V. K. Zworykin and J. A. Rajchman, "The Electrostatic Electron Multiplier," *I.R.E. Proc.*, 27 (1939), 558–566; C. C. Larson and H. Salinger, "Photocell Multiplier Tubes," *R.S.I.*, 11 (1940), 226–229; J. S. Allen, "Recent Applications of Electron Multiplier Tubes," *I.R.E. Proc.*, 38 (1950), 346–358.

[37] For early work along these lines, see: H. Gerdien, United States Patent, 1,004,012 (September 26, 1911); A. W. Hull, "The Effect of a Uniform Magnetic Field on the Motion of Electrons between Coaxial Cylinders," *Phys. Rev.*, 18 (1921), 31–57; E. G. Linder, "Description and Characteristics of the End-Plate Magnetron," *I.R.E. Proc.*, 29 (1936), 633–653.

[38] G. B. Collins, Editor, *Microwave Magnetrons*, Massachusetts Institute of Technology Radiation Laboratory Series, Vol. 6 (New York: McGraw-Hill Book Company, Inc., 1948); K. R. Spangenberg, *Vacuum Tubes* (New York: McGraw-Hill Book Company, Inc., 1948).

cylindrical anode and a coaxial cylindrical cathode inside an evacuated chamber as shown in Fig. 18a. The electrons emitted by the cathode are attracted to the anode by the radial electric field set up by an external source of voltage connected between the anode and cathode. If a sufficiently strong longitudinal magnetic field is impressed parallel to the axis, the paths of the electrons are curved, and they can be prevented from reaching the anode. The magnetic field is therefore a means

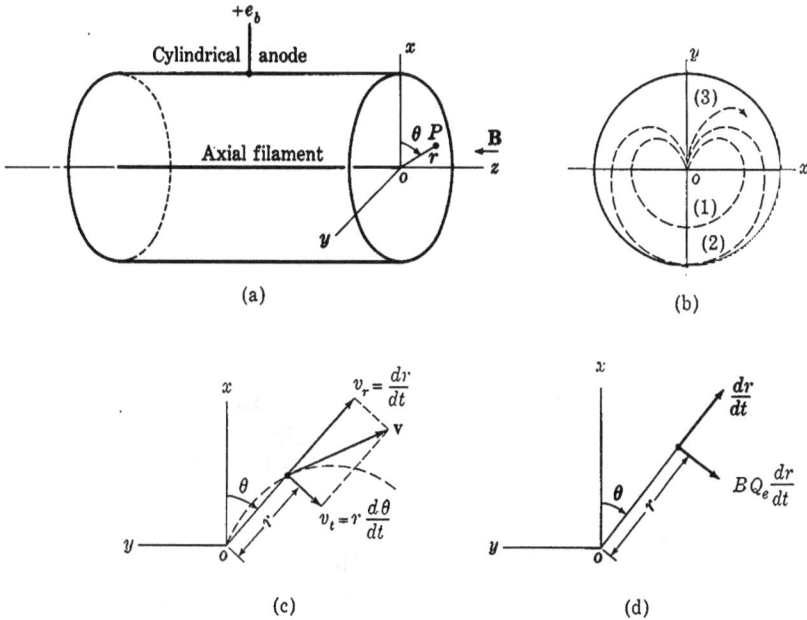

Fig. 18.   Electron trajectories in a magnetron.

of current control. Furthermore, under certain conditions mentioned subsequently, the tube may exhibit negative resistance, and the magnetron may then be employed in conjunction with a resonant circuit to generate oscillations of a very high frequency.

Electrons that leave the cathode of the magnetron follow curved trajectories such as paths 1, 2, or 3 in Fig. 18b for decreasing strengths of the magnetic field. The trajectory 2 corresponds to the condition that the electrons graze the anode. For magnetic flux densities greater than the critical value that corresponds to trajectory 2, the electrons miss the anode and are turned back to the filament; for magnetic flux densities smaller than this critical value, all the electrons reach the anode. A determination of this critical value of magnetic flux density, denoted by B_0, is therefore of interest.

In the derivation of the relation involving the plate voltage e_b, the critical magnetic flux density B_0, and the dimensions of the tube, it is necessary to make certain simplifying assumptions in order to obtain a solution with reasonable effort. The tube has cylindrical symmetry, and cylindrical co-ordinates can be used to advantage if the fringing of the electric field at the ends of the cylinder is considered negligible. Accordingly, it is assumed that (a) the effect of the fringing of the electric field is negligible, (b) the electrons are emitted from the cathode with zero initial velocity, (c) the radius of the cathode is small compared with the radius of the plate, and (d) the voltage drop along the cathode is small compared with the plate voltage e_b. The relations derived on the basis of these assumptions are found to be in good agreement with the results of experimental measurement of the characteristics of such tubes.

To utilize the cylindrical symmetry, it is convenient to express the position of an electron between the filament and anode, say at point P in Fig. 18a, in terms of the radius r and the angle θ. The solution is independent of z, since the fields are oriented so that the electron experiences no component of force along the z axis. The radius of the plate is denoted by r_p, and that of the cathode by r_k, where r_k is small compared with r_p. A solution for the trajectory might be obtained by employment of Eq. 108 in cylindrical–co-ordinate form and substitution of boundary conditions. However, in the following analysis the relations are built up as they are needed from fundamental physical principles.

For the critical magnetic flux density, the electron grazes the plate, and the radial component of its velocity dr/dt is therefore zero when r is equal to r_p. In order to utilize this important boundary condition, the first objective of this analysis is to obtain an expression involving dr/dt; and the second objective is to evaluate each term of this expression for the point on the grazing trajectory at which it touches the plate. A critical relation involving the plate voltage e_b, the critical flux density B_0, and the dimensions of the tube for the grazing condition is thereby obtained.

The velocity of the electron at any point may be resolved into a radial component

$$v_r = \frac{dr}{dt} \tag{131}$$

and a tangential component

$$v_t = r\frac{d\theta}{dt}, \tag{132}$$

as is shown in Fig. 18c, and the square of the speed of the electron is then

$$v^2 = \left(\frac{dr}{dt}\right)^2 + \left(r\frac{d\theta}{dt}\right)^2. \qquad [133]$$

Each term of Eq. 133 may be evaluated for the condition of an electron grazing the plate. Because of the boundary condition already stated,

$$\frac{dr}{dt} = 0 \qquad [134]$$

for a grazing electron at the plate, and, from Eq. 30, if the initial emission velocity of the electron is neglected,

$$v^2 = 2\frac{Q_e}{m_e}e_b \qquad [135]$$

when the electron reaches the plate. Evaluation of the third term of Eq. 133 is then the only remaining step.

This third term may be determined through equating the torque tending to rotate the electron about the axis of symmetry of the tube to the rate of change of the angular momentum of the electron about the same center. The torque is the result of the radial velocity component dr/dt and the magnetic flux density B producing the tangential force $BQ_e(dr/dt)$ and hence the torque $rBQ_e(dr/dt)$. The angular momentum is the product of the moment of inertia $m_e r^2$ and the angular velocity $d\theta/dt$, and therefore

$$\frac{d}{dt}\left(m_e r^2 \frac{d\theta}{dt}\right) = rBQ_e\frac{dr}{dt}, \qquad [136]$$

which, upon integration, gives

$$m_e r^2 \frac{d\theta}{dt} = \frac{1}{2}r^2 BQ_e + C. \qquad [137]$$

The constant of integration C may be evaluated from the initial conditions. When

$$t = 0, \, r = r_k \text{ and } \frac{d\theta}{dt} = 0; \qquad [138]$$

thus

$$C = -\frac{1}{2}r_k{}^2 BQ_e, \qquad [139]$$

and

$$m_e r^2 \frac{d\theta}{dt} = \frac{1}{2}BQ_e(r^2 - r_k{}^2). \qquad [140]$$

For an electron to graze the plate, B must have its critical value B_0 and r must be r_p. Since r_k is assumed to be much smaller than

r_p, r_k^2 may be neglected in comparison with r_p^2. Thus, from Eq. 140 is obtained

$$r\frac{d\theta}{dt} = \frac{1}{2}B_0\frac{Q_e}{m_e}r_p \qquad [141]$$

as the value to be used in the third term of Eq. 133 when the electron just grazes the plate.

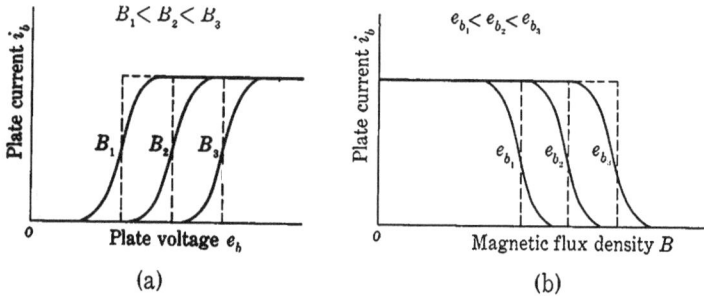

Fig. 19.   Characteristic curves of a magnetron.

The values from Eqs. 134, 135, and 141 may be substituted into Eq. 133 to obtain the relation

$$2\frac{Q_e}{m_e}e_b = \frac{1}{4}B_0^2\left(\frac{Q_e}{m_e}\right)^2 r_p^2 \qquad [142]$$

or

$$B_0 = \frac{1}{r_p}\sqrt{8\frac{m_e}{Q_e}e_b}\ . \qquad [143]$$

Substitution of the numerical value of $\frac{Q_e}{m_e}$ from Eq. 3 yields, in the mks system,

$$B_0 = 6.74 \times 10^{-6}\frac{\sqrt{e_b}}{r_p}\ \text{webers per square meter.} \qquad \blacktriangleright[144]$$

Equations 143 and 144 give the critical value of the magnetic flux density at which the anode current is interrupted as the flux density is increased. They show that the current in a tube with a large anode radius and small plate voltage may be interrupted by a smaller magnetic flux density than for the opposite conditions. Fig. 19 presents the general form of the characteristic curves of a cylindrical-anode magnetron. The dotted lines indicate the idealized characteristics based on Eq. 143; the solid lines the experimental ones. The difference between the idealized and the experimental curves results from the influence of

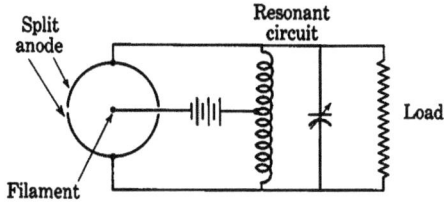

Fig. 20. Magnetron oscillator circuit.

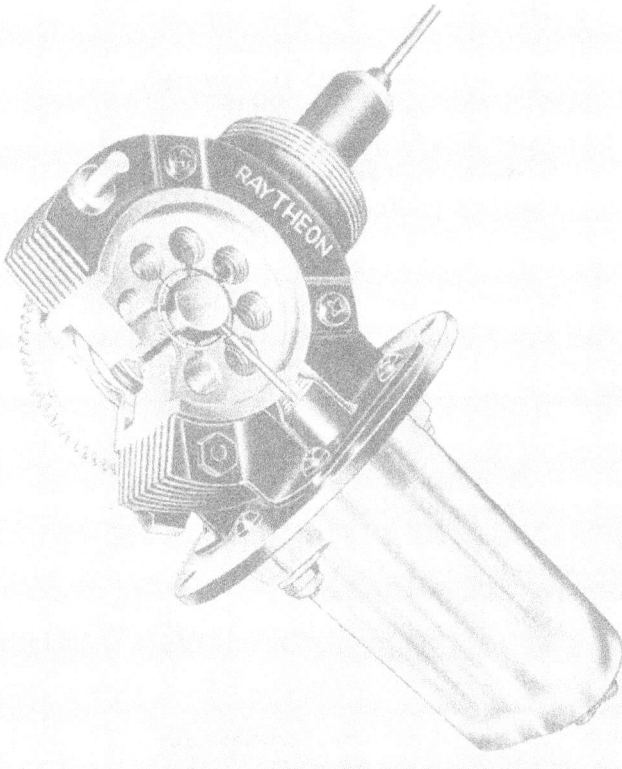

Fig. 21. Cutaway view showing internal construction of a magnetron capable of generating 50 kilowatts of peak power at 3200 megacycles per second under pulsed conditions. (*Courtesy Raytheon Manufacturing Company.*)

such causes as the initial velocities of the electrons at the cathode, the fringing of the electric field, the voltage drop along the cathode, and deviations of the cathode from true axial alignment.

An early form of the magnetron for use as an oscillator to generate high-frequency power involves separation of the anode into two segments[39] as is shown in Fig. 20. The two semicylinders are connected to the opposite ends of a resonant circuit with a return path through a direct-voltage source to the cathode. Frequencies of the order of 300 megacycles per second and higher can be generated in this way with relatively high efficiency.

Placing the resonant circuit inside the tube,[40] and increasing the number of pairs of anode segments, have made possible production of frequencies as high as 30,000 megacycles per second and instantaneous power outputs of several million watts. The internal structure of a typical magnetron for high-frequency generation is shown in Fig. 21. Each slot and associated hole forms a resonant cavity that is excited to oscillation by the passage of electrons across its mouth.

At the high frequencies produced by magnetron oscillators the electric field changes in magnitude during the passage of an electron from the cathode to the anode. Hence the conditions are no longer those of static fields, and the analysis of the behavior of the tubes requires that the time variation of the electric field be taken into account. Other tubes in which this time variation is important are discussed in Ch. IV.

PROBLEMS

1. What is the speed of an electron in miles per second after it has moved through a potential difference of 10 volts?

2. An electron and another particle, starting from rest, fall through a potential difference E in an electrostatic field. The particle has the same charge as the electron and 400 times the mass of the electron.

   (a) Is the time of flight of the particles affected by the distribution of potential which they encounter on their journey?
   (b) For a given potential distribution, do the times of flight of the two particles differ? If so, by how much?
   (c) What is the kinetic energy of each of the particles at the end of the journey?

3. Assume the parallel plane plates of Fig. 2 to be large and spaced 1 cm apart in vacuum. Instead of the battery shown, an impulse generator with polarity as

[39] G. R. Kilgore, "Magnetron Oscillators for the Generation of Frequencies between 300 and 600 Megacycles," *I.R.E. Proc.*, 24 (1936), 1140–1157; L. Rosen, "Characteristics of a Split-Anode Magnetron Oscillator as a Function of Slot Angle," *R.S.I.*, 9 (1938), 372–373.

[40] E. G. Linder, "The Anode-Tank-Circuit Magnetron," *I.R.E. Proc.*, 27 (1939), 732–738; J. B. Fisk, H. D. Hagstram, and P. L. Hartman, "The Magnetron as a Generator of Centimeter Waves," *B.S.T.J.*, 25 (1946), 1–188.

indicated supplies a potential difference which increases at a uniform rate from zero to one volt in 10^{-8} sec. At the end of this time the plates are short-circuited.

(a) If an electron is at rest very near the negative plate at the beginning of the voltage pulse, where is it at the time of short circuit?

(b) What is the total electron transit time to the positive plate?

4. Two large plane plates separated by a distance of 1 cm in a high vacuum are arranged as shown in Fig. 2. Instead of the battery, an alternating square wave of voltage having an amplitude of 1 volt and a period of 2×10^{-8} sec is impressed between the plates.

If an electron is at rest near the negative plate at the beginning of the first cycle, where will it be at the end of that cycle?

5. Two large plates separated by a distance d in a high vacuum are arranged as shown in Fig. 2. Instead of the battery a voltage source of instantaneous value

$$e_b = E_m \cos{(\omega t + \psi)}$$

is applied between the plates. The electric field between the plates is essentially uniform at any instant of time. Electrons are emitted from the left-hand plate and may be assumed to have zero initial velocity.

(a) Derive a general expression for the position of an emitted electron with respect to the left-hand plate as a function of time.

(b) Is it possible that some of the emitted electrons may never reach the right-hand plate? Either show that this condition cannot occur, or derive the conditions for which it can occur.

6. An electron is initially at rest at the surface of the left-hand plate of the parallel-plate configuration shown in Fig. 2. Instead of the battery a sinusoidal voltage source having a 1-volt peak value and a frequency of 50 megacycles per second is applied to the plates. The voltage is passing through its zero value at the instant of application and is increasing in the direction which makes the right-hand plate positive. The plates are infinite in extent and are placed 1 cm apart in vacuum.

(a) What is the maximum speed attained by the electron, and at what positions does this speed occur?

(b) What is the speed of the electron at time t equals 2×10^{-8} sec?

(c) Describe completely the motion of the electron.

7. The screen of an electron oscilloscope is 20 cm from the anode. Two electrons pass through the aperture in the anode traveling in parallel paths 0.1 mm apart.

(a) The electrons are 0.11 mm apart when they reach the screen. What is their transit time from anode to screen?

(b) If it is assumed that the electrons started from rest at the cathode, what is the accelerating voltage of the anode?

(c) Determine the anode accelerating voltage corresponding to an electron separation of 0.101 mm at the screen.

8. The voltages E_1 and E in the diagram of Fig. 22 each have a magnitude of 100 volts. An electron starts from rest at the left-hand electrode, passes through the hole in the center diaphragm, and finally reaches the anode on the right. The left-hand and right-hand electrodes are each 1 cm from the centrally located diaphragm.

(a) With what energy in electron volts does the electron strike the electrode at the right?

(b) On the assumption that the electrodes are infinite parallel plates, what is the time of transit of the electron between the left-hand electrode and the diaphragm?

(c) On the same assumption, what is the time of transit of the electron between the diaphragm and the right-hand electrode?

(d) If the voltage E is changed to 200 volts, what then is the answer to (c)?

9. An electron is moving at a speed of 10^8 cm per sec. Through how many volts of potential difference must it pass to double its speed?

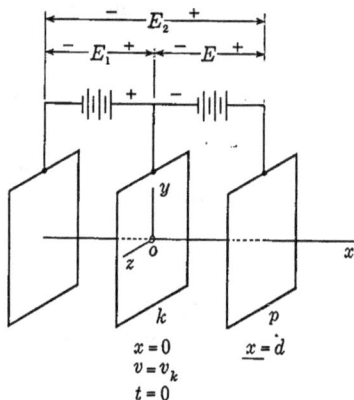

Fig. 22. Electrode configuration for Prob. 8.

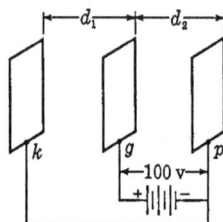

Fig. 23. Electrode configuration for idealized Barkhausen oscillator of Prob. 11.

10. A cathode-ray oscilloscope is built with a fluorescent screen deposited on a metallic conducting plate. A lead is available from the conducting screen. Three different connections are tried as follows:

(1) The screen is left without connection, that is, insulated from the anode.

(2) The screen is connected through a high resistance to the anode.

(3) The screen is connected directly through a low-resistance wire to the anode.

If secondary emission from the screen is neglected:

(a) How do the velocities of the electrons as they reach the screen differ in the three cases?

(b) Which of the connections should be employed in practical use of the tube?

11. In a Barkhausen oscillator, the grid of a triode is maintained at a positive potential, and the plate is maintained at essentially the cathode potential. The transit time of the electrons in such an electrode configuration is to be found.

In order to obtain an approximate solution, consider that the vacuum triode has parallel plane electrodes as shown in the diagram, Fig. 23, with interelectrode spacings d_1 and d_2, each equal to 1 cm. The plate p and cathode k are connected together, and the grid g is maintained at a positive potential of 100 volts with respect to them. The space charge is negligible, so that the potential distribution is an inverted $\mathbf{V}$ with its apex at the grid. An electron leaves the cathode with zero velocity. It may be assumed that the electron passes freely through the grid structure.

(a) What time elapses before the electron reaches the anode?

(b) On the assumption that the grid may be moved with respect to the plate and cathode, show what effect the location of the grid has on the transit time of an electron from cathode to anode. Is there a position of the grid that gives a minimum transit time? If so, where is it?

12. Find the electrostatic deflection sensitivity of the cathode-ray tube of Fig. 5 in volts per centimeter, if d is 1 cm, l is 4.5 cm, and l_s is 33 cm, for accelerating voltages E_b of 300, 1,000, 5,000, and 10,000 volts.

13. Electrons accelerated by a potential difference of 1,000 volts are shot tangentially into the interspace between two concentric semicylinders of radii 4 cm and 5 cm respectively. What potential difference must be applied between the plates to make the electron paths circular and concentric with the plates?

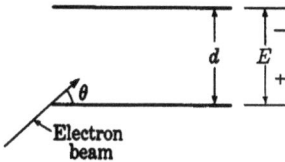

Fig. 24.   Configuration for   Fig. 25.   Diverging plate arrangement
        Prob. 14.   of Prob. 15.

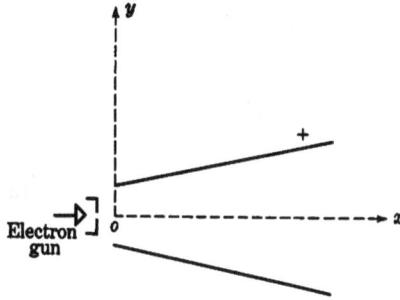

14. Two large capacitor plates are spaced a distance d apart and charged to a potential difference E as shown in Fig. 24. It may be assumed that the field intensity between the plates is uniform and that there is no fringing. A stream of electrons is projected at a velocity corresponding to an accelerating voltage E_0 into the electric field at the edge of the lower plate and at an angle θ with the plates.

(a) Find the relation among the voltages and θ necessary in order that the electrons may just miss the upper plate.

(b) If θ is 30°, d is 1 cm, and E is 50 volts, what initial kinetic energy should the electrons in the beam have in order that they may just miss the upper plate? Express the result in electron volts and joules.

15. An electron gun shoots an electron horizontally into an electric field established by a pair of diverging plane electrodes, with a velocity corresponding to a kinetic energy of 400 electron volts, as shown in Fig. 25. At any distance x from the origin, the electric field intensity is assumed to be given by the following approximate expression

$$\mathcal{E}_y = \frac{200}{x + 4} \text{ volts per cm,}$$

where x is in centimeters, and the field may be considered to be entirely in the vertical direction and to be directed upward within the space under consideration.

Using the axes indicated in the figure, find the equation of the path of the electron in the form

$$y = f(x).$$

16. The mass of an alpha particle is 7,344 times that of an electron. Its charge is twice that of an electron. If its velocity corresponds to a fall through a potential of 5×10^6 volts, what magnetic field is required to bend its path into a circle of radius 1 meter?

17. Find the magnetic deflection sensitivity of the cathode-ray oscillograph tube of Fig. 11 in centimeters per gauss for accelerating voltages of 100, 1,000, 5,000, and 10,000 volts. The distance l is 5 cm, and l_s is 35 cm.

18. An electron and an ionized hydrogen atom are projected into a uniform magnetic field of 10 gausses. Their velocities correspond to 300 volts, and they enter in a direction normal to the magnetic field.

(a) Find the ratio of the diameters of the circular paths traced by the particles.
(b) Find the time required for each of the particles to make a complete revolution.

19. An electron oscilloscope (see Fig. 5) is located with its axis parallel to and spaced 2 meters from a long conductor carrying direct current. The return conductor for the direct current is very far away, so that the magnetic field from it is negligible. The distance l_s from the anode to the screen is 50 cm, and the accelerating voltage E_b is 1,500 volts.

(a) What current in the conductor will cause a deflection of 1 cm?
(b) Is it possible to reorient the tube so that the deflection is zero for all values of current in the conductor?

20. Electrons are projected into a uniform magnetic field of 100 gausses. The speed of the electrons corresponds to 300 volts, and the beam makes an angle of 30° with the direction of the magnetic field. Find the diameter and pitch of the spiral path described by the electrons.

21. It is desired to find some of the relations among the size of a cyclotron and the voltage, frequency, and magnetic flux density applied to it.

In the cyclotron a proton is whirled in a succession of semicircular paths of increasing radius, and is alternately under the influence of an electric field which imparts linear acceleration and under that of a magnetic field which bends its path of flight. As indicated in the plan view of Fig. 12, the electric field lies in the plane of the paper and is normal to the magnetic field, which is perpendicular to and into the paper. The path of the proton is also in the plane of the paper. The magnetic field may be considered uniform in this plane, and its effect within the slit between the dees may be neglected because of the narrowness of the slit. The electric field, however, may be assumed to exist only in this slit and to be produced by an alternating potential difference

$$e = E \cos \omega t.$$

Owing to the narrowness of the slit, the time of flight of the proton through the electric field is small compared to the time it is acted upon by the magnetic field alone, so that if the proton is assumed to cross the slit at the times t equals 0, π/ω, $2\pi/\omega$, etc., the corresponding values of e are $+E$, $-E$, $+E$, etc.

(a) Derive the analytic relation between the angular frequency ω and the magnetic flux density B which must exist in order that the proton may cross the slit at the specified time intervals.

(b) If the speed of the proton in its final semicircular path is to correspond to 10^6 volts, and the magnetic flux density is 1,000 gausses, what will be the radius of the final path and what must be the angular frequency ω of the voltage e?

(c) Is the direction of travel of the proton counterclockwise as shown or clockwise?

22. Electrons emerge from a small hole in the anode of a cathode-ray tube in all directions contained in a cone of small angle. The accelerating potential difference

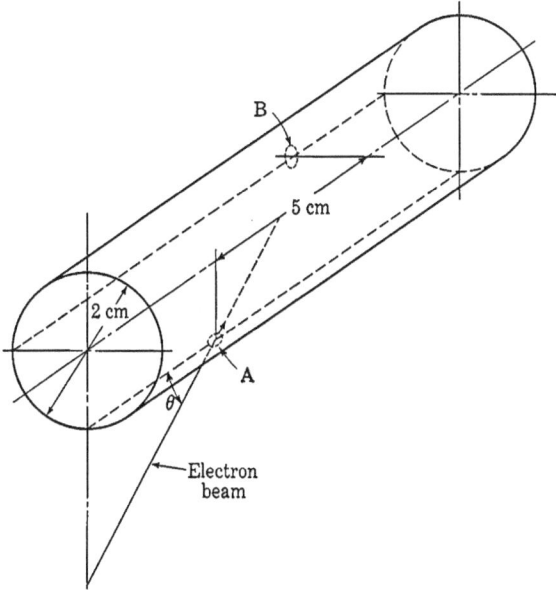

Fig. 26. Hollow cylinder of Prob. 23.

between cathode and anode is 200 volts. The distance from the anode to the screen is 20 cm, and the cathode is 5 mm from the anode.

The tube is placed in a long solenoid of 10 cm diameter having six turns per centimeter along its entire length, the tube and solenoid axes coinciding.

What is the smallest value of current in amperes in the coil that will produce a magnetic effect on the beam of electrons sufficient to cause them to focus on a spot the same size as the hole but at the screen?

23. An electron gun directs a beam of 1,000-volt electrons through a hole A in the wall of a hollow cylinder 2 cm in diameter, as indicated in Fig. 26. The axis of the gun intersects that of the cylinder at an angle θ. The cylinder wall has another hole B, removed 90° circumferentially and 5 cm axially from the hole A.

Determine the angle θ, and the density B and direction of the uniform axial magnetic field required to cause the electron beam to emerge from the hole B.

24. A region in space is permeated by a constant electric field parallel to the x axis and a constant magnetic field parallel to the y axis in a system of rectangular co-ordinates.

If an electron is projected into this space with a finite velocity, what are the necessary and sufficient conditions that must be fulfilled if is it to remain undeflected?

25. An electron leaves the origin with an initial velocity $\mathbf{v_0}$ in the positive direction along the x axis. There are a uniform electric field of intensity $\mathcal{E}$ in the positive x direction and a uniform magnetic field of flux density B in the positive z direction.

   (a) What differential equations determine the motion of the particle in the x–y plane?
   (b) How can these equations be solved?

26. A magnetron contains an axial tungsten filament of 0.03 cm diameter. The plate is a 1-cm diameter right-circular cylinder concentric with the filament, and it is 100 volts positive with respect to the filament. A magnetic flux density of 50 gausses is impressed parallel to the filament. Consider that the potential of all parts of the filament is the same and that the lengths of the filament and plate are equal and are large compared with the diameter of the plate.

   (a) What is the tangent of the angle between the velocity vector and the radius at the point where an electron strikes the plate?
   (b) What is the ratio of the speed with which an electron strikes the plate when the magnetic field is on and the speed with which it strikes when the field is off?

27. Find the approximate path of an electron emitted from the filament of a cylindrical-anode magnetron. The filament radius is r_k and is very small compared to the anode radius r_p. Assume no space charge and assume that the initial velocity of the electron at the filament surface is negligible.

CHAPTER II

Electron Emission from Metals

The dominant concept presented in Chapter I is the relative ease and precision with which the motion of a charged particle in free space may be controlled. To the imaginative mind this concept, together with a recognition of the many and varied trajectories that a particle may be made to follow by appropriate location of electric and magnetic fields, abounds with engineering possibilities. Since a charged particle in motion constitutes an electric current, control of the motion of the particle provides a new and effective means of controlling current. The development of many new and unusual devices thus is possible through different applications of controlling fields to the region traversed by the particles.

A matter that until now has been given only passing mention is the provision of a sufficient supply of charged particles to be controlled. The electron oscilloscope, the electron microscope, and a great number of other electrostatically or electromagnetically controlled devices, to be discussed subsequently, depend for their effectiveness, first, on relatively simple means for controlling the trajectory of the charged particles and, second, on an adequate supply of particles at the electrode where the trajectory begins.

The current in the devices previously mentioned is seldom greater than a few milliamperes, and the electric power involved is usually relatively small. In order that the features of the electrostatic or electromagnetic control may be applied to large amounts of power, a means for liberating charged particles from an electrode at a rapid rate is necessary. The factor often limiting the amount of power that can be controlled is the liberation of charged particles at an adequate rate. Usually the particles of major importance are electrons, and the liberation of them is known as *electron emission*.

1. -STRUCTURE OF SOLIDS

The several processes by which electrons are liberated from metals may be explained in terms of the atomic theory of matter. The essentials of this theory are discussed in this and following articles to assist in fitting into an orderly pattern the great number of diverse experimental data about the electrical conductivity of metals and electron emission from them.

According to present concepts, a metal is regarded as a conglomerate of crystals of various sizes and shapes. The crystals are usually small, though it is possible to grow single ones to dimensions of several inches. Within each crystal the atoms are held by interatomic forces in a regular pattern or space lattice whose dimensions can be measured by means of x-rays.

Each atom, in turn, consists of a nucleus surrounded by a dynamic array of electrons. The nucleus is a complex particle[1] which, though not much larger than an electron even in the heaviest atom, has most of the mass of the atom. The lightest nucleus so far discovered is that of the common form of hydrogen, which consists of a single proton, approximately 1,837 times the mass of an electron and of equal and opposite charge. Comparatively stable nuclei exist throughout the table of natural chemical elements up to the heaviest, uranium, which has an atomic number 92 (indicating that the atom contains 92 electrons) and a nuclear mass approximately 240 times that of a proton, or 441,000 times that of an electron. Heavier nuclei, such as plutonium, are produced in nuclear reactions between uranium nuclei and elementary particles, but they are not known to exist in nature.

The electrons in the atom are grouped in a more or less spherical distribution around the nucleus. The outermost ones, called the *valence electrons*, determine the chemical behavior of the atom, for they can interchange from one atom to another when the atoms are brought close together. In a metal, the atoms are closely spaced in a crystal lattice, and the valence electrons may move from atom to atom. Experiments show that, for every atom in the metal, approximately one electron may move about in this fashion; hence some 10^{22} electrons in every cubic centimeter are free to move. An impressed electric field causes them to move in the direction of the field and constitute an electric current. Because of the large concentration of the free electrons, the drift velocities, or average velocities of the electrons in the direction of the field, do not exceed a few centimeters per second, even when very large electric currents are carried by conductors of small cross section.

2. ELECTRON GAS IN A METAL

The study of the behavior of the free electrons in a metal and the methods by which they may be liberated constitutes a large field of investigation in physics at the present time. Though the theories

---

[1] E. C. Pollard and W. L. Davidson, *Applied Nuclear Physics* (2nd ed.; New York: John Wiley & Sons, Inc., 1951); R. E. Lapp and H. L. Andrews, *Nuclear Radiation Physics* (New York: Prentice-Hall, Inc., 1948).

cannot be given in detail **here, a few of the fundamental concepts** are helpful in understanding the operation of electronic devices.

Consider a single electron in the field of an isolated nucleus. **The** nucleus has a positive charge, and, when the electron is brought near

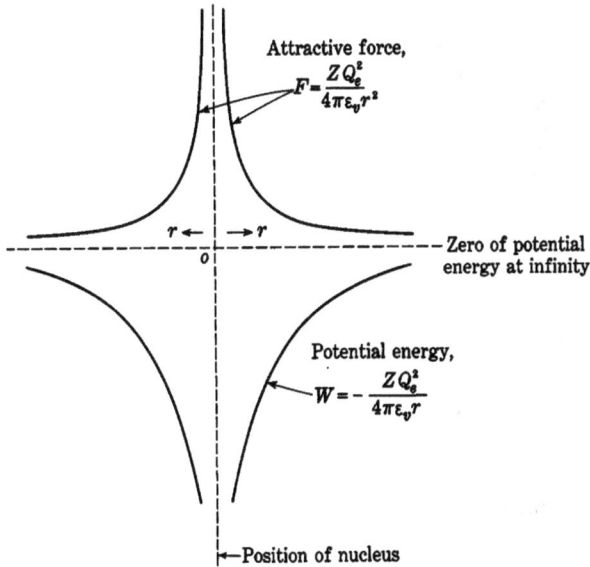

Fig. 1. Force on and potential energy of an electron in the vicinity of an isolated nucleus.

it, an attractive force occurs between the two in accordance with Coulomb's law,

$$F = \frac{ZQ_e \cdot Q_e}{4\pi\varepsilon_v r^2},$$ [1]

where

F is the attractive force,

Q_e is the charge on an electron,

Z is the atomic number of the nucleus or atom, which is unity for hydrogen,

r is the distance between the nucleus and the electron,

ε_v is the dielectric constant of free space.

This force has radial symmetry about the nucleus and can be represented as in Fig. 1. The potential energy of an electron at any point, with respect to an arbitrarily assumed zero of potential energy at infinity, is the work required to move the electron from infinity to that point, or

$$W - \int_{\infty}^{r} \frac{ZQ_e^2}{4\pi\varepsilon_v r^2}\, dr = - \frac{ZQ_e^2}{4\pi\varepsilon_v r}, \qquad [2]$$

where W is the potential energy of the electron at a distance r from the nucleus. This equation also may be plotted as in Fig. 1. The forces near a more complex atom can be represented in a similar manner, but the mathematical formulation is more complicated, because the atom consists not only of a small nucleus but also of a distribution of electrons in the near-by space.

Next consider two similar atoms close together. The force on an electron at any point is the sum of the forces due to the separate

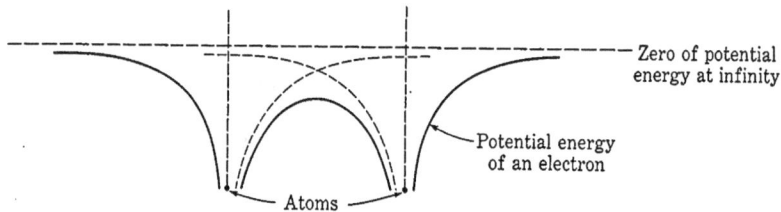

Fig. 2. Potential energy of an electron along a line through two adjacent atoms.

charges. Between the atoms, the forces on an electron tend to cancel. The net force is smaller than before, and for this reason the potential energy along a line through the two atoms is also smaller and can be represented qualitatively as in Fig. 2. The maximum of the potential energy is less than the potential energy at the same distance from an isolated atom.

In a metal, the atoms are spaced closely in a three-dimensional crystal lattice, and the potential energy is determined by the charges of all the electrons and nuclei. A two-dimensional diagram hence cannot show the potential energy of an electron at every point in the crystal; the potential energy along a line through a row of the nuclei of these atoms can, however, be represented as in Fig. 3. Within the metal, the potential-energy distribution is a series of humps between the nuclei, and at the boundary there is a potential-energy rise. If the net charge on the metal is zero, there is no electric field outside the metal, and the potential energy rises to a plateau flat to infinity. In view of this picture it is difficult to define the "boundary" of the metal to an accuracy smaller than the distance between nuclei. Since the forces acting on an electron extend some distance beyond the plane defined by the nuclei of the surface atoms, the boundary is not sharply defined from an electronic point of view.

The electrons in an atom have both kinetic and potential energy, and they tend to revert to the position of lowest potential energy, which is close to the nucleus. Atomic theories predict, however, and experiments confirm that only a restricted number of electrons can have any particular value of possible total energy. The atom may therefore be described as having a number of *energy levels* which electrons can occupy. All the energy levels have certain features governing the maximum number of electrons that can be accommodated in them. The lower the energy of the level, the more tightly bound is the electron. The lowest energy level accommodates only two electrons, which are considered to be closest to the nucleus. These two electrons

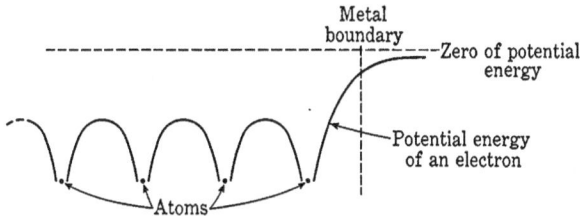

Fig. 3.   Potential energy of an electron along a line through a row of atoms in a metal.

form a *shell*, which is often described as the innermost or K shell of an atom. The next higher group of energy levels, usually known as the L shell, accommodates eight electrons, and so on.[2] In the normal state of the atom, the electrons always fill the energy levels in the order of their energy value until all the electrons are accounted for. This rule requires that some levels in one shell be filled before higher energy levels in shells interior to it. The outermost shell is often not completely filled; the extent to which it is determines the chemical valence ascribed to the atom. The electrons in this outermost shell are the *valence electrons*.

Normally the electrons in an atom cannot leave it—they do not have enough energy to surmount the potential-energy barrier shown in Fig. 1. In a metal, however, the potential-energy barriers between the atoms are lowered, owing to their close proximity; and the energy of some electrons is sufficient to enable them to move beyond the point where they are attracted to what might be called their "parent" nucleus. Some of these valence electrons therefore move very freely through the lattice in such a way as always to maintain the time

[2] W. G. Dow, *Fundamentals of Engineering Electronics* (2nd ed.; New York: John Wiley & Sons, Inc., 1952), 564–567.

average of the total charge within a region equal to zero. These are the *free electrons*; those that lie in energy levels well below the top of the humps and cannot leave their parent atoms are called the *bound electrons*. A superposed electric field caused by a potential difference applied between two parts of the metal can cause the free electrons to drift through the metal and constitute an electric current.

The very numerous free electrons move about in the metal with a random, chaotic motion, exchanging energy with each other and with the atoms in the lattice, which vibrate around their average positions. The random velocities of the free electrons have a wide range of values and on the average are very much larger than any velocities that may be imparted to the electrons by means of an external field. In fact, the *random current density*, or current crossing a unit surface in the metal because of random electron velocities, found by counting the electrons crossing in one direction and neglecting the equal number crossing in the reverse direction, is of the order of 10^{13} amperes per square centimeter. The exact value depends upon the metal and, to a slight degree, upon its temperature. The drift velocities acquired by the electrons because of an external electric field constitute only a very minor modification of their random velocities.

Just as the electrons within a single atom must have energies in certain energy levels, and cannot have intermediate values of energy, the energies that may be attained by an electron within a metal are limited. As a group of atoms is brought together to form a metal crystal, the energy levels of the individual atoms expand into bands of allowed energies separated by forbidden energy intervals.[3] The lower energy levels are only slightly affected and become very narrow bands separated by wide intervals. The bands nearer the tops of the humps in the potential-energy curve are wider; the one in which the energies of the valence electrons lie extends from a little below the tops of the potential-energy humps to a level several electron volts above them. Allowed bands of still higher energy exist, but ordinarily only a negligible number of electrons have sufficient energy to occupy levels in them. Only a limited number of electrons may have energies lying within each allowed energy band. The band is said to be *filled*, or all the energy levels in it are said to be occupied, if as many electrons as possible have energies within it.

The random motions of the valence electrons from atom to atom are possible because the band in which their energies lie is in part

[3] N. F. Mott and H. Jones, *The Theory of the Properties of Metals and Alloys* (Oxford: Oxford University Press, 1936); F. Seitz, *The Modern Theory of Solids* (New York: McGraw-Hill Book Company, Inc., 1940); W. Shockley, *Electrons and Holes in Semiconductors* (New York: D. Van Nostrand Company, Inc., 1950).

above the humps in the potential-energy curve. Since in a *metal* this band is not filled, an external field applied to the metal is capable of causing a shift in the energy distribution among the electrons in such a way that the electrons drift through the metal in the direction of the field; thus the electrical conductivity of a metal is obtained.

In a crystal of an *insulator*, the band of energy levels corresponding to the valence electrons is completely filled at a temperature of absolute zero, and a relatively wide gap of forbidden energy values exists above it. The net current associated with electron motion is zero despite any externally applied electric field because in a filled band each electron moving in one direction is accompanied by another moving in the opposite direction and with exactly the same energy. As the temperature of the insulator is increased from absolute zero, the thermal energy tends to induce some of the valence electrons to jump the gap of forbidden energy levels and occupy allowed unfilled levels above. Under the influence of an electric field these electrons could drift and constitute a conduction current. But the conductivity of such an insulator is small because the forbidden region is so wide that with practical temperatures only a negligible fraction of the thermally excited electrons can escape across it.

A *semiconductor* is a material with an energy-level distribution similar to that of an insulator, but having either a relatively narrow forbidden energy region or additional allowed energy levels lying in the otherwise forbidden region. These additional levels are associated with the presence of impurities or with imperfections in the crystal lattice of a pure substance. Semiconductors generally fall into one of three classes. In an *intrinsic* semiconductor, the forbidden energy region is so narrow that appreciable numbers of thermally excited electrons can jump across it. Electric conduction through drift of these electrons can then take place if an electric field is applied. Furthermore, the vacancies, left in the formerly filled band, called *holes*, can also contribute to the conduction by drifting in an electric field just as though they were electrons with a positive instead of a negative charge. In an *n-type extrinsic* semiconductor, also called a *donor* type, impurity or imperfection levels which are filled at a temperature of absolute zero lie near the top of the otherwise forbidden region. Upon elevation of the temperature, thermal excitation of electrons from these levels into the unfilled band above can cause appreciable conductivity. Finally, in a *p-type extrinsic* semiconductor, also called an *acceptor* type, impurity or imperfection levels which are unoccupied at a temperature of absolute zero lie near the bottom of the otherwise forbidden region. Electrons thermally excited from the filled band into these unoccupied levels leave vacancies or holes which can drift in an electric field and constitute a current. A distinguishing feature of the p-type and n-type semiconductors is that the sign of the Hall effect for one is opposite to that for the other. The Hall effect is the development of a potential gradient across a current-carrying conductor situated in a magnetic field. The direction of the potential gradient is normal to the directions of both the current and the component of magnetic field perpendicular to the current. Most semiconductors have a large and negative temperature coefficient of resistivity.

The picture of the behavior of electrons in a metal presented so far is entirely qualitative. For quantitative results, statistical methods must be used. In the study of the behavior of a single planet in motion about the sun, the positions of both bodies can be fairly closely specified at any instant in terms of a system of co-ordinates, and other properties such as velocity and momentum can be specified in terms of the co-ordinates and time. But when a problem involves a large number of particles, such as the electrons in a metal, and the experimental measurements concern the average behavior of very large numbers of them, an analytical method is appropriate which leads to average values of the properties of the particles and does not even consider exact positions and velocities as functions of time. Statistical methods are particularly well adapted to the calculation of such average values of the properties of great numbers of elementary particles. Statistical theories that utilize such methods are characterized by distribution functions. These functions make possible the calculation of various distributions—the distribution of particles in space, of velocities among the particles, of energy among the particles, and of other quantities.

An accurate determination of such distributions for electrons moving in a metal would be exceedingly difficult because of the complications introduced by the nonuniform potential energy of the electron within the metal, and because of the restrictions upon the values of energy that an electron may possess. An approximation that yields very useful results may, however, be obtained through calculating instead the distribution of kinetic energies among the electrons moving in a space of constant potential. This approximate distribution has been obtained by Fermi and Dirac; it takes into account the restrictions upon energies of the electrons, as prescribed by the Pauli exclusion principle, for electrons in a field-free space. The distribution is

$$N(W)\,dW = \frac{8\pi\sqrt{2}\,m_e^{3/2}}{h^3}\frac{\sqrt{W}}{1+\epsilon^{\frac{W-W_i}{kT}}}\,dW, \qquad [3]$$

where

W is the kinetic energy of an electron,

W_i is the so-called *Fermi level* of energy,

$N(W)$ is the number of electrons per unit volume per unit range of energy,

T is the absolute temperature of the electrons,

k is the Boltzmann constant,[4] $(1.38047 \pm 0.00026) \times 10^{-23}$ joule per degree Kelvin,

h is the Planck constant,

m_e is the mass of an electron,

ϵ is the Naperian base.

At T equals zero, W_i is given by

$$W_i = \frac{h^2}{8m_e} \left(\frac{3n}{\pi}\right)^{2/3}, \qquad [4]$$

where n is the concentration, or number per unit volume, of the electrons. If n is expressed as the number of electrons per cubic meter,

$$W_i = 3.64 \times 10^{-19} n^{2/3} \text{ electron volts.} \qquad [5]$$

The Fermi level W_i varies with temperature, but the variation is small, and the value in Eq. 5 holds to a high degree of accuracy for all temperatures below the melting point of metals. In Eqs. 3 and 4 h, m_e, and k are all universal constants that may be determined by appropriate experiments. For example, spectroscopic experiments serve well for a determination of h, and k can be determined from the study of ideal gases. The concentration of electrons n, however, is a specific property of the particular metal and varies from metal to metal. It may be determined by experiments on the diffraction of electrons from crystals, but the results have only a fair degree of precision.

It is well to examine the meaning of the kinetic-energy distribution. The quantity $N(W)\, dW$ signifies the number of electrons in a unit volume that have kinetic energies between the values W and $W + dW$. Thus the distribution function $N(W)$ is the number of electrons per unit volume per unit range of energy. To give the number that lie in a small range of kinetic energies ΔW, $N(W)$ at the middle of that range may be multiplied by ΔW. If the range ΔW is not small, the result must be found by integration of $N(W)$ between the limits of the range ΔW.

---

[4] R. T. Birge, "A New Table of Values of the General Physical Constants," *Rev. Mod. Phys.*, *13* (October, 1941), 233–239.

Although this energy distribution is a relatively complicated expression, certain facts regarding it when T is zero may be deduced rather simply. Inspection of Eq. 3 shows that, when T is zero, the exponential term is infinite for W greater than W_i, and zero for W less than W_i. Also, when W is less than W_i, all quantities in the expression except the term $\sqrt{W}$ in the numerator are constant as W varies. A plot of the distribution function is therefore parabolic, as in Fig. 4, which also shows the shape of the curve for a typical temperature, 1,500 degrees Kelvin. The number of electrons per unit volume that have kinetic energies between any pair of values of W is the integral of the function, or the area under the curve, between those values. The area under the whole curve is necessarily equal to n, the total number of electrons per unit volume.

Thus, according to the Fermi-Dirac statistics, electrons moving in a field-free space have kinetic energies because of the velocities associated with their temperature, and the distribution of energy among the electrons depends upon their temperature. Even at the absolute zero of temperature, the energy of the electrons is not zero but is distributed according to the curve of Fig. 4, and the maximum kinetic energy possessed by any electron is then W_i. At higher temperatures, more heat energy is stored in the electrons, and their average kinetic energy is increased. Some of the electrons attain energies greater than W_i, as is indicated by the second curve in Fig. 4.

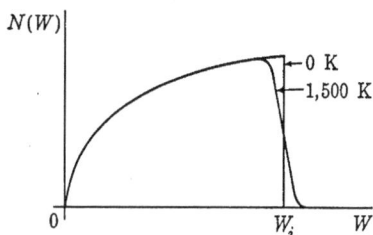

Fig. 4. Distribution function for the kinetic energies of electrons in a space of uniform potential.

The Fermi-Dirac distribution has been applied by Sommerfeld and others to the distribution of kinetic energy among electrons in a metal. The result must be regarded as only an approximation to the actual conditions, since it is derived on the assumption of a space of constant potential, whereas the potential in a metal is known to vary periodically in space because of the lattice structure of the crystal. Though the Fermi-Dirac distribution does not take into account the alternate bands of allowed and forbidden energies existing in a metal, the energy distribution of valence electrons, whose energies lie in a band extending above the potential-energy humps, is very similar to that of electrons in the field-free space assumed by the Fermi-Dirac distribution. Electrons that have energies well above the tops of the humps are particularly little affected in their motions as they pass over the peaks and valleys of potential energy. The Fermi-Dirac theory hence makes possible a good approximation to the kinetic-energy distribution of those electrons in a metal that occupy the range of energies well above the tops of the potential-energy humps. It is a poor approximation in the range of energies near the tops of these humps, and it is entirely inapplicable in lower energy ranges, where the motions of the electrons are very nearly what they would be in isolated atoms. Since, fortunately, the phenomena of interest in electron emission depend upon the energy distribution in the upper ranges, the Fermi-Dirac distribution is useful in studying them.

3. WORK FUNCTION; ELECTRON ESCAPE FROM A METAL

The fact that at normal room temperatures a metal does not lose electrons in appreciable quantities is evident, because a metal does not become positively charged when standing idle and well insulated. According to the theory of behavior of free electrons, the reason for the failure of the electrons to escape is the potential-energy "hill" or "barrier" at the boundary of the metal as shown in Fig. 3. Practically none of the electrons have total energies sufficient to enable them to surmount this barrier by changing their kinetic energy to potential energy. The barrier occurs because of the lack of symmetry in the arrangement of the charges about the positively charged nuclei at the boundary of the metal. An electron deep in the metal is surrounded by a symmetrical array of positive charges and, on the average, experiences no net force in any direction. On the other hand, an electron that starts through the boundary of the metal finds behind it positive charges pulling it back but none pulling it onward.

The way the potential-energy barrier holds the free electrons in the metal despite the fact that they move in all directions with considerable energy may be compared to the behavior of elastic balls placed in a dish and agitated with a stirrer at the center. In the numerous collisions that occur, some of the balls acquire considerable kinetic energy and start up the sloping sides of the dish. However, they are held in unless their kinetic energy is greater than the change in potential energy represented by the climb up the side.

In the metal, the number of electrons capable of passing over the potential-energy barrier is indicated by the distribution of energies associated with the component of velocity directed perpendicular to the boundary of the metal. If a rectangular co-ordinate system is set up with the x axis perpendicular to the metallic surface, an electron has values of kinetic energy W_x, W_y, and W_z associated with its three components of velocity along the three axes. For brevity, W_x is called the x-associated kinetic energy of an electron, and W_y and W_z are similarly named. For an electron to have a total kinetic energy greater than the potential-energy barrier is not enough; W_x, the kinetic energy associated with its component of velocity perpendicular to the barrier, must be greater than the potential energy of the barrier to make escape possible.

From the kinetic-energy distribution, Eq. 3, the following expression can be derived for the number of electrons $N_x(W_x)\,dW_x$ that arrive per unit time at a unit surface in the y–z plane with x-associated energy in the range from W_x to $W_x + dW_x$:

$$N_x(W_x)\,dW_x = \frac{4\pi m_e kT}{h^3} \ln\left(\epsilon^{\frac{W_i - W_x}{kT}} + 1\right) dW_x. \qquad [6]$$

The distribution function $N_x(W_x)$ is sketched in Fig. 5 for temperatures of zero and 1,500 degrees Kelvin.

For further discussion of the liberation of electrons from metals, the concept of the space distribution of the potential energy of an electron in a metal, or near the surface of a metal, may be combined with the concept of the x-associated kinetic-energy distribution among the electrons within the metal. The combination is made graphically in Fig. 6, where the curves of Fig. 3 and Fig. 5 are combined. To make this combination, a common reference level of energy in the two concepts must be found. In this connection it should be observed that at absolute zero all the energy levels in the band of the valence electrons below a certain level are occupied, and all the higher levels in this band are unoccupied. This situation corresponds to the Fermi-Dirac distribution at absolute zero, where electrons possess all allowable values of kinetic energy below W_i but none above. Accordingly, the diagrams of Fig. 3 and Fig. 5 are aligned so that, at a temperature of absolute zero, the W_i level of the Fermi-Dirac distribution coincides with the highest occupied level in the band of allowed energies occupied by the valence electrons. This alignment specifies the correct distribution of energies in the range well above the tops of the humps—the only range of energies in which the Fermi-Dirac distribution is a good approximation.

The energy difference between the zero level of kinetic energy and the potential-energy plateau outside the metal is called W_a, and the band of energies between this zero and the Fermi energy level W_i is

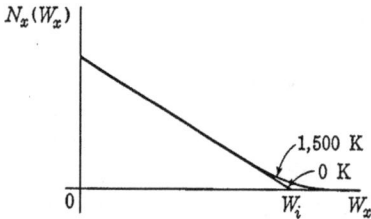

Fig. 5. Distribution function for the x-associated energy among the electrons that arrive at unit area of a boundary in unit time.

known as the *Fermi band*. The exact location of the kinetic-energy zero level is indefinite. It cannot be determined directly, since it lies in the poor-approximation range of the Fermi-Dirac distribution. Hence no specific potential in the metal can be said to correspond to the constant potential of the field-free space. Moreover, the kinetic-energy zero level cannot be located definitely by reference to W_i, since the height of W_i above the zero level depends upon the effective concentration of the electrons, and this concentration is known only very approximately. In Fig. 6 the zero level is placed a little below the

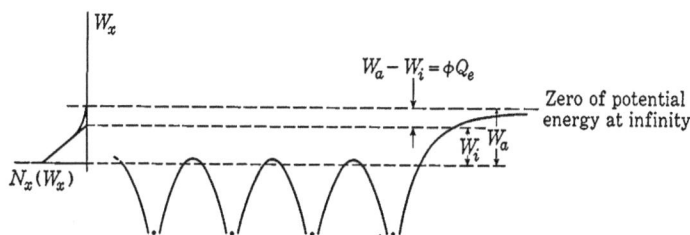

Fig. 6.   Superposition of the potential-energy diagram of Fig. 3 and the kinetic-energy distribution function of Fig. 5.

tops of the potential-energy humps in order that the band of occupied levels above it may correspond to the levels actually occupied by the valence electrons. Fortunately, the location of this level is not important in the analysis of the phenomena of electron emission, for, as is apparent from the remainder of this chapter, the phenomena of electron emission depend not on it but only on the energy difference between the W_a and the W_i levels and the electron energies near and above the W_i level.

In order that an electron may escape from a metal, the electron must have an x-associated kinetic energy equal to or greater than W_a, as Fig. 6 indicates. At zero degrees Kelvin, the maximum x-associated energy of any electron is W_i, which is less than W_a; hence none of the electrons escape at this temperature. The amount of energy $W_a - W_i$ is called the *work function* of the metal. It may be a function of the temperature, although experiments indicate that its dependence on the temperature is comparatively small. At a temperature of absolute zero, the work function is the smallest amount of energy that can enable an electron to escape from the metal; at other temperatures the work function is not the smallest amount. If the *voltage equivalent of the work function* is denoted by ϕ, then

$$\phi Q_e = W_a - W_i ,\qquad [7]$$

and ϕ in volts is numerically equal to $W_a - W_i$ expressed in electron volts. The experimentally measured values of the work function of various metals generally lie in the range from one to seven electron volts.

4. CONTACT DIFFERENCE OF POTENTIAL[5]

The concept of the work function is essential for an understanding of the process of liberation of electrons from metals. It is also useful in the explanation of the contact difference of potential. If two metals of different work function are brought into contact, an electric field is

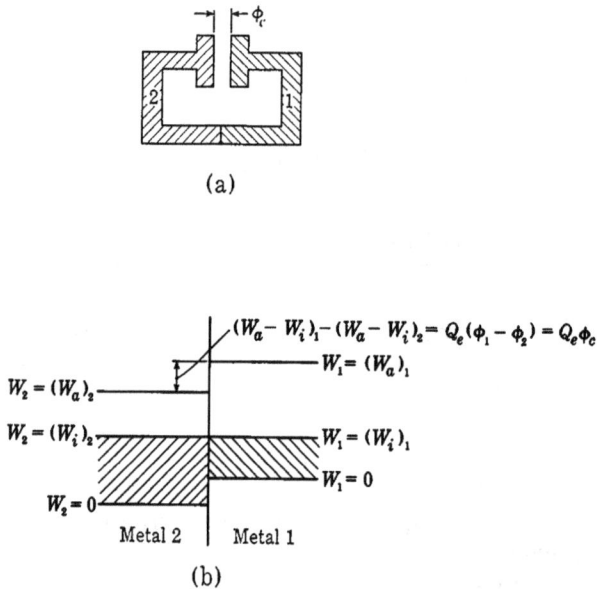

(a)

(b)

Fig. 7.   Contact potential difference between two unlike metals.

observed between the surfaces not in contact. The magnitude of the field is that which would be produced if the two metal surfaces differed in potential by an amount equal to the difference in the work functions of the metals. This apparent difference in potential, called the *contact potential difference* of the metals, may be explained in terms of an equilibrium among the electrons of the metals in contact.

When two metals are joined, as shown in Fig. 7a, the potential-energy barriers of each disappear at the surface of contact, and electrons flow from one metal into the other because of their kinetic energies. If the metals are dissimilar, the rate of flow across the surface

[5] W. G. Dow, *Fundamentals of Engineering Electronics* (2nd ed.; New York: John Wiley & Sons, Inc., 1952), 252–254.

in one direction will be greater initially than that in the other, even though the metals are uncharged. The metal which receives electrons at a greater rate, however, quickly acquires a negative charge which reduces its potential and sets up a retarding field that decreases the tendency for electrons to move in this direction. This reduction in potential continues until the rates of electron flow in the two directions are equal. At the absolute zero of temperature, this equilibrium condition occurs when the W_i energy levels in the two metals are aligned, as indicated in Fig. 7b. From this figure it is apparent that points just outside the surface of the two metals differ in potential energy by the difference in the work functions and hence differ in potential by an amount ϕ_c given by

$$\phi_c = \phi_1 - \phi_2 , \qquad\qquad \blacktriangleright[8]$$

where ϕ_1 and ϕ_2 are the voltage equivalents of the work functions in the two metals. The potential difference ϕ_c gives rise to an electric field between the surfaces of the metals and is the contact potential difference. Notice that an electron just outside a metal of high work function has an excess of potential energy and will move toward the metal of low work function. Thus the contact potential difference is in such a direction as to make the metal of low work function positive with respect to the other. At higher temperatures, the equilibrium condition of the electrons requires that the contact potential difference have a value only slightly different from its value at absolute zero.

5. THERMIONIC EMISSION

Several different means of giving energy to electrons are available. If, for example, the temperature of the metal is increased, the kinetic energy of some of the electrons increases, as is indicated by the distribution of kinetic energies for 1,500 degrees Kelvin in Figs. 5 and 6. A sufficient increase of temperature may give some electrons enough x-associated kinetic energy to enable them to surmount the potential barrier and escape. The liberation of electrons by this process is called *thermionic emission*. Electrons in a metal can also secure the additional energy required for liberation from photons of light impinging on the surface or from electrons or ions that strike the metal surface from the outside. The corresponding names for the electron liberation by these two processes are *photoelectric emission* and *secondary emission*. A fourth means for liberation of the electrons differs from the other three in that additional energy is not given to the electrons. Instead, an intense surface electric field produced by an adjacent positively charged electrode is used to modify the confining potential-energy barrier to such an extent that electrons with an energy near the W_i level penetrate it and escape. The resulting liberation of electrons is known as *field emission*. Since these four modes of electron emission are basic in the operation of many electronic devices, they and their engineering significance are discussed in more detail in this article and in those that follow.

Although it has been known for some 200 years that a negatively charged metallic body loses its charge more rapidly when heated than when not, the phenomenon of thermionic emission was not studied intensively or put to practical use until economic demand arose for efficient sources of electrons in radio communication. Edison[6] discovered the emission of electrons in a vacuum; Elster and Geitel,[7] Preece,[8] Hittorf,[9] and others made experimental studies of the effect; and Fleming[10] applied the principle to a wireless-telegraphy device. Richardson,[11] however, took the first important steps toward a clarification of the mechanism of thermionic emission from a theoretical standpoint. By application of the principles of thermodynamics and quantum mechanics, he, and later Dushman, derived a relationship, Eq. 13, between the emission current and the temperature of the metal. This result has since been shown to agree essentially with one derivable from the Fermi-Dirac statistics as applied to the electrons in a metal. The derivation from the Fermi-Dirac kinetic-energy distribution in Eq. 6 follows.

The discussion of Fig. 6 points out that only those free electrons with an x-associated kinetic energy W_x greater than W_a can escape from the metal. Analyses which take into account the wave nature of the electrons predict that not all these electrons will escape. Rather,

[6] T. A. Edison, "A Phenomenon of the Edison Lamp," *Engineering* (December 12. 1884), 553.

[7] J. Elster and H. Geitel, "Über die Elektrizität der Flamme," *Ann. d. Phys., 16* (1882), 193–222; "Über Elektrizitätserregung beim Contact von Gasen und glühenden Körpern," *Ann. d. Phys., 19* (1883), 588–624; "Über die unipolare Leitung erhitzter Gase," *Ann. d. Phys., 26* (1885), 2–9; "Über die Elektrisierung der Gase durch glühende Körper," *Ann. d. Phys., 31* (1887), 109–126; "Über die Elektrizitätserregung beim Contakt verdünnter Gase mit galvanisch glühenden Drähten," *Ann. d. Phys., 37* (1889), 315–329.

[8] W. H. Preece, "On a Peculiar Behavior of Glow-Lamps When Raised to High Incandescence," *Roy. Soc. Proc. (London), 38* (1885), 219–230.

[9] W. Hittorf, "Über die Elektrizitätsleitung der Gase," *Ann. d. Phys., 21* (1884), 119–139.

[10] J. A. Fleming, "On Electric Discharges between Electrodes at Different Temperatures in Air and in High Vacua," *Roy. Soc. Proc. (London), 48* (1890), 118–126; "A Further Examination of the Edison Effect in Glow Lamps," *Phil. Mag., 42* (1896), 52–102; United States Patent 803,684 (April 19, 1905).

[11] O. W. Richardson, "On the Negative Radiation from Hot Platinum," *Camb. Phil, Soc. Proc., 11* (1901), 286–295; *Emission of Electricity from Hot Bodies* (2nd ed.; New York: Longmans, Green & Co., 1921).

the electrons when approaching an irregularity behave as waves; some go on unimpeded, while others are reflected. Measurements of the energy distribution of the escaping electrons confirm these predictions. However, the probability of escape is very nearly unity whenever W_x exceeds W_a by more than a few hundredths of an electron volt. Since only a few electrons approach the surface with such a small excess of energy, it is a sufficiently good approximation to assume that all electrons for which W_x is greater than W_a *will* escape. On the basis of this approximation and Eq. 6, the number n_x of electrons that escape per unit area per unit time is

$$n_x = \int_{W_a}^{\infty} N_x(W_x)\, dW_x$$

$$n_x = \int_{W_a}^{\infty} \frac{4\pi m_e kT}{h^3} \ln \left(\epsilon^{\frac{W_i - W_x}{kT}} + 1 \right) dW_x \qquad \text{electrons per unit} \qquad [9]$$
$$\text{time per unit area.}$$

For values of $W_x - W_i$ much greater than kT, which are the conditions of importance since the usual value of $W_a - W_i$ is of the order of $20kT$, the quantity $\epsilon^{(W_i - W_x)/(kT)}$ is very small compared with unity, and an approximate form of Eq. 9, valid for the entire range of integration, is

$$n_x = \frac{4\pi m_e kT}{h^3} \int_{W_a}^{\infty} \epsilon^{\frac{W_i - W_x}{kT}} dW_x . \qquad [10]$$

Since each electron carries a charge Q_e, the electric current density through the surface of the metal is $Q_e n_x$, or

$$J_s = Q_e n_x = \frac{4\pi m_e kTQ_e}{h^3} \int_{W_a}^{\infty} \epsilon^{\frac{W_i - W_x}{kT}} dW_x \qquad \text{current units} \qquad [11]$$
$$\text{per unit area}$$

$$= \frac{4\pi m_e Q_e k^2}{h^3} T^2\, \epsilon^{-\frac{W_a - W_i}{kT}} \qquad \text{current units per unit area.} \qquad [12]$$

Equation 12 has the form

$$J_s = AT^2 \epsilon^{-\frac{b}{T}}, \qquad \blacktriangleright[13]$$

where A is a constant but b may vary slightly with the temperature. This relationship is one of two derived by Richardson and is known as *Richardson's equation.*

The quantity J_s is the *emission current density*; it is determined by the rate of electron escape through the surface of the metal. Later it is explained that the current actually conducted from the cathode across the space does not always equal the emission current because a fraction of the escaping electrons may return directly to the cathode.

Experiments purporting to determine the quantities A and b are described subsequently. A word of caution is necessary at this point, however. Because of the assumptions made in deriving Richardson's equation (Eq. 13) and because of the fact that b may vary with the temperature, none of the experimental measurements actually determines the terms in Eq. 12 that correspond to A and b in Eq. 13. Consequently, in the following discussion, Eq. 13 should be regarded as an

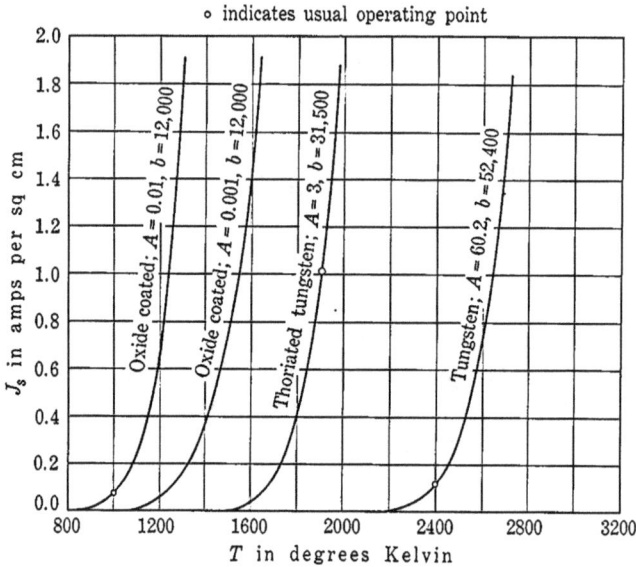

Fig. 8.   Emission current density from various thermionic emitters.

empirical equation that when both A and b are considered constant will fit, within the limits of experimental error, the data obtained from measurements. The derivation of Eq. 13 is, then, a guide to the form of the empirical equation, and the values of A and b determined by experiment are merely numerical constants that make the empirical equation fit the experimental data.

Since the constant A in Eq. 13 corresponds to a factor in Eq. 12 involving only quantities independent of the particular metal, A should apparently be the same for all metals. But substitution of numerical values gives for A, 120.4 amperes per square centimeter per degree Kelvin squared, and the values of A found to fit Eq. 13 to experimental data for pure metals when A and b are considered constant are usually about 60, or one-half of the theoretical value. This discrepancy may be attributed to the assumptions made in deriving Eq. 12 and to the fact that b is considered constant in making the empirical fit.

The current-temperature characteristic curve described by Richardson's equation is shown in Fig. 8 for several pairs of values of the constants. Because of the exponential form of the relation a small change in the temperature produces a large change in the current. For the same reason, at a given temperature the emission current depends markedly on the constant b, while the constant A plays a relatively unimportant part.

The foregoing derivation of Richardson's equation shows that the constant b is related to the work function and its voltage equivalent ϕ by the expression

$$b = \frac{W_a - W_i}{k} = \frac{Q_e \phi}{k} \text{ degrees Kelvin,} \qquad [14]$$

where

Q_e is the electronic charge in coulombs,

k is the Boltzmann constant in joules per degree Kelvin,

ϕ is the voltage equivalent of the work function.

When numerical values are substituted, this becomes

$$b = 11,605 \, \phi \text{ degrees Kelvin.} \qquad \blacktriangleright[15]$$

Representative values of the experimentally determined emission constants for several different pure metals are given in Table I. These values are taken from the International Critical Tables; the measurements were made on pure metals in the best obtainable vacuum with

TABLE I*

REPRESENTATIVE VALUES OF THERMIONIC-EMISSION CONSTANTS
FOR PURE METALS

$$J_s = A T^2 \epsilon^{-(b/T)}$$

Metal	A in amp/cm²/deg K^2	b in deg K	Melting point in deg K
Calcium	60.2	26,000	1,083
Carbon	60.2	46,500	3,773
Cesium	16.2	21,000	299
Molybdenum	60.2	51,500	2,893
Nickel	26.8	32,100	1,725
Platinum	60.2	59,000	2,028
Tantalum	60.2	47,200	3,123
Thorium	60.2	38,900	2,118
Tungsten	60.2	52,400	3,643

* The constants given in this table are adapted from the *International Critical Tables* (New York: McGraw-Hill Book Company, Inc., 1929), Vol. 1, pp. 103–104; Vol. 6, pp. 53–54; with permission.

careful, refined, and precise methods. Even with these precautions, considerable disagreement remains among the results of different experimenters—a matter in part explained by studies indicating that the work function[12] $W_a - W_i$ may differ at different temperatures. Furthermore, electron emission does not take place uniformly over the surface of the metal[13] but depends on the type of crystal face exposed; hence to establish identical conditions for different investigations on the same metal is difficult.

6. MEASUREMENT OF THERMIONIC EMISSION

To determine the emission constants for a particular metal, an experiment is usually performed in which a cathode in the form of a filament F, Fig. 9, surrounded by a cylindrical anode A fitted with guard rings G to eliminate end effects, is heated in the best vacuum obtainable. The temperature of the filament can be measured by an optical pyrometer sighted on it through hole H. The emission current I_s is drawn to the anode by the field set up by the voltage e_b, which is raised to a value large enough to collect all the electrons emitted. The current under these conditions is called the *saturation current*. From the measured values of current and the dimensions of the filament, the current density J_s may then be found for any particular filament temperature T. In order to determine the constants A and b from these data, it is convenient to plot* $\log (J_s/T^2)$ as a function of $1/T$. The logarithm of Eq. 13 is

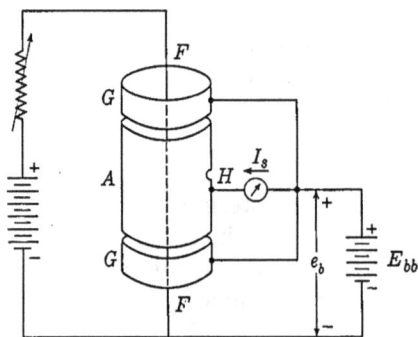

Fig. 9. Experimental measurement of the thermionic-emission constants.

$$\ln \frac{J_s}{T^2} = \ln A - \frac{b}{T}, \qquad [16]$$

or

$$\log \frac{J_s}{T^2} = -0.434\, b\, \frac{1}{T} + \log A. \qquad \blacktriangleright[17]$$

[12] J. A. Becker and W. H. Brattain, "The Thermionic Work Function and the Slope and Intercept of Richardson Plots," *Phys. Rev., 45* (1934), 694–705.

[13] R. P. Johnson, "Simple Electron Microscopes," *J. App. Phys., 9* (1938), 508–516.

* In this book, the symbol *log* is used to indicate the *logarithm to the base 10*, and the symbol *ln* to indicate the *logarithm to the base* ϵ.

When the data are plotted in this manner, therefore, the result should be a straight line with a slope of $-0.434b$, and an intercept on the axis of ordinates of $\log A$, as is shown in Fig. 10. The fact that most

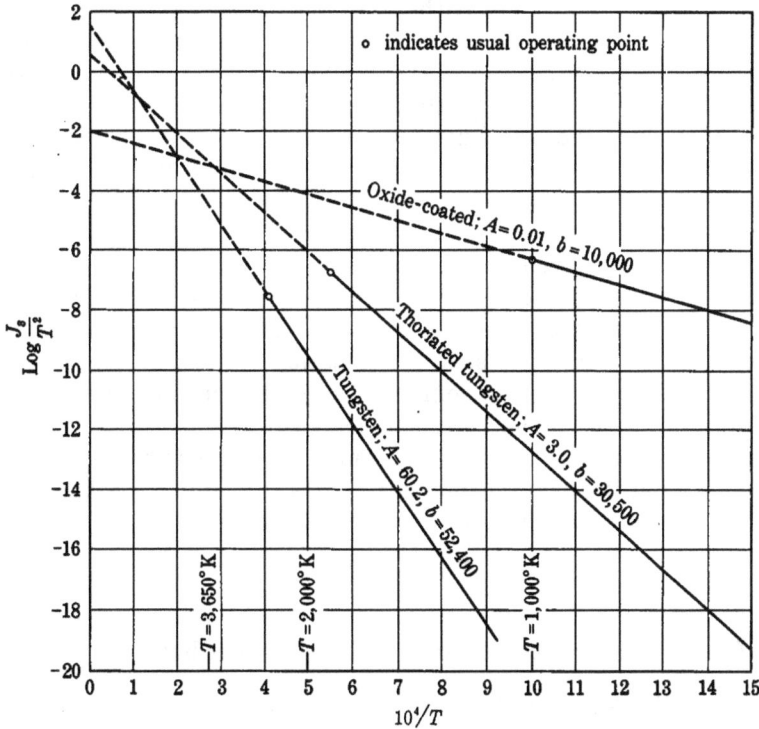

Fig. 10. Determination of the constants in Richardson's equation. The curves are plotted with J_s in amperes per square centimeter and T in degrees Kelvin.

measured data, when plotted in this way, give a straight line is taken as evidence for the validity of Richardson's equation.

The accuracy of the resulting values of A and b depends on the measurements of current, dimensions, and temperature. As an indication of the relative importance of these measurements in determining the constants, the temperature coefficient of the emission current may be derived from Eq. 13 as

$$\frac{dJ_s/J_s}{dT/T} = 2 + \frac{b}{T}.$$
[18]

The ratio b/T for tungsten has a value of about 23 at its usual operating temperature. Thus the measurement of the absolute temperature should be some 25 times as precise, on a percentage basis, as that of the current, to yield a result for A of the same precision as that of the measurement of J_s.

The temperature of a hot cathode may be obtained most precisely by means of an optical pyrometer. This instrument utilizes the ability of the eye to match the brightness of the image of the cathode, at a given wavelength of light, to that of the calibrated filament in the instrument. Since the brightness varies much more rapidly than the temperature, the method gives rather precise results. But because the brightness of a body at a given temperature depends not only upon the temperature but also upon the spectral radiation emissivity of the surface for the given wavelength, this emissivity must be known and must be properly used in the interpretation of the pyrometer indications. Otherwise, the instrument is useless for accurate temperature determinations. While the spectral radiation emissivity of tungsten is accurately known, that of some of the other thermionic emitters is not definite because of their varying composition.

Another method of depicting the emission characteristics of thermionic cathodes is by power-emission charts.[14] These are based on the Stefan-Boltzmann law of radiated power, which is

$$P = Ke_tT^4, \qquad\qquad \blacktriangleright[19]$$

where

P is the power per unit area radiated from a hot body,

T is the temperature in degrees Kelvin,

K is the Stefan-Boltzmann constant[15]—$(5.673 \pm 0.004) \times 10^{-8}$ watt per square meter per degree Kelvin fourth,

e_t is the total radiation emissivity (*see* Table II).

When Richardson's equation, Eq. 13, and the Stefan-Boltzmann law, Eq. 19, are combined to eliminate the temperature T, an expression for J_s in terms of P is obtained.[16] By skewing the co-ordinates, Davisson devised a cross-section paper upon which a plot of log J_s as a function of log P for an emitter is a straight line. Such a diagram is shown in Fig. 11. This type of plot has the advantage that measurements of the

[14] R. W. King, "Thermionic Vacuum Tubes and Their Applications," *B.S.T.J., 2* (1923), 31–37.

[15] R. T. Birge, "A New Table of Values of the General Physical Constants," *Rev. Mod. Phys., 13* (October, 1941), 233–239.

[16] W. G. Dow, *Fundamentals of Engineering Electronics* (2nd ed.; New York: John Wiley & Sons, Inc., 1952), 210–211.

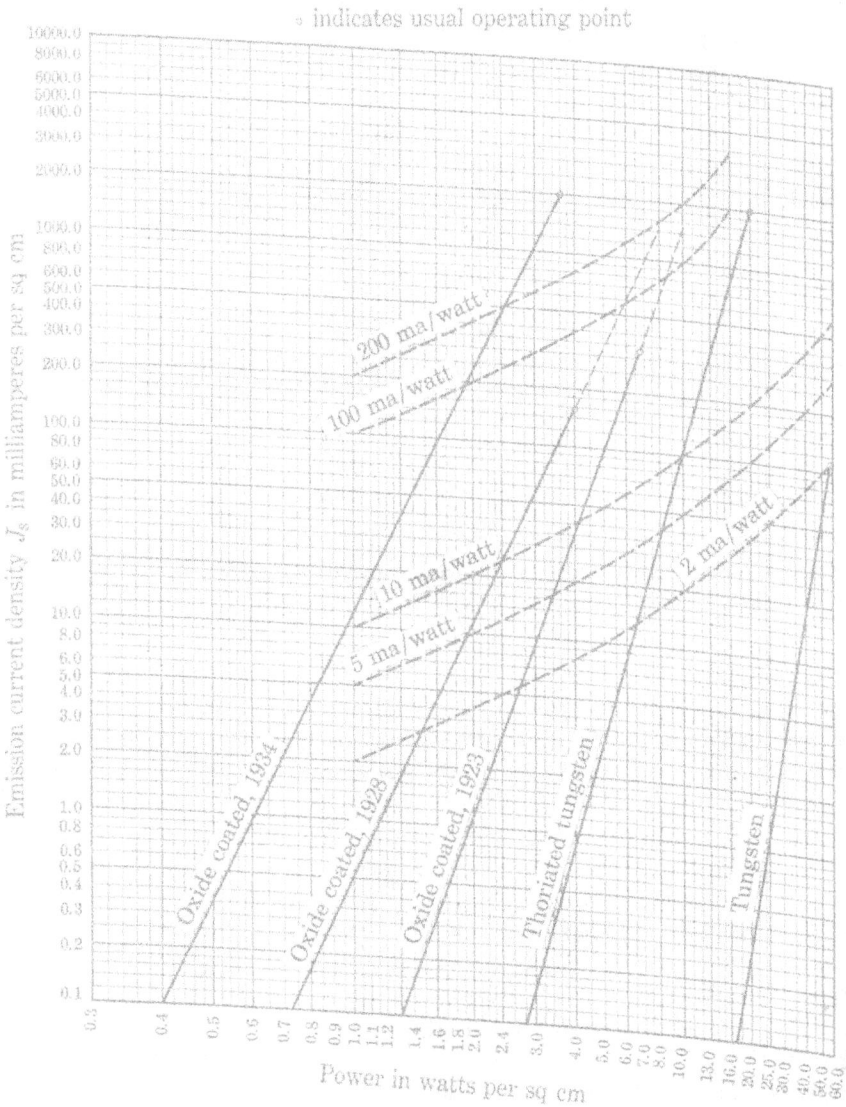

Fig. 11.   Power-emission chart.*

* The data for oxide-coated cathodes are from S. Dushman, "Electron Emission," *A.I.E.E. Trans.*, *53* (1934), Fig. 2, p. 1059, with permission. Those for thoriated tungsten are from R. W. King, "Thermionic Vacuum Tubes," *B.S.T.J.*, *2* (1923), Fig. 2, p. 4, with permission. Those for tungsten are computed from Table III.

emission at a few values of low power, where the current from the cathode and the heating of the anode are small, may be extrapolated to give the emission at larger values of power. It is to be noted, however, that the straightness of the line depends upon the assumption that the total radiation emissivity of the surface is constant. Table II, which gives values for the total radiation emissivity of several metals, shows that it varies appreciably with temperature.

TABLE II*

Metal	Total radiation emissivity e_t at		
	1,000 deg K	1,400 deg K	2,000 deg K
Tungsten	0.114	0.174	0.260
Platinum	0.134	0.182	
Molybdenum	0.096	0.145	0.210

* The constants of this table are taken from the *International Critical Tables* (New York: McGraw-Hill Book Company, Inc., 1929), Vol. 5, p. 243, with permission.

In general, a metal with a small value of b has a large emission current at a low temperature and therefore provides electrons at high emission efficiency. The value of b is not, however, the sole criterion of merit of an emitter—the maximum practical operating temperature must also be considered. This temperature is the highest value consistent with a rate of metal evaporation that will yield a reasonable life of the cathode. The rate of evaporation may be described by an equation containing an exponential term similar to that in Richardson's equation, and this rate usually becomes excessive at a temperature considerably below the melting point. Tantalum and tungsten, because of their low vapor pressures and high melting points, can be operated at a sufficiently high temperature to give an emission greater than that from any other pure metals despite their relatively large values of b. The metals with low values of b and consequent high emission at a given temperature are not practical emitters, because the maximum temperature to which they can be raised in a vacuum without excessive evaporation is low. Tantalum is seldom used as an emitter because of its tendency to become brittle as a result of recrystallization into large crystals. Also, the action of residual gases in the tube, such as oxygen and water vapor, may cause a great reduction in the emission from tantalum. Tungsten, then, is the only pure metal that is practical as an emitter at present.

7. THERMIONIC EMISSION FROM PURE TUNGSTEN

The emission current from a pure tungsten cathode is given by Richardson's equation with the values of A equal to 60.2 amperes per square centimeter per degree Kelvin squared, and b equal to 52,400 degrees Kelvin. Figures 8, 10, and 11 show the emission current in different ways. A tungsten cathode usually takes the form of a wire filament heated by an electric current passed through it in the manner illustrated in Fig. 9. The practical life[17] of the filament ends when its cross section decreases by about 10 per cent, because burn-out usually occurs soon thereafter. Economic considerations indicate that the temperature should be adjusted to produce this reduction in cross section in about 1,000 to 2,000 hours. Since the rate of evaporation depends on the circumferential area or first power of the diameter of the filament, while the amount of material to be evaporated depends on the cross section or second power of the diameter, the operating temperature for the same life is different for filaments of different sizes. Large-diameter filaments may be operated at higher temperatures than may those of small diameter, because their cross section is greater in proportion to their circumferential area.

Often an optical pyrometer is not available or cannot be used to determine the temperature of a filament, because an unobstructed view of the filament is not possible. In these circumstances the tables of Jones and Langmuir[18] are particularly useful if the filament is pure tungsten. An extract from these tables is given in Table III.

The Jones and Langmuir tables give, for a wide range of temperatures, the values of many properties of a *unit filament* of tungsten; that is, of a filament one centimeter long and one centimeter in diameter, having a radiating area of π square centimeters. The tabulated values apply only to an ideal filament defined in the tables and are not applicable if the filament is not straight or if the filament is cooled by gas in the tube. A correction for loss of heat through the leads is necessary for accurate results. The characteristics of a straight cylindrical tungsten filament of any known dimensions may be found with the help of these tables by simple methods explained in the tables and illustrated in the following example.

An ideal tungsten filament has a length l of 10 centimeters and a diameter d of 0.01 centimeter (about 0.004 inch) and operates at 2,400

---

[17] Y. Kusunose, *Calculations on Vacuum Tubes and the Design of Triodes: Researches of the Electrotechnical Laboratory, No. 237* (Tokyo, Japan), 135–139; J. J. Vormer, "Filament Design for High-Power Transmitting Valves," *I.R.E. Proc.*, 26 (1938), 1399–1407.

[18] H. A. Jones and I. Langmuir, "Characteristics of Tungsten Filaments as Functions of Temperature," *G.E. Rev.*, 30 (1927), 310–319, 354–361, 408–412.

TABLE III*

SPECIFIC CHARACTERISTICS OF IDEAL TUNGSTEN FILAMENTS

(For a wire 1 cm long and 1 cm in diameter.)

T	$W' = \dfrac{W}{ld}$	$P' = \dfrac{Rd^2}{l} \times 10^6$	$A' = \dfrac{A}{d^{3/2}}$	$V' \times 10^3 = \dfrac{V\sqrt{d}}{l} \times 10^3$	$I' = \dfrac{I}{ld}$	$M' = \dfrac{M}{ld}$	$R'_T/R'_{293°}$	$R_T/R_{293°}$
$\deg K$	$watts\ cm^{-2}$	$ohms\ cm$	$amp\ cm^{-3/2}$	$volts\ cm^{-1/2}$	$amp\ cm^{-2}$	$gram\ cm^{-2}sec^{-1}$		
273	0.0	6.37	0.0	0.0			0.911	
293	0.000100	6.99	3.727	0.02683			1.00	
300	0.00624	7.20	24.67	0.2530			1.03	
400		10.26					1.467	
500	0.0305	13.45	47.62	0.6404			1.924	
600	0.0954	16.85	75.25	1.268			2.41	
700	0.240	20.49	108.2	2.218			4.93	
800	0.530	24.19	148.0	3.581			3.46	
900	1.041	27.94	193.1	5.393			4.00	
1,000	1.891	31.74	244.1	7.749	3.36×10^{-15}	1.16×10^{-33}	4.54	
1,100	3.223	35.58	301.0	10.71	4.77×10^{-13}	6.81×10^{-30}	5.08	
1,200	5.210	39.46	363.4	14.34	3.06×10^{-11}	1.01×10^{-26}	5.65	
1,300	8.060	43.40	430.9	18.70	1.01×10^{-9}	4.22×10^{-24}	6.22	
1,400	12.01	47.37	503.5	23.85	2.08×10^{-8}	7.88×10^{-22}	6.78	

1,500	17.33	51.40	580.6	29.85	2.87×10^{-7}	7.42×10^{-20}	7.36
1,600	24.32	55.46	662.2	36.73	2.91×10^{-6}	3.92×10^{-18}	7.93
1,700	33.28	59.58	747.3	44.52	2.22×10^{-5}	1.31×10^{-16}	8.52
1,800	44.54	63.74	836.0	53.28	1.40×10^{-4}	2.97×10^{-15}	9.12
1,900	58.45	67.94	927.4	63.02	7.15×10^{-4}	4.62×10^{-14}	9.72
2,000	75.37	72.19	1,022	73.75	3.15×10^{-3}	5.51×10^{-13}	10.33
2,100	95.69	76.49	1,119	85.57	1.23×10^{-2}	4.95×10^{-12}	10.93
2,200	119.8	80.83	1,217	98.40	4.17×10^{-2}	3.92×10^{-11}	11.57
2,300	148.2	85.22	1,319	112.4	1.28×10^{-1}	2.45×10^{-10}	12.19
2,400	181.2	89.65	1,422	127.5	0.364	1.37×10^{-9}	12.83
2,500	219.3	94.13	1,526	143.6	0.935	6.36×10^{-9}	13.47
2,600	263.0	98.66	1,632	161.1	2.25	2.76×10^{-8}	14.12
2,700	312.7	103.22	1,741	179.7	5.12	9.95×10^{-8}	14.76
2,800	368.9	107.85	1,849	199.5	11.11	3.51×10^{-7}	15.43
2,900	432.4	112.51	1,961	220.6	22.95	1.08×10^{-6}	16.10
3,000	503.5	117.21	2,072	243.0	44.40	3.04×10^{-6}	16.77
3,100	583.0	121.95	2,187	266.7	83.0	8.35×10^{-6}	17.46
3,200	671.5	126.76	2,301	291.7	150.2	2.09×10^{-5}	18.15
3,300	769.7	131.60	2,418	318.3	265.2	5.02×10^{-5}	18.83
3,400	878.3	136.49	2,537	346.2	446.0	1.12×10^{-4}	19.53
3,500	998.0	141.42	2,657	375.7	732.0	2.38×10^{-4}	20.24
3,600	1,130	146.40	2,777	406.7	1,173	4.86×10^{-4}	20.95
3,655	1,202	149.15	2,838	423.4	1,505	7.15×10^{-4}	21.34

* The values given are taken from H. A. Jones and I. Langmuir, "The Characteristics of Tungsten Filaments as Functions of Temperature," *G.E. Rev.*, *30* (1927), Table I, pp. 312–313, with permission. The notation of Jones and Langmuir is retained in this table and in the example of its use.

degrees Kelvin. The electrical operating conditions and the life of the filament are to be found.

Solution: From the Jones and Langmuir Table I, $W' = 181.2$, $R' = 89.65 \times 10^{-6}$, $A' = 1,422$, $V' = 127.5 \times 10^{-3}$, $I' = 0.364$, $M' = 1.37 \times 10^{-9}$, $R'_T/R_{293°} = 12.83$. Thus:

$$\text{Power radiated} = W = W'ld = 181.2 \times 10 \times 0.01 = 18.2 \text{ watts,} \qquad [20]$$

$$\text{Resistance} = R = R'(l/d^2) = 89.65 \times (10/10^{-4}) \times 10^{-6}$$
$$= 8.965 \text{ ohms,} \qquad [21]$$

$$\text{Heating current} = A = A'd^{3/2} = 1,422 \times 0.01^{3/2} = 1.422 \text{ amp,} \qquad [22]$$

$$\text{Voltage drop} = V = V'(l/\sqrt{d}) = 127.5 \times (10/0.1) \times 10^{-3}$$
$$= 12.75 \text{ volts,} \qquad [23]$$

$$\text{Emission current} = I = I'ld = 0.364 \times 10 \times 0.01 = 0.0364 \text{ amp,} \qquad [24]$$

$$\text{Rate of evaporation} = M = M'ld = 1.37 \times 10^{-9} \times 10 \times 0.01$$
$$= 1.37 \times 10^{-10} \text{ gram per sec,} \qquad [25]$$

Ratio hot to cold resistance $= R_T/R_{293°} = R'_T/R_{293°} = 12.83.$ $\qquad$ [26]

From the rate of evaporation, the life—or time for 10 per cent evaporation—may be computed, if the effects of temperature and area variation on the evaporation are assumed to be negligible. The life is the ratio of the weight of tungsten to be evaporated to the rate of evaporation, or, since the density of aged tungsten is 19.35 grams per cubic centimeter,

$$\text{Life} = \frac{\pi d^2 l \times 19.35}{4 \times 10 \times 3,600 M} = \frac{\pi d \times 19.35}{4 \times 10 \times 3,600 M'} \qquad [27]$$

$$= 4.222 \times 10^{-4} \frac{d}{M'}, \text{ hours.} \qquad [28]$$

For the filament of this example,

$$\text{Life} = 4.222 \times 10^{-4} \frac{0.01}{1.37 \times 10^{-9}} = 3,080 \text{ hours.} \qquad [29]$$

Since the emission current and the power radiated are directly proportional to the surface area of the filament, their ratio—the *emission efficiency*—is given by the ratio I'/W' from the Jones and Langmuir tables. The emission efficiency found in this way is 2.01, 4.27, and 8.56 milliamperes per watt at temperatures of 2,400, 2,500, and 2,600 degrees Kelvin, which cover the normal operating range for the common sizes of filaments. In comparison with the emission efficiency of other practical emitters discussed subsequently, that of tungsten is very low. Tungsten is, therefore, an expensive source of electrons because of the power required, and is not used unless some other compensating advantage makes it desirable.

An outstanding advantage of tungsten over other types of emitters is that it is much less subject to loss of emitting properties as a result of the bombardment by positive ions occurring because of the small amount of residual gas always present in commercial vacuum tubes. While mechanically fragile because of recrystallization, tungsten is electrically rugged and finds its greatest application in high-voltage x-ray and power tubes where the positive ions have the greatest energy and the bombardment is therefore most severe. Tungsten, when hot, is effectively attacked even by traces of some gases,[19] particularly water vapor. Emission from tungsten is impaired by other gases, especially nitrogen, but in the presence of mercury vapor and the noble gases, neon, helium, argon, krypton, and xenon, it is unaffected.

8. THERMIONIC EMISSION FROM THORIATED TUNGSTEN

During the early stages of the development of tungsten filaments for incandescent lamps, operation on alternating current shortened the life of filaments because of the tendency of the material to "offset"; that is, the tendency for one part of the filament to slip sideways with respect to the remainder across a transverse crystal face. The addition of a small percentage of a foreign substance, such as thoria, to the tungsten was found to inhibit crystal growth and prevent offsetting. Later, measurements of the emissive properties of thoriated tungsten revealed that under certain conditions its electron-emission currents were enormously greater than those from pure tungsten. This increased emission is now thought to occur because thorium has a lower rate of evaporation from tungsten than from itself,[20] and thus a thin, possibly monatomic, layer of thorium may remain on the surface of hot tungsten at a temperature that would cause rapid evaporation from thorium metal. The emission properties of thorium may therefore be utilized at temperatures that would otherwise be impossibly high because of rapid evaporation. The work function of thoriated tungsten proves to be lower than that of either thorium or tungsten. This reduction is attributed to the fact that the adsorbed layer of thorium on the filament surface is electropositive with respect to the tungsten. Consequently, it constitutes a dipole layer with its positive side away from the tungsten, and thus lowers the potential barrier at the surface.

In the preparation of thoriated tungsten, 1 to 2 per cent of thorium oxide (thoria) is added to the tungsten powder before it is sintered, swedged, and drawn into wire form. After being mounted in the tube, the filament is usually carburized[21] by being heated to a temperature of about 2,000 degrees Kelvin in a low pressure of hydrocarbon gas

---

[19] I. Langmuir, "The Effect of Space Charge and Residual Gases on Thermionic Currents in High Vacuum," *Phys. Rev.*, Series 2, 2 (1913), 461–476.

[20] I. Langmuir, "Electron Emission from Thoriated Tungsten Filaments," *Phys. Rev.*, 22 (1923), 357–398.

[21] L. R. Koller, *The Physics of Electron Tubes* (2nd ed.; New York: McGraw-Hill Book Company, Inc., 1937), 34–38; C. W. Horsting, "Carbide Structures in Carburized Thoriated-Tungsten Filaments," *J. App. Phys.*, 18 (1947), 95–102; H. J. Dailey, "Designing Thoriated Tungsten Filaments," *Electronics*, 21 (January, 1948), 107–109.

or vapor, such as acetylene, naphthalene, or benzol, until its resistance increases by 10 to 25 per cent. This process allows the reduction of the thoria to metallic thorium at lower temperatures and is necessary if the filament is to withstand appreciable positive-ion bombardment in a high-voltage, high-power vacuum tube. The life of the filament as an emitter is increased, because the rate of evaporation of thorium from the carburized surface is several times smaller than from a surface of pure tungsten.

The activation of a thoriated-tungsten filament involves two steps, as follows:

(a) During the evacuation of the tube, the filament is heated to a temperature above 2,800 degrees Kelvin for about one minute. Thorium oxide decomposes at this temperature, the oxygen is pumped out, and thorium metal is left in the tungsten. At this temperature, the thorium evaporates rapidly as it diffuses to the surface of the tungsten.

(b) After the evacuation process is complete, or during it if the tube is operated on the pumping system, the temperature of the filament is held at about 2,400 degrees Kelvin for a few minutes. At this temperature, the thorium diffuses to the surface more rapidly than it evaporates and gradually builds up a thin layer there. However, the temperature of the filament must be lowered to 1,900 to 2,000 degrees Kelvin for continuous operation, in order that the rate of evaporation of thorium may be low enough to give a reasonable life.

Because the electric field used to draw the electrons to the anode has a marked effect on the emission from thoriated tungsten, the emission constants in Richardson's equation are difficult to obtain, and they yield information of little value for the practical use of the material. Values of A and b ranging from 3 to 59 amperes per square centimeter per degree Kelvin squared, and from 30,500 to 36,500 degrees Kelvin, respectively, have been published for fully activated thoriated tungsten. The abnormally great effect of the electric field vitiates any attempt to use these data in predicting the performance of a particular thoriated-tungsten filament in a vacuum tube employing considerable anode voltage, for a doubling of the anode voltage may increase the emission current by as much as 50 per cent. The same uncertainty applies to the power-emission plot for thoriated tungsten shown in Fig. 11, since the position of the curve shifts up or down considerably for different values of anode voltage. Possibly it is for this reason that such emission plots given by different authorities do not agree.

Fortunately this uncertainty regarding the emission properties of thoriated tungsten is not of great engineering importance, because this type of cathode is almost always used under conditions where the current is not limited by the emission itself. The filament is usually designed to furnish ten or more times the current normally required. For practical purposes, carburized thoriated tungsten may often be operated at a temperature as high as 2,000 degrees Kelvin with a reasonable life, and at that temperature it emits about 2.75 amperes per square centimeter and requires a heating power of 24 to 29 watts per square centimeter, depending on the degree of carburization. The total radiation emissivity increases with the degree of carburization, and the power required to maintain the given temperature of 2,000 degrees Kelvin increases accordingly. The emission efficiency of thoriated tungsten in practical use is therefore about 100 milliamperes per watt.

Despite the improved performance obtained by carburization of thoriated-tungsten filaments, the limitation on the use of them remains their susceptibility to deactivation by the action of positive ions. Although, as is discussed in Art. 2b, Ch. V, the deactivation is negligible for anode voltages below a low critical value, in most high-vacuum tubes the anode voltage is higher than this critical value, and a trace of residual gas pressure too small to affect the emission from a pure tungsten filament can cause rapid deactivation of a thoriated-tungsten filament. The action of even a few ions is severe at high anode voltages, and, because of the difficulty of maintaining a very high vacuum in a commercial tube, it was believed for many years that the practical limit of anode voltage with a thoriated-tungsten filament was about 5,000 volts. Refined techniques, however, now permit use of thoriated·tungsten filaments in tubes with anode voltages as high as 17,00ʋ volts.[22] Pure tungsten, on the other hand, is used in x-ray tubes with voltages of 350,000 and more.

9. Thermionic emission from oxide-coated cathodes

Wehnelt[23] discovered in 1903 that, when coated with certain oxides of the rare earths, a platinum ribbon used as a hot filament is a source of copious electron emission. However, because the technique of making durable coated filaments was undeveloped, most of the early vacuum tubes employed comparatively inefficient pure-tungsten

[22] R. B. Ayer, "Use of Thoriated-Tungsten Filaments in High-Power Transmitting Tubes," *I.R.E. Proc.*, *40* (1952), 591–594.

[23] A. Wehnelt, "Über den Austritt negativer Ionen aus glühenden Metalverbindungen und damit zusammenhangende Erscheinungen," *Ann. d. Phys.*, *14* (1904), 425–468.

emitters. Even now, when oxide-coated cathodes are used almost exclusively, and coating methods are highly developed, difficulties with emission are common in the manufacture of vacuum tubes.

Platinum was first used as a base, or core material, upon which to put the coating of oxides but because of its high cost other materials have been substituted. Pure nickel, nickel with a few per cent of either silicon or cobalt, and Konel metal, which consists of nickel, cobalt, iron, and titanium, have all been found satisfactory.

Numerous methods of coating the base are used, but most of them have as a final object a surface coating of barium oxide and strontium oxide. The emission from a mixture of the two oxides is higher than that from either alone. Because the oxides tend to form hydroxides (which are poor emitters on nickel) in humid air, either the carbonates or the nitrates are usually applied originally. The application may be made either through dipping the filament in a water suspension of the salts or through spraying the cathode with a suspension of them in amyl acetate or other suitable vehicle containing a little nitrocellulose. After the vehicle dries, the nitrocellulose serves as a binder to hold the carbonates together on the surface.

During the evacuation process, the coated cathode is heated to a temperature of 1,200 to 1,500 degrees Kelvin. At this temperature, the salts decompose into the oxides, and gas is evolved. This gas is removed by the pumps, and the oxides are left on the cathode surface. Unless the decomposition is carried to completion before the tube is sealed, the slow evolution of gas will eventually destroy the vacuum.

After the tube is sealed and separated from the pumping system, the cathode must be "activated" to build up the emission. Activation usually consists of the application over a period of time of 100 to 200 volts through a protective resistor to the tube. The cathode is made negative, and all other elements of the tube are connected together as a collector and made positive. The activation is continued until the emission reaches a stable value; the process may take several hours.

The action of the oxide-coated cathode is not fully understood at present, although many investigations of it have been made.[24] A thin film of barium metal is believed to be formed on the cathode surface by electrolysis, and, in addition, small particles of barium are thought to be distributed throughout the oxide. The oxide is a semiconductor and has a relatively high conductivity at the operating temperatures. The rather complex structure of the resulting composite

[24] J. P. Blewett, "Properties of Oxide-coated Cathodes," *J. App. Phys.*, *10* (1939), 668–679, 831–848; "Oxide Coated Cathode Literature, 1940–1945," *J. App. Phys.*, *17* (1946), 643–647; A. M. Bounds and T. H. Briggs, "Nickel Alloys for Oxide-Coated Cathodes," *I.R.E. Proc.*, *39* (1951), 788–799.

surface is particularly effective in reducing the potential-energy barrier, and hence the work function, of the surface. The supposition of a surface film of barium is borne out by the fact that in gas tubes oxide-coated cathodes exhibit a tendency to become deactivated[25] under positive-ion bombardment as do thoriated-tungsten cathodes. They are, however, somewhat more rugged in this respect than thoriated-tungsten cathodes.

Two factors make it difficult to determine experimentally the emission constants in Richardson's equation for oxide-coated cathodes. The first factor is that the total radiation emissivity varies widely from an average value of 0.35 and the spectral radiation emissivity also varies with the preparation of the cathode.[26] Temperature measurements on different cathodes require separate determinations of the emissivity unless the resistance-temperature characteristic of the core material is accurately known. When this characteristic is known, the temperature of the filament can be determined from the ratio of hot-to-cold resistance if proper correction is made for the effect of the leads. This method is of no avail with indirectly heated cathodes, which are commonly used. The second factor is that the emission current varies markedly with the anode voltage. As the anode voltage is raised, the emission current continues to increase even though the cathode temperature is fixed. This effect is much more pronounced with oxide-coated cathodes than with thoriated-tungsten cathodes, which, in turn, exhibit the effect to a much higher degree than do pure-tungsten cathodes. Consequently, a true saturation current does not appear with oxide-coated cathodes.

Despite these difficulties, the characteristics of oxide-coated cathodes have been measured, and values of A and b are often stated. The work function is apparently about one volt, which corresponds to b equal to 11,605 degrees Kelvin, and A has a value of the order of 0.01 ampere per square centimeter per degree Kelvin squared.

The representation of the characteristics of oxide-coated cathodes on a power-emission chart, Fig. 11, is also subject to the uncertainty caused by the lack of saturation of the emission current. As with thoriated tungsten, the curves given by different investigators do not agree, and they cover a wide range. The variations among the cathodes may be explained to some extent by the fact that continual improvement in coatings has been made, as is shown by the dates on the different curves of Fig. 11.

---

[25] A. W. Hull, "Gas-Filled Thermionic Tubes," *A.I.E.E. Trans.*, *47* (1928), 753–763; "Hot-Cathode Thyratrons," *G.E. Rev.*, *32* (1929), 213–223, 390–399.

[26] C. H. Prescott, Jr. and J. Morrison, "The True Temperature Scale of an Oxide-Coated Filament," *R.S.I.*, *10* (1939), 36–38.

The usual operating temperature for oxide-coated cathodes is about 1,000 degrees Kelvin, although some types of core materials permit lower values.[27] If the temperature of the cathode is raised above the normal value, the life of the cathode is materially shortened. During the life of an oxide-coated cathode, the emission sometimes rises at first but eventually drops off, despite the fact that coating material is still left on the surface. Evidence suggests[28] that the loss of emission is caused by the preferential evaporation of barium from the coating. Only the strontium oxide is left, and it is relatively inactive.

As with thoriated-tungsten filaments, oxide-coated cathodes are almost always designed to furnish considerably more emission than they are called upon to supply. If an attempt is made to draw saturation current from them by application of high plate voltage, hot spots usually develop that sometimes give off enough gas to initiate an arc in the tube. Lack of accurate knowledge of the maximum current that an oxide-coated cathode will furnish is therefore not a serious engineering difficulty from the standpoint of design. In general, such a cathode may be expected to supply about 100 to 200 milliamperes per watt when operated at a temperature to give an economical life.

Since oxide-coated cathodes are subject to deactivation by positive-ion bombardment, they are ordinarily used only in tubes with relatively low plate voltage. Because of their tendency to evolve gas, it is difficult to maintain an extremely high vacuum when they are employed in a tube, and so more positive ions are present. In addition, the evaporation of barium to other electrodes in the tube may lead to difficulties both from primary and from secondary emission. An oxide-coated cathode is not practical in high-power tubes where the anode itself is allowed to operate at high temperature, for back radiation from a high-temperature anode causes the cathode temperature to vary with load.

The fact that oxide-coated cathodes may be indirectly heated is one of their important advantages. Figure 12 shows the construction of a cathode of this type. It consists of a nickel cylinder coated on the outside with the oxides and heated from the inside by a tungsten or other heater wire, which may be insulated from the cylinder. With insulated heaters, several cathodes at different potentials may be heated from the same power source. Often, however, the refractory insulation cannot withstand a voltage greater than 100 volts without excessive

[27] E. F. Lowry, "Rôle of the Core Metal in Oxide-coated Filaments," *Phys. Rev., 35* (1930), 1367–1378.

[28] M. Benjamin and H. P. Rooksby, "Emission from Oxide-Coated Cathodes," *Phil. Mag., 15* (1933), 810–829.

leakage current, since the resistivity of the insulator decreases at high temperatures.

In tubes filled with gas or mercury vapor, a different form of indirectly heated cathode can be, and usually is, utilized. Such cathodes are discussed with gas tubes in Ch. V.

All measurements of the foregoing quantitative data for emission from oxide-coated cathodes were made with a continuous emission

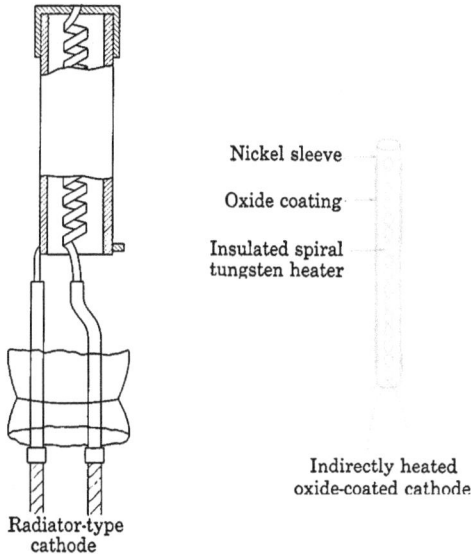

Nickel sleeve

Oxide coating

Insulated spiral
tungsten heater

Indirectly heated
oxide-coated cathode

Radiator-type
cathode

Fig. 12.   Indirectly heated oxide-coated cathodes.*

current. In some applications, however, particularly radar, large momentary pulses of emission current separated by relatively long intervals of zero current are required. For pulses approximately one microsecond long, emission current of the order of 100 amperes per square centimeter with satisfactory life is obtained.[29] This value is some 100 times larger than that for continuous emission. The maximum pulsed emission current is limited by the accompanying temperature rise of the cathode and by sparking or sputtering at the cathode, which results in destruction of the active coating. Consequently, the permissible pulsed emission current decreases with increase of the pulse width and its ratio to the interval of zero current in a repeated cycle.

* The diagram of the radiator-type cathode is after L. R. Koller, *The Physics of Electron Tubes* (2nd ed.; New York: McGraw-Hill Book Company, Inc., 1937), Fig. 15, p. 50, with permission.

[29] E. A. Coomes, "The Pulsed Properties of Oxide Cathodes," *J. App. Phys.*, *17* (1946), 647–654.

10. The Schottky Effect

As is mentioned in the preceding articles, the magnitude of the electric field at the surface of a thermionic emitter affects the rate of emission of the electrons. This influence is exerted through two mechanisms. One, discussed in Art. 11, depends upon the wave nature of the electron; the other, known as the Schottky effect, is most readily observed as a modification of the thermionic emission from a heated cathode. As a consequence of it, the current does not truly saturate as the voltage e_b across the tube in Fig. 9 is increased; rather it continues to increase slowly as long as e_b is increased. Schottky[30] was the first to give a plausible explanation of this behavior in terms of a change which the external field produces in the height of the potential-energy barrier at the surface of the metal.

If the surface of the thermionic emitter is assumed to be perfectly smooth except for the irregularities having atomic dimensions caused by the crystal lattice itself, the force acting on an electron removed from the metal surface by a distance large compared with the spacing of the atoms in the crystal lattice may be computed as though the surface of the emitter were a conducting plane surface. From this force it is possible to compute the shape of the potential-energy curve at the right of Fig. 3, except in a region close to the boundary of the metal. The force between an electron with a charge $-Q_e$ and a plane conducting surface may be found through computing the force between it and its electrical image—a charge $+Q_e$ at an equal distance on the other side of the plane, as shown in Fig. 13. According to Coulomb's law the attractive force is

$$F = \frac{Q_e{}^2}{16\pi\varepsilon_v x^2}, \qquad [30]$$

where x is the distance from the surface to the electron. This force function is plotted in Fig. 14 over the range of x in which it applies. Notice in particular that the equation does not apply when x equals zero, where according to the equation the force would be infinite. The actual value of the force for very small values of x depends upon whether the electron approaches an atom in the lattice or enters the metal between the lattice points of the crystal.

---

[30] W. Schottky, "Über den Einfluss von Strukturwirkungen, besonders der Thomsonschen Bildkraft, auf die Elektronenemission der Metalle," *Phys. Zeits.*, *15* (1914), 872–878; "Weitere Bemerkungen zum Elektronendampfproblem," *Phys. Zeits.*, *20* (1919), 220–228; "Über den Austritt von Elektronen aus Glühdrähten bei verzögernden Potentialen," *Ann. d. Phys.*, *44* (1914), 1011–1032; "Über kalte und warme Elektronenentladungen," *Zeits. f. Phys.*, *14* (1923), 63–106.

If an electric field $\mathcal{E}$ is set up at the surface of the emitter in such a direction as to aid the electrons in their escape from the metal, the electron is still attracted to the metal when close to it where the potential-barrier force predominates, but is pulled away from the metal at large distances where the force $-\mathcal{E}Q_e$ due to the electric field predominates. At a critical distance x_m, the two components of force are

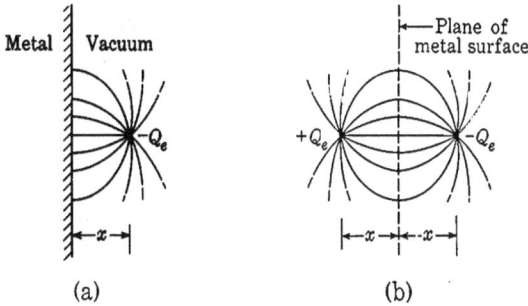

(a)   (b)

Fig. 13.   Electric field between an electron and a metal and between an electron and its image.

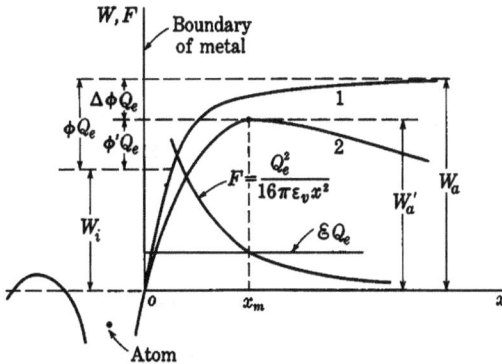

Fig. 14.   Image force and resulting potential-energy barrier.

equal and opposite, and the force on an electron there is zero. For larger distances, the net force on the electron tends to draw it away from the metal, since the force due to the electric field is then larger than the attractive image force. Thus the electron escapes in the presence of the electric field if it has sufficient kinetic energy on leaving the metal to reach the point where x equals x_m. Unless the external field is extraordinarily strong, the critical distance x_m is very large relative to interatomic distances, and the point at which the force on an escaping electron is zero lies well within the range in which the image-force law applies. Thus, at the critical distance,

$$\frac{Q_e^2}{16\pi\varepsilon_v x_m^2} = \mathcal{E}Q_e,$$ [31]

and

$$x_m = \frac{1}{2}\sqrt{\frac{Q_e}{4\pi\varepsilon_v\mathcal{E}}}.$$ [32]

The potential-energy barrier which exists in the absence of an external field is shown by curve 1 in Fig. 14, and the barrier, as modified by the field, by curve 2 in that figure. The reduction in the height of the barrier brought about by the field is the decrease in the work which must be done upon an electron to remove it from the metal. This reduction may be calculated in two parts: first, the reduction in the work needed to move an electron from inside the metal to the critical point; and, second, the reduction in the work needed to move the electron from this point to infinity. Although the force which opposes the outward flight of an electron as it moves from within the metal to the critical point depends upon the path by which the electron leaves the metal, the external field reduces this force at each point by an amount $\mathcal{E}Q_e$, whatever the path of the electron. Thus the field reduces the work required to move the electron to the critical point, where x equals x_m, by an amount $\mathcal{E}Q_e x_m$. Since when the field is present an electron at the critical point is free to escape, the external field reduces the work necessary to move the electron out from the critical point by the total energy which would be required for the electron to escape, starting from the critical point, in the absence of the field. The second part of the reduction in the potential-energy barrier is therefore

$$\int_{x_m}^{\infty} \frac{Q_e^2}{16\pi\varepsilon_v x^2}\,dx = \frac{Q_e^2}{16\pi\varepsilon_v x_m}.$$ [33]

Thus the total reduction in kinetic energy required for the escape of an electron brought about by the external field is

$$W_a - W_a' = \mathcal{E}Q_e x_m + \frac{Q_e^2}{16\pi\varepsilon_v x_m},$$ [34]

where W_a is the height of the potential-energy barrier in the absence of an external field, and W_a' is the height in the presence of the field. If the value of x_m from Eq. 32 is substituted into Eq. 34, the result is

$$W_a - W_a' = \frac{1}{2}Q_e\sqrt{\frac{Q_e\mathcal{E}}{4\pi\varepsilon_v}} + \frac{1}{2}Q_e\sqrt{\frac{Q_e\mathcal{E}}{4\pi\varepsilon_v}}$$ [35]

$$= Q_e\sqrt{\frac{Q_e\mathcal{E}}{4\pi\varepsilon_v}}.$$ [36]

Since the kinetic energy of the electrons within the metal is not affected by the external field, the energy W_i remains unchanged, and so the change in the work function is equal to the change of the height of the potential-energy barrier. In terms of the voltage equivalent of the work function,

$$\Delta\phi = \sqrt{\frac{Q_e\mathcal{E}}{4\pi\varepsilon_v}},$$ [37]

and

$$\phi' = \phi - \sqrt{\frac{Q_e\mathcal{E}}{4\pi\varepsilon_v}},$$ [38]

where

$\Delta\phi$ is the voltage equivalent of the reduction in the work function brought about by the external field,

ϕ' is the voltage equivalent of the effective work function of the metal in the presence of the field.

It follows that the work function is reduced in direct proportion to the square root of the field strength.

To find the effect of the field on the emission current, let J_s be the emission current density in the absence of an external electric field, and J be the corresponding current density in the presence of the field. Then,

$$J_s = AT^2\epsilon^{-\frac{\phi Q_e}{kT}}$$ [39]

and

$$J = AT^2\epsilon^{-\frac{\phi' Q_e}{kT}}.$$ [40]

If ϕ' is replaced by its value from Eq. 38, then

$$J = AT^2\epsilon^{-\frac{\phi Q_e}{kT}}\epsilon^{\frac{Q_e}{kT}\sqrt{\frac{Q_e}{4\pi\varepsilon_v}}\sqrt{\mathcal{E}}},$$ [41]

or

$$J = J_s\epsilon^{\frac{Q_e}{kT}\sqrt{\frac{Q_e}{4\pi\varepsilon_v}}\sqrt{\mathcal{E}}}.$$ [42]

When numerical values for Q_e, k, and ε_v are substituted, Eq. 42 becomes

$$J = J_s\epsilon^{0.44\sqrt{\mathcal{E}}/T}.$$ [43]

where $\mathcal{E}$ is in volts per meter and T is in degrees Kelvin. If Eq. 43 is multiplied by the area of the cathode, the total current is obtained in place of the current density, and

$$I = I_s\epsilon^{0.44\sqrt{\mathcal{E}}/T}.$$ [44]

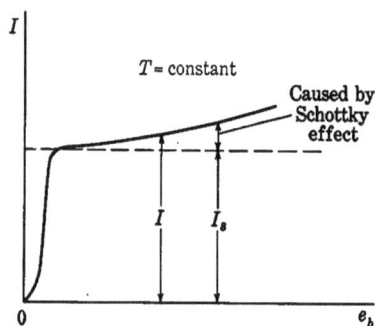

Fig. 15. Typical shape of the volt-ampere characteristic illustrating the Schottky effect.

Equation 44 gives the variation of the emission current with electric field strength. Since the electric field strength at the cathode is directly proportional to the anode voltage e_b for small currents, the Schottky effect becomes apparent in experimental measurements in the manner shown in Fig. 15. To find the true saturation current, I_s, the curve may be extrapolated to intersect the axis at e_b equals zero. This extrapolation is conveniently made on a semilogarithmic plot as in Fig. 16 because Eq. 42 multiplied by the area of the cathode yields

$$\log I = \log I_s + \frac{Q_e}{2.301kT} \sqrt{\frac{Q_e}{4\pi\varepsilon_v}} \sqrt{\mathcal{E}}, \qquad \blacktriangleright[45]$$

which is in the form of an equation of a straight line with a slope of $[Q_e/(2.301kT)]\sqrt{Q_e/4\pi\varepsilon_v}$ and an intercept on the log I axis of log I_s. Actually the extrapolation to find I_s can be carried out with $\sqrt{e_b}$ as the independent variable, since $\mathcal{E}$ is proportional to e_b. The plot should then be a straight line, but the slope is dependent on the factor that relates $\mathcal{E}$ and e_b.

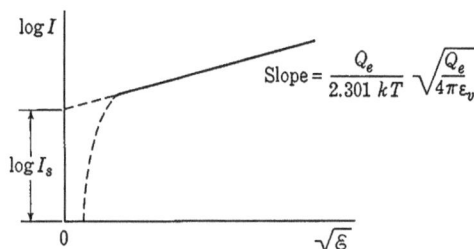

Fig. 16.   Semilogarithmic extrapolation for the determination of the true saturation current.

It must be remembered that this discussion of the Schottky effect is not a complete picture of the effect of an external field upon the emission of electrons by a metal. It does, however, provide a very satisfactory explanation of the major phenomena observed when an electric field is applied at the surface of a heated pure-metal cathode. A different effect of an external field is described in the next article.

11. Field emission

Appreciable amounts of current can be drawn from metals even at low temperatures by the application of very high electric fields at the surface. The magnitude of this effect was formerly thought to be predictable by the Schottky theory of Art. 10, but this theory is now known to be inapplicable, since, when experimentally measured, such currents often appear to be independent of temperature within the limits of the experimental error; whereas, according to the Schottky theory, there would be an appreciable dependence on temperature.[31] This type of emission is sometimes spoken of as pulling of electrons out of metals and is often called the *cold-cathode effect,* or *auto-electronic emission.* It may take place in a vacuum tube at a relatively low anode potential when the field is concentrated at sharp points. As the anode voltage is increased, more and more attention must be paid to the internal construction of vacuum tubes to prevent the concentration of the electric field at the edges or ends of electrodes and the consequent pulling of electrons out of the metal. Such emission is believed to play an important role at the cathode spot in arcs—for example, at the mercury-pool cathode of a rectifier—since the temperature is not thought to be high enough to provide the current by thermionic emission alone. Field-emission current densities of the order of thousands of amperes per square centimeter are possible.[32] Ordinarily such enormous current densities can be maintained only a short time if destruction of the cathode by the heat they produce is to be avoided.

The phenomena of field emission may be explained in terms of the wave nature of electrons. Electrons which approach the potential-energy barrier at the surface of the metal are in part reflected and in part transmitted through the metal surface. If no external field exists at the surface of the metal, the barrier is infinitely thick, as shown by curve 1, Fig. 14, and no electrons with an x-associated kinetic energy less than the top of the barrier can escape. If, on the other hand, an external field is applied, the thickness of the barrier is reduced, as shown by curve 2, Fig. 14, and of all the electrons which approach this barrier with less energy than that needed to surmount it, some will be reflected and some will penetrate it. The stronger the field, the thinner the barrier and the greater the number of electrons that will penetrate it. Since electrons having energies in the upper levels of the Fermi band may escape from the metal in this fashion, and since the number of

[31] S. Dushman, "Thermionic Emission," *Rev. Mod. Phys., 2* (1930), 470–473; A. L. Reimann, *Thermionic Emission* (New York: John Wiley & Sons, Inc., 1934), 64–66.

[32] C. M. Slack and L. F. Ehrke, "Field-Emission X-Ray Tube," *J. App. Phys., 12* (February, 1941), 165–168.

these electrons is tremendously greater than the number excited to higher energy levels by temperature, field emission is substantially independent of temperature. It is the tremendous rate at which electrons of energies in the Fermi band arrive at the metal surface that makes possible the enormous field-emission current densities. Even a value of thousands of amperes per square centimeter is small relative to the random current density of 10^{13} amperes per square centimeter.

The use of wave mechanics in analyses of the penetration of electrons through the potential-energy barrier shows that the field-emission current density from a clean pure-metal surface is given by a relation[33] which may be placed in the following form:

$$J = A_f \mathscr{E}^2 \epsilon^{-(b_f/\mathscr{E})},\qquad [46]$$

where

    J is the density of the field-emission current,

    $\mathscr{E}$ is the electric field intensity at the cathode surface,

    A_f is an approximately constant coefficient,

    b_f is approximately constant, and is determined mainly by the work function of the metal.

The striking similarity between Eq. 46 and Richardson's equation for thermionic emission should be noticed.

Experimental data can be fitted to Eq. 46 very well if A_f and b_f are regarded as constants to be determined to fit the data; in fact, an empirical equation of essentially this form was obtained before the wave-mechanical analysis was made. Experimental measurements[34] of field emission from a pure-metal point are very difficult because of differences in the emission from different crystal faces of a single-crystal point, because of nonuniformities in the field distribution produced by microscopic irregularities in the point, and because of the extreme difficulty of maintaining a clean point of a pure metal at low temperatures, even in the best vacuum at present obtainable.

[33] R. H. Fowler and L. W. Nordheim, "Electron Emission in Intense Electric Fields," *Roy. Soc. Proc. (London),* Series A, *119* (1928), 173–181; L. W. Nordheim, "The Effect of the Image Force on the Emission and Reflection of Electrons by Metals," *Roy. Soc. Proc. (London),* Series A, *121* (1928), 626–639.

[34] T. E. Sterne, B. S. Gosling, and R. H. Fowler, "Further Studies in the Emission of Electrons from Cold Cathodes," *Roy. Soc. Proc. (London),* Series A, *124* (1929), 699–723; E. W. Müller, "Die Abhängigkeit der Feldelektronenemission von der Austrittsarbeit," *Zeits. f. Phys.,* 102 (1936), 734–761; "Weitere Beobachtungen mit dem Feldelektronenmikroskop," *Zeits. f. Phys.,* 108 (1938), 668–680; M. Benjamin and R. O. Jenkins, "Distribution of Autoelectronic Emission from Single-Crystal Metal Points," Part I, *Roy. Soc. Proc. (London),* Series A, *176* (1940), 262–279.

12. SECONDARY EMISSION

A moving particle striking a solid surface may impart sufficient energy to an electron in the solid to enable the electron to escape through the potential-energy barrier at the surface. This process of escape, which was identified in 1902,[35] is called *secondary emission* because a primary particle must first strike the material before the secondary electron can escape. Although it is an incidental and often detrimental effect in some electron devices, secondary emission is

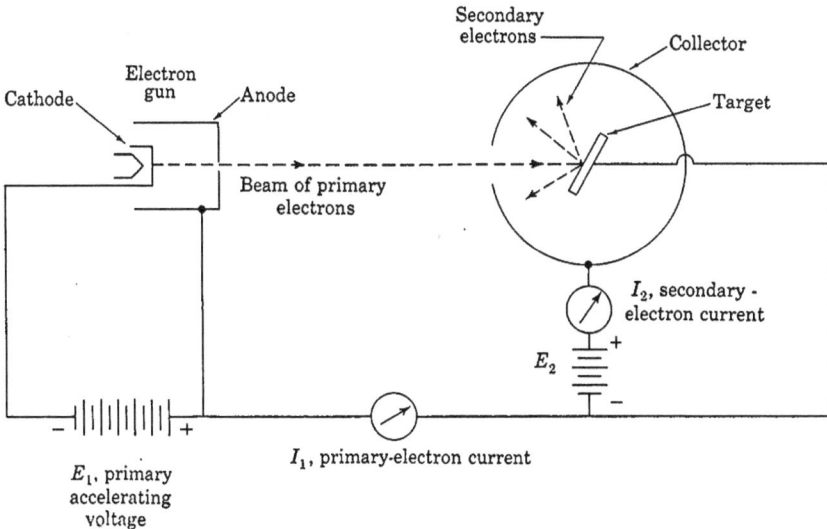

Fig. 17.   Measurement of secondary-electron emission.

responsible for the successful operation of others, including cathode-ray tubes for television and laboratories, electron multiplier tubes,[36] dynatrons,[37] some magnetrons,[38] certain high-performance amplifier tubes,[39] and storage tubes[40] in radar and electronic computers.

[35] L. Austin and H. Starke, "Über die Reflexion der Kathodenstrahlen und eine damit verbundene neue Erscheinung sekundarer Emission," *Ann. d. Phys.*, 9 (1902), 271-292.

[36] J. S. Allen, "Recent Applications of Electron Multiplier Tubes," *I.R.E. Proc.*, 38 (1950), 346-358.

[37] A. W. Hull, "The Dynatron," *I.R.E. Proc.*, 6 (1918), 5-35.

[38] G. Hok, "The Microwave Magnetron," *Advances in Electronics, Vol. II*, L. Marton, Editor (New York: Academic Press, Inc., 1950), 219-250.

[39] C. W. Mueller, "Receiving Tubes Employing Secondary Electron Emitting Surfaces Exposed to the Evaporation from Oxide Cathodes," *I.R.E. Proc.*, 38 (1950), 159-164.

[40] R. A. McConnell, "Video Storage by Secondary Emission from Simple Mosaics," *I.R.E. Proc.*, 35 (1947), 1258-1264; A. V. Haeff, "A Memory Tube," *Electronics, 20* (September, 1947), 80-83; Hans Klemperer, "Repeller Storage Tube," *Electronics, 21* (August, 1948), 104-106.

The method illustrated in Fig. 17 is one of several used for the experimental study of secondary emission. An electron gun directs a beam of primary electrons having an energy of E_1 electron volts through a hole in a collector against a target made of the material to be studied. The secondary electrons leaving the target are drawn to the collector by a voltage E_2 and cause the current I_2, while the primary electrons are responsible for the current I_1. The number of

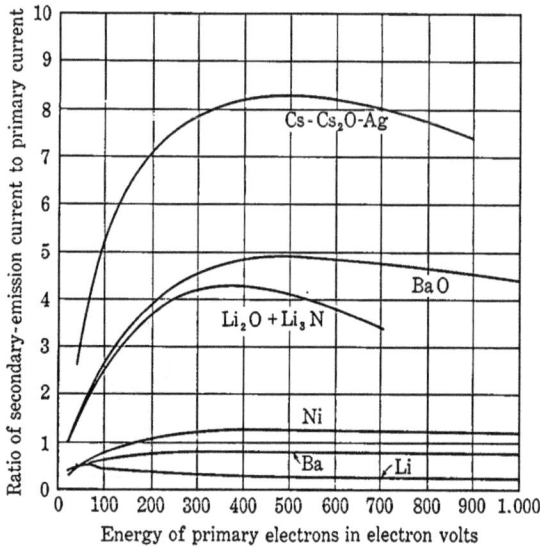

Fig. 18. Secondary-emission ratio as a function of the energy of the primary electrons for various surfaces.*

secondary electrons emitted per primary electron, called the yield or *secondary-emission ratio*, is thus I_2/I_1. This ratio depends on the target material and its surface condition, and on the type of primary particle, its energy, and its angle of incidence at the target.

The secondary-emission ratio is shown as a function of energy of the primary electrons in Fig. 18 for three pure metals and three composite surfaces. These curves are illustrative of the typical behavior of such materials. Secondary emission occurs from most surfaces whenever the energy of the incident electrons is much in excess of the work function of the surface. In many practical tubes, however, it does not become appreciable until the energy of the incident electrons is of the order of

* The curve for Cs-Cs$_2$O-Ag is taken from V. K. Zworykin, G. A. Morton, and L. Malter, "The Secondary Emission Multiplier—A New Electronic Device," *I.R.E. Proc.*, *24* (1936), Fig. 2, p. 355, with permission. The other curves are taken from H. Bruining, "Secondary Electron Emission," *Philips Tech. Rev.*, *3* (1938), Figs. 4 and 5, p. 82, with permission.

20 electron volts. As the energy of the primary electrons is increased, the secondary-emission ratio increases to a maximum at several hundred electron volts, and then decreases gradually. The maximum is less than 2 for pure metals, and less than 10 for most composite surfaces and alloys, although values as high as 16 have been found. It is increased by layers of gas or electropositive metals on the surface of the target, and is larger for many alloys than for any of their constituent materials alone.[41] Because of the sensitivity of the ratio to surface conditions of the target, reproducible results are difficult to obtain, and there is considerable lack of agreement among the numerous investigators who have contributed to the extensive literature on the subject.[42] The secondary emission from some composite surfaces consisting of a thin film of cesium on an intermediate film of cesium oxide covering a silver electrode, such as is used in the phototubes discussed in Art. 13, approaches 60 milliamperes per watt of incident primary-electron energy, a value that compares favorably with the more efficient thermionic emitters.[43]

. 　The energy values of the secondary electrons as they leave the surface are distributed over a wide range. This distribution in energy has been examined by several methods, one of which involves reversing the polarity of E_2 from that shown in Fig. 17. The electric field at the surface of the target is then in a direction to return the secondary electrons to the target; hence only electrons with values of initial energy above that corresponding to E_2 can reach the collector. Variation of the magnitude of this retarding voltage E_2 gives data from which the distribution in energy of the secondary electrons can be found. The curves shown in Fig. 19 are typical of the results obtained when the primary-electron energy is below about 1,000 electron volts. When tne primary-electron energy is 123 electron volts, for example, the curves show that most of the secondary electrons have energies less than 25 electron volts. A smaller group have energies only slightly less than that of the primary electrons. This second group presumably comprises primary electrons that are elastically reflected from the surface. The remaining few secondary electrons have energies spread throughout the range between the two larger groups, and are interpreted as being inelastically reflected primary electrons. As the

　[41] L. R. Koller, "Secondary Emission," *G.E. Rev.*, *51* (April, 1948), 33–40; *51* (June, 1948), 50–52.

　[42] H. Bruining, *Die Sekundar-Elektronen-Emission fester Korper* (Berlin: Springer-Verlag, 1942); K. G. McKay, "Secondary Electron Emission," *Advances in Electronics*, *Vol. I*, L. Marton, Editor (New York: Academic Press, Inc., 1948), 65–130.

　[43] V. K. Zworykin, "Electron Optical Systems and Their Application," *I.E.E.J.*, *79* (1936), 1–10; V. K. Zworykin, G. A. Morton, and L. Malter, "Secondary-Emission Multiplier—A New Electronic Device," *I.R.E. Proc.*, 24 (1936), 351–375.

primary-electron energy is increased to many thousands of electron volts, the distribution in energy of the secondary electrons changes from that shown in Fig. 19 to one with an increased percentage of high-energy electrons.[44]

Secondary electrons are not a factor in the operation of a simple two-electrode vacuum tube because there is no electrode situated near the anode which has a potential nearly the same as or higher than

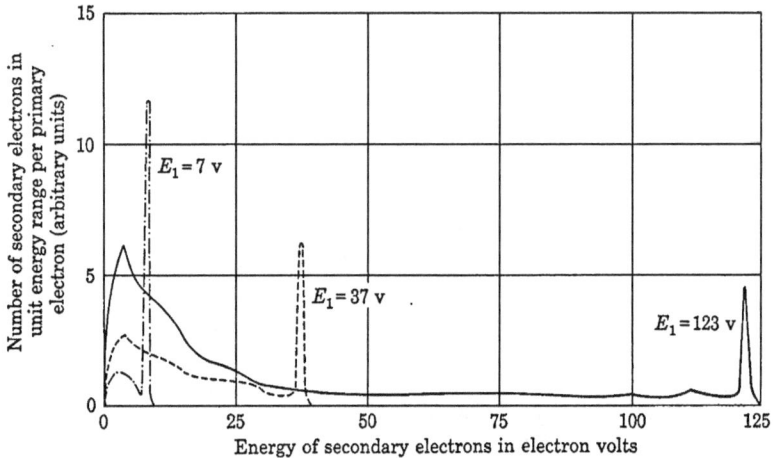

Fig. 19. Curves of the distribution in energy of the secondary electrons from a molybdenum target for three values of primary-electron energy.[*]

that of the anode, and can therefore collect or attract them; hence they do not leave the anode. Secondary emission from electrodes may become a factor in the operation of tubes having three or more electrodes if at least two are positive with respect to the cathode so that the current from the cathode divides between them. In these circumstances, the part of this electron stream that strikes one electrode, such as the target in Fig. 17, causes secondary electrons that can travel to another electrode of nearly the same or higher potential. The net current from the external circuit to the first electrode is reduced by the secondary emission; that to the second is increased. The reduction imposes an important limitation on the operation of tetrode tubes, and elimination of it is a major consideration in the design of many other multi-electrode tubes that are discussed in Ch. IV. To

[44] J. G. Trump and R. J. Van de Graaf, "Secondary Emission of Electrons by High Energy Electrons," *Phys. Rev.*, 75 (1949), 44–45; J. S. Allen, "Recent Applications of Electron Multiplier Tubes," *I.R.E. Proc.*, 38 (1950), 349.

[*] The curves are adapted from L. J. Haworth, "The Energy Distribution of Secondary Electrons from Molybdenum," *Phys. Rev.*, 48 (1935), Fig. 2, p. 90, with permission.

make the secondary emission small, the electrodes in some tubes are coated with carbon or are made of other materials having small values of secondary-emission ratio. Evaporation of the oxide coating from the cathode and subsequent condensation of the active material on the other electrodes often cause difficulty in such tubes because the oxide increases the secondary-emission ratio of a surface that otherwise has a small value. In certain tubes, on the other hand, secondary emission is utilized to enhance the total current to the anode.[45] In these, transfer of the oxide has a deleterious effect because it reduces the secondary-emission ratio at a surface that normally has a high value.

Secondary emission occurs not only from conductors, but also from insulators such as the glass walls and electrode supports of a vacuum tube and the fluorescent screen of a cathode-ray tube. For low-anode voltages, the inner surfaces of the glass walls tend to become charged negatively because of the stray electrons that strike them. At higher voltages, bombardment of the inner surface of the glass by stray electrons may cause an appreciable current of secondary electrons to leave the glass and go to the positive anode. If the rate at which the secondary electrons leave a spot on the wall becomes greater than the rate at which the primary electrons arrive, the spot becomes positively charged and attracts more primary electrons.[46] In high-voltage tubes, this process sometimes becomes cumulative, the wall becomes hot in one spot, the glass softens, and a puncture occurs, spoiling the vacuum.[47] Protecting shields and focusing electrodes are often included to prevent this action in high-voltage tubes with glass walls, such as rectifier or x-ray tubes.

Secondary emission is utilized in most cathode-ray tubes to keep the phosphor used as a fluorescent screen from acquiring a negative charge when struck by the beam. The phosphor used is an insulator. Were it not for secondary emission, its surface would charge negatively until it reached the potential of the cathode, or a slightly lower value, whereupon no further electrons could reach the screen and no visible spot would form. The manner by which secondary emission overcomes this difficulty in a typical cathode-ray tube is illustrated in Fig. 20. For anode-to-cathode voltages in the electron gun smaller

[45] C. W. Mueller, "Receiving Tubes Employing Secondary Emitting Surfaces Exposed to the Evaporation from Oxide Cathodes," *I.R.E. Proc.*, *38* (1950), 159–164; S. Nevin and H. Salinger, "Secondary-Emitting Surfaces in the Presence of Oxide-Coated Cathodes," *I.R.E. Proc.*, *39* (1951), 191–193.

[46] I. Langmuir, "Fundamental Phenomena in Electron Tubes Having Tungsten Cathodes, Part II," *G.E. Rev.*, *23* (1920), 589–590.

[47] E. L. Chaffee, *Theory of Thermionic Vacuum Tubes* (New York: McGraw-Hill Book Company, Inc., 1933), 45, 91.

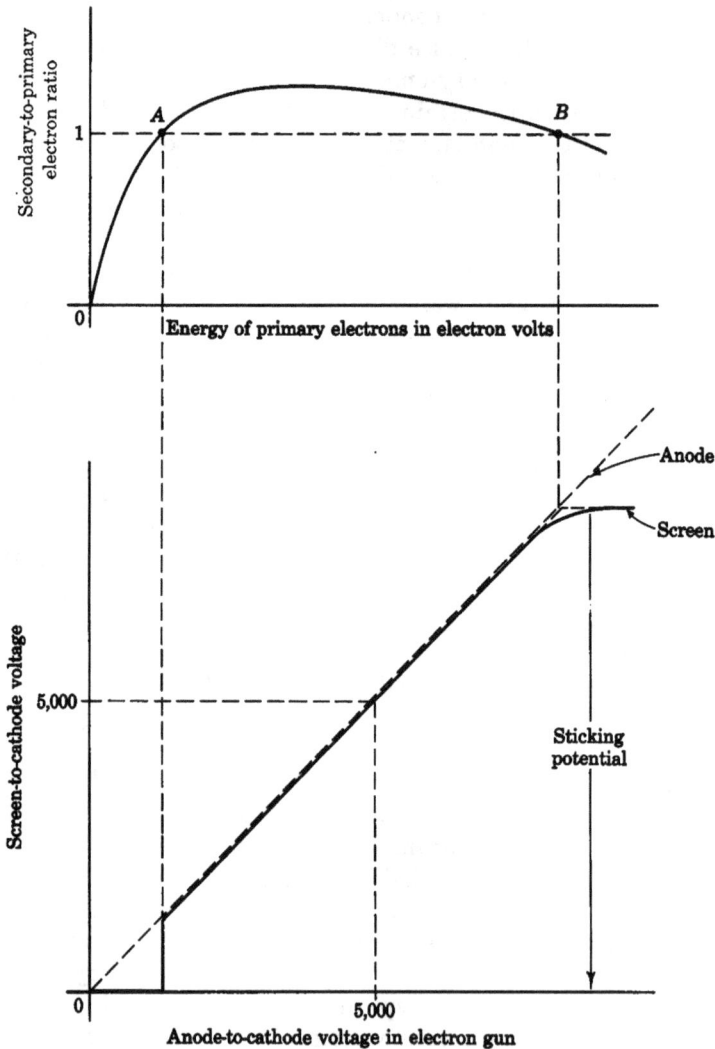

Fig. 20.   Screen-to-cathode voltage in a cathode-ray tube.

than the value corresponding to point A, the first *cross-over point* on the upper curve of the figure, the number of primary electrons exceeds the number of secondary electrons and the screen does charge negatively to nearly the cathode potential, as explained above. Once having reached that value, the screen potential tends to remain there regardless of any increase of electron-gun voltage because the only electrons that can reach the screen have negligible energy.

If, on the other hand, the electron-gun voltage is increased to a value between those corresponding to the first and second cross-over voltages, at points A and B, respectively, there is evidently a possibility that the primary electrons at the screen can have an energy corresponding to some voltage in the range between A and B. Consequently, the screen can lose more secondary electrons than it gains primary electrons, and hence can charge positively. For example, if such a value of electron-gun voltage is applied while the electron beam is turned off by a control means, leakage across the inner surface of the glass bulb tends to bring the screen potential to the same value as that of the anode of the electron gun. When the beam is turned on again, the first electron to arrive at the screen causes emission of more than one secondary electron, and the screen gains a net positive charge. The screen cannot continue to gain positive charge without limit, however. An equilibrium condition will necessarily be reached at which the screen loses secondary electrons at the same rate that it receives primary electrons, if the effects of leakage and other processes by which the screen might lose charge are negligible in comparison with the effect of secondary emission. To produce such an equilibrium when the primary electrons have an energy corresponding to voltages in the range between the cross-over values, the excess secondary electrons must be made to return to the screen. Thus the screen must acquire a potential relative to its surroundings that results in a retarding electric field at its surface sufficient to allow escape of only one secondary electron for each primary electron, and to turn back all others. This retarding field may be produced in either of two ways: either the screen may charge to a potential that is a few volts positive with respect to the potential of the anode of the electron gun, or the space charge of the secondary electrons, discussed in Ch. III, may produce the retarding field. When the beam is concentrated at a small area on the screen, the current density is usually large enough to make the effect of the space charge predominate,[48] and the equilibrium screen potential is a few volts below that of the anode, as is indicated in Fig. 20.

For values of electron-gun voltage above the value that corresponds to the second cross-over point at B in Fig. 20, the screen potential remains at essentially the potential of B because any higher value would result in fewer secondary electrons leaving than primary

[48] W. B. Nottingham, "Electrical and Luminescent Properties of Willemite under Electron Bombardment," *J. App. Phys.*, *8* (1937), 762–778; "Electrical and Luminescent Properties of Phosphors under Electron Bombardment," *J. App. Phys.*, *10* (1939), 73–83.

electrons reaching the screen and a corresponding accumulation of negative charge sufficient to return the screen potential to that of *B*. This limiting value of screen-to-cathode potential, called the *sticking potential*, is a characteristic of the phosphor and lies in the range of 4 to 10 kilovolts for the more common ones. This sticking potential places a limit on the brightness of spot that can be obtained in a cathode-ray tube unless a conducting screen with a direct connection, or other means different from secondary emission, is used for ridding the phosphor of the charge corresponding to the primary electrons.

Compared with electrons, positive ions are relatively inefficient in causing secondary-electron emission. Of the inert gases, the ions of helium are the most efficient,[49] and the efficiency decreases with increasing atomic weight. Only about one in ten 100-electron-volt helium ions produces a secondary electron from a nickel surface, but the efficiency increases with energy and, on the average, each 1,500-electron-volt helium ion produces a secondary electron.[50] Secondary-emission ratios considerably greater than unity are obtained from ions of higher energy.[51] The number of mercury ions required to free one electron is about one hundred times as great as for helium ions of the same energy. The same electric field that accelerates the positive ions toward the cathode surface accelerates the secondary electrons toward the anode; consequently, the secondary electrons always become a part of the space current. Fast neutral atoms also cause emission of secondary electrons, often with about the same efficiency as ions of the same gas.[52]

13. PHOTOELECTRIC EMISSION

Hertz discovered[53] in 1887 that the potential difference necessary to cause a spark between electrodes in air was reduced if the gap was

[49] F. M. Penning, "Liberation of Electrons from a Metal Surface by Positive Ions, Part II," *K. Akad. Amsterdam Proc.*, *33* (1930), 841–857; M. L. E. Oliphant, "The Action of Metastable Atoms of Helium on a Metal Surface," *Roy. Soc. Proc. (London)*, Series A, *124* (1929), 228–242.

[50] M. Healea and C. Houtermans, "Relative Secondary Electron Emission Due to He, Ne, and A Ions Bombarding a Hot Nickel Target," *Phys. Rev.*, *58* (1940), 608–610.

[51] A. G. Hill, W. W. Beuchner, J. S. Clark, and J. B. Fisk, "Emission of Secondary Electrons Under High Energy Positive Ion Bombardment," *Phys. Rev.*, *55* (1939), 463–470.

[52] H. W. Berry, "Secondary Electron Emission by Fast Neutral Molecules and Neutralization of Positive Ions," *Phys. Rev.*, *74* (1948), 848–849.

[53] H. Hertz, "Ueber sehr schnelle electrische Schwingungen," *Ann. d. Phys.*, *31* (1887), 421–448; "Ueber einen Einfluss des ultravioletten Lichtes auf die electrische Entladung," *Ann. d. Phys.*, *31* (1887), 983–1000.

illuminated by light from a second spark gap. Later, Hallwachs[54] found that a negatively charged body loses its charge rapidly when exposed to ultraviolet light. These phenomena, attributed to the emission of electrons under the action of the radiation, were explained by Einstein, who proposed in 1905 that Planck's quantum theory of radiation be applied to photoelectric studies. The philosophical reasoning of Planck indicated, and the photoelectric effect now demonstrates, that light behaves as though it travels in discrete units called quanta. The energy of a *quantum* is equal to hf, where h is the Planck constant and f is the frequency of the radiation. The term *photon* is applied to one quantum of light energy. When a photon of light impinges on a metallic surface, it may transfer enough energy to an electron at the surface of the metal to enable the electron to escape through the potential-energy barrier. For this escape to occur, if the metal temperature is zero degrees Kelvin, the energy of the photon must be equal to or greater than the work function, since the work function is then the least amount of energy that an electron needs to escape. Photoelectric emission from a metal at absolute zero occurs only when

$$hf > \phi Q_e. \qquad [47]$$

The frequency
$$f_0 = \frac{\phi Q_e}{h} \qquad [48]$$

is called the *threshold frequency*. Since

$$f = c/\lambda, \qquad [49]$$

where c is the speed of light and λ is the wavelength of the light, photoelectric emission occurs only if the incident light is such that

$$\lambda < \frac{hc}{\phi Q_e}. \qquad [50]$$

If the wavelength is represented in Angstrom units, denoted by the symbol Å—one Angstrom equals 10^{-8} centimeter—this relation becomes

$$\lambda < \frac{12{,}395}{\phi} \text{ Å}, \qquad [51]$$

where ϕ is in volts. The wavelength

$$\lambda_0 = \frac{12{,}395}{\phi} \text{ Å} \qquad \blacktriangleright[52]$$

is called the *threshold wavelength*.

[54] W. Hallwachs, "Über den Einfluss des Lichtes auf elektrostatisch geladene Körper," *Ann. d. Phys., 33* (1888), 301–312.

If the metal is not at a temperature of absolute zero, a few electrons are excited to higher energy levels by the temperature, and it is possible for a photon of light to give energy to one of these excited electrons and cause emission, even though the wavelength of the incident light is greater than the threshold wavelength computed by Eq. 52. However, the number of the electrons with high energies is very small relative to the total number of electrons in the metal, so that the emission current obtainable in this manner is exceedingly small. The observed emission from metals at room temperature or higher temperatures does not fall suddenly to zero at the threshold wavelength but decreases very rapidly to a minute value in a short range of wavelengths near the threshold.

The transfer of energy from a photon to an electron must occur in the force field represented by the potential-energy barrier at the surface of a metal. The difficulties of satisfying the conditions necessary for the transfer of energy without the presence of the force field are so great that the transfer cannot occur to an appreciable extent in the interior of the metal. Thus photoelectric emission is to be regarded as essentially a surface phenomenon.

In order that appreciable photoelectric emission may occur for all light in the visible spectrum—for which λ lies between about 4,000 and 7,600 Angstroms—the work function must be less than 12,395/7,600 or 1.63 volts. When the work function is less than this value, the energy of the radiation in the infrared portion of the spectrum may also cause photoelectric emission. Composite surfaces having a work function less than this value are available and make possible a great many applications of the photoelectric effect.

The considerations given thus far determine the *possibility* of photoelectric emission. The *amount* of photoelectric emission, however, is dependent on the intensity of the radiation, that is, on the number of light quanta that strike the surface per second. Experiment shows that the photoelectric current emitted from a surface is proportional to the light intensity over a range of at least 10^8 to 1, provided only that the color, or distribution of power with frequency, of the radiation is held constant.

The fundamental mechanism of photoelectric emission may therefore be said to be governed by two laws:

(a) Appreciable photoelectric emission can occur only provided the wavelength of the incident radiation is less than the threshold wavelength, or the frequency greater than the threshold frequency.

(b) The amount of photoelectric emission, or the emission current, is proportional to the intensity of the incident radiation for a fixed frequency distribution of the radiant power. This proportionality does not imply, however, a linear relation between the current and intensity in all practical applications of the photoelectric effect, which is a matter that is discussed in more detail subsequently.

The photoelectric current is a function of the distribution of the intensity with frequency of the light that strikes the surface. Because of the complex nature of the surfaces at which emission occurs, adequate

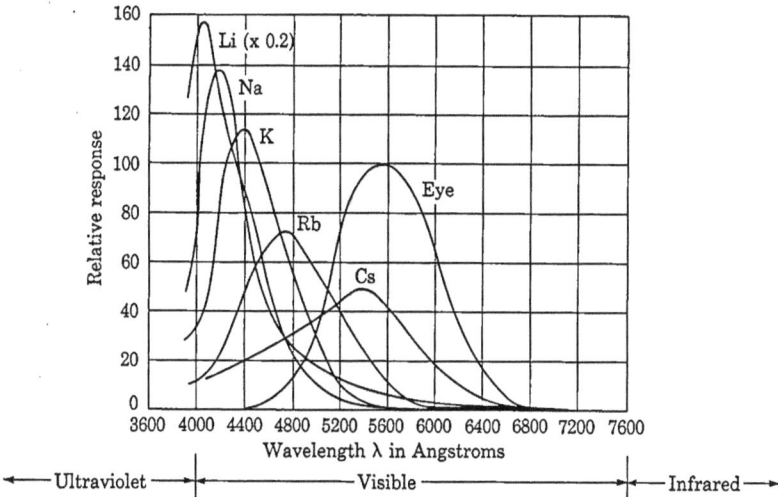

Fig. 21. Relative spectral-sensitivity curves for photoelectric emitters in arbitrary units, and relative-luminosity curve for the eye in per cent.*

quantitative explanations of the observed variations of the photoelectric current with the frequency of the light have not yet been given. For most surfaces, a given amount of radiant power in a small band at the blue end of the visible spectrum—that is, near 4,000 Angstroms—gives a greater photoelectric current than an equal amount of power distributed in a narrow band near the red end—that is, near 7,600 Angstroms. Because of their low work functions, the alkali metals, which have the approximate photoelectric work functions given in Table IV, are the most effective photoelectric emitters when exposed to visible light. Cesium, with the lowest work function of all the alkali metals, responds to the greatest range of the visible spectrum and is therefore much used in phototubes.

* The curves for the alkali metals are adapted from E. F. Seiler, "Color-sensitiveness of Photo-electric Cells," *Astrophys. J.*, 52 (1920), Fig. 4a, p. 143, with permission. The curve for the eye is a plot of data from *Recueil des travaux et compte rendu des séances, Commission Internationale de L'Éclairage, Sixième Session, Genève-Juillet*, 1924 (Cambridge, England: The University Press, 1926), p. 67.

The manner in which the emission current from a surface varies with the color or frequency of the incident radiant power is commonly represented as a spectral-sensitivity characteristic such as is shown in Fig. 21. The photoelectric response $R(\lambda)$ resulting from a given amount of radiant power confined to a narrow wavelength band and incident on the phototube window is plotted as the ordinate, and the mean wavelength λ of the band is plotted as the abscissa. Such curves are ordinarily obtained through passing light from an incandescent source through a prism, selecting a narrow band of wavelengths by means of

TABLE IV*

Metal	*Melting Point, degrees centigrade*	*Work Function, electron volts*	*Threshold Wavelength, Angstrom units*
Lithium	186.0	2.28	5,440
Sodium	97.5	2.46	5,040
Potassium	62.3	2.24	5,530
Rubidium	38.5	2.18	5,680
Cesium	26.0	1.91	6,490
Calcium	810.0	2.70	4,590
Barium	850.0	2.51	4,940

* The melting points are from *International Critical Tables* (New York: McGraw-Hill Book Company, Inc., 1929), Vol. *1*, pp. 103–105; the work functions are from V. K. Zworykin and E. G. Ramberg, *Photoelectricity and Its Application* (New York: John Wiley & Sons, Inc., 1949), Table 2.2, p. 30, with permission, and are selected values representative of those in the available literature; the threshold wavelengths are computed from Eq. 52 and the work functions.

a slit that can be moved through the spectrum, and plotting the ratio of the current given by the phototube when exposed to light from the slit to the current given by a blackened thermopile when exposed to the same light. The blackened thermopile absorbs all radiation equally and therefore gives a measure of the energy at that part of the spectrum. If the thermopile is calibrated to indicate the radiant power, the ordinates of the curves may be indicated in amperes per watt. In Fig. 21, however, the ordinates are to an arbitrary scale. As is indicated by the figure, the alkali metals exhibit a spectral sensitivity curve having a peak at some wavelength in the visible spectrum.

The spectral sensitivity of a phototube is of great importance in determining its suitability for use with a given light source. For example, the radiation from an incandescent tungsten lamp is distributed throughout the spectrum, as is shown in Fig. 22. Much of the power is radiated as heat in wavelengths beyond the visible part of the spectrum—hence the luminous efficiency of such lamps is low. Very little of the power is radiated in the blue end of the spectrum; thus, a blue-sensitive phototube gives relatively little response to an incandescent lamp. The total power from the lamp corresponds to the total area under its power-distribution curve. The total response of a phototube to such a power-distribution curve corresponds to the integral of the product of the spectral-sensitivity function and the power-distribution function, that is, to the area under the curve given by the product of the ordinates at each abscissa. In the same way the response

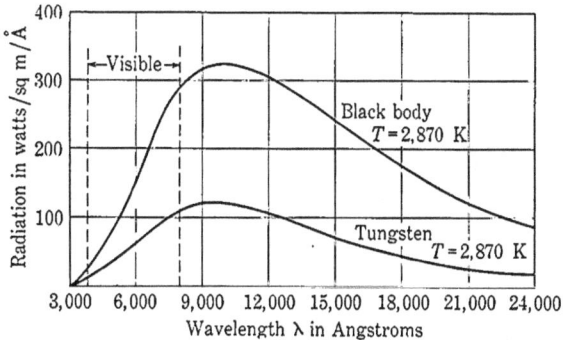

Fig. 22.   Radiation from a black body and from a tungsten-filament lamp.*

of the eye may be obtained from its spectral-sensitivity, or relative-luminosity, curve as given approximately in Fig. 21. When the area under this product curve is multiplied by the proper constant, the result is the lumen output of the lamp. A cesium surface matches the characteristics of the eye more closely than does one of any of the other pure alkali metals.

To increase the over-all sensitivity of phototubes to radiation from commonly used light sources such as the incandescent lamp, efforts have been made to increase the spectral sensitivity in the visible spectrum and extend it into the infrared region. Certain thin films of alkali metals on surfaces of other metals are found to have lower photoelectric work functions than either metal alone. This effect is similar to the reduction of the thermionic work function obtained with a thin film of thorium on tungsten. Combinations comprising a thin film of cesium on an oxidized backing surface of silver, on an antimony surface, or on a bismuth surface have received the greatest

* These curves are adapted from V. K. Zworykin and E. G. Ramberg, *Photoelectricity and Its Application* (New York: John Wiley & Sons, Inc., 1949), Fig. 2.4, p. 18, with permission.

commercial development.[55] Spectral-sensitivity curves for three standard thin-film emitters are shown in Fig. 23. The curve for cesium on cesium oxide on silver exhibits a double peak and extends far into the

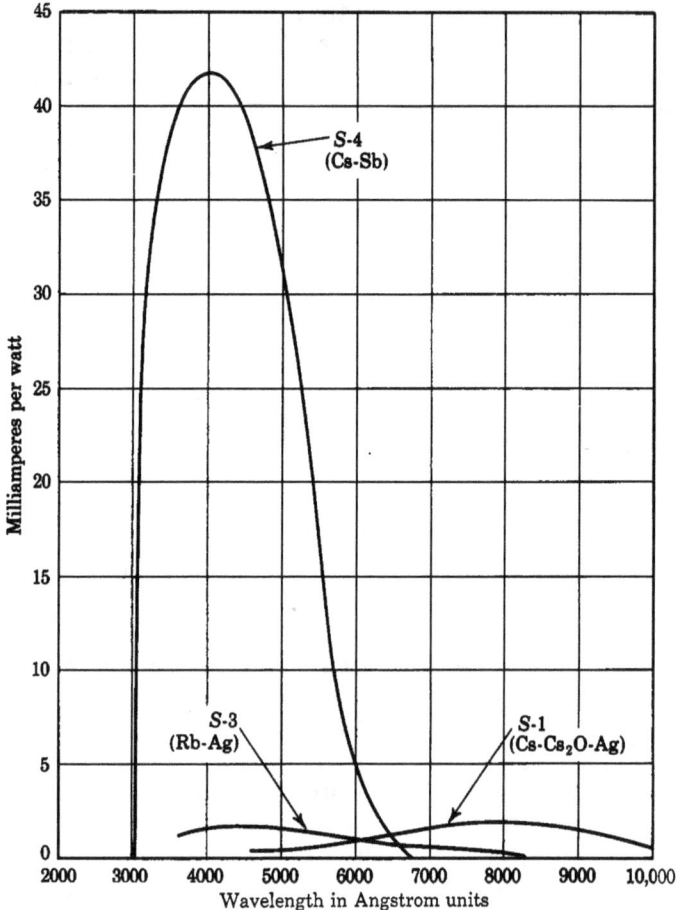

Fig. 23. Absolute spectral-sensitivity curves for three standard thin-film photocathodes.*

infrared region of the spectrum. This surface is therefore well suited for use with an incandescent source. Because of its higher absolute sensitivity, however, the cesium-on-antimony surface gives greater

[55] V. K. Zworykin and E. G. Ramberg, *Photoelectricity and Its Application* (New York: John Wiley & Sons, Inc., 1949), Ch. 3.

* These curves are taken from V. K. Zworykin and E. G. Ramberg, *Photoelectricity and Its Application* (New York: John Wiley & Sons, Inc., 1949), Fig. 20.1, p. 467, with permission.

total response to light from the usual incandescent lamp even though the peak of its spectral-sensitivity curve is at the blue end of the spectrum. The cesium-on-antimony surface is widely used as a semitransparent photocathode on a glass wall. Such a transparent cathode permits design of photoelectric devices in which light penetrates the cathode and releases electrons from the inner side.

A typical phototube, shown in Fig. 24, consists of a sensitive cathode surface of large area sealed in an evacuated bulb with an anode, or electron collector, in the form of a wire or ring. The anode is made small in order that it may not obstruct the passage of light to the cathode. The many ways of preparation of cathode surfaces constitute an art in themselves.[56] In potassium tubes, the alkali metal is often distilled into the bulb while the bulb is attached to the vacuum system. The metal coats the inner face of the glass, except for a region left for the entrance of light, and forms the active surface. Sodium is sometimes deposited on the inner surface of the bulb by electrolysis of sodium ions directly through the glass. Cesium is usually introduced by being formed from a chemical salt. A pellet of a mixture of the salt and a powdered metal—for example, cesium chloride and calcium, or cesium dichromate and silicon—is held in a small metallic container

Fig. 24.
A typical
phototube.
*(Courtesy
Radio
Corporation of
America.)*

in the bulb.[57] When the cesium is to be freed, the pellet is heated by induction, and the cesium, replaced in the salt by the powdered metal, boils out and condenses to form the active surface. The sensitivity may vary over a range of a thousand to one, or more, depending on the method of preparation.

The characteristics of the phototube for variable voltage and variable light flux are usually similar to the curves shown in Figs. 25 and 26, respectively, where i_b is the current through and e_b the voltage across the phototube, as indicated in Fig. 27. With the light flux held constant as in Fig. 25, a small current occurs even when e_b is equal to zero, because of the excess kinetic energy given to the electrons by the light quanta and the resulting initial velocities of the electrons as they leave the cathode. If the anode is made positive with respect to the cathode, the electrons are drawn from the cathode, and the current

[56] A. M. Glover, "A Review of the Development of Sensitive Phototubes," *I.R.E. Proc.*, 29 (August, 1941), 413–423.

[57] W. H. Nickless, "Manufacture of Caesium-Silver-Oxide Photocells," *Electronics*, 4 (August, 1932), 255–256; C. H. Prescott, Jr., and M. J. Kelly, "The Caesium-Oxygen-Silver Photoelectric Cell," *B.S.T.J.*, 11 (1932), 334–340.

increases. In the usual type of vacuum phototube, anode voltages of
50 volts or less draw all the electrons to the anode as rapidly as they
are emitted, and additional voltage does not result in an appreciable
increase of current above the resulting fixed "saturation" value. For
voltages below the value that draws all the electrons across, many of
the emitted electrons miss the small anode in their flight and return to

929

Average anode characteristics

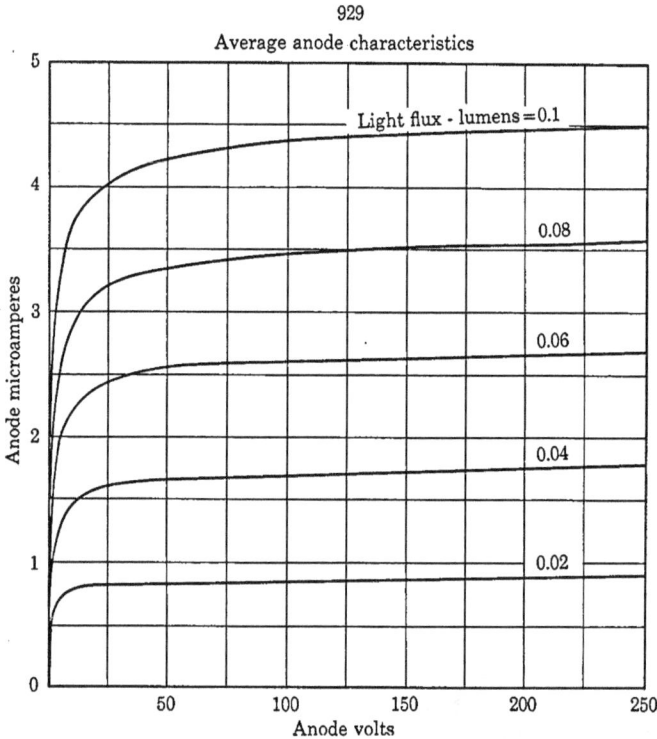

Fig. 25. Current through the vacuum phototube of Fig. 24 as a function
of its terminal voltage. (*Data by courtesy of Radio Corporation of
America.*)

the cathode. This is one example of a situation in which the space or
conduction current does not equal the emission current. Space charge,
which plays an important part in the thermionic tube and is discussed
in Ch. III, is comparatively unimportant in the phototube, because
the photoelectric currents are small, often less than a microampere.
For voltages above the minimum value that produces the saturation
current, an increase of light flux increases the saturation current in
direct proportion, as is illustrated by the curves of Fig. 26.

In the utilization of a vacuum-type phototube, a direct voltage E_{bb}
of the proper polarity to draw electrons from the cathode is connected

across a high-resistance R and the phototube in series, as shown in Fig. 27. The voltage drop across the resistance is generally used to excite the grid of a vacuum tube. When the phototube voltage is held constant and the light flux varied, the current is directly proportional to the light flux, as shown in Fig. 26. For all phototube voltages in excess of the saturation value, the curves lie essentially one on another. For phototube voltages below the saturation value, however, the current yield per lumen, while constant, is reduced in magnitude. The sensitivity of the phototube expressed in microamperes per

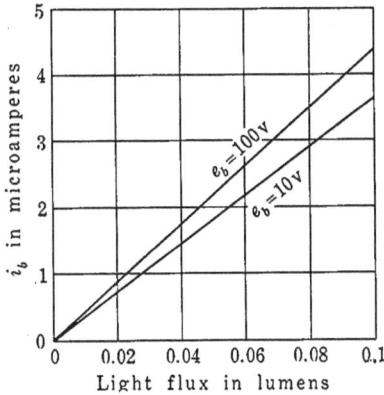

Fig. 26. Current from the vacuum phototube of Figs. 24 and 25 as a function of light flux.

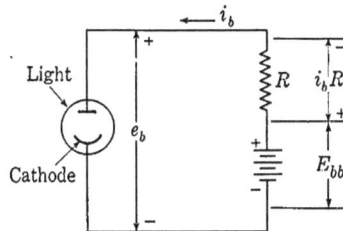

Fig. 27. Phototube connected in a circuit.

lumen is the slope of the upper curve in Fig. 26. Note that the value of the resistance R does not affect the current appreciably as long as the voltage of the battery is sufficient to keep the voltage across the phototube higher than the saturation value for all conditions of operation. Usually R is of the order of 5 to 25 megohms, the current is of the order of a microampere, and the applied battery voltage is of the order of 90 volts.

Gas is deliberately introduced in some phototubes to increase the yield of current for a given luminous flux. The increase of current, from fivefold to tenfold, results from ionization by collision. The amount of the increase is a critical function of the voltage applied to the phototube, as is explained in considerable detail in Art. 7, Ch. III. The gain in effective sensitivity resulting from the use of gas is obtained only at some sacrifice of linearity and speed of response.

Electron multiplication through secondary emission may also be used to increase the yield of current in a phototube. One device utilizing this principle is illustrated in Fig. 16, Ch. I. To eliminate the requirement of an auxiliary magnetic field inherent in that method, various types of electrostatic electron-multiplier phototubes have been originated.[58] A cross section of a common type is shown in Fig. 28. The electrodes in this tube are a photocathode 0, a series of secondary emitters 1 to 9, called *dynodes*, and an anode 10. It is designed for connection to a voltage source and load resistance R as shown in Fig. 29. The dynodes are maintained at increasing potential in sequence from cathode to anode by taps on a resistance voltage divider. Dynode 1, being at higher potential than photocathode 0, attracts photoelectrons released by light incident on it. Upon reaching dynode 1 each photoelectron produces, say, δ secondary electrons —the electron paths are drawn in Fig. 28 to illustrate the condition when δ equals two. Attracted to dynode 2 because of its still higher potential, these secondary electrons produce δ^2 new secondary electrons for each original photoelectron, and so on. Thus δ^9 electrons reach anode 10 for each

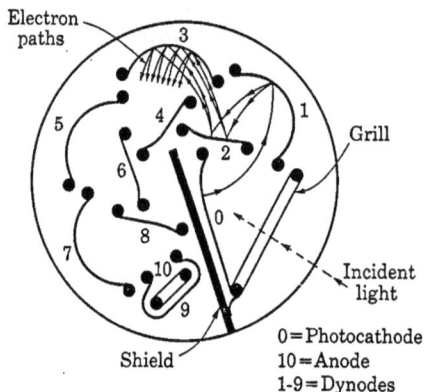

Fig. 28. Schematic cross section of a circular electrostatic multiplier phototube. (*Adapted by courtesy of Radio Corporation of America.*)

0 = Photocathode
10 = Anode
1–9 = Dynodes

photoelectron from the cathode—the multiplication, or current amplification, is δ^9. This total multiplication depends on the voltage impressed between the dynodes because the multiplication δ in each stage is a function of the energy of the incident electrons in the manner shown in Fig. 18. Actually, in a practical tube, not all the secondary electrons from one dynode reach the next. Some are lost in each stage, and the chief design problem in this type of tube is to shape the electrodes so as to reduce such escape and thereby make the current amplification large. For 50, 75, and 100 volts per dynode, the current amplification in the multiplier phototube of Fig. 28 is 8,000, 150,000, and 1,000,000, respectively.

Since the final dynode 9 almost surrounds the anode 10, a relatively small voltage e_b from dynode 9 to the anode is sufficient to collect

[58] H. E. Iams and B. Salzberg, "The Secondary Emission Phototube," *I.R.E. Proc.*, *23* (1935), 55–64; V. K. Zworykin, G. A. Morton, and L. Malter, "The Secondary Emission Multiplier—A New Electronic Device," *I.R.E. Proc.*, *24* (1936), 351–375; G. Weiss, "On Secondary Electron Multipliers," *Zeit. f. tech. Phys.*, *17* (1936), 623–629; R. W. Engstrom, "Multiplier Photo-Tube Characteristics: Application to Low Light Levels," *J.O.S.A.*, *37* (1947), 420–431.

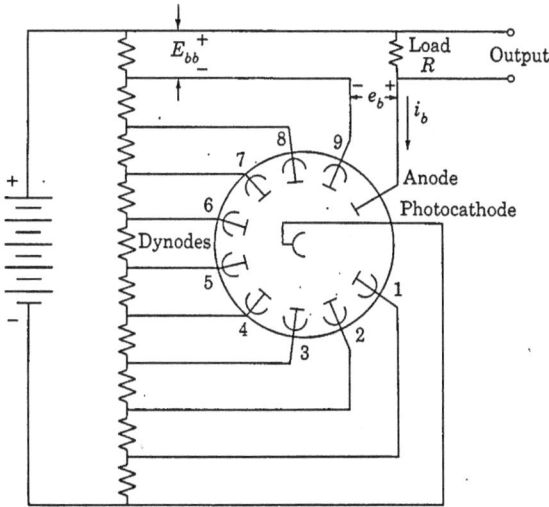

Fig. 29. Electrode connections for a multiplier phototube and load.

931-A
Average anode characteristics

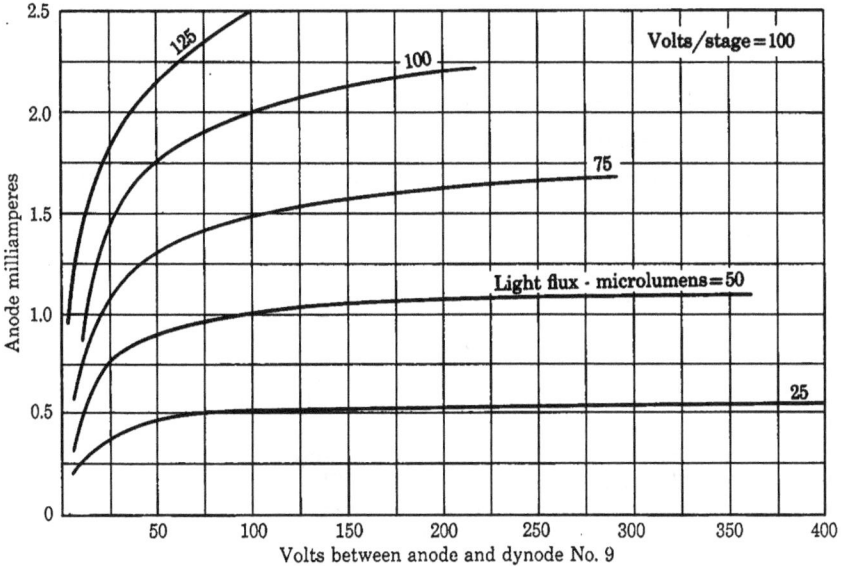

Fig. 30. Characteristic curves for a multiplier phototube. (*Data by courtesy of Radio Corporation of America.*)

essentially all the electrons from this dynode. Hence the output current i_b and the voltage e_b for a particular constant voltage between dynodes are related in the manner shown in Fig. 30.

PROBLEMS

1. By how many volts must the voltage equivalent of the work function change to reduce the emission of tungsten at a temperature of 2,400 K by 10 per cent?

2. The saturation current from two filaments of the same dimensions is measured for a temperature of 2,620 K. One filament is tungsten, the other is a material having an equal emission constant A but a work function half as great. What is the ratio of the emission currents?

3. Design a tungsten filament for a rectifier tube to deliver $\frac{1}{2}$ amp emission current. The filament voltage is to be 10 volts. Assume a temperature of 2,500 K. Give:

   (a) dimensions of filament,
   (b) heating current,
   (c) power to heat filament.

4. If all linear dimensions of a cylindrical tungsten wire filament are doubled, and the temperature is held constant, what changes occur in the emission current, heating power, life, heating voltage, and heating current?

5. Calculate the temperature consistent with 1,000 hours of life at which a tungsten filament should be operated if it is 32 mm long by 0.003 in. in diameter.

6. A vacuum diode has a 0.005-in. diameter tungsten filament of 2-in. effective length. The filament current is adjusted to maintain the filament at an initial temperature of 2,400 K. By what percentage will the emission have changed at the end of the life of the filament,

   (a) if the filament current is held constant?
   (b) if the filament voltage is held constant?

The filament may be assumed to be ideal.

7. A tungsten-filament vacuum tube is to furnish a continuous steady thermionic-emission current. The filament is 32 mm long and 0.003 in. in diameter. Assume that the tube is to be replaced after 10 per cent of the filament mass has evaporated.

   (a) If the cost of the tube is $1.00, and the cost of filament energy is 10 cents per kilowatt-hour, what is the minimum cost per ampere hour of saturation emission current over an extended period of time?
   (b) What will be the corresponding life in hours?
   (c) How does the cost of the tube compare with the total cost of the filament energy when the total cost of both is a minimum?
   (d) If approximations are used in the solution, state their nature and discuss any possible errors introduced by them. What additional considerations would determine the actual choice of filament current to be used?

8. At what temperature will a pure tungsten filament give an emission one-fiftieth as much as that from a thoriated-tungsten filament of the same dimensions if their temperatures are the same? The emission constants for thoriated tungsten may be taken as A equals 3.0 amp per cm^2 per (deg K)^2 and b equals 31,500 K.

9. (a) Approximately what emission could be expected from a tungsten filament 0.006 in. in diameter and 4 cm long when heated to a temperature of 2,400 K?

(b) What would be the temperature of a thoriated-tungsten filament of the same dimensions to give the same emission as that obtained in (a)? For thoriated tungsten, the emission constant A may be taken as 3.0 amp per cm^2 per (deg K)^2 and b as 31,500 K.

10. Construct a curve for thoriated tungsten on the power-emission chart, Fig. 11, using the data of Table III for the heating power, and the constants A equals 3.0 amp per cm^2 per (deg K)^2 and b equals 31,500 K in Richardson's equation for the emission current. Is it a straight line? Why?

11. Calculate the ratios of the emissions from pure-tungsten, thoriated-tungsten, and oxide-coated surfaces at the respective normal operating temperatures of 2,400 K, 1,900 K, and 1,000 K.

12. A typical vacuum triode has a cylindrical, oxide-coated, unipotential cathode with the following dimensions:

$$\text{Outside diameter} = 0.045 \text{ in.}$$
$$\text{Total length} = 26 \text{ mm}$$
$$\text{Coated length} = 20 \text{ mm.}$$

The rated operating conditions given by the manufacturer are:

$$\text{Cathode-heating voltage } E_f = 6.3 \text{ volts}$$
$$\text{Cathode-heating current } I_f = 0.3 \text{ amp}$$
$$\text{Normal plate current } i_b = 0.010 \text{ amp.}$$

(a) On the assumptions that none of the power is radiated from the ends of the cylindrical cathode, and that the radiation is uniform along the whole of the total length, calculate the total cathode emission, using the curve labeled 1934 on the power-emission chart, Fig. 11.
(b) What is the emission efficiency?
(c) What is the ratio of the normal plate current i_b to the saturation current I_s?
(d) Could you measure the saturation current? If so, how?
(e) Repeat the calculations for a triode which has a ribbon filament with a cross section of 0.020 in. by 0.0024 in. and a length of 110 mm, and rated values of filament voltage E_f equal to 2.5 volts, filament current I_f equal to 1.5 amp, and a normal plate current i_b of 0.035 amp.

13. What types of thermionic cathodes can be best employed in each of the following tubes? Give reasons.

(a) X-ray tubes.
(b) Kenotrons.
(c) Radio receiving tubes.
(d) Cathode-ray tubes.
(e) Gas-filled rectifier tubes.
(f) Radio transmitter tubes (plate voltage below 17,000).

14. Although the three common types of thermionic emitters have widely different emission efficiencies (milliamperes emission per watt of heating power), all three are nevertheless in general commercial use. Why is this true?

15. Determine the multiplication per stage in the multiplier phototube of Fig. 28 when it is operated at 50, 75, and 100 volts per dynode.

CHAPTER III

Electrical Conduction through Vacuum, Gases, and Vapors

In all the electronic devices discussed in Ch. I, the space through which the charged particles move is very highly evacuated. The number of gas molecules present is negligible, and their effects may therefore be ignored. Moreover, the current conducted across the interelectrode space by the charged particles is so small that the current density is not more than a few microamperes per square centimeter at any point. In these circumstances, the effect on the electric field of the charged particles in the interelectrode space and the effect of the gas molecules on the motions of the charged particles are negligible. Analysis of the operation of these devices hence is properly made on the assumption that the forces acting on the charged particles are only those caused by the electric and magnetic fields that may be present, and that these are in no way influenced by the presence of the particles.

Not all electronic devices may be analyzed so simply. In a large class—that of vacuum tubes, in which effects of the charges in the interelectrode region are important—the current density is relatively large, the electric field is influenced materially by the presence of the charged particles, and the study of the devices therefore must include the effect of space charge. The analysis is more involved, for the current and the electric field become interdependent quantities. Neither can be specified as a starting point in the analysis; rather, both must be sought together. For this reason, among others, the analysis of the behavior of many such devices is too difficult to be carried through to completion, and in their design resort to experimental methods is often made. The approximate behavior of certain vacuum electron tubes of relatively simple geometry can nevertheless be predicted and explained analytically, as is shown in Art. 1 following. This analysis is applicable to any electronic device in which the current is carried only by charged particles of one sign.

In still other electronic devices, the number of gas molecules in the interelectrode space is not so small that they have a negligible effect upon the operation of the tube. Because of the gas, many new phenomena occur in such tubes, and the analysis of their behavior is more complex. Sometimes the gas is present because of imperfect evacuation of an electron tube intended to function as a vacuum device, and the effect of the gas is an undesired modification of the characteristics of the tube. At other times the gas is deliberately introduced to produce a device with characteristics entirely different from those of vacuum tubes. The phenomena that occur in electron tubes because of the presence of gas are discussed in the later articles of this chapter.

1. LIMITATION OF CURRENT BY SPACE CHARGE

In general, current is conducted between the electrodes of electronic devices by both positively and negatively charged particles, that is, by positive ions, negative ions, and electrons. Sometimes, however, the

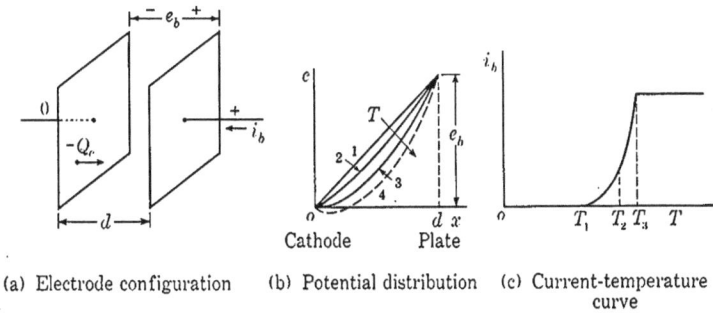

(a) Electrode configuration   (b) Potential distribution   (c) Current-temperature curve

Fig. 1. Potential distribution and current-temperature curve of a vacuum diode with infinite, parallel plane electrodes.

current is carried essentially by particles of one sign only, and the analysis is then greatly simplified. An example of such conduction is that found in a vacuum tube with a thermionic or photoelectric cathode. Because most of the gas is removed from the tube, positive and negative ions are not present in appreciable quantities, and the current is conducted by electrons only.

In the vacuum tube, the repulsive force between the electrons as they cross the interelectrode space places an upper limit on the magnitude of the current that can be conducted for a given applied voltage and geometrical configuration of the electrodes. If the plate voltage e_b in a diode such as that shown in Fig. 1, Ch. I, is held constant, and the temperature of the cathode is increased, the current through the tube increases at first in accordance with Richardson's equation but ultimately reaches a maximum value. Further increase of the cathode temperature does not then result in an appreciable further increase of current. This limitation of the current results from the repelling effect that the electrons in the space exert on those about to leave the cathode.

The effect may be visualized with the aid of Fig. 1, where for the sake of mathematical simplicity the cathode and plate are assumed to be parallel equipotential planes of dimensions large compared with the distance d between them. If the cathode temperature T is so low that no electrons are emitted, the electric field is uniform throughout the region between the plates, and for a given potential difference e_b the potential distribution in the space may be represented by a straight line, as shown by curve 1 in Fig. 1b. As the temperature of the cathode is increased and electrons are drawn across the space by the electric field, their negative charge reduces the potential in the space to some such curve as curve 2. The current then carried by the electrons might be that corresponding to T_2 in Fig. 1c. With increasing cathode temperature, the plate current i_b increases, and the curve of the potential distribution is forced lower and lower. When the charge in the space becomes so great that the potential distribution curve is depressed until its slope at the cathode is zero, as shown by curve 3, the electric field, or force on the electrons, at the cathode is zero. Still further increase of the cathode temperature results in such a curve as 4 in Fig. 1b, and the electric field at the cathode is then reversed. The force on the electrons at the cathode is back toward the metal, but some of them are emitted from the cathode with sufficient initial velocity to overcome the retarding force and pass the point of minimum potential. They are then drawn on to the anode by the electric field to the right of that point. Electrons having smaller values of initial velocity, however, are turned back by the retarding force and re-enter the cathode. The space current conducted from electrode to electrode is hence smaller than the emission current, and further increase of the cathode temperature beyond a value illustrated by T_3 in Fig. 1c does not result in an appreciable increase of the current i_b. *The current is said to be limited by the space charge of the electrons.*

In most practical applications of vacuum tubes, the plate voltage e_b is large compared with the retarding voltage represented by the minimum point on curve 4 in Fig. 1b. Under these conditions, the current for curve 3, for which the electric field at the cathode is zero, is essentially the same as the true space-charge-limited current that corresponds to curve 4. The following discussion therefore assumes that the condition of zero field at the cathode is the critical condition for the space-charge limitation of the current. When the effect of the initial velocities and the potential minimum is included, the mathematical treatment of the problem is much more difficult and results[1] in the appearance of correction terms involving the cathode

---

[1] T. C. Fry, "Thermionic Current between Parallel Planes; Velocities of Emission Distributed according to Maxwell's Law," *Phys. Rev.*, *17* (1921), 441–452; I. Langmuir, "Effect of Space Charge and Initial Velocities on the Potential Distribution and Thermionic Current between Parallel Plane Electrodes," *Phys. Rev.*, *21* (1923), 419–435.

temperature and the ratio of the actual current to the saturation current.

Another qualitative view useful in illustrating the effect of the space charge of the electrons is shown in Fig. 2. In Fig. 2a, the electrons are pictured in transit to the anode, with lines of electric force drawn between them and the positive charges at the anode. The magnitude of the electric field at any point between the plates equals the slope of the potential distribution curve in Fig. 1b and equals the density of the lines of force in Fig. 2. The density of the lines at the cathode is reduced by the presence of electrons in the space; and if the emission is increased sufficiently, the number of electrons in the space increases

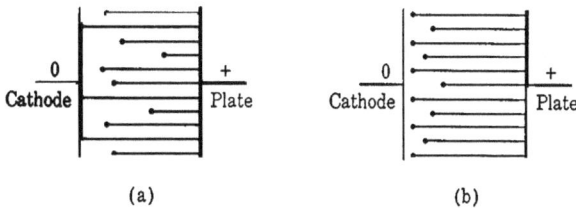

0   Cathode   +   Plate   0   Cathode   +   Plate

(a)   (b)

Fig. 2. Pictorial representation of the electrons and the lines of force in a vacuum diode.

until all the lines of force extending from the anode end on electrons, as in Fig. 2b. Then no additional electrons are drawn from the cathode except to replace those that arrive at the anode, since the electric field at the cathode is zero. The potential distribution corresponding to Fig. 2b is that of curve 3 in Fig. 1b; consequently, Fig. 2b illustrates the conditions in the diode when the current is limited by space charge and the effect of the initial velocities is neglected. For a more nearly exact picture of the field in the tube when the current is limited by space charge, the initial velocities must be taken into account and Fig. 2b must be slightly modified to show a small retarding field at the cathode surface and a position of zero field at a short distance from the cathode.

On the basis of the following five assumptions, a quantitative analysis of the problem can be made:

(a) The electrodes are infinite, parallel, plane, equipotential surfaces.

(b) The number of electrons emitted at the cathode exceeds the demand.

(c) The electrons start from rest at the cathode.

(d) Electrons only are present in the space.

(e) The anode voltage is constant and has been constant sufficiently long for the current to reach its steady-state value.

The first step in the analysis is to derive Poisson's equation, which may be developed from Gauss's theorem[2] that the surface integral of the normal component of the electric displacement leaving any closed surface is equal to the total charge enclosed by the surface. Gauss's theorem may be expressed in the form

$$\int_{\substack{\text{closed} \\ \text{surface}}} D_n \, ds = \int_{\substack{\text{enclosed} \\ \text{volume}}} \rho \, dv, \qquad [1]$$

where

D_n is the component of electric displacement normal to the surface,

ds is an element of area on the closed surface,

ρ is the charge density in the volume enclosed by the surface,

dv is an element of the volume enclosed by the surface.

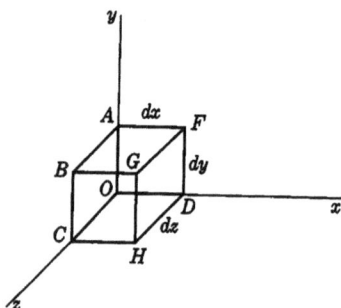

Fig. 3.   Element of volume for the derivation of Poisson's equation.

By means of this theorem, Poisson's equation in rectangular coordinates is derived as follows. If the cubic element of volume in Fig. 3 lies in an electric field, and the rectangular components of the electric displacement are D_x, D_y, and D_z, the flux of displacement inward through the face $OABC$ is

$$D_x \, dy \, dz,$$

and the flux of displacement outward through the opposite face $DFGH$ is

$$\left(D_x + \frac{\partial D_x}{\partial x} dx \right) dy \, dz.$$

The net flux outward through the faces perpendicular to the x axis is then

$$\frac{\partial D_x}{\partial x} \, dx \, dy \, dz.$$

The addition of similar expressions for the other directions gives the total outward flux as

$$\left(\frac{\partial D_x}{\partial x} + \frac{\partial D_y}{\partial y} + \frac{\partial D_z}{\partial z} \right) dx \, dy \, dz,$$

[2] N. H. Frank, *Introduction to Electricity and Optics* (2nd ed.; New York: McGraw-Hill Book Company, Inc., 1950), 41–45.

which, according to Gauss's theorem, equals $\rho \, dx \, dy \, dz$. Furthermore,

$$D_x = \varepsilon \mathcal{E}_x = -\varepsilon \frac{\partial e}{\partial x}, \qquad [2]$$

where ε is the dielectric constant, $\mathcal{E}_x$ is the electric field intensity in the x direction, and e is the electric potential* at any point in the field. Since relations equivalent to Eq. 2 hold for the other field components,

$$\frac{\partial^2 e}{\partial x^2} + \frac{\partial^2 e}{\partial y^2} + \frac{\partial^2 e}{\partial z^2} = -\frac{\rho}{\varepsilon}. \qquad \blacktriangleright[3]$$

Equation 3 is known as *Poisson's equation*. It, together with boundary conditions, governs the electric field in any region in which a charge is distributed in the space.

In order to carry further the analysis of the diode with parallel plane electrodes, shown in Fig. 1, assume that

x is the distance from the cathode to any point in the tube,

e is the potential,

ρ is the charge density,

J is the current density,

v is the velocity of the electrons.

The quantities e, ρ, and v are measured at a distance x from the cathode of the tube and are therefore functions of x. In addition, suppose that

e_b is the potential of the plate,

d is the distance from the cathode to the plate.

The quantities e and e_b are the potentials measured with respect to the potential of the cathode. Alternatively, e_b is the potential rise from the cathode to the plate, or the potential drop from the plate to the cathode. Since the geometry of the tube requires that the vectors representing the velocity of the electrons and the current density at each point in the tube be directed perpendicularly to the electrode planes, only the x components of these vectors need be considered. These are therefore represented here by the symbols v and J, respectively. Since the electrons actually move from the cathode to the plate, it is convenient to call velocities in this direction positive. Similarly, because the charge of the electron is negative, the actual

---

* In this article, the lower-case letters, e, e_b, and i_b are used despite the fact that the voltages and current they represent are supposed not to vary with time. The expressions derived are valid if these quantities vary, provided the variations are so slow that the displacement current in the interelectrode region is negligible. The effects of time variations of electrode voltages are discussed in Ch. IV.

direction of the current in the interelectrode space is from the plate to the cathode, and currents in this direction may conveniently be called positive.

The geometry of the tube requires also that the electric field be parallel to the x axis and that

$$\frac{\partial^2 e}{\partial y^2} = \frac{\partial^2 e}{\partial z^2} = 0. \tag{4}$$

Thus Poisson's equation reduces to the one-dimensional form

$$\frac{d^2 e}{dx^2} = -\frac{\rho}{\varepsilon_v}, \tag{5}$$

where the dielectric constant of free space ε_v is used since the space in the electron tube is a vacuum.

The current density J is given by

$$J = -\rho v, \tag{6}$$

where the minus sign appears because the positive directions of the current density and the velocity are chosen opposite to each other. This relationship may be shown by consideration of a volume v meters long and one square meter in cross section. The total charge in the volume is ρv coulombs. If the charges in the volume are moving with a velocity v meters per second toward the end of the volume, all of them will pass through the end in one second. The total charge transferred in one second across the end of the volume in the direction of the velocity—or the current density in this direction—is thus equal to ρv, the total charge in the volume. But the current density in the direction of the velocity is the negative of the current density that has been called J, so that J is given by Eq. 6.

From Eq. 30, Ch. I,

$$v = \sqrt{\frac{2Q_e}{m_e} e}. \tag{7}$$

Substitution of Eqs. 6 and 7 in Eq. 5 to eliminate ρ and v gives

$$\frac{d^2 e}{dx^2} = \frac{J}{\varepsilon_v \sqrt{\dfrac{2Q_e}{m_e} e}} = K \frac{J}{\sqrt{e}}, \tag{8}$$

where

$$K = \frac{1}{\varepsilon_v \sqrt{2 \dfrac{Q_e}{m_e}}}. \tag{9}$$

To integrate this differential equation relating J, e, and x, both sides may be multiplied by $2(de/dx)\,dx$. Then

$$\int 2 \frac{de}{dx} \frac{d^2 e}{dx^2}\, dx = 2KJ \int \frac{1}{\sqrt{e}} \frac{de}{dx}\, dx. \qquad [10]$$

The current density J appears outside the integral sign because it does not vary with x; it must be the same at every point in the tube. Since

$$\frac{d}{dx}\left(\frac{de}{dx}\right)^2 = 2 \frac{de}{dx} \frac{d^2 e}{dx^2}, \qquad [11]$$

Eq. 10 may be written

$$\int \frac{d}{dx}\left(\frac{de}{dx}\right)^2 dx = 2KJ \int \frac{de}{\sqrt{e}}, \qquad [12]$$

and

$$\left(\frac{de}{dx}\right)^2 = 4KJ\sqrt{e} + C_1. \qquad [13]$$

The boundary conditions for the evaluation of the constant C_1 are that when x is zero, e is zero, and, as discussed previously, de/dx is zero. Therefore C_1 is zero, and

$$\frac{de}{dx} = 2\sqrt{K}J^{1/2}e^{1/4}. \qquad [14]$$

The variables may be separated in this equation; thus

$$\int e^{-1/4}\, de = 2\sqrt{K}J^{1/2} \int dx; \qquad [15]$$

whence

$$\tfrac{4}{3} e^{3/4} = 2\sqrt{K}J^{1/2} x + C_2. \qquad [16]$$

Since e is zero when x is zero, the constant C_2 is zero, and

$$J = \frac{4}{9K} \frac{e^{3/2}}{x^2} = \frac{4\varepsilon_v}{9}\sqrt{2 \frac{Q_e}{m_e}} \frac{e^{3/2}}{x^2}. \qquad [17]$$

Since this equation applies for any value of x, it applies for conditions at the plate, where x equals d and e equals e_b. Thus

$$J = \frac{4\varepsilon_v}{9}\sqrt{2 \frac{Q_e}{m_e}} \frac{e_b^{3/2}}{d^2}. \qquad [18]$$

Any self-consistent system of units may be used for the quantities in Eq. 18. If the constants in the equation are evaluated in the mks system, the following important numerical form is obtained:

$$J = 2.33 \times 10^{-6} \frac{e_b^{3/2}}{d^2} \text{ amperes per unit area.} \qquad [19]$$

In this form of the equation, e_b is in volts and J in amperes per area unit associated with whatever linear unit is used for d. For example, if d is in centimeters, J is in amperes per square centimeter.

If the area of the electrodes is s_p, the total current crossing the tube from plate to cathode is Js_p. This is equal to the plate current i_b, or current entering the plate from the external circuit. Thus

$$i_b = 2.33 \times 10^{-6}\frac{s_p}{d^2}e_b^{3/2} \text{ amperes,} \qquad \blacktriangleright[20]$$

where e_b is in volts. Since s_p/d^2 is a nondimensional ratio, the only requirement is that s_p and d^2 be expressed in terms of the same unit of area. This relation, known as Child's law, Langmuir's equation, or the three-halves power equation,[3] shows that when the five assumptions previously listed are fulfilled, the current to the anode varies directly with the three-halves power of the anode voltage, and inversely with the square of the distance between the electrodes.

Additional information about the distribution of the potential, field strength, space charge, and electron velocity throughout the inter-electrode space may be secured from this derivation. It is readily shown that, under the assumed five conditions,

$$e = K_1 x^{4/3} \qquad [21]$$

$$\mathcal{E} = K_2 x^{1/3} \qquad [22]$$

$$v = K_3 x^{2/3} \qquad [23]$$

$$\rho = K_4 x^{-2/3}, \qquad [24]$$

where the K's are different constants. The expression for ρ gives an infinite charge density at the surface of the emitter, which is of course physically unattainable. This absurdity may be laid to the assumption that the electrons start from rest at the cathode.

A corresponding relation between the anode current and the anode voltage may be derived when the electron current is limited by space charge in a tube in which the anode and cathode are coaxial right-circular cylinders. Simplifying assumptions similar to those made for the parallel-plane problem are used; Poisson's equation is transformed into the form for cylindrical co-ordinates; an equation for the current in terms of the dimensions, charge density, and velocity, and an equation for the velocity in terms of the space potential are formulated;

[3] C. D. Child, "Discharge from Hot CaO," *Phys. Rev.*, Series I, *32* (1911), 498–500; I. Langmuir, "The Effect of Space Charge and Residual Gases on Thermionic Currents in High Vacuum," *Phys. Rev.*, 2 (1913), 450–486.

and the three are solved simultaneously, subject to boundary conditions.[4] The result when the outer cylinder is the anode is

$$i_b = 14.68 \times 10^{-6} \frac{l}{r_p} \frac{e_b^{3/2}}{\beta^2} \text{ amperes,} \qquad \blacktriangleright[25]$$

where

e_b is the anode voltage in volts,

r_p is the radius of the plate,

l is the length of the plate or cathode,

i_b is the plate current,

β^2 is a numeric dependent on the ratio r_p/r_k,

r_k is the radius of the cathode.

The quantity β is given by

$$\beta = y - \left(\frac{2}{5}\right) y^2 + \left(\frac{11}{120}\right) y^3 - \left(\frac{47}{3,300}\right) y^4 + \cdots , \qquad \blacktriangleright[26]$$

in which

$$y = \ln \frac{r_p}{r_k} . \qquad [27]$$

Since l, r_p, and r_k appear only in nondimensional ratios, they may be expressed in terms of any unit, provided the two quantities in each ratio are written in terms of the same unit. Table I gives values of β^2

TABLE I*

VALUES OF β^2 FOR VARIOUS VALUES OF r_p/r_k

r_p/r_k	β^2	r_p/r_k	β^2	r_p/r_k	β^2
1.0	0.000	6.0	0.836	20.0	1.072
1.5	0.119	7.0	0.887	30.0	1.091
2.0	0.279	8.0	0.925	50.0	1.094
2.5	0.412	9.0	0.955	100.0	1.078
3.0	0.517	10.0	0.978	500.0	1.031
4.0	0.667	12.0	1.012	5,000.0	1.002
5.0	0.767	14.0	1.035	30,000.0	0.999

* The data for this table are taken from I. Langmuir, "Currents Limited by Space Charge between Coaxial Cylinders," *Phys. Rev.*, *22* (1923), Table III, p. 353, with permission.

[4] K. T. Compton and I. Langmuir, "Electrical Discharges in Gases; Part II, Fundamental Phenomena in Electrical Discharges," *Rev. Mod. Phys.*, *3* (1931), 245–248.

for various values of r_p/r_k. For conditions where r_p/r_k is large, β^2 may often be taken as unity with sufficient precision for engineering use. A correction for the effect of initial velocities of the electrons in modifying Eq. 25 may be made.[5]

As the dimensions of a cylindrical tube are changed, the current changes in a way that depends on the ratio of the anode length to the anode radius, except for a small change in β^2. If all dimensions, including those of the filament, are modified in the same ratio, the volt-ampere relation when the current is limited by space charge is not altered; but, since the emitting area is changed in proportion to the square of the ratio by which the linear dimensions are changed, the saturation current also is modified by the square of this ratio.

By a process of dimensional reasoning it has been shown[6] that the current should be proportional to the three-halves power of the anode voltage for electrodes of any shape, provided the effect of the initial velocities of the emitted electrons may be neglected. Analytical derivation of the value of the proportionality constant, called the *perveance* of the tube,[7] is not practicable, however, except when the problem may be expressed in terms of a single dimension; that is, when the electron paths follow the lines of electric force, as they do in the cases of rectangular, cylindrical, and spherical symmetry.

2. OCCURRENCE OF GAS IN ELECTRONIC DEVICES

Although the effect of the gas is considered negligible in the operation of the electronic devices discussed previously, actually it is not physically possible to remove all the gas in the process of evacuation. Furthermore, gas is deliberately introduced in many electronic applications. In the remainder of this chapter, consideration is given to some of the fundamental aspects of the internal behavior, as well as the over-all electrical characteristics, of electronic devices in which the effect of the gas is of primary importance.

The gas pressures encountered in electronic phenomena range from several times atmospheric pressure (in high-pressure gaseous lamps and in the arcs of circuit breakers on high-voltage transmission lines), down through atmospheric pressure (in lightning, in corona discharges on high-voltage transmission lines or in high-voltage machines, and in arcs of various types), on down through reduced pressures in

---

[5] See Compton and Langmuir, reference 4, 252-255.

[6] See Compton and Langmuir, reference 4, 251-252.

[7] Y. Kusunose, *Calculations on Vacuum Tubes and the Design of Triodes: Researches of the Electrotechnical Laboratory, No. 237* (Tokyo, Japan, 1928), 2-15; "Standards on Electron Tubes; Definitions of Terms, 1950," *I.R.E. Proc.*, 38 (1950), 436.

the neighborhood of one-tenth of an atmosphere (in such familiar devices as neon signs, fluorescent lamps, and Tungar rectifiers), down further through pressures of about one hundred-thousandth of an atmosphere (in mercury-arc rectifiers and thyratrons), and finally down to pressures of below 10^{-8} atmosphere, called high vacua (in devices such as radio tubes, whose operation is only negligibly affected by the gas). This last class embraces all the devices properly classified as vacuum tubes, and the pressure that may be permitted in them depends on their dimensions, the type of gas, and the particular conditions under which they are operated. In no electronic device does the gas have absolutely no effect; the best condition possible to achieve is to make its effect negligible. Often the effect of the gas is harmful, but as the electrical properties of gases have become better understood, and gas pressures have become subject to better control, more use has been made of gas-filled tubes, with the result that, in present-day engineering practice, tubes containing gas occupy a position of great importance.

One of the important effects of the presence of gas molecules in electronic devices is their chemical reaction at the electrodes, which can destroy the useful properties of the device. This harmful action is often avoided by the use of mercury vapor or of an inert gas such as argon, neon, helium, krypton, or xenon. Another important effect of the presence of gas molecules results from collisions between the charged particles and the molecules. Several new phenomena are thus produced, the understanding of which requires some knowledge of atomic processes. The structure of atoms is touched upon in Arts. 1 and 2, Ch. II; some additional properties of atoms are described in the next few articles of this chapter.

3. Physical properties of atoms[8]

In electronic applications, the atoms of a gas are important chiefly as a result of their ability to absorb, transport, and give up energy. The energy may be acquired from or given up to other atoms of the gas. In addition, the electrodes and walls of the electronic device, and such particles as electrons and photons, may supply or receive energy. Since the atoms of a gas are free to move in the gas, and since they possess mass, they may acquire and transport kinetic energy. In addition, they may acquire and transport potential energy in internally stored form. The internal energy is associated with the energy levels in which the electrons of the atoms are located, as discussed in Ch. II.

---

[8] L. Tonks, "Electrical Discharges in Gases—Ionization and Excitation," *E.E.*, *53* (1934), 239–243.

Not all the electrons occupy the same energy levels. Those in the lower energy levels require more energy for separation from the nucleus than do those in the upper levels. An electron that normally occupies an energy level W_1 can move to another and higher energy level W_2, provided an amount of energy $W_2 - W_1$, called a *quantum* of energy, is received by the atom. After reaching the level W_2, the electron may go to a still higher level W_3, provided the atom receives an additional quantum of energy $W_3 - W_2$. This two-stage process can also be accomplished in a single "transition" with a quantum of energy $W_3 - W_1$. Since the various amounts, or *quanta*, of internal energy that an atom can absorb are therefore dependent on the differences in energy between the various levels in the atom, the atom can absorb only certain specific amounts of internal energy.

The electrons in the atom have a tendency to revert to the lowest possible energy levels. However, experimental studies of spectra show that the number of electrons that can exist at a given energy level in a particular atom is limited. When all the electrons occupy the lowest permitted levels, the atom is said to be in its *normal state.* An atom that has one or more of its electrons raised to a higher-than-normal energy level is said to be "excited," or in an *excited state*, and the energies required for the transitions from the normal to the excited state, usually expressed in electron volts, are called *excitation energies.*

In most atoms the excited state lasts for only about 10^{-8} second and terminates with the spontaneous return of the excited electron to a lower level. This termination is accompanied by the release of a photon (quantum of electromagnetic radiation, or light) whose energy is exactly equal to the quantum corresponding to the particular transition that the electron makes. The emitted radiation, which may or may not be in the visible spectrum, has a frequency proportional to the energy released by the atom; the relation is

$$f = \frac{\text{energy of quantum}}{h}, \qquad [28]$$

where h is the Planck constant and f is the frequency of the radiation. Most of the light from electric discharges originates by this process. The light is emitted at discrete frequencies corresponding to the differences in energy of certain of the various levels in the atom, and each atom therefore has its characteristic line spectrum. The frequencies of the emitted lines may lie in all regions of the spectrum, including the regions of x-ray, ultraviolet, visible, and infrared radiation. Certain gaseous-discharge tubes, constructed and operated so that these processes are accentuated, find extensive application as efficient sources of illumination.[9]

Some atoms have certain energy levels to which an internal electron may be raised but from which it apparently cannot revert to the normal state by giving up its energy in the form of a photon.[10] An atom that is so excited is said to be in a *metastable state*. It ordinarily gives up its energy to another atom in the gas, or to the walls or electrodes of the container; and the time that elapses before this surrender occurs, often referred to as the *life* of the metastable state, may be of the order of 0.1 second. It is therefore possible for metastable atoms to carry their internal energy for considerable distances before releasing it; this transportation of energy can, and often does, play an important part in conduction through gases.

The process by which an electron is completely separated from the atom is known as *ionization*, and the first, or minimum, *ionization energy* is the amount of energy which the atom must receive to separate the most easily removed of its electrons. Other ionization energies, required for the removal of other electrons, also exist. The part of the atom that remains after ionization is called a positive ion; it has a charge equal in magnitude and opposite in sign to that of the freed electron or electrons. The mass of the positive ion is practically that of the original atom, since the mass of the electron is negligible in comparison with that of the nucleus. After ionization has occurred, both the electron and the positive ion are free to move independently; and, if they are in an electric field, they are accelerated and acquire kinetic energy. It is to be emphasized, however, that the ion possesses ionization energy as well as its kinetic energy, and will release the former whenever any electron recombines with it to form a normal atom. The energy can be released in the form of radiation, but the absence of an appreciable amount of light of the proper character from the spectra of ordinary low-pressure gaseous discharges indicates that this type of spontaneous recombination does *not* occur in them to an appreciable extent.[11] Rather, the ion and electron combine at the surface of the electrodes or walls of the electron tube, and the energy is released as heat to these surrounding bodies.

Negative ions, consisting of an electron attached to a neutral atom, also exist. Such ions have the charge of one or more electrons and the

[9] S. Dushman, "Low-Pressure Gas-Discharge Lamps," *E.E.*, *53* (1934), 1204-1212, 1283-1296; W. E. Forsythe and E. Q. Adams, *Fluorescent and Other Gaseous Discharge Lamps* (New York: Murray Hill Books, Inc., 1948).

[10] G. P. Harnwell, *Principles of Electricity and Electromagnetism* (2nd ed.; New York: McGraw-Hill Book Company, Inc., 1949), 64.

[11] K. T. Compton and I. Langmuir, "Electrical Discharges in Gases; Part I, Survey of Fundamental Processes," *Rev. Mod. Phys.*, *2* (1930), 191-204.

mass of an atom. They differ from positive ions, however, in that they have much less internal energy. They form but rarely in the noble gases or in mercury vapor. On the other hand, oxygen is so avid for electrons that in ordinary discharges in gases containing as little as one

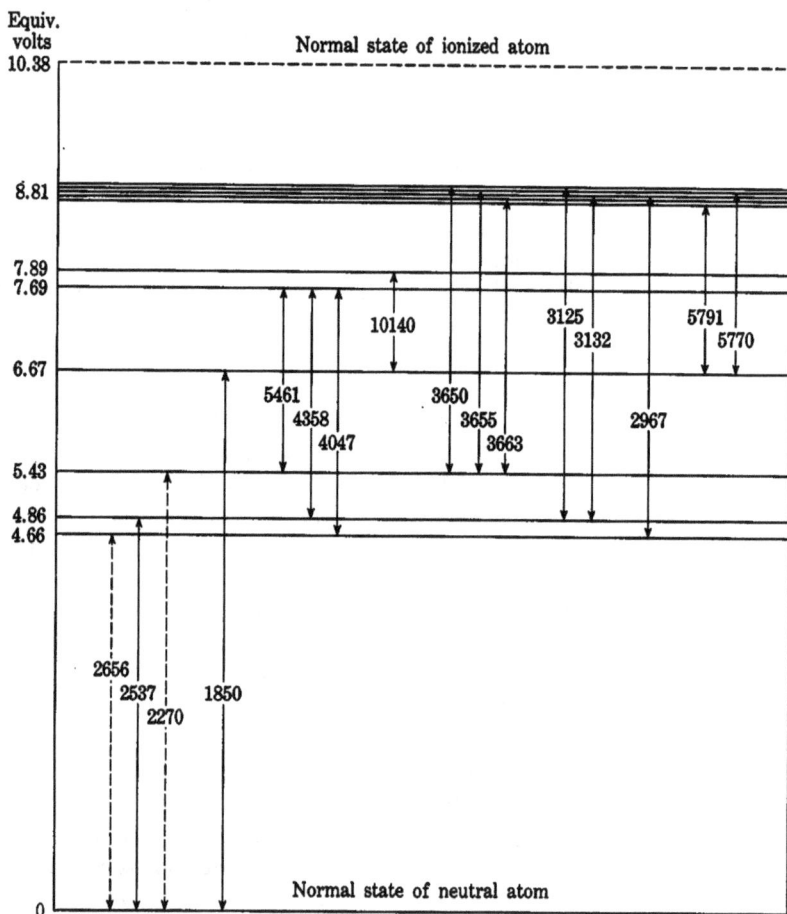

Fig. 4.   Lower energy levels of the mercury atom.* The numbers on the transition arrows give the wavelength of the corresponding radiation in Angstrom units (10^{-8} cm).

part in ten thousand of oxygen, most of the electrons become attached very quickly to oxygen atoms. For this reason, where it is desired that the conduction be mostly by electrons, the gases for use in electron tubes must be carefully freed from oxygen.

* This figure is taken from L. Tonks, "Electrical Discharges in Gases," *A.I.E.E. Trans.*, 53 (1934), Fig. 2, p. 241, with permission.

Figure 4 shows graphically the lower energy levels of the mercury atom. The first and third lowest levels shown are metastable levels, and the corresponding frequencies resulting from transitions from one of these to a lower energy level are "forbidden" or nonexistent lines in the mercury spectrum. Table II gives the first excitation energy, in every case that of a metastable state, and the ionization energy for a few other gases.

Thus far in this treatment, the atom has been discussed only from the standpoint of its ability to receive, hold, and spontaneously give

TABLE II*

FIRST EXCITATION ENERGY AND MINIMUM IONIZING
ENERGY FOR THE MONATOMIC GASES AND MERCURY

Gas	Excitation energy (volts)	Ionization energy (volts)
Helium (He)	19.73	24.48
Neon (Ne)	16.60	21.47
Argon (A)	11.57	15.69
Krypton (Kr)	9.9	13.3
Xenon (Xe)	8.3	11.5
Mercury (Hg)	4.66	10.39

* The data for this table are taken from A. W. Hull, "Fundamental Electrical Properties of Mercury Vapor and Monatomic Gases," *A.I.E.E. Trans.*, *53* (1934), Table I, p. 1436, with permission.

up energy in the form of radiation. No mention has been made of the methods by which energy can be imparted to atoms, or of other methods by which the atoms can give up internal energy. A discussion of these methods involves the atom not by itself but in relation to other atoms, electrons, ions, and photons that may exist in the discharge. The collision phenomena discussed in the following articles are of primary importance in the interactions that occur.

4. COLLISION PROCESSES IN A GAS—MEAN FREE PATH

The collisions that occur in a gas, among the gas molecules and between electrons and molecules, are of great importance in producing the phenomena associated with the conduction of electricity through a gas. These phenomena depend upon the type and number of collisions. The type of a collision, discussed in the next article, is determined by the way in which the energy of the colliding particles is redistributed. The number of collisions—the subject of this article— may be determined with the aid of an approximate picture of the motion of the gas molecules afforded by the *kinetic theory of gases*. According to this theory, the molecules may be regarded as small spheres in constant motion; each molecule continually collides with its neighbors and with the walls of the container, rebounds as a perfectly elastic ball might rebound, and follows a tortuous path made up of short jumps of varying length. An electron projected into the gas at high speed is pictured as a similar but much smaller sphere that

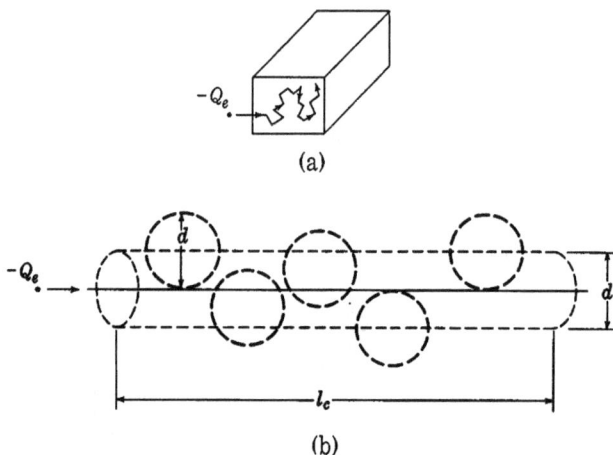

(a)

(b)

Fig. 5. Actual and equivalent-straight-line paths of an electron through a gas.

rebounds from the molecules and moves in a path similar to that shown in Fig. 5a.

This concept must be regarded as only an approximation to the conditions in a gas. Gas molecules do not always have spherical symmetry, and they never have definite material boundaries as do the balls of the kinetic-theory picture. Collisions between particles, such as molecules and electrons, are more accurately described in terms of interactions between the fields surrounding one of the colliding particles and the charges constituting the other particle. A complete determination of the forces between such particles is exceedingly difficult and has not yet been made. It is known, however, that such forces are very small if the distance between the particles is appreciable, and that they are very strong repulsive forces when the centers of the particles approach very closely. Sometimes strong attractive forces exist at separations slightly greater than those at which the repulsive forces exist; this possibility may result in the formation of chemical compounds. If the repulsive forces at close distances serve to deflect the approaching particles from one another without bringing about any changes in the internal energies of the particles, the collision is called *elastic*, and the effect is very similar to the process pictured by the kinetic theory. However, since the forces involved may depend upon the surrounding particles and the velocities of the colliding particles, as well as upon their nature, the diameter of the spheres of the kinetic theory cannot be regarded as an invariable quantity. If, because of the collision, changes occur in the internal energy of one of the colliding particles, a corresponding change occurs in the kinetic energy of the particles, and the collision is *inelastic*. Such collisions are not accounted for by the simple picture of the kinetic theory of gases.

The concept of gas molecules as spheres provides a means of calculating the *mean free path* λ_e of an electron projected into the gas. The mean free path of the electron is the average value of the different distances the electron moves between successive collisions. To obtain a value of the mean free path, assume that the gas molecules have an effective diameter d (about 10^{-8} centimeter) and are stationary, and that the electrons have a diameter (about 10^{-13} centimeter) so small compared with d that they may be taken as points. Actually, the gas molecules move in all directions at a variety of speeds. However, since the mass of a molecule is several thousand times that of an electron, and since at ordinary room temperatures the average kinetic energy of the molecules is about 1/30 electron volt, the speed of the molecules is generally negligible in comparison with the speed of an electron, even though its kinetic energy may be only a few electron volts.

Division of the total length of the broken-line path of an electron through the gas by the number of segments into which the path is split by the collisions gives the mean free path. The number of segments of the path may be taken to be the same as the number of molecules that the electron strikes in the course of its motion. To facilitate the calculation of this number, the actual path may be replaced by a straight path of equal length. This replacement is allowable, provided the length of path considered is sufficiently great, because the large-scale properties of the gas are supposed to be uniform, so that the same number of molecules are encountered along a path of the specified length, irrespective of its shape. If a cylinder of diameter d is drawn about this line as an axis, as shown in Fig. 5b, every molecule whose center lies in this cylinder will represent a collision, and the number of molecules having centers within a cylinder of length l_c equals the number of collisions the electron makes in moving the distance l_c. Thus

$$n_c = \frac{\pi d^2}{4} l_c n, \qquad [29]$$

where

n_c is the number of collisions of the electron in the distance l_c,

n is the number of molecules per unit volume.

Hence the average distance between collisions, or the mean free path of an electron, is

$$\lambda_e = \frac{l_c}{n_c} = \frac{4}{\pi d^2 n}. \qquad \blacktriangleright[30]$$

When the mean free path of one molecule moving among other molecules of the same gas is to be found, the diameter of the cylinder must be doubled, since the projectile molecule then has appreciable dimensions. In addition, the appreciable speed of the other molecules relative to that of the projectile molecule must be taken into consideration. This effect may be shown to introduce an additional factor of $\sqrt{2}$ into the expression for the mean free path.[12] The net result is

$$\lambda_m = \frac{1}{4\sqrt{2}} \lambda_e, \qquad \blacktriangleright[31]$$

where λ_m is the mean free path of the molecule.

The mean free paths for electrons and molecules in several gases are given in Table III. To convert these values to other conditions of pressure and temperature, the following relation for a gas is applicable:[13]

$$p = nkT; \qquad \blacktriangleright[32]$$

where

p is the pressure,

n is the number of molecules per unit volume, or the concentration of the molecules,

k is the Boltzmann constant,

T is the temperature in degrees Kelvin.

Consequently, since the concentration n is proportional to p/T, and the mean free path is inversely proportional to n, the mean free path λ_e or λ_m is directly proportional to the temperature T and inversely proportional to the pressure p.

As may be expected from the discussion of the nature of the collisions, the mean free path of an electron is not determined solely by the concentration of the gas; in fact, experimental measurements show that the mean free path of an electron depends on its speed. For

---

[12] J. H. Jeans, *The Dynamical Theory of Gases* (4th ed.; Cambridge: The University Press, 1925), 35–37.

[13] J. D. Cobine, *Gaseous Conductors* (New York: McGraw-Hill Book Company, Inc., 1941), 5.

gases with symmetrical molecules, such as the noble gases, it has a minimum value at a low speed corresponding to an energy of a few electron volts,[14] and rises to very high values at still lower speeds. This variation of the mean free path is called the *Ramsauer effect*.

Through reduction of the pressure the mean free path can be made greater than the distance between the electrodes in a tube, but this fact does not imply that there will then be no collisions between the

TABLE III*

KINETIC THEORY MEAN FREE PATHS λ AND NUMBER OF
COLLISIONS PER CM PATH, $\nu = 1/\lambda$, IN SEVERAL GASES AT
A PRESSURE OF 1 MM OF MERCURY AND A
TEMPERATURE OF 25 C

Gas	Mean free path in cm		Number of collisions per cm	
	Electron λ_e	Molecule λ_m	Electron ν_e	Molecule ν_m
Mercury	0.0149	0.00263	67.0	380.0
Argon	0.0450	0.00795	22.2	125.9
Neon	0.0787	0.01390	12.7	72.0
Helium	0.1259	0.02221	7.95	45.0
Hydrogen	0.0817	0.01444	12.2	69.1
Nitrogen	0.0425	0.00751	23.5	133.0
Oxygen	0.0455	0.00805	22.0	124.2
Carbon monoxide	0.0420	0.00743	23.8	136.5

* The data for this table are taken from K. T. Compton and I. Langmuir, "Electrical Discharges in Gases; Part I, Survey of Fundamental Processes," *Rev. Mod. Phys., 2* (1930), Table XV, p. 208, with permission.

electrons and molecules. The mean free path is a statistical average and indicates only the average distance traveled by a particle, or each of a group of particles, between successive collisions. To deduce the distribution of the free paths among a group of particles, suppose that N of the particles are projected into a gas-filled region. The fraction f_1 of the total number of particles that penetrates a distance of 1 centimeter without collision is, say, u, where u depends on the kind of gas and its concentration and must be positive and less than unity. By the nature of this process, the fraction f_1 of the remaining particles that penetrates the ensuing 1 centimeter of path without colliding is also u. Hence the fraction f_2 of the original particles that penetrates a distance

[14] W. G. Dow, *Fundamentals of Engineering Electronics* (2nd ed.; New York: John Wiley & Sons, Inc., 1952), 81–82.

of 2 centimeters without colliding is u^2, or the fraction f_x that penetrates a distance x centimeters without colliding is u^x. Since u is less than unity, a positive constant a may be so chosen that

$$\epsilon^{-a} = u, \qquad [33]$$

where the value of a, like that of p, depends on the nature and concentration of the gas. Thus

$$f_x = u^x = \epsilon^{-ax}. \qquad [34]$$

If N particles are projected into the region, the number of them that go x centimeters *or more* without collision is $N\epsilon^{-ax}$. The rate at which this number is diminished in the succeeding increment dx is

$$-\frac{d}{dx}(N\epsilon^{-ax}).$$

Thus,

$$-d(N\epsilon^{-ax}) = Na\epsilon^{-ax}\,dx \qquad [35]$$

is the number of particles that go a distance x without colliding but that do collide in the distance dx beyond x. The total distance traveled before collision by all those particles that collide between x and $x + dx$ is

$$xNa\epsilon^{-ax}\,dx.$$

The total distance l_t traveled before collision by all particles is therefore

$$l_t = \int_0^\infty xNa\epsilon^{-ax}\,dx = N/a. \qquad [36]$$

The average distance traveled by the particles before collision is then equal to l_t/N. By definition this distance is equal to the mean free path λ; thus

$$\lambda = 1/a. \qquad [37]$$

Consequently the fraction f_x of the particles that penetrates a distance x *without* colliding is, from Eqs. 34 and 37,

$$f_x = \epsilon^{-\frac{x}{\lambda}}. \qquad \blacktriangleright[38]$$

A curve of this distribution of the free paths is shown in Fig. 6. It shows that in going a distance equal to the mean free path, 63 out of 100 particles do collide, while only 37 out of 100 do not collide with gas molecules.

According to this analysis, collisions in a vacuum tube cannot be entirely prevented. The probability of a collision may, however, be reduced to a negligible value by reduction of the gas pressure. What constitutes a negligible collision probability depends on the purpose of the vacuum tube, but in general the pressure should be reduced until the mean free path is hundreds or thousands of times greater than the tube spacing before the gas can be said to play no appreciable part in the action of the tube. A consideration of the consequences of a collision, given in the next article, clarifies this requirement regarding the degree of perfection of the vacuum.

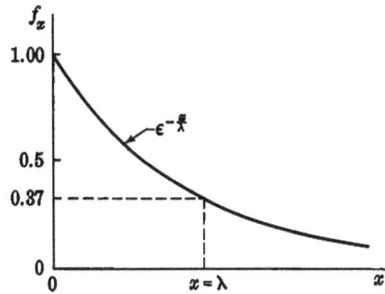

Fig. 6. Distribution of the free paths of a particle. The quantity f_x is the fraction of the total number of particles that penetrates a distance x *without* colliding.

5. CONSEQUENCES OF COLLISIONS[15]

From the discussion of atoms and collisions in the two preceding articles it appears that when an electron or other particle collides with an atom or molecule, the collision may be elastic or inelastic. If the internal energy of the atom does not change, the collision is elastic, and the kinetic energy of the particles is conserved at the collision. However, if a quantum of energy is absorbed internally by the atom in a process of excitation or ionization, the collision is inelastic, for the kinetic energy of the particles is then not conserved.

As the speed of the projectile electron is increased from zero, elastic collisions and inelastic collisions of several different types occur in different speed ranges. When the speed is less than that corresponding to the minimum excitation energy of the atom, the electron rebounds elastically from the atom or molecule, and in general its velocity suffers a change in direction. Because of the large mass of the molecule compared with that of the electron, and because momentum as well as kinetic energy must be conserved, the atom acquires on the average only a small fraction of the kinetic energy of the electron, and the electron rebounds with practically the whole of its original energy. In certain inert gases, electrons of very low energy sometimes seem to pass through the atom without deflection or loss of energy.

[15] A. W. Hull, "Fundamental Electrical Properties of Mercury Vapor and Monatomic Gases," *E.E.*, *53* (1934), 1435–1442.

If the energy of the projectile electron is increased to a value sufficient to excite the atom, inelastic collisions may occur. In such an event the atom absorbs a quantum of energy from the electron, and the electron retains the remainder of its kinetic energy. However, even when the energy of the electron is sufficient for excitation to occur, not all the collisions are inelastic. Measurements of the probability of excitation show that a maximum of about 1 to 3 out of 1,000 such collisions result in excitation, and that the excess of energy over the quantum involved must be small if the probability of excitation is to be large.

TABLE IV*

TOTAL IONIZING POWER OF ELECTRONS; IONS PER
PRIMARY ELECTRON

Gas	Energy of primary electron in electron volts			
	30	50	75	100
Mercury	1.1	1.4	. . .	2.7
Argon	0.45	0.9	. . .	1.6
Neon	. . .	. . .	1.2	2.0
Helium	. . .	1.2	. . .	2.9
Hydrogen	. . .	. . .	. . .	1.4
Nitrogen	. . .	. . .	1.3	1.6

* The data for this table are taken from I. Langmuir and H. A. Jones, "Collisions between Electrons and Gas Molecules," *Phys. Rev.*, *31* (1928), Table VIII, p. 402, with permission.

A larger increase of the energy of the projectile electron makes ionization of the atom possible. Again there is a fractional probability of this type of collision. The chance of ionization is nearly zero for electrons having exactly the ionization energy, increases to a maximum for energies about twice this critical value, and decreases for higher energies. Some idea of the relative susceptibility to ionization by electron collision of some of the more common gases may be obtained from Table IV, which gives the average number of ions produced by a single electron starting through a field-free gas with the initial energy indicated and continuing to ionize until its energy falls below the ionizing energy. The higher-energy electrons sometimes remove more than one electron from the atom, thus producing an ion that is multiple charged.

Electrons moving in a gas under the influence of an electric field produced, for example, by a pair of electrodes make numerous collisions with the molecules. The electrons are accelerated, and their energy is increased between collisions by an amount that increases with the mean free path and with the electric field. The electron rebounds with almost undiminished energy at each collision, so that at each collision an electron transfers only a small fraction—about 0.000005 for Hg, 0.00003 for A, 0.00005 for Ne, and 0.0003 for He—of its total energy to the molecule with which it collides. As the electrons move through the gas, they tend to acquire energy from the electric field until their mean energy is such that, on the average, they lose as much energy at a collision as they gain between collisions. Thus the mean energy of the electrons may become many times that of the molecules, and in gaseous discharges the electrons have a mean energy of several electron volts. This energy corresponds to a temperature of tens of thousands of degrees. Electrons having these energies may make inelastic as well as elastic collisions and may lose energy to the molecules, thus causing excitation and ionization of the gas.

The positive ions formed by the collisions of electrons with molecules behave similarly to the electrons, in that they receive energy from the electric field as they move through the gas. But, in making elastic collisions with gas molecules, the ions lose a considerable fraction of their energy—about half, on the average—with the result that their total energy is still small when they are able to give up as much energy at a collision as they receive between collisions. In addition to having a mean energy that is much lower than that of the electrons though larger than that of the molecules, the ions are inherently inefficient ionizing agents. Ionization of the gas molecules by positive-ion impact is therefore extremely small.

One of the more important effects of ionization in gases is the *neutralization of electronic space charge.* Because of their relatively large mass and consequently low speed for a given energy, positive ions remain much longer in the space between the electrodes than do the electrons and, therefore, for a given current, are much more effective in producing space charge. Consequently a small positive-ion current may neutralize the space charge of a relatively large electron current in a gas. The practical use of this neutralization is discussed in Art. 2, Ch. V. When the gas pressure is considerable, ionizing collisions may be so frequent that the resulting electrons and positive ions augment the current appreciably. It is this sort of amplification of current that is used to advantage in increasing the current output per photon in gas-filled phototubes. This use of gas is discussed in Art. 7.

In addition to their effect in the space between the electrodes, the positive ions have an important effect at the electrode and wall surfaces. The positive ions are neutralized when they strike the charged surfaces, and their internal ionization energy, together with part of their kinetic energy, is given up there. The energy may be used in heating the surface; it may actually remove cathode material by the process known as sputtering; it may be given off as radiation; or it may release or help to release an electron from the cathode. When the current through a gas tube is not limited by external agencies, the heating and sputtering processes may quickly destroy the electrodes.

Photons, which participate in gaseous conduction, are not influenced by the electric field between the electrodes. They are able to give up their energy to atoms and cause excitation of the atoms; the photons then cease to exist. After about 10^{-8} second the atom emits another photon, which can excite still another atom, and so on. This process, by which the excitation energy is passed from atom to atom, is known as *imprisonment of radiation*. Photons emitted by one kind of atoms in a mixture of gases can ionize other atoms in the mixture if these other atoms have an ionization potential lower than an excitation potential of the atoms of the first kind. In addition, photons can take part in multistage ionization and are able to give up their energy to the cathode and help to release electrons from it.

Excited atoms and metastable atoms can give up their energy directly, either to the gas atoms or to the cathode upon collision. Collisions between such extra-energy particles and ordinary gas atoms not only may result in the excitation of another atom but also may constitute a part of a multistage ionization. Collisions of such particles with the cathode may heat the cathode or may release or help to release an electron. Metastable and excited atoms are uninfluenced by the electric field and therefore move in the same manner as the other neutral molecules. Because of their long life the metastable atoms may transport their energy for considerable distances, independently of the electric field. Also, because of the relatively long life of the metastable state, there is an appreciable probability that metastable atoms may receive sufficient energy from a second collision to complete the detachment of an electron and thus to become ionized. Since this process requires two collisions, the effect is proportional to the square of the current. This type of ionization is of considerable importance in low-pressure arcs, and may account for the fact that the total voltage drop is lower than the ionizing potential in some arcs.

6. GASEOUS DISCHARGES

With the foregoing explanation of the electrokinetic properties of gases, it is possible to give an approximate explanation of the more common types of conduction through gases. The name given to such conduction is *gaseous discharge*, which perhaps has its origin in the fact that in the early experiments on electricity a capacitor or battery was frequently discharged through a gas. Ionization of the gas by some means is the dominant feature of all gaseous discharges. Since ions always tend to revert to normal atoms, a second important feature is the process by which this reversion takes place, that is, the de-ionization process. The different types of gaseous discharges are characterized and distinguished by the types and relative importance of the ionization and de-ionization processes that take place in them. In particular, gaseous discharges may be divided into two broad classes, *self-maintaining* and *nonself-maintaining*, depending on whether a source of ionization external to the discharge is necessary for the continuance of them.

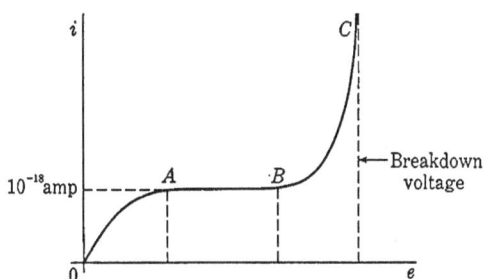

Fig. 7. Typical current-voltage characteristic of a discharge between two electrodes in a gas. Over the region *BC* it is named a Townsend discharge.

7. Townsend discharge

Perhaps the simplest example of gaseous conduction is the Townsend discharge.[16] If two metal electrodes of an area of a few square centimeters are located a few centimeters apart in a gas, and if a voltage is applied between the electrodes, the typical shape of the current-voltage characteristic that results is as shown[17] in Fig. 7. The current through the gas increases with the voltage at first, in the region *OA* of the figure; then becomes independent of voltage, or saturates, in the region *AB* at a minute value of current that is generally too small to be measured with an ordinary galvanometer; and finally increases rapidly as the voltage is increased further in the region *BC*.

The reason for the current in this gaseous discharge is that any gas is always in a state of partial ionization, because ions and electrons

[16] J. S. E. Townsend, *Electricity in Gases* (New York: Oxford University Press, 1915); *Electrons in Gases* (London: Hutchinson's Scientific and Technical Publications, 1947).

[17] K. K. Darrow, *Electrical Phenomena in Gases* (Baltimore: The Williams and Wilkins Company, 1932), 272–273. See footnote, p. 273 therein.

are continually produced in it at the electrodes and at the walls by
the action of ever-present cosmic rays, of radiation from radioactive
substances, and of similar radiation reaching the region. Air at atmospheric pressure, for example, although ordinarily considered to be a
good insulator, is ionized by these causes and does conduct electricity
to a very small extent. A voltage applied between two electrodes in a
gas, then, always causes the transportation of a few ions and electrons
and results in a minute electric current. As the voltage is increased in
the region OA of Fig. 7, the current increases, because more and more
of the ions are drawn to the electrodes before they recombine with

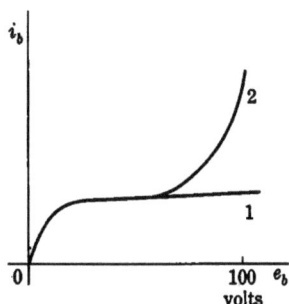

Fig. 8. Typical current-
voltage characteristic of a
gas-filled phototube. Curve
1 shows the current in the
absence of gas.

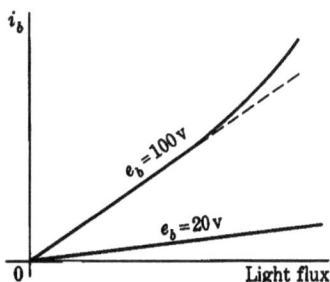

Fig. 9. Typical variation of
current with light flux in a gas-
filled phototube.

electrons at the walls of the container or drift out of the discharge
space. The current reaches a limit, shown in region AB, when the ions
and electrons are drawn to the electrodes by the applied electric field
as rapidly as they are produced in the interelectrode region, and the
current is then independent of the voltage over a considerable range.
The second increase in current, shown in the region BC, occurs
because the electric field is then sufficiently strong to impart enough
energy to those electrons produced by the external source of ionization to cause them to ionize molecules of the gas. New ions, excited
atoms, metastable atoms, and photons all begin to appear in the discharge, although at this stage it may remain invisible. Finally, as discussed in the next article, the current tends to increase without limit
for an infinitesimal increase in voltage. However, until this increase
happens, the discharge current is dependent on a supply of ions from
an external source and is nonself-maintaining.

A device in which the Townsend discharge plays a dominant role is
the gas-filled phototube. In a phototube having the current-voltage
characteristic shown as curve 1 in Fig. 8, addition of the proper

amount of a chemically inert gas will change the characteristic to the form shown as curve 2. Here the current that is increased by ionization resulting from collision of electrons with the gas molecules is that emitted by the light-sensitive cathode; this current is so much larger than the current produced by other external sources that the latter can be neglected. Figure 9 shows the typical dependence of the amount of current on the total light flux incident on the cathode when

5581
Average anode characteristics

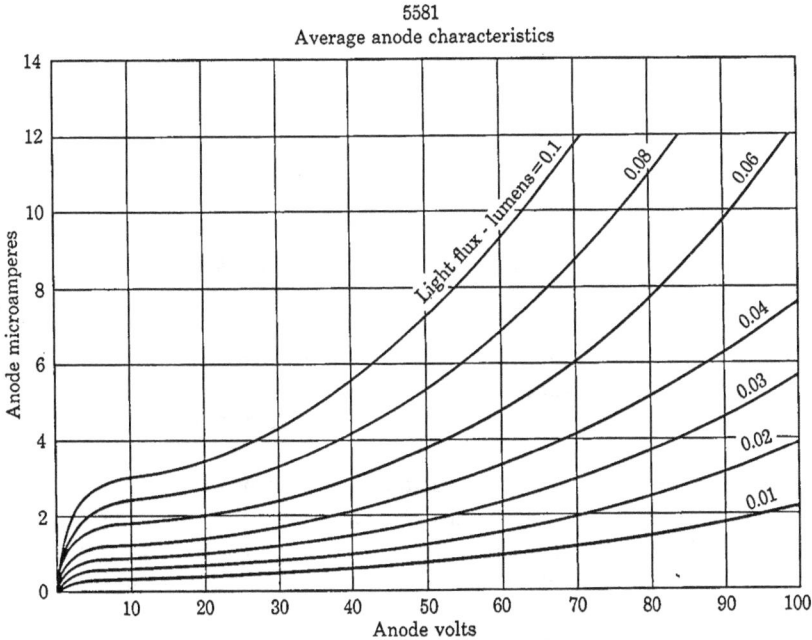

Fig. 10. Characteristic curves for a typical gas-filled phototube.
(*Data by courtesy of Radio Corporation of America.*)

the applied voltage is constant. For voltages below about 20, the current in a typical phototube is substantially linear with light flux; but for higher constant voltages, including the normal supply value of 100 volts, it departs appreciably from linearity and rises more and more steeply as the total light flux is increased. Because of the ionization, the addition of gas increases the sensitivity of the phototube —that is, the number of microamperes per lumen—by an amount dependent on the applied voltage, but it does so at some sacrifice in linearity and in speed of response of the current to sudden changes or rapid fluctuations of light. The loss in speed of response results from the time required for the ionization process to produce the equilibrium value of current. The increase of sensitivity, or the gas amplification factor, is limited practically to a factor of about ten by the fact that, if higher values are attempted, breakdown of the gas may occur in the manner described in the next article. A self-sustained discharge then occurs, the cathode is bombarded by positive ions, and damage to the light-sensitive surface usually results. Breakdown can be caused by an increase of either the applied voltage or the light flux; it is avoided through adhering to the limitations that the manufacturers set on the maximum permissible applied voltage, maximum incident light flux, and minimum series load resistance for each type of phototube. Curves for a particular gas phototube are shown in Fig. 10. It has the same dimensions and type of photocathode as the vacuum phototube of Figs. 24 and 25 of Ch. II.

8. BREAKDOWN[18]

When the voltage across a Townsend discharge is sufficient to cause the electrons to ionize some of the gas molecules upon collision, new electrons are liberated in the discharge region. These, in turn, are accelerated by the electric field and produce other new electrons. Thus an electron produced near the cathode by an external source of ionization liberates new ones in a geometric progression as the original electron and those liberated travel toward the anode. This cumulative process is known as an *electron avalanche*. The positive ions resulting from the ionization travel toward the cathode and may produce new electrons at or near it. Ultimately a voltage is reached at which, on the average, the products of one electron avalanche are able to produce another electron in a position such that it in turn can start a second electron avalanche as large as the first. Under these conditions the discharge becomes self-maintaining, for its continuance ceases to depend on an external source of ionization. The processes by which the products of the avalanche produce the new electron near the cathode may include positive-ion impact in the gas or at the cathode, the action of metastable atoms at the cathode, photoelectric action in the gas or at the cathode, and other means, but the details and relative importance of the phenomena are not clearly understood.

A gas is said to *break down* when the voltage across a Townsend discharge is increased until the condition of self-maintenance is reached. The voltage at which the discharge becomes self-maintaining is known variously as the breakdown, sparking, ignition, initial, or beginning voltage. Since this *breakdown voltage* depends on the number of ions,

---

[18] A survey of the facts known about breakdown appears in L. B. Loeb, *Fundamental Processes of Electrical Discharges in Gases* (New York: John Wiley & Sons, Inc., 1939), 408–595.

photons, or other ionization products formed in an electron avalanche, and on the effectiveness of these products in regenerating the original electron, it is a function of the particular gas; the concentration, or pressure, at a particular temperature; and the configuration and material of the electrodes. For parallel plane electrodes, as in Fig. 11a, the breakdown behavior may be explained qualitatively, provided any concentration of electric field caused by fringing at the edges of the electrodes is prevented by some means. As the pressure is decreased

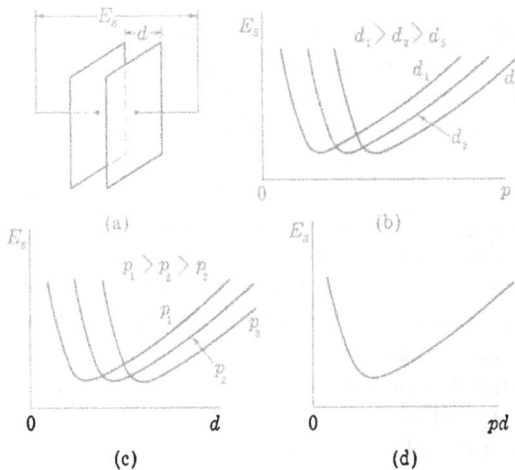

Fig. 11. Breakdown voltage between parallel plane electrodes.

from, say, atmospheric pressure, the breakdown voltage E_s decreases almost linearly for a time, as in Fig. 11b, then reaches a minimum value and rises again as the pressure is decreased further. An explanation of this behavior may be made on the basis of the change of the mean free path of the electrons with pressure. At a high pressure, the mean free path is so small and the collisions are so frequent that the electric field must be large in order to give the electrons ionizing speed in the short distance between collisions. At very low pressures, the mean free path is large, and an electron does not make enough collisions in the distance between electrodes to produce enough positive ions to reproduce itself, unless the voltage is increased in order to increase the ionizing ability of the electrons and positive ions. At some intermediate pressure, optimum conditions for ionization exist and the breakdown occurs most readily.

If the distance between electrodes is varied while the pressure is held constant, a similar variation in breakdown voltage, shown in Fig. 11c, occurs. When the spacing is short, only a few positive ions can be produced because of the limited number of collisions; and, to produce breakdown, the voltage must be high in order to accelerate the electrons rapidly and make them more effective as ionizing agents. As the spacing is increased, the number of collisions increases faster than the probability of ionization is decreased by the decreasing voltage per collision, so that the breakdown voltage is decreased. At some optimum spacing, the breakdown voltage is a minimum. At larger spacings, more mean free paths are included between the plates, and the voltage must be increased to produce the field strength necessary for ionization.

If the temperature, gas, and electrode material in a discharge tube are not changed, but the pressure and spacing are varied in such a way that their product remains constant, then, since the mean free path is inversely proportional to the pressure, the ratio between the spacing and the mean free path remains the same. That is, the number of mean free paths contained in the distance between the electrodes is not changed. Changes of pressure and spacing that keep this number constant do not change the number of collisions an electron makes in crossing the interelectrode space, since the distance between collisions is changed in the same ratio as the total distance. In addition, if the electrodes are parallel planes, such changes do not affect the average potential difference through which an electron moves between collisions, since the potential gradient and the distance between collisions change in inverse ratio. Thus the electron avalanche produced by one electron, and the resulting products of ionizing and excitation-producing collisions, are unchanged. In the same way the number of electrons produced near the cathode by these products remains unchanged, and thus the voltage required across the tube for breakdown remains the same. This result is *Paschen's law*, which states that for parallel plane electrodes in a particular gas at a particular temperature the breakdown voltage is a function only of the product of the pressure and the electrode separation. The typical form of the function is shown in Fig. 11d. The minimum breakdown voltage for air is about 350 volts, and at room temperatures it occurs for a product pd equal to about 0.06, where p is the pressure in centimeters of mercury and d is the electrode spacing in centimeters.

After a gas breaks down, the self-maintained discharge that is established may exhibit a variety of physical and electrical characteristics. The more common types of self-maintained discharges that result are the spark, glow, arc, corona, brush, and point discharges. The *spark* is essentially a transition phenomenon of short duration that is associated with a sudden release of a considerable amount of energy. It is characterized by intense ionization along the path of the discharge and intense excitation of the outer electrons of the atom, which results in radiation at characteristic frequencies of the atomic spectrum that are different from the characteristic frequencies in the spectra of the continuous discharges. A lightning stroke and the discharge of a capacitor through a spark gap are typical examples of the spark. A spark is frequently followed by a glow or an arc discharge, discussed in some detail in the following articles. The name *corona* is applied to a glow discharge on a curved electrode, and the name *point discharge* is applied to a glow discharge on a pointed electrode. A *brush discharge* is closely related to a glow. It has a branching structure that the glow does not have, although the electrical characteristics and the physical processes involved are essentially the same in both.

9. GLOW DISCHARGE[19]

The *glow discharge* is characterized visually by a soft luminosity in the gas, and electrically by a low current density and a voltage drop of the order of the minimum breakdown voltage of the gas, which is many times the ionization voltage. The voltage drop varies little with the current over a considerable range; hence the discharge is best controlled by external regulation of the current rather than the voltage. The familiar neon signs and certain voltage-regulator tubes are examples of commercial applications of this form of discharge.

Figure 12 shows the appearance, the potential distribution, and the distribution of other quantities in a glow discharge. Under the influence of the electric field, the positive ions in a glow discharge drift toward the cathode, where their presence tends to mask the effect of the negative charge on the cathode. The field from the cathode then reaches only a short distance into the gas before all the electric lines of force terminate on positive charges. The larger part of the voltage drop in the glow discharge is concentrated across this short distance, which is frequently composed of a series of relatively light and dark regions, called the Aston dark space, the cathode glow, the Crookes dark space, the negative glow, and the Faraday dark space. Near the anode appear the anode glow and the anode dark space. The remainder of the discharge, called the *positive column*, is a region of small voltage gradient because of the presence of electrons and positive ions in amounts that produce a small net space charge. The positive column behaves as a good conductor; it has the effect of establishing a virtual anode close to the cathode and is very important in the functioning of numerous gas tubes that utilize heat-shielded cathodes, as explained in Art. 2c, Ch. V. The voltage drop across the light and dark spaces

[19] K. G. Eméleus, *The Conduction of Electricity through Gases:* Methuen Monograph Series (3rd ed.; London: Methuen & Company, Ltd., 1951).

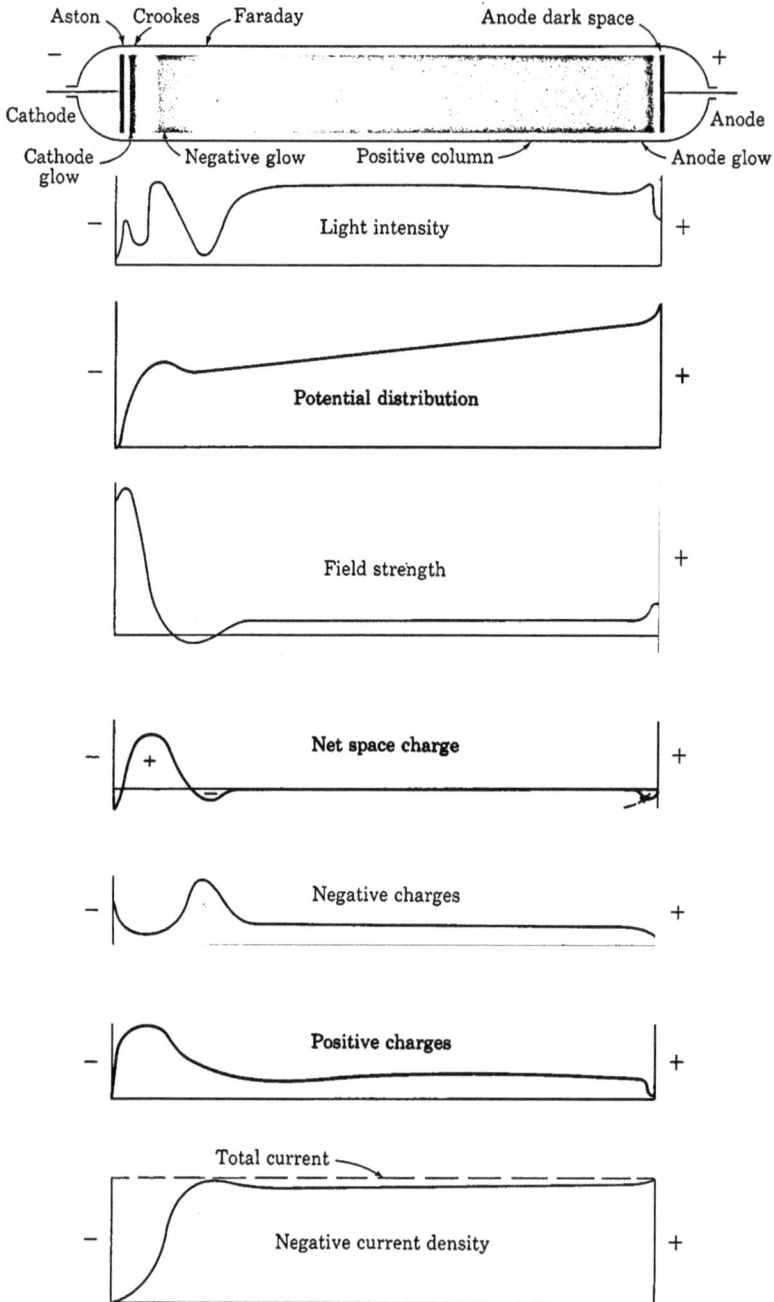

Fig. 12.   Appearance, nomenclature, and distribution of quantities in a
glow discharge.*

* This figure is reprinted from L. B. Loeb, *Fundamental Processes in Electrical Discharges in Gases* (New York: John Wiley & Sons, Inc., 1939), Fig. 269, p. 566, with permission.

near the cathode, sometimes called the *cathode fall of potential*, is nearly independent of the pressure, because the total thickness of these light and dark spaces tends to adjust itself so as to maintain the product of the pressure and distance approximately equal to the value at the minimum point of the breakdown-voltage curve, Fig. 11d.

TABLE V*

CATHODE DROP IN VOLTS

Electrode	Oxygen	Hydrogen	Nitrogen	Helium	Neon	Argon	Mercury vapor
Sodium	...	185	178	80	75	...	...
Gold	...	247	233	...	...	...	...
Magnesium	310	153	188	125	94	119	...
Mercury	...	270	226	142.5	...	...	340
Aluminum	...	171	179	141	120	100	...
Tungsten	...	...	...	...	125	...	...
Iron	343	198	215	161	...	131	389
Nickel	...	211	197	...	...	131	...
Platinum	364	276	216	160	152	131	340

* The data for this table are taken from J. Slepian, *Conduction of Electricity in Gases* (Pittsburgh: Westinghouse Electric & Manufacturing Company, 1933), 136, with permission.

When the current is small, the glow does not cover the whole of the cathode surface but concentrates on a part of it. As the current is increased, the area of the cathode covered by the glow increases linearly with the total current, so that the current density at the cathode remains constant. Under these conditions, the discharge is called a *normal glow*, and the voltage drop across the light and dark spaces near the cathode remains constant, as previously explained. Values of this cathode voltage drop for different pure-metal cathode materials and different gases are given in Table V. When composite surfaces, such as an oxide-coated surface, are used as cathodes with the noble gases, it is possible to obtain cathode voltage drops lower than those in the table, and values in the neighborhood of 50 to 100 volts are reached in commercial lamps and voltage-regulator tubes.[20]

In the normal glow, the current density at the cathode, called the *normal current density*, is a function of the cathode material, the gas, and the pressure. Experimental measurements show that the normal

[20] T. E. Foulke, United States Patents 1,965,582 and 1,965,583 (July 27, 1929); 1,965,585 (October 7, 1929); 1,965,587 (November 21, 1931); 1,965,588 (May 13, 1932).

current density varies with pressure according to the empirical relation

$$J_n = ap^b,$$ [39]

where

J_n is the normal current density in milliamperes per square centimeter,

a and b are constants dependent upon the gas and the electrode material,

p is the pressure of the gas in millimeters of mercury.

Values of a and b for several gas-electrode combinations are listed in Table VI. The constant a is numerically equal to the normal current

TABLE VI*

CONSTANTS a AND b, IN EQ. 39, FOR DETERMINING NORMAL
CURRENT DENSITIES FOR VARIOUS GAS-ELECTRODE
COMBINATIONS

Constant	Gas	Electrode material						
		Al	Zn	Cu	Fe	Ag	Au	Pt
a	H_2	0.140	0.120	0.125	0.140	0.125	0.150	0.125
	N_2	0.225	0.240	0.350	0.325	0.260	0.225	0.290
	Ne	0.008	0.006	0.024	0.026	0.021	0.019	0.011
b	H_2	2.05	1.94	1.86	1.89	1.86	1.80	1.90
	N_2	2.02	1.91	1.75	1.77	1.75	1.87	1.85
	Ne	1.50	1.83	1.06	1.38	1.00	1.14	1.30

* The data for this table are taken from W. Wien and F. Harms, *Handbuch der Experimentalphysik* (Leipzig: Akademische Verlagsgesellschaft M.B.H., 1929), Part III, Vol. *13*, 373.

density in milliamperes per square centimeter when the pressure is one millimeter of mercury. From the values of b listed, it is seen that for many gas-electrode combinations the normal current density varies approximately as the square of the pressure.

When the cathode becomes covered with the glow, the current density can no longer remain constant for further increase of total current, and the voltage drop and the current density at the cathode increase with the current. The discharge for these conditions is called an *abnormal glow*.

The volt-ampere characteristic of a glow discharge is of the general form shown in Fig. 13. The total voltage is made up of the sum of the voltages across the light and dark spaces near the cathode and the positive column, and, since the drop in the latter decreases only gradually with increased current, the total voltage is almost constant over a considerable range of current in which the discharge has the normal form. Beyond the limiting current for the normal form, the total voltage drop rises with an appreciable slope.

Experimentally determined volt-ampere characteristics of several glow-discharge devices are given in Fig. 14. During the experimental

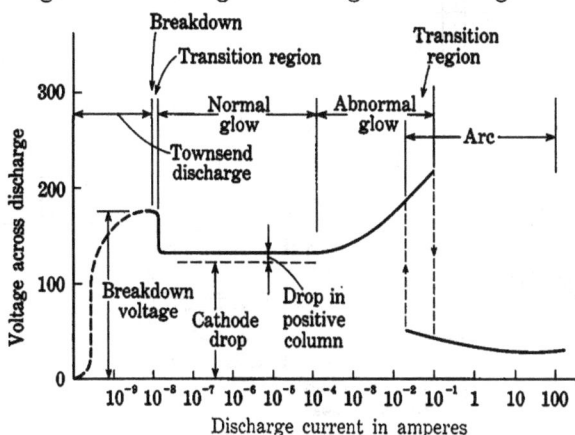

Fig. 13. Typical volt-ampere characteristic curve of a gaseous discharge through the Townsend, glow, and arc regions.

study, the electrodes were seen to become covered with glow at relatively small values of current. For the tubes in which this condition could be observed with certainty, the values of current corresponding to it are shown by an x in the figure. The data of Fig. 14 show that, even though the voltage drop increases when the total current is increased beyond the critical value that caused the glow to cover the electrodes, the increase in the voltage drop is small for values of total current up to several times this critical amount.

The essentially constant total voltage drop throughout a wide range of current variation is a property of the glow discharge frequently used in such industrial applications as the voltage-regulator tube, which is discussed in Art. 8, Ch. V, and Art. 14, Ch. VI. Another application of the glow-discharge phenomena is made in the Autovalve lightning arrester, which is designed to protect high-voltage electrical apparatus against damage due to lightning.[21] In this arrester a great many separate glow-discharge sections are connected in cascade, and the breakdown voltage of each section is made large compared with the

[21] J. Slepian, "Theory of the Autovalve Arrester," *A.I.E.E. Trans.*, *45* (1926), 169–177.

voltage required to maintain operation as a glow discharge. A large voltage is therefore required to start the discharge, but a relatively small voltage is required to maintain it after it is started. Thus, when lightning strikes a transmission line, and the sum of the breakdown

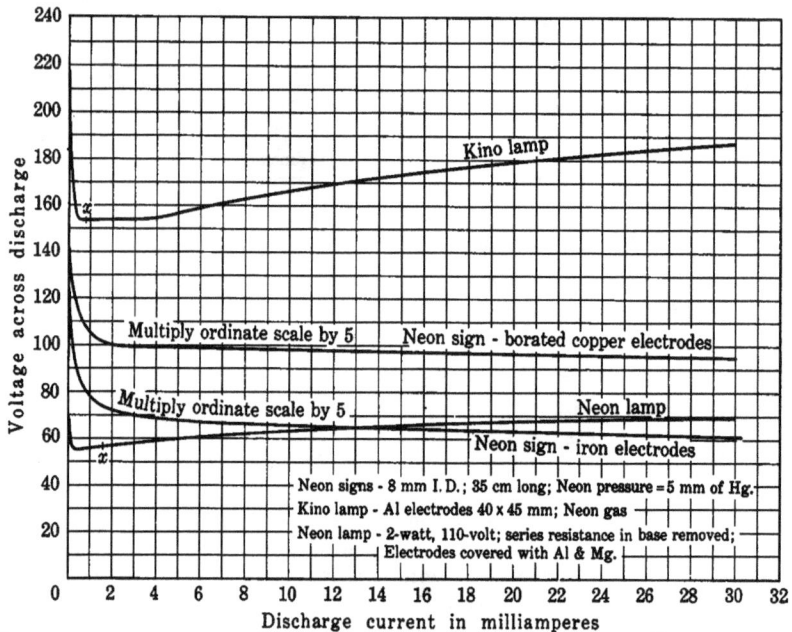

Fig. 14. Volt-ampere characteristic curves of several glow-discharge tubes. The point at which the cathode becomes covered with glow is marked by *x*. The breakdown voltage of the neon signs is greater than 1,500 volts.

voltages of the separate sections is exceeded, a multisection glow discharge is initiated in the arrester; the discharge drains the charge in the lightning stroke from the line and prevents excessive voltages from appearing across the terminal apparatus.

10. ARC DISCHARGE

Increase of the current in a gas discharge in the abnormal-glow region finally results in a sudden transition to a new type of discharge known as an *arc*. The transition stage, shown as an overlap of the abnormal-glow and arc regions in Fig. 13, is not clearly understood, and the current at which it occurs is not readily predictable. As the current in the abnormal-glow discharge is increased, the current density at the cathode increases, accompanied by increased heating and bombardment of the cathode by positive ions. At some point in the transition region, the current density at the cathode increases suddenly, and the

voltage across the discharge decreases suddenly to a much lower value
—the discharge becomes an arc. In the arc, the current is observed to
concentrate on a small spot on the cathode, and high temperatures of
the gas and electrodes may follow; violent sputtering or melting of the
electrodes frequently results if the current is allowed to increase to a
high value. The arc conducts a heavy current with a voltage drop that
is often less than the ionizing potential of the gas. This fact indicates
that some new process of reproduction of the electrons at the cathode
by the ionization and excitation products—more efficient than the
process effective in the glow discharge—must occur in the arc.

Since the physical and electrical properties of arcs vary markedly
with such factors as the gas pressure, electrode materials, and length,
a simple description suitable for all types of arcs is not readily made.
One definition[22] is: "An arc is a discharge of electricity, between electrodes in a gas or vapor, which has a voltage drop at the cathode of
the order of the minimum ionizing or minimum exciting potential of
the gas or vapor." An arc is generally characterized by a downward-
sloping volt-ampere characteristic. Thus the arc exhibits a negative
resistance to changes in current. The current in the arc therefore
depends primarily on the current-limiting properties of the circuit
supplying the power. Since most power-system circuits operate at high
voltage and have relatively little impedance, the current in an arc that
results from the breakdown of insulation or short circuits on such
systems is frequently of large magnitude and may be very destructive.
When a current is interrupted by the drawing apart of electrodes, as
in a circuit breaker, an arc is usually started, and its extinction before
the apparatus is damaged or destroyed has presented many difficult
problems for electrical engineers. The methods employed to quench
the arc in high-voltage circuit breakers[23] are good illustrations of the
fact that a sound knowledge of fundamental concepts of discharge
phenomena is essential for the successful design of apparatus. Although arcs are often harmful and annoying phenomena, there are
numerous devices other than the circuit breaker in which their
properties are used to advantage in engineering. An important
example is the mercury-arc rectifier, discussed in Art. 4, Ch. V, which
finds extensive application as a device to convert alternating-current
power into direct-current power with high efficiency and good
reliability.

[22] K. T. Compton, "The Electric Arc," *A.I.E.E. Trans.*, *46* (1927), 868.

[23] J. Slepian, "Some Physical Problems in the Electrical Power Industry," *J. App.
Phys.*, *8* (1937), 152–159; D. C. Prince, "Circuit Breakers for Boulder Dam Line," *E.E.*,
54 (1935), 366–372; D. C. Prince, J. A. Henley, and W. K. Rankin, "The Cross-Air-Blast
Circuit Breaker," *A.I.E.E. Trans.*, *59* (1940), 510–517.

This discussion of gaseous discharges can be summarized with reference to Fig. 13. As the current through a discharge tube is increased from zero, the voltage across the tube increases from zero to a few hundred volts in the region of the Townsend discharge. Further increase of the current results in breakdown, a lowering of the voltage, and a glow discharge. The voltage drop across the glow is at first nearly independent of the current but in the region of abnormal glow begins to increase. At some point in the abnormal glow region, the voltage suddenly drops to a few tens of volts or less, and an arc is formed. Thereafter the voltage decreases with increase of current, and destruction of the electrodes may result because of intense heating. The particular kind of discharge obtained in such a tube depends on the kind of gas, the spacing and material of the electrodes, the current and voltage capabilities of the source of power, and the previous history of the discharge region.

In all the forms of gaseous discharge described in the preceding articles, except the Townsend discharge, most of the electrons and ions that participate in the discharge are supplied by the discharge itself. If the cathode is made a copious source of electrons by an agency external to the discharge, a different form of gaseous conduction is obtained. For example, if the cathode is heated so that it becomes a thermionic emitter, as is done in gas-filled diodes used as rectifiers and in thyratrons, the discharge obtained resembles in some respects a glow discharge, but its current is usually higher and its voltage lower, and it is not self-maintained. Devices using gaseous conduction of this kind are discussed in Ch. V.

PROBLEMS

1. A tungsten-filament cathode 10 cm long by 1 mm in radius is concentric with a cylindrical molybdenum plate in an evacuated bulb. When the filament has a temperature of 2,400 K, and the molybdenum cylinder is made 100 volts positive with respect to the filament, the plate current i_b is 150 ma.

If end effects are neglected, what will be the plate currents corresponding to plate potentials of 200 and 400 volts?

2. A vacuum diode has a circular cylindrical plate concentric with an axial tungsten filament. The filament temperature is 2,400 K, and the tube dimensions are as follows:

Filament length = 10 cm
Filament diameter = 0.03 cm
Plate length = 10 cm
Plate diameter = 2 cm.

Effects associated with fringing of electric or magnetic fields near the ends of the structure may be neglected.

(a) What is the saturation current I_s from the filament?

(b) What plate voltage e_b is required to produce a plate current i_b equal to half the current in (a)?

(c) What value of uniform magnetic flux density directed parallel to the filament will reduce the plate current to zero if the plate voltage is 200 volts?

3. The electrode structure of a typical vacuum rectifier tube consists of a unipotential oxide-coated indirectly heated cathode of 0.0625 in. outside diameter by $\frac{7}{8}$ in. long, and a concentric cylindrical plate of 0.125 in. inside diameter by $\frac{7}{8}$ in. long.

(a) Calculate the approximate value of the plate voltage e_b when the plate current i_b is 100 ma.

(b) Discuss the factors that may be expected to produce deviations from the result of (a) in an actual tube.

4. A vacuum diode has a circular cylindrical plate concentric with an axial tungsten filament. The dimensions are:

$$\text{Filament length} = 5 \text{ cm}$$
$$\text{Filament diameter} = 0.02 \text{ cm}$$
$$\text{Plate length} = 5 \text{ cm}$$
$$\text{Plate diameter} = 2 \text{ cm.}$$

The initial filament temperature is 2,400 K.

(a) What direct plate voltage will be just sufficient to draw the total saturation current if end effects are neglected?

The filament power is now increased by 10 per cent.

(b) What is the per cent change in the direct plate voltage required to draw the total saturation current? Indicate whether this change is an increase or decrease.

5. Two vacuum diodes have right-circular cylindrical plate structures and round axial tungsten filaments. All linear dimensions of diode A are double those of diode B.

Show by sketches drawn approximately to scale how the electrical characteristics of the tubes differ. Include sketches of plate current i_b as a function of plate voltage e_b and sketches of plate current i_b as a function of filament temperature T, assuming that the plate current is determined primarily by the plate voltage or the filament temperature.

6. If the gas in an electron tube is nitrogen at 25 C:

(a) To what pressure must the gas be reduced in order that the chance of an electron's reaching the plate without collision with a molecule shall be at least 9,999 in 10,000 when the distance from the filament to the plate is 1 cm?

(b) What is the chance of an electron's colliding at least once if the pressure of nitrogen is: (1) 1 micron, that is, 10^{-3} mm of Hg? (2) 1 mm of Hg?

7. Consider electrons that are moving from the filament to the plate in a discharge tube through argon gas at low pressure. The mean free path of an electron in argon at this pressure is 4 cm. The distance between the filament and plate is 2 cm. What fraction of the number of electrons leaving the filament will probably collide at least once with argon atoms before reaching the plate?

CHAPTER IV

Vacuum Tubes

The preceding chapters deal primarily with the physical phenomena involved in electronic conduction. In this chapter and the next, the ways in which these phenomena combine to determine the characteristics and limitations of some of the more important electron tubes are discussed, and the resulting characteristics are described. In the remaining chapters these characteristics are used in analysis of the over-all behavior of electron-tube circuits in which the tube is associated with other circuit elements to form a working system, such as a rectified-power supply, an amplifier, or an oscillator.

As a guide to the material that follows, the various methods already mentioned for classifying electronic devices may be reviewed and expanded. A large class of electronic devices comprises *electron tubes*, in which electronic conduction takes place in an evacuated, or partially evacuated, enclosure. Within this group, the tubes may be divided into two classes on the basis of the electrical effect of the gas present in them after evacuation. The first class, which is discussed in this chapter, includes tubes in which the gas pressure is so low after evacuation that it has no appreciable effect on the tube operation. These are called *vacuum tubes* or *high-vacuum tubes*. The second class, which is discussed in Ch. V, includes tubes in whose operation gas, purposely inserted, plays a major role. These are called *gas tubes*.

Another method of classifying electron tubes is in accordance with the type of cathode used. Vacuum tubes include, among others, *thermionic tubes* (having a heated, thermionically emitting cathode) and *phototubes* (having a photoelectrically emitting cathode). Gas tubes include thermionic tubes, phototubes, and *cold-cathode tubes*, the last having an unheated cathode from which the electron emission is perhaps obtained by the process of secondary emission or field emission.

Still a further method of classifying electron tubes is in accordance with the number of electrodes that have an essential part in the operation for which the tube is intended. On this basis, as is stated in Art. 2, Ch. I, a tube containing two electrodes—one a cathode and the other an anode or plate—is named a *diode*; and a tube containing three electrodes—a cathode, a control electrode, or grid, and a plate—is named a *triode*. In a similar manner, tubes containing a cathode, a plate, and more than one grid are named *tetrodes, pentodes, hexodes, heptodes, octodes*, and so forth, in accordance with the total number of electrodes. The cathode in any one of these tubes may be an unheated electrode, an incandescent filament, or a combination of an oxide-coated sleeve and an internal heater element. This nomenclature is usually applied only to tubes in which the elements other than the cathode and anode are control electrodes. Multi-element tubes having additional electrodes that are not grids, such as the tubes with several anodes used in rectification, and special-purpose tubes, such as the cathode-ray tube, are usually distinguished by a name based on their function.

1. CHARACTERISTICS OF THERMIONIC VACUUM DIODES

The thermionic vacuum diode is a relatively simple and important electron tube. It consists of a plate and a heated cathode, as shown by the symbol for it in Fig. 1a, located in a region so well evacuated that the residual gas has a negligible effect on the operation of the tube. Thermionic vacuum diodes for high-voltage applications are often called *kenotrons*. The behavior of the thermionic vacuum diode may be described roughly as follows: Because of the substantial thermionic emission resulting from the high temperature of the cathode, and because of the close spacing of the electrodes, current is conducted through the tube with a relatively small voltage drop, provided the plate has a positive potential with respect to that of the cathode. On the other hand, if the applied voltage is reversed so that the plate is given a negative potential with respect to that of the cathode, a negligible current is conducted through the tube for moderate voltages, because electrons are not emitted from the unheated plate at an appreciable rate, and the vacuum is a good insulator. The tube thus has unidirectional conductivity. If an alternating-voltage source is connected across a series combination of the tube and an electrical load, such as a resistor, current conduction takes place only in one direction, and the current is zero during alternate half-cycles. Under these conditions, the tube causes the alternating-voltage source to send a unidirectional pulsating current through the load. The tube is then said to act as a *rectifier*. Since a voltage drop occurs across the tube while current is conducted, power is dissipated in the tube. This power loss is often undesirable, because it decreases the efficiency of power transfer from the alternating-voltage source to the load, and because it heats the plate of the tube. In order to determine the tube loss, the circuit efficiency, and other related quantities, the current-voltage relationship of the tube is needed. The factors governing this relationship are discussed in this article, and the behavior of the tube as a rectifier is analyzed in Chs. VI and VII.

Previous chapters describe two electronic phenomena of major importance among those that govern the current-voltage characteristic of a thermionic vacuum diode. These phenomena are thermionic emission and the limitation of current by space charge. If they were the only governing phenomena and were accurately describable by the theoretical relationships developed in the previous chapters, the graphical relations among the three variables—cathode temperature, plate voltage, and plate current—would have the shapes shown qualitatively in Figs. 1b and 1c.

In these figures, i_b and e_b are the plate current and plate voltage with assigned positive reference directions as indicated in the schematic circuit diagram of Fig. 1a. Lower-case letters are used for electrode voltages and currents in this chapter, as is done in the derivation of Child's law in Art. 1, Ch. III, in order to conform to the general scheme of symbols used in this book for vacuum-tube circuits, which is summarized in Art. 20, Ch. VIII. Capital letters, however, are frequently used for the variables on the axes of such characteristic curves in other literature on electron tubes. The electrode currents specified by the characteristic curves, such as those of Figs. 1b and 1c, are strictly correct only when the electrode voltages change so slowly that displacement currents are negligible. If the voltage variations are more rapid, the currents through the interelectrode capacitances effective when space charge is present must be added to the currents specified by the curves, in order to give the total electrode currents, as is discussed in Ch. XII for diodes and in Ch. VIII for triodes. If the voltage variations are exceedingly rapid, the cathode-to-plate transit time of the electrons becomes important, and the effects discussed in Art. 14 of this chapter must be considered.

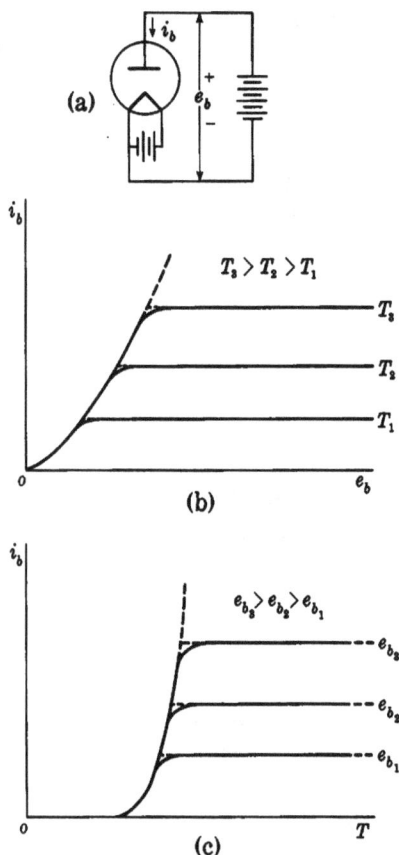

Fig. 1. Characteristics of an ideal vacuum diode.

The characteristic curves of actual diodes have, roughly, the shape of those in Figs. 1b and 1c, but noticeable departures from the idealized shapes exist. For example, the curves for a vacuum diode with a directly heated tungsten filament are shown in Fig. 2. In the regions roughly defined as *cd* along the curves corresponding to the intermediate values of the parameters T and e_b in Figs. 2a and 2b, respectively, the current is limited principally by the amount of the thermionic emission from the cathode. Because of the Schottky effect, however, the plate current increases with the plate voltage in these regions; consequently, the curves have an upward slope in the region *cd* of Fig. 2a and are separated slightly in the region *cd* of Fig. 2b. In

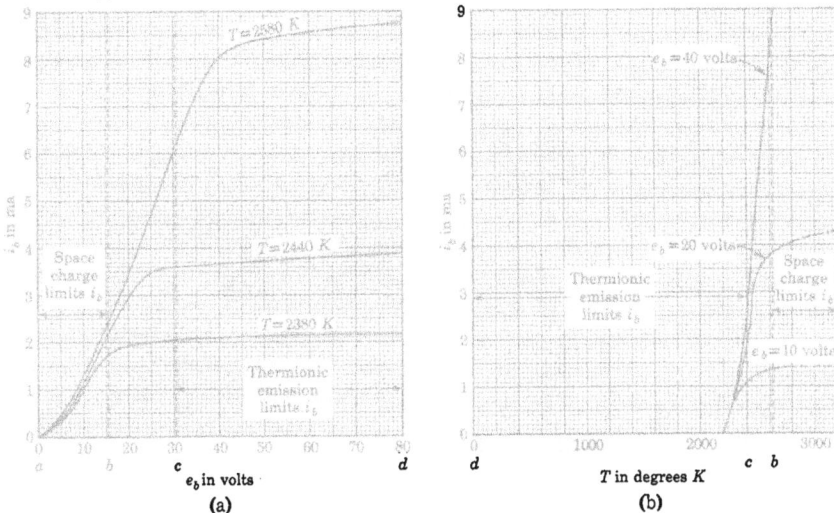

Fig. 2.   Characteristics of an actual vacuum diode with tungsten cathode.

the region roughly defined as *ab* along the same curves in Figs. 2a and 2b, the plate current is limited principally by space charge. However, the amount of the space-charge-limited current depends on the initial velocities of the electrons at the cathode, which in turn depend on the cathode temperature. Accordingly the plate current increases slightly with the temperature; the curves hence slope upward in the region *ab* of Fig. 2b and are separated slightly in the region *ab* of Fig. 2a. The separation of the curves at the smaller values of plate current is somewhat exaggerated in this figure.

A characteristic curve for an experimental vacuum diode having an indirectly heated oxide-coated cathode is shown in Fig. 3. Space charge is the only current-limiting factor in this tube. Limitation by thermionic emission is generally not effective in such tubes because heating of the plate, discussed in Art. 2, sets a practical limit on the space current before it equals the full thermionic emission. Even when the cathode is operated at a subnormal temperature, the abnormally

large Schottky effect of the oxide-coated cathode obscures any limitation by thermionic emission. The diode of Fig. 3 has a basic structure
to which in Arts. 5 and 8 one and two grids are added for illustration
of triode and tetrode behavior. The curves in Figs. 3, 7, 8, 9, and 17
thus show the comparative effect of addition of grids. The diameter of
the anode shown in Fig. 3 is larger than would ordinarily be employed,

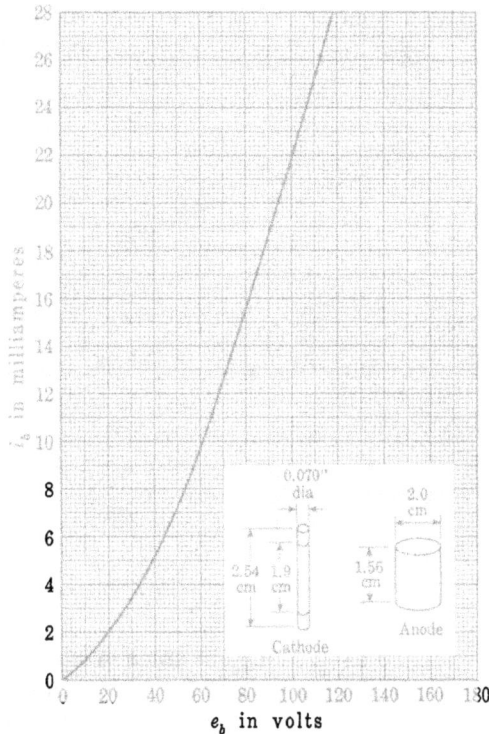

Fig. 3. Characteristic curve of an experimental vacuum diode
having an oxide-coated cathode and internal dimensions as
indicated.

and the voltage drop at a particular current is correspondingly greater.
Appendix C contains curves for commercial diodes, Types 6AL5,
5U4-G, and 5Y3-GT, which have closer spacing between cathode
and anode.

Several phenomena in addition to space charge and thermionic
emission have a minor influence on the shape of the characteristic
curves. One such effect is the contact difference of potential. As is
explained in Art. 4, Ch. II, differences in work function of two metals
cause an electrostatic field to appear between the metals when they

are joined at one point. Consequently, when a voltage is applied between the cathode and the plate of a diode, the field appearing between the electrodes in the tube is that which is set up by the applied voltage plus the contact potential difference. The contact potential difference is equal to the difference in the voltage equivalents of the work functions of the materials of the plate and the cathode; it is not modified if a third metal is used in the external connections of plate to cathode. The contact potential difference has the effect of shifting the curves of Figs. 1b, 2a, or 3 to the right or left by an amount of the order of one volt. If the work function of the plate is greater than that of the cathode, the shift is to the right; if it is less, the shift is to the left.

A second minor effect that influences the shape of the characteristic curve is that of the initial velocities associated with the thermal energies of the electrons as they leave the cathode. As a result of these velocities, the electrons are able to travel to the plate even though the plate voltage e_b is zero or slightly negative. Thus the curves in Figs. 1b, 2a, and 3 actually intersect the current axis above and the voltage axis to the left of the origin.[1] These intersections are too small, however, to show on the scales used. In many applications the failure of the curve to intersect the origin is negligible, but in others it cannot be neglected, and in some the negative-voltage part of the volt-ampere characteristic curve is the only portion that is utilized.

A relation between the plate current and the negative plate voltage that is applicable to parallel-plane diodes may be found from the principles given in Chs. II and III. As the plate voltage is reduced from positive values through zero to negative values, the potential minimum shown in Fig. 1b, Ch. III, moves away from the cathode and finally reaches the plate. For negative plate voltages having still larger magnitudes, no potential minimum exists, and space charge therefore no longer limits the current. Instead, the current is limited by the fact that, to reach the plate, an electron starting from inside the cathode must have sufficient kinetic energy not only to surmount the potential-energy barrier W_a at the surface of the cathode shown in Fig. 6, Ch. II, but also to overcome the retarding effect of the electric field established by the negative plate voltage. In other words, the electron must have an amount of initial energy $Q_e |e_b|$ in addition to the amount W_a required for escape from the cathode. Addition of $Q_e |e_b|$ to W_a in Eq. 12, Ch. II, therefore gives a relation between the retarding voltage and the current density at the plate instead of at the cathode. Expressed in terms of the saturation emission-current density

---

[1] E. L. Chaffee, *Theory of Thermionic Vacuum Tubes* (New York: McGraw-Hill Book Company, Inc., 1933), 79–82.

J_s at the cathode, the plate current is hence

$$i_b = s_p J_s \epsilon^{-\frac{Q_e |e_b|}{kT}}, \qquad [1]$$

where s_p is the area of the cathode or plate, and the dimensions of the parallel plane electrodes are assumed to be large compared with the separation. The factor $Q_e/(kT)$ is larger than 10 volt^{-1} for the normal operating temperatures of all ordinary thermionic cathodes. Consequently a negative plate voltage of one to two volts is generally sufficient to reduce the plate current to a negligibly small value.

Equation 1 is not exactly applicable to diodes having cylindrical electrodes because it does not account for the possibility that an electron may have a changing tangential component of velocity as well as radial component during its travel to the plate. Nevertheless, the plate current in cylindrical diodes is found[2] to change almost exponentially with the magnitude of the negative plate voltage over many decades of current ratio, and the logarithm of the plate current is essentially a linear function of the plate voltage within that range.

A third factor that influences the shape of the characteristic curves in a tube having a filament-type cathode is the potential distribution along the filament. This affects the tube characteristics in a way that is not included in the simple space-charge theory, since the filament is there assumed to be an equipotential surface. If the filament is considered to be made up of infinitesimal elements, and if their individual contributions to the plate current are summed by integration, the over-all effect of the filament voltage drop is determined. The plate current is thus found[3] to increase with the five-halves power of the plate voltage when the plate voltage is smaller than the filament voltage, and gradually to merge with the values given by the three-halves power relation, Eq. 25 of Ch. III, as the plate voltage becomes large compared with the filament voltage.

Finally a fourth factor that influences the shape of the characteristic curves is the fact that the temperature is never uniform along the cathode of a vacuum tube. Consequently the temperature-limited value of current for one part is reached at a lower plate voltage than for another. This effect influences principally the shape of the "knee" of the curve and accounts for the rounding of the knee. The non-uniform temperature distribution along a filament-type cathode is due

[2] W. Schottky, "Über den Austritt von Elektronen aus Glühdrähten bei verzögernden Potentialen," *Ann. d. Phys.*, *44* (1914), 1011–1032; L. H. Germer, "The Distribution of Initial Velocities among Thermionic Electrons," *Phys. Rev.*, *25* (1925), 795–807; T. S. Gray and H. B. Frey, "Acorn Diode Has Logarithmic Range of 10^9," *R.S.I.*, *22* (1951), 117–118.

[3] H. J. Van der Bijl, *The Thermionic Vacuum Tube and Its Applications* (New York: McGraw-Hill Book Company, Inc., 1920), 64–70.

to a number of factors, among which the cooling effect of the leads is important. This cooling effect makes the effective length of the filament less than the actual length.[4]

A nonuniform temperature distribution along the filament also results from the fact that in the filament the emission current is superposed on the current I_f supplied by the filament battery. When, as is customary, the negative end of the plate-voltage supply is connected to the negative end of the filament, the current in the filament at that end is $I_f + i_b$, and at the other end it is I_f. This excess current makes the negative end hotter than the positive end. To reduce the effect, a filament in a vacuum diode is often designed so that its heating current I_f is more than thirty times its rated plate current i_b. In a particular 300-kilowatt vacuum tube with a tungsten filament[5] and an available emission current of 200 amperes, which is several times the rated plate current, the heating voltage is 17 volts and the heating current has the large value of 1,800 amperes.

An additional effect of filament voltage drop consists of conduction from the negative end of the filament across the space to the positive end. This conduction is particularly large when the two ends of the filament are close together, as in the case of a **V**, **W**, or "hairpin" form. In fact, this type of conduction first brought to the attention of Edison what is now recognized as electronic conduction. The conduction current may become so large in a gas tube, where the space charge is neutralized by positive ions, that an arc will form between the ends of the filament and destroy it. Such arcs occurred in some of Edison's early lamps, because inferior vacuum pumps left considerable gas in the bulb.

The heat of evaporation of the electrons produces a cooling effect, which tends to counteract the heating effect caused by the conduction of the emission current in the filament. Each electron, as it leaves the filament, carries with it an amount of energy equal to ϕQ_e, where ϕ is the voltage equivalent of the work function of the filament, and the power required to replace this energy as the electrons evaporate into the vacuum is ϕi_b. In the large filament mentioned above, the work function ϕ has a value of about 4.5 volts, and the amount of heating power that would have to be supplied merely to evaporate all the emitted electrons is 4.5 times 200, or 900 watts. The cooling of the filament is often visible when the plate voltage is applied, even with small tubes. Upon entering the anode, the electrons give up an amount

[4] H. A. Jones and I. Langmuir, "The Characteristics of Tungsten Filaments as Functions of Temperature," *G.E. Rev.*, *30* (1927), 356.

[5] F. Banneitz and A. Gehrts, "Wassergekühlte Senderöhren (Grossleistungsröhren)," *E.N.T.*, *11* (1934), 214–231.

of heat that may be calculated in a similar manner if the work function of the anode is employed. However, as is discussed in Art. 2, this is only a small fraction of the total heating effect at the anode.

2. Maximum ratings and average characteristics of vacuum tubes

The maximum ratings of a diode—that is, the maximum anode current and voltage to which it should normally be subjected—are set by one or more of a number of factors. As is shown in the previous chapters, a more or less definite limitation is placed on the anode current by the cathode-heating power, the emission efficiency, and life considerations. On the other hand, heating of the anode or the glass bulb may impose a smaller limit. With a constant anode current, the total power dissipated as heat inside the tube is

$$P = e_b i_b + I_f E_f ,\qquad [2]$$

where I_f and E_f are the cathode heating current and voltage, respectively, and the contact difference of potential is considered negligible in comparison with e_b. In the steady state, all this power must leave the bulb through the walls, and because of absorption of some of the energy the walls are heated. To avoid evolution of gas, and, further, to avoid softening of the glass and consequent collapse of the walls under the pressure of the atmosphere, the walls should not operate at a temperature as high as that to which they were raised in the out-gassing process during evacuation; the outgassing temperature is about 350 degrees centigrade for soft or lime glass, and 550 degrees centigrade for hard, or Nonex, glass. Because of this limitation, glass bulbs are not generally used to enclose vacuum tubes of more than about one kilowatt rating.

Often the heat-radiating ability of the anode is the limiting factor in the tube rating. The anode is heated by two processes. First, the kinetic energy that the electrons possess because of their velocity is converted into heat when they strike the anode. Second, a portion of the energy radiated by the cathode is intercepted by the anode. If the contact difference of potential between the anode and the cathode and the heating effect caused by the work function of the anode, discussed in Art. 1, are neglected, the energy dissipated by each electron on arrival at the anode is $e_b Q_e$, and thus the power that the electrons carry to it is $e_b i_b$. Therefore the anode must radiate an amount of power

$$P_B = e_b i_b + f E_f I_f ,\qquad \blacktriangleright [3]$$

in which f is the fraction of the cathode-heating power intercepted as radiation by the anode. The factor f has a value less than unity, depending on the extent to which the anode surrounds the cathode. The temperature reached by a radiation-cooled anode is approximately that given by the Stefan-Boltzmann law, previously stated in Art. 6, Ch. II,

$$P_B = s_p e_t K T^4, \qquad \blacktriangleright[4]$$

where

s_p is the area of the radiation surface,

e_t is the total radiation emissivity (see Table II, Ch. II),

K is the Stefan-Boltzmann constant,

T is the surface temperature in degrees Kelvin.

The permissible temperature rise of the anode is limited in some devices by the melting point of the material of which it is made, and in others by the allowable emission current from the anode. This current becomes a reverse conduction current in a rectifier when the plate-voltage polarity reverses. For nickel, molybdenum, and tungsten, the permissible power dissipation per unit area is given by Kusunose[6] as 3, 5, and 8 watts per square centimeter, respectively. Graphite anodes, which are often used,[7] have a greater radiation emissivity and consequently a lower temperature for a given amount of power radiated. For the same reason, nickel anodes are often coated with carbon. Tantalum has the important advantage as an anode that it tends to absorb gas and thereby improves the vacuum.

For power ratings greater than about one kilowatt, dependence on cooling by radiation from the anode is generally considered undesirable because of the large dimensions required and the consequent bulk of the tube. Larger tubes are often constructed with water-cooled anodes, part of the tube wall being made of metal. The electrons are collected on the inner surface of the wall while cooling water is circulated on its outer surface. Cooling fins with forced-air circulation are also common.[8]

---

[6] Y. Kusunose, *Calculations on Vacuum Tubes and the Design of Triodes: Researches of the Electrotechnical Laboratory, No. 237* (Tokyo, Japan, 1928), 132; "Calculation of Characteristics and the Design of Triodes," *I.R.E. Proc., 17* (1929), 1742. The second reference is an abbreviated form of the first.

[7] E. E. Spitzer, "Anode Materials for High-Vacuum Tubes," *E.E., 54* (1935), 1246–1251; W. Espe and M. Knoll, *Werkstoffekunde der Hochvakuumtechnik* (Berlin: Julius Springer, 1936).

[8] M. van de Beek, "Air-Cooled Transmitting Valves," *Philips Tech. Rev., 4* (1939), 121–127; E. M. Ostlund, "Air Cooling Applied to External-Anode Tubes," *Electronics, 13* (June, 1940), 36–39. "Water-Cooling versus Air-Cooling for High-Power Valves," *I.E.E. Proc., 96, Part III* (1949), 220–221.

The maximum anode-voltage rating of a vacuum tube may be limited by electrolysis in the glass or by flashover outside the tube, rather than by heating. When the cathode and anode leads are brought through the glass side by side, and a high voltage is impressed between them, appreciable conduction between the leads and accompanying electrolysis of the glass may take place, especially if the glass is hot. This electrolysis results in deterioration of the glass and eventual air leakage, with consequent loss of the vacuum. The highest voltage permissible between adjacent leads depends upon the spacing, temperature, and kind of glass; for example, it is about 500 to 1,000 volts for the usual receiver-type tube in which the leads are only one or two millimeters apart.

In rectifier circuits, the polarity of the anode voltage reverses during a cycle, and, as is discussed in Ch. VI, the tube must be able to withstand an *inverse-peak anode voltage* (a voltage that makes the anode negative with respect to the cathode) of one to two times the peak value of the alternating voltage it is to rectify. For high-voltage rectification, a tube design is therefore required in which the anode and cathode leads are brought out at opposite ends and are sufficiently separated to prevent a spark from passing between them. A study of the dimensions of some commercial tubes indicates that they are usually made long enough so that the average gradient does not exceed about 7,000 volts per inch over the surface of the glass between the leads. To enable them to withstand a greater inverse-peak voltage, tubes are sometimes immersed in oil, the danger of flashover thus being lessened.

In tubes with grids, which are discussed in Art. 2, Ch. I, and in the remainder of this chapter, the rated values of the anode voltage and current are governed by cathode emission and anode heating as for diodes. Except for a fraction of the cathode power absorbed as radiation, the anode heating is the average product of the anode voltage and current, which may be determined by the methods given in Chs. VIII and X. The tube ratings must, however, take into account grid heating as well as anode heating. Computation of this grid heating is not always so straightforward as computation of anode heating, for two reasons. First, since the grids are ordinarily located between the cathode and the anode, and one grid is often very close to the cathode, a major fraction of the total grid heating may result from radiation from the cathode and anode. Second, grids often emit electrons; hence the product of the grid-to-cathode voltage and the current in the external circuit supplying the grid may not correspond to the kinetic energy of the electrons striking it. Generally, grid emission of several types may contribute to the grid current, but the types that can be significant for a particular grid depend on the potential of the grid relative to that of the adjacent electrodes. Grid-emission current from a grid having a potential more negative than that of all other electrodes, such as a control grid, can result from thermionic emission from the grid when heat radiation from other electrodes raises its temperature. Since appreciable current at a control grid is usually not tolerable, as is explained in Ch. VIII, limitation of the temperature of the other electrodes in a tube is often necessary to reduce such radiation. In some tubes the design provides for conduction of heat from the grids by materials having high heat conductivity, or by radiation of the heat by fins attached to the grid.

Grid-emission current from a grid having a potential intermediate between that of the cathode and another electrode adjacent to the grid can result not only from thermionic emission but also from secondary emission from the grid surface, in the manner described in Art. 12, Ch. II. Since the grid-emitted electrons reduce the net circuit current below the value corresponding to the primary electrons, and may, in fact, cause its direction to reverse, the heating caused by the primary electrons striking the grid may be much larger than the product of the grid-to-cathode voltage and the current in the external circuit. Limitation of the grid-to-cathode voltage is then necessary to prevent overheating of the grid. Both thermionic and secondary emission from grids are aggravated by evaporation of a film of oxide-coating material or thorium onto the grid surface when a composite cathode of the oxide-coated or thoriated-tungsten type is used. This evaporated film lowers the thermionic work function of the grid and increases its yield of secondary electrons.

Ratings based on the foregoing considerations are known as *absolute maximum ratings*. Damage to the tube or reduction of its life is likely if they are exceeded. Alternative ratings called *design-center maximum ratings* are commonly stated by manufacturers of tubes. These ratings are values about which variations associated with normally encountered fluctuations of the voltage of the alternating-current or direct-current line, storage battery, or other source supplying the power are expected and permissible. They are purposely chosen to allow for such fluctuations.

Manufacturers customarily publish characteristic curves, such as the current-voltage curve for diodes and other similar curves for tubes comprising grids, which are called *average characteristics*. Because the chemical and physical phenomena on which the behavior of the tube depends are complicated and the internal geometrical distances are small, control of the characteristics of individual tubes in manufacturing is difficult. The manufacturer inspects tubes as they are produced, and rejects all having characteristics that fall outside a standard specified range, which spans the average values desired. But the tolerance, or permitted range of variation, is larger than is generally realized. One prominent manufacturer states that "in general the designer should consider a probable plus or minus variation of not less than thirty per cent." Engineering design of equipment should take into account such probable variations, so that the equipment will function with replacement tubes having characteristics that fall at the extremes of the range covered by the specifications, as well as at the average point in the range. Furthermore, the design should anticipate that the characteristics of a particular tube will change with use. For example, manufacturers ordinarily consider that many common tubes have not reached the end of their useful life until the transconductance, defined in Art. 6, has decreased to 60 per cent of its rated average value.

3. CONTROL OF CURRENT IN VACUUM TUBES BY MEANS OF GRIDS

The insertion of a third electrode, known as a grid, between the cathode and plate of a vacuum diode for the purpose of controlling the plate current has proved to be of great importance to science and engineering. In the usual conditions of operation of such a vacuum triode, the grid is maintained at a negative potential with respect to the cathode; hence electrons are repelled, few reach the grid, and the electron current to the grid is very small. Little power, consequently, is dissipated in the grid circuit even though changes in the grid potential result in the control of considerable power in the plate circuit. Thus *amplification* is accomplished, in the sense that a comparatively large amount of power supplied by a plate-voltage source is controlled at the expense of a very small amount of power supplied by the circuit connected to the grid. As those engaged in the electronic art became more familiar with the characteristics of triodes, it was foreseen that the insertion of additional grids in the space between the cathode and plate, in a manner such as is shown in Fig. 4, would alter the tube characteristics in ways advantageous in certain applications. The properties of the multigrid devices as circuit elements were then carefully and thoroughly investigated. Tubes with as many as five grids, each grid having a definite and different function, are now not uncommon.

If the grid of a triode is maintained at a constant negative potential with respect to the cathode, few if any electrons can reach it, and the plate current may be considered to be the same as the current of electrons that leaves the cathode. Thus the effect of the grid voltage on the plate current may be determined from the effect of this voltage on the cathode current. When the plate and grid voltages are such that the

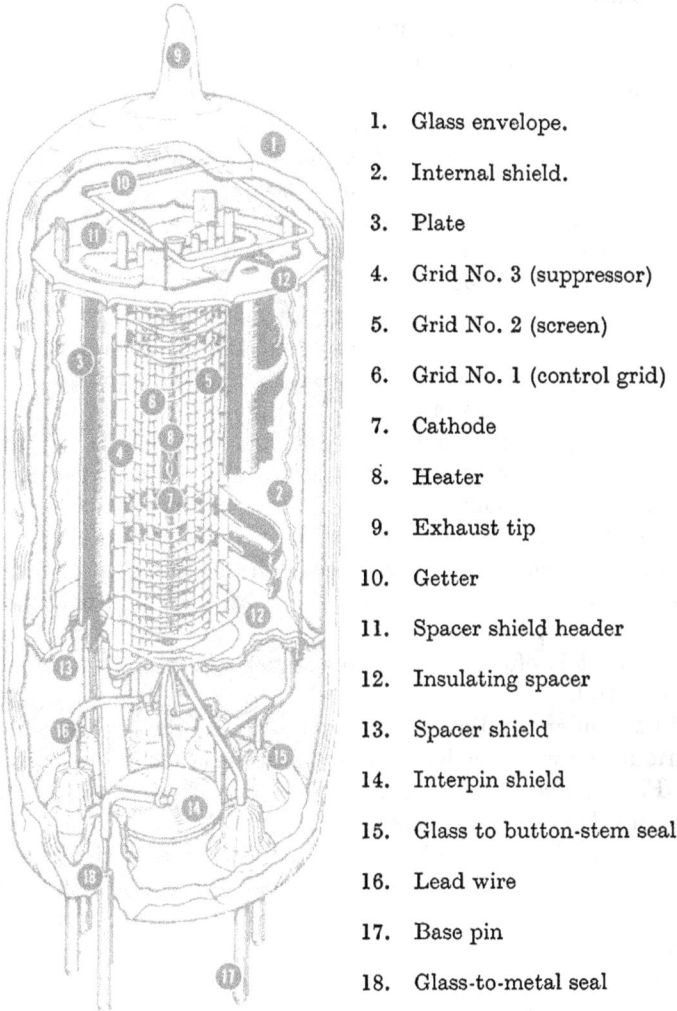

1. Glass envelope.

2. Internal shield.

3. Plate

4. Grid No. 3 (suppressor)

5. Grid No. 2 (screen)

6. Grid No. 1 (control grid)

7. Cathode

8. Heater

9. Exhaust tip

10. Getter

11. Spacer shield header

12. Insulating spacer

13. Spacer shield

14. Interpin shield

15. Glass to button-stem seal

16. Lead wire

17. Base pin

18. Glass-to-metal seal

2½ times actual size

Fig. 4.   Structure of a multi-grid miniature-type tube (2⅛ inches long).
(*Courtesy Radio Corporation of America.*)

cathode current is limited by the electron-emitting properties of the cathode, its magnitude is determined essentially by the material and temperature of the cathode, and variations of the grid voltage can affect it only slightly. On the other hand, when the current is limited by space charge, the grid voltage may affect the space charge and control the current effectively. A qualitative explanation of the effect of the grid may be made as follows. If, for simplicity, the initial velocities of emission are neglected, the analysis of Art. 1, Ch. III, shows that, for the current to be limited by space charge in a diode, the electric field at the surface of the cathode must be zero. If a grid is inserted, and if its potential is maintained at a value negative with respect to the potential at the point in the diode at which it is inserted, it serves to lower the potential distribution curve in its neighborhood. It acquires a negative charge, which tends to increase the total negative charge between the plate and cathode. Thus the potential gradient at the cathode surface tends to become negative, and the current drawn from the copiously emitting cathode tends to stop. The space charge therefore decreases throughout the interelectrode space until the gradient at the cathode surface is again zero, and the current establishes itself at a new and smaller value consistent with the smaller space charge then in the interelectrode space. In a similar manner, if the inserted grid is given a positive potential with respect to the potential at the point at which it is inserted, a larger electron space charge is needed to maintain the zero-field condition at the cathode, and the space current increases. In the ordinary use of triodes, however, a negligible grid current is desired, and the grid potential is therefore always kept negative with respect to the cathode. The control of current is effected through varying the magnitude of this negative potential.

As an aid to understanding the action of a grid, plots of the electric field in a triode, showing the hypothetical lines of force in the tube, are useful. In Fig. 5, the cathode and plate are drawn as parallel plane electrodes, and the grid is shown as a grating of round wires in a plane parallel to them. The plus and minus signs placed near the grid and plate indicate the sign of the potential of these electrodes with respect to the cathode, and the dots represent a few of the electrons in the space between the cathode and grid where the space charge is densest. This diagram is to be compared with that for a diode in Fig. 2b, Ch. III. The lines of force represent the direction and intensity of the electric field at each point; they may be considered to emanate from positive charges on the plate and to terminate on negative charges, either on the grid or on the electrons. Since the initial velocities of emission are neglected, the field intensity at the cathode surface is zero in the presence of the space-charge-limited current, and no lines of force from the anode reach all the way to the cathode.

In Fig. 5a, the grid potential is positive and modifies only slightly the field that would exist in the diode. It intercepts only a few of the

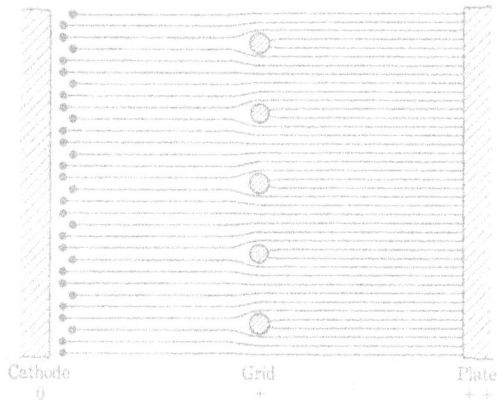

(a) With positive grid voltage

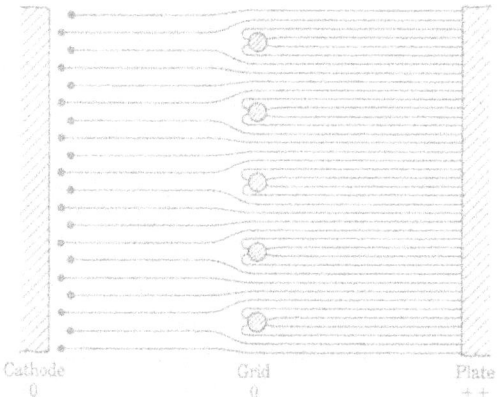

(b) With zero grid voltage

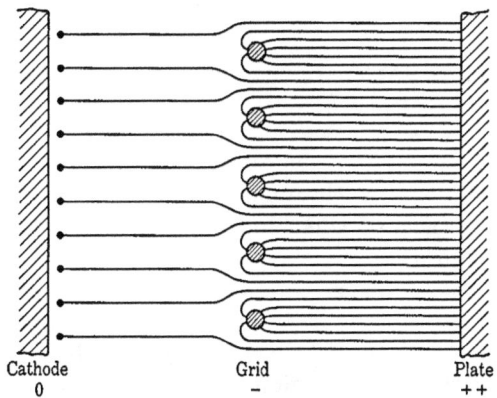

(c) With negative grid voltage

Fig. 5.   Electrostatic field in a triode with space-charge-limited current.

lines of force from the plate. The remainder terminate on the moving electrons that constitute the space charge. In Fig. 5b, the grid potential is reduced to that of the cathode, and its negative charge is therefore increased in magnitude. It intercepts more of the lines of force from the plate and causes greater distortion of the field. Fewer lines reach through to terminate on electrons and draw them to the plate. A smaller space charge and space current result. In Fig. 5c, the grid potential is made negative, so that the magnitude of the negative charge on the grid and the field distortion are further increased. Similarly a further decrease of space charge and current occurs.

If the grid potential is made negative and of sufficient magnitude so that all the lines of force that leave the plate terminate on the grid, none reach through to terminate on electrons. The field at the cathode and the charge in the interelectrode space are then simultaneously zero, and there is hence no plate current. The negative grid voltage of smallest magnitude required to make the plate current zero is called the *cut-off grid voltage*. Its value depends on the plate voltage used.

When the previously neglected initial velocities of the electrons are included in this qualitative picture, a minimum of potential exists near the cathode. The electric field near the cathode is in a direction to draw the electrons back to the cathode, and the lines of force from electrons in this region reach back to the cathode rather than to the plate. Electrons with sufficient initial velocity to pass through the potential minimum, however, can pass on to the plate. As the grid is made increasingly negative, the plate current decreases gradually toward zero in accordance with the discussion of Eq. 1 for cut-off in a diode. The plate current does not reach zero, however, because the initial velocities have a Maxwellian distribution, and a single discrete value of cut-off grid voltage therefore does not exist in an actual tube.

4. Approximate Analysis of Triode Current-Voltage Relations

The concepts of Art. 3 may be applied to obtain an approximate quantitative relation of the plate current in a triode to the grid and plate voltages. Analysis of this kind, carried through in a manner sufficiently thorough to permit reasonably accurate calculation of current-voltage relations from tube dimensions, is of importance to the tube designer; to the engineer interested primarily in the engineering application of vacuum tubes, however, it is not of sufficient importance to warrant its inclusion here. An exact analysis is very difficult because of the complications due to the presence of space charge and the irregular geometry of tube structures. An approximate treatment, however, is helpful as an aid to understanding the operation of the tube and interpreting the measured tube characteristics.

An approximate analysis based on an analogy between the diode and triode can be performed if a simple tube geometry is assumed and a sufficient number of approximations are included. Numerous analyses have been made,[9] and the discussion given here embodies basic ideas common to several of them. In Art. 1, Ch. III, it is shown that, provided the initial velocities of the electrons are neglected, the plate current in a diode is given by a relationship of the form

$$i_b = Ke_b{}^{3/2},$$ [5]

where

i_b is the plate current in the diode,

K is a constant, called the *perveance* of the tube, which depends only on the tube geometry,

e_b is the plate voltage of the diode.

If the plate voltage is applied to this same diode with the cathode unheated so that there is no current or space charge, the cathode and plate act as the electrodes of a capacitor, and a charge is induced on the cathode, having the value given by

$$Q_k = e_b c_{pk},$$ [6]

where

Q_k is the magnitude of the charge on the cathode,

c_{pk} is the capacitance* between plate and cathode of the diode with an unheated cathode.

Since Q_k is the charge on the surface of the cathode itself, c_{pk} is the

---

[9] H. J. van der Bijl, *The Thermionic Vacuum Tube and Its Applications* (New York: McGraw-Hill Book Company, Inc., 1920), 42–46, 155, 226–236; "Theory of the Thermionic Amplifier," *Phys. Rev.*, 12 (1918), 180–183; W. H. Eccles, *Continuous Wave Wireless Telegraphy* (London: The Wireless Press, Ltd., 1921), 333–342; F. B. Vodges and F. R. Elder, "Formulas for the Amplification Constant for Three-element Tubes in Which Diameter of Grid Wires Is Large Compared to the Spacing," *Phys. Rev.*, 24 (1924), 683–689; W. G. Dow, *Fundamentals of Engineering Electronics* (2nd ed.; New York: John Wiley & Sons, Inc., 1952), 81–173; and G. B. Walker, "Theory of the Equivalent Diode," *Wireless Engr.*, 24 (1947), 5–7; of particular interest are the concept of an equivalent grid plane, W. G. Dow, "Equivalent Electrostatic Circuits for Vacuum Tubes," *I.R.E. Proc.*, 28 (1940), 548–556; and the analysis involving the transit time of an electron, F. B. Llewellyn, *Electron-Inertia Effects: Cambridge Physical Tract* (Cambridge: The University Press, 1941).

* The lower-case letters are used for the capacitances discussed in this article to indicate that they are capacitances between the active portions of the electrodes in the tube with the cathode unheated. They should be distinguished from the interelectrode capacitances denoted by capital letters and used in subsequent discussions; these are the capacitances effective when the tube is operating as part of a circuit.

capacitance associated with electric displacement in the interelectrode region that would be traversed by electrons if the cathode were heated, and that does not include capacitance between lead-in wires and electrode supports. If the value of e_b given by Eq. 6 is substituted in Eq. 5,

$$i_b = \frac{K}{c_{pk}^{3/2}} Q_k^{3/2},$$ [7]

or

$$i_b = K_1 Q_k^{3/2},$$ [8]

where K_1 is a new constant dependent only on the tube geometry. Equation 8 states that the current in a diode is directly proportional to the three-halves power of the charge that would exist on the unheated cathode if, with the cathode cold, a plate voltage were applied to the diode equal to that applied when the cathode is hot. Since the charge on the unheated cathode is directly proportional to the electric field at its surface, an alternate interpretation of this equation is that the space-charge-limited plate current is directly proportional to the three-halves power of the electric field at the cathode surface in the tube with an unheated cathode.

The next step in the approximate analysis involves the assumption that, when the initial velocities of the electrons are again neglected, the current in a triode bears this same relation to the charge on its unheated cathode, provided the grid and plate voltages that induce this charge are equal to those that cause the current when the cathode is heated. That is,

$$i = K_2 Q_k^{3/2},$$ [9]

where

> i is the space-charge-limited current from the cathode of the triode,
>
> K_2 is a constant dependent only on the geometry of the triode,
>
> Q_k is the magnitude of the charge on the cathode of the triode when it is unheated.

It should be remembered that Eq. 9 is the result of an assumption regarding the similarity of the phenomena in the diode and the triode. More nearly exact analyses[10] show that under the usual operating conditions of triodes the assumption is approximately but not exactly true. An understanding of the significance of this assumption and the reasons for it can be gained from a comparison of the potential distribution diagrams for the diode and the triode given in Fig. 6. In Fig. 6a,

---

[10] W. G. Dow, "Equivalent Electrostatic Circuits for Vacuum Tubes," *I.R.E. Proc.*, *28* (1940), 548–556; F. B. Llewellyn, *Electron-Inertia Effects: Cambridge Physical Tract* (Cambridge: The University Press, 1941).

the upper curve is a plot of the potential at each point in a diode as a function of the distance from the cathode when there is no space charge in the tube, that is, when the cathode is unheated. Since in the diagram the cathode and plate are assumed to be parallel plane electrodes, this curve is a straight line. The lower curve indicates the effect of space charge when the cathode is heated to give an electron emission greater than the space-charge-limited current. The negative charges in the space depress the potential at each point sufficiently to make the slope of the curve just zero at the cathode surface if initial velocities are neglected. Along this curve the potential is directly

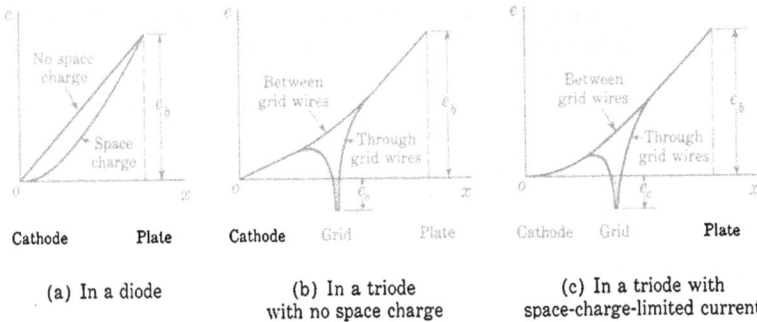

| (a) In a diode | (b) In a triode with no space charge | (c) In a triode with space-charge-limited current |

Fig. 6.   Potential distributions in diodes and triodes.

proportional to the four-thirds power of the distance from the cathode, as is shown in Art. 1, Ch. III.

In Fig. 6b, the potential distribution in a triode is shown under the conditions of no space charge. The figure is drawn for the triode of Fig. 5, in which the cathode and plate are in parallel planes. Because of the distortion of the field produced by the irregular geometry of the grid, it is not possible to represent the entire potential distribution by means of a single curve. The electric field varies along two dimensions. It can be determined by measurements on models,[11] flux-plotting methods,[12] or conformal transformations.[13] The upper curve in Fig. 6b, which represents the potential distribution along a line from the cathode to the anode midway between the grid wires, and the lower curve, which represents the potential distribution along a line through the center of a grid wire, are the limits between which the potential

[11] E. D. McArthur, "Determining Field Distribution by Electronic Methods," *Electronics, 4* (1932), 192–194.

[12] E. L. Chaffee, *Theory of Thermionic Vacuum Tubes* (New York: McGraw-Hill Book Company, Inc., 1933), 175–177.

[13] W. G. Dow, *Fundamentals of Engineering Electronics* (2nd ed.; New York: John Wiley & Sons, Inc., 1952), 81–127.

lies everywhere in the triode. In Fig. 6b, the two curves coincide and are straight lines in the regions near the plate and the cathode; they are separate and curved only in the distorted field region near the grid. This condition of uniform field everywhere except near the grid occurs in the more usual forms of triodes under normal operating conditions.

In Fig. 6c, the potential distribution resulting from a space-charge-limited current is drawn. Just as in the diode, the potential at each point is depressed by the space charge until the slope of the potential distribution curve at the cathode surface is zero. It is to be expected that the four-thirds power relation between the potential and the distance that was found for the diode should also be found in the triode in the region near the cathode where the field is uniform when the cathode is unheated. The similarity between the field in the regions near the cathode in the diode and triode, together with the fact that charges in regions far removed from the cathode have relatively less influence on the field there, substantiate the assumption that the currents in each tube are related in the same way to the charge or field at the cathode when no current flows.

The charge on the unheated cathode of the triode is the result of the electrostatic field set up in the tube by the voltages applied to the grid and plate. Since the charge induced on the cathode is a linear function[14] of the two voltages, its magnitude Q_k may be written as

$$Q_k = c_{pk}e_b + c_{gk}e_c, \qquad\qquad [10]$$

where

e_b is the plate voltage of the triode,

e_c is the *grid voltage* of the triode,* or potential of the grid measured with respect to the potential of the cathode,

c_{pk} and c_{gk} are the constant coefficients of the linear relation between charge and voltage.

The grid voltage e_c is measured in the same manner as the plate voltage. It may be expressed as the voltage rise from cathode to grid, or voltage drop from grid to cathode. Coefficients such as c_{pk} and c_{gk} are known in the general electrostatic theory of multiple-conductor systems as *direct capacitances* and are equal to the negatives of the corresponding coefficients of electrostatic induction or mutual capacitance operators.[15] The relation of the charge on the unheated cathode

---

[14] N. H. Frank, *Introduction to Electricity and Optics* (2nd ed.; New York: McGraw-Hill Book Company, Inc., 1950), 63–64; W. R. Smythe, *Static and Dynamic Electricity* (New York: McGraw-Hill Book Company, Inc., 1939), 25–26, 35–38.

* A small letter is used for the grid voltage as well as for the plate voltage, as explained in Art. 1, even though in this article each voltage is considered to be a constant.

[15] *American Standard Definitions of Electrical Terms—A.S.A. No. C42* (New York: American Institute of Electrical Engineers, 1941), 28–29.

to the electrode voltages may be found from an equivalent circuit in which capacitance elements equal to c_{pk} and c_{gk} are connected from plate to cathode and from grid to cathode, respectively. Just as the capacitance c_{pk} of the diode differs from the ordinary interelectrode capacitance of the tube measured under operating conditions, these capacitances also differ, in that the ordinary interelectrode capacitances involve contributions from electric displacement in regions not traversed by electrons, and are affected by the space charge in the tube.

If the charge Q_k in Eq. 9 is replaced by its value from Eq. 10, there results for the cathode current in the tube

$$i = K_2(c_{gk}e_c + c_{pk}e_b)^{3/2}, \tag{11}$$

or

$$i = K_3 \left(\frac{c_{gk}}{c_{pk}} e_c + e_b\right)^{3/2} \tag{12}$$

where K_3 is a new constant that depends only on the geometry of the tube. If the grid voltage e_c is negative, the grid current is a negligible fraction of the cathode current, which is therefore identical with the plate current i_b. Then

$$i_b = K_3 \left(\frac{c_{gk}}{c_{pk}} e_c + e_b\right)^{3/2}. \qquad \blacktriangleright[13]$$

Usually the capacitance c_{gk} is much larger than c_{pk}, because the grid surrounds the cathode or the plate sufficiently to shield one from the other. The shielding effect is large when the spacing of the grid wires is small compared with the distance from the grid to the plate, so that the field from the plate reaches through the grid to only a small extent. The coefficient of e_c in Eq. 13 is then larger than unity. Large values of the ratio c_{gk}/c_{pk} cause a given change in grid voltage to have a much greater effect on the plate current than does the same change in the plate voltage. This property of the triode makes possible its use as a voltage amplifier, for, as discussed in Ch. VIII, a small change in grid voltage may be made to cause a large change in plate voltage.

It is not necessary that the grid surround either the plate or the cathode; in fact, in some special-purpose tubes and in the original DeForest triode, the grid and the plate are placed on opposite sides of the cathode. However, the ratio c_{gk}/c_{pk} is then smaller than, or at most equal to, unity, and the relative effectiveness of the grid voltage in controlling the plate current is small.

The assumptions made in obtaining Eq. 13 should be reviewed. When the three-halves-power law of the diode is used, the initial velocities of emission, the contact potentials, the nonuniform cathode temperature or potential, and the possibility of emission limitation of current are neglected. In Eq. 9, the assumption that the current is directly proportional to the three-halves power of the charge on the cathode in the absence of space charge is made. To justify this assumption, even as an approximation, parallel-plane symmetry of the tube is assumed. To obtain Eq. 13 from Eq. 12, a negative grid voltage is assumed; and, finally, in the entire discussion the plate and grid voltages are assumed to be constant, so that displacement current may be neglected and the current caused by moving charges may be considered to be the entire current.

The effect of some of the neglected phenomena may be explained in a qualitative manner. For example, the initial velocities in a triode, just as in a diode, cause the space charge so to adjust itself that the potential gradient at the cathode surface is slightly negative instead of zero and a potential minimum is established at a short distance from the cathode. The net effect of these changes is an increase in the plate current. Relatively small if the plate and grid voltages are so adjusted that the plate current is a large fraction of the possible emission of the cathode, the increase is of importance if the grid voltage is near the cut-off value. In particular, the grid voltage needed for current cut-off becomes indefinite. As the grid potential is made more negative than the value that would otherwise produce cut-off, the plate current is reduced exponentially to an exceedingly small value, but because of the initial velocities it is not completely interrupted. In addition, since the initial velocities depend on the cathode temperature, the plate current is dependent to a small extent on the cathode temperature even though it is limited by space charge.

Another effect that has an influence on the plate current is the contact potential difference between the plate and cathode and between the grid and cathode. These differences of potential cause the effective potential differences between the pairs of electrodes to differ from the externally applied voltages. Since small changes of grid voltage may be much more important than equal changes of plate voltage, the contact potential difference between grid and cathode is of relatively greater importance. The assumption on which Eq. 9 is based leads to a constant value of c_{gk}/c_{pk} in Eq. 13. A more nearly exact analysis results in a similar equation in which the coefficient that replaces this ratio is very nearly but not exactly constant.[16] For simple tube

---

[16] Measurements by R. W. King, "Calculation of the Constants of the Three-Electrode Thermionic Vacuum Tube," *Phys. Rev.*, *15* (1920), 258, show that the relative effectiveness of the grid and plate voltages in controlling the space current is essentially constant throughout a range of cathode current of about 1,000 to one, indicating that the amount of the space charge has little effect on this quantity.

geometries, it is possible to calculate the value of the constant K_3, but, in general, this quantity is difficult to determine exactly.

5. CHARACTERISTIC CURVES OF TRIODES

The preceding discussion is intended primarily to impart an understanding of the underlying phenomena in a tube containing a grid, rather than to serve as an exact analysis of the tube behavior. To derive the exact functional relationship between i_b, e_b, and e_c from purely theoretical considerations is exceedingly difficult. The three-halves-power relation, Eq. 13, is never exactly applicable because of such complicating factors as the initial velocities of the electrons at the cathode, contact potentials among the electrodes, gas left in the tube after the evacuation process, and the Schottky effect. Adding empirically determined constants to the voltage symbols in Eq. 13 to account for these factors gives a closer approximation. With those constants included, changing the exponent to a value somewhat different from three halves gives an even closer approximation. But the effort involved in determining an exact functional relationship is hardly justifiable in view of the ease with which experimental tubes can be constructed and with which their performance can be determined by simple laboratory measurements. From a practical standpoint, representation of the tube characteristics by experimentally determined curves rather than by mathematical expressions is often preferable because all the complicating factors, which often vary among tubes of the same type, are thus taken into account. A graphical analysis using such characteristics to find the conditions of operation is then convenient, and is commonly employed.

In general, the plate current depends upon three independent variables, the plate voltage, the grid voltage, and the cathode heating voltage. When the heating voltage is more than sufficient to produce adequate cathode emission, however, the current is limited primarily by space charge, and is almost independent of the heating voltage. The plate current i_b is then essentially a function of the two independent variables e_b and e_c. This function cannot be represented by a single curve in a plane but can be represented by a three-dimensional surface. Such representation, although instructive,[17] is usually inconvenient; consequently the characteristics customarily are shown as three families of plane curves termed the *static characteristics*. The curve formed by intersection of the surface with a plane parallel to one through the axes for two of the three variables, called a principal

· [17] E. L. Chaffee, *Theory of Thermionic Vacuum Tubes* (New York: McGraw-Hill Book Company, Inc., 1933), 153–164.

co-ordinate plane, is projected onto the principal co-ordinate plane
to give each of the curves in each family of static characteristics. In
other words, the one of the three quantities i_b, e_b, or e_c whose axis is
perpendicular to the principal co-ordinate plane and whose value is
the same everywhere in the intersecting plane is a constant for a particular curve in the family.

A family of curves for which the principal co-ordinate plane is determined by the i_b and e_b axes is shown in Fig. 7. This family is commonly

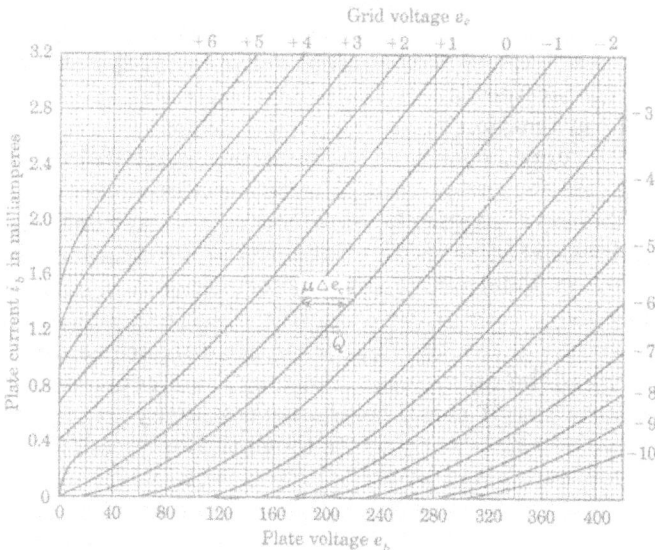

Fig. 7.   Plate characteristics of an experimental triode.

referred to as the *plate-current–plate-voltage characteristics*, or simply
the *plate characteristics*. The grid voltage is constant for each curve of
the family. The data of Fig. 7 are for a triode whose cathode and plate
have the same dimensions as have those in the diode of Fig. 3. The
triode differs from the diode only because of the addition of a helical
grid located between the anode and cathode. The grid consists of
wire 0.005 inch in diameter wound in a helix of 11 turns per centimeter for a length of 2.3 centimeters on a cylindrical mandrel 0.7
centimeter in diameter. The turns of the helix are fastened to two
wires, or "side rods," 0.020 inch in diameter parallel to the axis of the
mandrel. The degree to which the grid is effective in controlling the
space current is roughly evident from the fact that with the grid at
the potential of the cathode—that is, with e_c equal to zero—the plate
current of the triode is only about three per cent of the plate current of

the diode at the same plate voltage. Plate characteristics for representative commercial triodes are given in Appendix C.

Because of the approximations made in the analysis, Eq. 13 represents the experimental data only approximately. For e_c equal to zero, the plate current varies only approximately as $e_b^{3/2}$. When the grid voltage is made negative, an increase of plate voltage is necessary to bring the plate current back to the same value. To the extent that Eq. 13 holds, the magnitude of this required increase in plate voltage is $(c_{gk}/c_{pk})e_c$, which is usually written as μe_c, where μ is termed the *amplification factor*. The fundamental definition of the amplification factor appears subsequently in Art. 6. In so far as the amplification factor is constant, the curve for e_c equal to, say, -1 volt is the same as the curve for e_c equal to zero, moved in the *positive* direction along the e_b axis by an amount 1μ, as is indicated on the diagram by $\mu \, \Delta e_c$. Similarly a positive grid voltage tends to shift the plate characteristic in the other direction. However, a positive grid voltage also causes the shape of the curve to change, since Eq. 13 does not hold even approximately if the grid voltage is positive and the grid takes an appreciable fraction of the space current.

The data of Fig. 7, in effect, completely describe the tube, since all three variables are shown on it. However, the data are not continuously complete for the variable e_c, which is used as a parameter. In some applications, the projections on the other two principal coordinate planes are more convenient. A family of *plate-voltage–grid-voltage characteristics*, sometimes termed the *constant-current characteristics*, is shown in Fig. 8. This family may be obtained from Fig. 7, through plotting values of e_c and e_b from the curves that correspond to the same ordinate—that is, for constant plate current. From Eq. 13 and the foregoing considerations, along a curve for which i_b is constant,

$$e_b + \mu e_c = \left(\frac{i_b}{K_3}\right)^{2/3} = \text{constant}, \qquad [14]$$

or

$$e_b = \text{constant} - \mu e_c \, ; \qquad [15]$$

thus $-\mu$ *is the slope of the curves in* Fig. 8 and it is constant where the curves are straight and parallel. This condition is evidently true over a considerable range of values of e_c and e_b, even in the region where e_c is positive. Constant-current characteristics for a commercial triode are given in Fig. 9, Ch. X.

A family of *plate-current–grid-voltage characteristics*, often termed the *transfer characteristics*, is shown in Fig. 9. These too are obtained from Fig. 7 if values of i_b and e_c corresponding to several constant values of plate voltage are plotted. Along a horizontal line in Fig. 9, Eq. 14 is applicable; thus, for an increment Δe_b in the plate voltage, the grid voltage must be changed by an amount $\Delta e_b/\mu$ to maintain the same plate current, as is indicated on the diagram. In so far as μ

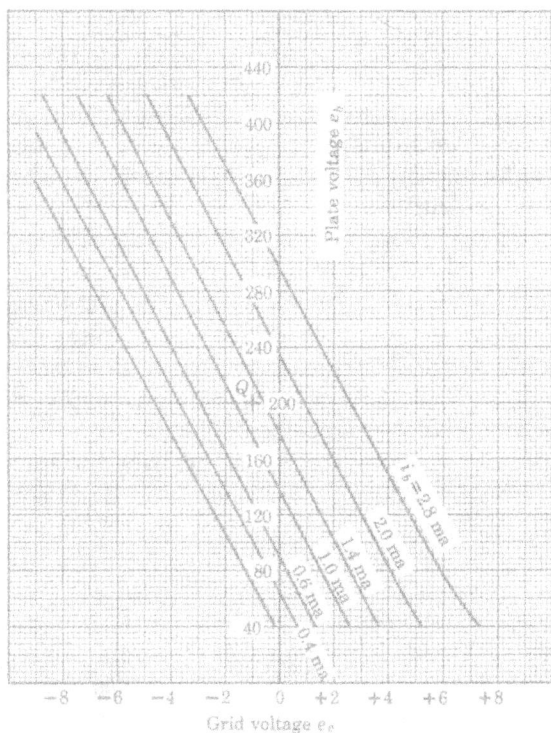

Fig. 8.   Constant-current characteristics for the tube of Fig. 7.

is a constant for this tube, all the curves in Fig. 9 are of the same shape in the negative grid-voltage region but are shifted horizontally from each other by the amount $\Delta e_b/\mu$, where Δe_b is the difference in the plate voltages for any two curves in question. Transfer characteristics for a pentode are shown in Fig. 20.

Curves of grid current i_c as a function of grid voltage e_c for several values of plate voltage are also given in Fig. 9. When the grid is positive with respect to the cathode it collects electrons, and the grid current is considerable. Increase of the plate voltage causes a decrease of the grid current because the plate then draws a larger fraction of the electrons from the cathode through the grid and on to itself. When, on the other hand, the grid is sufficiently negative to counteract the contact potential difference and the initial velocities of the electrons, the grid current due to electron collection is negligible. Ordinarily the grid current becomes zero at a negative grid voltage of one to two volts. At more highly negative grid voltages, the grid current is negative, as is explained in Art. 5, Ch. V, but is too small to show

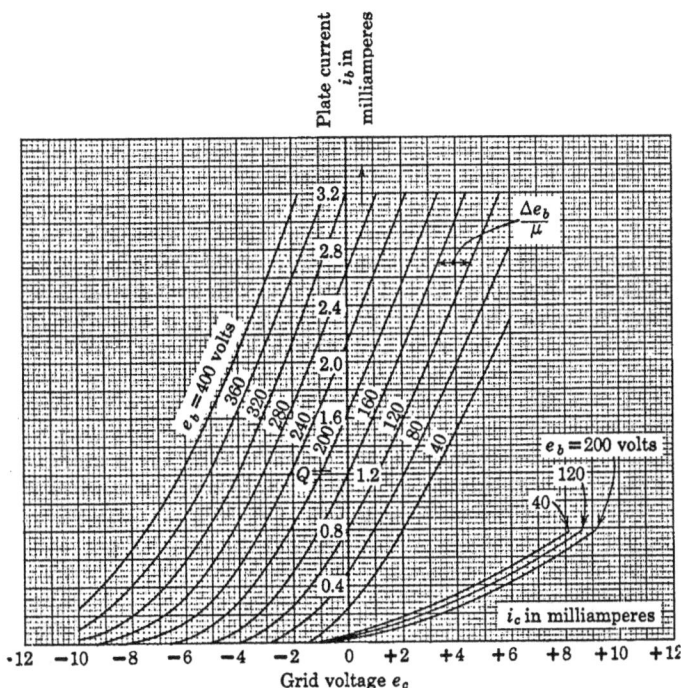

Fig. 9. Transfer characteristics for the tube of Fig. 7.

on the scale of Fig. 9, and is negligible in many electron-tube applications. However, if a voltage source having a very high internal resistance or capable of delivering only a very small current is to supply the grid voltage to be amplified, the grid current may be an important factor even though the control-grid voltage is negative. Vacuum tubes are sometimes used as sensitive electrometers or galvanometers, and in these applications the grid current is significant, for, although the resistance of the tube between grid and cathode is large, it is not then large in comparison with the resistance of the circuit that supplies the grid voltage. Consequently, the actual grid voltage at the tube terminals is appreciably smaller than the internal electromotive force of the voltage source by the amount of the voltage drop in the internal resistance of the source. Special electrometer tubes in which the grid current is of the order of magnitude of 10^{-15}

ampere are available for such applications.[18] The grid current for the usual negative-grid operation of standard receiving tubes is of the order of magnitude of 10^{-8} ampere, with the normal rated voltages applied to the tube electrodes, but may be reduced to the order of magnitude of 10^{-12} ampere when smaller electrode voltages are used.[19]

Tubes with more than one grid are often connected as triodes. The extra grids are ordinarily connected to the plate or to the control grid to form such a triode connection.

6. VACUUM-TUBE COEFFICIENTS

Although a graphical method using the characteristic curves is suitable for analysis of many vacuum-tube applications, it often is cumbersome and unwieldy. Hence an alternative but approximate method that permits application of linear circuit analysis is commonly used. This method involves as a first step the approximation of portions of the characteristic curves in terms of three coefficients, which, as is shown later, serve as figures of merit for many vacuum tubes. These coefficients describe the slopes of the three families of curves in Figs. 7, 8, and 9.

To appreciate the usefulness of the three coefficients, one should have a qualitative knowledge of the function usually expected of a triode, namely, to regulate the magnitude of the current in the circuit element connected in the plate circuit. The details of the circuit operation are given in Ch. VIII; only a brief qualitative discussion is given here. A common method of connection of a vacuum triode is shown in Fig. 10. The plate current is supplied by a battery or other direct-voltage source E_{bb}, and the load element connected in the plate circuit is frequently either a network for coupling this tube to a second tube or is a current-sensitive device such as an ammeter, relay, loud speaker, or telephone receiver. In the analysis of the circuit, the volt-ampere characteristics of the load element as a function of the current and frequency are assumed to be determinable and known. As the current i_b in the load element varies as a result of changes in the grid

[18] G. F. Metcalf and B. J. Thompson, "A Low-Grid-Current Vacuum Tube," *Phys. Rev. 36* (1930), 1489–1494; J. M. Lafferty and K. H. Kingdon, "Improvements in the Stability of the FP-54 Electrometer Tube," *J. App. Phys., 17* (1946), 894–900; J. A. Victoreen, "Electrometer Tubes for the Measurement of Small Currents," *I.R.E. Proc., 37* (1949), 432–441.

[19] W. B. Nottingham, "Measurement of Small D-C Potentials and Currents in High-Resistance Circuits by Using Vacuum Tubes," *J.F.I., 209* (1930), 287–348; R. D. Bennett, "An Amplifier for Measuring Small Currents," *R.S.I., 1* (1930), 466–470; C. E. Nielsen, "Measurement of Small Currents: Characteristics of Types 38, 954, 959 as Reduced Grid Current Tubes," *R.S.I., 18* (1947), 18–31; J. R. Prescott, "The Use of Multi-Grid Tubes as Electrometers," *R.S.I., 20* (1949), 553–557.

voltage e_c, the voltage drop across the load also varies, causing the plate voltage e_b of the tube to vary. Hence, in a circuit application, the changes in plate current caused by changes in grid voltage react upon the plate voltage in a manner that tends to neutralize the effectiveness of the grid. No one of the three variables i_b, e_b, and e_c is constant, and the tube operation does not follow any of the curves described in Art 5. However, these mutually dependent effects can be conveniently expressed in terms of the coefficients that give the relative variations among the three pairs of the variables i_b, e_b, and

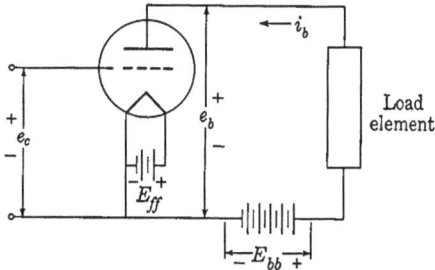

Fig. 10. Basic circuit diagram for use of a vacuum triode.

e_c, while the third quantity for each pair is held constant.

The first of the three coefficients, the *amplification factor*, denoted by the Greek letter μ, is defined as

$$\text{Amplification factor } \mu \equiv -\frac{\partial e_b}{\partial e_c} = -\frac{de_b}{de_c}\bigg|_{i_b \text{ constant}} \qquad \blacktriangleright[16]$$

This quantity may be interpreted as the negative of the slope of the curves in Fig. 8, and is approximately the ratio of a small change in plate voltage to the change in grid voltage required to bring the plate current back to the value it had before the plate voltage was changed. Since, when the reference directions of the voltages are assigned as in Fig. 10, the signs of the voltage changes must be opposite to maintain i_b constant, the slope or the ratio above is inherently a negative number, and the negative signs in Eq. 16 are needed if μ is to be a positive number. Much confusion can result from a lack of understanding of this matter of definition.

To determine the relation of μ to quantities previously defined, Eq. 13 may be written in the form

$$e_b = -\frac{c_{gk}}{c_{pk}}e_c + \left(\frac{i_b}{K_3}\right)^{2/3}. \qquad [17]$$

Thus

$$\frac{\partial e_b}{\partial e_c} = -\frac{c_{gk}}{c_{pk}}, \qquad [18]$$

and

$$\mu = \frac{c_{gk}}{c_{pk}}. \qquad [19]$$

Therefore, in so far as Eq. 13 is a sufficiently accurate approximation to the actual tube characteristics, the capacitance ratio of Art. 4 is the same as the coefficient μ defined by Eq. 16, and the use of μ for this ratio in Art. 5 is in accordance with the fundamental definition stated by Eq. 16. In regions where the measured curves differ from those given by Eq. 13, the amplification factor is still given by Eq. 16 but bears no special relation to the tube capacitances.

The second of the three coefficients is called the *plate resistance*, and is defined as

$$\text{Plate resistance } r_p \equiv \frac{\partial e_b}{\partial i_b} = \frac{de_b}{di_b}\bigg|_{e_c \text{ constant}} . \qquad \blacktriangleright[20]$$

It is the reciprocal of the slope of the plate characteristic curves in Fig. 7, and is approximately the ratio of a small change in plate voltage to the change in plate current accompanying it when the grid voltage is held constant. To distinguish it from the ratio e_b/i_b, the coefficient r_p is often called the *dynamic*, or *incremental*, or *variational*, *plate resistance*. The ratio e_b/i_b is then called the *static*, or the *d-c*, *plate resistance*.

The third of the three coefficients is the *transconductance*, which is defined as

$$\text{Transconductance } g_m \equiv \frac{\partial i_b}{\partial e_c} = \frac{di_b}{de_c}\bigg|_{e_b \text{ constant}} . \qquad \blacktriangleright[21]$$

It is the slope of the transfer characteristic curves in Fig. 9, and closely approximates the ratio of a small change in plate current to the change in grid voltage accompanying it when the plate voltage is held constant. Its name arises from the facts that it has the dimensions of conductance and that it relates a quantity in the plate circuit to one in the grid circuit. It is also known as the *mutual conductance*. In multigrid tubes there are interelectrode transconductances among all the grids and the plate.[20] Each of these coefficients is defined in a manner analogous to that of Eq. 21, but a descriptive name, such as *control-grid–to–plate transconductance*, is needed to distinguish one from the others. Since the control-grid–to–plate transconductance is the most important one, however, it is commonly called merely the transconductance even for multigrid tubes.

The three coefficients in the form of partial derivatives in Eqs. 16, 20, and 21 may be recognized as the slopes measured in the three directions parallel to the principal co-ordinate planes of the surface that represents the plate current as a function of the grid and plate

---

[20] "Standards on Electron Tubes: Definition of Terms, 1950," *I.R.E. Proc.*, *38* (1950), 428, 435, 437.

voltages, $i_b = f(e_b, e_c)$. A relation among the three coefficients may be found from the expression for the total differential of the plate current:

$$di_b = \frac{\partial i_b}{\partial e_b} de_b + \frac{\partial i_b}{\partial e_c} de_c. \qquad [22]$$

It expresses approximately the small change in plate current caused by simultaneous changes in the grid and the plate voltages. If the voltages are changed in such a way that their effects on i_b counteract each other, so that i_b remains constant, di_b is then zero, and, from Eqs. 20, 21, and 22,

$$\frac{1}{r_p} de_b + g_m de_c = 0, \qquad [23]$$

and

$$g_m r_p = - \frac{de_b}{de_c}\bigg|_{i_b \text{ constant}} . \qquad [24]$$

But the right-hand side of Eq. 24 corresponds to Eq. 16; hence,

$$\mu = g_m r_p . \qquad \blacktriangleright[25]$$

The three coefficients, amplification factor μ, plate resistance r_p, and mutual conductance g_m, define the behavior of a tube at one pair of values of e_c and e_b. Since the slopes of the curves in Figs. 7 and 9 vary considerably, r_p and g_m can be considered to be constants over only a relatively small range of variation; but the slope of the curves in Fig. 8 is almost uniform over a wide range; thus μ is a constant over a relatively large range of voltage and current variations. The curves in Fig. 11 show the dependence of the three coefficients on the plate current for a fixed plate voltage in a typical triode. Within the region where the curves of the families of Figs. 7, 8, and 9 are *straight, parallel,* and *equidistant for equal increments of* the parameter of the family, the coefficients are constants. To the extent that they are, the tube may be considered a linear circuit element for small variations, and the quantities μ, r_p, and g_m may be used as constants in a mathematical analysis of the tube's behavior. Furthermore it follows that under these conditions the three-dimensional surface representing the tube behavior is a plane. Although this condition never holds exactly in practice, it serves as a workable approximation over a small region of operation. These matters are better appreciated from a study of the tube as a circuit element, and their consideration in greater detail is given in Chs. VIII through XII.

When the grid voltage has a value such that appreciable grid current exists, expressions similar in form to those above can be derived for the grid circuit. Normally the grid current is caused primarily by electrons striking the grid when e_c is positive, and by electrons emitted and positive ions collected by the grid when e_c is negative, although there are a number of other sources of grid current. These are discussed in more detail in Art. 6a, Ch. V. When the operation is such that the grid current is appreciable, a convenient method for predicting the circuit operation is desirable. An experimental study of the grid-current properties of the tube shows that the grid current is

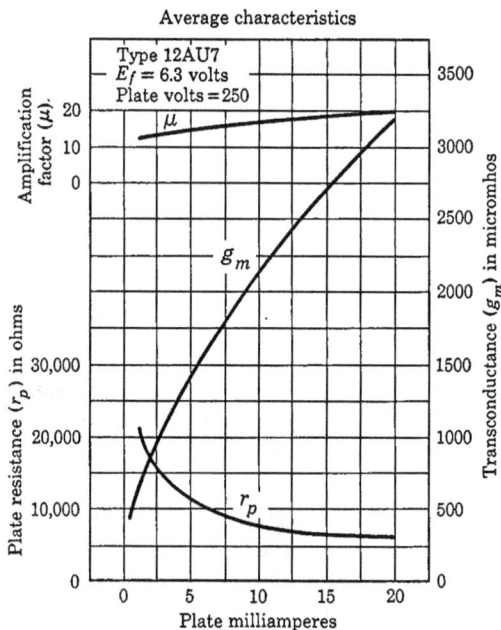

Average characteristics

Fig. 11. Curves showing the dependence of μ, g_m, and r_p on the plate current. (*Data by courtesy of Radio Corporation of America.*)

also a function of the grid and plate voltages. As shown by Chaffee,[21] expressions for the reflex amplification factor, reflex transconductance, and grid resistance can be derived for the grid circuit in a manner similar to that followed above for the plate circuit. However, these reflex coefficients need be used only when the grid current is appreciable.

As an example of the relations between the curves and the coefficients, the coefficients for the tube of Figs. 7, 8, and 9 may be found. At the point Q in Fig. 8, e_b is 200 volts, e_c is —1 volt, and i_b is 1.25 milliamperes. The slopes of the constant-plate-current lines on either

[21] E. L. Chaffee, "Equivalent Circuits of an Electron Tube and the Equivalent Input and Output Admittances," *I.R.E. Proc.*, *17* (1929), 1633-1648; *Theory of Thermionic Vacuum Tubes* (New York: McGraw-Hill Book Company, Inc., 1933), 164-166.

side of this point are about -38; thus μ has the value 38 for this tube.

The plate resistance r_p is the reciprocal of the slope of the curves in Fig. 7. At the point Q on that diagram for the same values of e_b, e_c, and i_b as before, the ratio $\Delta e_b / \Delta i_b$ of the curve through the point is about 94,000 ohms. Similarly, the mutual conductance g_m found from the

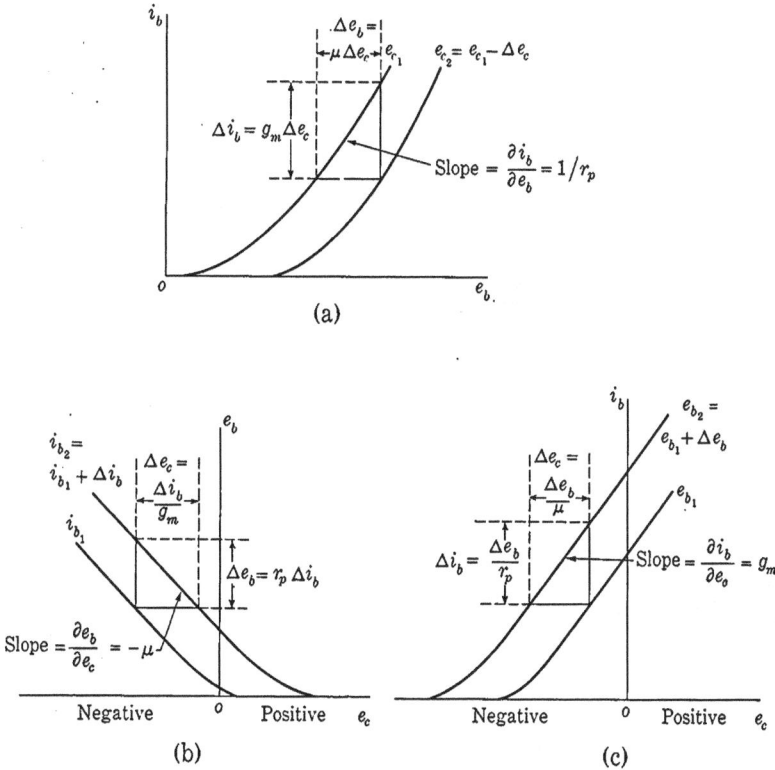

(a)

(b)

(c)

Fig. 12. Relationships among the tube coefficients and the characteristic curves.

slope of the curve through the corresponding point Q on Fig. 9 is approximately 0.000400 mho. The product

$$r_p g_m = 94,000 \times 0.000400 = 37.6 \qquad [26]$$

is in good agreement with the value of μ found above in accordance with Eq. 16, which serves as a check on the determination.

The three diagrams of Fig. 12, which are based on the assumption that the curves are straight and parallel throughout small regions, show the relationships among the various quantities. The *three* coefficients may be determined approximately from *any one* of the three plots by the methods suggested on the diagrams. Where the data of the curves are not available, electrical methods[22] for measuring the coefficients directly are in common use.

7. LINEAR ANALYTICAL APPROXIMATIONS OF TRIODE CHARACTERISTICS

Operation of a vacuum tube over small ranges of plate current and electrode voltages near any chosen operating point may be approximated satisfactorily for many practical purposes by assumption that the three coefficients μ, g_m, and r_p are constants within that range. For operation over a larger range such that the coefficients themselves are variables, a Taylor's-series expansion for the current as a function of two independent variables may be used. It has the form[23]

$$i_b + \Delta i_b = i_b + \underbrace{\frac{\partial i_b}{\partial e_c} \Delta e_c + \frac{\partial i_b}{\partial e_b} \Delta e_b}_{\text{linear terms}}$$

$$+ \frac{1}{2} \underbrace{\left[\frac{\partial^2 i_b}{\partial e_c{}^2} (\Delta e_c)^2 + 2 \frac{\partial^2 i_b}{\partial e_c \partial e_b} \Delta e_c \Delta e_b + \frac{\partial^2 i_b}{\partial e_b{}^2} (\Delta e_b)^2 \right]}_{\text{quadratic terms}} + \ldots \quad [27]$$

The infinite series expresses the change Δi_b from the current i_b at the point of expansion that accompanies changes Δe_c and Δe_b from the values e_c and e_b, respectively, at the point of expansion. The partial derivatives are to be evaluated at the point from which the changes take place. The series contains a pair of linear terms, a set of quadratic terms, and so on as indicated. A different form of Taylor's series for the current is discussed in Art. 9, Ch. XII. It is based on the assumption that μ is a constant, however, and is therefore less generally applicable than Eq. 27.

For many applications the higher-order terms in the Taylor's series may be neglected and only the linear terms retained. The linear approximation for increments in the current and voltages is thus

$$\Delta i_b = \frac{\partial i_b}{\partial e_c} \Delta e_c + \frac{\partial i_b}{\partial e_b} \Delta e_b, \quad [28]$$

[22] E. L. Chaffee, *Theory of Thermionic Vacuum Tubes* (New York: McGraw-Hill Book Company, Inc., 1933), 228–241; F. E. Terman and J. M. Pettit, *Electronic Measurements* (2nd ed.; New York: McGraw-Hill Book Company, Inc., 1952), 162–178; "Standards on Electron Tubes: Methods of Testing, 1950," *I.R.E. Proc.*, *38* (1950), 939–943.

[23] See P. Franklin, *Methods of Advanced Calculus* (New York: McGraw-Hill Book Company, Inc., 1944), 63–64; W. A. Granville, P. F. Smith, and W. R. Longley, *Elements of the Differential and Integral Calculus* (Boston: Ginn and Company, 1941), 489; or other similar textbooks.

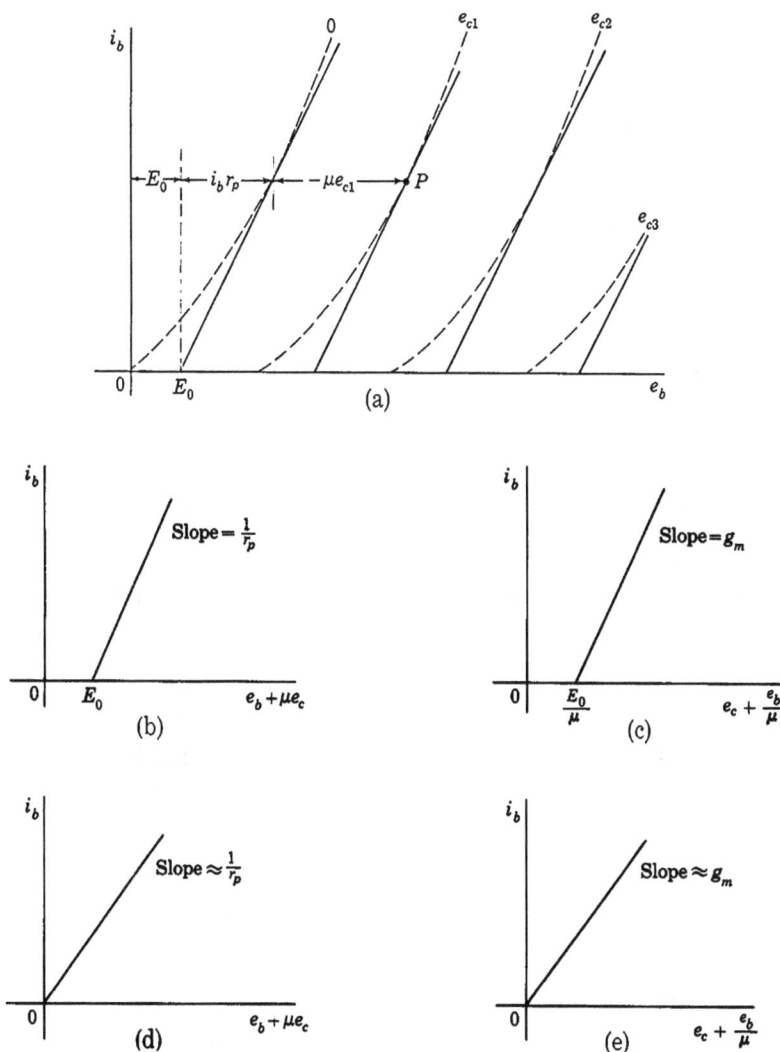

Fig. 13.   Linear approximations of triode characteristics.

or, in terms of the definitions in Eqs. 20 and 21,

$$\text{Linear approximation:} \quad \Delta i_b = g_m \Delta e_c + \frac{1}{r_p} \Delta e_b . \qquad \blacktriangleright [29]$$

This relation is of fundamental importance and serves as a basis for much of the analysis of circuit applications in the chapters that follow.

An analytical approximation for triode characteristics in terms of the total current and voltages, rather than their increments, may be obtained by the method illustrated in Fig. 13. For operation in the vicinity of point P in Fig. 13a, a tangent to the plate characteristic is drawn at P and tangents to the other curves in the family are drawn parallel to it. The slope of these curves is then $1/r_p$ and the horizontal spacing between them $\mu \Delta e_c$. The tangent to the curve for zero grid voltage intersects the axis for e_b at E_0. The plate-voltage co-ordinate of point P is therefore the sum of the three components E_0, $i_b r_p$, and $-\mu e_{c1}$ as is shown on the diagram. Thus the family of straight lines is described by

$$e_b = E_0 + i_b r_p - \mu e_c, \qquad \text{for } e_c \leq 0, \text{ and } i_b \geq 0, \qquad [30]$$

or by the equivalent expressions

$$i_b = \frac{1}{r_p} (e_b + \mu e_c - E_0), \qquad \text{for } e_c \leq 0, \text{ and } e_b \geq E_0 - \mu e_c, \quad [31]$$

and

$$i_b = g_m \left(e_c + \frac{e_b}{\mu} - \frac{E_0}{\mu} \right), \quad \text{for } e_c \leq 0, \text{ and } e_b \geq E_0 - \mu e_c. \qquad [32]$$

If the combination of e_c and e_b included in either Eq. 31 or 32 is interpreted as a single independent variable, the whole family of plate or transfer characteristics may be approximated by a single straight line as in Figs. 13b and 13c.

For some very approximate analyses it is sufficient to represent the plate characteristic for zero grid voltage by a straight line through the origin and the other curves of the family as straight lines parallel to it. Then E_0 is zero, and the family of curves is described by

$$i_b \approx \frac{1}{r_p} (e_b + \mu e_c) = g_m \left(e_c + \frac{e_b}{\mu} \right), \qquad \text{for } e_c \leq 0, \text{ and } e_b \geq -\mu e_c. \quad [33]$$

Note, however, that no true tangent to the curve for zero grid voltage in Fig. 13a can intersect the origin because the curve is concave upward and passes through the origin. Consequently, a family of curves represented by Eq. 33 is only a gross approximation of the tube behavior for any region of operation. On the other hand, a tangent at point P as in Fig. 13a is a curve that is correct at only one point. A family of straight lines having the same slope but a somewhat smaller value of E_0 might well be considered a better approximation for operation over a considerable range of current and voltages.

8. Tetrodes

The addition of a second grid was a natural development after the recognition of the effectiveness of one grid in a vacuum tube. Schottky[24]

[24] W. Schottky, "Über Hochvakuumverstärker, III Teil: Mehrgitterröhren," *Arch. f. Elek.*, *8* (1919), 299–329.

was one of the first to examine the characteristic of the double-grid tube, called a *tetrode*, and to point out certain aspects of its usefulness. Though, for reasons given subsequently, the tetrode has been practically superseded by the pentode, discussion of it is a useful preliminary to consideration of the pentode.

With the introduction of the second or auxiliary grid, the plate current becomes a function of the voltages of the three electrodes indicated in Fig. 14. When more than one grid is included, the voltages and currents of the grids are distinguished by integer numerical subscripts, beginning with 1 for the grid nearest the cathode and increasing in order for grids away from the cathode.[25] Generally the voltage of only one of the grids, called the *control grid*, is allowed to vary.

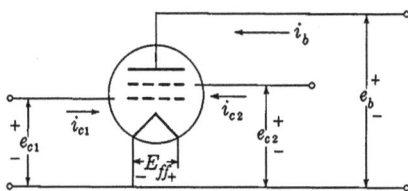

Fig. 14. Electrode voltages and currents in a tetrode.

One of the earliest uses of the auxiliary grid, but not a very common one, was placing it inside the control grid next to the cathode and making its voltage positive with respect to the cathode to help overcome the effect of the space charge. An increase of the plate current and mutual conductance of the tube is thereby accomplished. When so used, the grid nearest the cathode is called a *space-charge grid*.[26]

The second and more general use of tetrodes is that in which the grid nearest the cathode is used to control the plate current, as in the triode, and the voltage of the second grid is held constant. This disposition of the elements exhibits characteristics with considerable advantage over those of the triode for voltage amplification. The disadvantage of the triode is discussed quantitatively in Art. 8, Ch. VIII, which shows that the interelectrode capacitance between the plate and the grid places a limitation on the effectiveness of triodes as amplifiers in some circuits, because it constitutes a path for alternating current between the circuits connected to the grid and the plate. When the grid voltage is increased in the typical circuit of Fig. 10, the plate current increases and the plate voltage therefore decreases. The change in voltage that occurs across the interelectrode capacitance between the grid and the plate is the sum of the changes in the grid

[25] *Standards on Abbreviations, Graphical Symbols, Letter Symbols, and Mathematical Signs* (New York: The Institute of Radio Engineers, 1948), 3.

[26] G. F. Metcalf and B. J. Thompson, "Low-Grid-Current Vacuum Tube," *Phys. Rev.*, *36* (1930), 1489–1494; W. S. Brian, "Experimental Audio Output Tetrode," *Electronics*, *20* (August, 1947), 121–123.

and plate voltages, and the current through the capacitance is much larger than that which would result if the grid voltage alone changed. Under certain conditions, this displacement current from the grid to the plate may result in a tendency for sustained oscillations to develop, as is discussed in Art. 7, Ch. XI. The effect is roughly proportional to the product of the amplification factor and the grid-to-plate inter-electrode capacitance; and, since both these factors increase as the grid is made of finer mesh, there is a practical limit to the amplification that may be obtained with triodes.

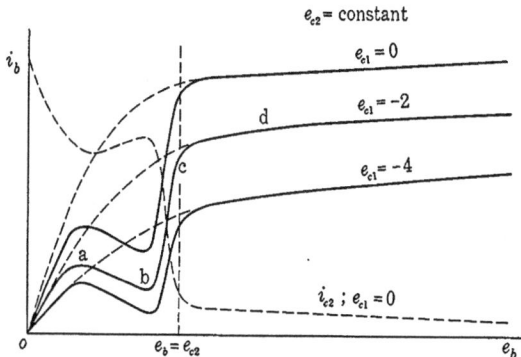

Fig. 15.   Typical plate characteristics of a tetrode.

It was not for some time after the development of tetrodes that the effectiveness of the second grid as a means for lowering the interelectrode capacitance was fully recognized and the use of tetrodes became common.[27] If the potential of the second grid e_{c2}, which is interposed between the control grid and the plate, is held fixed with respect to the cathode, the grid acts effectively as an electrostatic shield or screen between the control grid and plate and thereby reduces the capacitance between them to a very low value. When used in this manner, the second grid is called a *screen grid*.

Since the plate current is dependent on three voltages in the tetrode, a four-dimensional model is needed to represent its characteristics; or alternatively, several two-dimensional diagrams, each with one of the variables held fixed and another used as a parameter, may be used to describe graphically the tube behavior. The diagram most widely used is the one that shows the relation between plate current and plate voltage with the screen grid held at a fixed voltage and the control-grid voltage as the parameter of the family. Since the screen grid effectively shields the control grid from the plate, it also effectively

[27] A. W. Hull and N. H. Williams, "Characteristics of Shielded-Grid Pliotrons," *Phys. Rev.*, 27 (1926), 432–438; A. W. Hull, "Measurements of High-frequency Amplification with Shielded-Grid Pliotrons," *Phys. Rev.*, 27 (1926), 439–454.

shields the cathode from the plate. Most of the electrostatic lines of force from the plate therefore terminate on the screen grid, and few of them penetrate to the cathode. Thus the field at the cathode, or the charge on the cathode, under conditions of no space charge, is only slightly dependent on the plate voltage for fixed grid voltages. As a consequence, on the basis of the reasoning of Art. 4, the space current from the cathode is little influenced by the plate voltage.

The typical shapes of the plate characteristics of a tetrode with constant screen-grid voltage are shown in Fig. 15. Over that considerable portion of the curves where the plate voltage is higher than the screen-grid voltage, the plate current is almost independent of the plate voltage, for the reason stated above.

For plate voltages lower than the screen-grid voltage, several factors become important in causing the plate current to vary considerably with the plate voltage. The behavior may be described on the basis of the approximate potential-distribution diagrams of Fig. 16. In this figure, the tube is considered to comprise a cathode and an anode in parallel planes, and the grids are considered to be parallel wires located in planes. The potential distribution of Fig. 16 should be compared with the distri-

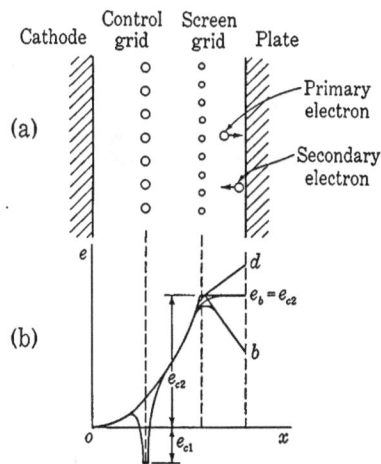

Fig. 16.　Approximate potential distributions in an idealized tetrode for several values of plate voltage.

bution of potential in a triode, indicated by Fig. 6. Under the conditions of the curve of Fig. 15 for e_{c1} equal to -2 volts, the control-grid voltage is negative, the screen-grid voltage is at a fixed positive value, and the plate voltage e_b is a variable. Potential distributions that correspond to conditions at points b and d in Fig. 15 are drawn in Fig. 16. When the plate voltage has a value higher than about 10 volts, the electrons that strike the plate release an appreciable number of secondary electrons at the plate surface. When the plate voltage has a value higher than the constant screen-grid voltage, as at d in Fig. 15, however, the electric field between the plate and the screen grid is then in such a direction as to force the electrons back toward the plate, and the secondary electrons play no part in the action of the tube. As the voltage of the plate is decreased from a high value and approaches that of the screen grid, the electric field between the screen grid and the plate approaches zero, and the secondary electrons, which have a small initial velocity upon emission, can travel against the slight retarding field and thus cross the space to the screen grid. The net plate current in the external circuit is the difference between the primary and secondary electron currents. Thus the plate current decreases appreciably before the decreasing plate voltage reaches the constant value of the screen-grid voltage.

Upon further reduction of the plate voltage to provide operation at a point below the constant screen-grid voltage at c in Fig. 15, an accelerating field is set up for the secondary electrons, and each escapes from the plate as soon as it is emitted. The speed of the primary electrons is also reduced as the plate voltage is decreased, however, and the efficiency of secondary emission, defined as the number of secondary electrons emitted per incident primary electron, also decreases. For plate voltages that provide operation near b, the secondary emission has its maximum effect. For high screen-grid voltages and plate surfaces that have a high secondary-emission efficiency, the curve may even dip into the region of negative plate currents. As the plate voltage is further reduced from its value at b, the speed of the primary electrons is reduced until they become relatively ineffective in producing secondary electrons; and for plate voltages below the value that provides operation at point a, a negligible number of secondary electrons leave the plate.

Since, because of the shielding effect of the screen grid, the plate voltage has little effect on the electric field near the cathode, the *space current* from the cathode is practically constant for the whole range of plate voltage. The negative control grid receives a negligible electron current; hence the sum of the plate and screen-grid currents is practically constant. Accordingly, the screen-grid current varies with plate voltage in the manner shown by the dotted curve i_{c2} in Fig. 15.

For ordinary operation in amplifiers when linearity is desired between the grid-voltage and plate-current variations, the considerations of Art. 6 apply and operation must be restricted to the region in Fig. 15, where the curves are straight, parallel, and equidistant for equal increments of the parameter of the family, that is, to the region where the plate voltage is higher than the screen-grid voltage. Thus, although the tetrode is an improvement over the triode, in that a relatively large amplification factor accompanied by a low grid-plate capacitance is obtained, the effects of secondary emission limit the applicability of the tube. The instantaneous plate voltage on the tetrode must not decrease to a value below that of the screen-grid voltage; otherwise the curvature of the characteristic curves causes an alteration of the waveform, and the output current variation is not a reproduction of the input voltage waveform. Since the maximum power output is a function of the maximum allowable variations in plate current and plate voltage, the secondary emission from the plate limits the power output of the tetrode for a given plate-supply voltage by restricting the allowable variation in plate voltage. Use of a third

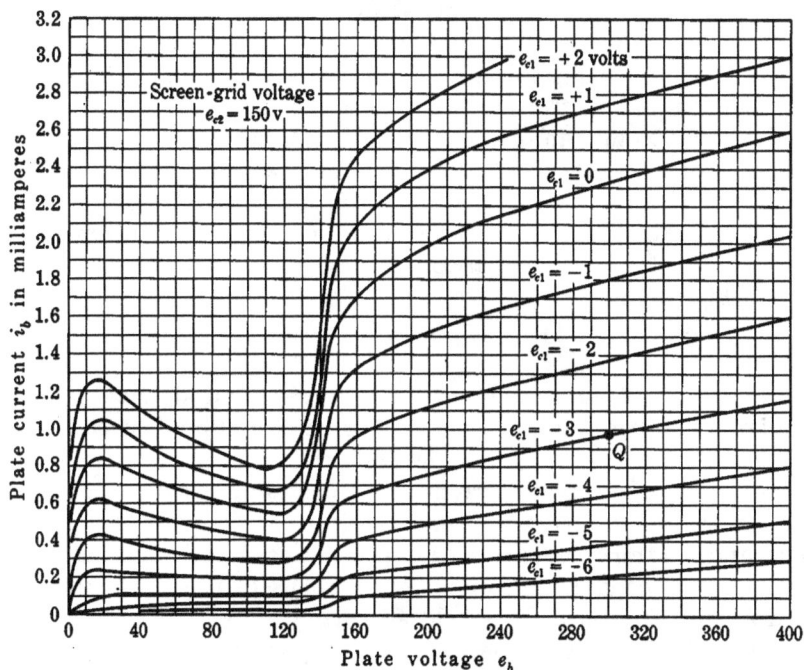

Fig. 17. Plate characteristics of an experimental screen-grid tetrode for a screen-grid voltage of 150 volts.

grid overcomes the effects of secondary emission in the manner discussed in Art. 9, and hence permits operation over a much larger range of plate voltage. Pentodes have therefore practically superseded tetrodes for most applications.

The region of lower plate voltages does have one feature that makes it desirable for certain purposes. Between points a and b on the curve of Fig. 15, the plate-characteristic curve has a negative slope. A positive increment of voltage is accompanied by a negative increment of current; in other words, the dynamic plate resistance r_p of the tube is *negative*. When operated in this region, the tube is not a passive circuit element but becomes an active element capable of supplying power to a load, and it can be used to support oscillations in a tuned circuit.

In this application, the tube is known as a *dynatron*.[28] The negative dynamic resistance is not a property of the tetrode alone; it is found in the characteristics of triodes or any multi-electrode tube when a grid is operated at a voltage higher than that of an adjacent positive electrode.

As an illustration of the effect of the screen grid, the plate characteristics of an experimental tube are shown in Fig. 17. This tube has a cathode, anode, and inner or control grid similar in dimensions to those of the tube to which Figs. 3 and 7 apply, but a screen grid is interposed between the control grid and the plate. The screen grid consists of wire 0.005 inch in diameter wound in a helix of 7.5 turns per centimeter for a length of 2.13 centimeters on a cylindrical mandrel 1.5 centimeters in diameter. The shapes of the curves in Fig. 17 are those of a typical screen-grid tetrode, and the effect of the screen grid is clearly evident from a comparison of these shapes with those of the curves in Figs. 3 and 7.

9. PENTODES

Use of three grids in a vacuum tube to form a pentode makes possible a large number of circuit connections for operation. The connection in widest use is that in which the first grid, nearest the cathode, is used for control of the plate current, the second grid is held at a fixed positive potential to screen the control grid from the plate, as in the tetrode, and the third grid, nearest the plate, is used solely to eliminate the effects of secondary emission. When so used, the third grid is termed a *suppressor grid*. It is held at a fixed voltage (usually zero) with respect to the cathode. Because of its improved performance as a result of the suppression of secondary emission, the pentode has almost superseded the tetrode for purposes of amplification. Much of the discussion of the tetrode in the previous article is pertinent to the pentode as well, however, and is of interest as an introduction to it.

Figure 18 shows schematically the electrodes of a pentode and an approximate potential-distribution diagram drawn for a tube having parallel plane plate and cathode and grids made of parallel wires in planes. The control grid is held at a negative voltage to minimize grid current, and the screen grid is held at a positive voltage to overcome the space-charge effects near the cathode. The screen grid is constructed so that it approximates an electrostatic shield between the control grid and the plate, but its mesh is so coarse as to permit the electrons to flow through it with little impediment. The suppressor grid ordinarily has an even coarser mesh.

[28] A. W. Hull, "The Dynatron, a Vacuum Tube Possessing Negative Electric Resistance," *I.R.E. Proc.*, *6* (1918), 5–35.

With the suppressor grid at the cathode potential, as shown in Fig. 18, an attracting electric field is always maintained near the anode to prevent secondary electrons from leaving the plate. As a result, secondary electrons are forced back to the plate even when the plate potential is below the screen-grid potential, and the pronounced dips and inflections in the plate character-istics of the tetrode shown in Fig. 15 do not appear in those of the pentode. The plate characteristics of a pentode are similar to the dotted curves in Fig. 15 instead of the solid ones.

Typical plate characteristics of the pentode are shown in Fig. 19. Because of the shielding effect of the screen grid, the potential of the plate has little effect on the electric field at the surface of the cathode. Hence, in accordance with the considerations of Art. 8, the potential of the plate has little influence on the plate current over the upper part of the range shown. Despite the elimination of second-ary-emission effects, however, there is still considerable rounding of the knees of the curves. Since for a linear relation between changes in the output current and the input voltage, the curves must be straight, parallel, and uniformly spaced, as is explained in Art. 6, this rounding results in a limitation of the power output of pentodes for a prescribed amount of harmonic distortion in the tube (see Arts. 13 and 16, Ch. VIII). Lack of sharpness

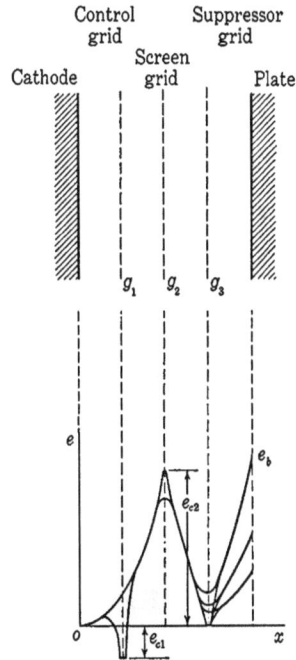

Fig. 18. Approximate poten-tial distributions in an idealized suppressor-grid pen-tode for several values of plate voltage.

of the knees of the curves is primarily a result of dispersion[29] of electrons as they pass between the suppressor-grid wires.

The characteristic curve of the screen-grid current as a function of plate voltage for one value of control-grid voltage is also shown in Fig. 19. The screen grid prevents changes of plate voltage from affecting the electric field near the cathode, and hence from affecting the current at the cathode. The sum of the plate and screen-grid

[29] O. H. Schade, "Beam Power Tubes," *I.R.E. Proc., 26* (1938), 137–181; W. G. Dow, *Fundamentals of Engineering Electronics* (2nd ed.; New York: John Wiley & Sons, Inc., 1952), 52–190; K. R. Spangenberg, *Vacuum Tubes* (New York: McGraw-Hill Book Company, Inc., 1948), 289–297.

currents, which equals the cathode current for negative values of the control-grid voltage, is therefore essentially independent of the plate voltage, and depends only on the control-grid voltage.

When only one of the family of curves for the screen-grid current is available, as in Fig. 19, the essentially constant value of the current that corresponds to the part of any other curve for which the plate voltage is well above the value at the knee of the plate characteristics

6AU6
Average plate characteristics
Pentode connection

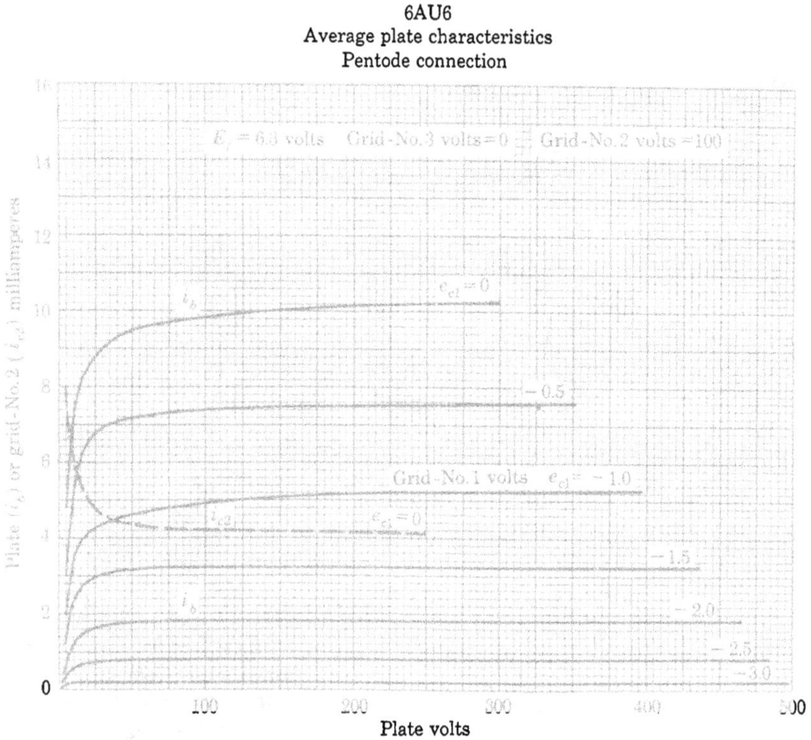

Fig. 19. Plate characteristics of a typical pentode tube with the suppressor grid connected to the cathode. (*Data by courtesy of Radio Corporation of America.*)

may be closely estimated on the assumption that the ratio of the screen-grid current to the plate current is independent of the control-grid voltage. This ratio is the fraction of the electrons passing the control grid that are intercepted by the screen grid divided by the fraction that are not so intercepted, and depends chiefly on the geometry of the tube structure. It is essentially the projected area of the screen-grid wires divided by the area between those wires. The numerical value of the ratio may be taken from any pair of screen-grid-current and plate-current curves for the same control-grid voltage, or it may be taken from any one set of typical operating data given by the manufacturer.

Transfer characteristic curves for the pentode of Fig. 19 are shown in Fig. 20. Here the plate and screen-grid voltages are held constant.

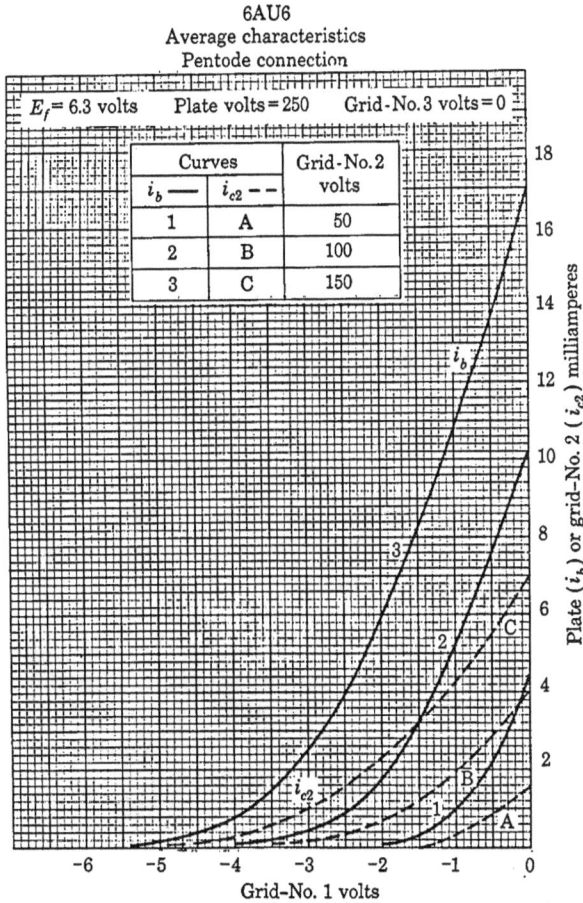

Fig. 20. Transfer characteristics for the pentode of Fig. 19. Curves for three different screen-grid voltages are shown. (*Data by courtesy of Radio Corporation of America.*)

Curves are given for three different values of screen-grid voltage, the intermediate value being the one at which the screen grid is held constant in Fig. 19. The effect of different values of plate voltage on the transfer characteristics is not shown, but the flatness of the plate characteristics makes it apparent that such an effect is small.

The curves of Fig. 19 are drawn for only one pair of values of the suppressor-grid and screen-grid voltages. Usually these are held fixed at definite values in tubes used in amplifiers. However, sometimes grids other than the first are used as additional control grids. Examples of such applications are use of the pentode as a suppressor-grid modulator,[30] which is discussed in Art. 5, Ch. XII, and as a mixer or a frequency converter, which is discussed in Art. 17, Ch. II. In general for each different connection of the pentode, including the triode connection mentioned in Art. 5, a different family of characteristic curves is needed fully to describe the operation. Because the possibilities for connection and operation of the tube are so numerous, however, the curves desired are often not published, and resort must be made to experimental measurements or to use of published coefficients that are applicable for stated quiescent operating conditions.

Since the screen and suppressor grids are ordinarily operated at fixed voltages, three coefficients μ, g_m, and r_p, as defined in Art. 6, usually serve to describe the operation of the tube if the variable e_c is replaced by e_{c1} for the pentode. From a consideration of typical curves for commercial pentodes and triodes of about the same over-all dimensions and operating voltages, it is evident that the vertical distance between curves, which is a measure of g_m, is of the same order of magnitude for the pentode as for the triode. However, the horizontal distance between the curves for the pentode in the region above the knee is much greater than for the triode, since the curves for the pentode are almost horizontal. Thus μ for the pentode is very large— as large as 1,000 to 2,000 and higher—yet this large value introduces no serious problem of instability, because of the low grid-to-plate interelectrode capacitance. Since the reciprocal of the slope of the curves is the plate resistance, and the curves are almost horizontal, the plate resistance also has a very high value in pentodes—it is of the order of one to two megohms for some types.

Some pentodes, designed so that the control grid is close to the cathode and the spacing between the control-grid wires is small, have a large ratio of mutual conductance to plate current.[31] A large value of this ratio is particularly important in tubes for application in wide-band amplifiers, such as those for television and radar service discussed in Art. 11, Ch. IX.

Linear representation of the characteristics of a pentode may be made in a manner similar to that discussed in Art. 7 for a triode. Thus, a tangent to the characteristic curve is drawn at any chosen

---

[30] C. B. Green, "Suppressor-Grid Modulation," *Bell Lab. Rec.*, *17* (1938), 41–44; K. R. Spangenberg, *Vacuum Tubes* (New York: McGraw-Hill Book Company, Inc., 1948), 270–272.

[31] A. P. Kauzmann, "New Television Amplifier Receiving Tubes," *RCA Rev.*, *3* (1939), 271–289.

point P in Fig. 21, and tangents to the other curves are constructed parallel to it. The tangent to the curve for zero grid voltage intersects the axis for i_b at, say, I_0. The vertical separation between this curve and any other is $g_m e_c$, where e_c is the grid voltage pertaining to the second curve. The current at any point in the neighborhood of P may therefore be expressed as

$$i_b = I_0 + \frac{e_b}{r_p} + g_m e_c, \quad \text{for } e_b \geqq 0, \text{ and } \left(-\frac{I_0}{g_m} - \frac{e_b}{\mu}\right) \leqq e_c \leqq 0, \quad [34]$$

or

$$i_b = g_m \left(e_c + \frac{e_b}{\mu} + \frac{I_0}{g_m}\right), \quad \text{for } e_b \geqq 0, \text{ and } \left(-\frac{I_0}{g_m} - \frac{e_b}{\mu}\right) \leqq e_c \leqq 0.$$
$$[35]$$

Since μ and r_p are so large, the terms involving them may often be neglected.

In pentodes, and tetrodes as well, the control grid is sometimes purposely made nonuniform. The spacing of the grid wires is made closer at the ends of the structure than at the center, often through the omission of one helical turn at the center of the grid. In tubes containing such control grids, the amplification factor decreases markedly when the plate current decreases. This change occurs

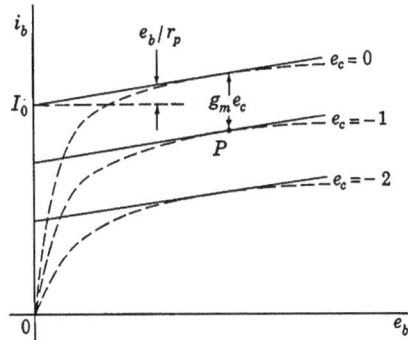

Fig. 21. Linear approximation of pentode characteristics.

because the tube behaves as a parallel combination of several small tubes, each having the amplification factor associated with one part of the grid structure. As the grid voltage is made increasingly negative, cut-off is reached for the high-amplification-factor sections of the grid before it is reached for the remaining sections. Thus, when the plate current is small, the electrons move through the low-amplification-factor section only, and the apparent amplification factor of the tube is nearly the same as that of this section of the grid. At smaller magnitudes of negative grid voltage, current flows through a larger portion of the grid structure, and the apparent amplification factor is greater.

The effect of this grid structure, in addition to causing the amplification factor of the tube to vary, is to require a very large negative grid voltage to produce cut-off. For tubes with nonuniform grids, the various curves of the family of Fig. 19, drawn for equal intervals of grid voltage, are spaced much more closely for highly negative values of grid voltage than for positive or slightly negative voltages. Such tubes are known as *remote-cut-off*, *variable-μ*, or *supercontrol tubes*; tubes having uniform grids are called *sharp-cut-off tubes*. The primary use of remote-cut-off tubes is in automatic-volume-control circuits. In these circuits, the direct component of the grid voltage is varied in such a way as to make the amplification of the tube smaller when a large signal voltage is applied to the grid than when a small signal voltage is applied.

10. BEAM POWER TUBES

After the development of tetrodes and pentodes along the lines described in the previous articles, it was realized that space charge in the region between the screen grid and the plate could be utilized to eliminate, or to aid in eliminating, the effects of secondary emission from the anode. *Critical-distance tubes*[32] and *beam power tubes*[33] utilize the space charge for this purpose. The objective in their design is to cause the space charge in the region between the screen grid and the plate to depress the space potential at some point in the region to a value far enough below the plate potential to prevent plate-emitted secondary electrons from reaching the screen grid. To accentuate the effect of the space charge, the anode is separated from the screen grid by a relatively large distance in critical distance tubes, and the electrons are concentrated in beams in beam power tubes.

Figure 22 shows the potential distribution inside an idealized beam power tube. This is similar to that shown for a tetrode in Fig. 16, but the space charge causes a dip in the potential distribution between the screen grid and the plate. Over a considerable region of operation, this dip can be made large enough to give a retarding field at the plate surface sufficient to eliminate the effects of secondary emission. The effect is that of a virtual suppressor grid, and no limitation caused by electron dispersion around suppressor-grid wires is introduced.

One method by which the electrons are concentrated in beams to enhance the effects of space charge is shown diagrammatically in Fig. 23. The pitches of the helical control grid and of the screen grid are made equal and their wires are aligned to confine the electrons to sheets or flat beams. Focusing electrodes, or beam-confining electrodes, at cathode voltage are included to aid in maintaining uniform current density over a cross section of the beam and to prevent the return of secondary electrons by paths outside the beam. In some beam power

---

[32] J. H. O. Harries, "The Anode to Accelerating Electrode Space in Thermionic Valves," *Wireless Engr.*, *13* (1936), 190–199.

[33] O. H. Schade, "Beam Power Tubes," *I.R.E. Proc.*, *26* (1938), 137–181.

tubes an actual suppressor grid is used, but space charge associated with the beam formation is an essential factor in the tube operation.

Plate characteristics of a typical beam power tube are shown in Fig. 24. The feature of these curves that constitutes the major advantage of the beam power tube over the pentode is the fact that the knees of the curves are sharper, and the relatively straight and almost horizontal part of the curves covers a wider range of plate voltage. As is shown in Arts. 13 and 16, Ch. VIII, this permits a greater power output and efficiency for a given percentage of harmonic generation than would be possible if the knees of the curves were rounder. A second important advantage is that the alignment of the control and screen grids in a beam power tube makes the screen-grid current smaller in proportion to the plate current than it is in a pentode.

The phenomena that occur in the neighborhood of the knee of the curves are somewhat complex.[34] As the plate voltage is varied in this region, a discontinuity and "hysteresis" effect occurs, the curves become double-valued functions, and the minimum space voltage in the region between the screen grid and the plate changes suddenly from one value to another. The effects of secondary emission are not completely suppressed in the region of low plate currents and voltages, but, since

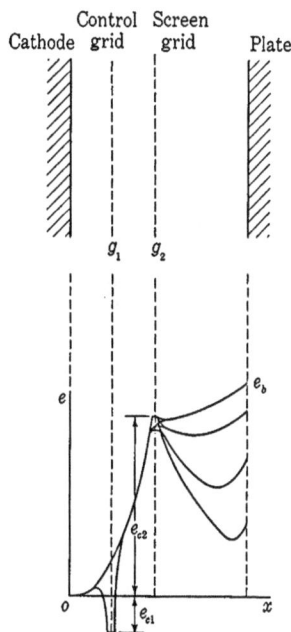

Fig. 22. Approximate potential distributions in an idealized beam power tube for several values of plate voltage.

operation almost never takes place in this region, they are not of serious consequence there.

The beam principle has also been applied to triodes. A water-cooled beam triode consisting of forty-eight paralleled beam-triode structures produces an output of 500 kilowatts in the medium- and high-frequency range.[35]

---

[34] B. Salzberg and A. V. Haeff, "Effects of Space Charge in the Grid-Anode Region of Vacuum Tubes," *RCA Rev.*, 2 (1938), 336–374; J. H. O. Harries, "Secondary-Electron Problems in Beam Power Tetrodes," *Electronic Engineering*, 14 (January, 1942), 586–587. K. R. Spangenberg, *Vacuum Tubes* (New York: McGraw-Hill Book Company, Inc., 1948), 245–265.

[35] "Super-Power Beam Triode," *Broadcast News*, 38 (March–April, 1950), 8–9; *Electronics*, 23 (June, 1950), 120.

11. Gated-beam Tube

The gated-beam tube[36] is essentially a pentode in which additional fixed-potential electrodes are included to focus the electrons in beams and thereby to give an unusually sharp transition between cut-off and a maximum-plate-current condition. A schematic diagram of its

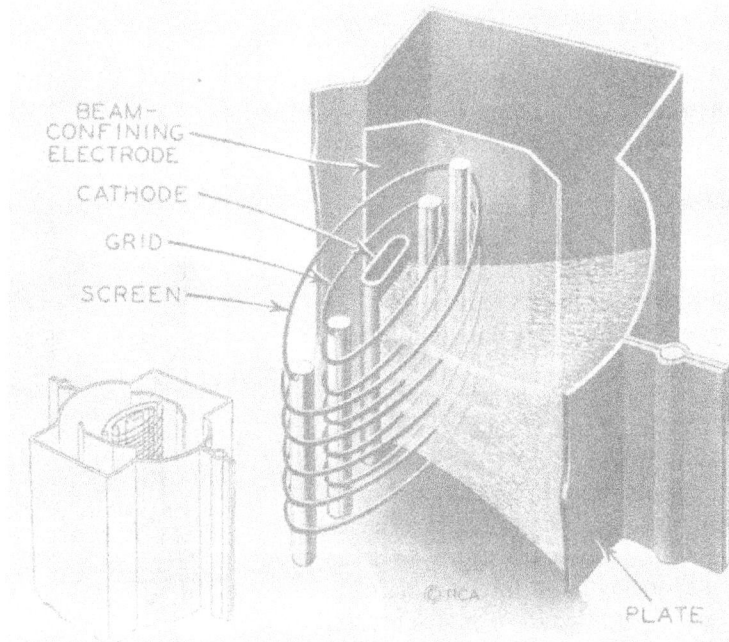

Fig. 23.   Diagram of the elements and electron paths in a beam power tube.   (*Courtesy Radio Corporation of America.*)

structure is shown in Fig. 25a. Electrons traveling from the cathode to the anode in it pass through the first control grid, the screen grid, and the second control grid in turn. These are also known as the limiter grid, accelerator grid, and quadrature grid, respectively. The other electrodes shown are designed to focus the electrons in sheet beams through the grid wires, and also to shield the second control grid and the cathode from variations in electric fields caused by changes in potential of the first control grid. The accelerator electrode surrounding the first control grid is connected to the screen grid and a voltage source that holds it at a positive potential with respect to

[36] R. Adler and A. P. Haase, "The 6BN6 Gated Beam Tube," *Proceedings of the 1949 National Electronics Conference, Vol. V* (Chicago: National Electronics Conference, Inc., 1950), 408–426; R. Adler, "A Gated Beam Tube," *Electronics, 23* (February, 1950), 82–85.

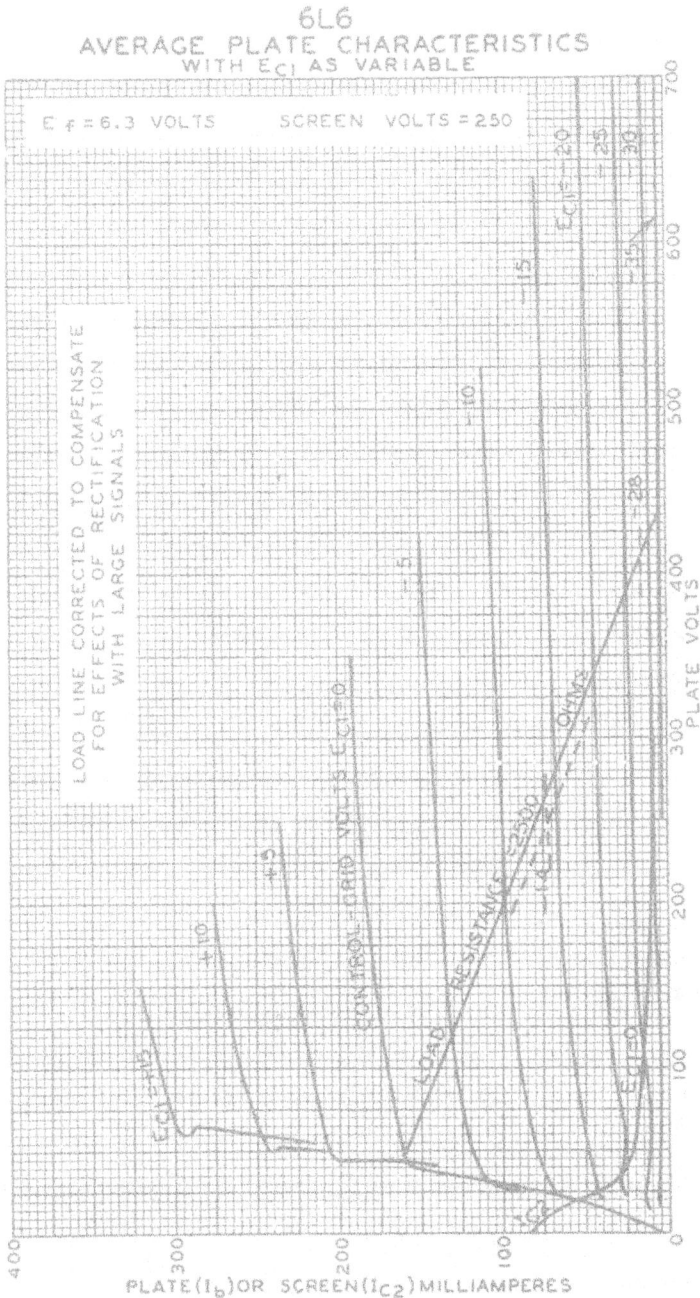

Fig. 24. Plate characteristics of the beam power tube of Fig. 23.
The manufacturer uses capital-letter symbols instead of lower-case
symbols. (*Courtesy Radio Corporation of America.*)

the cathode. The electrodes designated as focus, lens, and shield, are internally connected to the cathode.

Transfer characteristic curves for the tube, which are given in Figs. 25b and c, show that a relatively small change of either control-grid voltage is capable of switching on or off—that is, gating—the full

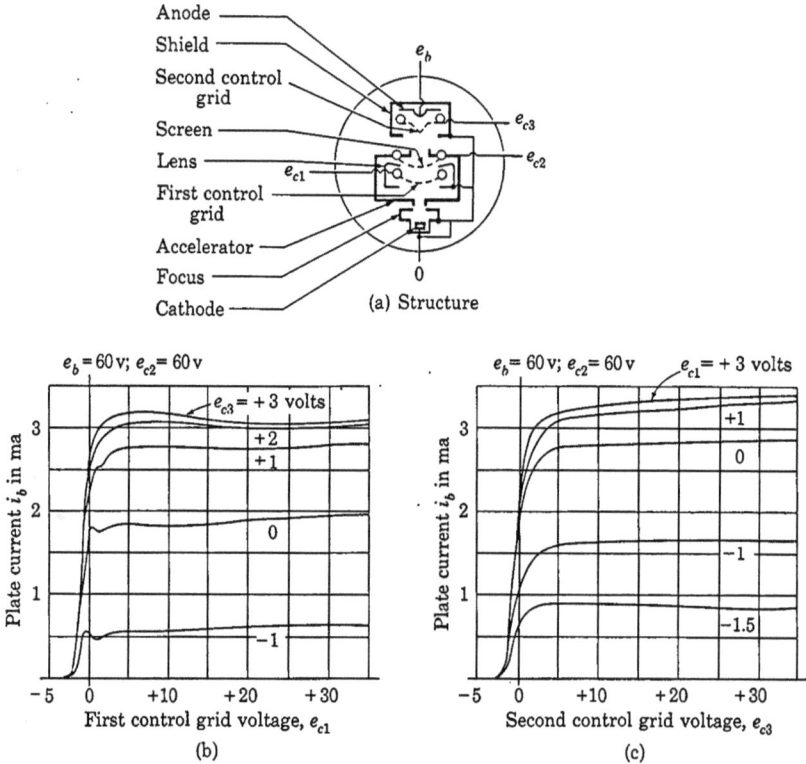

(a) Structure

$e_b = 60$ v; $e_{c2} = 60$ v

Plate current i_b in ma

First control grid voltage, e_{c1}

(b)

$e_b = 60$ v; $e_{c2} = 60$ v $e_{c1} = +3$ volts

Plate current i_b in ma

Second control grid voltage, e_{c3}

(c)

Fig. 25.   Gated-beam-tube structure and transfer characteristics.
(*Adapted by courtesy of General Electric Company.*)

anode current. The steepness of the curves during the transition indicates a high control-grid–to–anode transconductance within that range. Unlike the current in a simple pentode, however, the anode current reaches an abrupt maximum as the voltage of the control grid is increased. The beam current at the cathode remains essentially constant as the control-grid voltages are changed because the cathode is shielded from the control grids. Thus a decrease of anode current is accompanied by an equal increase of accelerator-electrode current, and the transconductance from either control grid to the accelerator electrode is negative—a feature applicable in some oscillator circuits discussed in Ch. XI.

The plate characteristics that give the anode current as a function of anode voltage for various values of the first control grid and constant values of the other electrode voltages are not shown here, but they resemble those of an ordinary pentode given in Fig. 19. The plate characteristics with the second control-grid voltage as a parameter, however, resemble those of a triode for negative values, and tend toward those of a pentode for positive values of the control-grid voltage.

Like the anode current, the current to either of the control grids also levels off at a maximum value as they are driven positive. The maximum current to the first control grid is about one-half milliampere, and that to the second about one-fifth milliampere. Positive-voltage excursions of the control grids therefore do not cause excessive grid current. They are hence permissible, and are practical when the sources supplying the control-grid voltages are capable of furnishing such small currents.

Designed originally for application in frequency-modulation receivers, the gated-beam tube has unique properties that give it a wide range of usefulness in operations known as clipping, limiting, discriminating, and coincidence detection in television, radar, nuclear measurements, and electronic computing machines.

12. SECONDARY-EMISSION TUBES

Amplification by secondary emission is employed in some tubes to supplement amplification by grid action.[37] Such use of secondary emission permits a several-fold increase of over-all transconductance without a corresponding increase in tube capacitance—an important improvement for wide-band and pulse amplifiers[38] (see Art. 11, Ch. IX). Short life and instability of the secondary-emitting properties of the dynode have been the major deterrents to greater production and use of such tubes.

Secondary emission is also employed to provide trigger action in

[37] J. L. H. Jonker and A. J. Overbeck, "Application of Secondary Emission in Amplifying Valves," *Wireless Engr.*, *15* (1938), 150–156; H. M. Wagner and W. R. Ferris, "The Orbital-Beam Secondary Electron Multiplier for Ultra-High Frequency Amplification," *I.R.E. Proc.*, *29* (1941), 598–602; C. W. Mueller, "Receiving Tubes Employing Secondary Electron Emitting Surfaces Exposed to the Evaporation from Oxide Cathodes," *I.R.E. Proc.*, *38* (1950), 159–164; G. Diemer and J. L. H. Jonker, "Secondary-Emission Valve," *Wireless Engr.*, *27* (1950), 137–143. *See also* manufacturer's data on Amperex Type EFP60 and National Union Type NU-5857.
[38] N. F. Moody and C. J. R. McLusky, "Secondary Emission Tubes in Wideband Amplifiers," *Wireless Engr.*, *26* (1949), 410–411.

some vacuum tubes.[39] In this application use is made of circuit instability associated with negative dynamic resistance between a dynode and the cathode. Such a negative resistance appears in the plate characteristics of the tetrode shown in Fig. 17.

13. OTHER MULTI-ELECTRODE VACUUM TUBES

Combinations of diodes, triodes, tetrodes, and pentodes in one envelope, known as multiple-unit tubes, constitute many of the types of multi-element tubes, and the characteristics of their groups of electrodes do not differ essentially from those discussed in previous articles of this chapter. However, some multi-element tubes comprise one cathode and one plate and more than three grids. In such tubes, two of the grids function as control grids and serve to control the plate current in accordance with the combined effect of their voltages. The remaining grids generally function as screen or suppressor grids to shield the control grids and the plate from one another and to suppress secondary emission from the plate, exactly as in tetrodes and pentodes. Sometimes a grid operated at a positive potential is used as an auxiliary anode whose current depends upon the voltage of one of the control grids.

The plate current in these tubes is proportional to the product of some function of the voltage of one control grid and some function of the voltage of the other control grid.[40] If the voltages of both control grids are varied, a component of the change in plate current may be obtained that is proportional to the product of the changes in the two voltages. This multiplication process makes these tubes applicable in circuits where some form of modulation (see Ch. XII) occurs. Because of the large number of electrodes in multi-element tubes, it is not practical to give static curves that completely specify the plate current as a function of all the electrode voltages. Accordingly, static characteristics are not often published; instead, the optimum operating conditions and the performance of the tubes in specific applications are generally given.

Multi-element tubes are particularly useful for frequency conversion,[41] that is, for changing a voltage of one frequency to one of another frequency. In this application, voltages of different frequencies are applied to the two control grids, and, because of the multiplication in

[39] C. F. Miller, "New Design for a Secondary-Emission Trigger Tube," *I.R.E. Proc.*, *37* (1949), 952–954.

[40] A. H. Wing, "On the Theory of Tubes with Two Control Grids," *I.R.E. Proc.*, *29* (March, 1941), 121–136.

[41] E. W. Herold, "The Operation of Frequency Converters and Mixers for Superheterodyne Reception," *I.R.E. Proc.*, *30* (February, 1942), 84–103.

the tube, new frequencies are produced in the plate circuit as is discussed in Art. 17, Ch. XII. The *pentagrid mixer*, a tube having five grids, is frequently used in this manner. In the *pentagrid converter*, the voltage for one grid is obtained through the use of that grid, together with the cathode and one of the additional electrodes, in an oscillator circuit (see Ch. XI); the voltage for the other control grid is obtained from an external source. Multi-element tubes are also used widely in circuits, such as automatic-volume-control circuits and volume expanders, in which the amplification of a voltage applied to one control grid is changed by the voltage applied to the other control grid.

14. USE OF VACUUM TUBES AT VERY HIGH FREQUENCIES

The diodes, triodes, and multigrid tubes described in the preceding articles have found increasing use in the production and amplification of electrical power at frequencies ranging from direct current to about 100 megacycles per second. Considerable attention has been devoted to extending the upper limit of this range; and, by careful design, triodes have been developed that may be used at frequencies as high as 3,000 megacycles per second.

As is discussed in Arts. 5 and 8, Ch. VIII, the behavior of a vacuum triode operated at moderate frequencies may be determined from an incremental equivalent circuit containing the resistances specified by the static characteristics and the proper interelectrode capacitances. Correct results are not obtained from the use of this equivalent circuit at frequencies of 100 megacycles or more, however, because such an equivalent circuit is not complete. The current in the plate circuit of electron tubes depends not only upon the rate of arrival of electrons at the plate but also upon the rate of change of the electric flux density at the plate. Each of these current components is influenced by the motion of electrons throughout the entire interelectrode space. Analyses[42] that take into account the total current caused by varying voltages have demonstrated that the simple equivalent circuit is correct only if the *transit time*, or time required for an electron to cross from the cathode to the plate, is short compared with the period of the voltage cycle. If the transit time is comparable to a period of the voltage cycle, the distribution of electrons in the interelectrode

---

[42] W. E. Benham, "Theory of Internal Action of Thermionic Systems at Moderately High Frequencies," *Phil. Mag.*, 5 (1928), 641–662; *11* (1931), 457–517; F. B. Llewellyn, "Vacuum-Tube Electronics at Ultrahigh Frequencies," *I.R.E. Proc.*, 21 (1933), 1532–1573; *Electron-Inertia Effects: Cambridge Physical Tract* (Cambridge: The University Press, 1941); W. R. Ferris, "Input Resistance of Vacuum Tubes as Ultrahigh-Frequency Amplifiers," *I.R.E. Proc.*, 24 (1936), 82–107; D. O. North, "Analysis of Effects of Space Charge on Grid Impedance," *I.R.E. Proc.*, 24 (1936), 108–136.

space depends upon the variations in voltage that occur while the
electrons are crossing the tube. The motion of this nonuniform distribution of electrons causes at the surface of the electrodes a varying
field, and a correspondingly varying surface charge, that contributes
to the current reaching the electrodes. The most important conclusions
of these analyses are:

(a) In a diode, the resistive part of the plate impedance varies with
frequency and is negative in certain frequency ranges.

(b) In a triode operated at a sufficiently low frequency with a resistive plate load, the plate current is in phase with the grid
voltage and 180 degrees out of phase with the plate voltage. At
a frequency for which the transit time is of importance, the
phase angle between plate current and plate voltage differs
from 180 degrees, and, as a result, an increase occurs in the
power dissipated at the plate for a given power output.

(c) In a triode, at high frequencies, power is required by the grid
because of a transit-time effect known as *electron loading* of the
grid. Electron loading is a loss at the input of the tube that must
be added to any losses that may result from electrons striking
the grid or from the grid-plate capacitance (see Art. 8, Ch. VIII).
Since the electron loading increases with frequency, it places an
upper limit on the frequency at which a triode can deliver
more power than it receives.

In order to avoid the undesirable effects of the transit time, tubes
are constructed in which that time is made very short through the use
of small interelectrode spacings and high plate voltages. However,
since the dimensions of the electrodes themselves must be small to
prevent excessive interelectrode capacitances, tubes suitable for use
at very high frequencies are necessarily small in size.[43] Hence the
power that may be dissipated in them and, consequently, the power
output obtainable from them are small.

Inductance of the lead connections to the electrodes is a further
limiting factor in the performance of high-frequency tubes.[44] The inductance of the cathode lead in particular is important for it tends to
resonate with the grid-to-cathode capacitance and to cause feedback
coupling between the grid and plate circuits. This coupling contributes

[43] C. E. Fay and A. L. Samuel, "Vacuum Tubes for Generating Frequencies above
One Hundred Megacycles," *I.R.E. Proc.*, *23* (1935), 199–212; B. Salzberg and D. G.
Burnside, "Recent Developments in Miniature Tubes," *I.R.E. Proc.*, *23* (1935), 1142–
1157.

[44] M. J. O. Strutt and A. van der Ziel, "The Causes for the Increase of the Admittances
of Modern High-Frequency Amplifier Tubes on Short Waves," *I.R.E. Proc.*, *26* (1943),
1011–1032.

to the loss at the input of the tube, as does the electron loading effect. To overcome the effects of lead inductance, resort is made to special disc-seal, or lighthouse, construction[45] so that the connections to the electrodes can become parts of the concentric transmission lines or cavity resonators that constitute the external circuits. Operation at frequencies as high as 3,000 megacycles per second is then possible.

Tubes virtually free from the effects that in triodes cause electron loading and limit the power output at high frequencies have been developed on the basis of an unique principle called *velocity modulation* of an electron beam.[46] These tubes operate effectively as oscillators and amplifiers in a range of frequencies extending upward from 600 megacycles per second called *microwave* frequencies. The Klystron shown in Fig. 26 is representative of such tubes, and is one of a class that involves interaction of an electron beam with an electromagnetic field.

A diagram of part of a tube employing a possible method of velocity modulation is shown in Fig. 27. Electrons emitted from an electron gun pass through two grids, between which a high-frequency voltage is impressed. As this voltage alternates, some of the electrons in the beam are accelerated and others are decelerated; that is, their velocities are modulated. Thus the electrons are given velocities that are functions of the time at which they passed through the grids. The grids are placed very close together, so that the time of electron transit between them is small compared with the time of a cycle of the impressed voltage. The crest value of the high-frequency voltage applied to the grids is much smaller than the direct voltage through which the electrons are accelerated in the gun. Therefore the number of electrons leaving the grids per unit time is approximately constant; that is, essentially no density modulation of the beam is produced. Moreover, little power is taken from the high-frequency source, since as many

[45] E. D. McArthur, "Disk-Seal Tubes," *Electronics*, 18 (February, 1945), 98–102; A. M. Gurewitsch and J. R. Whinnery, "Microwave Oscillators Using Disk-Seal Tubes," *I.R.E. Proc.*, 35 (1947), 462–473; D. R. Hamilton, J. K. Knipp, and J. B. H. Kuper, Editors, *Klystrons and Microwave Triodes*, Massachusetts Institute of Technology Radiation Laboratory Series, Vol. 7 (New York: McGraw-Hill Book Company, Inc., 1948).

[46] A. A. Heil and O. Heil, "Eine neue Methode zur Erzeugung kurzer ungedampfter elektromagnetischen Wellen von grosser Intensität," *Zeit. fur Phys.*, 95 (1935), 752–773; R. H. Varian and S. F. Varian, "A High-Frequency Oscillator and Amplifier," *J. App. Phys.*, 10 (1939), 321–327; W. C. Hahn and G. F. Metcalf, "Velocity-Modulated Tubes," *I.R.E. Proc.*, 27 (1939), 106–116; A. V. Haeff, "An Ultrahigh-Frequency Power Amplifier of Novel Design," *Electronics*, 12 (February, 1939), 30–32; D. R. Hamilton, J. K. Knipp, and J. B. H. Kuper, Editors, *Klystrons and Microwave Triodes*, Massachusetts Institute of Technology Radiation Laboratory Series, Vol. 7 (New York: McGraw-Hill Book Company, Inc., 1948).

electrons are accelerated as are decelerated, and the amounts of the acceleration and deceleration are the same. In this way, the process of velocity modulation avoids the loss in the input circuit that occurs in triodes at high frequencies.

A velocity-modulated beam of electrons is not directly usable but must first be converted to a density-modulated beam. The usual method of making this conversion is to allow the velocity-modulated beam to travel through a space free of all electric fields except that

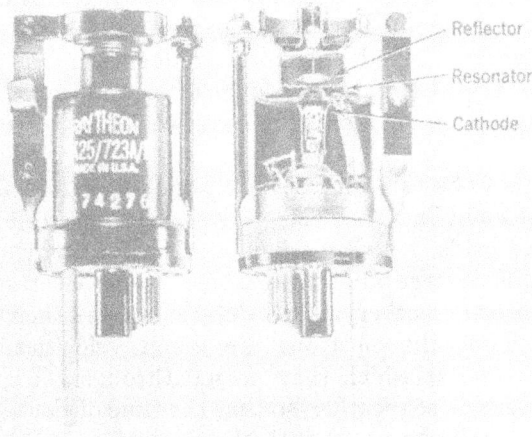

Fig. 26. External and cut-away views of a mechanically tunable reflex Klystron having an average power output of 35 milliwatts over a frequency range of 8,500 to 9,660 megacycles per second. (*Courtesy Raytheon Manufacturing Company.*)

produced by the electrons themselves. The electrons that are accelerated in passing through the grids during the positive half-cycle of the alternating-voltage wave travel faster than the electrons that are decelerated in passing through the grids during the negative half-cycle of the wave. As a result, the fast electrons overtake the slow electrons that left the grids one-half cycle earlier, and at some distance beyond the grids the electron beam is made up of alternate regions of high and low electron density, moving along the beam at the velocity imparted to the electrons in the electron gun. Thus density modulation is obtained from the original velocity modulation. The regions of high electron density are often referred to as *electron bunches*. Since this conversion is a result of the time required for electrons to move through the field-free space, transit time—the very phenomenon that leads to undesirable effects in a triode—is made useful in obtaining density modulation from velocity modulation.

In the design of the plate of a triode intended for use at very high frequencies, two conflicting requirements are encountered. The plate must be small if the tube is to operate at high frequencies; yet it must be large if it is to dissipate the power associated with the large electron current necessary for a large power output and the high plate voltage necessary for a short transit time. In tubes that use the process of induction of power from density-modulated electron beams, this difficulty is removed by the use of two separate electrodes—one to extract the high-frequency power from the beam and deliver it to an external circuit, and the other to dissipate the direct-current power that

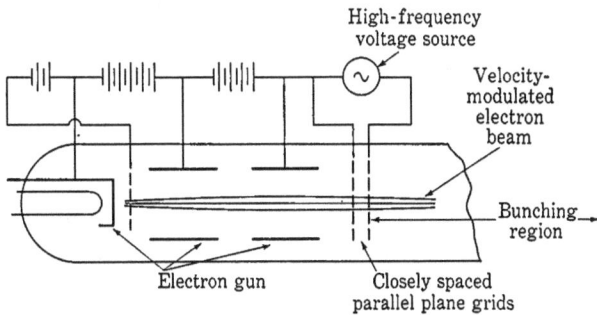

Fig. 27.   Method of producing a velocity-modulated electron beam.

remains in the electron beam after the high-frequency power has been removed. The mechanism by which the power transfer to the load takes place may be explained with reference to Fig. 28. As a bunch of electrons in the beam moves along the axis, it induces a positive charge on the inside surface of the output electrode, which is in the form of a coaxial-line or cavity resonator. This induced positive charge moves along the electrode abreast of the electron bunch. When the bunch passes the gap in the electrode, the positive charge must transfer from one side of the gap to the other. This transfer of charge constitutes an impulse of current in the resonant circuit. If the intervals between the times at which the induced charges flow through the resonant circuit are equal to the resonant period of the circuit, and if the losses of the circuit are small, the current impulses, occurring once each cycle, will sustain oscillations in the resonant circuit in a manner somewhat analogous to that by which a pendulum is maintained in oscillation by impulses delivered to it from an escapement mechanism in a clock. The electrons of the beam, after transferring some of their energy to the resonant circuit through interaction with the electromagnetic field at the gap, move along the center of the tube and strike the collector electrode at the end.

At the high frequencies used with such tubes, the ordinary resonant circuits of inductance and capacitance are replaced by cavity resonators[47] and resonant sections of transmission lines, because of the smaller losses in resonators of the latter form and their convenient size. In Fig. 28, a quarter-wavelength section of a coaxial transmission line[48] is shown as the resonant circuit. The two conductors of the line are the two concentric cylinders shown in section in the figure; the tube and the electron beam are placed in the hollow center of the inner conductor. The slot at one end of the inner conductor allows the

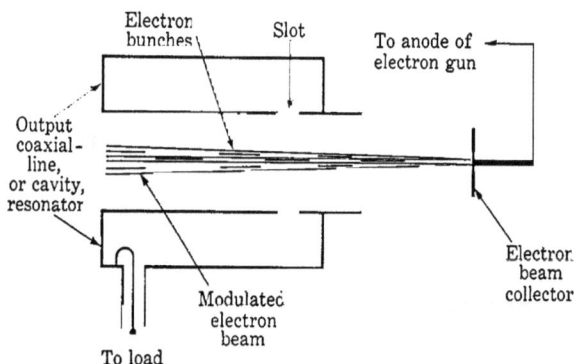

Fig. 28.   Method of obtaining power by induction from a density-modulated electron beam.

transfer of power from the electron beam to the resonator, and moreover constitutes an open circuit at that end of the line. The other end of the line is short-circuited. If the frequency of the current impulses is such that the wavelength associated with them is approximately four times the length of the line, standing waves are set up, just as standing sound waves are set up in an organ pipe driven at one end and closed at the other. Power may be withdrawn from the resonator to a load by a coupling method such as the loop shown. The loop is located in a region of high magnetic flux density, and a high-frequency voltage is induced in it by the changes in the magnetic flux it encloses. Since more power may be obtained from the resonant circuit than is required to produce the velocity modulation, which is converted to density modulation and then to power in the output circuit, a tube involving all these processes may serve as an amplifier or an oscillator.

[47] W. W. Hansen, "A Type of Electrical Resonator," *J. App. Phys.*, 9 (1938), 654–662; W. L. Barrow and W. W. Mieher, "Natural Oscillations of Electrical Cavity Resonators," *I.R.E. Proc.*, 28 (1940), 184–191.

[48] F. E. Terman, "Resonant Lines in Radio Circuits," *A.I.E.E. Trans.*, 53 (1934), 1046–1053.

The reflex type of Klystron illustrated in Fig. 26 is useful as an oscillator and is the type in widest use.[49] In it the electron stream is velocity-modulated during its passage through a gap in a cavity resonator as previously explained. Instead of continuing through the gap in a second and different resonator, however, the stream is directed toward an electrode having a potential below that of the electron-gun cathode. This reflects the electrons so that they again pass through the original resonator gap, but in the reverse direction. During this second passage, energy is delivered to the resonator by the bunched

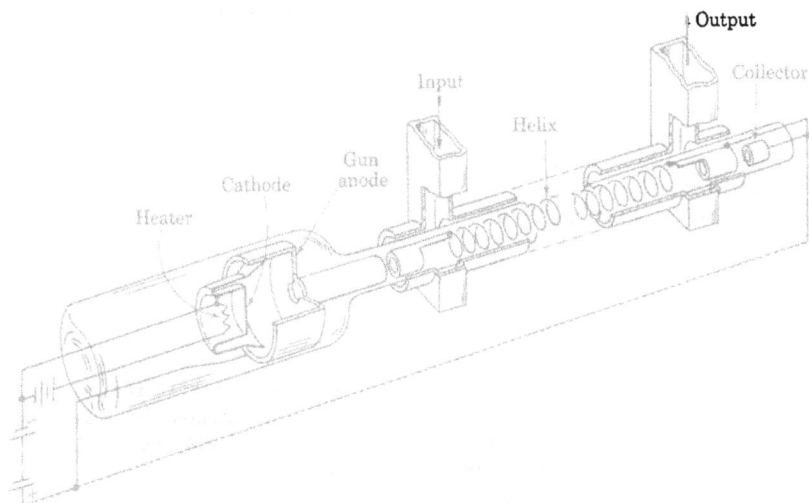

Fig. 29.   Schematic diagram of a traveling-wave tube.*

electrons, and oscillations are sustained at frequencies as high as 30,000 megacycles per second.

Other high-frequency tubes that depend on interaction of an electron beam with an electromagnetic field are the cavity magnetron[50] and the traveling-wave tube.[51] In the cavity magnetron, which is

[49] J. R. Pierce, "Reflex Oscillators," *I.R.E. Proc.*, *33* (1945), 112–118; E. L. Ginzton and A. E. Harrison, "Reflex-Klystron Oscillators," *I.R.E. Proc.*, *34* (1946), 97–117.

[50] J. B. Fisk, H. D. Hagstram, and P. L. Hartman, "The Magnetron as a Generator of Centimeter Waves," *B.S.T.J.*, *25* (1946), 1–188; G. B. Collins, Editor, *Microwave Magnetrons*, Massachusetts Institute of Technology Radiation Laboratory Series, Vol. 6 (New York: McGraw-Hill Book Company, Inc., 1948).

[51] J. R. Pierce, *Traveling-Wave Tubes* (New York: D. Van Nostrand Company, Inc., 1950); S. E. Webber, "1,000-Watt Traveling Wave Tube," *Electronics*, *23* (June, 1950), 100–103.

* This figure is reproduced from *Traveling-Wave Tubes*, J. R. Pierce, Bell Telephone Laboratories, Inc., copyright 1950, D. Van Nostrand Company, Inc., New York, Fig. 2.1, with permission.

discussed in Art. 8d, Ch. I, electrons are whirled past gaps in multiple resonators around the periphery of the tube and induce resonant oscillations in them. In the traveling-wave tube, which is illustrated in Fig. 29, energy is imparted to the electromagnetic wave traveling along a helical wire electrode by a beam of electrons directed along the axis of the helix. The mechanism of operation may be likened to that by which waves on water receive energy from the wind.

PROBLEMS

1. A vacuum diode has a circular cylindrical plate and an axial filament with the following specifications:

Plate

    Material: graphite
    Length = 7 cm
    Outside diameter = 5.5 cm
    Inside diameter = 5.0 cm
    Total radiation emissivity = 0.9.

Filament

    Material: thoriated tungsten
    Length = 7 cm (effective)
    Diameter = 0.03 cm.

Assuming that (1) the effective length of the filament is equal to that of the plate, (2) only the outside cylindrical surface of the plate is effective in radiating heat, (3) the plate current is limited by space charge, and (4) the fraction of the filament power dissipated by the plate is negligible,

(a) Find the temperature of the plate when the plate voltage is 700 volts.
(b) If the residual gas is nitrogen at a temperature of 500 K, to what pressure must it be reduced in order that, on the average, 99.99 per cent of the electrons do not collide with a gas molecule in traveling from the filament to the plate?

2. Because of the slow de-ionization of the mercury vapor, mercury-vapor gas tubes will not rectify a high-frequency current. As a means of increasing the allowable frequency it is proposed to remove the gas and convert the tube to a vacuum type. The mercury-vapor rectifier tube in question has the following ratings:

    Voltage drop while conducting = 10 to 15 volts
    Maximum allowable peak plate current = 2.5 amp.

For analysis, consider the tube as having cylindrical symmetry with an axial filament and the following specifications:

Plate

    Circular cylinder: 2 cm in diameter by 6 cm long
    Material: nickel sheet 0.010 in. thick
    Maximum allowable temperature = 600 K
    Total radiation emissivity = 0.1.

Filament

Oxide-coated; directly heated
Heating voltage = 2.5 volts
Heating current = 5.0 amp
Emission efficiency = 200 ma per watt
Length = same as plate length.

It may be assumed that:

(1) The Schottky effect is negligible.
(2) The plate radiates heat only from the outside surface.
(3) The filament power escapes through the open ends of the plate—none of it reaches the plate.
(4) $\beta^2 = 1.00$.

(a) What plate voltage is required to produce a plate current equal to the saturation current of the filament?

(b) What is the allowable continuous plate current expressed as a fraction of the saturation current?

3. A vacuum diode with a circular cylindrical plate and an axial tungsten filament is to be built. The tube is to have a saturation plate current of 1 amp, and the plate voltage required for this value of current when limited by space charge should be 600 volts. The plate must be capable of radiating continuously the power represented by the product of the foregoing current and voltage plus 75 per cent of the filament power, which it intercepts.

Assuming that the filament and plate are of the same length, that the filament temperature is 2,500 K, that end effects may be neglected, that the plate can safely radiate as heat 8 watts per sq in. of its outside surface area, and that β^2 is approximately 1.08, find:

(a) the plate diameter and length,
(b) the filament diameter and length,
(c) the filament heating current and voltage.

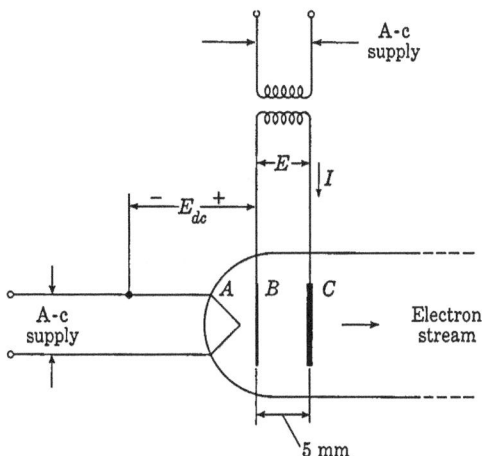

Fig. 30. Plane cathode heated by electron bombardment, Prob. 4.

4. In an experimental beam tube, a plane cathode is desired that will deliver about 1 amp of electron emission. The structure shown in Fig. 30 is proposed. The

tungsten filament A is a coil requiring 10 watts and operating at a temperature of 2,500 K. A direct voltage E_{dc} heats the plane nickel electrode B by electron bombardment. Electrode C is also plane and made of nickel, and is coated with barium and strontium oxides on both sides, whereas B is coated only on the side toward C as shown. After B becomes heated to a temperature of 1,000 K, a 60-cps alternating voltage having an effective value E is applied between B and C. Electrode C then becomes hot, and the voltage E_{dc} can be removed and the filament A disconnected. The right-hand side of C serves as a cathode for the main electron stream.

What effective voltage E and effective current I are required to maintain electrodes B and C at a temperature of 1,000 K if (1) B and C are 1-cm diameter discs spaced 5 mm apart, and (2) the heat losses from B and C are 4 watts per cm² at 1,000 K? State assumptions made in the solution.

5. Find the transit time for an electron in a vacuum diode having essentially parallel plane electrodes spaced 0.5 cm apart, a plate voltage of 100 volts, and current limited by space charge. Neglect initial velocities of the emitted electrons at the cathode. By what factor would the transit time be multiplied if the space charge were absent?

6. Determine the dynamic coefficients μ, r_p, and g_m from Fig. 7 for operation near the point where e_b is 280 volts and e_c is -4 volts. Show by a sketch how this was done. Use only this one family of curves.

7. Determine from the plate-characteristic curves for the Type 12AU7 tube the dynamic coefficients μ, r_p, and g_m for operation near the point where e_b is 245 volts and e_c is -10 volts (see manufacturers' curves in Appendix C). Show by a sketch how the determination was made. Compare the results with those given for the same plate current in Fig. 11.

8. From the plate characteristics for the Type 6SN7-GT triode, find values of μ, g_m, and E_0 that permit linear representation of the curves by Eq. 32 in the vicinity of the point at which e_b is 250 volts and e_c is -8 volts. Superpose the family of straight lines on the plate characteristics for comparison.

9. The characteristics of a certain triode are representable by the following relation:

$$i_b = 2 \times 10^{-6}(10e_c + e_b)^2, \qquad \text{for } e_c \leqq 0, \text{ and } e_b \geqq -10e_c \,,$$

where i_b is in amperes and e_c and e_b are in volts.

Determine mathematically, without using graphical methods, the amplification factor μ, dynamic plate resistance r_p, and mutual conductance g_m for operation near the point where e_b is 200 volts, and e_c is -10 volts.

10. Evaluate the coefficients K, n, and μ in the expression

$$i_b = K\left(e_c + \frac{e_b}{\mu}\right)^n$$

for the characteristics of the tube of Figs. 7, 8, and 9. A plot on log-log graph paper will be found helpful.

11. From the plate characteristics for the Type 6AU6 pentode shown in Fig. 19, find values for μ, g_m, and r_p that are applicable for operation in the vicinity of the point at which e_b is 200 volts, e_{c2} is 100 volts, and e_{c1} is -1 volt.

CHAPTER V

Gas Tubes

The more prominent physical processes that occur in gas tubes are discussed in Ch. III. The manner in which these phenomena determine the characteristics and limitations of various gas tubes important in engineering is described in this chapter. In some of these tubes, such as the gas phototube and the voltage-regulator tube, a gaseous discharge occurs between unheated metal electrodes. In others—the thermionic gas diode and triode, for example—a heated cathode is used as a source of electrons, and these participate in the discharge. Since their characteristics are very different from those of vacuum tubes, both cold- and hot-cathode gas tubes are suitable for entirely different circuit applications.

1. EFFECT OF GAS IN A THERMIONIC DIODE

If gas at a small fraction of atmospheric pressure is present in a thermionic diode, the characteristics of the tube are markedly altered from those discussed in the previous chapter. Such gas may be left by inadequate pumps and is likely then to contain oxygen or water vapor, which have chemical effects upon the cathode and hence decrease its emission. On the other hand, mercury vapor and the inert gases (helium, neon, argon, krypton, and xenon) have no chemical action on commonly used cathodes and are often introduced into tubes after exhaust for the specific purpose of neutralizing the electronic space charge.

An understanding of the action of a gas and the relative importance of the various effects discussed in Ch. III is obtained from an examination of the experimentally determined characteristics that result when increasing amounts of mercury vapor or an inert gas are introduced into a tungsten-filament vacuum diode. Tubes containing the small pressures of gas which will cause the behavior discussed here are not of practical importance at present. The tungsten-filament tube with a small amount of gas is mentioned because of the relative simplicity of its internal behavior, and because it illustrates the first stages of the transition that takes place as more and more gas is added to a tube with a thermionic cathode.

When no appreciable amount of gas is present, the characteristic current-voltage curve of such a tube with a constant filament temperature resembles the curve for p equal to zero in Fig. 1. With the

introduction of gas at a very low pressure p_1, the curve rises as shown
but reaches essentially the same saturation current. Further addition
of gas gives rise to a characteristic like that for p equal to p_2. Often the
curve contains discontinuities, and usually the current is different for
increasing and decreasing plate voltages; that is, the complete current-
voltage curve resulting when the voltage is increased from zero to a
large value and then is returned to zero may be a loop, as indicated in
the figure. For high voltages, the current frequently decreases as the
voltage increases. The major effect of the gas is evidently in the region
AB of Fig. 1, where the current in the absence of gas would be limited

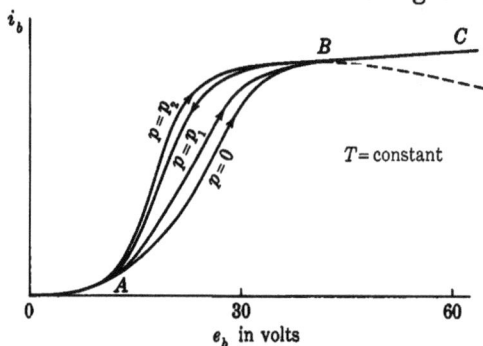

Fig. 1.   Change of the current-voltage curve of a diode having a
tungsten filament as the gas pressure is varied.

by space charge. The gas has little effect on the current-carrying
capacity of the tube, for the saturation current, indicated in the region
BC, is changed only slightly. Moreover, the gas does not have an
appreciable effect in the region OA, where the plate voltage is less than
a value of the order of the ionizing potential.

The statement that the effect of the gas is small in the region OA
does not mean that no departure from the high-vacuum curve occurs
for voltages below the ionizing potential of the gas, or that a departure
must occur at the ionizing potential. Below the ionizing potential, the
gas may impede the motion of electrons and tend to reduce the current
at a given voltage. In gases having metastable states, ionization and an
increase of current may occur for voltages below the ionizing potential.
Contact potential differences and initial velocities of the emitted
electrons also affect the voltage at which ionization becomes appreciable. As is pointed out in Art. 5 of Ch. III, ionizing collisions do not
occur frequently until the voltage exceeds the ionizing potential by a
considerable amount. Because of these considerations, it can merely be
said that, when the plate voltage is increased, the effect of the gas
becomes appreciable at a voltage of the order of the ionizing potential
of the gas.

As the plate voltage is increased above its value at point A, a few pairs of positive ions and associated electrons are produced by electrons from the cathode as they cross the tube. The additional electrons increase the current directly as they move in the electric field. The positive ions, on the other hand, have a twofold effect: First, because of their charge and motion, they contribute to the total current; and, second, because of their charge, they tend to neutralize the space charge of the electrons. The first effect, and that of the additional electrons as well, are not of importance here, however, as is evident from the experimental fact that the current in the region BC is not appreciably increased over the saturation value by the presence of the gas. At very high plate voltages—in a region beyond C —more effective ionization may be obtained if the gas pressure is sufficient, and the current may be increased by the ions just as in a Townsend discharge.

The second effect of the positive ions—the partial neutralization of the space charge—is of importance in the region AB. The filament can supply electrons at a rate corresponding to the current at point B, but, if the voltage is less than that at B, and if no positive ions are present, the electrons in the interelectrode space repel those about to leave the cathode, and the plate current is therefore limited to less than the filament emission. Positive ions in the interelectrode space partially neutralize the electron space charge and allow a greater portion of the available electrons to cross to the plate; thus the plate current is increased. If the positive ions are appreciably to neutralize the electronic space charge, the *number* of ions in the space must be comparable to the *number* of the electrons. This condition is possible even though the *rate* at which ions are produced in the space is very small compared to the *rate* at which electrons enter the space, for the electric field carries the electrons out of the interelectrode space much more rapidly than it does the positive ions.

The relatively sluggish movement of the positive ions is caused by their smaller charge-to-mass ratio, as is shown in the following analysis. According to Eq. 30, Ch. I, the speed attained by electrons or positive ions moving freely from rest in an electric field is

$$v = \sqrt{-2\frac{Q}{m}E}\,, \qquad [1]$$

where

    Q is the charge of the particle,

    m is the mass of the particle,

    E is the potential rise through which the particle moves.

If an electron and a positive ion move from rest through the *same* difference of potential, but in opposite directions,

$$\frac{v_e}{v_+} = \sqrt{\frac{Q_e m_+}{Q_+ m_e}},$$ [2]

where v_e, Q_e, and m_e are, respectively, the speed, the magnitude of the charge, and the mass of the electron; and v_+, Q_+, and m_+ are the corresponding quantities for the positive ion. For single-charged ions of mercury, the ratio v_e/v_+ is 605; for those of xenon, krypton, argon, neon, and helium, it is, respectively, 490, 391, 270, 192, and 85.5. The ratio of the current densities carried by moving space charges of electrons and ions is

$$\frac{J_e}{J_+} = \frac{\rho_e v_e}{\rho_+ v_+},$$ [3]

where ρ_e and ρ_+ are the space-charge densities of the electrons and the positive ions, respectively. Thus, in a region in which single-charged positive mercury ions and electrons have been accelerated by the *same* potential difference, the concentration of positive mercury ions can be equal to the concentration of electrons, so as to neutralize completely the electronic space charge, even though current conducted by the ions is only 1/605 that conducted by the electrons. The great agility of an electron in an electric field as compared to the agility of a positive ion, then, accounts for the effectiveness of positive ions in neutralizing electronic space charge without contributing appreciably to the space current.

If a vacuum tube is desired in which gas shall play no appreciable part in the over-all behavior, the pressure must be reduced until the number of electrons that make ionizing collisions in crossing the inter-electrode space is a very small fraction of the total number of electrons that cross. Because of the very great effect of positive ions upon the space-charge limitation of current, the fractional increase in the space-charge-limited current caused by ionization is much larger than the fraction of the electrons that ionize. To reduce the effect of the gas on the space-charge-limited current to, say, one per cent, it is therefore necessary to reduce the gas pressure until far fewer than one per cent of the electrons ionize gas particles. The pressure must often be reduced until only one electron in 10,000 or more ionizes; in practical vacuum tubes the pressure is frequently of the order of one billionth of an atmosphere.

2. Gas diodes with thermionic cathodes

The effects of gas on the characteristics of a diode that are discussed in Art. 1 are those occurring when only a very minute amount of gas is in the tube. Addition of a greater quantity of gas increases the number of positive ions produced by each electron and causes the volt-ampere curves of Fig. 1 to rise more steeply between points A and B; if the pressure of the gas is sufficient, the curve between these points becomes vertical or even slopes in a direction to indicate a decrease of voltage with increase of current. This characteristic is often represented by a volt-ampere curve, as in Fig. 2, where the positions of the current and voltage axes are reversed from those in Fig. 1. As the

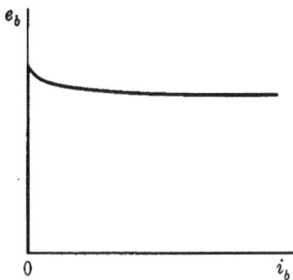

Fig. 2. Volt-ampere charac-
    teristic of a gas diode.

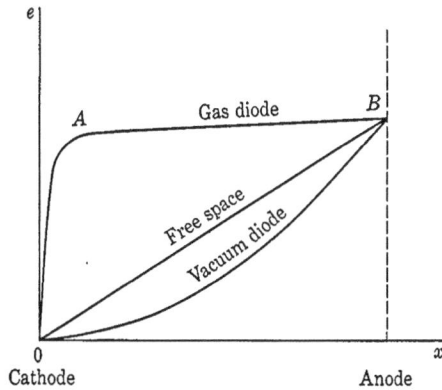

Fig. 3. Comparison of the potential distri-
    butions in vacuum and gas tubes.

voltage is increased from zero, the current in a typical tube is very small and the curve lies essentially on the vertical axis as long as the voltage is too low to cause appreciable ionization. When a critical value of current (near that of point A in Fig. 1) is reached, the current increases suddenly[1] to a value limited only by the external circuit resistance or by cathode emission, and a visible glow appears in the tube; the tube is said to *break down, fire,* or *start.*

Thermionic-cathode diodes into which an inert gas or mercury vapor is introduced to produce a characteristic curve of the form of that in Fig. 2 are known as *gas diodes.* In Fig. 2, the current corresponding to that in the range OA of Fig. 1 is shown as zero, since in the ordinary gas diode it is extremely small relative to currents shown on the scale of Fig. 2. Since oxide-coated cathodes are almost always used in gas diodes, the emission limitation of current indicated in region BC of Fig. 1 does not occur, and, as indicated in Fig. 2, the voltage across the

---

[1] I. Langmuir, "Electric Discharges in Gases at Low Pressures," *J.F.I., 214* (1932), 284–285; E. D. McArthur, *Electronics and Electron Tubes* (New York: John Wiley & Sons, Inc., 1936), 103–106.

gas tube—in contrast to that across a vacuum tube—is almost *independent* of the current over the entire useful current range. The voltage is of the order of the ionizing potential of the gas, although, because of the action of metastable atoms and double-step ionization, its exact value is often somewhat lower. In the usual gas diodes, it lies in the range of 10 to 20 volts. The constant drop is of great significance in the application of such devices, for *the current must be limited by the circuit external to the tube*; otherwise the tube will be destroyed. A gas tube placed across a source that supplies a constant voltage greater than the tube drop behaves as though it were essentially a short circuit. Because of their constant, low, voltage drop, gas diodes may carry large currents with small power loss, and therefore are efficient rectifiers in high-current circuits.

2a. *Potential Distribution in a Gas Diode.* The potential distribution inside a gas diode, shown in Fig. 3, is entirely different from that in the same tube with the gas removed. The figure is drawn for a diode in which the cathode and plate lie in parallel planes; thus the potential-distribution curve for free space is a straight line, and that for the vacuum tube is such that the potential at any point is proportional to the four-thirds power of the distance of the point from the cathode as given by Eq. 21, Ch. III. Poisson's equation, Eq. 3, Ch. III, shows the relation between the *rate of change of slope* of a potential-distribution curve with distance and the net *space-charge density*. The form of Poisson's equation suitable to a tube having parallel-plane symmetry is

$$\frac{d^2e}{dx^2} = -\frac{\rho}{\varepsilon_v},$$ [4]

where

    e is the potential at any point in the tube,

    x is the distance of the point from the origin, taken at the cathode,

    ρ is the space-charge density at the point,

    ε_v is the dielectric constant of free space.

According to this equation, the rate of change of the slope de/dx, with respect to x, is proportional to the net space-charge density. The minus sign in Eq. 4 indicates that a positive space charge causes the slope to decrease with increase of x; that is, it causes the curve to be concave downward. In accordance with this relation, the slope of the free-space curve in Fig. 3 is constant, since the space charge is everywhere zero; the curve for the vacuum diode is concave upward, corresponding to the presence of a negative space charge; and the curve for the gas diode corresponds to the presence of a positive-ion space charge in a region near A.

In order to explain the potential distribution in the gas diode, the changes that occur as a gas is admitted into the tube may be considered. If the gas pressure is very small, the few positive ions produced tend to raise and straighten the potential-distribution curve. The electrons emitted by the cathode then flow across to the anode at an increased rate in order to counteract this tendency, decreasing the potential near the cathode to maintain the field at the cathode equal to zero, as required for space-charge limitation of the current. Thus the few positive ions increase the space-charge-limited current, as indicated by the curves of Fig. 1. If the gas pressure is gradually raised, a rate of positive-ion production is finally attained at which a cumulative process sets in—the positive ions increase the current, and the greater current increases the rate of production of positive ions; and so on, until positive ions are produced at such a great rate that their charges tend to raise the potential at some point in the tube above the anode potential. When this situation exists, the current is no longer limited by space charge, and the volt-ampere characteristic of the tube is that typical of a gas tube—shown in Fig. 2. It is not possible for the potential at a point in the tube to attain a value appreciably greater than the anode potential,* for, as soon as the positive ions tend to form such a potential maximum, the electrons produced simultaneously with the ions are attracted to the point of highest potential (instead of to the anode) and neutralize the excess of positive ions. Thus the flat-topped potential-distribution curve shown in Fig. 3 for the gas diode is formed by the accumulation of positive ions and electrons in the region AB.

The interelectrode space of a conducting gas diode may be divided into two regions of very different characteristics. The first, called a *positive-ion sheath*, or *cathode sheath*, is the region OA of Fig. 3; and the second, called a *plasma*, is the region AB of that figure. In the positive-ion sheath, electrons emitted by the cathode are accelerated by the electric field and are projected into the plasma at high velocity. In a similar way, positive ions that reach the sheath from the plasma are drawn to the cathode. Because of the relatively constant potential of the plasma, the voltage drop across the sheath is essentially equal to that across the tube. This sheath voltage is often called the *cathode voltage drop*. Because of their greater mass, the positive ions attain velocities much smaller than those of the electrons, as indicated by Eq. 2. Thus, even though the positive-ion current is negligible in comparison to the electron current (because many electrons must cross

* See reference 1.

the tube for each positive ion produced, as discussed in Art. 1), the positive ions move so much more slowly than the electrons that in the region *OA*, except in the immediate vicinity of the cathode, the space charge is almost entirely that of the positive ions. Accordingly, the positive-ion current through the sheath is given approximately by the three-halves-power law, Eq. 20, Ch. III, modified because of the greater mass of the positive ions. This current must equal the rate at which positive ions reach the sheath from the plasma; the thickness of the sheath adjusts itself so that this equality results.

The accumulation of positive ions and electrons in the plasma results in nearly zero net space charge. Electrons are projected into the plasma from the sheath and, because of frequent collisions with the gas molecules, acquire a random distribution of velocities, like the velocity distribution among gas molecules. A few of these collisions produce the ions and new electrons that help to maintain the plasma. In addition, the electrons acquire a drift velocity toward the anode because of the small electric field along the plasma. The positive ions similarly have a range of velocities and drift toward the cathode and the walls of the tube, where they are neutralized by combination with electrons. The plasma may be thought of as a mixture of three gases: an electron gas, a positive-ion gas, and a molecular gas. Or it may be considered as very similar to a metallic conductor, in that the electrons move about it in random manner and conduct the current by acquiring drift velocities because of the electric field. The length of the plasma adjusts itself so as to fill the tube, except for the short space required by the positive-ion sheath about the cathode. Changing the anode-to-cathode spacing simply changes the length of the plasma and therefore, because of the small field in the plasma, has little effect on the voltage drop across the tube.

2b. *Cathode Disintegration by Positive-Ion Bombardment.* Because the region of steep potential rise near the cathode has a thickness of the order of a mean free path or less, nearly every positive ion that strikes the cathode does so with an energy corresponding to almost the full anode voltage. The resulting bombardment often destroys the coating of thorium or barium on composite cathodes and causes rapid loss of their efficient emitting properties. However, a threshold value[2] of ion energy, and hence of anode voltage, is found below which the cathode does not lose its activity. Figure 4 shows the results of an experiment on the disintegrating effect of mercury vapor in a diode with a thoriated-tungsten cathode. With increase of anode voltage, the

[2] A. W. Hull, "Gas-Filled Thermionic Tubes," *A.I.E.E. Trans.*, *47* (1928), 753-763; A. W. Hull, "Fundamental Processes in Gaseous Tube Rectifiers," *E.E.*, *69* (1950), 695-700.

current increases at first as the rate at which the emitted electrons are drawn to the anode approaches the rate at which they are emitted thermionically from the cathode. Above about 22 volts, however, the current decreases because of a reduction of the thermionic emission caused by destruction of the thorium coating by the impacts of the positive ions formed in the gas. The emission that occurs at these higher voltages is determined by the fraction of the surface that remains covered as an equilibrium is established between the rate at which the thorium is sputtered away by the bombarding ions and the rate at which it is replaced by the diffusion to the surface of thorium from within the metal. As long as the anode voltage is kept below the critical value of 22 volts with respect to the negative end of the cathode, the positive ions of mercury vapor have no deactivating effect, regardless of the rate at which they strike the cathode. The action of argon, neon, and mercury vapor at other pressures of the order of a fraction of a millimeter of mercury is similar to that in the example shown, except that the critical voltage, termed the *disintegration voltage*, depends on the kind of gas. The disintegration voltages are

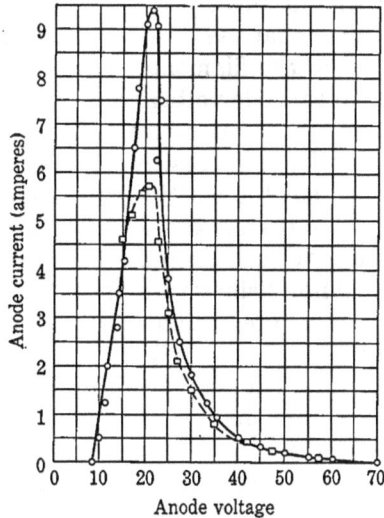

Fig. 4. Disintegrating effect of positive ions on a thoriated-tungsten filament at 1,900 degrees Kelvin in argon at 0.030 millimeter of mercury pressure, as a function of anode voltage. The full curve is for increasing voltage; the dotted curve for decreasing voltage.*

22 volts for mercury, 25 volts for argon, and 27 volts for neon; the positive ions have no injurious effect on the cathode, whether it be oxide-coated nickel or thoriated tungsten, when the anode voltage is kept below these values.

To have reasonable life, gas tubes with composite thermionic cathodes must be so designed and operated that the energy of the ions does not exceed the value corresponding to the disintegration voltage for the gas used. In general, the permissible current is considered to equal the thermionic emission of the cathode in a high vacuum. Particularly if the cathode is oxide-coated, much larger currents than this normal emission can be obtained, but only if the voltage drop is

* This figure is taken from A. W. Hull, "Hot-Cathode Thyratrons, Part I," *G.E. Rev.*, *32* (1929), Fig. 5, p. 215, with permission.

allowed to increase above the disintegration voltage, and the life of the cathode is then shortened. On this basis, the peak-current rating of a gas tube is determined by the thermionic emission of the cathode in the absence of gas. There is evidence,[3] however, that "the safe current rating of a cathode in a gaseous discharge is not a characteristic of the cathode alone but is also dependent on the gas in which the discharge takes place."

In tubes that use directly heated filaments as cathodes, the voltage drop along the filament produced by the heating current causes the anode-to-cathode voltage to vary along the length of the filament; the voltage from the negative end of the filament to the anode is greater than that from the positive end by the amount of the filament voltage. In order that the positive end furnish current to the arc as a cathode, the voltage from it to the anode must be at least as large as the minimum arc voltage drop for the gas used. In order that the negative end shall not disintegrate, the voltage from it to the anode must be less than the disintegration voltage. Thus, the maximum allowable filament voltage is the difference between the disintegration voltage and the minimum arc voltage drop. For example, for mercury-vapor tubes the minimum drop is about 10 volts, although it varies somewhat with the pressure of the mercury vapor, which is a function of the ambient temperature. Since the disintegration voltage is 22 volts, the filament voltage should not exceed 12 volts. If an alternating filament voltage is used, its effective value should not exceed $12/\sqrt{2}$, or 8.5 volts; it is seldom made greater than 5 volts. A second limitation on the filament voltage is set by the danger of an arc's striking between the ends of the filament and burning it out, but the first limitation usually predominates. Indirectly heated cathodes are not subject to these limitations, and with them a heating voltage having an effective value of 115 volts is sometimes used.

2c. *Emission Efficiency of the Cathode.* Although the addition of gas does not directly affect the current capacity of a thermionic tube, it does increase the practical efficiency of emission realizable by proper design of the cathode. In a vacuum tube, the surface of the emitter must be directly exposed and placed close to the anode in order that the space-charge effect be small; thus the cathode loses heat by radiation from the whole of the surface that emits electrons. These considerations apply to the indirectly heated cathodes in Fig. 12, Ch. II. In a gas tube, however, the space charge is neutralized, a thin sheath of positive ions exists near the cathode surface, and after passing through this sheath the electrons drift to the anode through the

[3] C. G. Found, "A New Method of Investigating Thermionic Cathodes," *Phys. Rev.*, *45* (1934), 524.

plasma. If an emitter is built in the form of a cavity with its internal surface oxide-coated, the cathode sheath covers the inner surface, and the plasma extends into the cavity. As a consequence, electrons can emerge from holes or crevices, and the spacing between anode and cathode becomes of little consequence, since it determines only the relatively unimportant drop through the plasma. Radiant heat, on the other hand, is propagated only in straight lines. If a cathode is constructed so that its effective heat-radiating area is small compared with its internal oxide-coated electron-emitting area, the power loss by radiation occurs over an area smaller than that available for electron emission, and the efficiency of thermionic emission is increased. Even coiling, corrugating, or crimping an ordinary oxide-coated ribbon filament, as is illustrated in Figs. 5a and 5b, increases its emission efficiency because of reflections between adjacent parts of the filament. Other types of high-efficiency cathodes are shown in Fig. 5. In Fig. 5f, heat-shielding cylinders are placed around the cathode, and fins inside the cylinders are coated to increase the emitting surface. The two external cylinders reflect heat inward and reduce the heat radiation, while the fins add much to the emitting area but little to the heat-radiating area. Heat-shielded cathodes having an emission current rating of 600 amperes and a heater rating of 80 amperes at 5 volts have been constructed. The corresponding emission efficiency is 1,500 milliamperes per watt, whereas a conservative rating of a plain oxide-coated ribbon cathode for use in gas tubes is 100 milliamperes per watt.

The cathode in a gas-filled thermionic tube *must be allowed to reach its normal operating temperature* before the plate circuit is closed, lest the voltage drop across the tube exceed the disintegration voltage because of the inability of the cathode to supply the emission current demanded by the load. The heating-time requirement makes necessary a considerable delay in putting tubes with large efficient cathodes into operation; for example, the heating time of the heat-shielded cathodes mentioned above is 30 minutes.

3. EFFECTS OF GAS PRESSURE ON THE CHARACTERISTICS OF A GAS DIODE

The pressure of the gas (strictly, the gas concentration) in a diode affects the operating characteristics of the tube in three ways. It influences the rate of evaporation of the cathode, the drop in the tube while conducting, and the breakdown voltage in the reverse direction. The practical range of operating pressure is set by these considerations.

If the pressure is high enough, it decreases the rate of evaporation

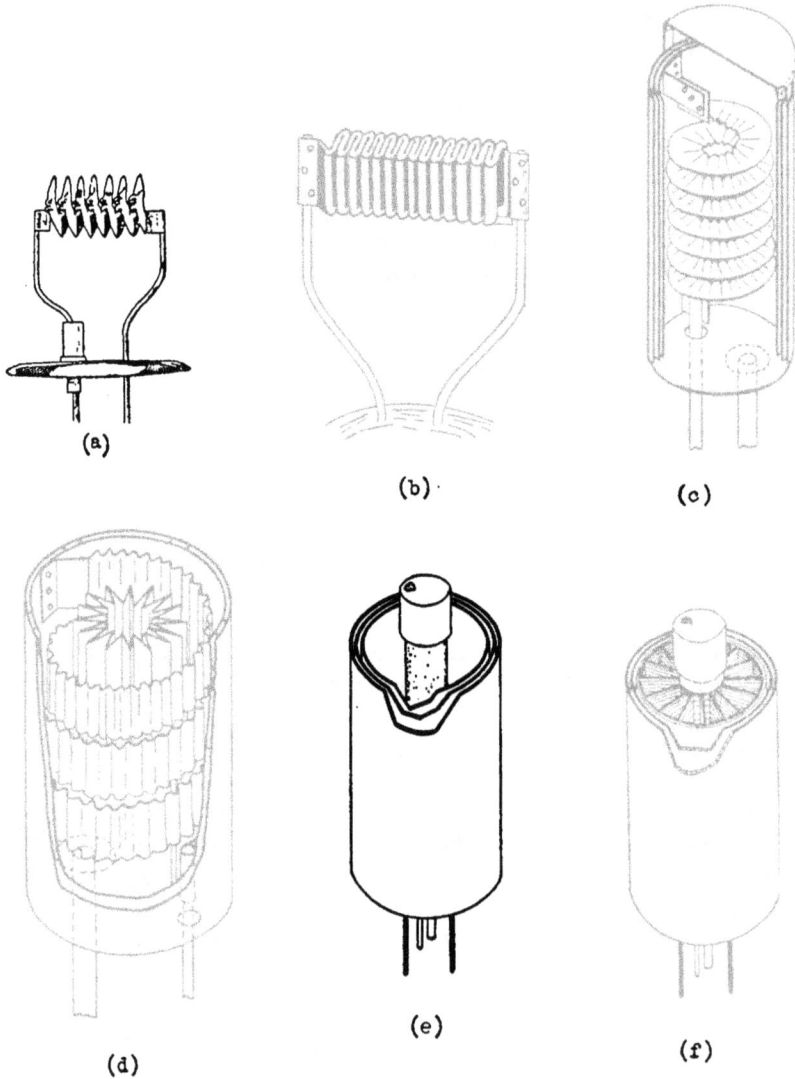

Fig. 5.   Heat-conserving cathodes for gas tubes.*

* Figures 5a, 5b, 5c, and 5d are adapted from E. F. Lowry, "Thermionic Cathodes for Gas-Filled Tubes," *Electronics*, 6 (October, 1933), Figs. 1, 2, and 3, pp. 280–281, with permission.   Figures 5e and 5f are adapted from A. W. Hull, "Gas-Filled Thermionic Tubes," *A.I.E.E. Trans.*, 47 (1928), Figs. 10 and 11, p. 757, with permission.

of the cathode and permits operation of the cathode at a higher temperature and emission efficiency. Commercial battery-charging rectifier tubes such as the Tungar[4] or Rectigon are of this type; they contain a thoriated-tungsten filament in argon gas at a pressure of about 5 centimeters of mercury. Because of the relatively high pressure, the temperature of the filament can be raised considerably above that which would give a reasonable life in vacuum, and the cathode efficiency is increased by a large factor. The increase of efficiency is accompanied by a tendency for the discharge to concentrate and heat the filament locally with resultant burn-out. In high-pressure mercury-vapor lamps, oxide-coated cathodes are used with no heating current. The arc concentrates on a small section of the coated surface and maintains its temperature, but the high pressure prevents evaporation or sputtering.

Although a high gas pressure is desirable in a thermionic rectifier tube because it reduces the rate of evaporation of the cathode, it is undesirable for the reason that it decreases the ability of the tube to withstand anode voltage of negative polarity. If the pressure is too high, breakdown followed by a glow or arc discharge may take place in a reversed direction during the inverse part of the cycle of alternating voltage across the tube. In these circumstances, the tube conducts in both directions and ceases to rectify. The inverse breakdown voltage follows Paschen's law and the breakdown-voltage curve, Fig. 11, Ch. III, and therefore decreases as the pressure is increased, until a minimum value is reached. In the Tungar and Rectigon tubes, the pressure is sufficient to cause an inverse breakdown voltage near the minimum value; these tubes consequently have a peak-inverse-voltage rating of only a few hundred volts and are not suitable for rectifying high voltages.

The pressure of the gas must not be reduced too far in an effort to increase the inverse voltage rating of a gas tube. If the pressure is too low, enough gas molecules may not be present to provide the positive ions needed to neutralize the space charge when the tube is conducting its rated current; a high voltage drop, instability,[5] and disintegration of the cathode will then result. At such low gas pressures, a lower voltage drop is obtained if the current is reduced, so that it is still practicable to operate the tube at a value of current smaller than its normal rating. Thus for every gas-filled thermionic tube there exists an operating range of pressure above which the allowable peak

[4] G. S. Meikle, "Hot-Cathode Argon-Gas-Filled Rectifier," *G.E. Rev.*, *19* (1916), 297–304; R. E. Russell, "The Tungar Rectifier," *G.E. Rev.*, *20* (1917), 209–215.

[5] A. W. Hull, "Fundamental Processes in Gaseous Tube Rectifiers," *E.E.*, *69* (1950), 698.

inverse voltage of the tube is lowered, and below which the allowable peak anode current is lowered.

The mercury vapor in a hot-cathode mercury-vapor tube is at an equilibrium pressure dependent upon the temperature of the liquid mercury. The pressure therefore depends upon the temperature of the coolest part of the tube, which is the place at which the mercury vapor condenses. The relation between the pressure in the tube and the temperature of the condensed mercury is shown in Fig. 6. Since the inverse

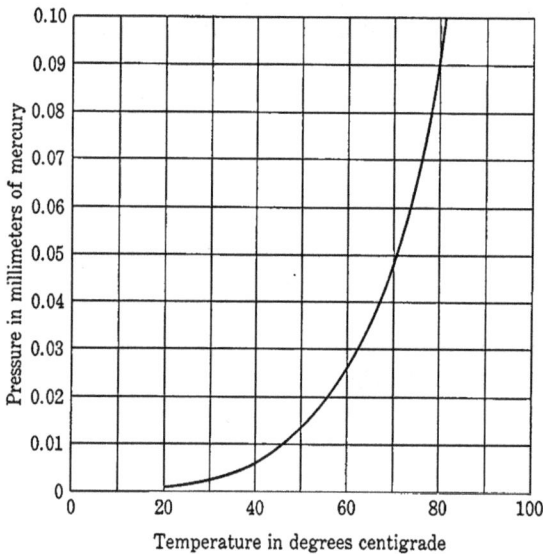

Fig. 6. Pressure of mercury vapor as a function of the condensed-mercury temperature.*

voltage that the tube will withstand depends upon the pressure of the mercury vapor, it depends also upon the temperature of the condensed mercury, in the manner shown for a particular tube in Fig. 7. For another tube, the curves of Fig. 8 show not only the dependence of the permissible inverse voltage on the mercury temperature but also the variation of the tube drop and the permissible peak anode current with this temperature. From these curves, it is apparent that the tube can withstand only a low inverse voltage at temperatures above 75 degrees centigrade, and that the lower the temperature of the mercury, the higher the voltage the tube can rectify. At temperatures below about 25 degrees centigrade, however, the peak current that may be drawn from the tube is less than the rated peak current of 600 amperes,

* Data for this curve are taken from *International Critical Tables* (New York: McGraw-Hill Book Company, Inc., 1929), Vol. 3, p. 206, with permission.

since otherwise the tube drop would exceed the disintegration value of 22 volts. Presumably, allowance of a safety factor is the reason why the curves of permissible peak plate current show current values some-what less than those that cause a tube voltage drop equal to the disintegration voltage.

Gas tubes in which the inert gases argon, neon, helium, krypton, and xenon are used possess an advantage over mercury-vapor tubes in that the concentration of the gas is not dependent upon temperature.

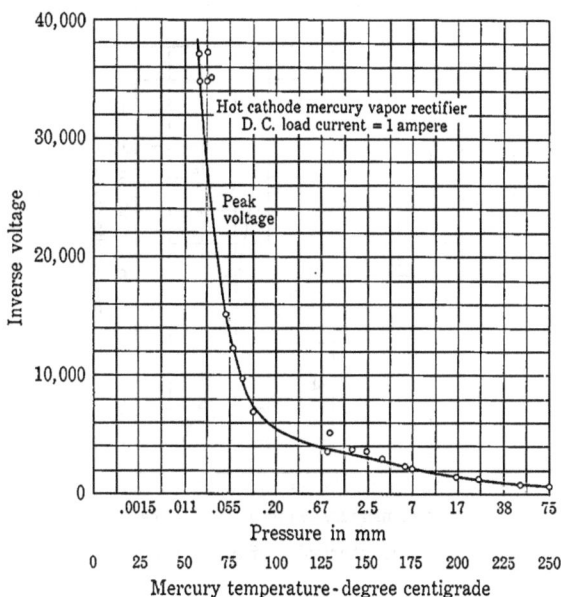

Fig. 7.   Inverse breakdown voltage of a particular gas diode as a function of the mercury-vapor pressure and temperature.*

However, the concentration of the gas in these tubes does change appreciably but slowly during the life of the tubes because of a *clean-up* action that occurs—the gas disappears by being trapped in the walls and electrodes of the tube as a result of the action of the discharge. The inert gases are not generally used in high-voltage gas tubes because the gas pressure in these tubes must remain within narrow limits, and the life of such a tube is therefore short because of the clean-up phenomenon. In low-voltage tubes, the inert gases are often used, for the pressure is not required to remain within such narrow limits, and enough gas to allow for clean-up can be introduced initially.

* This figure is taken from A. W. Hull, "Gas-Filled Thermionic Tubes," *A.I.E.E. Trans.*, 47 (1928), Fig. 13, p. 758, with permission.

4. TUBES COMPRISING MERCURY-POOL CATHODES

When large currents are to be conducted, mercury-pool cathodes are usually used in gas tubes. Since the mercury pool is not of itself a source of electrons, some means (such as those discussed subsequently in connection with special mercury-arc tubes) must be employed to start the arc. Once started, the arc in such a tube concentrates on a small bright spot that moves erratically over the surface of the mercury pool. The mercury-pool cathode then serves a double purpose; it supplies not only the electrons but also the mercury vapor through

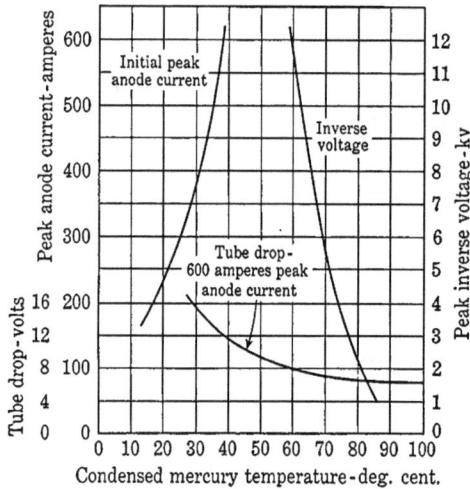

Fig. 8.   Characteristics of a gas diode.*

which the current is conducted. The cathode is self-restoring, for the tube can be designed so that the mercury evaporated from the cathode by the heat resulting from the arc is caused by gravity to run back to the pool after condensation. Disintegration of the cathode is therefore not a factor that limits the current, and heavy overload currents can be drawn momentarily without injury.

The *cathode spot* formed by the arc at the surface of the mercury pool is the source of a tremendous emission current density of electrons— it is estimated to be about 4,000 amperes per square centimeter. When the total current is large, several spots often appear, each carrying about 30 to 40 amperes.[6] The phenomena that give rise to such a high rate of emission of electrons are not clearly understood.

* This figure is taken from H. C. Steiner, A. C. Gable, and H. T. Maser, "Engineering Features of Gas-Filled Tubes," *E.E.*, *51* (1932), Fig. 2, p. 313, with permission.

[6] E. D. McArthur, *Electronics and Electron Tubes* (New York: John Wiley & Sons, Inc., 1936), 137.

Thermionic emission is believed to be a minor factor in accounting for the liberation of electrons, as the temperature of the spot on the surface of such a good heat conductor as mercury could hardly be high enough with the power input involved.[7] Furthermore, the rate of evaporation of the mercury is not sufficient to indicate a high temperature. Bombardment by the relatively low-voltage positive ions present is considered to be a process too inefficient to play an important role in causing the electron emission. However, a plasma and positive-ion sheath form in the arc just as in the discharge in the thermionic gas diode, and the potential distribution in the arc is as shown in Fig. 9.

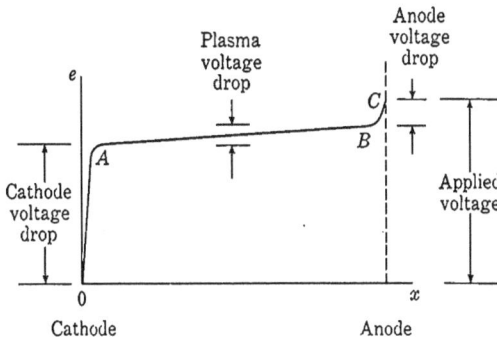

Fig. 9.   Potential distribution in the arc of a tube with a mercury-pool cathode.

It is thought that, as a result of this potential distribution, the electric field at the cathode surface is sufficiently great to pull the electrons out of the liquid metal by the process of field emission, discussed in Art. 11, Ch. II.

The arc in tubes with mercury-pool cathodes may be divided into three parts: the *cathode sheath* (or positive-ion sheath about the cathode), shown as region *OA* in Fig. 9; the *plasma*, shown as region *AB*; and the *anode sheath* (or sheath of electrons about the anode), shown as region *BC*. The cathode sheath is essentially the same as that in a thermionic gas diode (see Art. 2a); it consists of positive ions moving from the plasma to the cathode and electrons moving in the reverse direction. Though the electron current exceeds that of the positive ions, the electrons have a negligible effect on the space charge in the sheath. The cathode voltage drop is found[8] to be about 9.9 volts over a wide range of currents and pressures. Since the work function of mercury is 4.5 volts, and electrons liberated by field emission are

[7] K. T. Compton, "The Electric Arc (Abridgment)," *A.I.E.E.J.*, *46* (1927), 1194–1196.

[8] E. S. Lamar and K. T. Compton, "Potential Drop and Ionization at Mercury Cathode," *Phys. Rev.*, *37* (1931), 1069–1076; K. T. Compton, "On the Theory of the Mercury Arc," *Phys. Rev.*, *37* (1931), 1077–1090.

obtained from very nearly the W_i energy level in the metal, only 9.9 — 4.5, or 5.4, volts are available to accelerate the electrons after emission and give them the kinetic energy that results in the production of mercury ions. The fact that this voltage is lower than the minimum ionizing potential of 10.4 volts indicates that ionization takes place by multiple impact in the manner described in Art. 5, Ch. III. At low currents, however, the single-impact ionization predominates and the voltage drop rises as the current decreases.

The plasma in the mercury arc, like that in the thermionic gas diode, is a region of low potential gradient. In it, positive ions and electrons

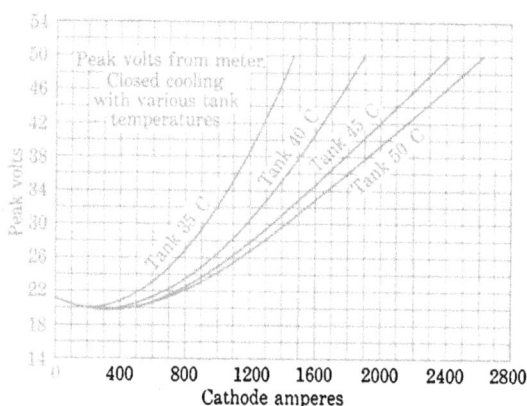

Fig. 10. Voltage drop in the arc of a mercury-arc rectifier.*

are present in approximately equal numbers, and the potential gradient has a value between about 0.05 and 0.2 volt per centimeter at the usual operating pressures and current densities. The plasma serves chiefly as a highly conducting path for the main length of the arc.

Owing to their random motion in the plasma, electrons tend to strike the anode at a rapid rate. If this rate is sufficient to supply the anode current to the external circuit, the plasma extends to the anode. If the anode current is greater than this value, however, an increased potential gradient develops in the region *BC* of Fig. 9 near the anode to draw more electrons from the plasma. This region of increased potential gradient is the anode sheath. Very few positive ions are found in the sheath, for the electric field opposes their motion from the plasma toward the anode. Accordingly, the anode sheath is a region of space-charge-limited flow of electrons. The *anode voltage drop*, which is the voltage across this sheath, is the principal reason for an increased

* This figure is taken from A. W. Hull and H. D. Brown, "Mercury-Arc Rectifier Research," *A.I.E.E. Trans.*, *50* (1931), Fig. 14, p. 750, with permission.

over-all arc voltage drop at high currents. Figure 10 shows typical curves of the total voltage drop (which is of the order of 25 volts) as a function of current and mercury-vapor pressure. The rise at low currents results for the most part from an increased cathode drop, and the rise at high currents is caused principally by an increase of the anode drop.

The mercury-pool cathode is an exceedingly rugged source of electrons, since the current that can be drawn from it is practically unlimited. Tubes in which it is employed generally have a large momentary overload capacity, and their ratings are usually set by the heating of the anode and other parts, rather than by the properties of the cathode. Since a large fraction of the power represented by the product of the cathode voltage drop and the current is converted to heat at the cathode spot, mercury is evaporated rapidly from the surface of the pool. The random motion of the spot over the surface of the pool is attributed to the high vapor pressure thus developed over the spot. To maintain a low pressure in the tube and a consequent high inverse-voltage strength, the mercury vapor must be condensed as rapidly as it is produced. This condensation is accomplished either by the air cooling of a large glass condensing chamber, such as that shown in Fig. 11, or by water cooling of the walls of a steel-walled

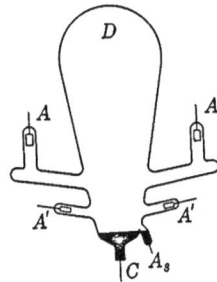

Fig. 11.   Glass-bulb mercury-arc rectifier.

tube, such as those shown in Figs. 12 and 13. The rate at which the mercury may be condensed is an important limitation on the allowable average current of the arc.

4a. *Mercury-Arc Rectifier.* In all tubes employing mercury-pool cathodes, some provision must be made for starting the arc, for it cannot be started simply by the application of normal anode voltage. Furthermore, with certain methods of starting, provision must also be made for maintaining the arc, for it will be suddenly extinguished if the current falls below about one ampere. Generally, some form of auxiliary starting electrode is provided that initiates an arc at the surface of the mercury pool, whereupon the arc transfers to the main anode. The rectifiers with mercury-pool cathodes described in this and the following article are characterized by the means employed to start the arc and to maintain it continuously when an alternating voltage is supplied to the anode.

Early mercury-arc rectifiers were built in glass bulbs. The advantage of glass is that, since the walls are insulating, the electrode leads may be sealed into the glass and a vacuum-tight tube may be obtained.

Thus the complete rectifier may be permanently evacuated. Figure 11 shows the construction of a two-anode glass-bulb rectifier. The excitation anodes A' and A' are connected to a separate transformer with a small fixed load and serve only to prevent extinction of the arc when the current to the main anodes A and A falls to zero. To start the arc, the bulb is tipped until the mercury of the cathode C makes contact with that in the arm of the ignition electrode A_s. A current is made to flow through this contact, and the bulb is tipped back again. The arc that results when the contact breaks is picked up and maintained by the excitation anodes.

Steel-tank rectifiers were developed to overcome the cooling and mechanical limitations of glass-bulb rectifiers. Figure 12 shows a cross section of a multi-anode steel-tank rectifier. The rectifier is kept at the proper temperature by a system of water jackets and internal coils as shown. The circulating water may be either cooled or heated, by external means, to maintain the proper temperature.

The construction of the main and exciting anodes is shown in Fig. 12. The ignition anode 17 consists of a rod extending to within a fraction of an inch of the mercury pool. A solenoid 3, external to the vacuum chamber, drives the ignition anode into the mercury pool, and a spring draws it back out. The ignition anode is supplied with a small direct voltage so that an arc results when the anode leaves the mercury pool. This arc is immediately transferred to the excitation anodes. The ignition anode is sometimes used as an excitation anode in small rectifiers. The tank is insulated from the cathode as well as from the anodes so that the arc cannot terminate on the tank and melt a hole through the steel. Pumping equipment shown at 26 and 35 in Fig. 12 is provided to remove gas that leaks through the insulating seals at the electrodes and evolves from the electrodes and internal wall surfaces.

Steel-tank rectifiers have been made in a wide range of ratings and are used for the rectification of large amounts of power. For example, the rectifier illustrated in Fig. 12 has a power rating of 3,000 kilowatts and an output voltage of 600 volts, and higher ratings are common. An output voltage rating of 20,000 is about the highest yet found practical. For the reasons explained in Ch. VI, steel-tank rectifiers are generally used with polyphase alternating-voltage supplies and frequently have six or twelve main anodes. Conduction takes place from the cathode to each of the anodes in turn during the successive parts of a cycle of the supply voltage. When conduction takes place to one anode all others are at large negative voltages with respect to it, and arc-back, or conduction from anode to anode, must be inhibited. For this purpose, each anode is surrounded by a shield and a baffle,

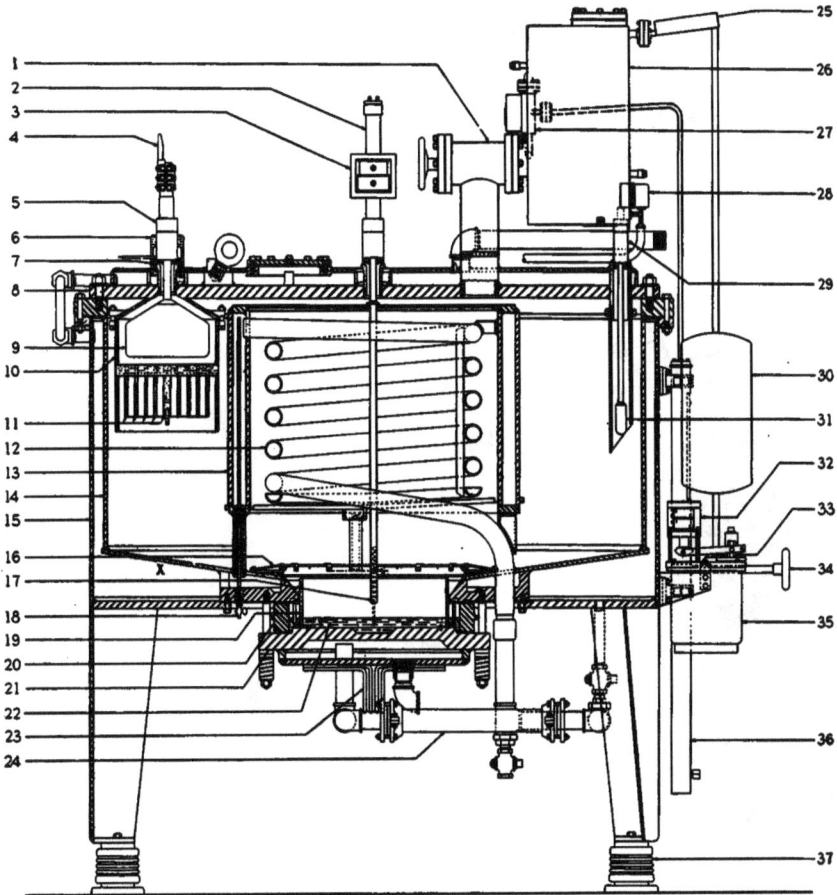

1.	Manual vacuum valve	14.	Vacuum tank	28.	Thermal relay for mercury
2.	Starting-anode armature	15.	Water jacket		condensation pump
	sleeve	16.	Mercury separator	29.	Excitation-anode
3.	Starting-anode solenoid	17.	Starting-anode tip		insulating seal
	winding and yoke	18.	Air vent	30.	Gas receiver tank
4.	Main-anode terminal	19.	Quartz arc shields	31.	Excitation-anode tip
5.	Main-anode insulating seal	20.	Cathode insulator	32.	Rotary-pump-valve
6.	Main-anode heater cover	21.	Cathode plate		solenoid
7.	Main-anode heater	22.	Cathode mercury	33.	Rotary-pump valve
8.	Tank cover plate	23.	Cathode terminal	34.	Vacuum gauge operating
9.	Main-anode tip	24.	Cathode insulating pipe		hand wheel
10.	Baffle cylinder	25.	Air-cooled mercury trap	35.	Rotary vacuum pump
11.	Baffle	26.	Mercury condensation	36.	Vacuum gauge
12.	Internal cooling coil		pump	37.	Rectifier insulators
13.	Internal cooling cylinder	27.	Vacuum detector		

Fig. 12.   Vertical section and list of parts of a twelve-anode steel-tank rectifier having an output of 3,000 kilowatts at 600 volts. (*Courtesy General Electric Company.*)

shown as 10 and 11, respectively, in Fig. 12. The problem of preventing arc-back is, however, not yet completely solved.[9]

Fig. 13. Ignitron tube, which employs a mercury-pool cathode. (*Courtesy Westinghouse Electric Corporation.*)

Fig. 14. Internal view of an ignitron tube similar to the one in Fig. 13. The ignitor appears inside a ring near the bottom. Fins on the anode lead cool the anode. (*Courtesy Westinghouse Electric Corporation.*)

4b. *Excitron.* The *excitron*[10] is a single-anode steel-tank rectifier having a mercury-pool cathode, a main anode, an excitation anode, a grid, and a solenoid-operated mercury-jet device for starting the arc. Energizing the solenoid directs a mercury jet upward from the pool

[9] A. W. Hull and H. D. Brown, "Mercury-Arc Rectifier Research," *A.I.E.E. Trans.*, *50* (1931), 744–756; J. Slepian and L. R. Ludwig, "Backfires in Mercury-Arc Rectifiers," *E.E.*, *50* (1931), 793–796; A. W. Hull, "Fundamental Processes in Gaseous Tube Rectifiers," *E.E.*, *69* (1950), 698.

[10] H. Winograd, "Development of Excitron-Type Rectifier," *A.I.E.E. Trans.*, *63* (1944), 969–978.

to the excitation anode located close to the mercury surface. When the jet stream breaks, an arc is formed and a cathode spot is maintained by the auxiliary source of direct current supplying the excitation anode. Current to the main anode then occurs under control of the grid much as in thyratrons, which are described in Art. 6.

4c. *Ignitron.* Figures 13 and 14 show an *ignitron*, which is a single-anode mercury-arc rectifier with a pool-type cathode in which the arc is started by a method different in principle from that described in the previous articles.[11] The tip of an ignitor, a rod made of a semiconductor such as boron carbide, which has a relatively high electrical resistivity, is immersed in the mercury-pool cathode. When a current of about 30 amperes is passed through the ignitor to the pool by an auxiliary source of about 200 volts, a cathode spot forms at the surface of the pool. If the anode is positive at the time, an arc starts between the cathode and the anode, and the tube conducts. It is, of course, necessary to initiate the arc at the beginning of each conducting period. Ignitrons are built with single anodes only—a construction that reduces the likelihood of arc-back, since the only unwanted conduction that can take place is that between the anode and cathode when the anode is at a negative potential. They are available in either glass or permanently evacuated metal envelopes, some of which are water cooled. Ignitrons are used extensively to control the heating current in resistance welders. They are especially suited to this application, since the tube must supply current in short pulses having an extremely large peak value and a much smaller average value.

Ignitrons and excitrons in large sizes supplant multiple-anode steel-tank rectifiers in many applications. Figure 15 shows a group of six ignitrons arranged for rectification of power from a six-phase supply. A cross-section view of one of these ignitrons is shown in Fig. 16. Ignitrons capable of withstanding high inverse voltage are constructed with several voltage-dividing grids or electrodes between the cathode and anode.

4d. *Capacitron.* An electrode separated from the mercury pool by a thin layer of insulating material is used for starting a mercury arc in a variety of gas tubes. Sudden application of a relatively high voltage between the electrode and the pool causes formation of a cathode spot.[12] Such a tube has been called a *capacitron.* A glass-covered metal electrode immersed in or floating on the pool is used for starting the

---

[11] J. Slepian and L. R. Ludwig, "A New Method for Initiating the Cathode of an Arc," *A.I.E.E. Trans.*, *52* (1933), 693–700; D. E. Marshall and W. W. Rigrod, "Characteristics of Resistance Ignitors," *Electronics*, *20* (May, 1947), 122–126.

[12] M. A. Townsend, "Cathode Spot Initiation on a Mercury Pool by Means of an External Grid," *J. App. Phys. 12* (1941), 209–215.

arc in one type of tube.[13] Another uses the glass wall of the tube as the insulation. The starting electrode is then outside the tube and is called a *band-igniter*. Band-igniter mercury-pool tubes are used for the control of resistance welding.[14] A band igniter is also suitable for control of the starting of an arc in a cold-cathode tube filled with an

Fig. 15. Six-phase ignitron mercury-arc rectifier rated at 3,000 kilowatts and 625 volts. (*Courtesy General Electric Company.*)

inert gas, and is employed in tubes for flash and stroboscopic photography.[15]

5. EFFECT OF GAS IN A THERMIONIC TRIODE

In the treatment of the triode in Ch. IV, it is assumed that, when the grid is at a negative potential with respect to the cathode, no

[13] K. J. Germeshausen, "A New Form of Band Igniter for Mercury-Pool Tubes (Letter)," *Phys. Rev.*, 55 (1939), 228; Hans Klemperer, "Dielectric Igniters for Mercury-Pool-Cathode Tubes," *Electronics*, 14 (November, 1941), 38–41.

[14] T. S. Gray and W. B. Nottingham, "Half-Cycle Spot-Welder Control," *R.S.I.*, 8 (1937), 65–68; T. S. Gray and J. Breyer, Jr., "An Electronic Control Circuit for Resistance Welders," *A.I.E.E. Trans.*, 58 (1939), 361–364.

[15] H. E. Edgerton and J. R. Killian, Jr., *Flash!—Seeing the Unseen by Ultra-High-Speed Photography* (Boston: Hale, Cushman & Flint, 1939), 194–203 (includes an extensive bibliography); P. M. Murphy and H. E. Edgerton, "Electrical Characteristics of Stroboscopic Flash Lamps," *J. App. Phys.*, 12 (1941), 848–855.

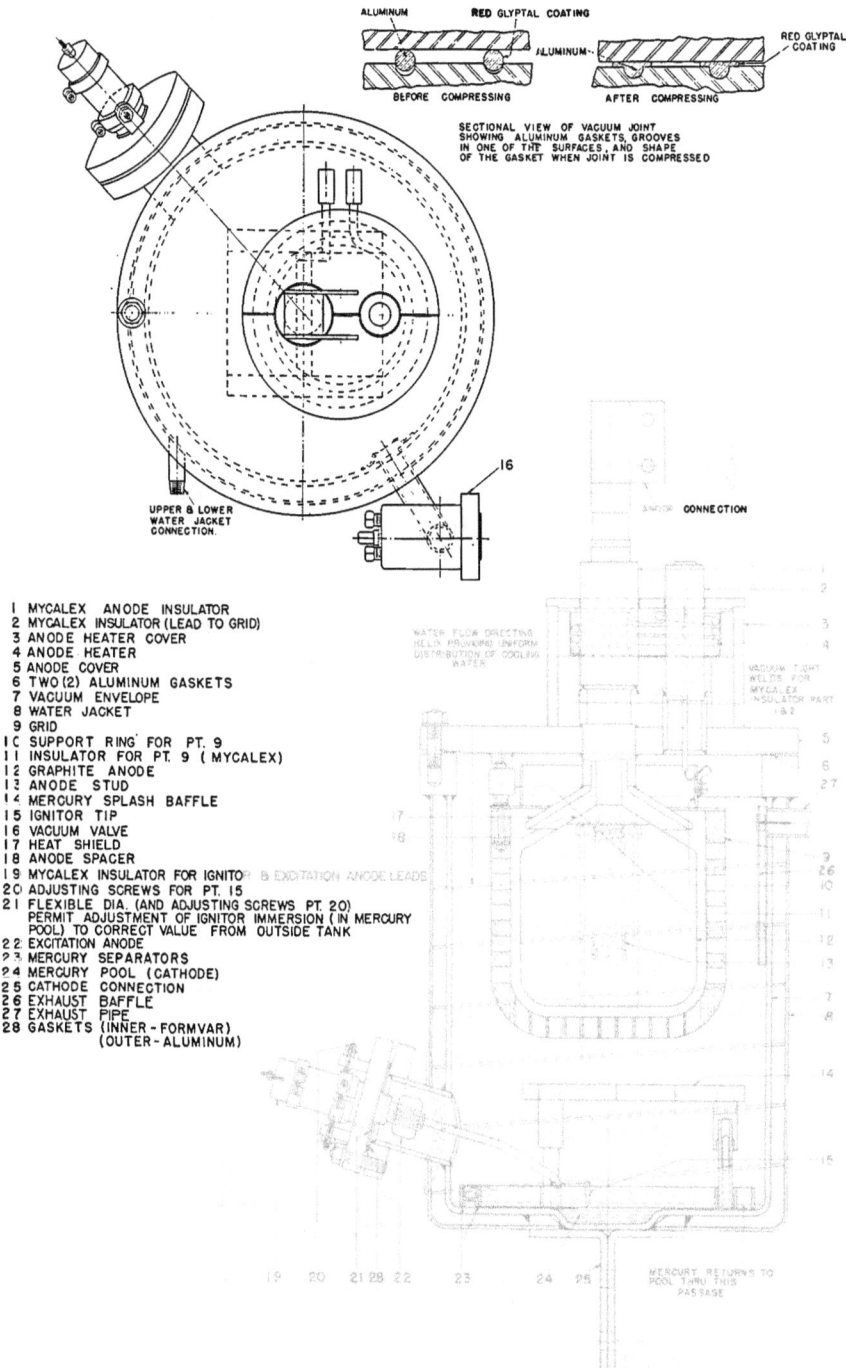

1 MYCALEX ANODE INSULATOR
2 MYCALEX INSULATOR (LEAD TO GRID)
3 ANODE HEATER COVER
4 ANODE HEATER
5 ANODE COVER
6 TWO (2) ALUMINUM GASKETS
7 VACUUM ENVELOPE
8 WATER JACKET
9 GRID
10 SUPPORT RING FOR PT. 9
11 INSULATOR FOR PT. 9 (MYCALEX)
12 GRAPHITE ANODE
13 ANODE STUD
14 MERCURY SPLASH BAFFLE
15 IGNITOR TIP
16 VACUUM VALVE
17 HEAT SHIELD
18 ANODE SPACER
19 MYCALEX INSULATOR FOR IGNITOR & EXCITATION ANODE LEADS
20 ADJUSTING SCREWS FOR PT. 15
21 FLEXIBLE DIA. (AND ADJUSTING SCREWS PT. 20)
    PERMIT ADJUSTMENT OF IGNITOR IMMERSION (IN MERCURY
    POOL) TO CORRECT VALUE FROM OUTSIDE TANK
22 EXCITATION ANODE
23 MERCURY SEPARATORS
24 MERCURY POOL (CATHODE)
25 CATHODE CONNECTION
26 EXHAUST BAFFLE
27 EXHAUST PIPE
28 GASKETS (INNER - FORMVAR)
         (OUTER - ALUMINUM)

Fig. 16.   Vertical section and list of parts of one tank of the ignitron mercury-arc
rectifier of Fig. 15.   (*Courtesy General Electric Company.*)

current flows to the grid, because the electrons are then repelled from it. However, a negative grid does attract positive ions, so that the presence of a slight amount of gas may give rise to a considerable grid current. Measurement of the current to a negative grid provides a sensitive test of vacuum conditions and is used for this purpose by tube manufacturers. In order that a triode may operate efficiently as an amplifier, a high degree of vacuum must be maintained to prevent the formation of positive ions.

Increased grid current, when the grid is negative, is the first manifestation of gas that becomes apparent as the pressure in a triode is increased. Further increase of pressure results in the occurrence of the same phenomena as in diodes—appreciable neutralization of space charge, with departures of the plate characteristics from their normal form, takes place. At higher pressures, the negative space charge becomes completely neutralized in most of the space—a plasma develops—and a positive-ion sheath forms around the cathode. When the plasma forms, the grid almost entirely loses its effectiveness in controlling the plate current, and the behavior of the tube is completely altered.[16] It becomes a *gas triode*, or *thyratron*, and is useful in a manner entirely different from that in which the vacuum triode is useful.

The total conduction grid current of any thermionic triode consists of a number of components,[17] of which the more important are current resulting from:

(a) electrons collected by the grid,

(b) positive ions collected by the grid,

(c) conduction along the insulation-resistance paths between the grid and other electrodes of the tube,

(d) secondary emission from the grid,

(e) photoelectric emission from the grid,

(f) thermionic emission from the grid,

(g) bombardment of the grid by the soft x-rays produced at the plate.

In gas tubes, the total grid current is often of the order of milliamperes. The first two of the above components of grid current predominate. In the operation of vacuum tubes, it is sometimes extremely important to keep the total grid current small—less than 10^{-15} ampere in certain

[16] A. W. Hull, "Hot-Cathode Thyratrons," *G.E. Rev.*, *32* (1929), 213–223, 390–399.

[17] G. F. Metcalf and B. J. Thompson, "Low-Grid-Current Vacuum Tube," *Phys. Rev.*, *36* (1930), 1489–1494; H. B. Michaelson, "Variations of Grid Contact Potential and Associated Grid Currents," *J.F.I.*, *249* (1950), 455–473.

applications—with the result that precautions must be taken in the design of the tube to keep any of the above components of grid current from becoming excessive.

The variation of the first three components of grid current in a typical vacuum tube may be illustrated as in Fig. 17. Both the plate current i_b and the grid current i_c are plotted for a constant plate voltage as functions of the grid voltage e_c. The scales for the two currents, however, are not the same; for, while the plate current may be of

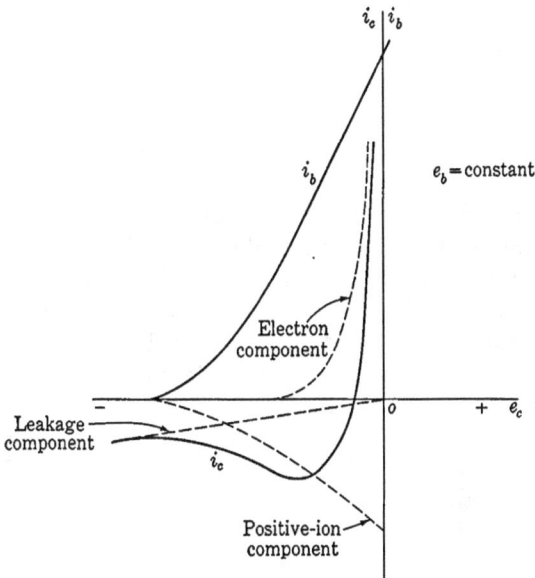

Fig. 17.   Grid current resulting from electrons, positive ions, and leakage.

the order of milliamperes, the grid current may be of the order of microamperes or less. The component of the grid current caused by the electrons that strike the grid is shown as a dotted curve in Fig. 17. Because of the initial velocities of the electrons as they leave the cathode and the contact potential difference between the grid and the cathode, electrons may reach the grid even though it is negative with respect to the cathode. The contact potential difference, which depends on the material and surface conditions of the two electrodes, has the effect of causing a shift of the whole curve to the right or left by a volt or so, as is discussed in Art. 1, Ch. IV. Since the initial velocities of the electrons are distributed over a wide range, the current approaches zero gradually as the grid potential is made increasingly negative with respect to the cathode. The *average* kinetic energy[18] of the therm-ionically emitted electrons associated with their component of velocity in a direction normal to the cathode surface is

$$\frac{kT}{Q_e} = \frac{T}{11,605}$$ [5]

electron volts, where T is the temperature of the cathode in degrees Kelvin. Consequently, the electrons emitted from a filament with a temperature as high as 2,400 degrees Kelvin have an average energy associated with their normal component of velocity of only about 0.2 electron volt. As is explained in Art. 1, Ch. IV, the number of electrons having higher energies drops off exponentially as the energy possessed by them increases, so that a negative grid voltage of one or two volts is generally sufficient to reduce the electron component of the grid current to a negligible value.

The component of the total grid current that passes through the insulation resistance in the vacuum tube, termed the leakage current, is shown as a second dotted curve in Fig. 17. The path of the leakage current is both inside and outside the tube across the glass and other insulating surfaces that separate the electrodes and separate their leads. In the diagram, the leakage is considered to exist only between the grid and the cathode, and the current is therefore approximately a linear function of the grid voltage. Under some conditions, leakage from the plate to the grid adds a relatively constant value to this component. The leakage currents would be measurable when the cathode is cold, were it not for the effect of changes in the conductivity of the glass and other insulators brought about by the higher temperatures used in operation. Unless special precautions are taken to diminish them, the leakage currents alone may be appreciable in some applications of vacuum tubes.

The component of the total grid current caused by the positive ions collected by the grid is shown as the third dotted curve in Fig. 17. This current is a function of the rate of ionization occurring in the tube. For small pressures and constant plate voltage, it is almost directly proportional to the electron current that causes the ionization and to the pressure. In fact, the linearity of the relation between grid current and pressure is utilized in the ionization gage[19] in which the electron current to a positive electrode in the form of a grid is held constant by

[18] L. H. Germer, "The Distribution of the Initial Velocities among Thermionic Electrons," *Phys. Rev.*, *25* (1925), 795–807; W. B. Nottingham, "Thermionic Emission from Tungsten and Thoriated-Tungsten Filaments," *Phys. Rev.*, *49* (1936), 78–87.

[19] O. E. Buckley, "An Ionization Manometer," *Nat. Acad. Sci. Proc.*, *2* (1916), 683–685.

adjustment of the cathode temperature, and the current to a collector at a fixed negative potential is assumed to be directly proportional to the pressure. The proportionality between the positive-ion component of the grid current and the electron plate current i_b, at constant pressure, is shown by the curve in Fig. 17, the positive-ion current being zero when the plate current is zero. Since an electron current to the grid is arbitrarily called a positive current, the positive-ion current is negative. It is therefore plotted downward, and is an image of the plate current about the horizontal axis. The total grid current is the algebraic sum of the three components and a typical curve of it has the form shown as i_c in Fig. 17.

An increase of the gas pressure in the tube causes the positive-ion component of grid current to increase but leaves the electron component and the leakage components the same. The positive-ion component would be directly proportional to the pressure, but for the fact that the plate current also increases for a given grid voltage because of the effect of the additional positive ions in neutralizing the space charge. Thus, when the pressure is increased from p_1 to p_2, as shown in Fig. 18, the curve of the plate current for p_2 lies higher than that for p_1, and the positive-ion component of the grid current increases more rapidly than the pressure. Although not shown on the diagram, the curves for both the plate current and the grid current for decreasing grid voltage generally depart from the curves for increasing grid voltage.

As the pressure is increased further, a value is finally reached at which the behavior of the tube is altered materially and the tube takes on the characteristics of a thyratron. If, at this pressure, the grid is made highly negative before the plate voltage is applied, and is then gradually made less negative, the plate current and the grid current increase continuously until a particular value of grid voltage is reached, at which the tube fires. Then a plasma forms in the tube and both currents increase suddenly; the plate current is limited only by the resistance of the external circuit, just as it is in a diode when a plasma forms. The grid and cathode of the triode are surrounded by positive-ion sheaths, and the grid voltage no longer controls the plate current effectively. This behavior is illustrated by the curves for the pressure p_3 in Fig. 18; the solid horizontal lines indicate the lack of control of the grid after the tube fires. *The grid acts as a trigger to control the starting of the discharge.*

This description of the effects of gas pressure includes only a few of the important phenomena that occur to some degree in all thermionic triodes. The transition from vacuum-tube behavior to thyratron behavior is largely dependent upon the gas pressure. The structure and materials of the tube elements, the type of cathode, the electrode voltages, the insulation used, and the particular gas used, all have definite effects upon the behavior of thermionic triodes containing gas.

6. THYRATRONS

Although it is applicable to a gas triode with a mercury-pool cathode, the name *thyratron* is ordinarily restricted to a gas triode that

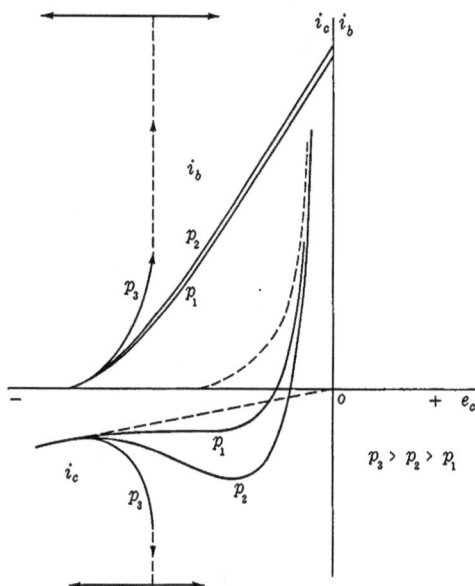

Fig. 18.   Effect of gas pressure on the grid and plate currents.

has a continuously emitting thermionic cathode, sufficient gas pressure to permit the formation of a plasma when normal plate voltage is applied, and a grid to prevent the formation of the plasma until a desired instant.[20] In such a tube, the *trigger action* of the grid described in the previous article is the dominant feature. The grid acts as a trigger, in that, when it is brought from a sufficiently negative value to a more positive value, it allows the anode current to start, or the tube to fire. Before the grid voltage reaches a critical value, dependent

[20] A. W. Hull, "Hot Cathode Thyratrons," *G.E. Rev.*, *32* (1929), 213–223, 390–399; A. W. Coolidge, Jr., "A New Line of Thyratrons," *A.I.E.E. Trans.*, *67* (1948), 723–727; H. de B. Knight, "Hot-Cathode Thyratrons: Practical Studies of Characteristics," *I.E.E. Proc.*, *96*, Part III (1949), 361–378. For accounts of early work in the year 1913 on this type of tube, refer to G. W. Pierce, United States Patents 1,087,180 (February 17, 1914); 1,112,549 (October 6, 1914); 1,112,655 (October 6, 1914); 1,450,749 (April 3, 1923).

upon the anode voltage, the anode current is negligible; but, after the critical voltage is once reached, the grid exerts no further appreciable control over the anode current, the anode voltage drops to a value of the order of 10 to 20 volts, and the anode current *must be limited by impedance in series with the anode* to prevent destruction of the tube. Furthermore, an *impedance is necessary in series with the grid* to prevent excessive grid current, especially if an alternating voltage is applied to the grid, so that it is positive during part of each cycle. Returning the grid voltage to the negative value it had before the tube fired cannot stop the anode current; in fact, the grid is incapable of stopping the anode current (except when the anode current is limited to a small value by the circuit; then a highly negative grid voltage may sometimes interrupt it). In order for the grid to regain control, the anode current must be stopped either by reduction of the anode voltage to zero, or by reversal of it.

A tube of this type will serve as a relay but not as an amplifier.[21] As a relay, it is extremely fast in its action but must be "reset" for each new cycle of operation by reduction of the anode voltage and current to zero. This reduction is most conveniently accomplished through the utilization of an alternating supply voltage for the anode circuit so that the anode becomes negative periodically and remains negative for a sufficient time to allow the plasma to disappear. The grid voltage determines the point in each positive half-cycle of the anode voltage at which the tube fires. During the following negative half-cycle, the current stops and allows the grid to regain control, so that it may determine the start of conduction in the next positive half-cycle. In this way the grid determines the average value of the rectified anode current.

Not only is the circuit behavior of a thyratron different from that of a vacuum triode, but also its internal structure requires a different design, and the factors in its physical behavior must be fully understood before it can be applied to advantage. Because of the formation of a plasma, it is possible to use high-efficiency heat-shielded oxide-coated cathodes, such as those described in Art. 2c, and thus a high peak-current rating relative to the heating power required may be obtained. However, adequate time must be allowed for the cathode to reach its operating temperature before current is drawn from it, and the peak anode current must be limited to the rated value. Destructive

---

[21] Gas tubes that amplify but operate on different principles are described by P. T. Weeks and J. D. LeVan, "A New Type of Gas-Filled Amplifier Tube," *I.R.E. Proc.*, *24* (1936), 180–189; E. O. Johnson, "Controllable Gas Diode," *Electronics*, *24* (May, 1951), 107–109; and E. O. Johnson and W. M. Webster, "The Plasmatron, a Continuously Controllable Gas-Discharge Developmental Tube," *I.R.E. Proc.*, *40* (1952), 645–659.

bombardment of the active surface by positive ions will otherwise occur. Furthermore, it is necessary to limit the average anode current to prevent overheating of the tube or any of its parts. The cathode in particular may be overheated by the kinetic energy and the heat of neutralization of the ions that strike it.

6a. *Grid Action before Starting.* Figure 19 shows the approximate manner[22] in which the grid current of a thyratron varies before starting for constant pressure and several fixed values of the anode voltage.

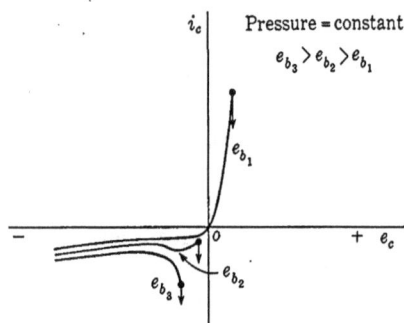

Fig. 19. Grid current prior to and at the starting condition.

When the anode voltage is low, as at e_{b1}, the number of ions formed by each electron traveling to the anode is small, a relatively high anode current is necessary to start a plasma, and the grid voltage must be made positive before this current can pass and a plasma can form. The starting condition is denoted by a dot and an arrow, to indicate a sudden change in grid current. As the grid is made positive, it draws an electron current from the cathode, and considerable positive grid current flows at starting. For higher values of positive anode voltage, the efficiency of ionization is greater, less current to the anode is required for starting, and the discharge can start for negative grid voltages and relatively small grid currents. In general, the scale of current for normal tubes in a diagram like Fig. 19 is such that the starting grid current for negative grid voltages is of the order of 0.001 microampere or less, and the critical anode current is of the order of one microampere. Thus the anode and grid currents that precede starting may be made small by proper tube design and can often be neglected. Some tubes, however, may require a grid current of as much as one-half ampere for starting, and thus require considerable power for grid excitation. In applications where the power available to excite the grid is small, the starting grid current must be considered, but in other applications the tube may often be considered as a voltage-operated device.

6b. *Starting Characteristics.* If a thyratron is operated so that the grid current may be neglected, its characteristic of importance is the

[22] W. B. Nottingham, "Characteristics of Small Grid-Controlled Hot-Cathode Mercury Arcs or Thyratrons," *J.F.I.*, *211* (1931), 271–301; H. W. French, "Operating Characteristics of Small Grid-Controlled Hot-Cathode Arcs or Thyratrons," *J.F.I.*, *221* (1936), 83–102.

starting characteristic,[23] or *control characteristic*—that is, the relation between grid and anode voltages for the starting condition. Families of such characteristic curves are shown in Fig. 20 for three different tubes and for various condensed mercury temperatures. The starting characteristic is the boundary between two regions—one to the right and above the characteristic, in which the tube will conduct; and the other

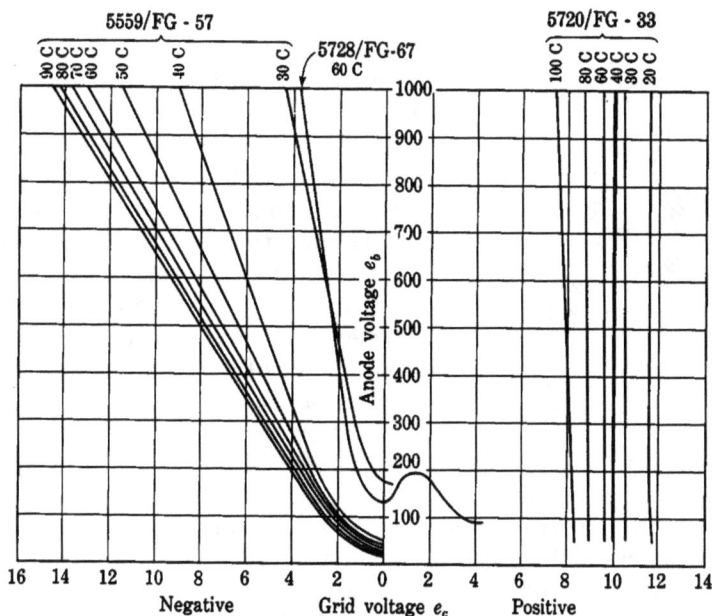

Fig. 20.   Starting characteristics of mercury-vapor thyratrons at various condensed-mercury temperatures.*

to the left and below the characteristic, in which it will not conduct. If for any fixed anode voltage the grid voltage is brought from a point to the left of the curve, conduction starts when the grid voltage reaches the curve; or, for a fixed grid voltage, conduction starts if the anode voltage is increased until it reaches the curve. It is to be noted, however, that the grid voltage can control only the start of conduction; conduction does not stop as the curve is crossed from the right and above.

[23] A. W. Hull, "Hot-Cathode Thyratrons, Part I," *G.E. Rev.*, 32 (1929), 218–219.

* The curves for the thyratron 5559/FG-57 are adapted from *Publication GET-437* (Schenectady: General Electric Company, September 20, 1932), with permission; those for the thyratron 5720/FG-33 from *Publication GET-435* (Schenectady: General Electric Company, September 20, 1932), with permission; and the curve for the thyratron 5728/FG-67 is experimentally determined.

Thyratrons are classified as *negative-control* or *positive-control* types, depending upon whether the starting grid voltage is negative or positive over substantially the whole operating range.[24] The two families of characteristics for the Type 5559/FG-57 at the left and the Type 5720/FG-33 at the right of Fig. 20 are representative of tubes of these two classes, respectively, while the characteristic for the Type 5728/FG-67 in the middle of the diagram is intermediate and falls into neither class.

Over the greater part of the range, the starting characteristic is practically a straight line, and for some negative-control tubes its projection passes through the origin. Under these conditions, the *grid-control ratio* of the tube, defined[25] as "the ratio of anode voltage to negative grid voltage that will just allow current to start," is constant and is a useful term in describing the tube. It is somewhat analogous to the amplification factor of a vacuum tube.

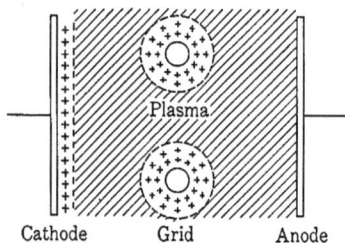

Cathode   Grid   Anode

Fig. 21.   Positive-ion sheaths around a negative grid.

6c. *Grid Action after Starting.* The explanation of the fact that the grid has practically no effect upon the anode current when it is again made negative after starting may be made in terms of the space-charge sheath of positive ions that forms around the negative grid. Figure 21 shows a thyratron anode and cathode and two cylindrical wires that form part of the grid structure. When conduction starts, a positive-ion sheath forms around the cathode, and the remaining space between the anode and cathode becomes filled with a plasma in which the concentrations of positive ions and electrons are approximately equal and in which the potential is constant and practically equal to that of the anode, as is discussed in Art. 2a. When the grid is made negative with respect to the space, it repels electrons and attracts positive ions so that the space adjacent to it is filled with positive ions all moving toward the grid under the influence of the electric field. When an equilibrium rate of flow of positive ions is reached, the total positive charge in the space outside the grid wire equals the negative charge induced on the grid wire by the applied voltage. Under these conditions, the net charge inside the boundary of the positive-ion sheath is zero, and the plasma outside the sheath is unaffected by the presence of the negative grid. Consequently, conduction between the

[24] H. C. Steiner, A. C. Gable, and H. T. Maser, "Engineering Features of Gas-Filled Tubes," *E.E.*, *51* (1932), 314.

[25] A. W. Hull, "Hot-Cathode Thyratrons, Part I," *G.E. Rev.*, *32* (1929), 219.

anode and cathode can take place through the remaining plasma between the grid sheaths, and the negative grid does not control or stop the anode current.

The foregoing discussion is based upon the assumption that the thickness of the sheaths around the grid wires is small compared with the distance between wires. Justification of that assumption requires further analysis, as follows. Since the current through the sheath is carried essentially only by positive ions, and electrons are not present in the sheath in appreciable numbers, the positive-ion current from the outer boundary of the sheath to the grid wire is limited by space charge in the same way as is the electron current through a vacuum (see Art. 1, Ch. III). If the sheath thickness is assumed to be small compared with the diameter of the grid wire, the equation for current between parallel planes can be applied, and subsequent verification of the assumption can be made. When the mass of the positive ion is used instead of that of the electron, Eq. 19, Ch. III, becomes

$$J_+ = \frac{2.33 \times 10^{-6}}{\sqrt{1,823M}} \frac{e_{ga}^{3/2}}{d^2}, \qquad [6]$$

where

J_+ is the positive-ion current density in amperes per square centimeter,

M is the atomic weight of the positive ions,

d is the thickness of the positive-ion sheath in centimeters,

e_{ga} is the voltage across the sheath, which equals approximately the voltage from grid to anode, since the plasma is approximately at the anode potential.

In a vacuum hot-cathode tube, the space-charge-limited current varies with the voltage across the space-charge region, and the thickness of the region filled with a space charge is equal to the distance between the electrodes and is hence fixed. In contrast, the space-charge-limited current through a positive-ion sheath surrounding a negative electrode in a plasma is determined by the number of positive ions that move across the boundary and into the sheath as a result of their random motion in the plasma. In the usual type of low-pressure mercury-vapor discharge, this current, called the *random positive-ion current*, has a density of the order[26] of one milliampere per square centimeter and varies[27] directly with the anode current. Consequently,

[26] T. J. Killian, "Uniform Positive Column of an Electric Discharge in Mercury Vapor," *Phys. Rev.*, *35* (1930), 1249.

[27] A. W. Hull, "Hot-Cathode Thyratrons, Part I." *G.E. Rev.*, *32* (1929), 215.

to the first approximation, the current density into the outer surface of the sheath surrounding a negative grid is constant and independent of the magnitude of the negative grid voltage. As the grid voltage is varied, the thickness of the positive-ion sheath varies, but the grid current remains practically constant as long as the sheath remains thin. The thickness of the positive-ion sheath, as found from Eq. 6, is

$$d = \sqrt{\frac{2.33 \times 10^{-6}}{\sqrt{1,823M}} \frac{e_{ga}^{3/2}}{J_+}} \text{ centimeters.} \qquad [7]$$

Under typical conditions, the anode voltage of a mercury-vapor tube is about 10 volts, and the grid voltage may be assumed to be, say, 10 volts negative with respect to the cathode, so that e_{ga} is about 20 volts. Since the random positive-ion current density is of the order of 10^{-3} ampere per square centimeter, and the atomic weight of mercury is 200.6, an approximate value of d is

$$d = \sqrt{\frac{2.33 \times 10^{-6}}{\sqrt{1,823 \times 200.6}} \times \frac{20^{3/2}}{10^{-3}}} = 0.018 \text{ centimeter.} \qquad [8]$$

Thus it is apparent that the sheath thickness is small under normal conditions, and that the application of the equation for the space-charge-limited current between parallel planes is justified as an approximation. Only through making the grid very negative or the current density in the plasma very small can the sheaths be made to overlap between the grid wires and to extinguish the arc.

If the grid voltage is made positive, the potential of the grid approaches that of the plasma; and, because of their random velocities, some of the electrons in the plasma are able to reach the grid. Since the random current density of electrons in the plasma is much greater than that of positive ions, the electron grid current is equal to the positive-ion current even when the grid potential is a few volts negative with respect to the plasma. At more positive grid potentials, the electron current predominates. When the grid voltage reaches the potential of the plasma, the positive-ion sheath disappears, the plasma extends to the grid surface, and the grid current becomes very large— comparable to, or even greater than, the anode current; the grid begins to take over the function of the anode.

The typical manner in which the grid current varies as a function of grid voltage and anode current after conduction has started is shown in Fig. 22. With the formation of a plasma and a positive-ion sheath around the grid, the grid current changes suddenly, as indicated by the arrow in Fig. 19 to much larger values along one of the family of curves of Fig. 22. The currents in Fig. 22 may be of the order of a billion times larger than those in Fig. 19. After conduction starts, changes of the grid voltage in the region of negative voltages produce changes in the thickness of the positive-ion sheath with small accompanying changes of the grid current, as shown. The almost linear changes of grid current with anode current, when the grid voltage is held constant, are noticeable. Because of the electron current that reaches the grid when it is at a potential near or above that of the plasma, the grid current may become very large if the grid is positive.

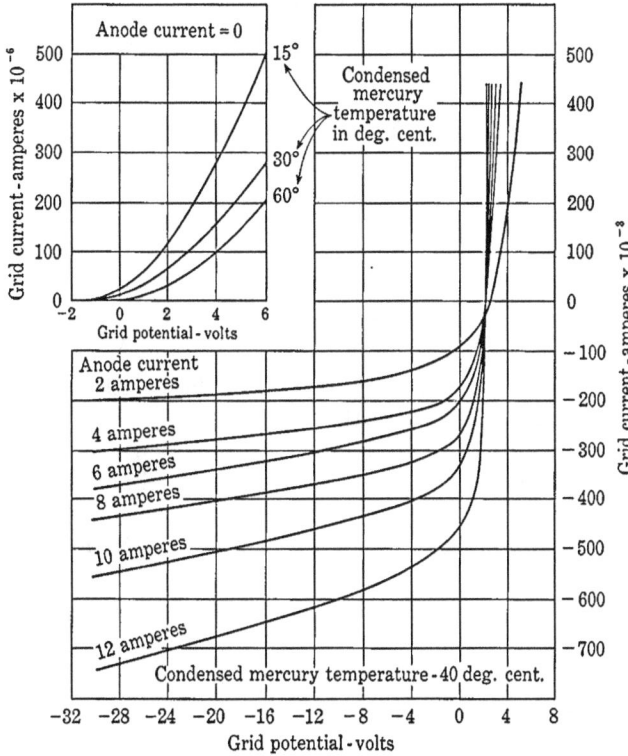

Fig. 22. Grid current in an inverter-type thyratron after conduction starts.*

For this reason, a current-limiting impedance is needed in series with the grid. The large currents that may occur at the positive peaks of the grid voltage cycle, and the sudden changes of current that occur when the tube fires, may cause surges of current and voltage in the grid circuit. The effect of these surges upon the source of grid voltage may require consideration.

* This figure is taken from H. C. Steiner, A. C. Gable, and H. T. Maser, "Engineering Features of Gas-Filled Tubes," *E.E.*, *51* (1932), Fig. 5, p. 314, with permission.

6d. *Ionization and De-ionization Times.* The speed with which a thyratron acts is of importance in determining the uses to which the tube may be put. This speed depends upon the *ionization time*, which is the time required for the plasma to form and the anode current to attain its final value after the grid or anode voltage is suddenly changed to a value that allows the tube to fire, and upon the *de-ionization time*, which is the time required for the plasma to disappear and allow the grid to regain control after the anode voltage is removed. For example, these times place an upper limit upon the frequency of the supply voltage for which the tube will serve as a grid-controlled rectifier, and limit the use of the tube in the inversion of direct current to alternating current.

The *ionization time* of the thyratron ranges from a fraction of a microsecond to several microseconds.[28] It depends upon the tube structure, the gas pressure, the electrode supply voltages, and the circuit parameters.

The *de-ionization time* depends upon the same factors as the ionization time, but is usually somewhat longer, ranging from a few microseconds to several hundred microseconds. It is a function not of the tube alone but of the tube and circuit as a unit.[29] The process of de-ionization[30] consists essentially of diffusion of the positive ions of the plasma into the positive-ion sheaths on the walls and electrodes, whereupon they are drawn to the surfaces and neutralized. As the positive-ion concentration decreases, the sheath thickness increases in accordance with Eq. 7; and, finally, when the sheaths around the grid surfaces overlap, the grid regains control, and the plate may be made positive again without the occurrence of conduction. Regaining of control occurs most quickly in tubes with small grid holes; hence tubes designed for short de-ionization time are ordinarily positive-control tubes (see Art. 6b) and require considerable starting power.

Hydrogen is used as the gas in some thyratrons to give a very short

[28] A. W. Hull, "Characteristics and Functions of Thyratrons," *Physics*, *4* (1933), 71; L. B. Snoddy, "Ionization Time of Thyratrons," *Physics*, *4* (1933), 366–371; J. B. Woodford, Jr., and E. M. Williams, "The Initial Conduction Interval in High Speed Thyratrons," *J. App. Phys.*, *23* (1952), 722–724.

[29] H. C. Steiner, A. C. Gable, and H. T. Maser, "Engineering Features of Gas-Filled Tubes," *E.E.*, *51* (1932), 314; H. de B. Knight, "The Deionization Time of Thyratrons: A New Method of Measurement," *I.E.E. Proc.*, **96**, Part III (1949), 257–261; H. H. Wittenberg, "Pulse Measuring of Deionization Time," *E.E.*, *69* (1950), 823–827; K. W. Hess, "Measuring the Deionization Time of Gas-Filled Diodes and Triodes," *Philips Tech. Rev.*, *12* (December, 1950), 178–184.

[30] A. W. Hull, "Characteristics and Functions of Thyratrons," *Physics*, *4* (1933), 71; L. Malter and E. O. Johnson, "Studies of Thyratron Behavior," *RCA Rev.*, *11* (June, 1950), 165–189.

de-ionization time.[31] Rapid de-ionization occurs because the hydrogen ions, being light in weight, diffuse out of the plasma faster than those of other gases. Hydrogen thyratrons are particularly suitable for applications involving pulse techniques, such as radar.

6e. *Thyratron Grids.* In contrast to its function in the vacuum triode, the grid in the thyratron must control the maximum current density instead of the average current density across its area. Starting of the gas tube will occur if the current density at some point reaches a critical value; hence, if there is one hole in the grid larger than the others, the current density through the largest hole tends to control the starting, and the remainder of the grid could as well be solid, instead of being perforated, without altering the starting characteristics materially. Also, because of this condition the critical current density may occur along a path around instead of through the grid. Consequently, in the gas tube the grid must surround either the anode or the cathode more completely than is necessary in the vacuum tube; the usual structure of the vacuum triode is unsatisfactory for a thyratron.

If the grid surrounds either the anode or the cathode alone, the space between the grid and the remaining electrode will be influenced by the electric field set up by charges on the glass walls.[32] The accumulation of charges there is subject to much fluctuation, and their influence gives rise to erratic starting characteristics. Hence it is desirable that the grid shield both the anode and the cathode from the glass walls. When grids are used in large polyphase rectifiers with several anodes and a single mercury-pool cathode, however, the potential of the steel wall remains practically constant with respect to the cathode, and individual grids are placed around each anode.

If the grid emits too many electrons, they may cause sufficient ionization during their passage to the anode to allow the main discharge to start. Since erratic starting may result from such grid emission, grid surfaces are usually treated to inhibit electron emission and to increase their total radiation emissivity, thus lowering the operating temperature. Sputtering of the oxide-coating material from the cathode onto the grid may lead to difficulty from increased grid emission.

In conformity with these principles, the internal structure of a typical single-grid thyratron appears as in Fig. 23a. The grid consists of

[31] K. J. Germeshausen, "The Hydrogen Thyratron," *Pulse Generators*, Massachusetts Institute of Technology Radiation Laboratory Series, Vol. 5, G. N. Glascoe and J. V. Lebacqz, Editors (New York: McGraw-Hill Book Company, Inc., 1948), Ch. 8; H. Heins, "Hydrogen Thyratrons," *Electronics*, 19 (July, 1946), 96–102.

[32] A. W. Hull, "Hot-Cathode Thyratrons, Part I," *G.E. Rev.*, 32 (1929), 219.

a solid metal cylinder surrounding both the anode and the cathode, so as to shield them from the glass walls, and a diaphragm, having a single hole between the two. The grid is made of a material having a high work function, such as nichrome, carbonized nickel, graphite, or iron, and because of its large radiating surface it remains cool and gives negligible electron emission. When the diaphragm has a large single hole, a negative-control tube is the result; when the diaphragm

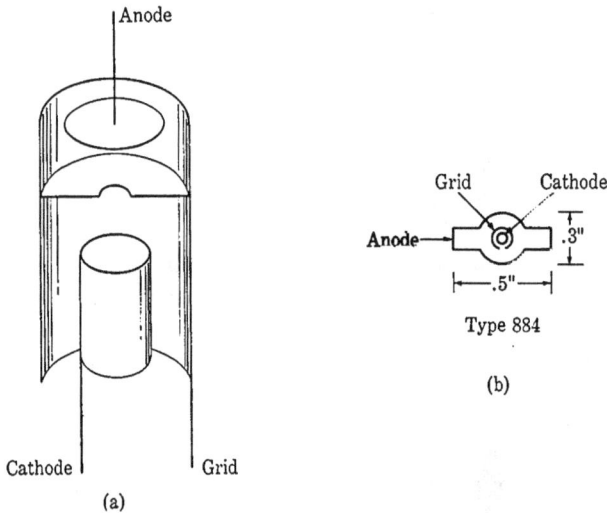

Fig. 23.   Internal structure of single-grid thyratrons.*

is perforated with a number of smaller holes, the tube is a positive-control thyratron.

The large area of the grid pictured in Fig. 23a gives rise to a large starting grid current and a high interelectrode capacitance with a resultant low grid input impedance, which is undesirable if the grid is to be excited from a source of high impedance and low power. To reduce the starting grid current and the capacitance, the various grid functions are divided in the double-grid tubes shown in Fig. 24. In these tubes the shield, or blocking grid, which is held at a fixed potential with respect to the cathode, performs most of the shielding function and collects most of the starting and de-ionizing grid current. The control, or trigger grid, which is a short cylinder, a wire ring, or a straight wire between the two diaphragms of the shield grid or in the slot of the blocking grid, serves to control the starting of the tube. Because of its small area, however, the control grid requires only a

* Figure 23b is taken from H. H. Wittenberg, "Pulse Measuring of Deionization Time," *E.E.*, *69* (1950), Fig. 1, p. 824, with permission.

low[33] starting current and has only a small capacitance. By variation
of the magnitude of the voltage between the shield grid and the
cathode, these tubes may be made to start at either positive or negative
grid voltages.

6f. *Effects of Gas Pressure on the Characteristics of a Thyratron.* In
a thyratron, the inverse voltage required for breakdown and the tube
drop while conducting are affected by the pressure of the gas in the

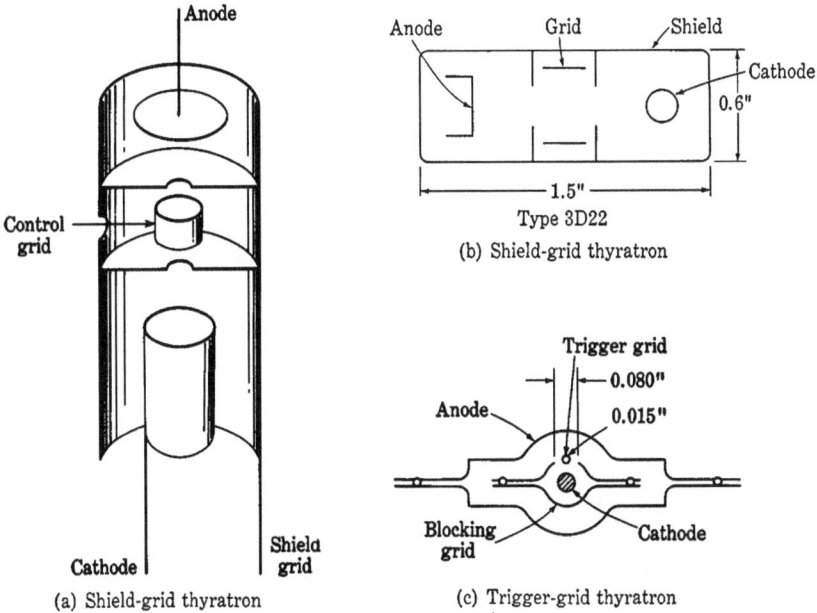

Fig. 24.   Internal structure of double-grid thyratrons.*

same way as in a gas diode (see Art. 3). Therefore the pressure must be
restricted to a definite operating range if the thyratron is to withstand
its rated inverse voltage and deliver its rated current without damage.
Furthermore, the gas pressure affects the starting characteristics, as
shown in Fig. 20. If the pressure is increased, the mean free path of the
electrons is decreased, so that each electron makes more collisions and
produces more ions in crossing the tube. Thus with increased pressure

[33] O. W. Livingston and H. T. Maser, "Shield-Grid Thyratrons," *Electronics*, 7 (1934);
114–116; L. Malter and M. R. Boyd, "Grid Current and Grid Emission Studies in
Thyratrons—The Trigger-Grid Thyratron," *I.R.E. Proc.*, *39* (1951), 636–643.

    * Figure 24b is taken from H. H. Wittenberg, "Pulse Measuring of Deionization Time,"
E.E., *69* (1950), Fig. 1, p. 824, with permission, and Fig. 24c is taken from L. Malter and
M. R. Boyd, "Grid Current and Grid Emission Studies in Thyratrons—The Trigger-Grid
Thyratron," *I.R.E. Proc.*, *39* (1951), Fig. 7, p. 639, with permission.

the critical condition for the formation of a plasma is reached at a lower value of initial electron current, and, for each value of anode voltage, starting occurs at more negative grid voltages. In mercury-vapor tubes, the pressure depends upon the temperature of the coolest spot on the wall of the tube—Fig. 6 shows that the mercury-vapor pressure approximately doubles for a ten-degree increase in temperature—and therefore the starting characteristics change rapidly with temperature. If the tube is to be used in a circuit in which the magnitude of the starting voltage is critical, the temperature of the tube must be controlled within close limits. Positive-control tubes in particular show great variations in the starting characteristics with temperature and exhibit anomalous effects[34] and double-valued characteristics, as shown for the Type 5728/FG-67 tube in Fig. 20. Often tubes of this type will not start until they have reached a temperature of 40 degrees centigrade or more. In tubes containing an inert gas, variation of the starting characteristics with temperature does not occur, since the concentration of the gas does not change with temperature. Because of the clean-up action, however, the gas pressure does decrease during the life of the tube; hence, for a given plate voltage, progressively more positive grid voltages are needed to start the tube.

Another effect of the gas pressure on the characteristics of thyratrons occurs because a cold-cathode glow discharge may take place between the grid and the anode if the voltage between these electrodes is allowed to exceed the breakdown voltage. The ionization resulting from such a glow discharge may cause the formation of a sufficiently thin positive-ion sheath around the grid to allow the main anode current to start and thus cause the grid to lose control. Accordingly, the voltage between grid and anode that a thyratron will withstand without starting is limited. Although the maximum permissible voltage between grid and cathode is sometimes specified, the rating commonly given to indicate this limitation is termed the *maximum forward anode voltage*. It is dependent upon the gas pressure in the tube, because the breakdown voltage decreases as the pressure is increased in accordance with a curve similar to the left branch of one of the curves of Fig. 11b, Ch. III. Thus the upper limit of the operating temperature of a mercury-vapor tube is set either by the maximum value of the forward anode voltage or by the peak inverse voltage the tube must withstand. Tubes filled with an inert gas have, in general,

---

[34] W. B. Nottingham, "Characteristics of Small Grid-Controlled Hot-Cathode Mercury Arcs or Thyratrons," *J.F.I.*, *211* (1931), 271–301; H. W. French, "Operating Characteristics of Small Grid-Controlled Hot-Cathode Arcs or Thyratrons," *J.F.I.*, *221* (1936), 83–102.

lower forward anode voltage ratings than do mercury-vapor tubes, because sufficient gas must be put into the tube initially to allow for some clean-up, and the breakdown voltage is therefore low.

7. PERMATRON

A magnetic field is used to control the starting of a gas discharge in a tube called a *permatron*.[35] The magnetic field is produced by a coil surrounding the tube and supplied by an auxiliary control circuit. The field deflects electrons from the hot cathode toward a collector electrode at low voltage and prevents the formation of an arc and plasma between the cathode and anode. Reducing the field below a critical value permits the arc to start. The tube has applications similar to those of a thyratron.

8. COLD-CATHODE TUBES

Tubes employing a gaseous discharge from an unheated cathode have a wide range of application, including rectification, voltage regulation, control, counting, and production of light.

A cold-cathode gas diode having one electrode much larger in area than the other can be used as a rectifier.[36] Figure 25 shows the volt-ampere characteristic curve of a tube utilizing this principle. The cathode for current in the positive direction is large in area and coated with a material of low work function; hence the discharge for this polarity is a normal glow discharge that has a low voltage drop. The cathode for current in the negative direction is small and of higher work function, however, and hence the discharge for this reversed polarity is an abnormal glow discharge, which has a much higher voltage drop for any particular value of current. When the tube is used in a circuit with alternating voltage applied, this asymmetrical characteristic results in rectification if the voltage has an amplitude larger than the value that causes appreciable current in the positive direction, but smaller than the value at the negative "knee" of the curve. Other rectifiers of the so-called cold-cathode type have a cathode that is coated with emitting material, which becomes hot under ionic bombardment and emits thermionically. The cathode voltage drop in such tubes is only about 25 volts even though no external source of cathode-heating power is required.

The approximate constancy of the voltage drop across a normal

[35] W. P. Overbeck, "The Permatron—a Magnetically Controlled Industrial Tube," *A.I.E.E. Trans.*, *58* (1939), 224–228.

[36] A. E. Shaw, "Cold-Cathode Rectification," *I.R.E. Proc.*, *17* (1929), 849–863.

glow discharge over a considerable range of current, discussed in Art. 9, Ch. III, is used for maintaining constant voltage across a load in the gas-filled voltage-regulator diode. Application of the tube for that purpose, which is explained in more detail in Art. 14, Ch. VI, depends upon the principle that, when connected in parallel with a load that is

Average anode characteristic

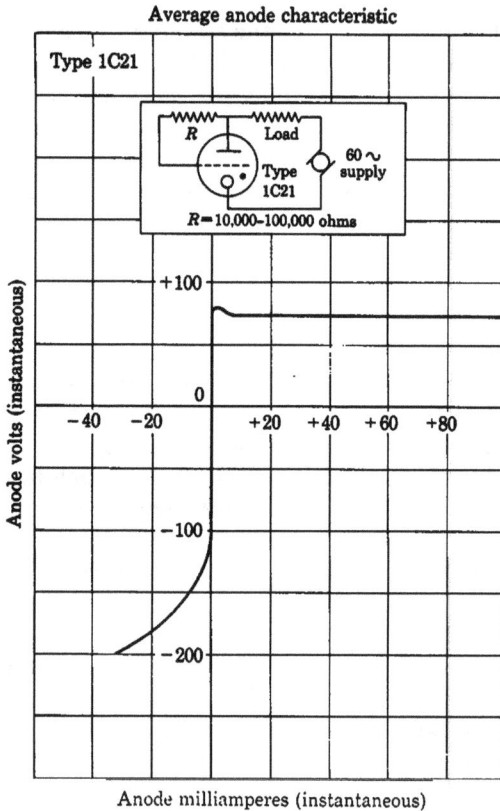

Fig. 25. Asymmetric volt-ampere characteristic curve of a cold-cathode rectifier tube. (*Adapted by courtesy of Radio Corporation of America.*)

fed from a source having considerable internal series resistance, the gas diode will draw a larger or smaller current to cause a compensating voltage drop in the series resistance if the source voltage changes. If, on the other hand, the load current changes while the source voltage remains constant, the current in the diode changes by an almost equal but opposite amount so as to keep the source current and the voltage drop in the series resistance essentially constant and thereby to maintain the load voltage almost constant.

The precision of voltage stabilization that can be obtained through use of a gas diode is limited by several properties of such tubes. Typical tubes have a voltage drop that changes by 2 to 4 per cent over the full rated range of current. The voltage drop is subject to gradual changes of as much as 0.3 per cent during operation,[37] and depends on the ambient temperature, with a coefficient of the order of 10 millivolts per degree centigrade.[38] Voltage jumps or discontinuities of as much as 0.2 per cent occur spontaneously in a random manner, and oscillations that are inherent in the tube are likely to occur at various points in the operating range of current.[39] Special preparation of the cathode can reduce some of these undesirable effects.[40]

The largest voltage that voltage regulator tubes employing the glow discharge will regulate is limited to about 150 volts by the properties of that type of discharge. For stabilization of higher voltages up to tens of kilovolts, the corona discharge is used.[41] The electrodes in the corona-discharge tube are usually coaxial cylinders or a point and a plane. If the voltage across them is increased from zero, negligible current results until a critical corona voltage is reached. Further increase of voltage is accompanied by a rapid increase of current. Thus the tube may be used for voltage regulation in the same manner as the glow-discharge type. The corona voltage is dependent upon the dimensions, and upon the gas and its pressure. One electrode is sometimes made movable by external means to provide for adjustment of the regulated voltage.

Control of the starting of a cold-cathode gaseous discharge is obtained if a third electrode, often called a *starter anode* or *control anode*,[42] is added to a tube containing a cathode and a main anode. The cathode of a typical cold-cathode gas triode is made of nickel

[37] G. M. Kirkpatrick, "Characteristics of Some Voltage-Regulator Tubes," *I.R.E. Proc., 35* (1947), 485–489.

[38] F. A. Benson, W. E. Cain, and B. D. Lucas, "Variations in the Characteristics of Some Glow-Discharge Voltage-Regulator Tubes," *J.S.I., 26* (1949), 399–401.

[39] E. W. Titterton, "Some Characteristics of Glow-Discharge Voltage-Regulator Tubes," *J.S.I., 26* (1949), 33–36.

[40] T. Jurriaanse, "A Voltage Stabilizing Tube for Very Constant Voltage," *Philips Tech. Rev., 8* (1946), 272–277. See also manufacturer's instructions for Type 5651 tube.

[41] I. H. Blifford, R. G. Arnold, and H. Friedman, "Corona-Tube Regulators for High Voltages," *Electronics, 22* (December, 1949), 110–111; S. W. Lichtman, "High-Voltage Stabilization by Means of the Corona Discharge Between Coaxial Cylinders," *I.R.E. Proc., 39* (1951), 419–424.

[42] S. B. Ingram, "The 313A Vacuum Tube," *Bell Lab. Rec., 15* (December, 1936), 114–116; W. E. Bohls and C. H. Thomas, "A New Cold-Cathode Gas Triode," *Electronics, 11* (May, 1938), 14–16, 72–74; S. B. Ingram, "Cold-Cathode Gas-Filled Tubes as Circuit Elements," *E.E., 58* (1939), 342–347; G. H. Rockwood, "Current Rating and Life of Cold-Cathode Tubes," *E.E., 60* (1941), 901–903.

coated with a mixture of barium and strontium produced from their oxides, and the gas is neon with one per cent of argon added.

The primary use of such tubes is based on the principle that production of ionization in one gap lowers the breakdown voltage across a nearby gap. Thus if a voltage almost sufficient to cause breakdown

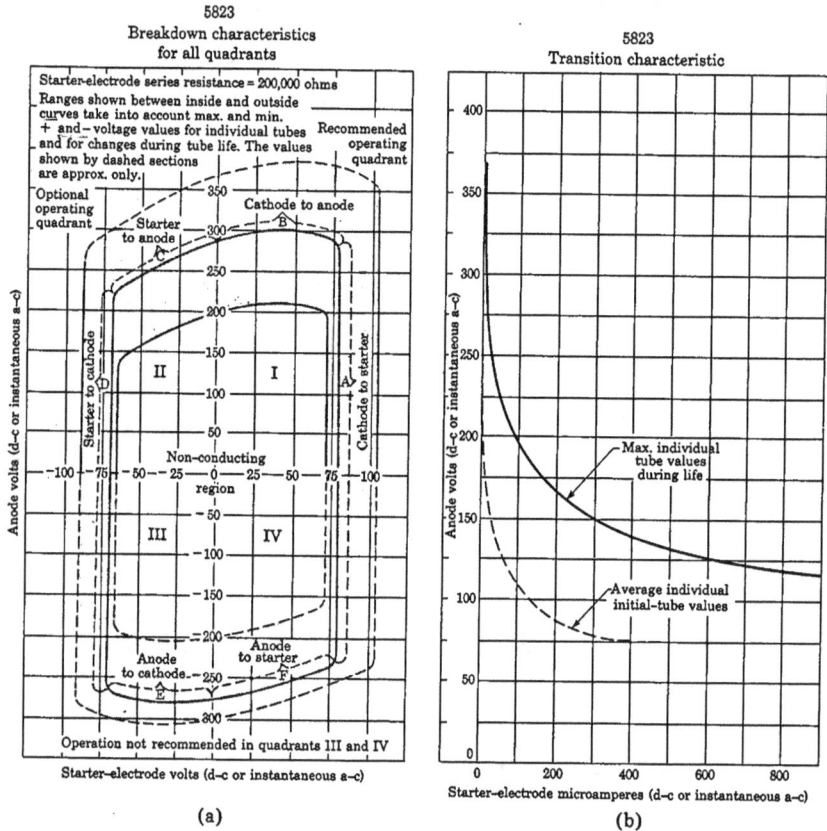

5823
Breakdown characteristics
for all quadrants

5823
Transition characteristic

Starter-electrode series resistance = 200,000 ohms
Ranges shown between inside and outside
curves take into account max. and min.
+ and − voltage values for individual tubes
and for changes during tube life. The values
shown by dashed sections
are approx. only.

Recommended
operating
quadrant

Optional
operating
quadrant

Cathode to anode

Starter
to anode

II   I

III   IV

Non-conducting
region

Anode
to cathode.

Anode
to starter

Operation not recommended in quadrants III and IV

Starter-electrode volts (d−c or instantaneous a−c)

Anode volts (d−c or instantaneous a−c)

Starter to cathode

Cathode to starter

Max. individual
tube values
during life

Average individual
initial-tube values

Starter-electrode microamperes (d−c or instantaneous a−c)

(a)
(b)

Fig. 26. Characteristic curves for a cold-cathode control tube.
(*Adapted by courtesy of Radio Corporation of America.*)

between the main anode and cathode is applied through a load, breakdown in this main gap will be induced if a discharge is started in the auxiliary or control gap between the starter anode and the cathode. Actually, there are three possible paths for conduction among the three electrodes and two directions for each, making six possible types of breakdown. The conditions for these six types of breakdown in a typical tube are shown in Fig. 26a. Simultaneous application of anode and starter-electrode voltages that correspond to points inside the closed curve does not cause conduction. Conduction begins, however, when the point corresponding to the applied voltages crosses the curve from inside to outside. The path and direction of the resulting conduction depend on the section of the curve that is crossed, as is indicated in the figure.

In the usual application of the tube, the operating point is caused to cross the curve at the upper part of section *A* of the closed curve. Conduction is thus initiated under control of the starter electrode, yet the anode voltage is high enough to cause transfer of the conducting path to the main gap from anode to cathode. Such transfer takes place, however, only if the current in the control gap is caused to exceed a critical value. The starter-electrode current at which the transfer takes place is given by Fig. 26b for various values of anode voltage. If, for example, an anode voltage of 150 volts is applied, a starter-electrode current of about 25 microamperes is required to cause transition of the discharge to the main gap in an average tube. The control circuit for such tubes must therefore be designed to furnish appreciable current and power.

Cold-cathode gas triodes are frequently used with alternating or pulsating anode supply voltage for selective ringing of four-party telephone instruments, and with a radio-frequency control signal on the starter electrode to permit remote control of equipment.

A class of cold-cathode glow-discharge tubes for specialized counting, stepping, or relay applications is composed of tubes having one anode and several cathodes, often ten or more, usually located in a ring around the anode.[43] One or two auxiliary electrodes are interposed between the cathodes. Application of pulses to the auxiliary electrodes causes the main discharge to transfer from cathode to cathode around the ring. Thus the position of the visible glow indicates the number of pulses that have occurred. Output pulses may be taken from one or more of the individual cathodes to adapt the tube to a variety of applications, particularly electronic digital computation.

The *Strobotron* is a cold-cathode gas discharge tube designed to produce intense flashes of light at controlled instants for stroboscopic applications.[44] It consists of an anode, a cathode, and two control

[43] J. J. Lamb and J. A. Brustman, "Polycathode Glow Tubes for Counters and Calculators," *Electronics, 22* (November, 1949), 92–96; R. C. Bacon and J. R. Pollard, "The Dekatron—a New Cold Cathode Counting Tube," *Electronic Engineering, 22* (1950), 173–177; M. A. Townsend, "Construction of Cold-Cathode Counting or Stepping Tubes," *E.E., 69* (1950), 810–813; G. H. Hough and D. S. Ridler, "Some Recently Developed Cold Cathode Glow Discharge Tubes and Associated Circuits," *Electronic Engineering, 24* (1952), 152–157, 230–235, 272–276.

[44] H. E. Edgerton and K. J. Germeshausen, "A Cold-Cathode Arc-Discharge Tube," *E.E., 55* (1936), 790–794, 809; R. C. Hilliard, "Gaseous Discharge Tubes and Applications," *Electronics, 19* (March, 1946), 122–127.

electrodes in a tube filled with neon at a pressure of about five milli-meters of mercury. Application of a suitable voltage between various pairs of electrodes starts a glow discharge that transfers immediately to the anode and the cathode. The cathode is a cesium compound that releases cesium under ionic bombardment and facilitates transition of the glow discharge to an arc. In the application of the Strobotron as a stroboscopic light source a capacitor is discharged periodically through the tube, and the discharge, which is of short duration and amplitude of one hundred amperes or more owing to the formation of the arc, produces short flashes of light of high intensity. The tube is also useful as a control device for producing current pulses. Its control characteristics are conveniently represented by a family of closed curves similar to the one in Fig. 26a, but having the two control-electrode voltages as co-ordinates and the anode voltage as a para-meter.[45]

A continuous light output of controlled intensity is produced by a cold-cathode glow discharge tube known as a *glow modulator lamp*.[46] It contains a hollow or crater-type cathode that produces a high ionization density. The light output as viewed from the open end of the crater is of high intensity and is approximately proportional to the current. The tube is therefore useful for converting current variations into light variations for such applications as recording of sound on film and facsimile recording.

[45] A. B. White, W. B. Nottingham, H. E. Edgerton, and K. J. Germeshausen, "The Strobotron—II," *Electronics*, *10* (March, 1937), 18–21.

[46] R. C. Hilliard, "Gaseous Discharge Tubes and Applications," *Electronics*, *19* (March, 1946), 122–127.

CHAPTER VI

Rectifier Circuits

An important field of engineering is the application of electronic circuit elements to perform useful functions in power, control, measurement, and communication systems. From the background of analysis of circuits containing nonlinear resistive circuit elements given in *Electric Circuits*[1] and that of the behavior of electronic devices given in the previous chapters of this book, it is possible to proceed to a study of the basic methods by which electronic devices may be applied in engineering, as is done in the remaining chapters.

In this chapter, the application of circuit elements possessing a certain type of nonlinearity to the rectification of alternating current is considered; and in Chs. VII and XII further aspects of rectification are discussed. The analyses given here apply not only to circuits containing electron tubes, such as vacuum and gas diodes, thyratrons, and mercury-arc rectifiers, but also to circuits containing many other devices that, though physically different, have the same type of nonlinearity. Among these devices are the barrier-layer rectifiers such as copper-oxide and selenium rectifiers, and the crystal diodes discussed in Ch. XIII, and certain electrolytic cells. In order not to restrict the discussion to any one device, a broad consideration is given to *rectifiers*, which are defined as follows:[2] "A rectifier is a device which converts alternating current into unidirectional current by virtue of a characteristic permitting appreciable flow of current in one direction only." According to this definition, the term "rectifier" signifies a device for controlling a current as a function of its *direction*. A mechanical analogy is found in the trap valve of a reciprocating pump, which allows the flow of liquid in one direction but blocks its flow in the opposite direction.

A nonlinear element, to be useful practically as a rectifier, must conduct substantially in only one direction. Since not all nonlinear circuit elements possess this particular characteristic, a rectifier is a special kind of nonlinear resistance element.

Rectifiers find practical application in the production of direct current from an alternating-current source—for example, in battery chargers, radio-receiver and radio-transmitter power supplies, x-ray

[1] E. E. Staff, M.I.T., *Electric Circuits* (Cambridge, Massachusetts: The Technology Press of M.I.T.; New York: John Wiley & Sons, Inc., 1940), 657–723.

[2] *American Standard Definitions of Electrical Terms—A.S.A. No. C42* (New York: American Institute of Electrical Engineers, 1941), 79.

machines, railway substations, direct-current power transmission, electrolytic processes, and rectifier-type alternating-current meters. They are used also as detectors in radio receivers, as dynamo field dischargers, and in many other ways.

1. ELEMENTARY RECTIFIER THEORY

Rectification, or conversion of alternating current to direct current, usually performed by devices that conduct current more readily in one direction than in the other, may be accomplished by other means as well. For example, such time-varying circuit elements as synchronous commutators and contacts actuated by vibrating armatures may be made to vary their resistance in synchronism with the alternations of the source voltage, and thus to produce a unidirectional current. Rotating machines, such as synchronous converters and motor-generator sets, may also be used to produce direct current from alternating current, but the process is not then called rectification nor are the devices known as rectifiers. In Fig. 1, a parallel exposition of rectification by means of a nonlinear and a time-varying circuit element* brings out the similarities and the differences between the above two methods of obtaining direct current better than can a long discussion. In the figure, Column I may be considered as referring to a simple type of thermionic vacuum-tube rectifier such as might be used for the high-voltage supply to a radio transmitter; the resistance of the diode when conducting is neglected here. Column II may represent a synchronous-switch rectifier such as is sometimes used in storage-battery charging. The so-called mechanical rectifiers operating on the principle of the time-varying element might be thought to have possibilities equal to or greater than rectifiers employing electron tubes, but so far they have not been developed to nearly the same state of perfection.

Figure 1 shows that the cycle of the alternating-voltage source may be divided into two parts—one in which the rectifying element conducts current and the other in which the element is nonconducting. If a time-varying element is used to obtain rectification, the lengths of its periods of current conduction and interruption are independent of the remainder of the circuit. On the other hand, if a nonlinear element is used, the lengths of its conducting and nonconducting periods are affected by the other elements of the circuit. To understand this essential difference in the behavior of rectifying elements of the two

---

* Strictly speaking, these are circuit elements whose parameters are nonlinear or vary with time. They are called "nonlinear parameter elements" and "time-varying parameter elements" in *Electric Circuits,* but here the abbreviated terms are used.

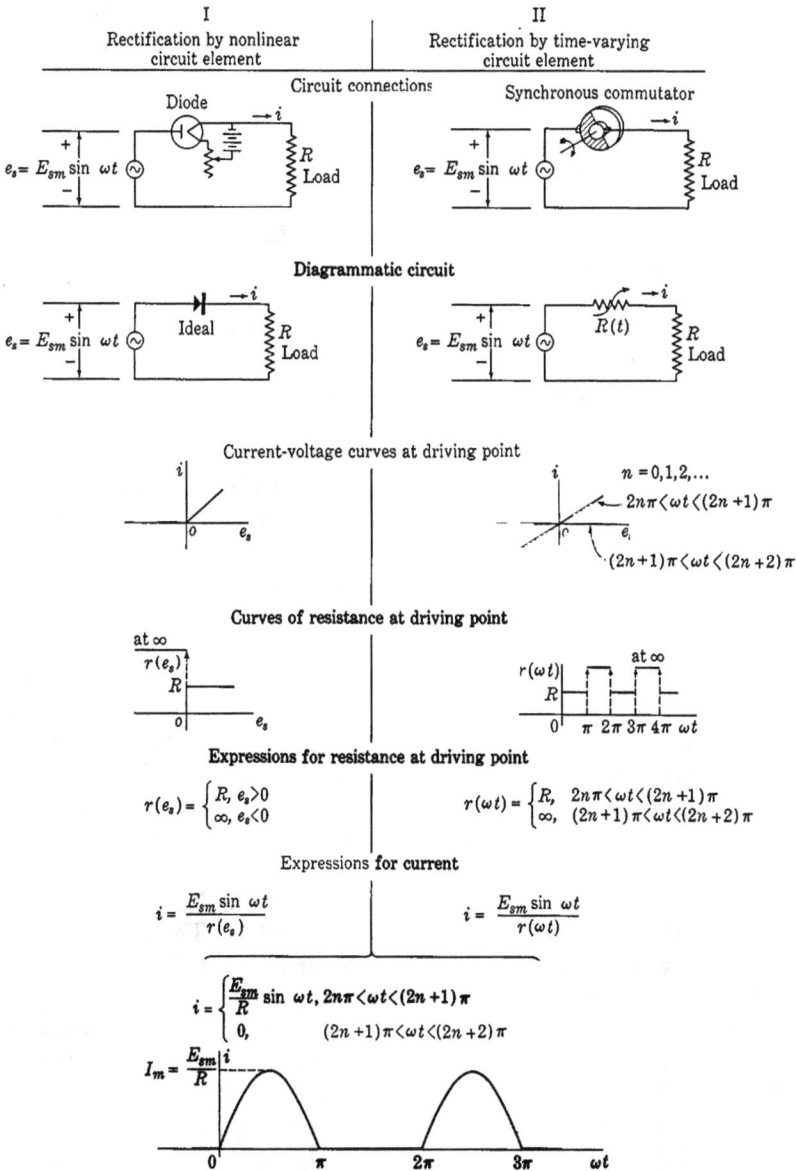

Fig. 1.   Comparison of rectification by nonlinear and time-varying circuit elements.

types, suppose a battery is inserted in series with the resistance R in each of the circuits of Fig. 1. The battery will have no effect upon the period of conduction of the synchronous commutator, for its time of conduction depends only on the speed of the driving motor and the length of the conducting segment. However, the battery will cause the diode to conduct current for a longer or shorter time, depending on its polarity, for its voltage adds to the voltage of the source and alters the time during which the total voltage applied to the diode is positive (see Art. 6). This dependence of a rectifying device of the nonlinear type on the circuit conditions, and of a rectifying device of the time-varying class on time alone, is characteristic of the distinction between nonlinear and time-varying elements.

Because of the difference between these two classes of circuit elements, the analyses of rectifier circuits in which they are used are very different. In this book only rectifier circuits using nonlinear elements are considered; in general, analysis of these circuits is much more difficult than that of circuits containing time-varying elements.

2. Graphical analysis of rectifier circuits

Since the rectifier element is nonlinear, a method of analysis suitable for nonlinear circuits must be used for determination of the instantaneous currents and voltages in a rectifier circuit. The algebraic or differential equation that describes the instantaneous equilibrium in the circuit is nonlinear, and its solution by straightforward methods is often impractical. Hence a graphical method, or one involving approximation of the nonlinear element by a linear resistance and a constant voltage source, is ordinarily used.

Analysis of the basic rectifier circuit in Fig. 2a by two different graphical methods is illustrated in Figs. 2b and 2c. The circuit comprises a resistance load R fed from an alternating voltage source e_s through a nonlinear rectifying element having a terminal voltage e_b. A thermionic vacuum diode is chosen for illustration, but the methods are equally applicable for any other nonlinear resistance element. The functional relation between the current through the nonlinear element and the voltage across it, which may be expressed in functional notation as $i_b(e_b)$, is represented graphically by the current-voltage characteristic curve labeled $i_b(e_b)$ in Figs. 2b and 2c. The sine wave of the supply voltage e_s is drawn so that instantaneous values of e_s may be projected upward from it to the voltage axis in each of the diagrams.

The solution in Fig. 2b is based on the principle that the series connection requires equality between the value of current given by the characteristic of the rectifier in terms of its terminal voltage and the value of current given by the characteristics of the circuit external to the rectifier, in terms of that same terminal voltage. The rectifier requires that the current satisfy the $i_b(e_b)$ curve. Likewise in terms of

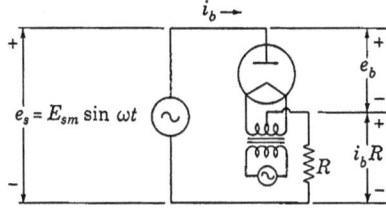

(a) Basic rectifier circuit

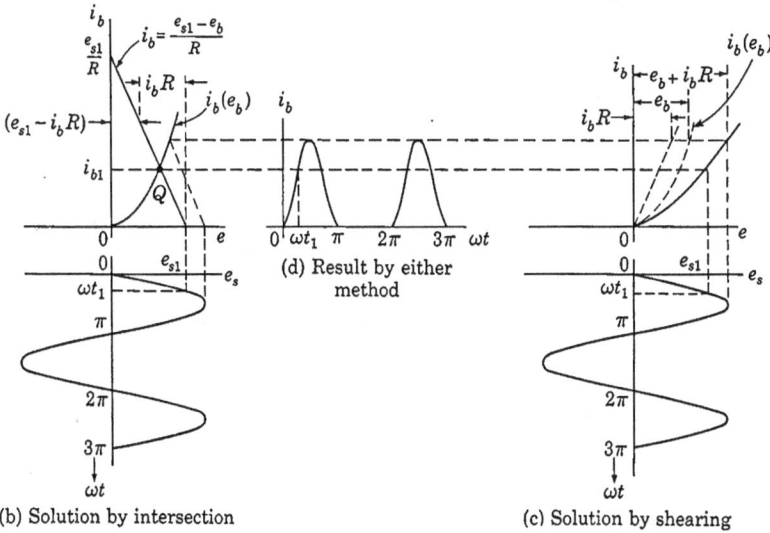

(b) Solution by intersection

(d) Result by either method

(c) Solution by shearing

Fig. 2.   Graphical analysis of rectifier operation.

e_b, the circuit external to the device requires that the current satisfy the relation

$$i_b = \frac{e_{s1} - e_b}{R},\qquad [1]$$

where e_{s1} is any particular instantaneous value of the source voltage. This relation describes a straight line through the points e_{s1}/R on the current axis and e_{s1}, as projected from the sine wave, on the voltage axis. The intersection of the $i_b(e_b)$ curve and the straight line, at Q in Fig. 2b, therefore gives the value of the current i_{b1} that satisfies the requirement of equality between the currents in the series-connected rectifier and circuit. Any other value of e_s gives another straight line parallel to the first that intersects the $i_b(e_b)$ curve at a new value of current. For example, the peak current corresponding to the peak voltage E_{sm} is determined by the dotted sloped line in the diagram.

The complete current wave in the circuit is found by repetition of the foregoing graphical construction for all values of e_s throughout the voltage cycle, and is shown in Fig. 2d. It is a distorted sine wave for positive values of source voltage and zero for negative values.

The principle used in the alternative graphical solution shown in Fig. 2c involves addition of the current-voltage characteristics of all the linear and nonlinear elements in the circuit to obtain an over-all characteristic for the circuit that is driven by the voltage source. In order to obtain the over-all voltage $e_b + i_b R$ existing across this circuit for any value of current i_b the voltage $i_b R$ across the resistance, which is the abscissa of the straight line that represents the current-voltage characteristic of the resistance at this value of i_b, is added graphically to e_b, which is the abscissa of the rectifier characteristic $i_b(e_b)$ at the same value of i_b. Repetition of this procedure gives the over-all current-voltage characteristic for the circuit that is shown as a solid curve in the figure. This graphical process is known as shearing the rectifier characteristic in the resistance line. Projection of instantaneous values of e_s onto the solid curve in Fig. 2c gives the same values of current in Fig. 2d as are obtained by the method of Fig. 2b.

The method of Fig. 2b has the advantage that no point-by-point graphical addition is required, whereas the method of Fig. 2c involves a simpler construction once the over-all curve is found. Both give exact results for a pure-resistance load. If appreciable inductance or capacitance is present, neither method is adequate, but an extension of either is of assistance in a step-by-step solution of the corresponding nonlinear differential equation. The method of Fig. 2c is also readily extended to solve the problem of a linear and nonlinear resistance element connected in parallel and fed from a current source or from a voltage source and series resistance.

3. ASSUMPTIONS FOR SIMPLIFYING ANALYSIS

Because the graphical methods described in the previous article are cumbersome, the use of an approximate method of analysis is often more convenient. A first major simplification that may be made is an idealization of the volt-ampere characteristic curve of the rectifier. The commercially important rectifiers may be divided into two classes according to the general shape of their characteristics:

▶Vacuum Type: Rectifiers such as the thermionic vacuum diode and the barrier-layer rectifier, whose volt-ampere characteristic in the conducting direction may be approximated by a straight line through the origin.

Gas Type: Rectifiers such as the mercury-arc rectifier and the thermionic gas diode, whose characteristic curve of current as a function of voltage in the conducting direction may be closely approximated by a vertical line.◀

The actual and the idealized forms of the volt-ampere characteristics for each of these types are shown in Fig. 3.

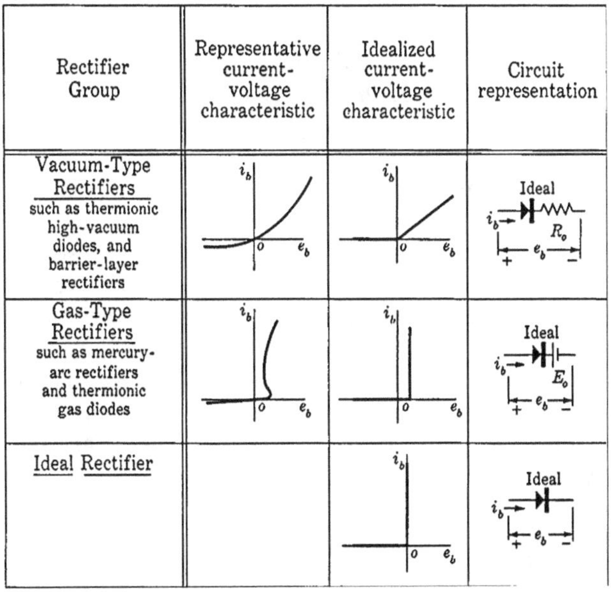

Rectifier Group	Representative current-voltage characteristic	Idealized current-voltage characteristic	Circuit representation
Vacuum-Type Rectifiers such as thermionic high-vacuum diodes, and barrier-layer rectifiers			
Gas-Type Rectifiers such as mercury-arc rectifiers and thermionic gas diodes			
Ideal Rectifier			

Fig. 3.   Characteristics of vacuum-type, gas-type, and ideal rectifiers.

The current-voltage characteristic of the *ideal rectifier*, shown also in Fig. 3, is obtained by a further simplification of the characteristic of rectifiers of either type. The voltage drop across the ideal rectifier when it is conducting is negligible; that is, the current-voltage characteristic of a vacuum-type rectifier is assumed to have such a steep slope in the conducting region that it coincides with the i_b axis, or the curve for a gas-type rectifier is assumed to be displaced from the i_b axis by such a short distance that it coincides with that axis. The ideal rectifier may be visualized as a switch that is closed and has zero resistance if the current flows through it in one direction, and is open and conducts no current if, because of a reversed voltage, current tends to pass through it in the opposite direction. Actual rectifiers may be replaced by the ideal in the analysis of rectifier circuits if only the salient features of the operation of the circuit are desired. Because there is no consumption of power by an ideal rectifier, it is not a sufficiently good approximation to an actual rectifier if such quantities as heating of the rectifier element, circuit efficiency, and voltage drop across the rectifier are to be found.

As shown in Fig. 3, a rectifier of the vacuum type may be approximately represented as an ideal rectifier in series with a resistance R_0, for the current-voltage curve of such a combination is identical with the idealized curve for rectifiers of this type if R_0 is chosen equal to the reciprocal of the slope of the idealized rectifier current-voltage curve in the conducting region. In a similar fashion, a gas-type rectifier may be approximately represented by a battery in series with an ideal rectifier, as in the figure. The voltage of the battery is E_0—the displamceent of the conducting portion of the idealized curve from the i_b axis—and the battery polarity is so chosen that the battery voltage opposes the flow of current in the conducting direction, just as does the rectifier voltage drop. If a vacuum diode is to be represented by an idealized characteristic, a better approximation than the idealized characteristic shown for that type in Fig. 3 is sometimes obtained through considering the diode as a series combination of an ideal rectifier, a battery of voltage E_0', and a resistance R_0'. The resulting characteristic is a combination of those for the vacuum-type and the gas-type rectifiers and, by proper adjustment of the magnitudes of E_0' and R_0', may be made to approximate reasonably well the actual characteristics of the diode. This idealization is used in Art. 12.

A second major simplification that can be made concerns the alternating-current source, which in most cases is connected to the rectifier through a step-up or a step-down transformer. Although a consideration of this transformer and of the internal impedance of the source is necessary for a complete solution of the problem, particularly in high-power rectifiers, the analysis is considerably complicated by their inclusion. In the circuits of this chapter, the transformer is considered to be ideal,[3] and the source is considered to have zero effective internal series impedance and to provide a sinusoidal voltage. Thus a sinusoidal voltage of constant amplitude, independent of the load, is applied to the rectifier circuit, as indicated in Fig. 4. In Fig. 4 are

[3] E. E. Staff, M.I.T., *Magnetic Circuits and Transformers* (Cambridge, Massachusetts: The Technology Press of M.I.T.; New York: John Wiley & Sons, Inc., 1943), Ch. X.

shown transformers suitable for use with half-wave and full-wave rectifier circuits, described in Arts. 4a and 4b. The transformer for the half-wave circuit furnishes a single sinusoidal voltage e_s and that for the full-wave circuit furnishes two sinusoidal voltages, e_s and e_s', which are 180 degrees out of phase on the basis of the reference directions assigned in the figure.

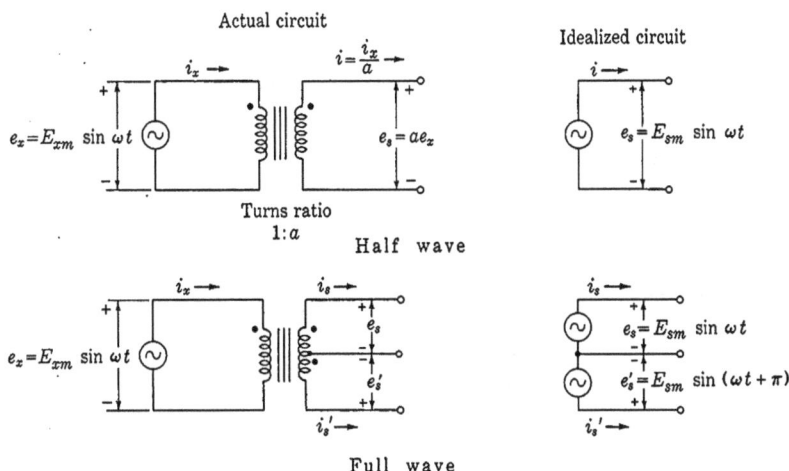

Fig. 4. Idealized transformer and power source for half-wave and full-wave rectifiers.

A rectifier operating with a resistance load in the simple circuit of Fig. 1 produces a current that, though *unidirectional*, is not *smooth*. Rather, it flows in pulses, as indicated by the current curve of Fig. 1. In a limited number of practical cases, such a pulsating unidirectional current is satisfactory. Most electrochemical applications, such as the charging of storage batteries and electroplating, are of this kind, and very simple rectifiers suffice. In many applications, however, as in radio-receiver and radio-transmitter power supplies and in direct-current power transmission, a substantially steady current is required. Special filter circuits are employed to smooth out the pulsations in the load current. Figure 5 shows the complete connections for a full-wave rectifier of the type often employed to supply direct-current power at 200 to 1,000 volts as a plate supply in radio apparatus. The rectifier elements commonly used are thermionic vacuum or gas diodes. The functional diagram of this figure represents symbolically the divisions that can conveniently be made in this circuit. The rectifying devices may be connected in a number of different ways, and the manner selected influences the form of the voltage applied to the smoothing-filter circuit. The filter circuit may be a simple series inductance or shunt capacitance, or a more complicated structure comprising many elements. The shunt capacitance tends to maintain the voltage across its terminal constant, and the series inductance tends to maintain the current through itself constant. Together, these elements tend to

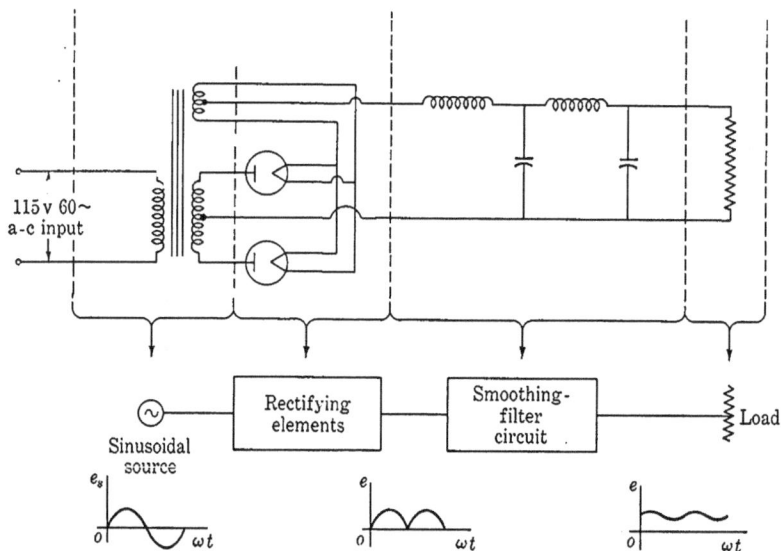

Fig. 5. Complete full-wave rectifier-and-filter circuit.

prevent the variations in current and voltage at the output of the rectifier, or input of the filter, from reaching the output of the filter. The choice of elements in a filter circuit is determined by the nature of the associated circuits and the degree to which the current pulsations must be smoothed.

4. Vacuum-type Rectifier with Resistance Load

The half-wave and the full-wave circuits are two very common connections of rectifiers. In this article, the simplifying assumptions introduced in Art. 3 are used to determine the performance of vacuum-type rectifiers when utilized in these circuits. As a basic problem, it is supposed that these rectifiers deliver power to a resistance load and that no filter circuit is used, a condition seldom encountered in practice; the modification of the performance and the complications of analysis introduced by use of filter circuits are reserved for consideration in subsequent articles.

Among the frequently important characteristics of a rectifier circuit are: (a) the waveform of the current delivered to the load, (b) the regulation of the output voltage, (c) the efficiency of the circuit in converting alternating- to direct-current power, (d) the heating of the rectifier element, (e) the peak value of the current in the rectifier element, and (f) the peak value of the voltage across the rectifier

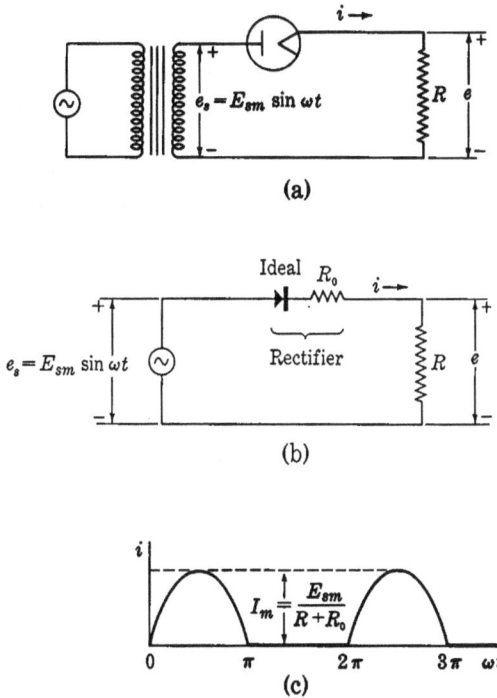

(a)

(b)

(c)

Fig. 6. Half-wave circuit with a vacuum-type rectifier and resistance load.

element in the inverse direction. These characteristics, as determined in the following articles, can be expected to deviate to some extent from those that may be determined by measurement on the circuits, because of the simplifying assumptions that are made. Furthermore, in practice, when the over-all losses and efficiency are to be found, it is necessary to take into account such things as the power loss in the transformers and the power needed to heat the filaments if thermionic rectifier tubes are used, and the power required to operate the vacuum pumps if a steel-tank mercury-arc rectifier is used; all these losses have been omitted from the analyses here.

4a. Half-Wave Circuit. The half-wave rectifier circuit with a resistance load and no filter is shown in Fig. 6. Actual connections of a diode used as a rectifier are shown in Fig. 6a, except that the connections for heating the cathode are omitted, and the idealized circuit appears in Fig. 6b. In the idealized circuit, the rectifier is shown as an ideal rectifier in series with a resistance R_0. The load resistance is denoted by R, the applied voltage is

$$e_s = E_{sm} \sin \omega t, \tag{2}$$

and the rectified load current is denoted by i.

Except for the ideal rectifier element, which behaves as a switch operated by the direction in which the current tends to flow, all the elements of the idealized circuit are linear. Therefore the waveform of the current is shown in Fig. 6c; it is composed of alternate half-sinusoids of maximum or peak value

$$I_m = \frac{E_{sm}}{R_0 + R} \qquad \blacktriangleright[3]$$

and current-free periods. The load voltage is

$$e = Ri \tag{4}$$

and therefore has the same waveform as the current.

The average or direct value of the current, I_{dc}, is

$$I_{dc} = \frac{1}{2\pi} \int_0^{2\pi} i \, d(\omega t) \tag{5}$$

$$= \frac{1}{2\pi} \int_0^{\pi} I_m \sin \omega t \, d(\omega t) + \frac{1}{2\pi} \int_\pi^{2\pi} 0 \, d(\omega t). \tag{6}$$

Thus

$$I_{dc} = \frac{I_m}{\pi} = \frac{E_{sm}}{\pi(R_0 + R)}. \tag{7}$$

The root-mean-square value of the load current, denoted by I, may be obtained from the expression

$$I - \sqrt{\frac{1}{2\pi} \int_0^{2\pi} i^2 \, d(\omega t)} = \sqrt{\frac{I_m{}^2}{2\pi} \int_0^{\pi} \sin^2 \omega t \, d(\omega t)} \, ; \tag{8}$$

whence,

$$I = \tfrac{1}{2} I_m = \frac{E_{sm}}{2(R_0 + R)}. \qquad \blacktriangleright[9]$$

The components of the load voltage e may be found in a similar way. By the use of Eqs. 4, 7, and 9, these components may be found directly from the current components. The peak value, average value, and effective value of the voltage are

$$E_m = E_{sm} \frac{R}{R + R_0} = \frac{E_{sm}}{(1 + R_0/R)}, \qquad [10]$$

$$E_{dc} = \frac{E_{sm}}{\pi\left(1 + \dfrac{R_0}{R}\right)}, \qquad \blacktriangleright[11]$$

$$E = \frac{E_{sm}}{2\left(1 + \dfrac{R_0}{R}\right)}, \qquad [12]$$

respectively. Examination of Eq. 11 shows that, unless the load resistance is always large compared with the resistance of the rectifier, the regulation of this rectifier may be poor; that is, the change of the direct voltage with change in load resistance may be great. The maximum value of the direct voltage occurs when R is very large compared with R_0, and is equal to E_{sm}/π.

The Fourier series for the current wave is found by the usual[4] analytical methods to be

$$i = I_m \left(\frac{1}{\pi} + \frac{1}{2} \sin \omega t - \frac{2}{3\pi} \cos 2\omega t - \frac{2}{15\pi} \cos 4\omega t - \cdots \right) \quad [13]$$

when the origin for the current and voltage waves is so chosen that the voltage wave is expressible as in Eq. 2. This expansion of the current wave into a Fourier series shows that the current may be considered to be the sum of an infinite number of current components. The first term of the series is the average or direct value determined in Eqs. 5, 6, and 7. The second term is a component of the same frequency as the supply voltage and has a peak value

$$\sqrt{2}I_1 = \tfrac{1}{2}I_m \qquad [14]$$

where I_1 is the effective value of the fundamental-frequency component. In a similar way, the third term is a second-harmonic component having an effective value I_2 and peak value

$$\sqrt{2}I_2 = \frac{2}{3\pi} I_m, \qquad [15]$$

and so on. It should be noticed that the frequency of the lowest-frequency alternating component of current is the same as that of the supply voltage.

[4] P. Franklin, *Differential Equations for Electrical Engineers* (New York: John Wiley & Sons, Inc., 1933), 69–74.

The power developed in the load resistance as a result of the direct component of the current, called the *direct-current power output*, is

$$P_{dc} = I_{dc}^2 R = \frac{E_{sm}^2 R}{\pi^2 (R_0 + R)^2}.$$ [16]

This power is not the total heat dissipated in the resistor, because the heating is caused by the total current, not merely by the direct component. However, there would be no need to rectify the current if the objective were only to heat the resistor. It is assumed that, whatever the load may be, it behaves as a resistance, but that the output power of interest is P_{dc}, the power associated with the direct component of the current.

With a given rectifier resistance R_0 and a given load resistance R, the power output can be controlled only by a change in the applied voltage e_s. Provision for this adjustment must be made in certain rectifier applications. For this purpose, tapped transformers, induction regulators and other similar schemes are used.

Since the actual rectifier element is represented by an ideal rectifier in series with a linear resistance R_0, the power dissipated in it is the sum of the power consumed by the ideal rectifier and that consumed by the resistance. But there is no power loss in the ideal rectifier. Hence, since R_0 is a *linear* resistance, the power dissipated in the rectifier in the form of heat is

$$P_p = I^2 R_0 = \frac{E_{sm}^2 R_0}{4(R_0 + R)^2}.$$ [17]

Note that this relationship for the power loss in the rectifier can be used only because the resistance R_0 is assumed to be linear. The temperature to which the tube or other rectifying device is raised in consequence of this heat is a major factor in determining the power rating of the device, as it is in practically all other forms of electrical power apparatus.

The total power drawn from the source equals the total power dissipated in the rectifier circuit. It is given by the expression

$$P_{in} = I^2 (R_0 + R) = \frac{E_{sm}^2}{4(R_0 + R)}.$$ [18]

From Eqs. 16 and 18, the rectifier efficiency for conversion of alternating-current power to direct-current power may be calculated as

$$\eta = \frac{P_{dc}}{P_{in}} = \left(\frac{2}{\pi}\right)^2 \frac{1}{1 + (R_0/R)} = \frac{40.6}{1 + (R_0/R)} \text{ per cent.}$$ [19]

The efficiency approaches a theoretical maximum of 40.6 per cent for this idealized half-wave rectifier circuit as R becomes large compared with R_0. The efficiency of this rectifier is less than 100 per cent for two reasons. First, only a portion of the power input to the rectifier circuit is delivered to the load. The remainder is lost as heat in the rectifier element. The factor $\dfrac{1}{1 + (R_0/R)}$ in Eq. 19 accounts for this loss in the rectifier. Second, only a portion of the power delivered to the load is direct-current power. The remainder is dissipated as heat associated with the alternating components of the load current. The factor $(2/\pi)^2$ in Eq. 19 accounts for this loss in the circuit.

Since both the current supplied to the load and the voltage developed across it have the half-sinusoidal waveform shown in Fig. 6c and represented by the Fourier series of Eq. 13, the ultimate purpose of the rectifier—to produce a steady output voltage and current from the sinusoidal voltage source—is but imperfectly achieved. A measure of the *ripple* voltage or current—that is, the waviness or lack of smoothness of the waveform—is given by the *ripple factor*. This quantity is defined as

$$\gamma \equiv \frac{\text{Effective value of the alternating components of voltage (or current)}}{\text{Direct or average value of the voltage (or current)}}. \qquad \blacktriangleright[20]$$

This definition may be put in a more convenient form for analysis by recognition of the fact[5] that the effective value of a wave made up of harmonic components is the square root of the sum of the squares of the effective values of the separate components. Thus, since the direct component is a harmonic component of zero frequency but is not an alternating component in the sense used here, for a current wave,

$$\gamma = \frac{\sqrt{I^2 - I_{dc}{}^2}}{I_{dc}} = \sqrt{\frac{I^2}{I_{dc}{}^2} - 1}. \qquad \blacktriangleright[21]$$

For the present problem of the half-wave rectifier with a resistance load and no filter,

$$\frac{I}{I_{dc}} = \frac{I_m/2}{I_m/\pi} = \frac{\pi}{2} = 1.57, \qquad [22]$$

$$\gamma = \sqrt{(1.57)^2 - 1} = 1.21. \qquad [23]$$

In all rectifier circuits, the rectifying element must withstand voltage in the reverse or inverse direction during a part of the cycle.

[5] E. E. Staff, M.I.T., *Magnetic Circuits and Transformers* (Cambridge, Massachusetts: The Technology Press of M.I.T.; New York: John Wiley & Sons, Inc., 1943), Art. 9e, Ch. VI.

The peak inverse voltage in this half-wave circuit with a resistance load is E_{sm} and is independent of the load current. Thus the rectifying elements must be rated for a

$$\text{Maximum peak inverse voltage} = E_{sm} = \pi E_{dc} \big|_{I_{dc}=0}. \qquad [24]$$

4b. *Full-Wave Circuit.* By means of a circuit that does not simply discard the lower half of the wave, as in Fig. 6, but uses it also to send current through the load in the proper direction, a smoother wave can

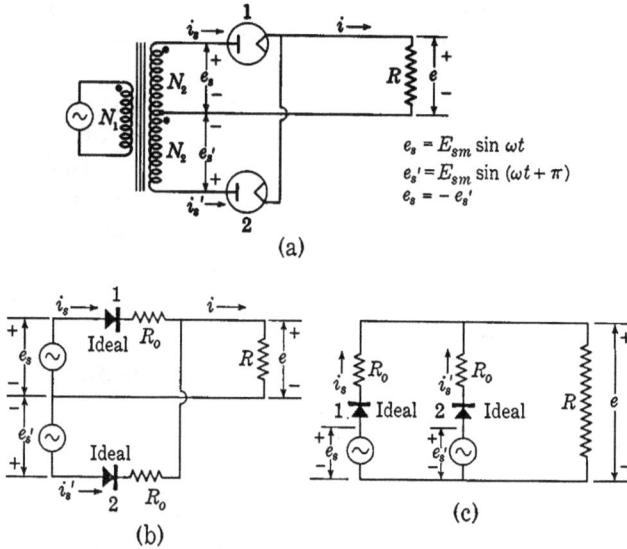

$$e_s = E_{sm} \sin \omega t$$
$$e_s' = E_{sm} \sin (\omega t + \pi)$$
$$e_s = -e_s'$$

(a)

(b)

(c)

Fig. 7.   Full-wave circuit with vacuum-type rectifiers and resistance load.

be supplied at a greater efficiency. The *full-wave connection*, also called the *diametric* and the *biphase connection*, is such a circuit. A large majority of single-phase rectifiers employ this connection of the rectifying devices (together with some smoothing-filter circuit). The full-wave connection with a resistance load and no filter is shown in Fig. 7. A center tap on the transformer secondary winding is used, as is shown in Fig. 7a, and the sinusoidal voltages e_s and e_s' are therefore equal in magnitude and 180 degrees out of phase when their positive reference directions are assigned as indicated; that is,

$$e_s = E_{sm} \sin \omega t, \qquad [25]$$

$$e_s' = -e_s, \qquad [26]$$

$$e_s' = E_{sm} \sin (\omega t + \pi). \qquad [27]$$

For analysis, the idealized sources in Figs. 7b and 7c having the characteristics discussed in connection with Fig. 4 are used. The two rectifiers are assumed to have identical characteristics—an assumption fairly well justified in practice. During one half of the cycle, the voltage across rectifier 1 only is in the conduction direction; hence this rectifier passes current while rectifier 2 is nonconducting. During the other half of the cycle, the conditions are reversed, and rectifier 2 carries current while rectifier 1 is nonconducting. Both the current i_s through rectifier 1 and the current i_s' through rectifier 2 have the same waveform as in a half-wave connection, but the two current waves are displaced from each other by 180 degrees. Both currents flow through the load resistance R in the same direction; hence the waveform of the resultant current

$$i = i_s + i_s' \qquad [28]$$

is that shown in Fig. 8. In the separate rectifiers, the currents have the waveform of the half-wave rectifier connection shown in Fig. 6c and already studied.

The maximum value of the instantaneous current in the full-wave rectifier is given by

$$I_m = \frac{E_{sm}}{R_0 + R} \qquad [29]$$

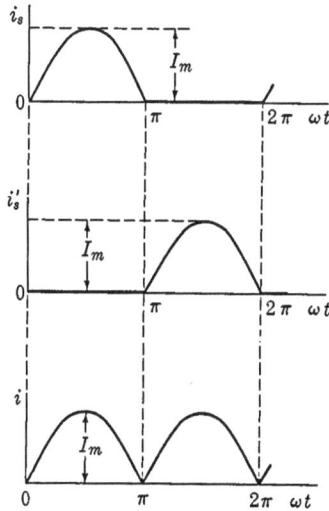

Fig. 8. Waveforms of the currents in the full-wave circuit with vacuum-type rectifiers, a resistance load, and no filter.

just as in the half-wave circuit, since each half of the circuit acts independently of the other. Note that E_{sm} in Eq. 29 is the maximum value of the voltage between one end terminal and the center tap of the transformer secondary winding, not the maximum value of the full end-to-end terminal voltage of the winding, and R_0 is the resistance of one rectifier.

The average or direct value of the load current determined by the method of Eqs. 5 and 6 is twice that of the half-wave circuit because the additional lobe in the current wave makes the second term on the right-hand side of Eq. 6 equal to the first, rather than to zero. Thus

$$I_{dc} = \frac{2}{\pi} I_m . \qquad \blacktriangleright[30]$$

Furthermore, the effective or rms value of the current determined by the method of Eq. 8 is

$$I = \frac{1}{\sqrt{2}} I_m,$$ [31]

since the term under the square root sign in Eq. 8 is also doubled.

Correspondingly, the maximum, average, and effective values of the load voltage are

$$E_m = \frac{E_{sm}}{1 + (R_0/R)},$$ [32]

$$E_{dc} = \frac{2}{\pi} \frac{E_{sm}}{1 + (R_0/R)},$$ ▶[33]

$$E = \frac{1}{\sqrt{2}} \frac{E_{sm}}{1 + (R_0/R)},$$ [34]

respectively.

The Fourier series for the load current can be found through taking the sum of the series for the individual rectifier currents as follows:

$$i_s(\omega t) = I_m \left(\frac{1}{\pi} + \frac{1}{2} \sin \omega t - \frac{2}{3\pi} \cos 2\omega t - \frac{2}{15\pi} \cos 4\omega t - \cdots \right),$$ [35]

$$i_s'(\omega t) = i_s(\omega t + \pi)$$ [36]

$$= I_m \left(\frac{1}{\pi} - \frac{1}{2} \sin \omega t - \frac{2}{3\pi} \cos 2\omega t - \frac{2}{15\pi} \cos 4\omega t - \cdots \right),$$ [37]

$$i = i_s + i_s'$$ [38]

$$= I_m \left(\frac{2}{\pi} - \frac{4}{3\pi} \cos 2\omega t - \frac{4}{15\pi} \cos 4\omega t - \cdots \right).$$ [39]

From Eq. 39, the lowest-frequency ripple component of the full-wave rectified current is seen to be of angular frequency 2ω; that is, it has twice the frequency of the applied voltage. The lowest frequency in the load current of the full-wave connection, then, is double that in the half-wave connection. This fact offers a decided advantage in smoothing, for smaller elements are needed in the filter circuit to reduce the ripple to an equally small value. It is interesting to consider that the addition of the two currents i_s and i_s', each having a term $\sin \omega t$, gives rise to a current without this term but containing all the other terms of i_s and i_s'.

Because of the relative directions of the windings and the currents, the separate currents i_s and i_s' in Figs. 4 and 7 tend to magnetize the core of the transformer in opposite directions. The load component of the current in the primary winding is therefore

$$i_x = \frac{N_2}{N_1}(i_s - i_s'),$$ [40]

where

i_x is the primary current,

N_1 is the number of turns in the primary winding,

N_2 is the number of turns in the secondary winding between the center tap and either end terminal.

Graphical addition of the currents i_s and $-i_s'$ similar to the addition of i_s and i_s' performed in Fig. 8, or the addition of the series in Eqs. 35 and 37, gives

$$i_x = \frac{N_2}{N_1} I_m \sin \omega t.$$ [41]

The full-wave rectifier circuit therefore draws a sinusoidal current from the line despite the fact that the individual rectifier and secondary currents are highly nonsinusoidal. The transformer losses are therefore small compared with those in the half-wave circuit, because the load component of the primary current is sinusoidal and because there is no net direct-current magnetization of the core to cause excessive exciting current. A smaller transformer can therefore be used, which is an important advantage.

The direct-current power output is

$$P_{dc} = I_{dc}{}^2 R = \left(\frac{2}{\pi}\right)^2 \frac{E_{sm}{}^2 R}{(R_0 + R)^2}.$$ ▶[42]

Compared with the half-wave connection, the full-wave connection gives a direct current twice as large and a power output four times as large when the peak values of the currents are the same in each circuit; that is, when the resistance of the rectifying devices and the magnitudes of the applied voltages (measured from center tap to one end of the supply transformer in the case of the full-wave rectifier) are the same for each connection. The power output depends on E_{sm}, R, and R_0 in the same way as in the half-wave circuit.

A further advantage of the full-wave circuit over the half-wave circuit may be deduced from a consideration of the power dissipated in the rectifying elements when identical elements are used in both circuits. Equations 7 and 30 show that, for the same direct output current, the maximum current I_m in a rectifying element in the full-wave circuit need be only one-half that in the half-wave circuit. Since the waveform of the current through the rectifying elements is the same in both circuits, and since the power loss in one of these elements is proportional to the square of the peak current, the power loss in one rectifying element of the full-wave circuit is one-fourth that in the rectifying element of the half-wave circuit for the same direct output current. Thus the total loss in the two rectifying elements of a full-wave circuit is one-half that in the rectifying element of a half-wave circuit when both circuits furnish the same direct output current, or the same direct output power, to identical loads. Consequently, the total power rating of the rectifying elements used in the full-wave connection can be less than that of the rectifying element used in the half-wave connection for equal outputs.

As a result of the use of both half-cycles of the supply voltage by a full-wave rectifier, the conversion of alternating current to direct current is more nearly complete than in the half-wave circuit, and a larger portion of the power dissipated in the load is the power associated with the direct component of current. For this reason, the efficiency of rectification is increased. The total power supplied to the circuit is

$$P_{in} = I^2(R_0 + R) = \frac{E_{sm}{}^2}{2(R_0 + R)}. \qquad [43]$$

From Eqs. 42 and 43, the efficiency is found to be

$$\eta = \frac{P_{dc}}{P_{in}} = 2\left(\frac{2}{\pi}\right)^2 \frac{1}{1 + (R_0/R)} = \frac{81.2}{1 + (R_0/R)} \text{ per cent.} \quad \blacktriangleright[44]$$

This value is just twice that for the half-wave connection. The theoretical maximum efficiency for the full-wave rectifier is 81.2 per cent, and it obtains when the load resistance R is very large compared with the internal rectifier resistance R_0. The maximum efficiency corresponds to a very small output, as it does for the half-wave rectifier. The factor $\dfrac{1}{1 + (R_0/R)}$ specifies the fraction of the input power delivered to the load, and the factor $2(2/\pi)^2$ is the fraction of the power delivered to the load that is converted to direct-current power.

It should be noticed that the factor $\dfrac{1}{1 + (R_0/R)}$, which depends upon the fraction of the *input power* lost in the rectifier elements, is the same for both half-wave and full-wave circuits, even though the total loss in both the rectifiers of a full-wave circuit is only half that in the rectifier of a half-wave circuit when the *output power* of both circuits is the same.

Since for the full-wave rectifier circuit,

$$\frac{I}{I_{dc}} = \frac{I_m/\sqrt{2}}{2I_m/\pi} = \frac{\pi}{2\sqrt{2}} = 1.11, \qquad [45]$$

the ripple factor is:

$$\gamma = \sqrt{(1.11)^2 - 1} = 0.48. \qquad [46]$$

A comparison of the full-wave with the half-wave connection shows that γ drops from 1.21 to 0.48.

As is evident from Fig. 7a, the cathodes of both rectifiers may be connected together in the full-wave circuit. This condition makes possible the use of mercury-arc rectifiers in which two anodes draw current from one cathode, as well as small thermionic vacuum or gas tubes constructed with both rectifier elements in the same bulb and with their cathodes heated from a common source.

Since the two rectifiers in the full-wave circuit are connected in series across the supply transformer, the voltage impressed on the nonconducting rectifier is the full transformer secondary voltage less the voltage drop in the conducting rectifier. The voltage drop in the conducting rectifier approaches zero with the load current; hence the maximum value of the peak inverse voltage to which the tubes in this circuit are subjected as the load is varied is

$$\text{Maximum peak inverse voltage} = 2E_{sm} = \pi E_{dc}\big|_{I_{dc}=0}, \qquad [47]$$

This value bears the same relation to the output voltage as does the value for the half-wave rectifier given in Eq. 24.

5. Gas-type Rectifier with Resistance Load

When gas-type rectifiers which have a constant voltage drop while conducting, are used in one of the circuits discussed in the preceding articles, the circuit operation is slightly different from that obtained with vacuum-type rectifiers, because the current does not begin to flow through the rectifier until the applied voltage e_s reaches the value E_0 of the constant voltage drop across the rectifier, and it ceases to flow as soon as e_s becomes less than E_0. Similarly, if a rectifier is used to charge a battery in series with a resistance, the voltage of the battery prevents the flow of current until the source voltage exceeds the battery voltage.

In Fig. 9 is shown a half-wave rectifier circuit containing a gas-type rectifier and a resistance load R. The rectifier element is represented by an ideal rectifier in series with a battery of voltage E_0, since the current-voltage characteristic of this combination, shown in Fig. 9b, is the same as that of the idealized gas-type rectifier. The waveform of the supply voltage is

$$e_s = E_{sm} \sin \omega t. \tag{48}$$

A series of curves showing the waveform of the current i through the load for different ratios of E_0 to E_{sm} is given in Figs. 9c and 9d. This

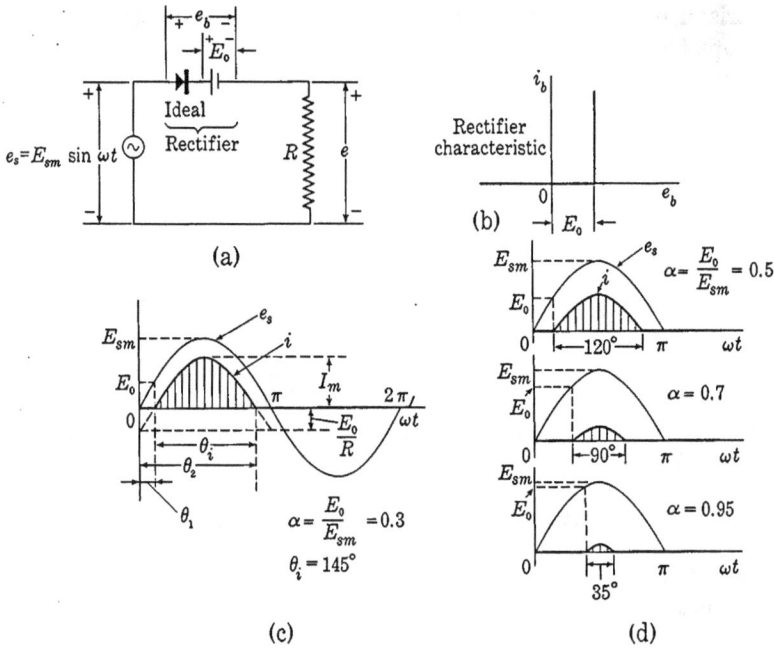

Fig. 9. Half-wave circuit with a gas-type rectifier and a resistance load.

series indicates the way in which the peak value of the load current

$$I_m = \frac{E_{sm} - E_0}{R}, \tag{49}$$

the waveform of this current, and the angle of conduction θ_i (that is, the angle in the supply-voltage cycle during which current flows) change with the ratio

$$\alpha = E_0/E_{sm}. \tag{50}$$

If E_0 were equal to zero, the current would pass through the rectifier during the entire half-cycle and would rise to the peak value

$$I_m{}' = E_{sm}/R. \tag{51}$$

Therefore the constant rectifier voltage drop E_0 may be thought of as producing two effects: first, a reduction of the conduction period; and, second, a reduction of the voltage effective in causing current.

The current through the rectifier may be expressed analytically by

$$i = \frac{e_s - E_0}{R} = \frac{E_{sm} \sin \omega t - E_0}{R} \qquad [52]$$

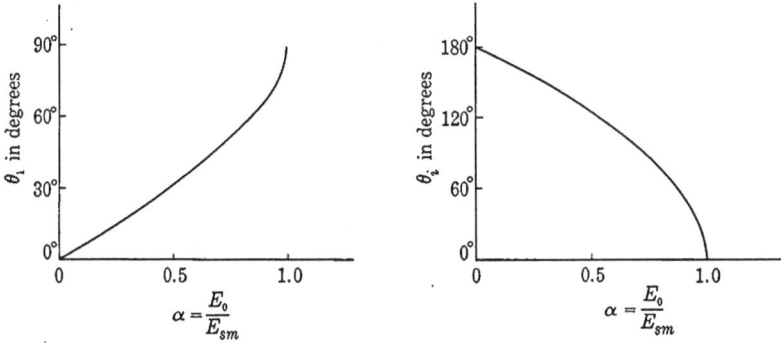

Fig. 10. Curves of ignition angle θ_1 and conduction angle θ_i as functions of α, or E_0/E_{sm}, in the circuit of Fig. 9a.

during the period of conduction. Substitution of Eqs. 50 and 51 in Eq. 52 gives

$$i = I_m{}' [(\sin \omega t) - \alpha], \text{ for } e_s \geqq E_0, \qquad [53]$$

$$i = 0, \qquad\qquad\qquad \text{ for } e_s \leqq E_0, \qquad [54]$$

where it is to be noted that $I_m{}'$ is *not* the actual peak value of the current but is the limit approached by the peak current as E_0 becomes small.

The ignition angle θ_1 and the conduction angle θ_i may be obtained through equating i, in Eq. 53, to zero:

$$I_m{}' [(\sin \theta_1) - \alpha] = 0; \qquad [55]$$

whence,

$$\theta_1 = \sin^{-1} \alpha. \qquad [56]$$

By symmetry,

$$\theta_2 = \pi - \theta_1, \qquad [57]$$

and it follows that

$$\theta_i = \theta_2 - \theta_1 = \pi - 2\theta_1 = \pi - 2 \sin^{-1} \alpha. \qquad [58]$$

Curves of θ_1 and θ_i as functions of α are shown in Fig. 10.

The graphical representation in Fig. 9d helps to make the physical situation clearer. If the arc drop E_0 is small compared with the applied voltage E_{sm}, current passes substantially during one half of the period; but, if E_0 is comparable with E_{sm}, the current is reduced to small pulses occupying only a small fraction of the period. Obviously, the

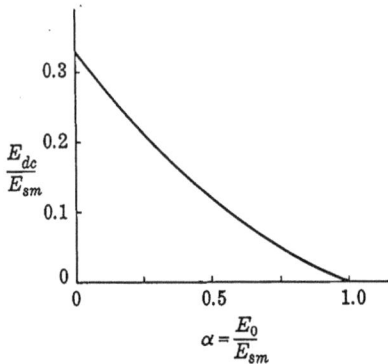

Fig. 11. Direct voltage across load per unit peak alternating supply voltage as a function of α, or E_0/E_{sm}, in the circuit of Fig. 9a.

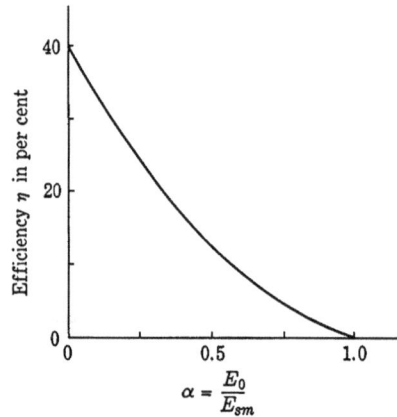

Fig. 12. Efficiency η as a function of α, or E_0/E_{sm}, for the rectifier of Fig. 9a.

value of the direct current through the load is reduced as α becomes larger. The direct current is given by

$$I_{dc} = \frac{1}{2\pi} \int_{0_1}^{\pi-0_1} i\, d(\omega t) = \frac{I_m'}{2\pi} \int_{0_1}^{\pi-0_1} [(\sin \omega t) - \alpha]\, d(\omega t) \qquad [59]$$

$$= -\frac{I_m'}{2\pi} \left[(\cos \omega t) + \alpha \omega t \right]_{\sin^{-1}\alpha}^{\pi-\sin^{-1}\alpha} \qquad [60]$$

$$= \frac{I_m'}{\pi} (\sqrt{1 - \alpha^2} - \alpha \cos^{-1}\alpha). \qquad [61]$$

The direct component of the voltage across the load is $I_{dc}R$, or

$$E_{dc} = \frac{E_{sm}}{\pi} (\sqrt{1 - \alpha^2} - \alpha \cos^{-1}\alpha). \qquad \blacktriangleright[62]$$

The ratio E_{dc}/E_{sm} is plotted as a function of α in Fig. 11. It is evident that the direct output voltage and current are reduced rapidly as the value of α, or E_0/E_{sm}, increases. In many practical applications, particularly in high-power and high-voltage rectifiers, E_0 is small compared with E_{sm}, and α is hence a small fraction.

The effective or rms value of the current is given by

$$I = I_m' \sqrt{\frac{1}{2\pi} \int_{\theta_1}^{\pi - \theta_1} [(\sin \omega t) - \alpha]^2 \, d(\omega t)} \tag{63}$$

$$= I_m' \sqrt{\frac{1}{2\pi} \left[(1 + 2\alpha^2) \cos^{-1} \alpha - 3\alpha\sqrt{1 - \alpha^2} \right]}. \tag{64}$$

The power loss in the rectifier is

$$P_p = \frac{1}{2\pi} \int_{\theta_1}^{\pi - \theta_1} E_0 i \, d(\omega t) = \frac{E_0}{2\pi} \int_{\theta_1}^{\pi - \theta_1} i \, d(\omega t), \tag{65}$$

or

$$P_p = E_0 I_{dc}. \qquad \blacktriangleright[66]$$

Equation 66 may be contrasted with Eq. 17.

The total power input is given by

$$P_{in} = I^2 R + I_{dc} E_0, \tag{67}$$

and the direct-current power output by

$$P_{dc} = I_{dc}^2 R. \qquad \blacktriangleright[68]$$

By the use of Eqs. 67 and 68, the efficiency is found to be

$$\eta = \frac{P_{dc}}{P_{in}} = \frac{I_{dc}^2 R}{I^2 R + I_{dc} E_0}. \tag{69}$$

This relation may be put in more useful form by means of Eqs. 50, 51, 61, and 64, as follows:

$$\eta = \frac{I_{dc}^2}{I^2 + \alpha I_{dc} I_m'} \tag{70}$$

$$= \frac{\left(\dfrac{I_m'}{\pi}\right)^2 (\sqrt{1 - \alpha^2} - \alpha \cos^{-1} \alpha)^2}{\dfrac{(I_m')^2}{2\pi} \left[(1 + 2\alpha^2) \, (\cos^{-1} \alpha) - 3\alpha\sqrt{1 - \alpha^2} \right]}$$
$$+ \frac{\alpha (I_m')^2}{\pi} \left[\sqrt{1 - \alpha^2} - \alpha \cos^{-1} \alpha \right] \tag{71}$$

$$= \frac{2}{\pi} \frac{\left(\alpha \cos^{-1} \alpha - \sqrt{1 - \alpha^2} \right)^2}{\cos^{-1} \alpha - \alpha\sqrt{1 - \alpha^2}} \tag{72}$$

$$= \frac{\pi}{2} \frac{\left(\alpha \cos^{-1} \alpha - \sqrt{1 - \alpha^2} \right)^2}{\cos^{-1} \alpha - \alpha\sqrt{1 - \alpha^2}} \times 40.6 \text{ per cent.} \tag{73}$$

Values of this quantity are plotted as a curve in Fig. 12, which shows that the efficiency drops rapidly as α increases. Or, stated differently, the efficiency increases with the applied voltage E_{sm} and approaches its maximum value when E_{sm} becomes large compared with E_0. Mercury-arc rectifiers demand high voltages if their operation is to be economical. The maximum theoretical value of η equals 40.6 per cent and coincides with that for the half-wave rectifier in Art. 4a. A second important fact to be read from Eq. 73 is that the efficiency is, to the degree of approximation contained in this analysis, independent of the load and dependent only on the ratio of tube voltage drop E_0 to applied alternating voltage E_{sm}. This behavior may be termed a characteristic of mercury-arc and thermionic-gas-diode rectifiers. Because of this dependence of efficiency on the value of α, gas-type rectifiers are most useful in applications where E_0 is small compared with E_{sm}.

The foregoing analysis may be extended to apply to the full-wave connection through recognition of the fact that there are two similar pulses of current and voltage per cycle of the supply voltage instead of one.

6. BATTERY-CHARGING RECTIFIERS

The analysis of the gas-type rectifier with a resistive load may be applied equally well to a rectifier used to charge a battery. In Fig. 9a, if the battery of voltage E_0 is thought of as the battery to be charged, and the resistance R as a resistance placed in series with the battery to control the charging current, the diagram applies to a half-wave battery-charging circuit. If a vacuum-type rectifier element is used, the total resistance consists of R_0 and R in series; if a gas-type rectifier is used, the total "battery voltage" in the circuit consists of the rectifier voltage drop E_0 added to the actual battery voltage. Provided E_0 and R are interpreted in this manner, most of the equations of Art. 5 may be applied to the battery-charging circuit.

7. POLYPHASE RECTIFIERS

For the rectification of large amounts of power, polyphase rectifier circuits are used in preference to the single-phase circuits of the preceding articles, for several reasons. First, most electrical power is at present generated and distributed as three-phase power; hence a large-power rectifier should be suitable for use with a three-phase supply. Second, as appears in the discussion that follows, the output voltage of a polyphase rectifier is smoother than that of a single-phase rectifier when no filter is used with either. A smooth output voltage can thus be produced with a simpler and less expensive filter with the poly-phase circuit; or, in applications where ripple in the output is unimportant and no filter is used, the efficiency of the polyphase rectifier is higher. Third, the utilization of supply transformers and other associated equipment is better with certain polyphase circuits than with single-phase circuits—that is, for a given rectifier output the

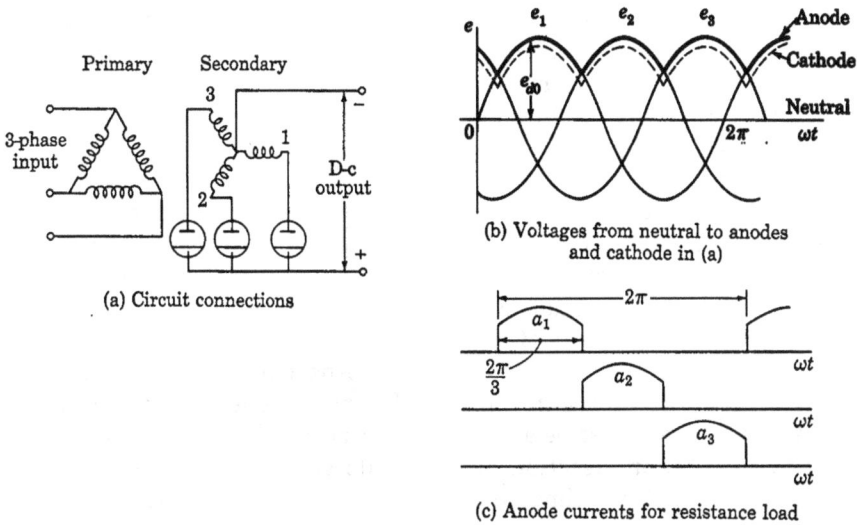

Primary   Secondary

3-phase input

D-c output

(a) Circuit connections

(b) Voltages from neutral to anodes and cathode in (a)

(c) Anode currents for resistance load

Fig. 13.   Basic three-phase rectifier circuit with a delta-wye-connected transformer.

rating of the associated equipment needed with the polyphase rectifier is smaller.

Most high-power rectifiers consist of either a number of anodes that operate from a common cathode or a number of single-anode rectifiers that have their cathodes connected together. The following discussion applies equally well to either form. Mercury-pool-type cathodes are illustrated in the diagrams. Hot cathodes are, however, equally applicable. Some means for starting the arc, such as an ignitor, is required with the mercury-pool cathode, but is omitted from the diagrams. A few special rectifier circuits do not have the common-cathode feature—one such type is discussed in Art. 8. The analysis of these circuits is an extension of the analysis given here.

Figure 13 shows a simple three-phase three-anode mercury-arc rectifier with a delta-wye-connected transformer. The transformer in this and other polyphase rectifier circuits serves two purposes. First, it provides the proper voltage for the rectifier from whatever supply is available; and, second, it provides a common connection to the load for the three anodes. With three, or any other number of anodes, there must be a common connection to the load through the transformer secondary windings; hence these windings must always have a neutral connection. *This neutral is used as a reference point in the following analysis, and voltages are given with respect to it.*

In this elementary discussion the effects of the transformer resistance and reactance are neglected; the anode voltages thus are assumed to be sinusoidal and equal to the primary voltages multiplied by the appropriate transformation ratio. The anode voltages are then equal to the secondary induced voltages e_1, e_2, and e_3 for the three-anode rectifier of Fig. 13a. These three anode voltages are shown plotted as functions of time in Fig. 13b, which shows that for part of the time more than one anode is positive with respect to the neutral. Since these voltages are taken with respect to the transformer neutral rather than with respect to the cathode, it *does not* follow that more than one anode will conduct at a time. If the arc voltage drop is assumed to be independent of the current and to be the same for all anodes, the potential of the cathode with respect to the neutral is as shown by the dotted wave in Fig. 13b. For two anodes to conduct at the same time, they must be at the same potential. Since this equality occurs only at the intersection of the anode-voltage curves, only one anode will conduct at a time, and this anode is the one that has the highest potential with respect to the neutral. The arc transfers from anode to anode in the order 1, 2, 3, 1, 2, 3, and so on. This transfer is termed *commutation* of the current, and the difference voltage, $e_2 - e_1$, $e_3 - e_2$, and so on, when positive, is called the *commutating voltage*.

For convenience, the voltage of the conducting *anode* with respect to the transformer neutral may be defined as the *rectified voltage*. The symbol e_{d0} indicates the instantaneous value of this rectified voltage, and the symbol E_{d0} its average value or direct component. In the three-phase rectifier of Fig. 13a, the rectified-voltage wave follows the positive caps or envelope of the sinusoidal anode voltages e_1, e_2, and e_3 in Fig. 13b. At every instant it is greater than the cathode-to-neutral voltage (shown as a dotted line) by the amount of the arc voltage drop in the rectifier, which is relatively constant. Thus the rectified voltage defined here is *not* the true voltage across the load, for it includes the arc voltage drop and the voltage drop in a smoothing reactor if one is used in series with the load. However, the selection of the rectified voltage as the quantity for analysis in the following discussion simplifies the relationships. The actual direct output voltage may be obtained through subtraction of the essentially constant arc voltage drop and the direct voltage drop in the resistance of the filter from E_{d0}, the average value of the rectified voltage.

Though most of the power generated at present is three-phase, optimum conditions of rectification are sometimes not realized unless a polyphase source having a number of phases greater than three is

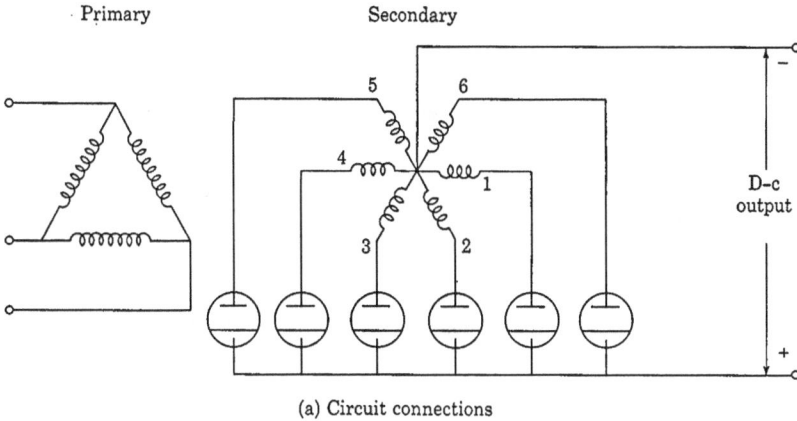

(a) Circuit connections

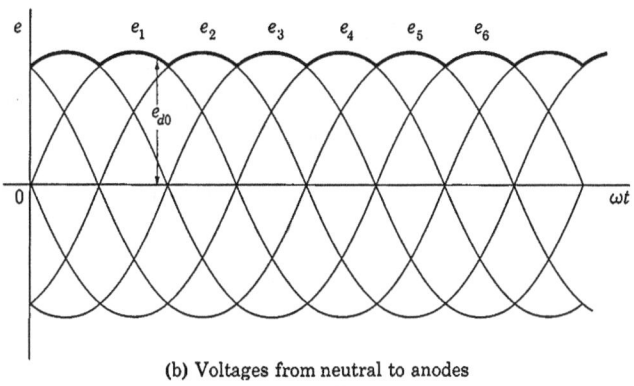

(b) Voltages from neutral to anodes

Fig. 14.   Delta, six-phase, star rectifier circuit.

used. Consequently transformation of the available three-phase power into six-phase, twelve-phase, or other polyphase power by means of transformers, called phase transformation, is commonly employed in the alternating-current supply circuit of the rectifier. An understanding of the methods of phase transformation discussed in

Magnetic Circuits and Transformers is therefore desirable as a background for this discussion.

Six-phase operation is obtained if three center-tapped secondary windings, each associated with one of the primary windings, are connected to six rectifiers as in Fig. 14a. This connection has many of the same advantages over the connection of Fig. 13 that the single-phase full-wave circuit of Fig. 7 has over the half-wave circuit of Fig. 6, for it resembles three single-phase full-wave rectifier circuits connected in parallel, but with their corresponding secondary voltages displaced from one another by 120 electrical degrees. Because of the interconnection, however, it differs from such an arrangement of three circuits in several important respects, as is discussed subsequently. The cathode potential of the rectifying devices follows somewhat below the envelope of the secondary voltage waves in Fig. 14b, and the lowest ripple frequency in the output voltage is consequently six times the supply frequency—twice that in the three-phase rectifier of Fig. 13.

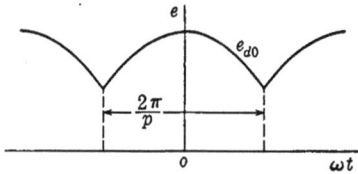

Fig. 15. Waveform of the instantaneous rectified voltage e_{d0} in a p-phase rectifier.

One or a number of rectifiers having p anodes may be connected to a p-phase system of voltages in a star circuit similar to that of the three-phase and six-phase rectifiers of Figs. 13 and 14. The operation of such a rectifier circuit is similar to that of the three-phase rectifier circuit except that the arc is commutated p times in each cycle instead of three times. Thus each anode conducts for $2\pi/p$ radians in each cycle of the supply voltage, and the rectified voltage e_{d0} consists of a series of sine-wave caps, each of which extends over that interval.

The rectified voltage e_{d0} for the general case of the p-phase rectifier may be expanded into a Fourier series and thus resolved into a steady direct voltage and a series of harmonic voltages. Since the number of phases affects the value of E_{d0} as well as the magnitude and frequency of the harmonic voltages, it must be considered in the choice of a rectifier. If the instant at which a peak of the rectified voltage occurs is chosen as the zero instant of time, as shown in Fig. 15, the voltage wave is an even function, and only cosine terms appear in the Fourier series. If the effective value of each of the secondary phase voltages $e_1, e_2, e_3, \cdots$ is E_s, the equation for the rectified voltage between the limits $-(\pi/p)$ and $+(\pi/p)$ is

$$e_{d0} = \sqrt{2}E_s \cos \omega t. \qquad [74]$$

The average value of e_{d0} is given by

$$E_{d0} = \frac{p}{2\pi} \int_{-(\pi/p)}^{+(\pi/p)} \sqrt{2}E_s \cos \omega t \, d(\omega t) = \sqrt{2}E_s \frac{p}{2\pi} \sin \omega t \Big|_{-(\pi/p)}^{+(\pi/p)} \quad [75]$$

$$= \sqrt{2}E_s \frac{\sin \dfrac{\pi}{p}}{\dfrac{\pi}{p}}. \quad \blacktriangleright[76]$$

Table I gives the average rectified voltage E_{d0} in terms of the effective value of the voltage E_s for various values of the number of phases p.

TABLE I

p	2	3	4	6	12	∞
$\dfrac{E_{d0}}{E_s}$	0.90	1.17	1.27	1.35	1.40	$\sqrt{2}$

If E_n represents the effective value of the nth harmonic component of the rectified voltage (that is, the component having a frequency n times the supply frequency), by Fourier analysis,

$$E_n = \left| \frac{p}{\pi} \int_{-(\pi/p)}^{+(\pi/p)} E_s \cos \omega t \cos n\omega t \, d(\omega t) \right|. \quad [77]$$

The magnitude sign is inserted because, since E_n is an effective value, it is positive, whereas the member of the equation within the magnitude bars is a coefficient of the Fourier series and may be either positive or negative. Upon integration and substitution of the limits, this relation gives

$$E_n = \left| E_s \frac{p}{\pi} \left[\frac{\sin\left(\dfrac{n\pi}{p}\right)\cos\left(\dfrac{\pi}{p}\right) - \cos\left(\dfrac{n\pi}{p}\right)\sin\left(\dfrac{\pi}{p}\right)}{n-1} \right. \right.$$
$$\left. \left. + \frac{\sin\left(\dfrac{n\pi}{p}\right)\cos\left(\dfrac{\pi}{p}\right) + \cos\left(\dfrac{n\pi}{p}\right)\sin\left(\dfrac{\pi}{p}\right)}{n+1} \right] \right| \quad [78]$$

Since the lowest value of n is equal to p, and higher values of n must be multiples of p, n/p may be replaced by m, where m is an integer;

sin $n\pi/p$ then becomes sin $m\pi$ and is equal to zero. Substitution of m for n/p in Eq. 78 therefore gives

$$E_n = \left| E_s \frac{p}{\pi} \left(\cos m\pi \right) \left(\sin \frac{\pi}{p} \right) \left(\frac{-1}{n-1} + \frac{1}{n+1} \right) \right| \qquad [79]$$

$$= \left| -2E_s \frac{p}{\pi} \left(\sin \frac{\pi}{p} \right) \frac{\cos m\pi}{n^2 - 1} \right|. \qquad [80]$$

From Eqs. 80 and 76, the peak value of the nth harmonic, $\sqrt{2}E_n$, expressed as a fraction of E_{d0}, is therefore

$$\frac{\sqrt{2}E_n}{E_{d0}} = \left| -\frac{2}{n^2 - 1} \cos m\pi \right|. \qquad [81]$$

Since

$$\cos m\pi = \pm 1, \qquad [82]$$

$$\frac{\sqrt{2}E_n}{E_{d0}} = \frac{2}{n^2 - 1}, \qquad \blacktriangleright[83]$$

which is independent of p. Table II gives $\sqrt{2}E_n/E_{d0}$ for several values of n.

TABLE II

n	2	3	4	6	8	9	12	18	24
$\dfrac{\sqrt{2}E_n}{E_{d0}}$	0.667	0.250	0.133	0.057	0.032	0.025	0.014	0.006	0.0035

It must be remembered that these harmonic voltages exist only when n is a multiple of p. For example, if p equals 6, n can have only the values 6, 12, 18, and so on. The magnitude of a harmonic voltage of a particular frequency is independent of the number of phases when expressed as a fraction of the average rectified voltage.

Table I shows that the direct component of the rectified voltage E_{d0} increases as the number of phases increases and that it approaches the peak value of the anode voltage as p approaches infinity. For all practical purposes, it is equal to the peak anode voltage for p equal to or greater than twelve. Table II shows that the magnitude of the ripple voltage decreases and its principal frequency increases as the number of phases increases. For example, a six-phase rectifier has the same percentage of sixth, twelfth, and eighteenth harmonics as a three-phase rectifier, but it has no third, ninth, and fifteenth harmonics, which are present in the three-phase rectifier. Tables I and II show two advantages of a large number of phases. That there are also some disadvantages will appear later.

Usually, a load reactor is connected in series with the transformer neutral of a power rectifier. This is a smoothing reactor and absorbs most of the ripple in the load voltage. In a few rectifier applications, the ripple may not be undesirable and the reactor is then omitted. If a reactor is used, it is appropriate to assume that the load current is a constant I_{dc}. Of course the load current can never be absolutely constant, but for six- and twelve-phase rectifiers the ripple voltage is very small, so that a reactor large enough to give substantially constant load current is not economically prohibitive. Because of the reactor, the load current is assumed to be essentially a constant in the following analysis, no matter how many phases are used.

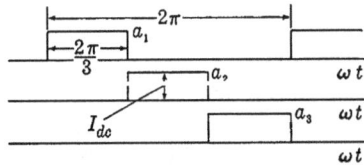

Fig. 16. Anode currents in a three-phase rectifier with a load reactor.

The symbols a_1, a_2, a_3, $\cdots$ are used here to denote the respective instantaneous anode currents. From previous considerations, it is clear that, if the transformer reactance is neglected and the load current has the constant value I_{dc}, the anode currents are equal to I_{dc} for $2\pi/p$ radians and are equal to zero during the remainder of a cycle. Figure 16 shows the anode currents for the three-phase rectifier of Fig. 13 when a load reactor is used. From this figure it may be seen that the average value of each anode current is $I_{dc}/3$ for this circuit, and I_{dc}/p in general. The root-mean-square or effective value of each anode current is denoted by I_a and given by

$$ I_a = \sqrt{\frac{1}{2\pi} \int_0^{2\pi} a^2 \, d(\omega t)} = \sqrt{\frac{1}{2\pi} \left(I_{dc}^2 \frac{2\pi}{p} \right)} = \frac{I_{dc}}{\sqrt{p}}, \qquad \blacktriangleright[84] $$

where a may be any one of the anode currents a_1, a_2, a_3, $\cdots$.

Since the losses in the transformer depend upon E_s and I_a, the required volt-ampere rating of the transformer is not equal to the output power rating of the rectifier unit and is different for different values of p. The product $E_s I_a$ is the required volt-ampere rating per phase of the transformer secondary windings, and the total required rating of the transformer secondary windings is given by

$$ P_2 = p E_s I_a = \sqrt{p} E_s I_{dc} . \qquad \blacktriangleright[85] $$

The direct-current power, including the loss in the rectifiers, is given by

$$P = E_{d0}I_{dc} = \sqrt{2}E_sI_{dc}\frac{p}{\pi}\sin\frac{\pi}{p}.$$ ▶[86]

The ratio of this power, which is approximately the output rating of the rectifier unit, to the required rating of the transformer secondary windings is

$$\frac{P}{P_2} = \sqrt{\frac{2}{p}}\,\frac{\sin\dfrac{\pi}{p}}{\dfrac{\pi}{p}}.$$ ▶[87]

This quantity, known as the *secondary utilization factor*, is similar to a power factor except that it results from a nonsinusoidal transformer current rather than a phase angle between the current and voltage. The secondary utilization factor has a maximum value when p is 2.69, but of course p must always be an integer. Table III gives the

TABLE III

Number of phases, p	2	3	4	6	12	∞
Secondary utilization factor	0.636	0.675	0.636	0.551	0.399	0

secondary utilization factor for several values of p. This table shows that, as the value of p increases above three, the required rating of the transformer secondary windings must increase for the same output, and therefore the transformer becomes more expensive. This fact is one of the chief disadvantages of a large number of phases.

All the foregoing results involve the assumption that the transformer secondary windings are connected in star. Through use of special connections, most of the advantages of both a large and a small number of phases can be obtained simultaneously. For example, in the delta, six-phase, double-wye connection shown in Fig. 17 the ripple voltage is the same as that of a star-connected six-phase rectifier, and yet the secondary utilization factor is equal to that of the three-phase rectifier. This connection is effectively two three-phase rectifiers operating in parallel, but with their sets of secondary voltages displaced 60 degrees from each other. The neutrals of the two wyes are connected together through a center-tapped reactor known as an *interphase transformer* (abbreviated I.P.T.), and the load, and smoothing reactor when used, are connected from the center terminal of the interphase reactor to the cathodes of the rectifiers.

Figure 17b shows the rectified voltages of the separate wyes that would result if their neutrals were connected to different loads. The solid sine waves are for the 1–3–5 set of rectifiers, and the dotted sine waves for the 2–4–6 set. The interphase transformer acts as an inductive potential divider across the difference of these two voltages so that the instantaneous rectified voltage measured from the center terminal of the interphase transformer is equal to the instantaneous

Primary   Secondary

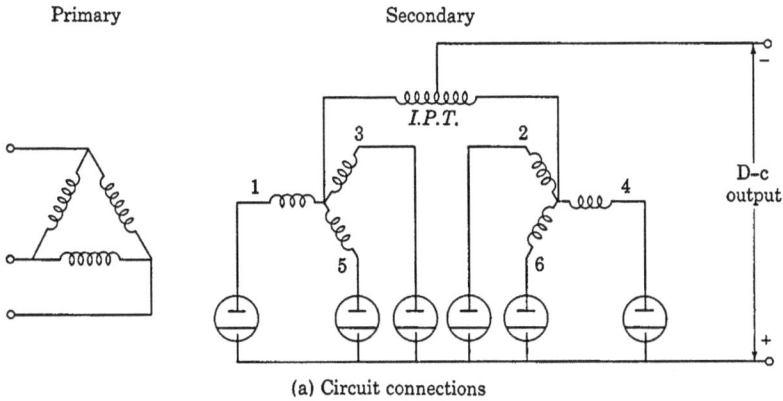

(a) Circuit connections

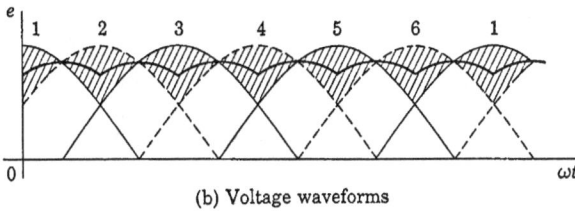

(b) Voltage waveforms

Fig. 17.   Delta, six-phase, double-wye rectifier circuit with interphase transformer.

average of the rectified voltages of the two separate wyes, as is shown by the heavy curve in Fig. 17b. From Tables I and II, the rectified voltage of the 1–3–5 wye is

$$e_{d1} = 1.17E_s[1 + 0.25 \cos 3\omega t - 0.057 \cos 6\omega t$$
$$+ 0.025 \cos 9\omega t - \cdots]. \qquad [88]$$

Since the rectified voltage of the other wye is shifted 60 degrees, it is

$$e_{d2} = 1.17E_s[1 + 0.25 \cos 3(\omega t + 60°) - 0.057 \cos 6(\omega t + 60°)$$
$$+ 0.025 \cos 9(\omega t + 60°) - \cdots] \qquad [89]$$

$$= 1.17E_s[1 - 0.25 \cos 3\omega t - 0.057 \cos 6\omega t$$
$$- 0.025 \cos 9\omega t + \cdots]. \qquad [90]$$

The instantaneous rectified voltage of the system, which is the instantaneous average of e_{d1} and e_{d2}, is

$$e_{d0} = 1.17E_s[1 - 0.057 \cos 6\omega t + \cdots]. \qquad [91]$$

The average value of the rectified voltage is the same as for a three-phase rectifier. That is, E_{d0} equals $1.17E_s$, whereas, without the inter-phase transformer—that is, in a six-phase star connection—E_{d0} equals $1.35E_s$. The ripple frequencies and the percentage ripple, however, are the same as for a six-phase rectifier. Since each anode conducts for one-third of a cycle, the secondary utilization factor is the same as for a three-phase rectifier, which is the maximum obtainable with a star connection. Also, the rectifiers work under better conditions than they would without the interphase transformer, since each anode current while flowing is only one-half the load current, whereas without the interphase transformer it is equal to the load current. This reduced current generally gives a smaller arc voltage drop and therefore a lower loss in the rectifiers.

When the transformer leakage inductance is neglected, one result of the approximate analysis is that the arc shifts instantly from one anode to the next. This apparent sudden commutation of the current would be accompanied by an infinite change of the anode current in a transformer winding, which is, of course, impossible. A more nearly exact analysis requires a study of the effect of the transformer leakage inductance upon the anode current and voltage and upon the rectified voltage.[6] Such an analysis shows that the leakage inductance prolongs the conduction period, reduces the direct output voltage, and reduces the effective value of the anode current.

8. Bridge, or double-way, rectifier circuits

Several different circuits in addition to the half-wave and full-wave connections have been devised for rectifying single-phase alternating current. One of these, the single-phase *bridge connection*, or diametric double-way circuit, drawn in alternate ways in Figs. 18a and 18b, is an arrangement of rectifiers that permit full-wave rectification of the current from a *two-terminal* source. A transformer without a center tap may be used as a supply, which may be an advantage in some applications. During the half-cycle in which the source voltage e_s is

[6] D. C. Prince and F. B. Vodges, *Principles of Mercury-Arc Rectifiers and Their Circuits* (New York: McGraw-Hill Book Company, Inc., 1927); L. B. W. Jolley, *Alternating-Current Rectification and Allied Problems* (3rd ed.; New York: John Wiley & Sons, Inc., 1928); O. K. Marti and H. Winograd, *Mercury-Arc Rectifiers—Theory and Practice* (New York: McGraw-Hill Book Company, Inc., 1930); H. Rissik, *The Fundamental Theory of Arc Converters* (London: Chapman and Hall, Ltd., 1939).

positive (relative to the reference direction shown on the figures), the rectifier elements 1 and 3 conduct current while the rectifiers 2 and 4 act as open circuits. Thus a current flows through the load resistance R in the positive direction of the current i. During the following half-cycle, the voltage e_s reverses, so that the rectifiers 2 and 4 conduct and

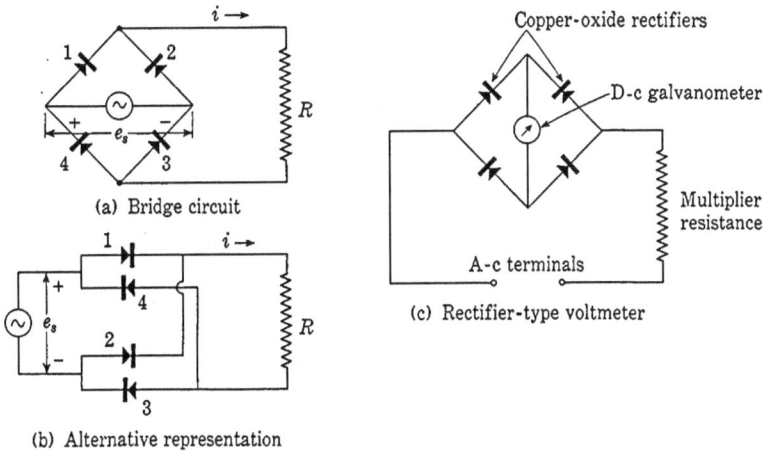

(a) Bridge circuit

(b) Alternative representation

Copper-oxide rectifiers

D-c galvanometer

Multiplier resistance

A-c terminals

(c) Rectifier-type voltmeter

Fig. 18.   Single-phase bridge, or diametric double-way, rectifier circuit and its application in the rectifier-type voltmeter.

rectifiers 1 and 3 do not conduct. The current through the load is again in the same direction; thus full-wave rectification is obtained. This type of circuit is called a *double-way rectifier*[7] because "the current between each terminal of the alternating-voltage supply and the rectifying elements connected to it flows in both directions." Both the primary and the secondary currents in the supply transformer are sinusoidal; consequently, for the same output, this transformer can be smaller than the transformer for the full-wave circuit.

The bridge circuit is at a disadvantage if a low voltage is to be rectified, not only because it requires four rectifiers instead of the one or two rectifiers used by the half-wave and full-wave circuits but also because two rectifiers in series are always carrying the current. For this reason, the voltage drop in the rectifiers while conducting causes a greater power loss than it would in other circuits. When not conducting, however, each rectifier is effectively connected directly across the alternating-current source, so that its required peak-inverse-voltage rating is approximately equal to the peak voltage of

[7] *American Standard for Pool-Cathode Mercury-Arc Power Converters—A.S.A. No. C34.1-1949* (New York: American Institute of Electrical Engineers, 1949), 7.

the source, and, for a given direct output voltage, is only half that required in the half-wave or full-wave circuits. This fact, together with the fact that the transformer required is smaller, of lower voltage, and of a simple standard two-winding type, makes the bridge circuit especially suited to the rectification of high voltages. Also the bridge circuit is particularly well adapted to application of selenium and copper-oxide rectifiers because they have relatively low peak inverse voltage ratings.

The bridge circuit is sometimes used, as shown in Fig. 18c, with small copper-oxide elements as rectifiers and a sensitive direct-current ammeter as a load, to form a rectifier-type ammeter, voltmeter, or galvanometer that may be used to measure alternating currents or voltages over a fairly wide range of frequencies.[8] Care should be exercised in interpreting the readings of rectifier meters, particularly when the unknown current or voltage contains harmonics, for the deflection of the instrument is not directly proportional to the root-mean-square value of the input wave.

A polyphase embodiment of the bridge, or double-way, principle is the rectifier connection shown in Fig. 19. It may be looked upon as two three-phase rectifiers supplied from the same three-phase voltages, and connected in series. The lower set of rectifiers produces positive output voltage with respect to the secondary neutral, just as in Fig. 13. Since the cathodes, rather than the anodes, of the upper set are connected to the transformer terminals, the upper set conducts current from the transformer in a direction opposite to that of the lower set. Hence the upper set produces negative output voltage with respect to the neutral. The average value of the total output voltage of the two in series is thus twice that of either alone. The neutral connection is not used as an output terminal and hence is shown dotted in Fig. 19.

The waveform of the output voltage may be determined from a consideration of the secondary voltage waves, which are drawn in Fig. 19b. The potential of the common cathode of the lower set of three rectifiers lies below the upper envelope of the voltage waves by the amount of the rectifier voltage drop. Similarly, the potential of the conducting anode of the upper set lies above the lower envelope by the same amount. The rectified output voltage inclusive of these voltage drops, e_{d0}, is hence measured between the upper and lower envelopes as indicated. The ripple in this rectified voltage is the same as in a six-phase star rectifier circuit.

[8] J. Sahagen, "The Use of the Copper-Oxide Rectifier for Instrument Purposes," *I.R.E. Proc.*, *19* (1931), 233–246.

The rectifiers conduct in pairs to provide a path for current through the transformer from the upper to the lower output terminal, just as in the single-phase rectifier of Fig. 18a. The conduction periods of the rectifiers within the upper and lower sets do not coincide in time,

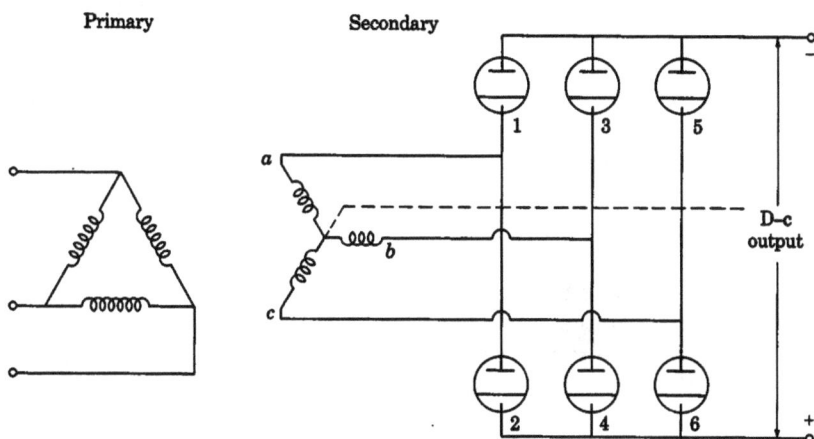

(a) Circuit connections

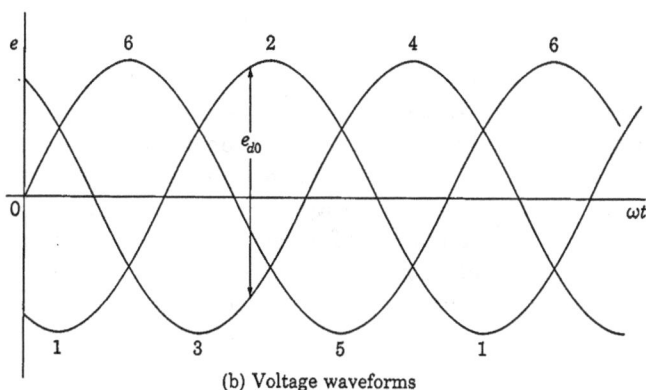

(b) Voltage waveforms

Fig. 19.   Delta, six-phase, wye, double-way (bridge) rectifier circuit.

however. Each rectifier conducts for 120 degrees per cycle corresponding to a lobe along the upper or lower envelope of the voltage waves as is indicated in Fig. 19b. When the phase order is *a–b–c*, the rectifiers conduct in pairs in the order 1–6, 3–6, 3–2, 5–2, 5–4, 1–4, 1–6, and so on. Since each rectifier conducts one third of the time, its required average current rating is one third of the average output current.

When one rectifier in a set is conducting, the inverse voltage across the other two is the line-to-line secondary voltage. Hence the maximum inverse peak voltage across any rectifier equals the peak value of the line-to-line secondary voltage.

Each secondary winding carries the current conducted by two rectifiers. These currents flow in opposite directions through the winding, however; hence the average winding current is zero. Direct-current magnetization of the core and consequent excessive exciting current are thereby eliminated. Furthermore, since the winding carries current in one direction or the other for two thirds of the time, the secondary utilization factor for this connection is relatively large. This large secondary utilization factor together with the facts that only three two-terminal secondary windings are needed and that their required voltage ratings are only half that needed for the same average output voltage in a three-phase wye connection makes the transformer relatively inexpensive for a particular power output. This connection is widely used, particularly in power supplies for radio transmitters.

9. Rectifiers with a smoothing capacitor

All the single-phase rectifier circuits previously considered have a resistance load and no filter for reducing the pulsations in the output voltage. The full-wave and bridge circuits give a smoother output wave than the half-wave circuit, and a substantial reduction in the ripple factor is gained by their use. Further reduction of the ripple factor can be achieved by the use of a polyphase source of voltage so connected that a pulse of current is obtained in the output from each of the phases during every cycle of the applied alternating voltage, as discussed in Art. 7. The production of a very smooth wave-form would, however, entail the use of a source having so many phases as to be prohibitive in most applications. Consequently, some sort of filter circuit is commonly used as a means of reducing the ripple where a very smooth output voltage is required.

One of the simpler methods for smoothing the current is the use of a capacitor connected in parallel with the load, in the manner shown in Fig. 20 for a half-wave rectifier. The capacitor stores energy obtained from the source during the period that the rectifier conducts, and delivers energy to the load during the nonconducting period. In this way, the time during which current passes through the load is prolonged considerably as compared with the time of conduction through the rectifier; in fact, the load current may be continuous, whereas the rectifier current flows in pulses.

To simplify the analysis of the behavior of the rectifier with a smoothing capacitor, the rectifier is assumed to be perfect. In other words, it is assumed to be the ideal rectifier described in Art. 3 and to act as a switch that is closed for current in the positive direction but opens when the current tends to reverse. A further assumption is that the internal impedance of the source and trans-
former used is zero. Though these two assumptions are often not justified in practice,[9] the analysis that follows may serve as a basis for estimating the behavior of an actual circuit.

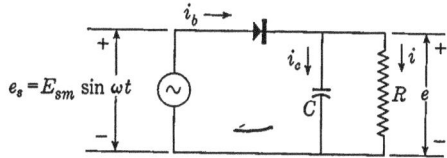

Fig. 20.   Half-wave rectifier circuit with a shunt capacitor for smoothing.

As a first step in the analysis, a qualitative understanding of some aspects of the circuit performance can be gained from a consideration of the simplified circuit of Fig. 21a, where the load resistance R is considered to be so large as to require a negligible current and is therefore

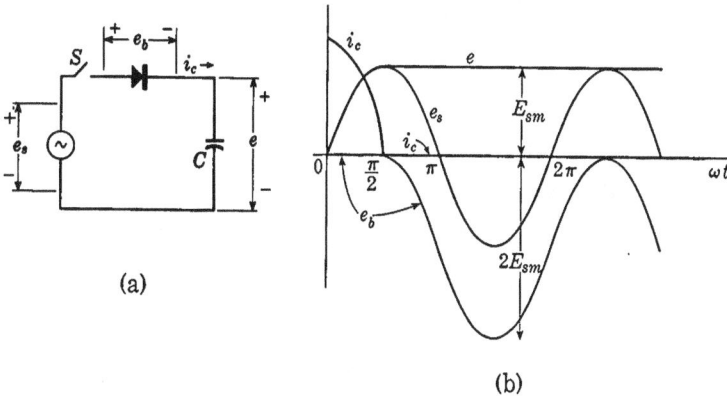

(a)

(b)

Fig. 21.   Half-wave rectifier circuit with a smoothing capacitor but no load.

omitted. Assume that the switch S is closed at a time t equal to zero with the capacitor uncharged. The voltage of the source is

$$e_s = E_{sm} \sin \omega t, \qquad [92]$$

and the current through the capacitor is then

$$i_c = C \frac{de_s}{dt} = \omega C E_{sm} \cos \omega t. \qquad [93]$$

Thus the current wave through the circuit is the pulse shown in Fig. 21b

[9] O. H. Schade, "Analysis of Rectifier Operation," *I.R.E. Proc.*, *31* (1943), 341–361.

as part of a cosine wave. Since the voltage across the rectifier, e_b, is zero during conduction, the voltage across the capacitor, e, which is the output voltage, equals the source voltage, e_s, while de_s/dt and the current are positive. These conditions obtain from the instant S is closed until the time at which ωt equals $\pi/2$. At this time, the current

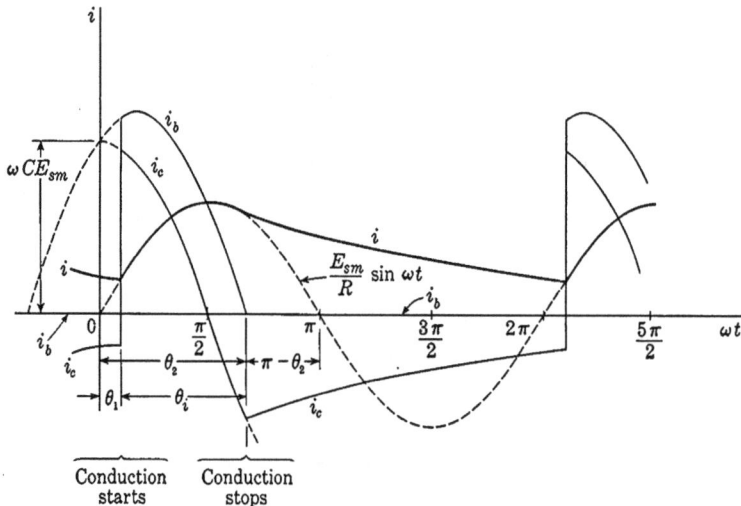

Fig. 22. Typical waveforms in the circuit of Fig. 20.

would reverse if it were not for the rectifier. The rectifier prevents such a current reversal, and the capacitor cannot discharge. The capacitor voltage therefore remains at its peak value E_{sm}, and the voltage across the tube, given by

$$e_b = e_s - e,\qquad [94]$$

has the waveform shown and is always in the inverse direction. It may be noted that the peak value of the inverse voltage across the tube is $2E_{sm}$.

When a resistance load R is added so that the circuit of Fig. 21a becomes that of Fig. 20, the capacitor voltage does not remain at E_{sm} once the capacitor is charged, because the capacitor discharges through the load resistance while the tube is not conducting. During each non-conducting period, the capacitor loses part of its charge; and, during every positive half-cycle of applied voltage, current flows through the rectifier to replace the charge drained off during the previous non-conducting period. The behavior of the circuit can then be described as a steady state consisting of a series of identical repeated transients, and a transient analysis is necessary to describe it quantitatively.

The curves of Fig. 22 serve to portray the general nature of the waveforms of the currents defined in Fig. 20 and to define the *ignition angle* θ_1 at the start of conduction, the *extinction angle* θ_2 at the end of the conduction period, and the *angle of conduction* θ_i. Two relations among the currents are apparent. First, during the conduction period,

$$i_b = i + i_c;$$ [95]

and, second, during the nonconduction period,

$$i_b = 0,$$ [96]

and therefore

$$i_c = -i.$$ [97]

During the conduction period, the rectifier is essentially a short circuit—it is assumed to be ideal—and the currents are given by

$$i = \frac{E_{sm}}{R} \sin \omega t,$$ [98]

$$i_c = C\frac{de_s}{dt} = \omega C E_{sm} \cos \omega t,$$ [99]

$$i_b = \frac{E_{sm}}{R} \sin \omega t + \omega C E_{sm} \cos \omega t$$ [100]

$$= E_{sm} \sqrt{(1/R^2) + (\omega C)^2} \sin (\omega t + \tan^{-1}\omega C R).$$ [101]

The extinction angle θ_2 may be determined by the condition that at the end of the conduction period,

$$i_b = i + i_c = 0,$$ [102]

and

$$\omega t = \theta_2 .$$ [103]

Therefore

$$\frac{E_{sm}}{R} \sin \theta_2 + \omega C E_{sm} \cos \theta_2 = 0,$$ [104]

and

$$\omega C R = -\frac{\sin \theta_2}{\cos \theta_2} = -\tan \theta_2;$$ [105]

thus, in terms of the dimensionless ratio of resistance to capacitive reactance, $\omega C R$, which is used as the independent variable in much of the subsequent analysis to make the results universally applicable for all combinations of resistance, capacitance, and frequency,

$$\theta_2 = \tan^{-1}(-\omega C R).$$ [106]

Since R, ω, and C are all positive quantities, θ_2 must lie in the second or fourth quadrant. Furthermore, as C approaches zero, θ_2 must approach the value for a simple resistance load, namely, 180 degrees. Hence

$$90° < \theta_2 < 180°. \qquad [107]$$

When ωt equals θ_2, conduction through the rectifier stops, and the capacitor begins to discharge through the resistor R until the time at which ωt equals $2\pi + \theta_1$ is reached. During this nonconduction part of the cycle, the rectifier acts as an open circuit, and the current through the load is described by

$$\frac{di}{dt} + \frac{1}{CR} i = 0 \qquad [108]$$

as the capacitor discharges. Consequently,

$$i = A\epsilon^{-\frac{t}{CR}} = A\epsilon^{-\frac{\omega t}{\omega CR}}, \qquad [109]$$

where the constant A may be determined from the condition that the capacitor voltage, and hence the current through the load resistor, must be continuous from the conduction period to the blocking period at the instant when ωt equals θ_2; that is, the currents given by Eqs. 98 and 109 must be equal for ωt equal to θ_2. Thus

$$\frac{E_{sm}}{R}\sin\theta_2 = A\epsilon^{-\frac{\theta_2}{\omega CR}}, \qquad [110]$$

whence

$$A = \frac{E_{sm}}{R}\sin\theta_2\,\epsilon^{\frac{\theta_2}{\omega CR}}. \qquad [111]$$

With the constant A evaluated, the expression for the load current during the blocking period becomes

$$i = \frac{E_{sm}}{R}\sin\theta_2\,\epsilon^{-\frac{\omega t - \theta_2}{\omega CR}}. \qquad [112]$$

The end of the blocking period, or the start of the next conduction period, is reached when the supply voltage e_s becomes equal to the voltage Ri obtainable from Eq. 112, because, from that time on, there is a tendency for the voltage across the tube to be positive, and the tube therefore conducts. At the end of the blocking period,

$$\omega t = 2\pi + \theta_1. \qquad [113]$$

Consequently, the ignition angle may be found from the condition that Ri from Eq. 112 equals e_s from Eq. 92 when ωt equals $2\pi + \theta_1$. Thus

$$E_{sm}\sin\theta_2\,\epsilon^{-\frac{2\pi + \theta_1 - \theta_2}{\omega CR}} = E_{sm}\sin(2\pi + \theta_1), \qquad [114]$$

or

$$\sin\theta_1 = \sin\theta_2\,\epsilon^{-\frac{2\pi - (\theta_2 - \theta_1)}{\omega CR}}, \qquad [115]$$

which can also be expressed as

$$\sin \theta_1 = \sin (\tan^{-1}\omega C R)\, \epsilon^{-\frac{2\pi-(\theta_2-\theta_1)}{\omega C R}}, \qquad [116]$$

where $\tan^{-1}\omega C R$ is an angle in the first quadrant and has the same sine as the angle $\tan^{-1}(-\omega C R)$ in the second quadrant. Also, since

$$\theta_i = \theta_2 - \theta_1, \qquad [117]$$

$$\sin (\theta_2 - \theta_i) = \sin \theta_2\, \epsilon^{-\frac{2\pi-\theta_i}{\omega C R}}. \qquad [118]$$

If $\sin (\theta_2 - \theta_i)$ is replaced by its equal, $\sin (\pi - \theta_2 + \theta_i)$, and $(\pi - \theta_2)$ in turn is replaced by $\tan^{-1}\omega C R$, an angle in the first quadrant, Eq. 118 takes the form

$$\sin [\tan^{-1} (\omega C R) + \theta_i] = \sin (\tan^{-1}\omega C R)\, \epsilon^{-\frac{2\pi-\theta_i}{\omega C R}}. \qquad [119]$$

Equations 115 and 118 are transcendental and cannot be solved explicitly, but graphical methods of solution may be used with success. Since θ_2 is determined by $\omega C R$ and Eq. 106, if the two members of Eq. 115 or 116 are plotted as functions of θ_1 for a particular value of $\omega C R$, the intersection of the two resulting curves determines θ_1 for that particular value of $\omega C R$. Repetition of this procedure for different values of $\omega C R$ allows a curve of θ_1 as a function of $\omega C R$ or of $\tan^{-1}\omega C R$ to be obtained point by point. The results of such a process of graphical solution are shown in Fig. 23, where curves of θ_1 and θ_2 are plotted as functions of $\tan^{-1}\omega C R$.

Since the operation of a full-wave rectifier with a smoothing capacitor is the same as that of the half-wave rectifier, except that two pulses of current instead of one are received by the capacitor in each cycle, this same analysis may be applied to the full-wave rectifier. However, in Eq. 113 and the equations that follow it, 2π must be replaced by π to account for the shorter time that the capacitor is required to store a charge. The curve for θ_1 in the full-wave rectifier is found through making this change, and is plotted in Fig. 23. The curve for θ_2 is the same for either rectifier connection.

The conduction angle θ_i, obtainable from Eq. 118 or 119, is the vertical distance between the curves for θ_1 and θ_2 in Fig. 23; a separate curve of it is plotted for the half-wave rectifier. Thus, if the load resistance R and the reactance of the capacitor $1/(\omega C)$ are known, this diagram gives the three angles θ_1, θ_2, and θ_i necessary for the remainder of the analysis of the circuit behavior. The ignition angle θ_1 is substantially zero for $\omega C R$ less than unity ($\tan^{-1}\omega C R$ less than 45 degrees) in the half-wave rectifier, but it increases rapidly for larger values of

ωCR and approaches 90 degrees as a limit. Clearly, one effect of increasing the capacitance is to shorten the conduction angle θ_i.

The foregoing determination of the expressions for the several currents appropriate both to the conduction and to the nonconduction

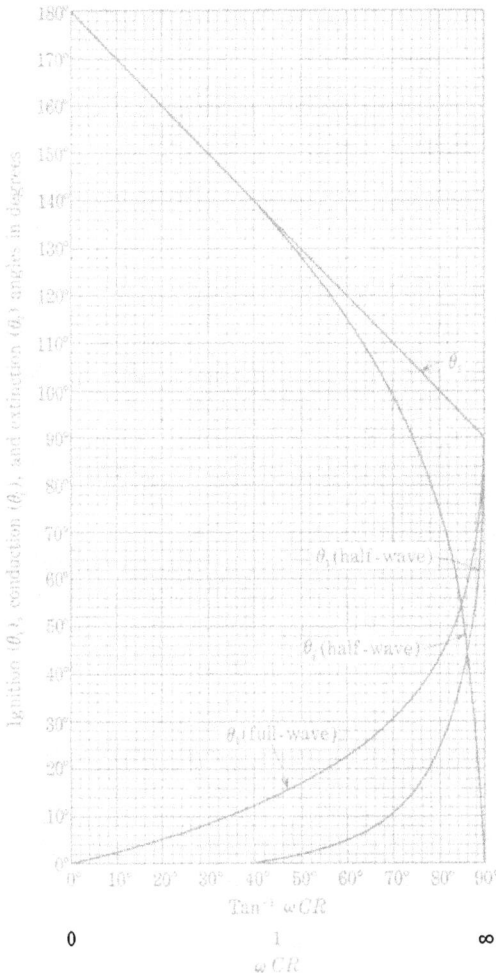

Fig. 23.   The ignition angle θ_1, the extinction angle θ_2, and the conduction angle θ_i as functions of $\tan^{-1} \omega CR$.

periods, and for the beginning and end of these periods, makes it possible to summarize the existing currents in the half-wave circuit. They are periodic and for one cycle are given by the following expressions:

Load current:

$$
i = \begin{cases} \dfrac{E_{sm}}{R} \sin \omega t, & \theta_1 \leqq \omega t \leqq \theta_2 & \blacktriangleright[120] \\[2em] \dfrac{E_{sm}}{\sqrt{R^2 + (1/\omega C)^2}} \, \epsilon^{\frac{\tan^{-1}(-\omega C R) - \omega t}{\omega C R}}, & \theta_2 \leqq \omega t \leqq 2\pi + \theta_1 . & \blacktriangleright[121] \end{cases}
$$

Capacitor current:

$$
i_c = \begin{cases} E_{sm} \, \omega C \cos \omega t, & \theta_1 \leqq \omega t \leqq \theta_2 & [122] \\[2em] -\dfrac{E_{sm}}{\sqrt{R^2 + (1/\omega C)^2}} \, \epsilon^{\frac{\tan^{-1}(-\omega C R) - \omega t}{\omega C R}}, & \theta_2 \leqq \omega t \leqq 2\pi + \theta_1 . & [123] \end{cases}
$$

Rectifier current:

$$
i_b = \begin{cases} E_{sm} \sqrt{(1/R)^2 + (\omega C)^2} \sin (\omega t + \tan^{-1} \omega C R), & \theta_1 \leqq \omega t \leqq \theta_2 & \blacktriangleright[124] \\[1em] 0, & \theta_2 \leqq \omega t \leqq 2\pi + \theta_1 . & \blacktriangleright[125] \end{cases}
$$

Corresponding expressions for the full-wave connection are obtained if 2π is replaced by π in Eqs. 120 to 125.

Curves of the load current i and the rectifier current i_b both multiplied by R/E_{sm}, are drawn for several values of the parameter $\omega C R$ in Figs. 24b and 24c. The relation between these curves and the curves of Fig. 23 is shown by the reproduction of Fig. 23 with axes interchanged in Fig. 24a. For each value of $\omega C R$, the angles of ignition and extinction are determined in Fig. 24a and projected into Figs. 24b and 24c. The horizontal lines in Fig. 24a show the extent of the conduction period, and the portion of a sine wave shows the relation of the conduction period to half a cycle. The current curves illustrate the change in waveform effected by an increase in the capacitance C when the resistance R and the peak voltage E_{sm} are held constant. Other curves —for example, those for various values of R with C held constant— may be obtained from this diagram if the currents are multiplied by the appropriate factor. The curves of Fig. 24b show the load current. When $\omega C R$ is zero, the waveform is the rectified half sinusoid of Art. 4a. An increase of the capacitance C increases the period of conduction through the load and reduces the minimum value to which the current drops, thereby smoothing the wave. In the limit, when $\omega C R$ is made infinite, this circuit would supply the load with a steady direct current with no ripple.

The magnitude of I_{dc}, the steady component of the load current, increases with increasing C, and it follows from Fig. 24b that the magnitude approaches E_{sm}/R as a limit when $\omega C R$ approaches infinity.

The expression for the direct current in the load is

$$I_{dc} = \frac{1}{2\pi}\left[\int_{\theta_1}^{\theta_2}\frac{E_{sm}}{R}\sin\omega t\, d(\omega t)\right.$$

$$\left. + \int_{0}^{2\pi+\theta_1}\frac{E_{sm}}{\sqrt{R^2+(1/\omega C)^2}}\,\epsilon^{\frac{\tan^{-1}(-\omega CR)-\omega t}{\omega CR}}\,d(\omega t)\right], \quad [126]$$

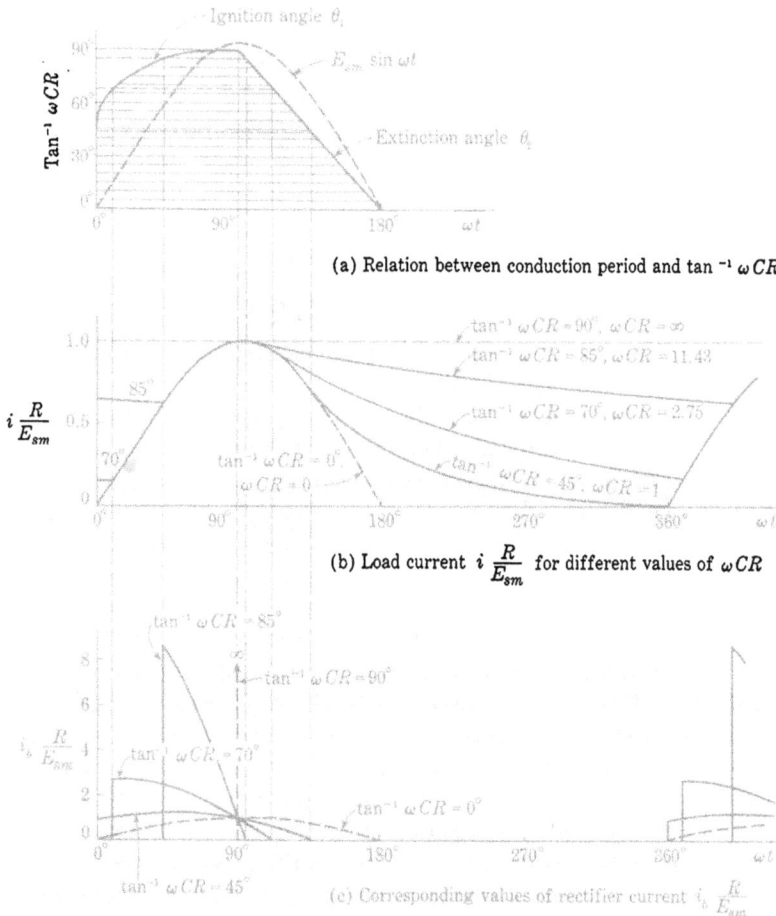

(a) Relation between conduction period and $\tan^{-1}\omega CR$

(b) Load current $i\dfrac{R}{E_{sm}}$ for different values of ωCR

(c) Corresponding values of rectifier current $i_b\dfrac{R}{E_{sm}}$

Fig. 24. Dimensionless curves of load and rectifier currents for the circuit of Fig. 20.

which, after integration, the use of Eq. 115, and certain trigonometric transformations, may be put in the form

$$I_{dc} = \frac{E_{sm}}{2\pi}\sqrt{(1/R)^2+(\omega C)^2}\,(1-\cos\theta_i). \qquad \blacktriangleright[127]$$

The effect of capacitance variations on the direct load current is shown by Eq. 127 written in the nondimensional form

$$I_{dc} \frac{R}{E_{sm}} = \frac{1}{2\pi} \sqrt{1 + (\omega C R)^2}\ (1 - \cos \theta_i). \qquad [128]$$

This equation, when plotted, yields the curve for $I_{dc}(R/E_{sm})$ in Fig. 25a. Since the capacitance C does not appear in the left-hand member of

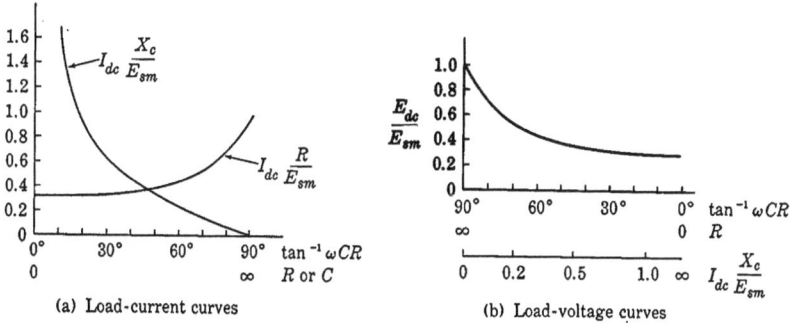

(a) Load-current curves   (b) Load-voltage curves

Fig. 25. Dimensionless curves showing the dependence of the direct components of load current and load voltage on load resistance and smoothing capacitance.

Eq. 128, this curve may be used to determine the variation of the direct current with the capacitance when the other circuit parameters are held constant. It shows quantitatively the increase in current produced by an increase in the capacitance, which is indicated qualitatively in Fig. 24b.

Another nondimensional form of Eq. 127,

$$I_{dc} \frac{X_c}{E_{sm}} = I_{dc} \frac{1/(\omega C)}{E_{sm}} = \frac{1}{2\pi} \sqrt{1 + \left(\frac{1}{\omega C R}\right)^2}\ (1 - \cos \theta_i), \qquad [129]$$

plotted as shown in Fig. 25a by the curve for $I_{dc}(X_c/E_{sm})$, shows the effect of load-resistance variations on the direct load current. Since R does not appear in the left-hand member of Eq. 129, the curve plotted from that equation illustrates the variation of I_{dc} with R when the remaining circuit parameters are constant. When R is zero, an infinite short-circuit current is indicated, because the rectifier loss and the internal impedance of the alternating-current source are neglected in this analysis. The open-circuit current is, of course, zero.

The waveform of the output voltage is the same as that of the output current, since the load is a resistor. If the load resistance is so large that the load current may be assumed to be negligible, the circuit is

equivalent to the simpler circuit of Fig. 21a; and a perfectly steady
output voltage equal to E_{sm}, the peak value of the supply voltage,
results, as shown in Fig. 21b. For lower values of the load resistance,
the capacitor is appreciably charged and discharged during each cycle,
so that a ripple voltage is produced and the average value of the output voltage is decreased. The average output voltage E_{dc} is obtained
if Eq. 127 is multiplied by R; it is

$$E_{dc} = \frac{E_{sm}}{2\pi} \sqrt{1 + (\omega C R)^2} \, (1 - \cos \theta_i). \qquad \blacktriangleright[130]$$

From this equation, the curve of Fig. 25b, which shows the variation
of E_{dc} with $\omega C R$, is plotted. The additional abscissa scale for
$I_{dc}(X_c/E_{sm})$ is obtained from the curve for this quantity in Fig. 25a.
This scale permits determining the variation of the output voltage
with output current when all parameters of the circuit except the load
resistance R remain constant. It is seen that the regulation of this
rectifier-and-filter circuit is very poor.

The curves of Fig. 24c show the instantaneous rectifier current for
several values of $\tan^{-1}\omega C R$ from 0 to 90 degrees. A comparison with
the load-current curves of Fig. 24b brings out the fact that as the load
current is smoothed, the rectifier current—and thus the supply current
—becomes more and more peaked and rises to a higher and higher
value. A shortening of the conduction period and an increase in the
peak value of the rectifier current are necessarily associated with
smoothing by the parallel-capacitor circuit. Under operating conditions, these large peak currents may damage the rectifying device
even though its average-current rating is not exceeded by the direct-
current output and its loss is below the permissible value. This fact
is true particularly of a gas tube having a thermionic cathode as
discussed in Art. 2b, Ch. V. The connection of a resistor or inductor in
series with the rectifier is sometimes used to reduce this peak current
to a tolerable value. In general, smoothing by the method of a parallel
capacitor is not good engineering practice when a gas tube is used, and
the smoothing is obtained preferably by the use of a series inductor, as
discussed in the next article.

Derivation of a complete expression for the ripple voltage in the
output of a rectifier with a smoothing capacitor for all values of load
is difficult. When the load current is so small that the load voltage
decays only a small percentage during the nonconduction interval,
however, the relatively simple approximate method[10] that follows
yields results of great utility.

[10] L. B. Arguimbau, *Vacuum-Tube Circuits* (New York: John Wiley & Sons, Inc.,
1948), 27–28.

The rectifier current may be represented by the general Fourier series:

$$i = I_{dc} + \sqrt{2}I_1' \sin \omega t + \sqrt{2}I_1 \cos \omega t + \sqrt{2}I_2' \sin 2\omega t$$
$$+ \sqrt{2}I_2 \cos 2\omega t + \cdots . \qquad [131]$$

The coefficients in this series are given by[11]

$$I_{dc} = \frac{1}{2\pi} \int_0^{2\pi} i \, d(\omega t), \qquad [132]$$

$$\sqrt{2}I_1' = \frac{1}{\pi} \int_0^{2\pi} i \sin \omega t \, d(\omega t), \qquad [133]$$

$$\sqrt{2}I_1 = \frac{1}{\pi} \int_0^{2\pi} i \cos \omega t \, d(\omega t), \qquad [134]$$

$$\sqrt{2}I_2' = \frac{1}{\pi} \int_0^{2\pi} i \sin 2\omega t \, d(\omega t), \qquad [135]$$

$$\sqrt{2}I_2 = \frac{1}{\pi} \int_0^{2\pi} i \cos 2\omega t \, d(\omega t), \qquad [136]$$

and so forth.

If the origin is shifted 20 degrees to the right in Fig. 24c, and the load resistance, or ωCR, is large, the current pulse persists for only a short time near $0, 2\pi, 4\pi, \cdots$. Then $\cos \omega t, \cos 2\omega t, \cdots$ are closely approximated by unity, and $\sin \omega t, \sin 2\omega t, \cdots$ by zero during the time that current exists. Consequently,

$$\sqrt{2}I_1 \approx \frac{1}{\pi} \int_0^{2\pi} i \, d(\omega t), \qquad [137]$$

or

$$\sqrt{2}I_1 \approx 2I_{dc} . \qquad [138]$$

Likewise,

$$\sqrt{2}I_2 \approx 2I_{dc} , \qquad [139]$$

and I_1' and I_2' are zero. Thus the limiting amplitude of each harmonic in the Fourier series for the rectifier current is twice the direct component.

The total current through the rectifying element divides between the load resistance and the parallel capacitor. The direct component passes through the resistance only, and produces the direct output voltage,

$$E_{dc} = I_{dc}R. \qquad [140]$$

[11] See, for example, P. Franklin, *Differential Equations for Electrical Engineers* (New York: John Wiley & Sons, Inc., 1933), 69.

The alternating components exist in both the capacitor and the resistor. But, as is shown by Fig. 24c, the current pulse can be considered short compared with one cycle only when ωCR is larger than, say, 20. In other words, the approximate analysis is applicable only when the resistance is at least 20 times as large as the capacitive reactance at the angular frequency ω. For circuit conditions satisfying that requirement, an error of not more than about five per cent is made if the whole of the fundamental component I_1 is assumed to pass through the capacitor. The effective value of the fundamental component of the output voltage is then

$$E_1 \approx \frac{I_1}{\omega C} \approx \frac{\sqrt{2}I_{dc}}{\omega C}. \qquad [141]$$

Since the capacitive reactance for the second harmonic is only half that for the fundamental, the approximation is even better for it, and

$$E_2 \approx \frac{I_2}{2\omega C} \approx \frac{I_{dc}}{\sqrt{2}\,\omega C}. \qquad [142]$$

In accordance with Eq. 20, the ripple factor for the half-wave rectifier is

$$\gamma = \frac{\sqrt{E_1{}^2 + E_2{}^2 + \cdots}}{E_{dc}} \approx \frac{\sqrt{2}\dfrac{I_{dc}}{\omega C}}{I_{dc}R}\sqrt{1 + \tfrac{1}{4} + \cdots}; \qquad [143]$$

thus

$$\gamma \approx \frac{1.58}{\omega CR}, \qquad \omega CR > 20. \qquad [144]$$

The foregoing method of analysis is also applicable to the full-wave circuit with a smoothing capacitor if the fact that the capacitor receives two short pulses of current during each cycle of the supply voltage instead of one is taken into account. The direct output current is given by Eq. 132. The fundamental component I_1 determined by Eq. 134 is zero because $\cos \omega t$ has equal values but of opposite sign for the two pulses. All higher odd harmonics are zero for the same reason. Only the second and higher even harmonics exist. The amplitude of the second harmonic expressed by Eq. 136 leads to the value in Eq. 139 because $\cos 2\omega t$ is approximately unity during both pulses of current. As in the half-wave circuit, $I_1{}'$ and $I_2{}'$ are zero. Accordingly, the expressions for the voltages, Eqs. 140 and

142, also apply to the full-wave circuit. Since E_1 is zero, however, the ripple factor is given by

$$\gamma = \frac{\sqrt{E_2{}^2 + E_4{}^2 + \cdots}}{E_{dc}} \approx \frac{\sqrt{2}\,\dfrac{I_{dc}}{2\omega C}}{I_{dc}R} \sqrt{1 + \tfrac{1}{4} + \cdots}, \qquad [145]$$

or

$$\gamma \approx \frac{1.58}{2\omega C R}, \qquad \omega C R > 20, \qquad\qquad [146]$$

just half the value for the full-wave circuit.

The peak inverse voltage across the rectifying element in the half-wave rectifier with a smoothing capacitor is shown in Fig. 21 to be $2E_{sm}$ for the limiting condition of zero load. The same value applies to the full-wave circuit with a smoothing capacitor because for zero load the output voltage is constant and each half of the circuit acts independently of the other. Presence of appreciable load current reduces the peak inverse voltage somewhat in either the half-wave or the full-wave rectifier with a smoothing capacitor, but in each connection, the tubes must be able to withstand a

$$\text{Maximum peak inverse voltage} = 2E_{sm} = 2E_{dc}\big|_{I_{dc}=0}. \qquad [147]$$

Comparison with Eqs. 24 and 47 shows that for the same direct output voltage, addition of a capacitor *reduces* the peak inverse voltage requirement on the rectifying elements.

Since the direct output voltage is essentially equal to the peak or crest value of the input wave for negligible load current, the rectifier with a smoothing capacitor may be used as the basis for a peak-indicating voltmeter.[12] An electrostatic voltmeter across the capacitor, or a sensitive direct-current measuring element in series with a large load resistance, gives an indication directly proportional to the peak value in one of the half-cycles of the input. Reversal of the input connections provides a measure of the peak value in the other half-cycle. Thus nonsymmetrical waveforms can be measured.

10. HALF-WAVE RECTIFIER WITH SMOOTHING INDUCTOR

The series inductor shown in the circuit of Fig. 26a performs the same function as does the capacitor in the circuit of Fig. 20, in that it smooths the output current by storing energy during one part of the supply-voltage cycle and delivering it to the load during the other

[12] C. H. Sharp and E. D. Doyle, "Crest Voltmeters," *A.I.E.E. Trans.*, *35* (1916), 99–107.

part, but the inductor performs this function in a different way. In Fig. 26a, the inductance L may be considered to be that of an inductor connected in series with the rectifier for the purpose of smoothing the current; or it may be thought of as the heretofore neglected inductance in the transformer and in the source of alternating current; or it may be considered to be part of a load comprising inductance and resistance.

In the analysis that follows, the rectifier is considered perfect, as in the analysis in Art. 9; that is, the voltage drop in the rectifier while it

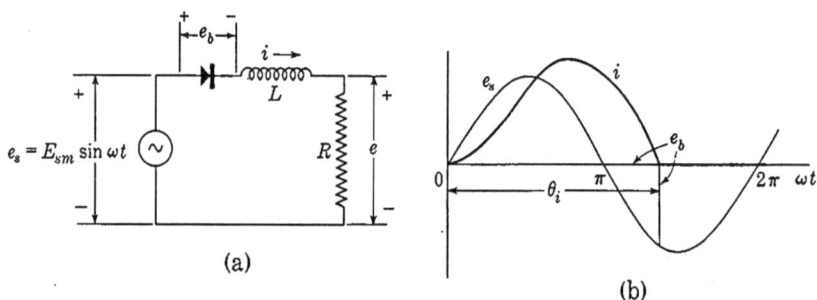

(a)

(b)

Fig. 26.   Half-wave rectifier circuit with a series inductor for smoothing.

conducts is assumed to be zero. The general nature of the current waveform is represented by the diagram of Fig. 26b, which serves to define the conduction angle θ_i. In this circuit, the ignition angle is zero and the extinction angle equals the conduction angle. During the conduction period, the current is determined by the differential equation

$$L\frac{di}{dt} + Ri = E_{sm} \sin \omega t, \qquad [148]$$

which has the solution,[13] subject to the condition that the initial current is zero,

$$i = I_m \sin (\omega t - \theta_z) - I_m \sin (-\theta_z)\epsilon^{-\frac{R}{L}t}, \qquad [149]$$

where I_m is the amplitude of the steady-state current given by

$$I_m = \frac{E_{sm}}{\sqrt{R^2 + (\omega L)^2}}. \qquad [150]$$

[13] Obtainable by the methods discussed in E. E. Staff, M.I.T., *Electric Circuits* (Cambridge, Massachusetts: The Technology Press of M.I.T.; New York: John Wiley & Sons, Inc., 1940), 341–345.

The angle θ_z is the impedance angle of the inductor and load—that is, the angle by which the voltage leads the current—given by

$$\theta_z = \tan^{-1} \frac{\omega L}{R};$$ [151]

and

$$I_m \sin(-\theta_z) = -E_{sm} \frac{\omega L}{R^2 + (\omega L)^2}.$$ [152]

In this solution, $I_m \sin(\omega t - \theta_z)$ is the steady-state current that the voltage $E_{sm} \sin \omega t$ would produce in the RL circuit, and $I_m \sin(-\theta_z)$ is the value of the steady-state component when t is equal to zero. As will appear in the following analysis, the conduction angle θ_i must be less than 2π radians; that is, the current must decrease to zero and tend to reverse before the beginning of the next cycle of the supply voltage. Since the rectifier does not permit a reversal of the current, the current must remain zero until the supply voltage again becomes positive, as shown in Fig. 26b. Therefore a steady-state condition consisting of repeated identical transients, each having the initial value zero, is reached by the current. This periodic current is given for one cycle by the expressions

$$i = \begin{cases} \dfrac{E_{sm}}{\sqrt{R^2 + (\omega L)^2}}\left[\sin\left(\omega t - \tan^{-1}\dfrac{\omega L}{R}\right) + \dfrac{\omega L}{\sqrt{R^2 + (\omega L)^2}}\,\epsilon^{-\frac{R}{\omega L}\omega t}\right], \\ \qquad\qquad\qquad\qquad\qquad\qquad 0 \leq \omega t \leq \theta_i \qquad \blacktriangleright[153] \\ 0, \qquad\qquad\qquad\qquad\qquad\;\; \theta_i \leq \omega t \leq 2\pi. \qquad \blacktriangleright[154] \end{cases}$$

The value of the conduction angle θ_i depends on the values of R and ωL, or on $\tan^{-1}(\omega L/R)$. It may be found from the requirement that the expression for i must be equal to zero when ωt is equal to θ_i. To this end, the bracketed term in Eq. 153 is equated to zero:

$$\sin\left(\theta_i - \tan^{-1}\frac{\omega L}{R}\right) + \frac{\omega L}{\sqrt{R^2 + (\omega L)^2}}\,\epsilon^{-\frac{R}{\omega L}\theta_i} = 0.$$ [155]

This transcendental equation can be solved graphically, but a new solution is not required, because by a simple change of variable the equation assumes the form of Eq. 119. The values plotted in Fig. 23 may then be used to secure θ_i. If θ_i is replaced by $2\pi - \theta_i'$, Eq. 155 may be put in the form

$$\sin\left(\tan^{-1}\frac{\omega L}{R} + \theta_i'\right) = \sin\left(\tan^{-1}\frac{\omega L}{R}\right)\epsilon^{-\frac{R}{\omega L}(2\pi - \theta_i')},$$ [156]

which is the same as Eq. 119 except that $\omega L/R$ replaces ωCR. The curve of θ_i as a function of $\tan^{-1}(\omega L/R)$, obtained in this manner, is shown with axes reversed in Fig. 27a, where the extent of the horizontal lines measures θ_i for the associated values of $\tan^{-1}(\omega L/R)$.

The effect of increasing the inductance L is to lengthen the conduction period to angles greater than 180 degrees. As $\omega L/R$ approaches

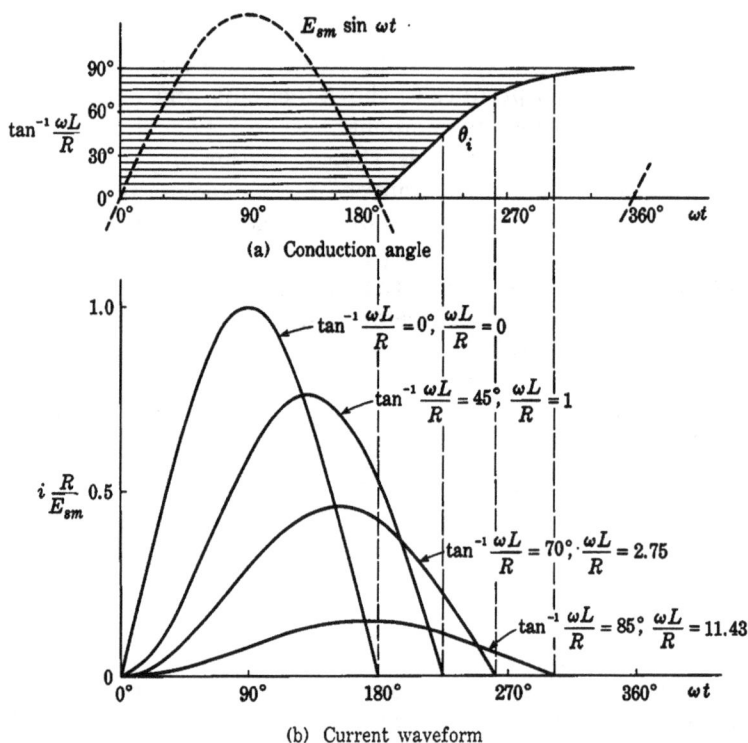

Fig. 27. Conduction angle and current waveforms in the circuit of Fig. 26 for various values of $\omega L/R$, or $\tan^{-1}(\omega L/R)$.

infinity, the conduction period approaches the entire cycle of 360 degrees as a limit. Curves of the instantaneous current multiplied by R/E_{sm} are plotted in Fig. 27b; they illustrate the lengthening of the conduction angle with increasing $\omega L/R$. When a smoothing inductor is used with a full-wave rectifier, the conduction period may be so extended that current flows in one rectifier until it begins in the other. In such circumstances, the current in the inductor never falls to zero. The analysis appropriate to such a circuit is markedly different from that given here, as is shown in Art. 12. Because of their ability to make the conduction continuous, series inductors are generally employed with mercury-arc rectifiers in full-wave or polyphase circuits to insure that the cathode current never falls to zero, and therefore that the arc does not extinguish.

Since the average value of the current is proportional to the area under one of the curves of Fig. 27b, these curves show qualitatively that the direct component of current decreases if, for a constant value of R, the inductance L is increased to increase the smoothing effect. This behavior is opposite to a corresponding result obtained for the capacitor smoothing circuit; in the latter, the direct component of current increases with the smoothing effect. The dependence of the direct current on the circuit parameters and source

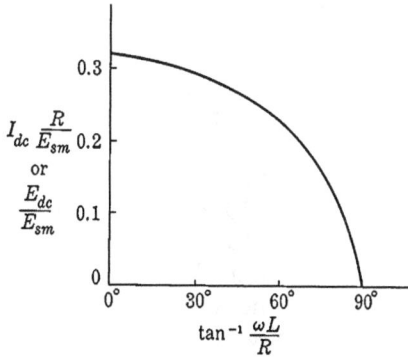

Fig. 28. Dimensionless curve showing the dependence of the direct component of the load current on $\tan^{-1}(\omega L/R)$.

voltage may be obtained quantitatively as follows. From Eq. 148,

$$i = \frac{E_{sm}}{R} \sin \omega t - \frac{\omega L}{R} \frac{di}{d(\omega t)}. \qquad [157]$$

Thus

$$I_{dc} = \frac{1}{2\pi} \int_0^{\theta_i} \left[\frac{E_{sm}}{R} \sin \omega t - \frac{\omega L}{R} \frac{di}{d(\omega t)} \right] d(\omega t) \qquad [158]$$

$$= \frac{E_{sm}}{2\pi R} (1 - \cos \theta_i), \qquad \blacktriangleright[159]$$

where the second term under the integral in Eq. 158 integrates to zero, because i is zero for both limits of the integral. From Eq. 159, the curve of $I_{dc}(R/E_{sm})$ as a function of $\tan^{-1}(\omega L/R)$ is plotted in Fig. 28. This curve shows the manner in which the direct current changes when L alone is varied. The direct component of the voltage across the load resistance R is

$$E_{dc} = I_{dc} R = \frac{E_{sm}}{2\pi} (1 - \cos \theta_i), \qquad \blacktriangleright[160]$$

and the function E_{dc}/E_{sm} is therefore represented by the same curve as $I_{dc}(R/E_{sm})$ in Fig. 28.

From Figs. 24b, 24c, and 27b, it may be observed that with either a capacitor or an inductor as a smoothing element *an increase of the conduction period is accompanied by a decrease in the direct component of the load current, and vice versa,* provided the smoothing element alone in each circuit is adjusted. This similarity of behavior of the two filter circuits appears in Eqs. 127 and 159, which give the direct current in the two circuits:

<table>
<tr><td align="center">*Capacitor Circuit*</td><td align="center">*Inductor Circuit*</td></tr>
<tr><td>$I_{dc} = \dfrac{E_{sm}}{2\pi}\sqrt{(1/R)^2 + (\omega C)^2}\,(1 - \cos\theta_i)$</td><td>$I_{dc} = \dfrac{E_{sm}}{2\pi}\dfrac{1}{R}(1 - \cos\theta_i)$</td></tr>
</table>

The rather striking likeness[14] of the two circuits is evident in another way. Equation 119 for the conduction angle of the capacitor circuit is the same as Eq. 156, from which the conduction angle of the inductor circuit is determined if ωCR and $\omega L/R$ are the same. From these equations and the relation $\theta_i{'}$ is equal to $2\pi - \theta_i$, it is apparent that if the capacitor circuit has a value of ωCR equal to $\omega L/R$ in the inductor circuit, the conduction angle in the first is less than 180 degrees (the conduction angle when no filter is used) by exactly the same amount as that angle in the second is greater than 180 degrees. This relation is seen graphically in Figs. 23 and 27a.

When conduction ceases in the circuit of Fig. 26, the voltage e_b across the rectifying element changes abruptly to the corresponding negative value of the supply voltage e_s, as is shown in the figure. Thus, for any value of load giving a conduction angle smaller than 270°, the peak inverse voltage is E_{sm}. As the load is varied, therefore, the

$$\text{Maximum peak inverse voltage} = E_{sm} = \pi E_{dc}\big|_{I_{dc}=0}, \quad [161]$$

just as in Eq. 24 for the half-wave circuit without a smoothing element.

11. Voltage-multiplying rectifier circuits

A rectifier circuit that has the property of delivering a direct output voltage in excess of the peak value of the applied alternating voltage is often of value—for example, in supplying x-ray tubes, or in furnishing high plate voltages to vacuum tubes when a transformer is prohibited, as in a-c–d-c radio receivers. The so-called voltage-doubling circuit shown in Fig. 29 has for large values of load resistance an output voltage almost twice the peak value of the voltage of the source. It is essentially two half-wave rectifiers with smoothing filters connected

[14] A. Glasser and K. Müller-Lübeck, *Theorie der Stromrichter* (Berlin: Julius Springer, 1935), 47–61.

in series but supplied from the same source. It operates by alternately charging each of the two capacitors to the voltage E_{sm}. The capacitors continually discharge in series through the load. They also act to smooth out the pulsations in the output voltage.

The waveform of the output voltage e may be deduced qualitatively from Fig. 29b. The capacitors have voltages corresponding approximately to the solid waves, one positive and the other negative with

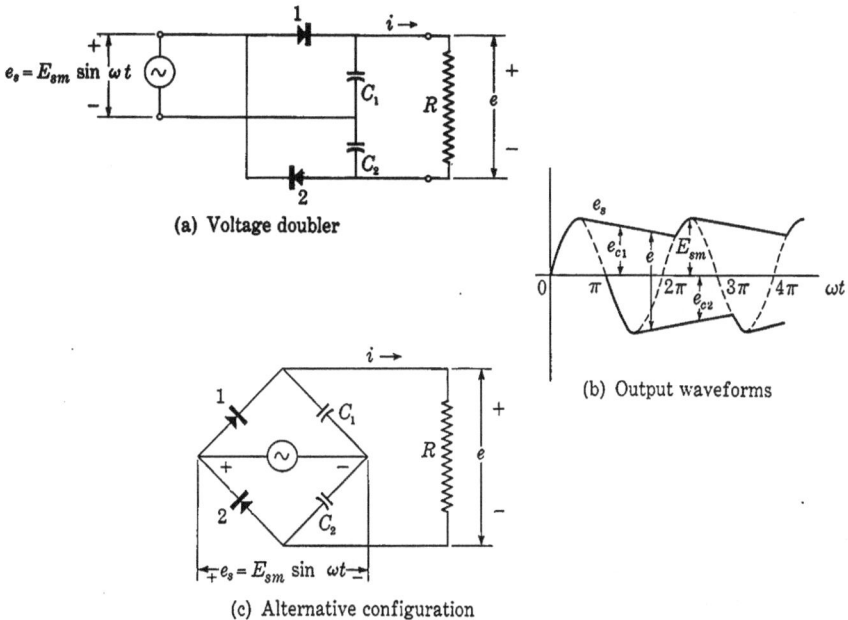

(a) Voltage doubler

(b) Output waveforms

(c) Alternative configuration

Fig. 29.   Voltage-doubling rectifier circuit.

respect to the common point between the capacitors. The output voltage across the load resistance is the voltage between the solid curves, as indicated. Since two pulses occur in the output for each cycle of the supply voltage, this connection is sometimes called a full-wave voltage doubler. It is also called the symmetrical voltage doubler because it can be drawn in the configuration of Fig. 29c. A disadvantage of this circuit is that neither terminal of the output may be connected to either terminal of the power source. Thus the source and the load may not simultaneously be connected to ground.

Voltage-multiplying rectifier circuits that permit a common connection between the alternating voltage supply and the direct voltage output are shown in Fig. 30. A qualitative explanation of the behavior of the voltage-doubler circuit of Fig. 30a is as follows. During the first negative half-cycle of e_s after the voltage source is connected, capacitor C_1 charges through rectifier 1 to the voltage E_{sm} with the polarity shown. During the next positive half-cycle, this capacitor voltage and the source voltage act in series and add to charge capacitor C_2 through rectifier 2. Capacitor C_1 is discharged by a like amount at the same time, but its charge is replenished in full during the next negative half-cycle. As the process repeats, the charge taken from C_1 and added to C_2 diminishes toward zero until finally C_2 is charged to $2E_{sm}$. Only one pulse of charge is fed to the output capacitor per cycle of the supply voltage; thus the lowest harmonic component in the

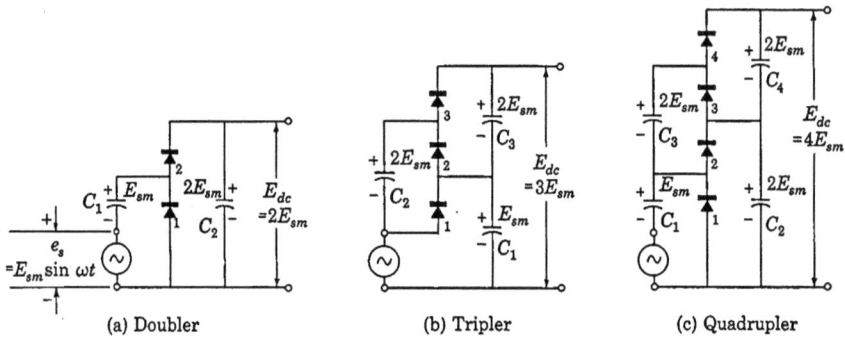

(a) Doubler   (b) Tripler   (c) Quadrupler

Fig. 30. Voltage-multiplying rectifier circuits having a common input and output terminal.

output voltage is the same as the power-supply frequency, and is half the lowest frequency in the output of the full-wave voltage doubler in Fig. 29. If the positive and negative parts of the input voltage wave are nonsymmetrical, rather than symmetrical as in a sine wave, the action of the circuit is to charge the output capacitor to a voltage equal to the numerical sum of the peak values of the two parts. Hence the connection in Fig. 30a, together with a means for measuring the direct output voltage without drawing appreciable current, is useful as a peak-to-peak indicating voltmeter.

The voltage-tripler circuit in Fig. 30b is essentially the voltage-doubler circuit of Fig. 30a in series with a half-wave rectifier. In the final state for negligible load current, capacitor C_1 is charged through rectifier 1 to the voltage E_{sm} during, say, the odd half-cycles. The source and C_1 in series charge C_2 to $2E_{sm}$ through rectifier 2 during the even half cycles, and the source and C_2 in series charge C_3 and C_1 in series to $3E_{sm}$ during the odd half-cycles.

The voltage-quadrupler circuit in Fig. 30c is effectively two voltage-doubler circuits in series. A final equilibrium state is reached in which C_1 tends to charge to the voltage E_{sm} through rectifier 1; C_2 to $2E_{sm}$

through C_1 and rectifier 2; C_3 to $2E_{sm}$ through C_2, rectifier 3, and C_1; and C_4 to $2E_{sm}$ through C_1, C_3, rectifier 4, and C_2 during successive half-cycles. Through application of the principles illustrated in Fig. 30, the direct output voltage can be brought to any desired multiple of the amplitude of the alternating source voltage.

12. FULL-WAVE RECTIFIER WITH INDUCTOR-INPUT FILTER

The treatment given in Art. 9 shows that in a filter circuit comprising only a smoothing capacitor, the ripple voltage decreases for a particular load current as the size of the capacitor is increased. Unfortunately, the peak and effective values of the rectifier current tend to increase as the size of the capacitor is increased, and, if the rectifier is of the gas type and the power source has low impedance, the peak current may rise to such a large value that the cathode of the rectifier is damaged by positive-ion bombardment unless means are employed to keep this peak current within reasonable limits. Also the increase in the effective value of the rectifier current with increased capacitor size results in an increase of the heating in the windings of the transformer or in the circuit used to supply power to the rectifier, and, for a unit having a particular transformer or supply circuit, decreases the efficiency and the over-all direct-current rating. As shown in Art. 10, however, the presence of series inductance tends to reduce both the peak value and the effective value of the rectifier current, in addition to reducing the ripple voltage. Thus the effect on the peak and effective values of the current in an inductive circuit as the inductance is increased is opposite to that in a capacitive circuit as the capacitance is increased, but the use of series inductance has the disadvantage of decreasing the voltage and current output of the rectifier. As a consequence, small or medium rectified-power supply units (that is, those with direct-current ratings of about 100 to 500 milliamperes, and 100 to 1,000 volts) whose output voltage must be very smooth frequently comprise a full-wave rectifier and a filter circuit having both inductance and capacitance.

The industrial demand for filter capacitors with ratings above 1,000 volts and filter inductors with ratings greater than 0.5 ampere is comparatively small, since the chief use of rectifier filters is in broadcast radio receivers, where the voltage and current ratings are relatively low. The increase in cost for larger ratings is hence more than proportional to the increase in rating. For this reason, considerable study is necessary when currents larger than about 0.5 ampere or voltages larger than about 1,000 volts are to be dealt with. Such matters as the relative cost of inductors and capacitors for the various currents and voltages involved play an important role in the determination of the most suitable combination of filter elements. In these higher-power units, a polyphase rectifier with a relatively simple filter often is practically desirable, since, as discussed in Art. 7, the rectifier itself produces a smoother voltage if the number of phases is increased.

The *inductor-input filter** shown in Fig. 31 utilizes an inductor to limit the peak and effective values of the rectifier and transformer currents, and a capacitor to smooth and support the voltage across the load. Though it is possible to analyze the full-wave rectifier with

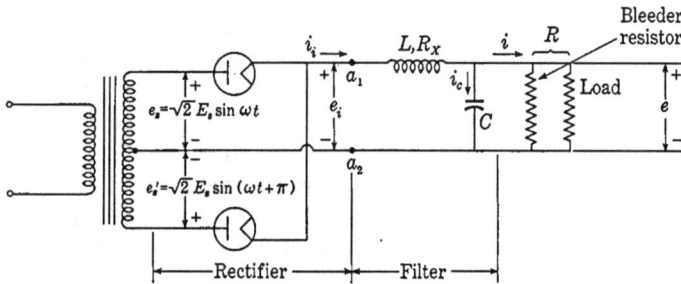

Fig. 31. Full-wave rectifier circuit with inductor-input filter.

an inductor-input filter by the methods used in Arts. 9 and 10, the procedure is relatively difficult. Since in the most usual conditions of operation, the current in the inductor never falls to zero, neither the current in the inductor nor the voltage across the capacitor at the beginning of each cycle is a known initial condition. The determination of the current and voltage waves as a series of repeated transients is therefore difficult. However, the very condition that current in the inductor is never zero—that is, that the instant one rectifier ceases to conduct the other begins—makes possible a relatively simple analysis suitable to this condition of operation.

Under the condition of continuous current in the inductor, and *only under this condition*, the waveform of the filter-input voltage (that is, the voltage at the terminals $a_1 a_2$ of the filter in Fig. 31) is independent of the waveforms elsewhere in the filter and of the constants of the filter circuit and the load. For example, if gas tubes are used, the potential of the terminal of the filter that connects to the cathodes of the tubes (terminal a_1) is always within ten volts or so of the potential of the anode of the *tube that is conducting*. The voltage between the anode of either tube and the terminal a_2 is established by the induced

---

* A filter inductor is frequently referred to as a *choke*, and the name *choke-input filter* is often used. Similarly, the name *condenser-input filter* instead of *capacitor-input filter* is frequently encountered.

voltage of the transformer; and the tubes, because of their uni-directional conduction property, act to commutate the applied voltage and thus to apply to the terminals a_1a_2 of the filter of Fig. 31 a rectified full-wave voltage e_i, as shown in Fig. 32. This voltage waveform occurs only when each tube conducts for an entire half-cycle—that is, when conduction in the inductor is continuous—and it differs from a rectified sinusoidal voltage only in that it is lowered at

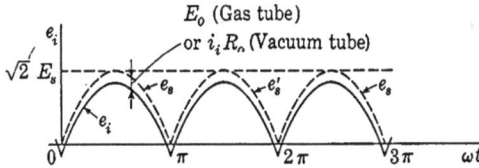

Fig. 32. Rectified sinusoidal voltage applied to the filter of Fig. 31 at terminals a_1a_2.

each point by the amount of the voltage drop across a tube during conduction.

For a waveform such as that shown in Fig. 32, the current and voltage in each branch of the filter circuit, including the output branch, can be calculated if the elements in the circuit are assumed to be linear. The solution involves the following steps: First, the rectified voltage waveform is resolved into its Fourier components; second, each component of voltage is considered to be impressed across the filter terminals at a_1a_2 and the current and voltage at the output caused by each component of voltage are computed; and, third, the component currents and voltages are then added in each branch of the circuit, with appropriate regard to phase and frequency, to give the total current in the branch.

If the voltage drop in the rectifier is temporarily neglected, and if a sinusoidal voltage of effective value E_s is applied between each anode and the center tap of the supply transformer, the waveform of the filter-input voltage e_i is a rectified sinusoid of peak value $\sqrt{2}E_s$. From Eq. 39, the Fourier series for this waveform is

$$e_i = \sqrt{2}E_s \left[\frac{2}{\pi} - \frac{4}{3\pi}\cos 2\omega t - \frac{4}{15\pi}\cos 4\omega t - \cdots \right], \qquad [162]$$

in which only a direct component and even harmonics of the supply frequency occur.

Figure 31 shows the load resistor paralleled by a second, or bleeder resistor, the use of which is discussed subsequently. For simplicity of analysis, R is used to denote the parallel resistance of the two, and hence denotes the true load resistance only if the bleeder resistor is omitted. The resistance of the inductor may be denoted by R_X, and the conductance of the capacitor and the effective impedance of the transformer may be neglected. Then, for the steady-state continuous-conduction condition, the direct output current is

$$I_{dc} = \frac{2\sqrt{2}}{\pi} \frac{E_s}{R + R_X} \qquad \blacktriangleright[163]$$

and the direct load voltage is

$$E_{dc} = \frac{2\sqrt{2}}{\pi} \frac{R}{R + R_X} E_s . \qquad \blacktriangleright[164]$$

Also, if the Q of the inductor is large,[15] so that $2\omega L$ is large compared with R_X, as is usual, the effective value of the second-harmonic component of i_i, the filter-input current at a_1, is

$$I_{i2} = \frac{4}{3\pi} \frac{E_s}{Z_i} \qquad [165]$$

where Z_i is the magnitude of the complex impedance

$$Z_i = j2\omega L + \frac{1}{(1/R) + j2\omega C} , \qquad [166]$$

which is the impedance at the input terminals of the filter. The effective value of the second-harmonic component of voltage across the load resistor is then

$$E_2 = \frac{4}{3\pi} \frac{E_s}{Z_i} \frac{1}{\sqrt{(1/R)^2 + (2\omega C)^2}} . \qquad [167]$$

In most applications of rectified-power supplies, it is desirable that the impedance as viewed from the load terminals back into the filter be small for all frequencies present in the load current. This condition is necessary in order that the alternating components of the load current produced by variations in the load itself may not cause corresponding alternating components of output voltage to appear. The load current often contains alternating components of frequencies as low as a few cycles per second. To satisfy the internal-impedance requirement, the filter-capacitor reactance at the lowest frequency of the alternating components of load current must be made small. Consequently, at the

[15] For the meaning and use of the constant Q, see E. E. Staff, M.I.T. *Electric Circuits* (Cambridge, Massachusetts: The Technology Press of M.I.T.; New York: John Wiley & Sons, Inc., 1940), 319–325.

frequency of the second-harmonic ripple component, which is many times higher, the capacitor reactance is usually negligible compared with the resistance of the load. In effect, the filter capacitor by-passes essentially all the alternating component of the current and eliminates it from the load. Furthermore, the capacitive reactance is almost always negligible compared with the inductive reactance at the ripple frequencies. In fact, the filter elements are purposely chosen so that this condition will exist, in order to make most of the ripple voltage appear across the inductor rather than across the load, or capacitor, and thereby to make the filter effective. For these reasons, it is usually true that

$$2\omega L \gg \frac{1}{2\omega C} \ll R, \qquad [168]$$

and, under these conditions, Eqs. 165 and 167 simplify to

$$I_{i2} \approx \frac{4}{3\pi} \frac{E_s}{2\omega L}, \qquad [169]$$

and

$$E_2 \approx I_{i2}/(2\omega C) \qquad [170]$$

$$\approx \frac{4}{3\pi} \frac{1/(2\omega C)}{2\omega L} E_s = \frac{4}{3\pi} \frac{E_s}{4\omega^2 LC}. \qquad \blacktriangleright[171]$$

For the fourth-harmonic frequency it is an even better approximation to write

$$4\omega L \gg \frac{1}{4\omega C} \ll R; \qquad [172]$$

hence the effective value of the fourth-harmonic component of the filter-input current is

$$I_{i4} \approx \frac{4}{15\pi} \frac{E_s}{4\omega L}, \qquad [173]$$

$$\approx 0.1 I_{i2}, \qquad [174]$$

and the fourth-harmonic voltage across the load resistor is

$$E_4 \approx \frac{4}{15\pi} \frac{E_s}{16\omega^2 LC}. \qquad [175]$$

From Eqs. 171 and 175, the ratio of the fourth-harmonic to the second-harmonic voltage across the load resistor is given approximately by

$$\frac{E_4}{E_2} \approx \frac{1}{20} = 5 \text{ per cent;} \qquad [176]$$

hence the ripple is predominantly second-harmonic, and the ripple factor, defined by Eq. 20, Art. 4a, may be written approximately as E_2/E_{dc} or, from Eqs. 164 and 171

$$\gamma = \frac{1}{6\sqrt{2}\omega^2 LC} \frac{R + R_X}{R},$$ [177]

which, for a 60-cycle-per-second supply voltage and for the condition of R large compared with R_X, reduces to

$$\gamma = \frac{0.83 \times 10^{-6}}{LC}.$$ ▶[178]

In the derivation of Eqs. 162 through 178, the current in the inductor is assumed to be continuous. This assumption, however, is not realized at all loads. The waveform of the current when the average value is larger than the amplitude of the ripple component is shown in Fig. 33a. If the harmonic currents of order higher than the second are neglected, this current into the filter at a_1 is representable by

$$i_i = I_{dc} + \sqrt{2}I_{i2} \cos 2\omega t.$$ [179]

The curve of Fig. 33a, a reproduction of an oscillogram taken on the circuit of Fig. 31, shows that the magnitudes of the harmonics of order higher than the second are small compared with the magnitude of the second harmonic, a fact that is in agreement with the condition indicated by Eq. 174. As the load resistance is increased from a value corresponding to an output current waveform such as in Fig. 33a, the amplitudes of the alternating components remain constant because, as is shown by Eqs. 171 and 175, they are independent of the load resistance. Consequently, when the load current is reduced to a critical value, the total current in the inductor reaches zero at one point in the ripple cycle, as is illustrated in Fig. 33b. Further reduction of the load current tends to make the current reverse for a short interval during the cycle.

However, the rectifiers conduct only in one direction; hence, if I_{dc} tends to become smaller than $\sqrt{2}I_{i2}$, conduction must cease during part of the cycle. The circuit then experiences a transient that repeats every half-cycle. The analysis in the derivation of Eqs. 162 through 178 is no longer applicable, because the filter-input voltage does not have the waveform indicated by Fig. 32. Once interrupted, the inductor current remains at zero until the instantaneous transformer voltage applied to the rectifier and circuit exceeds the instantaneous voltage on the capacitor. Since the load current is all supplied from the capacitor during the period when the inductor current is zero, the instant in the half-cycle at which conduction starts again depends upon the voltage on the capacitor when conduction ceases, the size of the capacitor, the magnitude of the load current, and the voltage and frequency of the source.

The calculation of the direct and ripple voltages at the load when conduction is not continuous involves a more complicated analysis than that indicated above. The general shape of the curve of E_{dc} as a function of I_{dc} may be estimated, however, as follows. As is discussed in connection with Fig. 21, for the condition of infinite load resistance, or zero load current, the steady-state voltage on the capacitor is equal to $\sqrt{2}E_s$. For I_{dc} equal to or greater than $\sqrt{2}I_{i2}$, the analysis

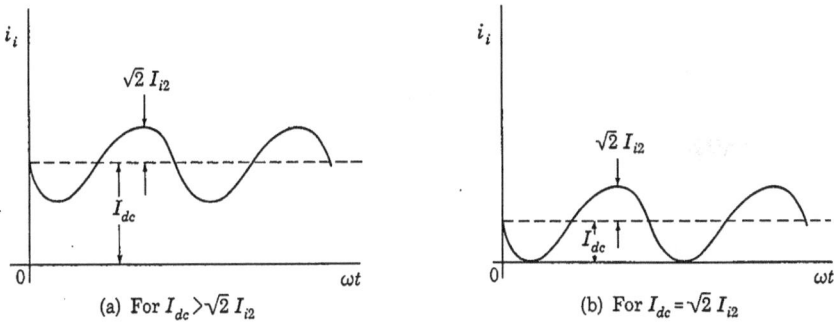

(a) For $I_{dc} > \sqrt{2}I_{i2}$ (b) For $I_{dc} = \sqrt{2}I_{i2}$

Fig. 33.   Waveform of current in inductor of inductor-input filter.

given above is applicable and Eq. 164 gives E_{dc}. As indicated by that equation, if the rectifier voltage drop and the resistance in the inductor are neglected, the direct output voltage is constant for all values of I_{dc} larger than $\sqrt{2}I_{i2}$. The curve of E_{dc} as a function of I_{dc}, then, is essentially horizontal for large currents, and rises to intersect the voltage axis at $\sqrt{2}E_s$, as is shown in Fig. 34, the dotted line being the idealized curve and the solid line an actual curve that differs from the ideal curve on account of the rectifier voltage drop and the rectifier and inductor resistances.

The approximate maximum value of load resistance at which the direct load voltage commences to increase rapidly is readily found through equating the value of I_{dc} in Eq. 163 to $\sqrt{2}I_{i2}$, the peak value of the second-harmonic current component obtained from Eq. 169; that is, from the equation

$$\frac{2\sqrt{2}}{\pi}\frac{E_s}{R+R_X} = \frac{4\sqrt{2}}{3\pi}\frac{E_s}{2\omega L},$$ [180]

which yields

$$R + R_X = 3\omega L.$$ ▶[181]

If the frequency of the supply voltage is 60 cycles per second, Eq. 181 gives

$$R + R_X = 1{,}131L, \qquad \blacktriangleright[182]$$

and, as long as the value of $R + R_X$ is less than $1{,}131L$, the direct voltage is essentially constant, since I_{dc} is then greater than $\sqrt{2}I_{i2}$. As $R + R_X$ is made larger than $1{,}131L$, however, the direct voltage increases and approaches $\sqrt{2}E_s$ as R approaches infinity, since I_{dc} is then smaller than $\sqrt{2}I_{i2}$.

Because of the uncertainty in the assumptions involved in calculations of this kind, it is usual practice to take the critical value[16] of R—

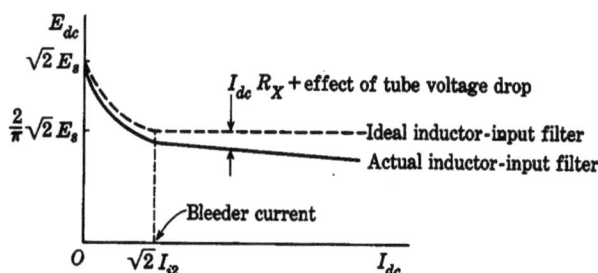

Fig. 34.   Direct output voltage as a function of direct current for a full-wave rectifier with an inductor-input filter.

that is, the *maximum* value for which the regulation is maintained small, or the value at which the break in the characteristic curve occurs—as

$$R = 1{,}000L \qquad \blacktriangleright[183]$$

for a rectifier with a 60-cycle-per-second supply. Frequently a resistor of this value, called a *bleeder resistor*, is permanently connected across the output of an inductor-input filter to keep the output voltage from increasing excessively and thereby possibly causing puncture of the dielectric of the capacitor if the load resistor is accidentally disconnected. This resistor is indicated in Fig. 31, and its current in Fig. 34.

In an actual circuit, the rectifier voltage drop may not always be negligible. If gas tubes are used as the rectifiers, the tube voltage drop may usually be assumed to be constant, and the average and instantaneous values of the voltage applied to the filter may be decreased by

---

[16] F. S. Dellenbaugh, Jr., and R. S. Quimby, "The Important First Choke in High-Voltage Rectifier Circuits," *QST*, *16* (February, 1932), 14–19; "The First Filter Choke—Its Effect on Regulation and Smoothing," *QST*, *16* (March, 1932), 26–30; "The Economical Design of Smoothing Filters," *QST*, *16* (April, 1932), 33–40.

the tube voltage drop E_0, where E_0 is 10 to 20 volts. Alternatively, if vacuum tubes are used, their effect is approximated if the rectifiers are represented by constant resistances R_0 in series with ideal rectifiers, as in Arts. 4a and 4b. A somewhat better approximation to the actual tube characteristic may be obtained if both a resistance R_0' and a battery of voltage E_0' are placed in series with each ideal rectifier. The resistance R_0' corresponds to the reciprocal of the mean slope of the volt-ampere characteristic of the tube over the working region, and E_0' is the intercept of this mean slope with the voltage axis.[17] On the basis of this approximation, Eqs. 163 and 164 become

$$I_{dc} = \frac{\frac{2\sqrt{2}}{\pi} E_s - E_0'}{R + R_X + R_0'} \qquad \blacktriangleright[184]$$

and

$$E_{dc} = \left(\frac{2\sqrt{2}}{\pi} E_s - E_0'\right) \frac{R}{R + R_X + R_0'} \qquad \blacktriangleright[185]$$

For this condition, the ripple voltage is not appreciably affected unless $2\omega L$ is not large compared with R_0 or R_X, and the ripple factor γ is given by

$$\gamma = \frac{R + R_X + R_0'}{\left(6\sqrt{2} - 3\pi \dfrac{E_0'}{E_s}\right) \omega^2 L C R}. \qquad [186]$$

Although only a single-section inductor-input filter is shown in Fig. 31, double and sometimes triple sections are used, each section consisting of a series inductor and shunt capacitor. From the foregoing analysis, and on the basis of the assumptions used for the derivation of Eqs. 162 through 178, the ripple voltage for a two-section filter is

$$E_2 \approx \frac{4}{3\pi} \frac{E_s}{(4\omega^2 L_1 C_1)(4\omega^2 L_2 C_2)}, \qquad \blacktriangleright[187]$$

where L_1 and L_2 are the inductances of the first and second inductors, and C_1 and C_2 the capacitances of the first and second capacitors, respectively. Note that the attenuation of the ripple voltage is proportional to the product of $(\omega^2)^n$ and all the LC products for the n filter sections. Each LC section of the filter reduces the ripple voltage by a factor $1/(4\omega^2 LC)$.

The analysis of rectifier-filter circuits is seldom performed with a

---

[17] This representation of the nonlinear tube characteristic is essentially the same as that of E. E. Staff, M.I.T., *Electric Circuits* (Cambridge, Massachusetts: The Technology Press of M.I.T.; New York: John Wiley & Sons, Inc., 1940), 686–687.

high degree of accuracy, because of the uncertainty of the values to be used for C and L with any particular type of capacitor and inductor. Most inexpensive electrolytic capacitors, and even oil-filled paper capacitors, are nonlinear; and their incremental capacitance—the parameter that determines their filtering effectiveness—is a function of their age, the average voltage, the temperature, and the like. Similarly, the value to be used for L is the incremental inductance, and this quantity varies with the output current, since the presence of the direct-current component in the winding of the inductor modifies the apparent or incremental permeability of the core by an amount depending upon the magnitude of the direct current.[18] Although air gaps are usually introduced into the magnetic core of the inductor, the inductance parameter is often appreciably nonlinear.

An economical filter frequently used employs an inductor whose incremental inductance decreases appreciably with an increase in direct current; this is known commercially as a *swinging choke*. If the filter is designed so that the values of L and C occurring with full-load operation give satisfactory filtering, the increase in L that occurs as I_{dc} is decreased is sometimes beneficial. The increase in L serves to decrease the ripple voltage and at the same time to cause the value of load current at which I_{dc} equals $\sqrt{2}I_{i2}$ to be smaller than that occurring with a fixed value of L. Thus the range over which I_{dc} can be varied without causing E_{dc} to increase above the value of $(2\sqrt{2}E_s)/\pi$ is extended. A larger value of bleeder resistance can then be used, and the power dissipation in the bleeder circuit is decreased.

When a swinging choke is used in a filter that uses more than one LC section, its properties are doubly useful only if it is connected directly to the cathodes of the rectifiers; that is, if it is the first inductor. This connection is necessary because the relation given by Eq. 181 applies only between R and the incremental inductance of the *first* inductor; the condition for continuous conduction in the first inductor is not affected by the addition of extra filter sections. The first inductor is doubly important, for it serves to control, first, the ripple voltage, and, second, the minimum value of I_{dc} for good voltage regulation. With the values of C generally used, the inductors in the other sections serve essentially only to decrease the ripple voltage. The extent to which any increase in their incremental inductance with decrease in I_{dc} aids the condition expressed by Eq. 181 is negligible.

---

[18] E. E. Staff, M.I.T., *Magnetic Circuits and Transformers* (Cambridge, Massachusetts: The Technology Press of M.I.T.; New York: John Wiley & Sons, Inc., 1943), Art. 15, Ch. VI.

As an example of the application of the foregoing analysis, the data given below illustrate the agreement between experimental and computed results in an inductor-input filter circuit similar to that of Fig. 31.

The inductor has a resistance of 3.3 ohms and an inductance of about 1.58 henries, and the capacitor a capacitance of about 53 μf. The impedance of the supply transformer is negligible. Since the values of the circuit elements are chosen to be in agreement with the criterion that $2\omega L$ is large compared with $1/(2\omega C)$, and to give values of I_{i2} and E_2 relatively easy to measure without disturbing the circuit, the values do not represent usual filter practice. The effective voltage E_s applied between each anode of the rectifier and the center tap of the transformer is 220 volts at 60 cps. Then, for the circuit using a mercury-vapor tube, and on the assumption that the tube voltage drop E_0' is 10 volts and its resistance R_0' is negligible, Eq. 185 gives

$$E_{dc} = \frac{2\sqrt{2}}{\pi} 220 - 10 \qquad [188]$$

$$= 188 \text{ volts}, \qquad [189]$$

since the direct voltage drop in the inductor is negligible. From Eq. 169,

$$I_{i2} = \frac{4}{3\pi} \frac{220}{2 \times 377 \times 1.58} \qquad [190]$$

$$= 0.078 \text{ amp}, \qquad [191]$$

and, from Eq. 171,

$$E_2 = \frac{I_{i2}}{2\omega C} = \frac{0.078}{2 \times 377 \times 53 \times 10^{-6}} \qquad [192]$$

$$= 1.95 \text{ volts}. \qquad [193]$$

Also, from Eq. 174,

$$I_{i4} = 0.1 I_{i2} = 0.1 \times 0.078 \qquad [194]$$

$$\approx 0.008 \text{ amp}. \qquad [195]$$

From Eqs. 191 and 195, the effective value of the capacitor current, which is essentially a second harmonic plus a fourth harmonic, is

$$I_c \approx \sqrt{0.078^2 \times 0.008^2} \qquad [196]$$

$$\approx 0.079 \text{ amp}. \qquad [197]$$

The maximum value of R for good regulation, from Eq. 183, is approximately

$$R = 1,000 \times 1.58 = 1,580 \text{ ohms}. \qquad [198]$$

The value of I_{dc} below which the voltage increases appreciably is then

$$I_{dc_{min}} = \frac{188}{1,580} \qquad [199]$$

$$= 0.119 \text{ amp}. \qquad [200]$$

The measured volt-ampere characteristic for the filter is shown in Fig. 35, the solid curves being obtained with a Type 83 mercury-vapor tube as a rectifier, and the dotted curves with a Type 5Z3 vacuum tube. The voltage is seen to begin to increase appreciably at about 0.120 amp (see Eq. 200). The measured effective value of the ripple voltage is approximately 1.85 volts (see Eq. 193), and the direct load voltage approximately 185 volts (see Eq. 189). The measured effective value of the capacitor current is 0.081 amp (see Eq. 197).

The measured and computed values are probably in better agreement than is ordinarily possible with filters assembled from the less expensive types of inductors and capacitors commercially available. The measured values of the ripple voltage, as well as the values of L

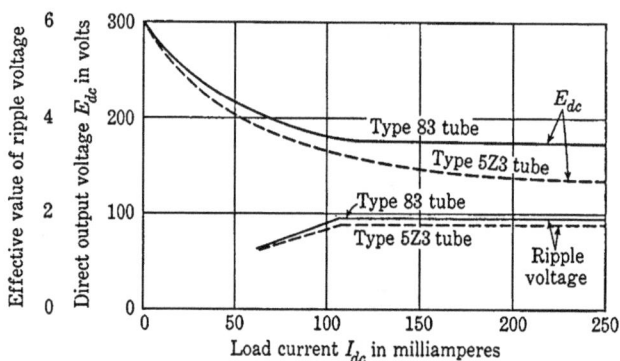

Fig. 35.   Experimental curves of direct output voltage and ripple voltage for an inductor-input filter.

and C, may be in error by as much as 5 per cent, on account of the difficulty of measuring these quantities. An equally satisfactory agreement between measured and computed values is found for the circuit using the Type 5Z3 tube if Eqs. 184 and 185 are used with the appropriate values of R_0' and E_0'.

13. FULL-WAVE RECTIFIER WITH CAPACITOR-INPUT FILTER

The filter circuit most commonly used in small power units is the capacitor-input filter shown in Fig. 36. It consists of a first capacitor C_1 to smooth and support the output voltage, followed by an inductor-capacitor combination, $L–C_2$, to reduce further the ripple voltage. Because of this additional filtering, the first capacitor can be much smaller and the peak and effective values of the rectifier current correspondingly smaller than would be required to provide the same ripple voltage for a particular direct current in the absence of the inductor. The presence of the inductor does little otherwise to decrease the peak and effective values of the rectifier current for a given direct current. This filter circuit should therefore not be used with gas tubes unless the impedance of the transformer or supply circuit is sufficient to keep the magnitude of the current peaks within permissible limits.

Analysis of the capacitor-input filter circuit for the ripple or direct voltage is somewhat more complicated than that given for the inductor-input filter. Note particularly that the capacitor-input filter behavior cannot be derived from the waveform given in Eq. 162 for the filter-input voltage e_i because the current i_i in Fig. 36 always falls to zero during some parts of the cycle, just as it does in the half-wave circuit with a smoothing capacitor, Fig. 20. Hence an interval exists during which the filter-input terminals are not effectively connected to

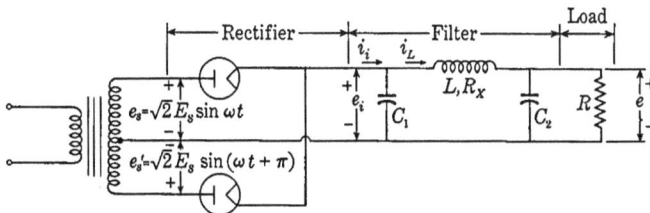

Fig. 36.   Full-wave rectifier circuit with capacitor-input filter.

either of the transformer terminals, and the filter-input voltage therefore does not have the waveform of Eq. 162 or Fig. 32. The waveform of the filter-input voltage depends on the filter parameters and the load current. Consequently, an adequate analytic attack is difficult; and, to find the direct component of the output voltage, resort is commonly made to experimental results.

As an aid to predicting the performance of rectifiers with capacitor-input filters, experimentally determined curves such as those in Fig. 37 are furnished by tube manufacturers. These curves show the direct component of the *input* voltage to the filter as a function of the direct output current for various magnitudes of the effective value of the supply voltage E_s, the input capacitance C_1, and the source impedance. To find the direct output voltage, it is merely necessary to subtract the direct component of the inductor voltage drop $R_X I_{dc}$ from the direct input voltage.

The ratings of the tube and the transformer frequently are such that the allowable peak current of the tube permits the first capacitor to be charged to a value that differs by only a negligible amount from the peak value of the applied voltage. Under these conditions, or when the amount Δe by which the peak voltage of the first capacitor does not reach peak supply voltage is small and can be estimated with reasonable accuracy, the direct load voltage and ripple voltage may

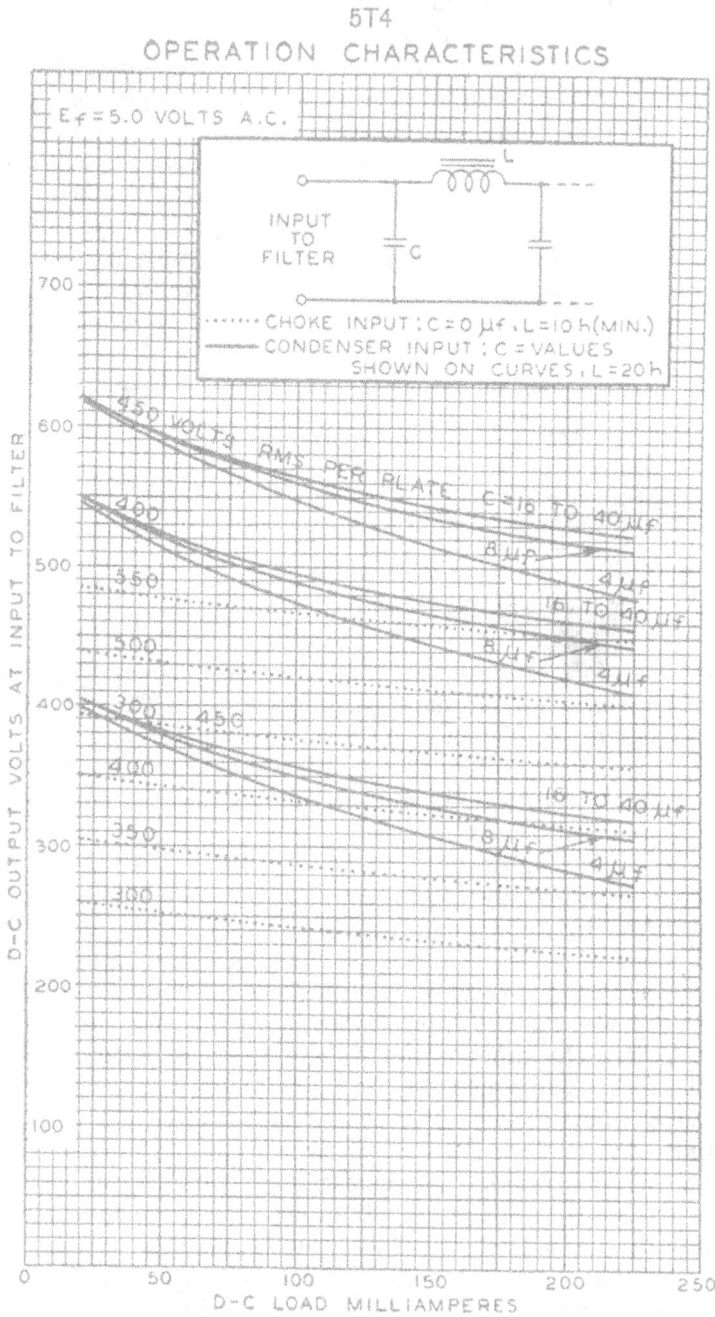

Fig. 37. Direct-current characteristics of a full-wave rectifier and filter with a Type 5T4 tube. (*Courtesy Radio Corporation of America.*)

be computed. On the assumption that Δe is known, the peak voltage $e_{i_{max}}$ on the first capacitor is

$$e_{i_{max}} = \sqrt{2} E_s - \Delta e. \qquad [201]$$

If it is assumed that the rectifier stops conducting when the capacitor voltage reaches this maximum value, the minimum voltage $e_{i_{min}}$ across this capacitor is less than $e_{i_{max}}$ by the amount the capacitor voltage decreases during the period when the tubes are not conducting.

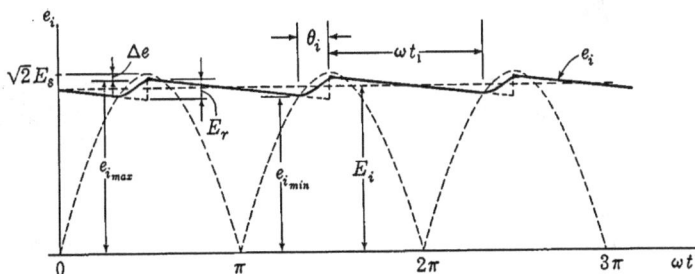

Fig. 38.   Approximate waveform of filter-input voltage e_i across the first capacitor of a capacitor-input filter.

If the current in the inductor is assumed to be constant, which assumption merely means that L is large enough to keep the ripple component of the current small, the decrease in voltage of the first capacitor is

$$e_{i_{max}} - e_{i_{min}} = I_{dc} t_1 / C_1 , \qquad [202]$$

where t_1 is the time during each half-cycle that the tubes do *not* conduct and C_1 is the capacitance of the first capacitor. The approximate waveform of e_i is shown by the solid line in Fig. 38. If both the rise and the fall of e_i are assumed to be linear with time, the direct voltage at the load is

$$E_{dc} = \frac{e_{i_{max}} + e_{i_{min}}}{2} - I_{dc} R_X = E_i - I_{dc} R_X, \qquad [203]$$

which, by means of Eq. 202, can be put in the form

$$E_{dc} = e_{i_{max}} - I_{dc} \left(\frac{t_1}{2C_1} + R_X \right), \qquad \blacktriangleright [204]$$

where R_X is the resistance of the inductor.

Because of the assumptions, the result given by Eq. 204 is a good approximation only when $e_{i_{max}}$ is nearly equal to $\sqrt{2} E_s$ and when $e_{i_{max}} - e_{i_{min}}$ is small. The result is also dependent upon the accuracy with which t_1 is known; and, for $e_{i_{max}} - e_{i_{min}}$ to be small, the conduction period of the tubes must be small and t_1 nearly equal to $1/(2f)$, where f is the supply frequency. The waveform of e_i may then be approximated by the dotted curve in Fig. 38 during the conduction period, and

$$E_{dc} \approx e_{i_{max}} - I_{dc}\left(\frac{1}{4fC_1} + R_X\right),$$ [205]

which, for small values of Δe, is

$$E_{dc} \approx \sqrt{2}\,E_s - I_{dc}\left(\frac{1}{4fC_1} + R_X\right).$$ [206]

Even when the rectifier voltage drop and the alternating-voltage source impedance are neglected, this rectifier-and-filter circuit behaves therefore, as if it had an inherent internal resistance of $1/(4fC_1)$ ohms to direct current, to which must be added the ohmic resistance in the inductor. This inherent resistance is to a large measure responsible for the downward slope of the curves in Fig. 37. Since the capacitance C_1 cannot be increased beyond a certain limit without exceeding the allowable peak and effective values of the rectifier current, and since vacuum rectifier tubes having a relatively large voltage drop are generally used, the direct voltage from a capacitor-input filter usually decreases appreciably as the load current is increased.

For a determination of the ripple voltage in the output, the approximate method leading to Eqs. 131 through 146 may be applied when the load is small enough that the input current i_i flows in short pulses near the peaks of the alternating supply voltage. As is mentioned in the discussion following Eq. 144, only the even harmonics exist in the voltage across the first capacitor, and the effective value of the second-harmonic voltage is

$$E_{i2} = \frac{\sqrt{2}I_{dc}}{2\omega C_1}.$$ [207]

This voltage is impressed on the L–C_2 filter. Generally the capacitive reactance of C_2 is made small compared with the load resistance to prevent cyclic changes in output current from affecting the output voltage, and the inductive reactance of L is made large compared with the capacitive reactance of C_2 to provide a large reduction in the ripple components of voltage. Thus,

$$2\omega L \gg \frac{1}{2\omega C_2} \ll R$$ [208]

as for the inductor-input filter. The same steps in analysis as are used for the inductor-input filter are therefore applicable. The effective value of the second-harmonic component in the inductor current is then

$$I_{L2} = \frac{E_{i2}}{2\omega L} = \frac{\sqrt{2}I_{dc}}{4\omega^2 LC_1},$$ [209]

and the effective value of the second-harmonic in the output voltage is

$$E_2 = I_{L2}X_{C2} = \frac{\sqrt{2}I_{dc}}{8\omega^3 LC_1C_2}.$$ [210]

For the fourth harmonic, the corresponding voltage components are

$$E_{i4} = \frac{\sqrt{2}I_{dc}}{2\omega C_1},$$ [211]

and

$$E_4 = \frac{\sqrt{2}I_{dc}}{64\omega^3 LC_1C_2}.$$ [212]

Thus the fourth harmonic in the output is only one-eighth as large as the second, and higher even harmonics are even more sharply attenuated. It is reasonable to neglect all harmonics except the second, therefore, and express the ripple factor in the output as

$$\gamma = \frac{E_2}{E_{dc}} = \frac{\sqrt{2}}{8\omega^3 LC_1C_2R}.$$ [213]

A comparison of Eq. 187 with Eq. 210 shows that the ripple suppression in the inductor-input filter is superior to that in the capacitor-input filter. From Eq. 187, the ripple voltage with the inductor-input filter is seen to be essentially independent of load current; whereas, from Eq. 210, the ripple voltage with the capacitor-input filter is to a first approximation proportional to load current.

Often sufficient filtering is obtained if a resistance is substituted for the inductance in the capacitor-input filter. The change of output voltage with load and the required total capacitance tend to be larger when a resistance is used, but the saving in cost, bulk, and weight that accompany elimination of the inductor may more than justify the substitution. Several resistance-capacitance sections in tandem are frequently used to obtain a desired reduction in the ripple voltage.

Where a high order of accuracy is essential, the analysis given here is inadequate, and experimental methods must be used or a more accurate analysis[19] must be performed. When the analysis must be

---

[19] O. H. Schade, "Analysis of Rectifier Operation," *I.R.E. Proc.*, *31* (1943), 341–361.

more accurate, the methods outlined in Arts. 9 and 10 and in *Electric Circuits*[20] can be combined to the degree necessary to achieve the accuracy desired. The task is not an impossible one by any means, the only problem being to maintain a reasonable balance between the value of the more accurate data and the cost of the engineer's time required to obtain these data.

14. Voltage stabilization by gas-discharge tubes

Stabilization of the output voltage as well as reduction of the ripple voltage from the rectifier may be accomplished by a glow-discharge or corona-discharge voltage-regulator tube used in the manner shown at

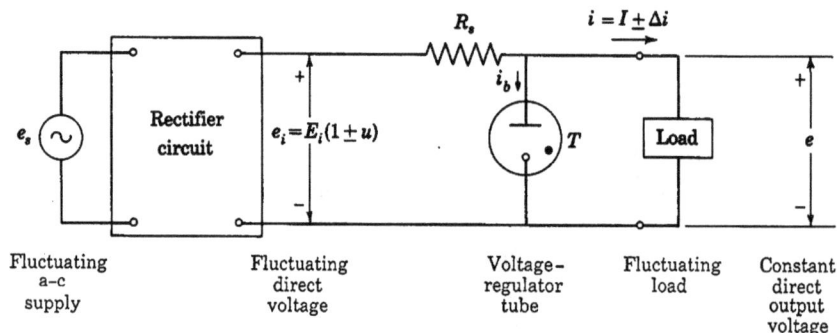

Fig. 39.   Voltage regulation by gas-discharge tube.

T in Fig. 39. Glow-discharge tubes, as is pointed out in Art. 9, Ch. III, and Art. 8, Ch. V, have an essentially constant voltage drop for current values over a wide range. Their volt-ampere characteristic is almost horizontal, as is shown for various glow-discharges in Fig. 14, Ch. III. Corona-discharge tubes have a similarly shaped characteristic curve but at a higher voltage level. Consequently, when connected in parallel with the load as at T in Fig. 39, the tube tends to by-pass around the load any current changes associated with fluctuations in the amplitude of the alternating supply voltage or with the ripple voltage. Furthermore, for a constant amplitude of supply voltage, the tube maintains the current in the series resistance R_s almost constant despite changes in the load current. The tube thereby maintains the output voltage essentially constant despite changes in both the supply voltage and the load.

Several factors must be taken into account in the application of a

---

[20] E. E. Staff, M.I.T., *Electric Circuits* (Cambridge, Massachusetts: The Technology Press of M.I.T.; New York: John Wiley & Sons, Inc., 1940), 676–712.

voltage-regulator tube. First, the circuit must supply sufficient voltage to start the discharge. A starting voltage 10 to 30 per cent larger than the constant operating voltage is required by most tubes. Second, the circuit must not allow the current i_b through the tube to exceed a specified maximum value, denoted hereafter by i_2; otherwise overheating and permanent change in the tube characteristics will occur. A current of 40 milliamperes is the maximum allowable in many glow-discharge tubes. Third, the circuit must not allow the current in the tube to fall below a specified minimum value, denoted hereafter by i_1; otherwise instability and pronounced change of the voltage drop with current are likely to occur. The minimum allowable current for many glow-discharge voltage-regulator tubes is 5 milliamperes. Since the tube maintains the voltage across the series resistance R_s almost constant when the amplitude of the supply voltage is constant, the sum of the tube and load currents is then essentially constant. The maximum permissible variation in the *load* current is hence established by the tube-current limitations, and equals $i_2 - i_1$.

The instantaneous current i_b through the tube in Fig. 39 is given by

$$i_b = \frac{e_i - E_b}{R_s} - i,$$ [214]

where E_b is the terminal voltage of the tube, which is the same as the load voltage, and is assumed constant. In ordinary applications of the voltage-regulator tube, the load current i varies between known limits. It may therefore be expressed as $I \pm \Delta i$, where I is the mean load current and Δi is half the total variation in i. Although the magnitude of the voltage e_i is not known, its random *fractional* variation is known to be the same as that of the supply voltage e_s, which may be as large as plus or minus 15 per cent in typical applications. Superposed on this random fluctuation is the cyclic fractional variation expressed by the ripple factor for the particular rectifier circuit used. Thus the voltage e_i may be expressed as $E_i(1 \pm u)$, where E_i is the mean output voltage of the rectifier circuit, and u is half the total fractional variation of e_i.

Since the fluctuations in e_i and i are random, the largest value of e_i and the smallest value of i may occur simultaneously. Under these conditions, the tube current in Eq. 214 is a maximum, which must be no larger than the maximum permissible value, i_2, specified by the manufacturer. Thus

$$\frac{E_i(1 + u) - E_b}{R_s} - (I - \Delta i) \leqq i_2 .$$ [215]

Similarly, for the other extreme

$$\frac{E_i(1-u) - E_b}{R_s} - (I + \Delta i) \geq i_1 , \qquad [216]$$

where, as is previously stated, i_1 is the minimum permissible value specified by the manufacturer.

To minimize the cost of the rectifier circuit, it may be desirable that E_i have the lowest permissible value. To find this value, note that as E_i is decreased, the required value of R_s expressed by Eq. 216 decreases also, and approaches zero as $E_i(1-u)$ approaches E_b. But, since $E_i(1+u)$ remains larger than E_b, the left-hand term of Eq. 215 then approaches infinity, so that Eq. 215 is violated. Thus there is a minimum value of E_i, and a corresponding minimum value of R_s, for which Eqs. 215 and 216 are satisfied. For these minimum values, the conditions of equality in the two equations occur together. Simultaneous solution of the equations gives for the minimum values

$$\text{Minimum permissible } E_i = \frac{E_b}{1 - u\,\dfrac{\dfrac{i_2 + i_1}{2} + I}{\dfrac{i_2 - i_1}{2} - \Delta i}} \qquad [217]$$

and

$$\text{Minimum permissible } R_s = \frac{E_b}{\dfrac{i_2 - i_1}{2u} - \dfrac{i_2 + i_1}{2} - \dfrac{\Delta i}{u} - I} \cdot \qquad [218]$$

For any selected value of E_i larger than that given by Eq. 217, there is a range of permissible values for R_s that is obtained through substitution for E_i in Eqs. 215 and 216. Conversely, for any selected value of R_s larger than that given by Eq. 218, a range for E_i is found.

Once values of E_i and R_s are selected by the foregoing procedure, the question of whether or not the minimum starting voltage requirement of the tube is satisfied should be examined. The voltage available to start the discharge is the fraction of E_i that appears across the load with the tube removed. The load and R_s then act as a voltage-divider network. This voltage available at the tube terminals must be greater than the rated starting voltage of the tube when e_i and the load resistance have their minimum values; otherwise the discharge will not start and no stabilizing action will occur. Adoption of higher values of E_i and R_s than originally chosen may thus be found necessary to provide adequate starting voltage.

Although e_i and R_s are shown as being external to the rectifier circuit in Fig. 39, the analysis is also applicable if e_i is interpreted as being the internal direct-voltage source in the rectifier circuit obtained by application of Thévenin's theorem. The resistance R_s then includes the internal series resistance in the Thévenin's equivalent as well as the externally added resistance. Consequently, it is not necessary that the rectifier circuit have good voltage regulation at the rectifier circuit terminals when a voltage-regulator tube is used. The internal series resistance that gives rise to the poor regulation is absorbed in the required value of R_s.

The filter elements in the rectifier circuit may be made smaller than would otherwise be required when a voltage regulator is used, because the tube reduces the ripple along with the supply-voltage variations. For high ripple frequencies and rapid fluctuations in the load and supply voltages, however, the volt-ampere relation for the tube ceases to follow an essentially horizontal line. It becomes a loop with a major axis having appreciable slope.[21] The effective impedance of the tube thus increases with the frequency or rate of rise of the fluctuations, and the voltage stabilization becomes poorer. Furthermore, even for slow changes or steady values of the tube current, the instabilities, lack of reproducibility, and discontinuities in the volt-ampere characteristic mentioned in Art. 8, Ch. V, impose limitations on the applicability of gas-discharge voltage-regulator tubes.

Gas-discharge voltage-regulator tubes are frequently operated in series to provide a regulated voltage higher than that afforded by one tube. A series arrangement with connections from each tube may be used as a voltage divider to give multiple regulated output voltages from one source. A parallel combination of voltage-regulator tubes for the purpose of obtaining increased current range is not practical, however. Since the starting voltages of two such tubes are never equal, one tube would tend to start first and prevent application of sufficient voltage for starting the other. Furthermore, since the volt-ampere characteristics are never identical and the operating voltages across the two tubes are constrained to be equal, even if both tubes did start, the current would always be much larger in one than in the other.

15. ELECTRONIC VOLTAGE STABILIZERS

Stabilization of the output voltage of a rectifier is accomplished by

[21] H. J. Reich, *Theory and Applications of Electron Tubes* (2nd ed.; New York: McGraw-Hill Book Company, Inc., 1944), 433–436; W. C. Elmore and M. Sands, *Electronics: Experimental Techniques* (New York: McGraw-Hill Book Company, Inc., 1949), 367–370.

a number of methods[22] that utilize amplifier tubes. These voltage-stabilizer circuits also serve to filter and reduce the ripple voltage. A representative arrangement in common use is shown in Fig. 40. A full explanation of the behavior of this circuit involves the principles of amplification and feedback discussed in Ch. IX. A qualitative description, however, is that tube T_1 serves as a variable resistance that is automatically adjusted by the circuit to maintain the output voltage E_{dc} approximately constant regardless of changes in the load current

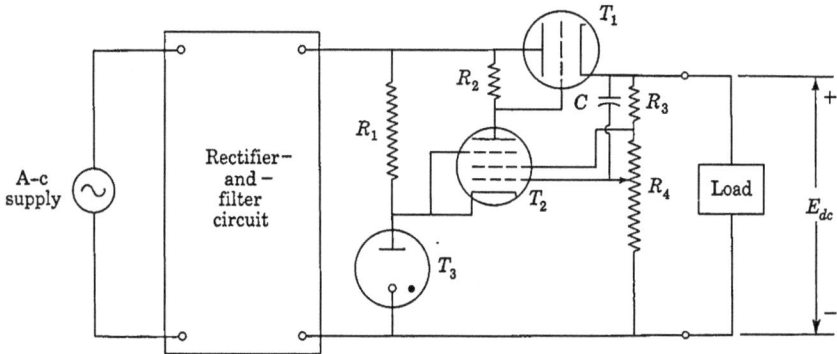

Fig. 40.   Degenerative-type electronic voltage-regulator circuit.

and alternating supply voltage. The voltage-regulator tube T_3 establishes a fixed voltage reference at its terminals in the manner described in Art. 14. A fraction of the output voltage, αE_{dc}, determined by the adjustable tap on resistor R_4, is compared with the fixed reference value by the amplifier tube T_2. Any change in E_{dc} causes a change in the difference between αE_{dc} and the voltage across T_3. This change in amplified by T_2 and applied to the grid of T_1 to increase or decrease its effective resistance to direct current so as to restore E_{dc} approximately to the value it had before it started to change.

Adjustment of the output voltage over a considerable range may be made by means of the tap on resistor R_4. The maximum current output is limited by heating in tube T_1; several tubes in parallel are used when a large current is needed. Capacitor C is added to make the circuit especially responsive to rapid changes in the load or supply

[22] F. V. Hunt and R. W. Hickman, "On Electronic Voltage Stabilizers," *R.S.I.*, *10* (1939), 6–21; A. B. Bereskin, "Voltage Regulated Power Supplies," *I.R.E. Proc.*, *31* (1943), 47–52; W. R. Hill, Jr., "Analysis of Voltage Regulator Operation," *I.R.E. Proc.*, *33* (1945), 38–45; A. Abate, "Basic Theory and Design of Electronically Regulated Power Supplies," *I.R.E. Proc.*, *33* (1945), 478–482.

voltage. A number of modifications of this basic stabilizer circuit have been made to gain further constancy of the output voltage. The ultimate limit in stability of the output voltage for all of them, it must be recognized, is the stability of the voltage across the voltage-regulator tube T_3. Several special voltage-reference tubes are available for applications of this kind. A representative tube has a stability of 0.1 volt in 90 volts when the current through it is maintained constant.

PROBLEMS

1. A vacuum diode is available whose volt-ampere characteristic can be closely represented by the relation

$$i_b = 10^{-4}e_b{}^{3/2},$$

where i_b is in amperes and e_b in volts.

(a) If this diode is connected in series with a resistance of 1,000 ohms across a 200-volt direct source, what will be the equilibrium value of the current?

(b) If the diode is connected in series with an inductance of 30 henries and the same source, derive a relation between the current and the time after closure of the switch.

(c) If the diode is connected in series with a capacitance of 50 μf and the same source, derive an expression for capacitor voltage as a function of time after closure of the switch.

2. The relation between the instantaneous values of the voltage across and the current through a hot-cathode fluorescent lamp 1 in. in diameter is given by

$$e = \frac{2.37l}{0.56 + i} + 12.5,$$

where l is the length of the tube in inches, e is in volts, and i is in amperes. The starting voltage may be assumed to be twice the operating voltage for 0.1 amp.

(a) What is the maximum length of lamp that can be operated from a 120-volt rms sinusoidal source?

(b) If the length is three-fourths that in (a), what value of series resistance should be used if the maximum instantaneous value of the current is to be limited to 1 amp?

(c) Sketch the waveform of the current and obtain its approximate effective value.

3. A circuit commonly used in rectifier-type indicating instruments is the bridge rectifying circuit shown in Fig. 18c. The instrument has a d'Arsonval movement that gives a deflection proportional to the average value of the current through it, but the scale is calibrated to indicate the rms value when a sinusoidal voltage is applied.

What does the instrument read and what is the true rms value when it is connected to

(a) a 115-volt d-c source?

(b) a 115-volt rms sinusoidal source?

(c) the sources of (a) and (b) in series?

4. Rectifier-type ammeters and voltmeters are sometimes constructed with the circuit arrangement of Fig. 41. What reasons can you give for the disposition of the two rectifying elements?

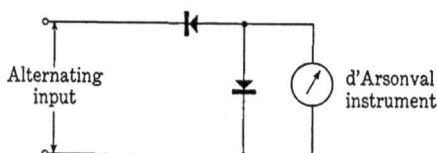

Fig. 41. Rectifier-type instrument circuit for Prob. 4.

5. (a) Derive Eq. 127 from Eq. 126.
   (b) Obtain an expression corresponding to Eq. 127 for a full-wave rectifier with a smoothing capacitor. What is its approximate form for large values of ωCR?

6. A high-voltage vacuum kenotron rectifier is used in a circuit similar to that shown schematically in Fig. 21a with a 1-μf capacitor as the plate load. The sinusoidal source voltage e_s has a peak value of 100,000 volts at a frequency of 60 cps. The switch S is closed when the applied voltage is passing through zero with a positive rate of change. The capacitor is initially uncharged.

The rectifier tube has a tungsten filament with a saturation current of 250 ma. The voltage drop required to overcome the space charge corresponding to this current is negligible in comparison with the peak applied voltage.

(a) Sketch the waveforms of the source voltage e_s, the voltage across the tube e_b, and the voltage across the capacitor e for cycles when the capacitor is uncharged, charged to approximately half its final voltage, and charged to its final voltage.
(b) What ultimate value does the voltage across the capacitor reach?
(c) Find the increments of voltage acquired by the capacitor during the first half-cycle and the half-cycle when it has acquired half its final charge. Assuming that over this period the average increment of voltage acquired per cycle is the average of these values, find the time required to charge the capacitor to half its final value.

7. Find the ripple factor of the load voltage for the full-wave rectifier shown in Fig. 7 when a capacitor is placed across the load. Compare this with the ripple factor without the capacitor.

The values to be used are:

$R_0 = 0$ $E_{sm} = 100$ volts
$R = 10,000$ ohms   Frequency $= 60$ cps.
$C = 10\ \mu$f

8. A full-wave rectifier that is supplied from a 60-cps source through a transformer having an effective secondary voltage of 350 volts from the center tap to each anode terminal utilizes a Type 5Y3-G vacuum rectifier tube. Each diode in the tube may be approximated by a constant resistance of 400 ohms during the time it conducts. The load is a resistance that varies from 1,000 to 100,000 ohms, and a smoothing inductor having an inductance of 5 henries and a resistance of 200 ohms is used. Plot the direct output voltage and the ripple factor as functions of the load resistance.

9. Determine the ratios of the maximum peak inverse voltage across the rectifiers to the direct output voltage at no load and to the peak value of the alternating supply voltage for

(a) the bridge rectifier circuit of Fig. 18 when used with a smoothing capacitor,

(b) the voltage-doubling circuit of Fig. 29, and

(c) the voltage-multiplying circuits of Fig. 30.

10. A gas diode having a voltage drop of 15 volts while conducting is used to charge batteries from a 115-volt 60-cps supply. Normally the rectifier is used in series with a control resistor to charge a 6-volt, 60-ampere-hour battery in about 10 hours. On a particular occasion, the operator is given 10 such batteries to charge. He places them in series and gradually decreases the current-limiting resistance until a d-c ammeter in the circuit indicates the usual charging rate. This has barely been accomplished when a 10-amp fuse in series with the tube blows out. What is the probable explanation? The explanation should be supported by diagrams and approximate computations.

11. A transformer, a gas diode, and a 1-ohm current-limiting resistor are used in a half-wave rectifier circuit for the purpose of charging a battery from a 60-cps sinusoidal source of alternating current. The battery is rated at 6 volts and 100 ampere-hours. The battery voltage may be assumed to be 6 volts during the entire charging time, and the battery may be assumed to become fully charged when 100 ampere-hours have been delivered to it. The voltage drop across the rectifier tube when it carries current is constant and equal to 10 volts, and the filament of the tube, which is supplied from a separate winding on the transformer, requires 35 watts.

(a) What alternating voltage is required across the secondary winding of the transformer to charge the battery in 10 hours?

(b) Calculate the over-all conversion efficiency as defined by

$$\text{Efficiency} = \frac{\text{D-c power delivered to the battery}}{\text{Total power consumed}},$$

when the transformer efficiency is taken as 90 per cent.

(c) What is the cost of charging the battery if the cost of energy at the primary winding of the transformer is 5 cents per kwhr?

12. The circuit diagram of Fig. 42 is that of a full-wave mercury-arc rectifier in which a center-tapped inductor L is used because of the lack of a center tap on the transformer. It is desired to find the value of L necessary to maintain the arc when the load resistance varies between 4 and 24 ohms.

The arc voltage drop is 20 volts, and the arc extinguishes if the instantaneous value of its current falls below 2 amp. If the coupling coefficient between the two halves of the inductor is 0.95, what value of L is needed?

13. A full-wave mercury-arc lamp for a-c operation is supplied from a center-tapped transformer secondary winding as in Fig. 7. An inductance, as well as the resistance shown, is included in series with the common cathode to stabilize the arc and prevent it from extinguishing during the cycle.

If the lowest instantaneous value of the current must be greater than one-half the average current, what is the limiting value of inductance needed? Neglect the voltage drop in the arc during conduction.

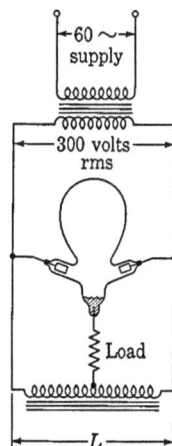

Fig. 42. Full-wave mercury-arc rectifier circuit for Prob. 12.

14. A delta, six-phase, star rectifier, illustrated in Fig. 14, delivers 1,000 amps at 600 volts to a d-c load. Connected in series with the load is a reactor that has a resistance of 0.01 ohm and an inductance sufficiently large to make the ripple in the output current negligibly small. The voltage drop in the arc of the mercury-arc rectifiers used is constant at 20 volts during conduction, and the transformer leakage reactance is negligible. Find the following:

(a) Required current and voltage ratings of the individual secondary windings.
(b) Required total volt-ampere rating of the secondary windings.
(c) Secondary utilization factor.
(d) Required volt-ampere rating of the reactor.

15. A delta, six-phase, double-wye rectifier, illustrated in Fig. 17, is used to supply the same current and power to the same load through the same reactor and rectifiers as in Prob. 14. Find the following:

(a) Required current and voltage ratings of the individual secondary windings.
(b) Required volt-ampere rating of the interphase transformer.
(c) Required total volt-ampere rating of the secondary windings.
(d) Secondary utilization factor for the transformer.
(e) Ratio of the output power to the sum of the volt-ampere ratings of the secondary windings and the interphase transformer.

16. Six hot-cathode mercury-vapor rectifier tubes are used in the delta, six-phase, wye, double-way (bridge) rectifier circuit shown in Fig. 19. Each tube has ratings as follows:

Average plate current = 10 amp max
Peak plate current   = 40 amp max
Peak inverse voltage = 10,000 volts max.

Connected in series with the load is a reactor having negligible resistance and sufficient inductance to make the ripple in the output current negligibly small. The voltage drop in the tubes while they conduct may be neglected. Find the following:

(a) Maximum rms secondary voltage to neutral permitted by the tube ratings.
(b) Maximum direct load voltage permitted by the tube ratings.
(c) Maximum direct load current permitted by the tube ratings.
(d) Minimum volt-ampere rating of the secondary windings required to produce the voltage and current of (b) and (c).
(e) Secondary utilization factor.

17. Construct the three voltage waveforms across the source, the rectifier, and the load for the conditions in Fig. 20.

18. Construct the four voltage waveforms across the source, the rectifier, the inductance, and the load for the conditions in Fig. 26.

19. Make an analysis of a series-inductor smoothing filter for a full-wave rectifier by a method similar to that employed in Art. 12a of this chapter. Neglect the voltage drop across the tube while it conducts and consider the load to be a pure resistance.

20. In the circuit of Fig. 31, the voltage drop across the tube is negligible while the tube conducts, E_s is 300 volts, f is 60 cps, L is 10 henries, C is 5 μf, and the maximum average current rating of each tube is 100 ma. Find:

(a) the direct output voltage,
(b) the value of "bleeder" resistance to be used,
(c) the minimum permissible value of load resistance,
(d) the rms value of the 120-cps component of the output voltage.

21. A full-wave rectifier employing mercury-vapor diodes is supplied by a 60-cps transformer with center-tapped secondary rated 300 volts rms on each side of the center tap. The load resistance R is coupled to the rectifier through a two-section inductor-input filter employing identical 10-henry inductors and 10-μf capacitors. The voltage drop in the diodes during conduction is 10 volts. The transformer may be considered ideal and the inductors assumed dissipationless.

(a) What is the maximum value of load resistance consistent with good voltage regulation? Explain your answer.
(b) What is the approximate rms ripple voltage at the load if R is 2,500 ohms?
(c) In what ways will the performance of the power supply be affected if vacuum diodes are used instead of mercury-vapor diodes?

22. A Type 5U4-G vacuum rectifier is used in a full-wave rectifier circuit. The filter circuit consists of a single inductor-input L section. The filter inductor has an inductance of 13 henries and a resistance of 121 ohms. The filter capacitor has a capacitance of 8 μf. The volt-ampere characteristic of the tube may be considered to be equivalent to that of a fixed voltage E_0' of 10 volts in series with a fixed resistance R_0' of 175 ohms. The supply frequency is 60 cps.

What are the direct load voltage and the ripple factor if the total rms voltage of the transformer secondary winding is 900 volts and the load resistance is 1,800 ohms?

23. A single-phase full-wave rectifier is to supply a 500-watt resistance load at a direct voltage of 500 volts from a 60-cps source. The rms ripple voltage of the lowest-order harmonic in the output is to be 0.001 times the direct voltage, and the maximum impedance as viewed from the load into the rectifier output terminals must be less than 100 ohms at 30 cps (the lowest important frequency at which the load might vary if the rectifier were supplying an audio amplifier). The voltage drop in the tube is constant at 15 volts during conduction.

(a) Design the rectifier for minimum cost, using an inductor-input filter of single L section, specifying the transformer voltage, filter inductance, and capacitance, and taking into account the cost of inductance as $2.50 per henry, the ratio L/R_X for the inductance as 1/15, and the cost of capacitors as $0.50 per μf. It may be assumed that the load will fall intermittently to 25 per cent of full load, but the current through the choke must be continuous even with the lighter load.
(b) If the transformer efficiency is 90 per cent, what is the over-all efficiency of conversion from a-c to d-c power in the rectifier?
(c) Using the above cost data, determine whether the filtering can be accomplished with least expense by a single-section or a double-section inductor-input filter. Specify the inductance and capacitance of each element, and calculate the cost of the filter.

24. A transformer with a center-tapped secondary winding rated 330 volts rms on each side of the center tap, a 60-cps supply, two capacitors each rated 8 μf, and a swinging choke rated 5 henries at 200 ma and 25 henries at 25 ma with 75 ohms

winding resistance are to be used for a full-wave single-phase rectifier and filter circuit using first a mercury-vapor tube and second a Type 5U4-G tube. The inductance of the swinging choke may be assumed to vary linearly with the average current through it between the limits given above.

(a) Plot for each rectifier tube the direct voltage and the ripple voltage obtained with an inductor-input filter that utilizes the two capacitors in parallel for load currents up to 200 ma.

(b) Plot the direct output voltage and the ripple voltage obtained with a capacitor-input filter and the Type 5U4-G tube for load currents up to 200 ma.

25. If the maximum permissible direct output current from a full-wave vacuum rectifier tube is 75 ma per plate, and the maximum permissible peak inverse voltage is 1,400 volts, what are the maximum direct output voltage and current that can be supplied by this tube when it is used in a full-wave rectifier circuit with an inductor-input filter? Neglect the voltage drop in the tube during conduction and the resistance of the inductor.

26. A full-wave rectifier comprises a Type 5T4 tube and a capacitor-input filter made up of two 8-μf capacitors and a 20-henry inductor. The transformer supplies 300 volts rms per anode to the tube, and the resistance of the inductor is 75 ohms. Plot curves of direct output voltage and ripple voltage for load currents up to 225 ma.

27. A full-wave rectifier for aircraft use has a 400-cps supply frequency, an rms transformer secondary voltage of 350 volts per anode, negligible voltage drop in the tubes while they conduct, and a load resistance of 7,000 ohms.

Find the direct output voltage and the 800-cps rms component of the load voltage when

(a) no filter is used,

(b) a filter composed of a 1-henry inductor having negligible resistance is used, and

(c) an inductor-input filter composed of the inductor in (b) and a 1-μf capacitor is used.

28. A Type OC3/VR105 voltage-regulator tube is used to stabilize the voltage across a load in which the current may vary between 0 and 20 ma. The alternating supply voltage may fluctuate plus and minus 10 per cent about its mean value. The direct starting voltage for the tube is 115 volts, its direct operating voltage is 105 volts, and the permissible range of its direct operating current is 5 to 40 ma.

Find the minimum permissible mean value of the direct output voltage of the rectifier circuit, E_i in Fig. 39, and the corresponding minimum series resistance, R_s. If the mean direct output voltage is increased to 300 volts, what is the range of permissible resistance values?

CHAPTER VII

Controlled-Rectifier Circuits

In rectifier circuits that employ diodes, the direct-current output to a fixed load is not subject to control except by a means such as an autotransformer or a similar device that adjusts the alternating supply voltage. If, however, a tube having an associated control means—for example, a thyratron, ignitron, or cold-cathode gas triode—is substituted for the diode, the direct-current output can be controlled by appropriate adjustment of the control means—grid potential or ignitor current. As is discussed in Art. 6, Ch. V, the control means in these tubes determines the instant at which conduction starts, but cannot stop conduction. These devices are therefore not generally suitable for use with a direct voltage supplied to the anode. However, when an alternating anode-voltage supply is used, as in rectifier circuits, de-ionization may take place during the negative half-cycles of the applied voltage, and the grid or other control means may thus regain control. By delay of the point at which conduction starts in each positive half-cycle, the average current in the anode circuit may be controlled. In contrast, the instantaneous current is the quantity controlled in the usual vacuum-tube amplifier circuit. In the articles that follow, the thyratron is used as an illustration, but the basic principles outlined for control of a rectified current are applicable to other controlled rectifiers as well.

The methods of control[1] by thyratrons may be classified in two ways: First, the control may be classed as *on-or-off* or as *continuous*; that is, (a) the grid voltage may be used to determine whether an anode current of a magnitude determined by the load flows or is interrupted, or (b) the grid voltage may be adjusted to vary the average anode current continuously over some range. Second, the control may be classed according to the form of the grid-voltage wave.[2] For example, the grid voltage may be a steady direct voltage, a sinusoidal voltage of the same frequency as the anode-supply voltage, a sharply peaked voltage wave, or any combination of these. The discussion that follows covers several of the numerous possible methods.

---

[1] W. R. G. Baker, A. S. Fitzgerald, and C. F. Whitney, "Industrial Uses of Electron Tubes," *Electronics*, 2 (1931), 467–469; "Electron Tubes in Industrial Service," *Electronics*, 2 (1931), 581–583.

[2] J. H. Burnett, "Thyratron Grid Circuit Design," *Electronics*, 24 (March, 1951), 106–111.

1. CRITICAL-GRID-VOLTAGE CURVE

A basic step in the analysis of any method of control by a thyratron is the construction of a curve that shows the critical grid voltage as a function of time. This curve may be determined from the control characteristic, which is discussed in Art. 6b, Ch. V, by the graphical method shown in Fig. 1. The sinusoidal anode supply voltage, e_s, for a

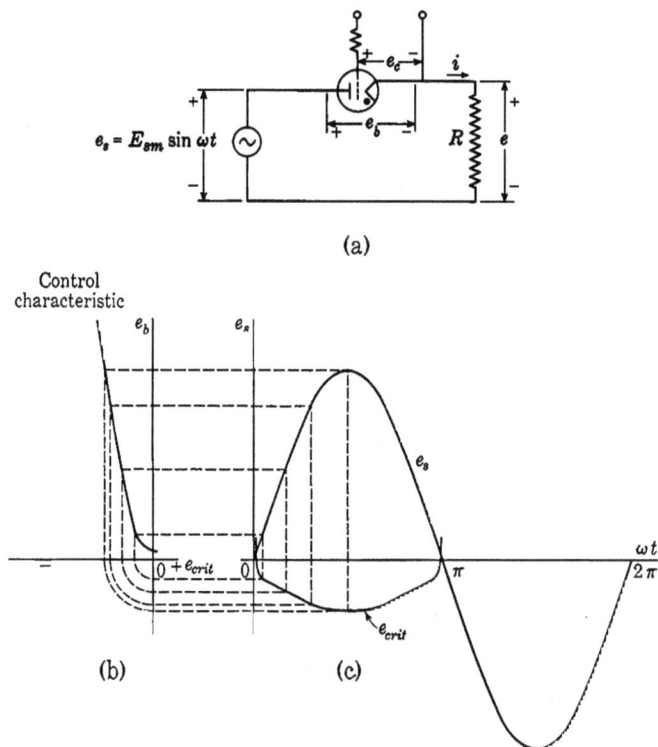

(a)

(b)   (c)

Fig. 1.   Construction of the critical-grid-voltage curve.

thyratron connected as in Fig. 1a is plotted in Fig. 1c so that instantaneous values of anode voltage may be conveniently projected onto the anode-voltage axis of the control characteristic in Fig. 1b. For each value of e_s, there is a value of grid voltage—the critical grid voltage given by the control characteristic—at which the tube will fire if the grid potential is changed from more negative values to that value. Projecting points on the curve of Fig. 1c onto the control characteristic, and then down and around as shown, gives the curve e_{crit} of Fig. 1c, called the *critical-grid-voltage curve*. This curve specifies as a function of time the grid voltage that will just allow the tube to fire if the anode voltage follows the curve of e_s. In a half-wave rectifier circuit in which the load is a resistor, as in Fig. 1a, the anode voltage drop e_b equals the anode-supply voltage e_s *before conduction begins.* Therefore, in such a circuit, the critical-grid-voltage curve derived from the anode-supply voltage may be used to determine the point in the cycle at which the tube fires. The point at which the curve of applied grid voltage e_c crosses the curve of the critical voltage e_{crit} *from below* is the firing point. Conduction begins at this point and continues until nearly the end of the half-cycle, where the anode voltage becomes less than the tube-voltage drop. In some circuits—for example, in those containing energy-storage devices and in polyphase circuits—the actual anode voltage that is applied to the tube itself may differ from the anode-supply voltage even before the tube fires. The critical-grid-voltage curve must then be derived from the curve of the actual anode voltage.

The critical-grid-voltage curve is drawn for a negative-control thyratron in Fig. 1c. For a positive-control tube, the curve would lie above the horizontal axis. In many applications, the critical grid voltage remains so small throughout the cycle that assuming the curve to lie on the axis does not introduce appreciable error into the analysis.

2. CONTROL BY DIRECT GRID VOLTAGE

Control of the average rectified current through variation of the applied direct grid voltage, sometimes called *bias control,* is illustrated in Fig. 2. In the circuit of Fig. 2a, a thyratron is employed as a half-wave rectifier to supply direct current to a resistance load. The grid of the thyratron is connected to a grid-bias-supply battery E_{cc} by means of a current-limiting resistor R_1 and a resistance voltage divider R_2 and R_3. Either part of the voltage divider may be a device (such as a phototube) whose resistance, often nonlinear, depends on a quantity (such as light intensity) whose value is to control the thyratron current. The current-limiting resistor must be large enough to prevent the flow of excessive grid current after the tube fires. Varying R_2, or R_3, or the ratio of the two, controls the rectified current in the anode circuit over a limited range.

The circuit behavior for one setting of the voltage divider is illustrated by the waveforms in Fig. 2b, in which e_s is a sinusoidal supply voltage, e_{crit} is the corresponding critical grid voltage, and e_c is the voltage applied to the grid. When e_c has the magnitude shown in the figure, the grid becomes more positive than the critical voltage at the point θ_1, and conduction to the anode takes place. The anode voltage

e_b drops to about 10 or 15 volts and remains almost constant as long as an anode current exists, and the anode current has the waveform indicated in the figure. During the negative half-cycle, the anode voltage equals the supply voltage, and thus the peak value of the inverse anode voltage equals that of the supply voltage.

During the conduction period, the grid current causes a voltage drop in the grid-circuit resistors that affects the waveform of the grid

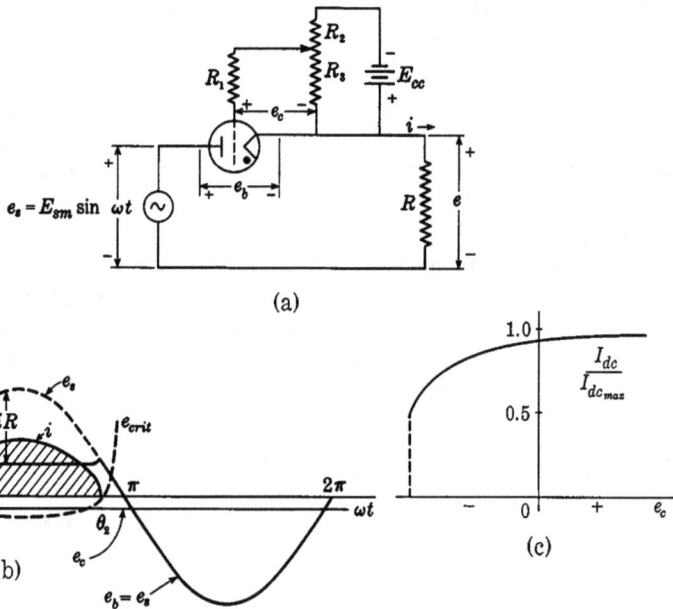

Fig. 2. Continuous control of the average load current by adjustment of the direct grid voltage.

voltage e_c. This effect, which is not indicated by the curve of e_c in Fig. 2b, has, however, no influence on the waveform of the anode current, because of the trigger action of the grid.

From Fig. 2b it is seen that if the direct grid voltage is changed by some means such as adjustment of the resistances in the voltage divider R_2 and R_3, the crossing point of the curves of applied and critical grid voltage is moved, and hence the ignition angle θ_1 is changed. Thus the portion of the cycle during which the anode current flows is controlled by the grid voltage. Since the average value of the anode current is proportional to the cross-hatched area shown, the direct output current of the rectifier is controlled in this way.

The average value of the current, I_{dc}, may be determined for any particular ignition angle θ_1 by a method similar to that used in Art. 5, Ch. VI. Thus, during conduction,

$$i = \frac{E_{sm} \sin \omega t - E_0}{R}, \qquad \theta_1 < \omega t < \theta_2, \qquad [1]$$

where E_0 is the voltage drop in the tube when it conducts. The average value of the current is therefore

$$I_{dc} = \frac{1}{2\pi} \int_{\theta_1}^{\theta_2} \frac{E_{sm} \sin \omega t - E_0}{R} d(\omega t) \qquad [2]$$

For analysis the small rise in e_b near the end of the conduction period may be neglected and E_0 considered constant until the current stops. Conduction then ceases at the point where e_s equals E_0, thus

$$\theta_2 = \pi - \sin^{-1}(E_0/E_{sm}), \qquad [3]$$

and

$$I_{dc} = \frac{1}{2\pi} \int_{\theta_1}^{\pi - \sin^{-1}(E_0/E_{sm})} \frac{E_{sm} \sin \omega t - E_0}{R} d(\omega t), \qquad [4]$$

which gives

$$I_{dc} = \frac{E_{sm}}{2\pi R} \left[\sqrt{1 - \left(\frac{E_0}{E_{sm}}\right)^2} - \frac{E_0}{E_{sm}}(\theta_2 - \theta_1) + \cos \theta_1 \right]. \qquad [5]$$

If the tube voltage drop E_0 is small compared with the peak supply voltage E_{sm}, Eq. 5 reduces to

$$I_{dc} = \frac{E_{sm}}{2\pi R} [1 + \cos \theta_1], \qquad E_0 \ll E_{sm}, \qquad [6]$$

which may be written in the form

$$\frac{I_{dc}}{I_{dcmax}} = \frac{1 + \cos \theta_1}{2}, \qquad E_0 \ll E_{sm}, \qquad [7]$$

where I_{dcmax} is the value $E_{sm}/(\pi R)$ that occurs when θ_1 is zero. Note that *these relations are basic for any method of control that causes ignition at the angle θ_1*; they apply not only to control by a direct grid voltage, but also to control by the other methods discussed in the articles that follow.

When the grid voltage in Fig. 2 has a positive value large enough to make the conduction period last for essentially the entire half-cycle, the direct current has its maximum value I_{dcmax}. For other values of grid voltage, the direct current is reduced as indicated by the curve of Fig. 2c. If the grid voltage is just equal to the maximum negative value of the critical grid voltage, θ_1 is 90 degrees, and conduction does not start until the middle of the positive half-cycle. Consequently the average rectified load current is then reduced to approximately half of its maximum value. For larger negative values of the grid voltage, conduction cannot occur at all, and therefore the curve of Fig. 2c drops suddenly to zero at this grid voltage.

Small changes in the control characteristic of the tube result in large changes of the direct load current, because of the smallness of the angle of intersection of the curves of applied grid voltage and critical grid voltage. Since the control characteristics depend upon the gas pressure, which changes both with the temperature and with the age. of the thyratron (see Art. 6f, Ch. V), this method of control inherently lacks precision and dependability. It is especially unsatisfactory if mercury-vapor tubes are used, for the temperature and vapor pressure are usually subject to wide fluctuation. The method of continuous control by a variable direct grid voltage that involves combining it with a constant alternating voltage as described in Art. 4 can be made relatively free from this objection and hence is to be preferred in most applications.

On-or-off control with satisfactory reliability may be obtained through variation of the direct grid voltage in the circuit of Fig. 2a if the total change in direct voltage is large enough to extend over a considerable range of e_c above and below the value that corresponds to the dotted vertical line in Fig. 2c. Thus, if the magnitude of E_{cc} in Fig. 2a is large compared with that of the maximum negative value of the critical grid voltage curve in Fig. 2b, switching the grid connection from one end to the other of E_{cc} turns essentially the full rectified current on or off with precision regardless of moderate changes of the control characteristic.

Various control circuits in which closing or opening a switch produces the necessary change in grid voltage are shown in Fig. 3. In these, the purpose of the resistor R_1 is to limit the current to the grid during the conduction period, and that of R_2 is to limit the current through the battery when the switch is closed. Since the resistors also limit the current through the switch, its current capacity may be small compared with that of the thyratron. Thus delicate contacts, or a low-current device such as a phototube, may be used as the switch to control a relatively heavy current by this means.

The waveform of the supply voltage e_s and the critical-grid-voltage curve e_{crit} derived from it are shown in Fig. 3b. With the switch S in Fig. 3a open, the grid is connected to the battery of voltage E_{cc} through R_1 and R_2, and is negative. If this negative voltage is greater in magnitude than the maximum negative value of the critical grid voltage, as shown in the figure, no anode current flows during the cycle, for the critical-voltage curve is never crossed by the curve of actual voltage. However, if the switch S is closed, the grid voltage e_c

and the anode voltage e_b are equal, and both rise with the supply voltage e_s. Conduction starts at the angle θ_1, where the applied grid voltage becomes equal to the critical grid voltage. At this instant, as is shown by the curves of e_b and of e_c with S closed, the current i rises suddenly, and both the anode and the grid voltages decrease to low values and remain almost constant during the conduction period; this lasts until ωt equals θ_2, at which angle the anode-supply voltage becomes too low to maintain ionization and the current stops suddenly. The resistor R absorbs the difference between the anode

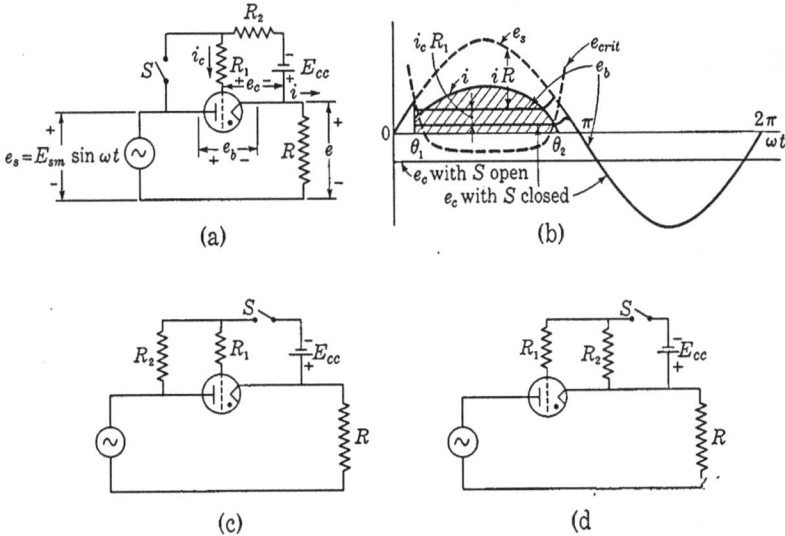

Fig. 3.   On-or-off control of the average load current.

voltage and the supply voltage during the conduction period; the resistor R_1 absorbs the difference between the grid and supply voltages. The load current i, shown by the curve that forms the upper boundary of the cross-hatched area, assumes such a value that the voltage drop across the load resistor equals the difference between the supply voltage and the tube voltage drop. As the end of the conduction period is approached, the anode voltage rises slightly and the current decreases rapidly, while the grid voltage also rises to the sine curve as the plasma in the tube disappears. During the negative half-cycle, the voltages of the anode and the grid are both equal to the anode-supply voltage. Hence the grid battery E_{cc} receives a net rectified charging current. The circuit of Fig. 3c is similar to that of Fig. 3a except that the positions of S and R_2 are interchanged, and conduction stops when S is closed. With the circuit of Fig. 3d, on-or-off control is obtained by use of a direct grid excitation only.

3. CONTROL BY PHASE SHIFT OF ALTERNATING GRID VOLTAGE

Continuous control of the average anode current through use of an alternating voltage of adjustable phase[3] as grid excitation is accomplished by the circuit of Fig. 4a. In this circuit, both the anode- and

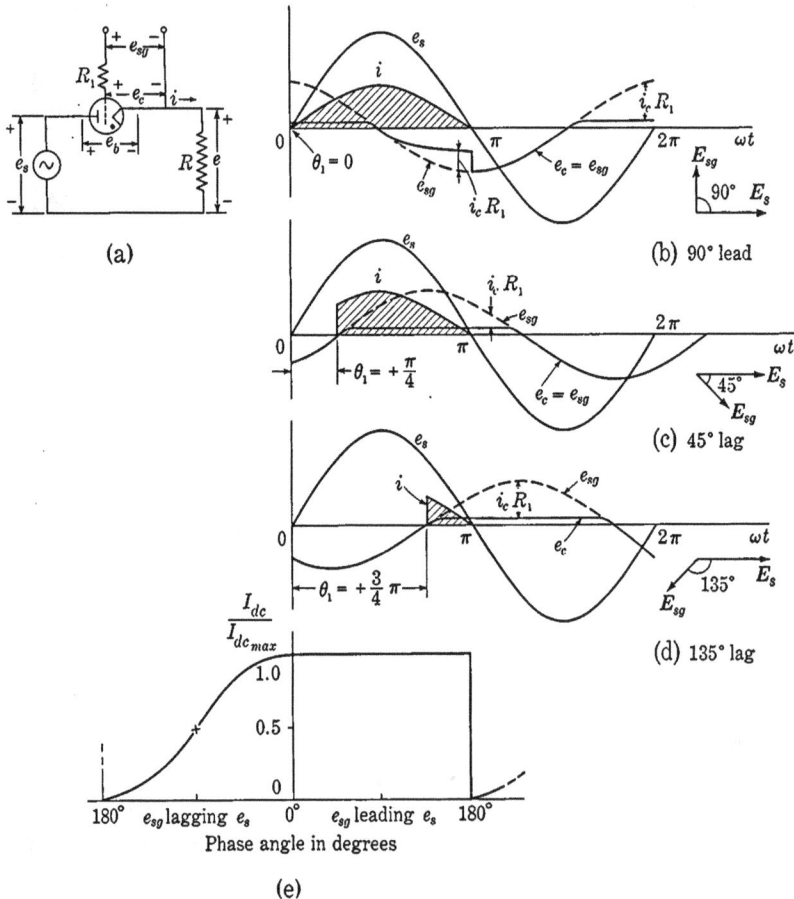

Fig. 4.   Control of the average load current by phase shift of an alternating grid voltage.

the grid-supply voltages e_s and e_{sg} are sinusoidal alternating voltages of the same frequency and ordinarily come from the same power source. However, a phase-shifting device or circuit not shown in the diagram is interposed between the power source and the grid to produce a phase angle between these voltages. Three diagrams of the

[3] A. W. Hull, "Hot-Cathode Thyratrons—Part II, Operation," *G.E. Rev.*, **32** (1929), 393–398.

current and voltage waveforms for different phase shifts of the grid voltage are given in Figs. 4b, 4c, and 4d. In each of these, the assumptions are made that (a) the critical grid voltage is zero for all values of the anode voltage and (b) the voltage across the tube when it is conducting is zero. A more nearly exact treatment would necessarily involve the instantaneous magnitudes of these two quantities, as does the discussion in Art. 2.

In Fig. 4b the grid-supply voltage e_{sg} leads the anode-supply voltage e_s by 90 degrees; thus the grid voltage e_c becomes positive before the anode voltage e_b does, and is still positive at the time the anode voltage reaches positive values. As a consequence, conduction starts at the beginning and lasts until the end of the positive half-cycle. This condition evidently is true not only for 90 degrees but also for any *leading phase angle* of the grid voltage. Variation of the phase angle of the grid voltage over a range of 0 to 180 degrees leading therefore produces no change in the average value of the load current I_{dc}, as is indicated in Fig. 4e; the ignition angle θ_1 remains zero, and the average current remains at its maximum value $I_{dc_{max}}$. The shape of the grid-voltage wave departs from a sinusoid in the manner shown because at the beginning of the positive half-cycle the tube is conducting, the grid draws a large electron current from the plasma, and most of the grid-supply voltage e_{sg} appears across the grid-current-limiting resistor R_1. After the grid-supply voltage becomes negative at 90 degrees, the grid draws a positive-ion current from the plasma, and a smaller but appreciable voltage drop in R_1 may occur until conduction stops and de-ionization takes place.

In Fig. 4c, the grid-supply voltage lags the anode-supply voltage by 45 degrees. In the initial part of the cycle, the grid voltage is negative; hence the grid current i_c is then negligible, and the grid voltage e_c equals the grid-supply voltage e_{sg}. Because of the negative grid voltage, the tube cannot conduct. When the grid becomes positive, at an angle of 45 degrees in the anode-supply-voltage cycle, ignition occurs and the anode current rises suddenly and varies sinusoidally for the remainder of the half-cycle. As in Fig. 4b, most of the grid-supply voltage is absorbed in the grid resistor as long as the grid-supply voltage is positive. When the grid-supply voltage becomes negative, however, the grid voltage follows the sine curve closely, since the grid current is then very small. The effect of the lagging phase angle of the grid voltage is to delay the start of conduction by a length of time corresponding to the angle of lag. The ignition angle θ_1 equals the magnitude of the lagging phase angle of the grid voltage. As is shown by Fig. 4d, large lagging phase angles reduce the average anode current toward zero as the lagging phase angle approaches 180 degrees.

From these considerations, it is evident that the curve of the average load current I_{dc} as a function of the phase shift of the grid-supply voltage has the general form shown in Fig. 4e. The universal relations expressed by Eqs. 1 to 7 are applicable to this phase-shift method of control, and describe this curve when the ignition angle θ_1 is expressed in terms of the phase angle between the grid- and anode-supply voltages. This ignition angle is equal to the magnitude of the phase angle for grid-supply voltages that lag the anode-supply voltage by angles up to 180 degrees, provided the critical grid voltage and the tube voltage drop during conduction are zero, as is assumed in Fig. 4. The ignition angle is zero for grid-supply voltages that lead by angles up to 180 degrees, as is explained in the foregoing paragraphs. Accordingly, Fig. 4e is plotted from Eq. 7.

If the critical grid voltage and the tube voltage drop during conduction are not negligible, the average current is represented by Eq. 5, but the ignition angle θ_1 no longer equals the magnitude of the phase angle for lagging grid voltages. Rather, it has a somewhat different value corresponding to the time at which the grid-voltage wave intersects the critical-grid-voltage curve. If the load in the anode circuit is not a pure resistance, but contains series inductance or parallel capacitance, the curve of Fig. 4e is not applicable, and the analysis for a half-wave grid-controlled rectifier must be made somewhat as outlined in Arts. 9 and 10, Ch. VI. For a full-wave or polyphase grid-controlled rectifier circuit with an inductor-input filter, the analysis of the conditions for continuous conduction, discussed in Art. 12a, Ch. VI, can be extended to determine the value of a *critical inductance*.[4]

Because of the sudden jump that occurs at the 180-degree position in Fig. 4e, a phase-shifting method that varies the phase of the grid voltage through a small angle about 180 degrees gives on-or-off control of the current.

Variations of an alternating grid voltage other than simple phase shift are sometimes useful in a grid-controlled rectifier. For example, the voltage actually applied to the grid often varies in both amplitude and phase. The changes in amplitude are sufficiently small to have little effect upon the starting of conduction, and the control is primarily the result of the variations in phase. It is to be noted that variation of an alternating grid voltage *in amplitude alone* is *not* an effective method of obtaining wide-range continuous control, as may be seen from consideration of the effect such a variation would have upon the firing point.

---

[4] W. P. Overbeck, "Critical Inductance and Control Rectifiers," *I.R.E. Proc.*, **27** (1939), 655–659.

It is ordinarily desired that the length of the conduction period and the average value of the anode current be as nearly independent of changes in the control characteristic as possible. Thus the curves of applied grid voltage and critical grid voltage should intersect at a relatively large angle. Such a condition can be obtained if the peak value of the grid-supply voltage is large relative to the critical grid voltage. Good design of thyratron circuits therefore includes ample amplitude of grid-supply voltage, and correspondingly adequate grid-current-limiting resistance.

Improved precision in the determination of the instant of ignition in the cycle is obtained if an alternating grid voltage is used that has a wave front that rises more steeply than that of a sine wave.[5] One method of producing a steeper wave front is to design the transformer so that the steel core through its windings operates well above the knee of the curve of flux density as a function of magnetizing force at the instant the primary current reaches its maximum value. If the magnetizing force is maintained nearly sinusoidal, the waveform of the flux through the secondary winding is then almost flat topped with steep sides. Short surges of voltage are induced in the secondary winding by the rapid flux changes corresponding to the steep sides, and almost zero voltage is induced by the flat tops of the flux wave. Thus the secondary voltage is a sharply peaked alternating wave that may be used to determine precisely the ignition angle of the thyratron. Phase shift of the primary supply voltage provides continuous control of the ignition angle. To assure that ignition will occur only on the steep front of the voltage wave, a negative direct bias voltage somewhat smaller than the amplitude of the peaked wave is ordinarily connected in series with the peaked voltage. The direct bias voltage is frequently obtained from a parallel combination of resistance and capacitance connected in series with the grid—a capacitor is connected across the resistor R_1 in Fig. 4a. The rectified grid current develops a direct voltage drop across the resistor. This voltage is smoothed by the capacitor in the manner described in Art. 9, Ch. VI.

A peaked wave of secondary voltage is produced if the primary winding of a transformer is supplied with an abnormally large exciting current from an alternating voltage source through a large resistance. The resistance should be large enough to govern the wave shape of the exciting current and maintain it essentially sinusoidal despite the changing primary inductance that accompanies the saturation of the core. Such a method is inefficient, however, because of the excessive

[5] M. M. Morack, "Voltage Impulses for Thyratron Grid Control," *G.E. Rev.*, *37* (1934), 288–295; O. Kiltie, "Transformers with Peaked Waves," *E.E.*, *51* (1932), 802–804.

power loss in the resistance. When efficiency is important, peaked-voltage transformers are used that have cores with multiple flux paths of different characteristics designed so that the core in the secondary winding saturates, but that in the primary does not.[6]

4. CONTROL BY MAGNITUDE OF A DIRECT GRID VOLTAGE SUPERPOSED ON AN ALTERNATING GRID VOLTAGE

Although, as is explained in Art. 2, a variable direct grid voltage is unsatisfactory for continuous control of the average current in a

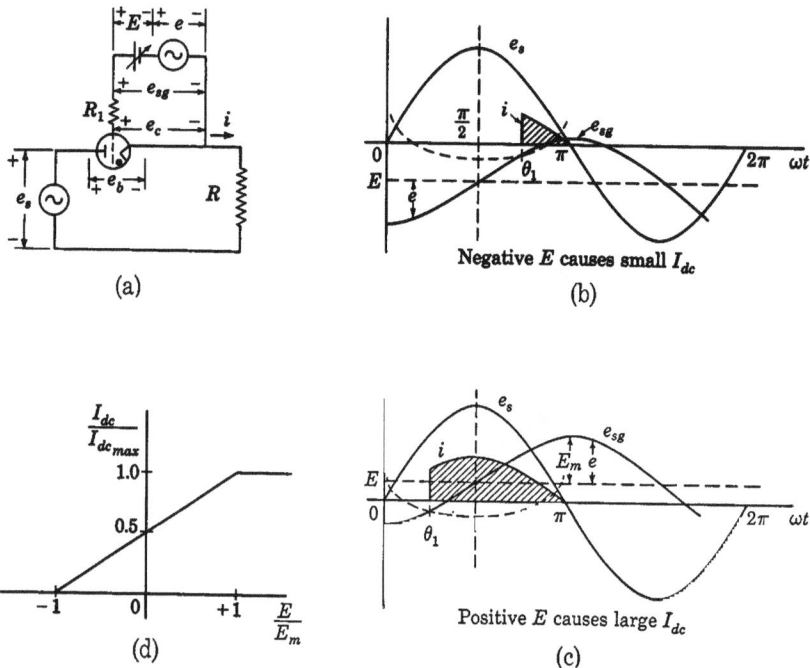

(a)

(b)

Negative E causes small I_{dc}

(d)

(c)

Positive E causes large I_{dc}

Fig. 5.   Varying direct grid voltage E gives phase control when superposed on lagging alternating bias voltage e.

thyratron when used alone, it can be utilized effectively if combined with an alternating component of grid voltage in the manner shown in Fig. 5. When the direct grid voltage E is negative, and the alternating voltage e lags the supply voltage e_s by 90 degrees, as in Fig. 5b, their sum, which is the grid-supply voltage e_{sg}, intersects the critical grid-voltage curve at an angle θ_1 greater than $\pi/2$. When the direct grid voltage is positive, however, the intersection is at an angle

[6] R. Lee, *Electronic Transformers and Circuits* (New York: John Wiley & Sons, Inc., 1947), 209–211.

smaller than $\pi/2$, and the average anode current is hence larger than for a negative direct grid voltage. Change of magnitude and sign of the direct voltage E thus give continuous control of the average anode current I_{dc}.

A relation between the average anode current and the direct grid voltage follows if θ_1 in Eqs. 1 to 7 is expressed in terms of E. For the particular conditions that the alternating component of grid voltage lags the anode-supply voltage by 90 degrees, and that the critical grid voltage is negligibly small, the ignition angle may be seen from Figs. 5b and 5c to be

$$\theta_1 = \frac{\pi}{2} - \sin^{-1}\frac{E}{E_m}, \qquad < \frac{E}{E_m} < +1, \qquad [8]$$

where E_m is the amplitude of the alternating component e of the grid voltage. Equation 7 then becomes

$$\frac{I_{dc}}{I_{dc_{max}}} = \frac{1}{2}\left[1 + \cos\left(\frac{\pi}{2} - \sin^{-1}\frac{E}{E_m}\right)\right], \qquad [9]$$

$$= \frac{1}{2}\left[1 + \sin\left(\sin^{-1}\frac{E}{E_m}\right)\right], \qquad [10]$$

$$\frac{I_{dc}}{I_{dc_{max}}} = \frac{1}{2}\left[1 + \frac{E}{E_m}\right], \qquad -1 < \frac{E}{E_m} < +1. \qquad [11]$$

Thus, I_{dc} varies linearly with E as is shown in Fig. 5d. If the phase angle between the alternating component of the grid voltage and the anode supply voltage is not 90 degrees, or the critical grid voltage is not negligible, continuous control may occur, but the relation between I_{dc} and E is no longer linear.

5. CONTROL BY AMPLITUDE OF AN ALTERNATING VOLTAGE SUPERPOSED ON A LAGGING VOLTAGE

An alternating voltage having a slowly changing amplitude is often available for control of a thyratron, particularly in various feedback-control systems. Combining the voltage of varying amplitude with a second alternating voltage of constant amplitude and phase, but lagging the anode-supply voltage E_s, as is illustrated vectorially in Fig. 6, gives continuous control of the average anode current. No direct bias voltage is used. When the amplitude of E_1 changes, the phase angle of the grid-supply voltage E_{sg}, which is the sum of E_1 and the lagging voltage E_2, varies and gives phase control. The amplitude of E_{sg}

also varies, but the effect of this variation is negligible if the critical-grid voltage is small in comparison with the amplitude.

As the amplitude of E_1 shrinks from the value shown in Fig. 6a, it may be considered to pass through zero and take on negative values as in Fig. 6b. Alternatively, the phase of E_1 may be considered to change through 180 degrees at the instant the amplitude decreases to zero and begins to increase to positive values again. This method,

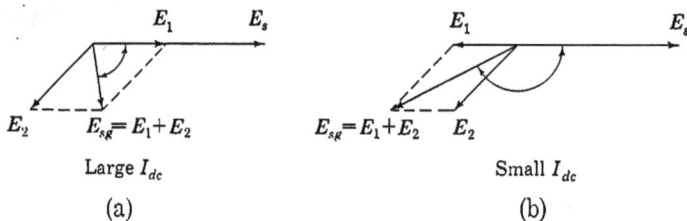

Large I_{dc} Small I_{dc}

(a)   (b)

Fig. 6. Alternating voltage with changing amplitude, E_1, gives phase control when superposed on a lagging voltage of constant amplitude and phase, E_2, to form the grid-supply voltage E_{sg}.

therefore, permits detection of phase reversal as well as amplitude change in a control system.

6. PHASE-SHIFTING METHODS

One of the more satisfactory methods of obtaining an alternating voltage of variable phase is use of a phase-shifting device[7] having a polyphase stator winding and a single-phase rotor winding. When the stator is supplied with polyphase currents and the rotor is prevented from turning, the phase angle of the induced voltage in the rotor winding with respect to the stator voltages changes with the position of the rotor, and a voltage of adjustable phase is obtained if provision is made for mechanical adjustment of the rotor position. However, such a phase shifter is often not the most economical method, and one of the phase-shifting circuits described below may be more suitable.

In the phase-shifting circuit of Fig. 7a, used with a thyratron, the sinusoidal supply voltages e_s and e_s', represented by the complex numbers E_s and E_s', are equal in magnitude and alike in phase and may come from a center-tapped transformer secondary, as shown in the figure. The voltage E_s supplies current through the thyratron rectifier to the load resistor R. The grid-supply voltage e_{sg}, represented by the complex number E_{sg}, is obtained between the junction

[7] When designed so that the rotor may turn or be turned freely, the device is given various trade names, such as *Selsyn, Synchro,* or *Autosyn;* see, for example, E. M. Hewlett, "The Selsyn System of Position Indication," *G.E. Rev.,* **24** (1921), 210–218.

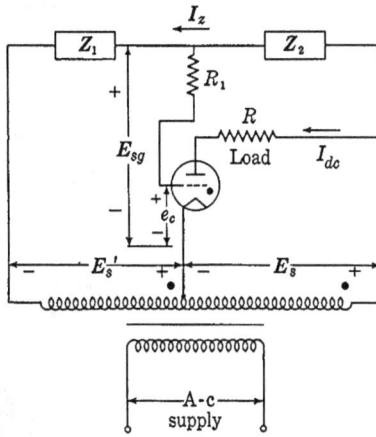

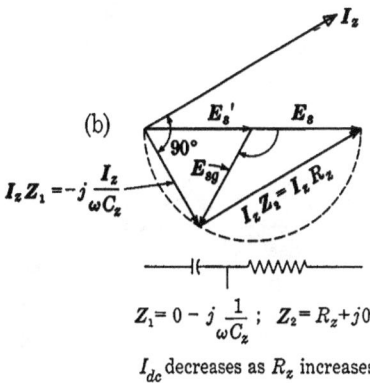

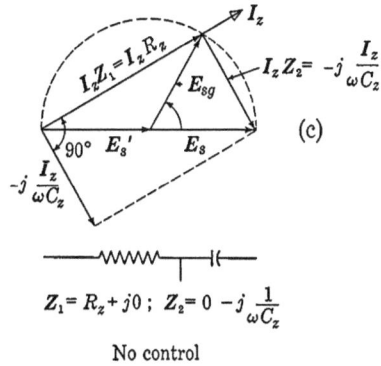

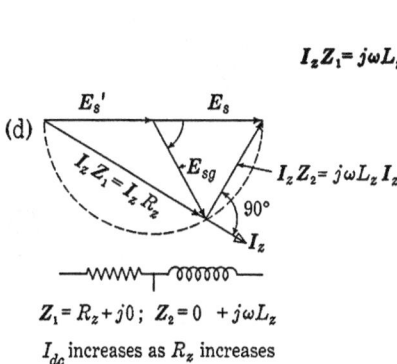

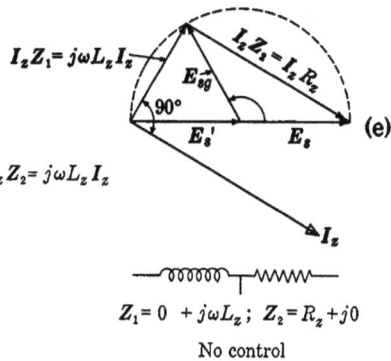

Fig. 7. Phase-shifting circuit and vector, or phasor, diagrams.

point of the two impedances Z_1 and Z_2 and the transformer center tap.

If a resistance is chosen for one of the two impedances and either inductance or capacitance is chosen for the other impedance, phase shift of the grid voltage may be accomplished by adjustment of the magnitude of either of the impedances. The vector, or phasor, diagram for the circuit may be constructed with the aid of the vector (complex-number) equations

$$I_z(Z_1 + Z_2) = E_s + E_s' \qquad [12]$$

$$E_{sg} = I_z Z_1 - E_s', \qquad [13]$$

where I_z is the complex number representing the current through the impedances Z_1 and Z_2. These equations follow from Kirchhoff's laws and the positive reference directions assigned in the figure, if R_1 is assumed to be so large that the grid current is negligible. For instance, if a capacitance C_z is chosen as the impedance Z_1, and a resistance R_z is chosen as the impedance Z_2, the vector diagram shown in Fig. 7b is applicable. The current I_z through the two impedances leads the voltage $E_s + E_s'$, and resolution of this voltage into the components across the two impedances results in the right triangle of vectors shown. The component across C_z lags the current vector by 90 degrees, and the component across R_z is parallel to the current vector. The grid-supply-voltage vector E_{sg} is drawn from the point that corresponds to the center of the transformer winding to the vertex of the triangle, and it lags the anode voltage E_s by the phase angle indicated. It may be shown that,[8] regardless of the value of R_z or C_z, the tip of the vector E_{sg} falls on the circle through the three vertices of the triangle; hence E_{sg} is constant in amplitude. Variation of either R_z or C_z results, therefore, in a variation of the phase angle but in no change in amplitude of the grid voltage. The phase angle of E_{sg} with respect to E_s is given by

$$\angle_{E_s}^{E_{sg}} = -2 \tan^{-1} \omega C_z R_z, \qquad \blacktriangleright [14]$$

and E_{sg} lags E_s for all values of the impedances. Consequently, in accordance with the discussion of Fig. 4, control over the average load current I_{dc} may be accomplished. If R_z is increased and C_z is held constant, the phase angle increases and the average load current therefore decreases. Likewise, if R_z is held constant and C_z is increased, a decrease in the load current results. This, then, is a satisfactory method of control; it is in wide use. If a phototube is substituted for the impedance Z_2, the control may be accomplished by a variation in light intensity, but the foregoing circuit analysis is not strictly

[8] E. E. Staff, M.I.T., *Electric Circuits* (Cambridge, Massachusetts: The Technology Press of M.I.T.; New York: John Wiley & Sons, Inc., 1940), 478–487.

applicable, because of the nonlinearity of the volt-ampere characteristic of the phototube.

If the resistor and the capacitor are interchanged, the vector diagram of Fig. 7c is obtained. The phase angle of E_{sg} with respect to
E_s is then given by

$$\angle_{E_s}^{E_{sg}} = 2 \tan^{-1} \frac{1}{\omega C_z R_z}. \tag{15}$$

With this combination of impedances, E_{sg} leads E_s for all values of
R_z and C_z; hence no control over the average load current I_{dc} is
obtained.

When inductance is chosen as one of the parameters and resistance
as the other, the vector diagram of either Fig. 7d or Fig. 7e results.
With one choice, control over the average load current results; with
the other, no control is obtained. If it is desired that the load current
increase when the resistance increases, the circuit to which the
diagram of Fig. 7d applies may be used. Thus, if a phototube is
substituted for R_z, the load current increases as the light intensity is
decreased.

These vector diagrams are applicable for a determination of the
phase angle at which conduction will start in the circuits of Figs. 7b
and 7d if the load is essentially resistive and the effects of grid current
are negligible, because the grid voltage e_c then equals the grid-supply-
voltage e_{sg} and the anode voltage equals e_s before conduction starts.
If, however, the load includes appreciable shunt capacitance, the
load voltage contains a transient component that persists from one
positive half-cycle into the next, as is explained in Art. 9, Ch. VI.
Consequently, the anode voltage is not truly sinusoidal in the nonconduction period, and differs from e_s at the instant conduction
starts. The vector diagrams hence are not applicable in determining
that instant. A similar difficulty is encountered if attempt is made to
extend this analysis and apply it to a grid-controlled full-wave
rectifier with an inductive load.

Even though the grid current is negligible when the tube is not conducting, the much larger grid current during the conduction part of
one cycle may cause a transient component of voltage that persists
across the capacitance C_z in Fig. 7b or the inductance L_z in Fig. 7d
until the instant conduction starts in the next cycle. During the conduction period the grid current charges C_z, or establishes a current in
L_z, through the grid resistance R_1. If, as is assumed earlier, a sufficiently
large value of R_1 can be included in the circuit, this effect of grid
current can be made negligible. But R_1 is limited because it must be
small enough to permit sufficient current for firing the tube. Although

the charge in C_z or the current in L_z is established through R_1, it must decay through R_z during the nonconduction period, because the grid current is then essentially zero. Consequently, if R_z is comparable with R_1 and, simultaneously, the time constant R_zC_z, or L_z/R_z, is comparable with or larger than the nonconduction period, the effect of the grid current may be appreciable even for the largest practical value of R_1. The vector diagrams are then not applicable in determining the instant conduction starts, because the grid voltage e_c is nonsinusoidal and does not equal e_{sg} prior to that instant. Furthermore, when the impedances Z_1 and Z_2 are large, the effects of the capacitances between the pairs of electrodes of the tube, including the stray capacitances due to the wiring, must be included in the analysis, and a modification of the vector diagram is required.

7. IGNITRON EXCITATION CIRCUITS

Ignitron tubes require an auxiliary ignitor-excitation circuit to send a pulse of current through the ignitor at the instants ignition of the main arc is desired. When the ignitron is to operate continuously from an alternating anode-supply voltage, as in rectifier service, these pulses must be periodic and synchronized with the alternating anode voltage so as to cause ignition at the same phase angle in successive cycles. In other applications only one, or a few, ignitor-current pulses are required at precisely controlled times. Various methods of supplying the ignitor-current pulses in a manner that satisfies the requirements of the particular ignitron application have been devised.[9] Several of the more commonly used methods are shown in Fig. 8.

The simplest type of ignitor excitation circuit is the so-called *anode firing* method shown in Fig. 8a. In this method an auxiliary hot-cathode gas tube T_2 is used to conduct the ignitor current. During the early part of each positive half-cycle prior to ignition of the main arc in the ignitron tube T_1, the alternating voltage source sends current through the load and auxiliary tube T_2 to the ignitor. When this current reaches a critical value, the main arc starts. The voltage applied to the auxiliary tube and ignitor thereupon decreases to a low constant value, and the ignitor current hence decreases to zero or a negligibly small value. The auxiliary tube serves the important function of preventing reverse current through the ignitor during the inverse half-cycle of anode voltage. Appreciable reverse current would materially shorten the life of the ignitor.

[9] Hans Klemperer, "A New Ignitron Firing Circuit," *Electronics*, *12* (December, 1939), 12-15; H. C. Meyers and J. H. Cox, "Excitation Circuits for Ignitron Rectifiers," *A.I.E.E. Trans.*, *60* (1941), 943-948.

Phase control of the ignitron load current may be obtained through use of a thyratron for anode firing, as in Fig. 8b, instead of the diode T_2 of Fig. 8a. An alternating grid voltage for the thyratron having the same frequency as the supply voltage for the ignitron but delayed in phase by one of the methods discussed in Art. 6 causes a correspondingly delayed ignition of the main arc in T_1.

Ignitor excitation from the anode of the ignitron by the methods of Figs. 8a and 8b has the advantage of extreme simplicity. One disadvantage is that the ignitor current must pass through the load.

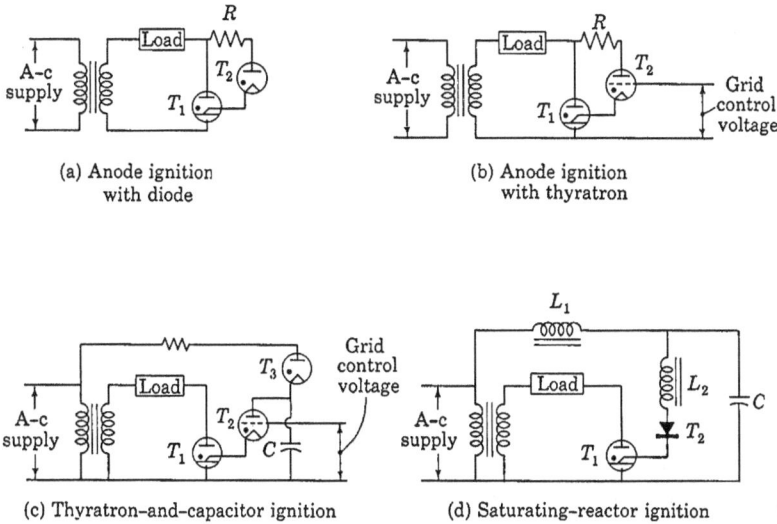

(a) Anode ignition
with diode

(b) Anode ignition
with thyratron

(c) Thyratron–and–capacitor ignition

(d) Saturating–reactor ignition

Fig. 8.   Ignitor-excitation systems.

The method is therefore suitable only for relatively heavy load currents; if the load varies and its resistance becomes too large, ignition cannot take place. A second disadvantage is that ignition cannot be made to take place at the beginning of the cycle. Time must elapse before the supply voltage becomes large enough to cause the necessary value of ignitor current.

Satisfactory ignition over all values of load current within the rating of the ignitron is obtained if a path other than through the load is provided for the ignitor current as in Figs. 8c and 8d. In each diagram, energy is accumulated in a capacitor C during part of the cycle, and is discharged suddenly through the ignitor at the desired instant. Capacitor C charges through a rectifier T_3 in Fig. 8c, and discharges through the thyratron T_2 and the ignitor to cause ignition when the control-grid voltage starts conduction in the thyratron.

In the method of Fig. 8d, capacitor C charges through reactor L_1 during the positive half-cycle of supply voltage. At the same time the current through reactor L_2, rectifier T_2, and the ignitor gradually increases, but is limited to a low value by the large inductance of L_2. The steel core of L_2 is designed, however, so that it saturates sharply when the current reaches a critical magnitude. The inductance then decreases and C discharges suddenly to cause ignition. The core of reactor L_1 is designed so that appreciable saturation does not occur. The rectifier T_2 is included to prevent reverse current through the ignitor. Phase-shifting equipment is ordinarily interposed between the alternating voltage supply and reactor L_1 to adjust the time of ignition in the cycle.

The saturating-reactor method of ignitor excitation has the important advantage that it does not require a hot-cathode gas tube, and is widely used. Hot-cathode tubes are not well suited to ignitor excitation because the ignitor current has a large peak value relative to its average value. A hot cathode adequate to supply the required peak current is capable of supplying much more than the required average current, and the hot-cathode tube is therefore unduly large and expensive for this service.

PROBLEMS

1. What is the average value of the load voltage in the thyratron circuit of Fig. 9 if the tube voltage drop is negligible during conduction, the critical grid voltage is zero, and the battery voltage is 81.4 volts?

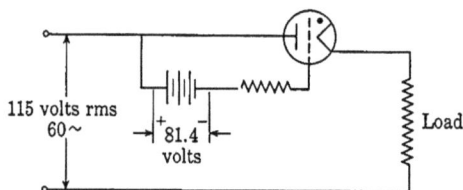

Fig. 9.   Thyratron circuit for Prob. 1.

2. Construct the waveform of the rectified charging current through the direct grid-voltage source E_{cc} in Fig. 2.

3. The phase-controlled thyratron circuit of Fig. 10 employs an anode load resistance R of 200 ohms, a grid-current-limiting resistor R_g of 50,000 ohms, a fixed capacitor C_z of 0.1 μf, and a phase-control resistor R_z that may be varied from zero to 10^5 ohms maximum. The supply transformer provides a secondary voltage of 115 volts rms at 60 cps on each side of the center tap. It may be assumed that the tube voltage drop is negligible during the conduction period, that grid current is negligible, and that the critical grid voltage is zero for all positive values of anode voltage.

(a) Find the average values of the anode current when R_z is 0, 5,000, 20,000, and 100,000 ohms. Plot the average value of the plate current as a function of the resistance R_z.

(b) Sketch the wave shapes of the instantaneous grid voltage e_{kg}, the anode voltage e_{ka}, and the load voltage e_{ab} as functions of time when R_z is 10,000 ohms.

(c) What should be the value of R_z to operate a 64-volt incandescent lamp at full brilliancy when it is substituted for R? Assume that the lamp resistance remains constant throughout each cycle.

(d) Determine the required power rating of R_z to prevent its overheating for *any* setting. (The power rating equals the maximum $I_z{}^2$ times the maximum R_z.)

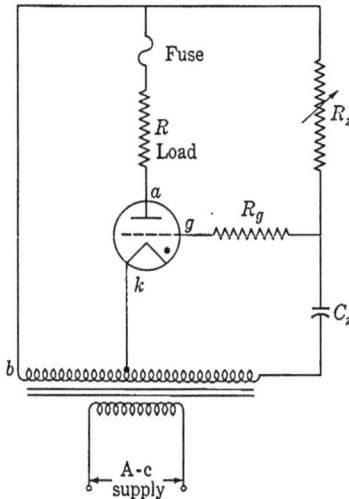

Fig. 10. Basic thyratron phase-shift circuit for Probs. 3 and 4.

4. In the thyratron circuit of Fig. 10, R is 100 ohms, R_g is 10^5 ohms, and R_z and C_z are so arranged as to give phase control over substantially the entire range of 180°. A fuse is used to protect the thyratron from excessive current. The transformer secondary furnishes a sinusoidal voltage of 115 volts rms on each side of the center tap at 60 cps. The tube voltage drop is negligible during the conduction period, the grid current is negligible, and the critical grid voltage is zero for all positive values of anode voltage.

Determine, using a factor of safety of two, the current rating of the fuse necessary to permit operation over the full range of phase control.

5. A Type 5559/FG-57 thyratron is connected as shown in Fig. 11 to regulate the average current in the load resistor R through variation of the resistance R_z. In this problem, however, the operation is to be considered for only one value of R_z, namely, R_z equal to 1,000 ohms. The tube is immersed in a constant-temperature bath at 60 C.

All transformer voltages indicated are rms values. R_z is 1,000 ohms, R_g is 10,000 ohms, C_z is 2.65 μf, and the frequency is 60 cps. Control characteristics of the Type 5559/FG-57 thyratron are given in Fig. 20 of Ch. V.

(a) Sketch as a function of ωt the anode voltage and the grid-supply voltage. Determine and indicate on this sketch the angle at which the tube fires.

(b) If the temperature is not controlled and may vary from 30 C to 60 C, between what limits will the firing angle vary?

(c) If the R_z-C_z combination is connected between taps a and b instead of between d and f, approximately (within 50 per cent) how much does the firing angle vary as the temperature varies between 30 C and 60 C?

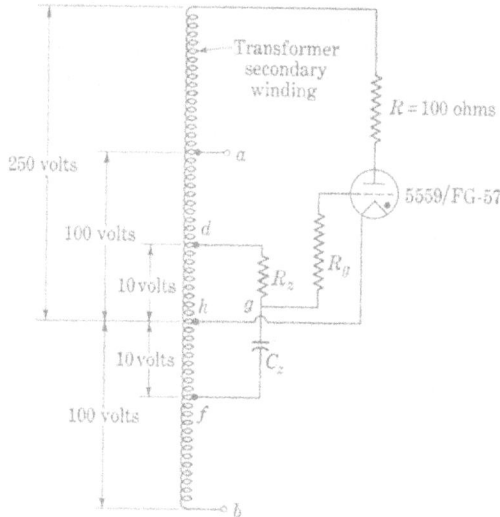

Fig. 11.   Thyratron phase-shift circuit for Prob. 5.

6. A thyratron is to be connected as shown in Fig. 12 to regulate the current flowing through the load R, which is 200 ohms. It is desired to regulate the load current by variation of the phase of the grid voltage with respect to the anode voltage by the method shown.

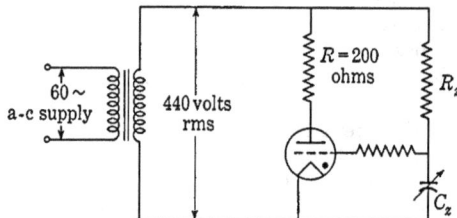

Fig. 12.   Thyratron phase-shift circuit for Prob. 6.

(a) If R_z is 10^6 ohms, specify the range of C_z needed to vary the average output current from 100 per cent of the maximum obtainable with half-wave rectification to 60 per cent of that value. Assume that the grid requires zero current, that the voltage drop in the thyratron is negligible while the tube conducts, and that the critical grid voltage is zero.

(b) Plot as a function of time for the maximum and minimum values of C_z:
   (1) anode voltage,
   (2) anode current,
   (3) grid voltage.
(c) Draw vector diagrams showing the anode voltage and the grid voltage corresponding to maximum and minimum C_z.

7. The Type 866-A mercury-vapor rectifier tube has the following ratings:

Peak inverse voltage (for condensed mercury temperature 25 C to 60 C) = 10,000 volts max

Peak anode current = 1 amp max
Average anode current = 0.25 amp max
Tube voltage drop (while conducting) = 10 to 20 volts.

First, two of these tubes are used in parallel in a half-wave single-phase rectifier circuit with a resistance load and no smoothing filter.

(a) What is the largest effective value of the transformer secondary voltage and the corresponding smallest value of the load resistance that can be used without overloading the tubes?

Second, two of these tubes are used in a full-wave single-phase rectifier circuit with a resistance load, a single-section inductor-input smoothing filter, and a 60-cps supply. Assume that the inductor has negligible resistance, the capacitance is 10 μf, and the tube voltage drop is negligible.

(b) If the inductor has a very large inductance, what is the largest effective value of the transformer secondary voltage and the corresponding smallest value of the load resistance that can be used without overloading the tubes?
(c) Which of the following criteria place the lower limit on the value of inductance that should be used in (b) above?
   (1) The peak-anode-current rating of 1 amp,
   (2) The requirement of continuous current in the inductor,
   (3) The requirement that the series resonant frequency of L and C must be less than 120 cps.

Third, a grid is introduced into each tube and phase control is added. The critical grid voltage may be assumed to be zero regardless of the anode voltage. In the circuit of (a) above:

(d) What maximum peak-forward-anode-voltage rating should the tube have to satisfy the circuit requirements?
(e) For what angles between the grid and anode voltages does the tube experience the maximum peak forward anode voltage in (d) above?
(f) If phase-shift grid control is introduced in the full-wave circuit of (b) above, can the maximum peak forward anode voltage become greater than the maximum peak inverse voltage as the inductance is varied? Why?

8. A Type 5728/FG-67 thyratron is connected as a relaxation oscillator as shown in Fig. 13. In the circuit R is 10 ohms, R_1 is 10,000 ohms, R_2 is 1,000 ohms, R_g is 5,000 ohms, and C is 0.1 μf. The inductance L is negligible. A direct voltage of 1,000 volts is applied at the input terminals. The starting characteristic of the thyratron (see Fig. 20, Ch. V) at a temperature of 60 C should be used and the further assumptions should be made that when the capacitor voltage has dropped to 5 per cent of its peak value the tube current suddenly stops, that the grid

current is zero at all times, and that the tube voltage drop is negligible during the conduction period.

Determine the frequency of oscillation of the capacitor voltage.

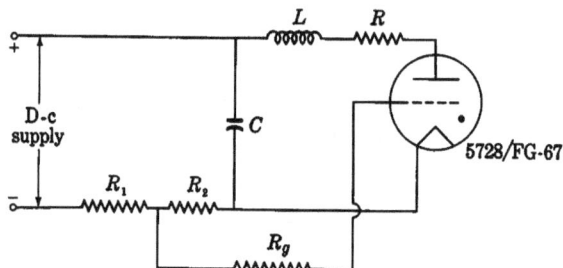

Fig. 13.   Thyratron relaxation oscillator circuit for Probs. 8 and 9.

9. The relaxation oscillator of Fig. 13 employs a Type 5728/FG-67 thyratron and is supplied by a direct voltage source of 1,000 volts. In that circuit R is 1 ohm, L is 5 mh, R_2 is 20,000 ohms, R_g is 10,000 ohms, C is 0.1 μf, and R_1 is 0. Use the starting characteristic for the Type 5728/FG-67 (see Fig. 20, Ch. V) at a temperature of 60 C.

(a) Calculate the value of the peak current through the thyratron.
(b) Determine the frequency of oscillation of the capacitor voltage.
(c) State the assumptions made in arriving at the above solutions and show that they are justified.

10. The circuit for a stroboscope shown in Fig. 14 is used to obtain pulses of current of very high magnitude and short duration at the frequency of a commercial 60-cps power line. One of the tubes is a vacuum diode used as a rectifier, and the voltage drop across it when it is conducting may be assumed to be zero. The other tube is a Strobotron. For analysis, the Strobotron may be assumed to behave as a thyratron with a critical grid voltage of $+200$ volts, a negligible drop when it conducts, and a negligible grid current.

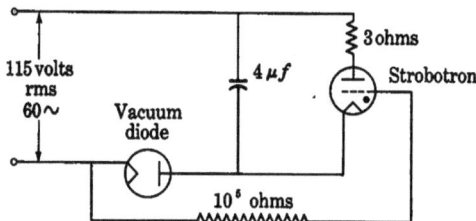

Fig. 14.   Strobotron flash circuit for Prob. 10.

In the operation of the circuit, the capacitor is charged through the rectifier during one half-cycle. The inverse voltage of the rectifier is applied to the grid of the Strobotron during the next half-cycle and causes the Strobotron to conduct. The capacitor then discharges quickly through the Strobotron and 3-ohm resistor, causing a pulse of light of short duration from the gaseous discharge. Since the light varies markedly with the current, the pulse may be assumed to end when the current through the Strobotron drops to 0.05 amp.

(a) At what instant in the cycle of the line voltage will the Strobotron fire?

(b) What are the peak value and the duration of the current pulse through the Strobotron?

(c) Sketch the waveforms of the currents and voltages in the circuit.

11. The diagram of Fig. 15 shows a type of thyratron circuit used as a so-called electronic switch for making two waves visible on a single electron-oscilloscope tube. The alternating voltage supplied to the grids fires the tubes alternately, and, when one tube becomes conducting, the capacitor C causes the anode of the other tube to remain negative with respect to its cathode long enough for de-ionization. The waveform of the voltage across ab is essentially rectangular, and this voltage is used to render one or the other of two amplifiers inoperative in synchronism with the a-c control voltage by making the corresponding grid highly negative with respect to its cathode.

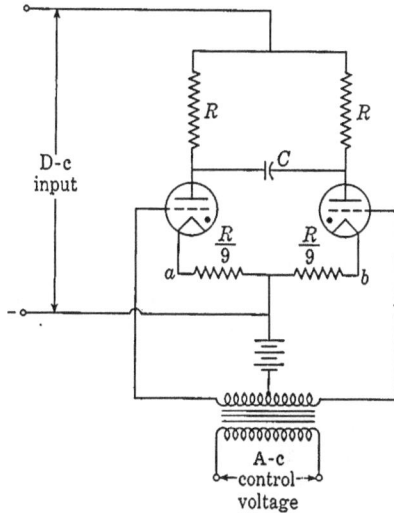

The capacitor charging current may be assumed to drop to zero during each charging period (half-cycle). The critical grid voltage and the tube voltage drop while the tube conducts may also be assumed to be zero.

(a) What should be the relationship among R, C, and the frequency of the control voltage to make 10 per cent of the time of a half-cycle available for de-ionization if the two resistors $R/9$ are short-circuited?

Fig. 15. Square-wave generator circuit for Prob. 11.

(b) Repeat (a) with the two resistors $R/9$ not short-circuited.

12. An ignitron is used with a resistive load and with anode firing for ignitor excitation as in Fig. 8a. The rms alternating secondary voltage is 600 volts at 60 cps, the peak load current is 1,600 amp, and the resistance R is 4 ohms. The arc voltage drops in the ignitron T_1 and the auxiliary diode T_2 during conduction may be assumed constant at 20 volts and 10 volts, respectively. Ignition may be assumed to occur when the ignitor current reaches 40 amp. The corresponding voltage across the ignitor is 150 volts. The maximum rated average current through the ignitor is 2 amp.

(a) At what angle in the cycle does ignition occur?

(b) What fraction of the maximum permissible value is the average ignitor current? For this analysis the ignitor resistance may be assumed constant.

CHAPTER VIII

Vacuum Tubes as Linear Circuit Elements; Class A Single-Stage Amplifiers

As is apparent from the earlier chapters of this book, grid-controlled vacuum tubes are but one class of the steadily growing group of devices described generally as electron tubes. This class is perhaps the most important one, particularly in the communications field, where such tubes are used in numerous different roles throughout the electrical system. The number of vacuum tubes manufactured each year now exceeds 350,000,000; their power ratings vary from a small fraction of a watt to as much as 500 kilowatts. The ability of the grid-controlled vacuum tube to control large amounts of power, together with its ability to do so at frequencies up to many millions of cycles a second, makes it one of the most important engineering devices of this age.

The applications of vacuum tubes may for convenience be classified according to the functions that they perform when associated with other elements in circuits. Basic among these functions is that of an *amplifier*, which is "a device whose output is an enlarged reproduction of the essential features of the input wave and which draws power therefor from a source other than the input signal."[1] In other fundamental functional operations that it performs, the tube is said to serve as a rectifier, oscillator, modulator, demodulator, mixer, detector, or discriminator. In still others, it serves as a so-called clipper, clamper, limiter, peaker, gate, multivibrator, trigger generator, blocking oscillator, or sweep generator. Though other more specialized functions may be enumerated, the foregoing lists contain the principal ones. In the remaining chapters of this book, the grid-controlled vacuum tube is considered as a circuit element, and the behavior of the complete circuit containing such tubes is examined with reference to the particular functional operation the circuit is desired to accomplish.

1. BASIC CONSIDERATIONS

As far as its ordinary operation in a circuit is concerned, the triode may be represented as a three–terminal network, or two–terminal-pair network, having certain electrical characteristics determined by the

---

[1] *Standards on Antennas, Modulation Systems, and Transmitters: Definitions of Terms, 1948* (New York: The Institute of Radio Engineers, 1948), 11.

physical phenomena taking place inside it. When the rate of change or frequency of the changes in the electrode voltages is low enough to make displacement currents through the interelectrode capacitances of the tube negligible, the electrical characteristics may be shown by a set of static characteristic curves that give the relations between the instantaneous currents and voltages for the pairs of terminals. Representative static characteristics of a triode are discussed in Art. 5, Ch. IV. If these characteristics and the way in which the three terminals are connected to other circuit elements and to voltage and current sources are given, the problem of analysis in the usual type of application becomes the determination of the transient or steady-state behavior of the circuit.

Since the behavior of the vacuum tube is fundamentally nonlinear, a method of analysis suited to the nonlinearity is necessary. In the material that follows, circuit behavior under three conditions is considered. The first holds when direct voltages only are applied, as in Art 2. The second condition is that occurring when increments such as pulses or alternations are superposed on the direct quantities, but the increments have amplitudes small compared with the corresponding direct quantities. For this condition, discussed in this and the next chapter, a reasonable approximation results from a superposition of a linear behavior for the incremental components on the nonlinear behavior for the direct components—and the so-called *linear* behavior of the tube occurs. The third condition, treated in the final chapters of the book, exists when the variations superposed on the direct components are so large that the linear approximation is not valid and non-linearity must be considered for both the varying and the direct components.

In this and the remaining chapters, the filament or heating current is considered to be maintained constant at its proper rated value (as specified by the manufacturer on the basis of life or other considerations), although the heating circuit (comprising two filament or heater leads, rheostat, and battery, transformer, or other power supply) is not drawn in the circuit diagrams. Instead, this part of the tube and circuit is usually represented by a single symbol as in Fig. 1a. Typical connections for heating the cathode are shown in Fig. 1b.

For clarity in a discussion of electron-tube performance, adoption of a logical consistent convention for letter symbols to represent various total and component voltages and currents in the circuit is highly desirable. Established principles governing selection of such symbols are:[2]

"Symbols for quantities in electrode circuits of electron tubes are developed from the proper quantity symbol and subscripts

[2] *Standards on Abbreviations, Graphical Symbols, Letter Symbols, and Mathematical Signs, 1948* (New York: The Institute of Radio Engineers, 1948), 2–3.

representing the electrodes concerned. When one of the electrodes concerned is the cathode, the subscript 'k' may be omitted and the single subscript understood to mean 'with respect to the cathode.'

"Instantaneous current and voltage values of a varying component may be represented by lower-case symbols with the subscripts 'g' and 'p' for grid and plate, respectively.

"Instantaneous total values of current and voltage (no-signal dc value plus varying-component value) may be represented by lower-case symbols and the subscripts 'b' for plate and 'c' for grid."

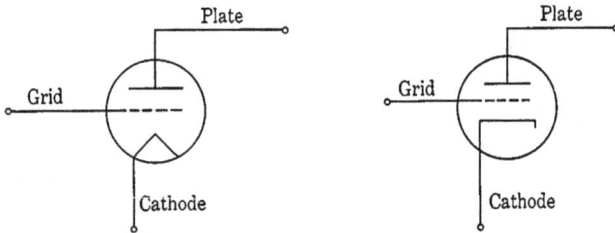

(a) Alternative symbols for triode used to represent the tube and its actual cathode–heating circuit

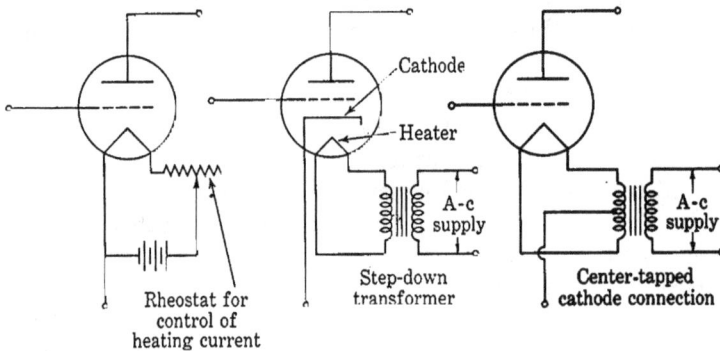

(b) Several methods of heating tube cathodes

Fig. 1.   Graphical symbols for triodes.

In accordance with these principles, Fig. 2 shows the letter symbols for electrode voltages and currents, with the reference directions assigned as positive in order to establish the algebraic signs in the succeeding analysis. Such a selection of directions is, of course, entirely arbitrary. It need not agree with the physical facts. Any discrepancy between the actual direction and the chosen reference direction is accounted for by the algebraic sign of the numerical value. The plus and minus sign shown near e_c, for example, mean that whenever the grid is at a higher potential than the cathode, e_c is a positive number, and whenever the grid is at a lower potential than the cathode, e_c is a negative number. The convention generally adopted is to refer all electrode voltages to the cathode or to the negative terminal of the filament, that is, to take the cathode as the zero reference of potential. Several possibilities for expressing in words the assigned reference direction are available. For instance, e_c may be expressed synonymously as the voltage of the grid with respect to the cathode, the voltage rise from the cathode to the grid, or the voltage drop from the grid to the cathode. Frequently e_c is merely called the grid voltage, the words "with respect to the cathode" being understood.

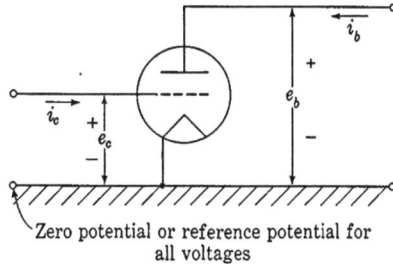

Zero potential or reference potential for all voltages

e_c ≡ instantaneous total voltage rise from cathode to grid

e_b ≡ instantaneous total voltage rise from cathode to plate

i_c ≡ instantaneous total current through the external circuit toward the grid

i_b ≡ instantaneous total current through the external circuit toward the plate

Fig. 2. Assigned positive reference directions for currents and voltages in a triode.

The assigned positive reference directions of the currents shown are selected on the basis of the fact that the conventional direction of current inside the tube is from the plate or grid to the cathode—electrons move from cathode to plate, and not in the opposite direction. The letter symbols used for the various components of the voltages and currents are defined in the following articles as the need for them appears, and a summary of the definitions is given in Art. 20.

As long as the potential of the grid is negative with respect to that of the cathode—that is, as long as e_c is negative—the magnitude of the conduction grid current in normal operation is very small, as is pointed out in Art. 5, Ch. V. When the impedance in the grid circuit is also small, the effect of the small grid current on the electrode voltages is negligible, and it is often neglected. For this reason, the tube is assumed in the next four articles to operate entirely as an electrostatically controlled device, and only negative values of e_c are considered. In the analysis given later in Art. 8, the displacement grid current through the capacitance of the tube is included, and in Ch. X, where positive values of e_c are considered, the conduction grid current is included.

2. QUIESCENT OPERATION; NO GRID-SIGNAL VOLTAGE

In the usual amplifier, the voltage to be amplified is introduced between the grid and the cathode so as to vary the plate current, and a load is connected in series with the plate. A direct-voltage supply is connected in series with the cathode, plate, and load to furnish the plate current, and another direct-voltage supply is connected in series with the cathode, grid, and voltage to be amplified, so as to keep the instantaneous potential of the grid with respect to the cathode negative at all times and thereby prevent the occurrence of

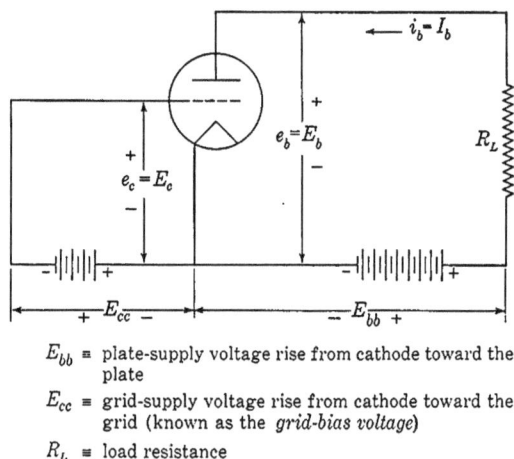

E_{bb} = plate-supply voltage rise from cathode toward the plate

E_{cc} = grid-supply voltage rise from cathode toward the grid (known as the *grid-bias voltage*)

R_L = load resistance

Fig. 3.   Schematic circuit diagram for a vacuum-triode amplifier with a resistance load before introduction of the grid-signal voltage.

grid current and the attendant power drain on the source of voltage to be amplified. The load may take various forms—for example, an indicating instrument, a loud speaker, a transformer with a loaded secondary winding, a relay, or a coupling network leading to the ultimate load. In the analysis of the circuit, it is necessary to consider the characteristics of both the load and the tube in order to determine the over-all electrical performance of the combination.

A problem of basic importance is to determine the steady value of the plate current that exists with the direct voltages applied, but prior to introduction of the voltage to be amplified. The circuit is then in the *quiescent* state, and conditions are shown as in Fig. 3. The load generally consists of resistance with or without inductance or capacitance. But when the electrode supply voltages are constant, only the resistance portion of the load affects the current; hence only a resistance R_L is indicated on the diagram. In this figure, the new symbols

E_{cc} for the grid-supply voltage and E_{bb} for the plate-supply voltage, each with an assigned positive reference direction, are introduced and defined.

Plus and minus signs near E_{cc} indicate that it is the voltage measured with respect to the cathode. Those on the battery symbol indicate that the battery is connected in a direction that actually makes the grid negative with respect to the cathode. Thus, for this condition, the actual polarity is opposite to the reference direction; hence E_{cc} *is a negative number*. But the algebraic sign of the letter symbol E_{cc} in literal equations pertaining to the circuit is determined by the reference direction. Much confusion can result from a failure to recognize the necessity for a reference direction for each letter symbol, and the distinction between it and the actual direction.

The voltage E_{cc} is called the grid-bias voltage, often abbreviated to *grid bias*, because it polarizes or biases the voltage of the grid with respect to the cathode so that values having one sign, usually negative, predominate.

Since the characteristics of the tube are generally given in graphical form, a graphical solution for the current and voltages in the circuit is convenient. By Kirchhoff's law in Fig. 3,

$$e_b + i_b R_L = E_{bb}, \qquad [1]$$

Thus, the current i_b is constrained by the circuit external to the tube to the value

$$i_b = \frac{E_{bb} - e_b}{R_L}. \qquad [2]$$

Likewise the current i_b is constrained by the tube alone to values given by the graphical plate characteristics, which may be mathematically described in functional notation as

$$i_b = f(e_c, e_b), \qquad [3]$$

wherein e_c equals E_{cc} for the particular conditions of Fig. 3.

Equations 2 and 3 are two expressions for i_b in terms of the variable e_b, and the series connection requires that i_b be equal in each expression; thus

$$f(E_{cc}, e_b) = \frac{E_{bb} - e_b}{R_L}. \qquad \blacktriangleright[4]$$

A graphical solution for the values of e_b and i_b that satisfy Eq. 4 may be had through superposing plots of the terms on both sides of the equation as functions of the variable e_b, and locating the intersection of the two curves. The problem and the graphical method of its solution are similar to those for a diode with a resistance load analyzed in

Art. 2, Ch. VI. Superposed plots of the two terms in Eq. 4 are shown in Fig. 4. The curve corresponding to the given value of the grid voltage (that is, to e_c equal to E_{cc}) is chosen from the family of plate characteristics. Note that, since the grid-bias battery in Fig. 3 makes the grid negative with respect to the cathode, the curve selected in Fig. 4 is one for which E_{cc} is a *negative* voltage. The ordinates of the

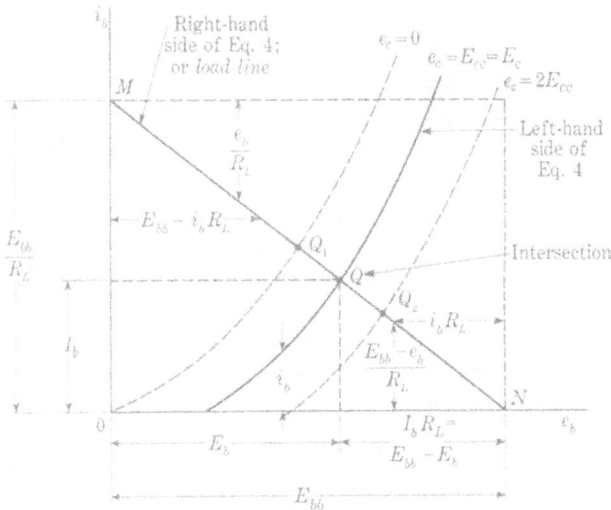

E_c = no-signal or quiescent value of the voltage rise from cathode to grid (when the varying component of grid voltage is zero)

E_b = no-signal or quiescent value of the voltage rise from cathode to plate (when the varying component of grid voltage is zero)

I_b = no-signal or quiescent value of the current through the external circuit toward the plate (when the varying component of grid voltage is zero)

Fig. 4.   Graphical construction for no-signal or quiescent operating conditions with resistance load.

straight line, called the *load line* or *resistance line* and drawn between the point M, for which the co-ordinates are

$$e_b = 0, \quad i_b = \frac{E_{bb}}{R_L}, \tag{5}$$

and the point N, for which the co-ordinates are

$$e_b = E_{bb}, \quad i_b = 0, \tag{6}$$

give the values of the current i_b in Eq. 2. The intersection of the curve and the straight line at the point Q determines the values of e_b and i_b that simultaneously satisfy Eqs. 2, 3, and 4. Since these values of

e_b and i_b are the particular ones that exist in the circuit in the absence of a varying component of the grid voltage e_c, they are denoted by capital letters. The point Q is called the *quiescent operating point*, and the values of i_b, e_c, and e_b at that point, called the no-signal or quiescent values, are I_b, E_c, and E_b, respectively, as indicated on Figs. 3 and 4. The quiescent plate voltage E_b differs from the plate-supply voltage E_{bb} because of the voltage drop in the load. Since the product of grid current and series resistance in the grid-supply voltage is assumed negligible, however, no corresponding voltage drop in the grid circuit exists. Hence the quiescent grid voltage E_c equals the grid-supply voltage E_{cc}. In other circumstances discussed subsequently, this condition is not true, and E_c must be distinguished from E_{cc}.

If the grid-supply voltage E_{cc} is changed to a new value, such as E_{cc} equal to zero, the operating point moves along the resistance line to the new position at the intersection of the load line and the corresponding new plate characteristic, as is shown by Q_1. If, on the other hand, the magnitude of the negative grid-supply voltage is increased— for example, to twice its value for point Q—the operating point moves to point Q_2. The advantage of the plate characteristics over either of the other two families of characteristic curves for the graphical solution with a resistance load is that the operating point moves in a straight line across the plate characteristics for changes in grid voltage.

If the plate-supply voltage E_{bb} is changed by an amount ΔE_{bb} and the grid-supply voltage E_{cc} is held constant, the load line is shifted horizontally by ΔE_{bb}, and the operating point Q moves to a new position on the plate-characteristic curve for the particular constant value of E_{cc}.

This graphical method is applicable to the solution of a circuit consisting of any two circuit elements connected in series, whether or not they are linear—for example, two vacuum tubes or phototubes connected in series.[3] It can also be extended to treat two circuit elements connected in parallel.

3. Quiescent Operation with a Cathode-Bias Resistor

Instead of a separate battery or power supply, the voltage drop across a resistor in the cathode circuit is commonly used as the grid-bias voltage because this method has the economic advantage of permitting supply of all voltages in an amplifier from a single rectified-power supply. Figure 5 shows the basic circuit for an amplifier with

---

[3] T. S. Gray, "A Photoelectric Integraph," *J.F.I.*, *212* (1931), 89; J. W. Horton, "The Use of a Vacuum Tube as a Plate-Feed Impedance," *J.F.I.*, *216* (1933), 749–762.

this biasing method. The voltage to be amplified is usually introduced as the input voltage at e_s. Under the no-signal or quiescent
conditions considered here, however, e_s is zero. The corresponding
steady component of plate current through the cathode resistor R_k
makes the cathode positive with respect to the negative end of the
plate power supply E_{bb}, shown as ground. It therefore makes the
cathode positive with respect to the grid or, alternatively, the grid
negative with respect to the cathode, which is the desired condition.

A capacitor C_k is included across R_k in the diagram. Its purpose,
which is further explained in Art. 7, is to provide a low-impedance

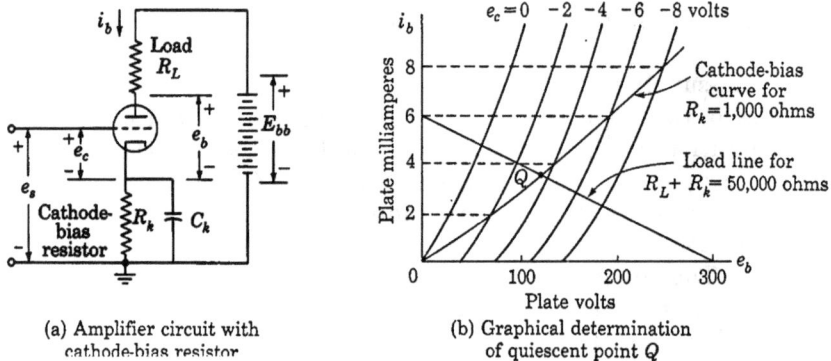

(a) Amplifier circuit with
cathode-bias resistor

(b) Graphical determination
of quiescent point Q

Fig. 5.   Grid-bias voltage obtained from a cathode-bias resistor.

path for varying components of current and thereby to by-pass them
around R_k. In this way it prevents the development of an appreciable
varying component in the voltage across R_k, which would tend to
counteract the input voltage and reduce the amplification. Capacitor
C_k has no effect on the steady currents and voltages existing during
the quiescent conditions under consideration here, however.

Determination of the quiescent conditions in this circuit must take
into account the fact that the plate current and the grid-bias voltage
are mutually dependent. A graphical method of solution is indicated
in Fig. 5b. In the plate circuit, R_L and R_k are in series; hence,

$$i_b = \frac{E_{bb} - e_b}{R_L + R_k}. \qquad [7]$$

Operation is therefore constrained by the circuit external to the tube
to lie on a load line constructed as discussed in Art. 2, but for
$R_L + R_k$ instead of R_L. The tube requires that the plate current
satisfy

$$i_b = f(e_c, e_b), \qquad [8]$$

but the grid connection imposes the additional constraint

$$e_c = e_s - i_b R_k ,$$ [9]

which, for the quiescent or no-signal condition, becomes

$$e_c = -i_b R_k .$$ [10]

Only one point on each plate characteristic curve corresponding to Eq. 8 also satisfies Eq. 10. A curve through these points, called the *cathode-bias curve*, and constructed as indicated on Fig. 5b, is therefore the locus of i_b as a function of e_b as constrained by the tube and its grid circuit for one particular value of R_k. The larger R_k, the lower lies the cathode-bias curve on the plate characteristics. The intersection of the load line with the cathode-bias curve is the quiescent operating point Q to which the quiescent current I_b and voltage E_b correspond.

In a pentode with a negative control-grid voltage and zero or negative suppressor-grid voltage, the cathode current is the sum of the plate and screen-grid currents. Hence this sum must be substituted for i_b in Eq. 10 when a pentode is used. When adequate data are not available, as in Fig. 19, Ch. IV, the screen-grid current for any value of negative control-grid voltage may be estimated through the principles given in Art. 9, Ch. IV. The cathode-bias curve constructed on the pentode plate characteristics proves to be almost horizontal in the normal linear range of operation. Consequently, the quiescent conditions for a pentode can often be determined most readily through a cut-and-try process of searching on the diagram for the pair of plate characteristics for which e_c most nearly equals the voltage across R_k caused by the sum of the plate and screen-grid currents, and interpolating between those curves. For a triode, however, plotting the cathode-bias curve is usually the most effective method of finding the quiescent conditions.

4. Operation with Grid-Signal Voltage in the Linear Region of the Characteristic Curves

The basic circuit for an amplifier with a resistance load is shown in Fig. 6. This circuit differs from that of Fig. 3 in that a voltage to be amplified, e_g, called the *grid-signal voltage*, is applied at the input in series with the source of grid-bias voltage. The amplified voltage appears at the output as a component of the voltage across the load resistor R_L. The objective of the analysis here is to find by graphical means the relation between the amplified voltage and the grid-signal voltage.

In Fig. 6, the total grid voltage e_c consists of the grid-signal voltage e_g superposed on the quiescent value E_c (which for these conditions equals E_{cc}); that is,

$$e_c = E_c + e_g .$$ [11]

The grid-signal voltage is thus an *increment* in e_c. It causes corresponding increments in the total plate current i_b and plate voltage e_b,

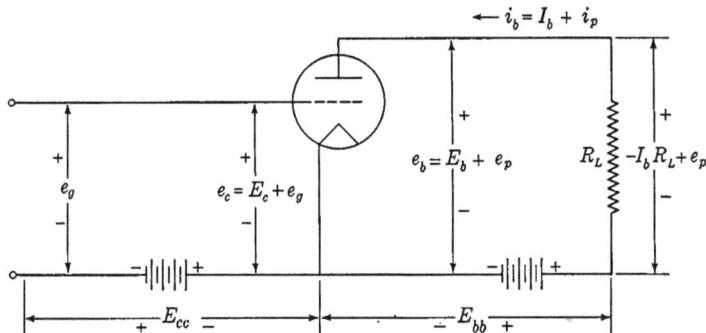

$e_g \equiv$ instantaneous value of the varying component of the voltage rise from cathode to grid; (*grid-signal voltage*)

$e_p \equiv$ instantaneous value of the varying component of the voltage rise from cathode to plate

$i_p \equiv$ instantaneous value of the varying component of current through the external circuit toward the plate

Fig. 6.   Basic circuit for a linear amplifier.

denoted by i_p and e_p, respectively. Thus i_b may be expressed as the sum of a quiescent component and an incremental component

$$i_b = I_b + i_p ,$$ [12]

and, similarly,

$$e_b = E_b + e_p .$$ [13]

The incremental grid-signal voltage e_g can have *any* waveform, either periodic or nonperiodic. It may, for example, be a periodic square wave, or merely a step voltage—that is, a direct-voltage increment. Determination of the response of the amplifier to complex periodic waveforms can be obtained through superposition of the separate responses to the components in a Fourier series that represents the grid-signal voltage, if linearity between the response and the input is first established. Consequently, analysis of the amplifier behavior with a sinusoidal grid-signal voltage

$$e_g = \sqrt{2}E_g \sin \omega t,$$ [14]

which follows, is of basic importance. The treatment here neglects displacement currents in the interelectrode regions, however, and hence is applicable only when the rate of change or the frequency of the grid-signal voltage is small enough to make such currents negligible.

At instants when the grid-signal voltage e_g is zero, conditions in Fig. 6 are identical with those in Fig. 3; hence the instantaneous plate current and voltage are found at point Q on a load line in Fig. 7, which

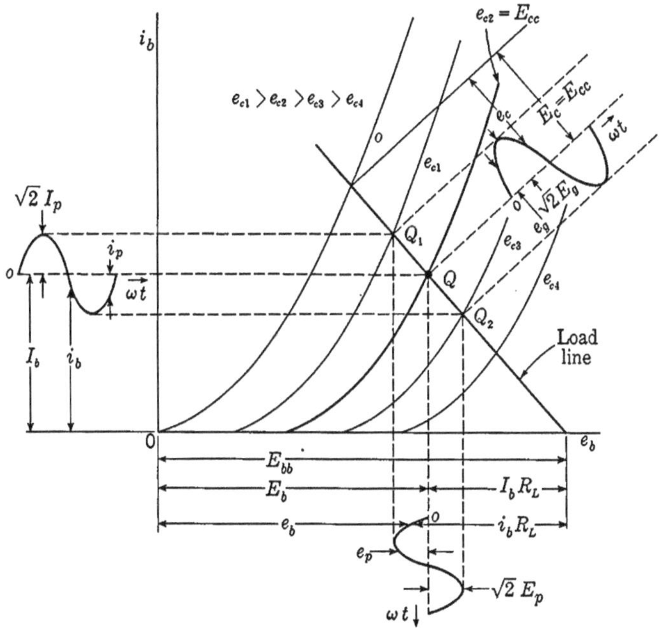

$E_g \equiv$ effective value of the sinusoidally varying component of voltage between grid and cathode

$E_p \equiv$ effective value of the sinusoidally varying component of the voltage between plate and cathode

$I_p \equiv$ effective value of the sinusoidally varying component of the plate current

Fig. 7.   Sinusoidal operation of a linear amplifier with a resistance load.

is constructed as in Fig. 4. As e_g takes on positive or negative values, operation shifts to new positions along the load line corresponding to the new values of e_c. At the instant when e_g has its maximum positive value $\sqrt{2}E_g$, for example, the total grid voltage is $\sqrt{2}E_g + E_c$, shown equal to e_{c1} in Fig. 7, and the operating point is at Q_1. Similarly, when e_g has its maximum negative value, operation shifts to Q_2 for e_{c3}, which equals $-\sqrt{2}E_g + E_c$. The grid-bias voltage E_c, of course, is numerically negative and is made larger in magnitude than $\sqrt{2}E_g$ so that e_{c1} is maintained negative.

If the plate characteristics are essentially straight, parallel, and equidistant for equal increments of the grid voltage, geometry indicates that the distance the operating point moves from Q along the load line is in direct proportion to e_g. Thus projection from a sine

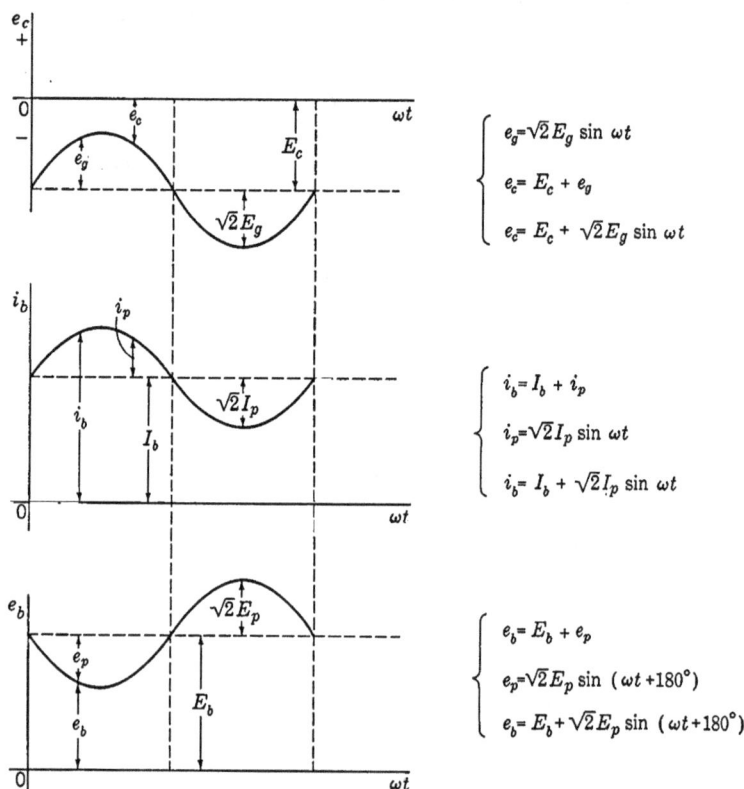

$$\begin{cases} e_g = \sqrt{2}\,E_g \sin \omega t \\ e_c = E_c + e_g \\ e_c = E_c + \sqrt{2}\,E_g \sin \omega t \end{cases}$$

$$\begin{cases} i_b = I_b + i_p \\ i_p = \sqrt{2}\,I_p \sin \omega t \\ i_b = I_b + \sqrt{2}\,I_p \sin \omega t \end{cases}$$

$$\begin{cases} e_b = E_b + e_p \\ e_p = \sqrt{2}\,E_p \sin (\omega t + 180°) \\ e_b = E_b + \sqrt{2}\,E_p \sin (\omega t + 180°) \end{cases}$$

Fig. 8.   Waveforms of the voltages and currents in Fig. 7 plotted on a common time scale.

wave having a time axis drawn perpendicular to the load line and representing e_g gives instantaneous values of i_b and e_b as shown in Fig. 7. The varying components i_p of i_b and e_p of e_b are therefore also sinusoidal with the relative values shown in Fig. 7. The operation is then said to be *linear*, or restricted to the linear region of the tube characteristics. For these conditions, the coefficients μ, g_m, and r_p are constants throughout the range of operation.

Waveforms of the total currents and voltages and their components are drawn in Fig. 7 and are shown on a common time scale in Fig. 8.

If I_p is defined as the effective value of the varying component of the plate current, and E_p is defined as the effective value of the varying **component of the voltage from the cathode to the plate, for linear operation,**

$$i_p = \sqrt{2}I_p \sin \omega t, \qquad [15]$$

and

$$e_p = \sqrt{2}E_p \sin (\omega t + 180°) = -\sqrt{2}E_p \sin \omega t, \qquad [16]$$

in which values of E_p and I_p for a particular value of E_g may be obtained from the graphical construction. Thus, on the basis of the positive reference directions assigned in Fig. 6, it is apparent that in this particular example, which involves a resistive load, the alternating component of the plate current is in phase with the alternating component of the grid voltage, but the alternating component of the plate voltage is 180 degrees out of phase with the alternating components of the plate current and the grid voltage.

When the plate characteristics are not a family of essentially straight parallel lines equally spaced for equal increments of the grid voltage, the assumption of linearity is not fulfilled, and the scale along the load line for e_g is nonlinear. Hence the projection from a sine wave of e_g indicated in Fig. 7 is then not valid even though e_g is sinusoidal in accordance with Eq. 14; and the voltage waveform across the load is not an exact reproduction of the grid-signal voltage. The coefficients μ, g_m, and r_p are not constant throughout the full range of operation. The effects of this nonlinearity in the tube are treated in Art. 13 and in later chapters. It results in waveform distortion and in modulation, with the introduction of new frequencies by the vacuum tube.

The degree of nonlinearity occurring in the operation of the tube serves as one basis for classification of amplifiers. Definitions of the standard classes are:[4]

▶ *"Class A Amplifier*—A Class A amplifier is an amplifier in which the grid-bias and alternating grid voltages are such that plate current in a specific tube flows at all times.

Note—To denote that grid current does not flow during any part of the input cycle, the suffix 1 may be added to the letter or letters of the class identification. The suffix 2 may be used to denote that grid current flows during some part of the cycle.

"Class AB Amplifier—A Class AB amplifier is an amplifier in which the grid-bias and alternating grid voltages are such that plate

[4] *American Standard Definitions of Electrical Terms—A.S.A. No. C42* (New York: American Institute of Electrical Engineers, 1941), 234.

current in a specific tube flows for appreciably more than half but less than the entire cycle.

Note—See note under Class A amplifier.

"*Class B Amplifier*—A Class B amplifier is an amplifier in which the grid-bias is approximately equal to the cut-off value so that the plate current is approximately zero when no exciting grid voltage is applied, and so that plate current in a specific tube flows for approximately one-half of each cycle when an alternating grid voltage is applied.

Note—See note under Class A amplifier.

"*Class C Amplifier*—A Class C amplifier is an amplifier in which the grid-bias is appreciably greater than the cut-off value so that the plate current in each tube is zero when no alternating grid voltage is applied, and so that plate current flows in a specific tube for appreciably less than one-half of each cycle when an alternating grid voltage is applied.

Note—See note under Class A amplifier."◀

In these definitions, the statements that grid current does or does not flow are not to be interpreted literally but rather are to be interpreted with regard for the effect of the grid current on the amplifier behavior. The grid current is never zero except at one particular value of grid voltage for any particular value of plate voltage, as is shown in Fig. 17, Ch. V. Furthermore, except at the very lowest frequencies, displacement currents exist in the tube that cause current in the grid circuit even though no electrons or ions reach the grid. However, the conduction grid current often has a negligible effect on the circuit operation, and the suffix 1 is then used. The grid current may have an appreciable effect on the circuit operation for either of the two following reasons. First, it produces a voltage drop across any series circuit element in the grid circuit, thus modifying the voltage actually applied to the grid; and, second, it may take some of the emission current from the cathode, thereby reducing the plate current from the value this current would otherwise assume. Both these effects are detrimental to the ideal operation of many types of circuits; the second may even endanger the life of the tube by causing the grid to overheat. If the external impedance in the grid circuit is large, the first effect may be of consequence even for negative grid voltages. Ordinarily, the tube and circuit are designed so that both the foregoing effects of grid current are negligible. The second effect becomes appreciable only when the grid becomes more positive than the plate—that is, in the "overexcited" condition—and is generally negligible for grid voltages smaller than the plate voltage. The first effect is important primarily in audio-frequency amplifiers, repeaters, and small tubes; the second occurs principally in high-power oscillators, amplifiers, and modulators.

When the effect of the grid current is appreciable because the grid voltage e_c reaches values that cause appreciable *electron* current to the grid, the suffix 2 is ordinarily used. When the effect of the grid current is appreciable because the series impedance in the grid circuit is very high but the electron current to the grid is made negligible by maintenance of a negative grid voltage throughout the cycle, neither of the subscripts is applicable, because these definitions are not intended to include those conditions, which are discussed in Art. 5, Ch. V.

On the basis of these definitions, the circuit of Fig. 6 is a Class A_1 amplifier when the operating conditions are restricted to the region in which μ, g_m, and r_p are essentially constants, as in Fig. 7. It belongs in Class A because the negative grid voltage is not sufficiently large in magnitude at any time to prevent the flow of plate current. It requires the subscript 1 because the grid current is kept negligible at all times in most applications by the fact that the grid voltage is always negative. Since, in addition to the foregoing, the input and output voltages are linearly related, the device is hereafter called a *linear Class A_1 amplifier*. Amplifiers of the other classifications are discussed in subsequent chapters.

Although operation with a triode is illustrated in Figs. 6 and 7, the principles of graphical analysis used and the conclusions reached are directly applicable to operation with tetrodes and pentodes as well.

5. INCREMENTAL EQUIVALENT CIRCUITS FOR LINEAR CLASS A_1 OPERATION

The preceding article shows that when the operation of the tube is restricted to the linear region of the characteristic curves, and the load is a resistance, the varying components of the grid voltage and of the plate current and voltage are linearly related. A relation among the varying components that holds for any type of load is expressed approximately by the total differential of the plate current, as is explained in Arts. 6 and 7 of Ch. IV. Since the plate current is a function only of the grid and plate voltages, the total differential is

$$di_b = \frac{\partial i_b}{\partial e_c} de_c + \frac{\partial i_b}{\partial e_b} de_b. \qquad [17]$$

If e_c is changed from the quiescent value E_c by a small increment Δe_c, i_b is changed by an increment Δi_b from its quiescent value I_b, and, because of the load, e_b is changed from its quiescent value E_b by an increment Δe_b. The relation among the increments is the same as that between the differentials in Eq. 17 if operation is confined to the region for which the plate characteristics consist of straight, parallel lines, spaced equidistant for equal increments of the grid voltage, because the partial derivatives are then constants within that region. Hence

$$\Delta i_b = \frac{\partial i_b}{\partial e_c} \Delta e_c + \frac{\partial i_b}{\partial e_b} \Delta e_b. \qquad [18]$$

This relation will also be recognized as the linear approximation to the increment in plate current expressed by the linear terms in the Taylor's series expansion for the plate current, Eq. 27, Ch. IV—a series representation that is especially useful when the characteristic curves are not straight.

The incremental quantities Δe_c, Δi_b, and Δe_b are equivalent to the previously defined varying components e_g, i_p, and e_p, respectively. Furthermore $\partial i_b / \partial e_c$ is g_m evaluated at the quiescent operating point, and $\partial i_b / \partial e_b$ is $1/r_p$ also evaluated at that point. Thus Eq. 18 becomes

$$i_p = g_m e_g + \frac{e_p}{r_p}. \qquad \blacktriangleright[19]$$

Solving for e_p gives

$$e_p = -g_m r_p e_g + i_p r_p, \qquad [20]$$

or, since $g_m r_p$ equals μ,

$$e_p = -\mu e_g + i_p r_p. \qquad \blacktriangleright[21]$$

Equations 19 and 21 apply to the tube, but they also suggest simple circuits, which must therefore behave as does the tube. Equation 19, for example, indicates that i_p is the same as the current from an ideal constant-current source $g_m e_g$ added to the current through a resistance r_p caused by a voltage e_p across it. It therefore applies to the circuit shown in Fig. 9a. In this diagram e_g is the terminal voltage at the input, and e_p and i_p the terminal voltage and current, respectively, at the output. Each is assigned the same reference direction as the total quantity e_c, e_b, or i_b of which it is a positive increment, and, to indicate that e_g and e_p are measured from a common point, the cathode, a connection is shown between the lower terminals. So far as increments are concerned, conditions at the terminals marked g, p, and k are then the same as those at the grid, plate, and cathode, respectively, of the actual tube. With these reference directions at the terminals established, a resistance r_p across the output terminals, and an ideal current source $g_m e_g$ with its reference direction downward as shown, account for the terms and their algebraic signs in Eq. 19. The input terminals are shown open because the grid is assumed to remain negative so that the conduction grid current is negligible.

The foregoing considerations indicate that, for analysis involving transient or steady-state increments from the quiescent values of the plate current and voltage, a grid-controlled vacuum tube viewed from its plate and cathode terminals is in fact *indistinguishable* from an ideal current source $g_m e_g$ shunted by a resistance r_p, provided the following two requirements are met. First, the increments must be small enough to restrict the operation to the linear region of the characteristic curves, and, second, the increments must have rates of change slow enough to make the displacement currents in the interelectrode capacitances negligible. When those conditions are fulfilled,

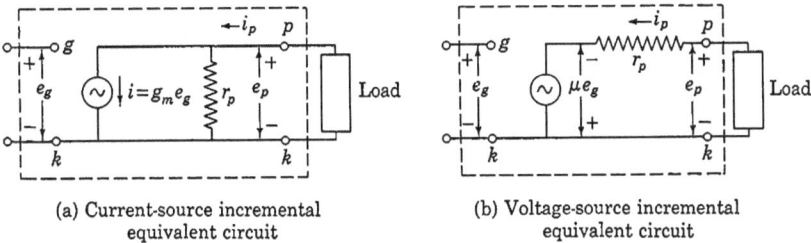

(a) Current-source incremental   (b) Voltage-source incremental
    equivalent circuit   equivalent circuit

Fig. 9. Alternative equivalent circuits for incremental components of voltage and current and for linear Class A_1 operation. The encircled symbol denotes an incremental time function having any waveform, sinusoidal or otherwise.

the representation is correct regardless of the waveform of the grid-signal voltage and the kind of load. Figure 9a is therefore called the *current-source equivalent circuit* for the vacuum tube. In the figure the symbol for the ideal current source denotes an increment having any wave-form, sinusoidal or otherwise. The current-source equivalent circuit is particularly useful in the analysis of tetrode and pentode circuits because the plate resistance of these tubes is very large. The plate resistance r_p can therefore often be neglected, and the tube assumed to behave as a simple current source.

Equation 21, on the other hand, suggests that for increments the tube behaves as if the output voltage e_p and current i_p are supplied by an ideal voltage source, or electromotive force, μe_g through a series resistance r_p. It applies to the circuit shown in Fig. 9b as well as to the tube. In this diagram, the terminal voltages and currents, reference directions, and interconnections are established on the same principles as are stated above for Fig. 9a. With the reference directions at the terminals established, an ideal voltage source μe_g with its reference direction downward and a series resistance r_p account for the terms and their algebraic signs in Eq. 21. Again, these considerations

indicate that, for analysis involving transients or steady-state increments from the quiescent values of the plate current and voltage, a
grid-controlled vacuum tube viewed from its plate and cathode terminals is *indistinguishable* from an ideal voltage source μe_g and series
resistance r_p provided the two requirements set forth in the previous
paragraph are fulfilled. Namely, the increments must be small enough
to restrict the operation to the linear region of the characteristic
curves, and the increments must have rates of change slow enough to
make the displacement currents in the interelectrode capacitances
negligible. Figure 9b is therefore called the *voltage-source equivalent*

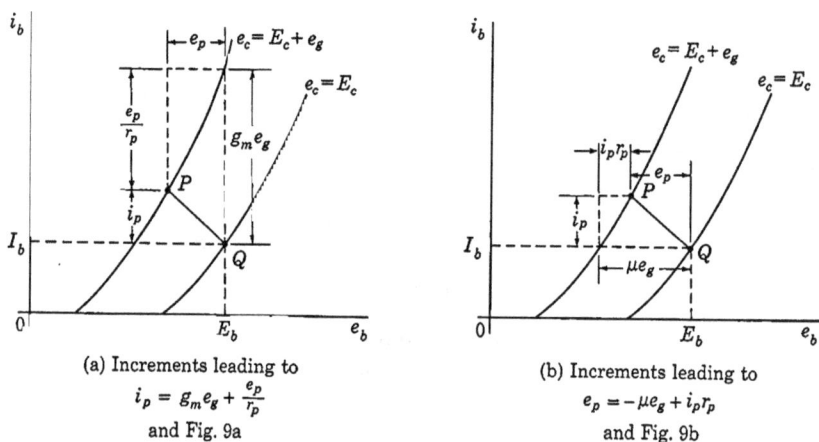

(a) Increments leading to

$$i_p = g_m e_g + \frac{e_p}{r_p}$$

and Fig. 9a

(b) Increments leading to

$$e_p = -\mu e_g + i_p r_p$$

and Fig. 9b

Fig. 10.   Graphical interpretation of incremental relationships.

circuit for the vacuum tube. Figures 9a and 9b are evidently equivalent to one another because they are both derived from Eq. 19.
Equivalence between them will also be recognized from basic network
theory.[5]

A graphical interpretation of the relationships among the terms in
Eqs. 19 and 21, the quantities in Figs. 9a and 9b, and the characteristic curves is given in Fig. 10. When an increment e_g is applied,
operation shifts from the quiescent point Q to some new operating
point P. Increments i_p and e_p corresponding to the shift occur as indicated. In accordance with the considerations of Art. 6, Ch. IV, the
increments $g_m e_g$ and e_p/r_p in Fig. 10a and μe_g and $i_p r_p$ in Fig. 10b may
next be identified. Algebraic addition of the increments on the diagrams, with due regard for the facts that e_p is a negative number for

[5] E. E. Staff, M.I.T., *Electric Circuits* (Cambridge, Massachusetts: The Technology
Press of M.I.T.; New York: John Wiley & Sons, Inc., 1940), 133–135, 389–390.

the conditions shown, but all other values are positive, leads to
Eqs. 19 and 21.

If the grid-signal voltage e_g is a small sinusoidal alternating voltage

$$e_g = \sqrt{2}E_g \cos \omega t, \qquad [22]$$

Fig. 9b shows that the triode then behaves for small sinusoidal *changes*
or *increments* of voltage and current exactly as does an alternating-
voltage generator of electromotive force.

$$\mu e_g = \mu\sqrt{2}E_g \cos \omega t \qquad [23]$$

having an internal series resistance r_p. The alternating current the
triode supplies to any load can therefore be calculated by any of the
methods that are applicable when any alternating-current generator
having those characteristics is the source of power.

If the load impedance is linear and passive, so that it is expressible
as

$$Z_L = R_L + jX_L = Z_L\underline{/\theta}, \qquad [24]$$

and, if the varying component of grid voltage is that given by Eq. 22,
the variations e_p and i_p are also sinusoidal and of the same frequency
as e_g, because the circuit is equivalent to that of an alternating-current
generator supplying a linear passive load. Thus all the incremental
components of the currents and voltages may be represented by complex quantities. The sinusoidal variations in e_c, e_b, and i_b may be
represented by the complex effective values, E_g, E_p, and I_p, respectively, where the assigned positive reference direction of each is
taken as the same as that of the corresponding instantaneous value in
Fig. 9b. Equation 21 then becomes, in complex form,

$$E_p = -\mu E_g + r_p I_p , \qquad [25]$$

and the equivalent circuit for complex quantities becomes that shown
in Fig. 11. A load comprising resistance and inductance in series has
been chosen for illustration in Fig. 11, but any other linear passive load
with a conduction path for the steady component of plate current is
permissible.

From the circuit of Fig. 11, all the currents and voltages may be
expressed in terms of the grid-signal voltage E_g. Thus

$$I_p = \frac{\mu E_g}{r_p + Z_L}. \qquad [26]$$

Also,

$$E_p = -I_p Z_L. \qquad [27]$$

Substitution of Eq. 26 in Eq. 27 gives

$$E_p = -\mu E_g \frac{Z_L}{r_p + Z_L}.$$
▶[28]

By means of Eqs. 24, 25, 26, 27, and 28, the vector, or phasor, diagram for the *sinusoidally varying incremental component quantities* in the circuit comprising the vacuum tube and load may be drawn as shown

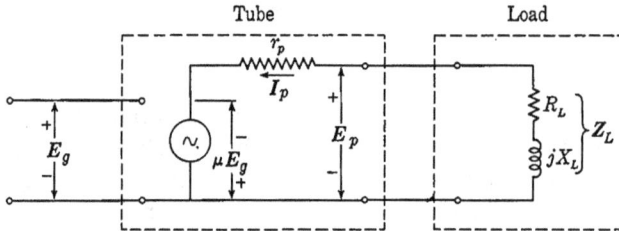

E_g = complex effective value of the sinusoidally varying component of the voltage rise from cathode to grid

E_p = complex effective value of the sinusoidally varying component of the voltage rise from cathode to plate

I_p = complex effective value of the sinusoidally varying component of the current through the load toward tne plate

Fig. 11. Voltage-source incremental equivalent circuit of the vacuum-tube circuit with complex representation of the sinusoidally varying incremental voltages and currents, and of the impedances.

in Fig. 12. In this figure, the grid voltage has been taken as the reference, although the construction of the diagram is perhaps most readily made if the plate current is taken as the starting point.

In Fig. 12, with E_g as the reference vector and positive angles measured counterclockwise from it,

$$E_g = E_g \underline{/0}, \qquad\qquad [29]$$

$$I_p = I_p \underline{/\phi} \quad (\textit{Note: In Fig. 12, } \phi \text{ is negative}), \quad [30]$$

$$E_p = E_p \underline{/\psi_p}. \qquad\qquad [31]$$

The following relationships are evident on the diagram:

Impedance angle of load,

$$\theta = \tan^{-1}\frac{X_L}{R_L}; \qquad\qquad [32]$$

Phase angle of current,

$$\phi = -\tan^{-1}\frac{X_L}{r_p + R_L}; \qquad\qquad [33]$$

Phase angle by which load voltage I_pZ_L leads the grid voltage,

$$\theta + \phi = \tan^{-1}\frac{X_L}{R_L} - \tan^{-1}\frac{X_L}{r_p + R_L};$$ [34]

Phase angle by which plate voltage leads the grid voltage,

$$\psi_p = 180° + \phi + \theta.$$ [35]

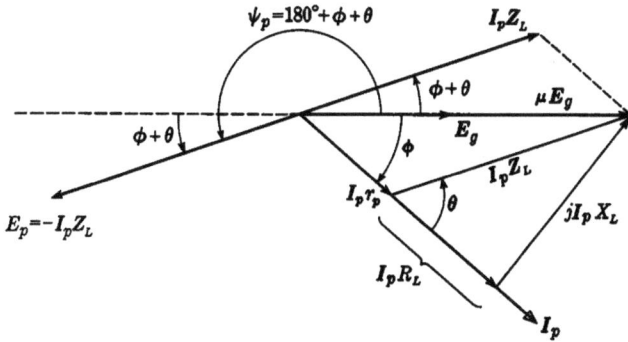

Fig. 12.   Vector, or phasor, diagram for the sinusoidal alternating components in the circuit of Fig. 11.

The relationships among the *scalar* effective values of the alternating-current quantities may be obtained from Eqs. 24, 26, and 28. Thus

$$I_p = \frac{\mu E_g}{\sqrt{(r_p + R_L)^2 + X_L^2}},$$ [36]

$$E_p = \mu E_g\frac{\sqrt{R_L^2 + X_L^2}}{\sqrt{(r_p + R_L)^2 + X_L^2}}.$$ [37]

Equations 24 and 29 through 37 completely define the relationships among the complex effective values in the equivalent circuit. From Eqs. 34 and 35, it is evident that there is a phase angle between E_g and E_p that approaches 180 degrees as the reactive component of the load impedance approaches zero.

The concept of the equivalence between the tube and the circuits in Fig. 9 is a valuable aid in the analysis of the operation of the tube. However, care must be exercised that the concept be used only when the assumptions upon which it is based are fulfilled. One assumption made is that the interelectrode capacitances in the tube have no effect on the circuit operation. The effect of these capacitances is discussed subsequently in Art. 8. Another assumption is that the operation is restricted to the linear region of the tube characteristics. This assumption is never fulfilled exactly; but, when the effect of the nonlinearity is so small that the waveform of the plate-current variation closely approximates that of the grid-signal voltage, the methods of analysis given in Art. 13 are applicable. When the effect of the nonlinearity on the waveform is greater, the methods given in Ch. X may be used. *Note that the incremental equivalent circuit cannot be used to determine correctly the quiescent operating point.*

It should be recognized, in a consideration of the linearity of the tube characteristics over the region of operation, that the region depends not only on the electrode-supply voltages but also on the type of load. Article 4 shows that, when the load consists of resistance only, the locus of the operating point Q on the plate characteristics is a straight line. However, if the load impedance includes a reactive component, the locus is not a straight line but becomes an ellipse, in accordance with the analysis given in many textbooks on electric circuits,[6] as long as the path of operation does not extend beyond the region where μ, g_m, and r_p are constants.[7] If the path does extend outside this region, waveform distortion occurs, and the principles of nonlinear circuit analysis must be used.

6. AMPLIFICATION, OR GAIN

Usually the purpose of an amplifier is to enlarge the voltage, current, or power of the input wave. The enlargement is conveniently described by the ratio of the increment in the output voltage, current, or power to the corresponding input quantity. *This ratio is called the voltage, current, or power amplification or gain.** In general, the term amplification may refer to specific features of the output and input waves such as their instantaneous, average, effective, or complex values; hence a further qualifying adjective is generally needed for clarity. When the amplifier circuit contains only resistances, however, as in the circuits considered thus far, the ratios for all these features of the voltages differ only as to algebraic sign, and the term voltage amplification is therefore used for all of them. When the amplifier

---

[6] E. E. Staff, M.I.T., *Electric Circuits* (Cambridge, Massachusetts. The Technology Press of M.I.T.; New York: John Wiley & Sons, Inc., 1940), 333–336.

[7] A. A. Nims. "Circle Diagrams for Tube Circuits," *Electronics, 12* (May, 1939), 23–26; W. G. Dow, *Fundamentals of Engineering Electronics* (2nd ed.; New York: John Wiley & Sons, Inc., 1952), 67–70; A. J. H. van der Ven, "Output Stage Distortion," *Wireless Engr., 16* (1939), 450 (Fig. 23).

* The term "gain" is sometimes reserved for the power ratio, but this distinction is not made here because the terms amplification and gain are now widely used as synonyms.

circuit contains inductance or capacitance, on the other hand, the ratios for some of the several features of the voltages are complex or vary with time, and must therefore be distinguished from one another.

When the input-signal voltage is sinusoidal, and operation is in the linear region of the tube characteristics, the incremental output voltage is also sinusoidal for any amplifier; hence the voltage ratio is conveniently expressed as the *complex voltage amplification* A, given by

$$A = \frac{\text{complex value of the output voltage}}{\text{complex value of the input voltage}} \qquad [38]$$

$$= A\underline{/\theta_A} \qquad [39]$$

where A is the ratio of the amplitudes or effective values of the output and input voltages, hereafter referred to merely as the *voltage amplification or gain*, and θ_A is the *phase angle* of the ratio. In many amplifiers, some of which are discussed subsequently, circuit elements lie between the tube and the input and output terminals; hence the input and output voltages involved in the voltage amplification are not the grid-to-cathode and plate-to-cathode voltages at the tube terminals. In the simple amplifier of Fig. 11, however, the output voltage at the load is E_p and the input voltage is E_g; thus, from Eq. 28, the complex voltage amplification for the amplifier with a load Z_L is

$$A = \frac{E_p}{E_g} = -\mu \frac{Z_L}{r_p + Z_L} = -\mu \frac{1}{1 + \dfrac{r_p}{Z_L}}, \qquad \blacktriangleright[40]$$

and, if

$$Z_L = R_L + jX_L, \qquad [41]$$

in accordance with Eqs. 29, 31, 34, 35, and 37,

$$A = \mu \frac{\sqrt{R_L^2 + X_L^2}}{\sqrt{(r_p + R_L)^2 + X_L^2}}, \qquad [42]$$

and

$$\theta_A = 180° + \tan^{-1}\frac{X_L}{R_L} - \tan^{-1}\frac{X_L}{r_p + R_L}. \qquad [43]$$

Although the angle θ_A is the same as the angle ψ_p in Fig. 12, where the reference vector is E_g, the angle is here given the new symbol θ_A, which is independent of the choice of the reference vector. The angle θ_A, then, is the phase angle by which the plate voltage E_p leads the grid voltage E_g. In general, the impedance Z_L may be any combination of resistance, capacitance, and inductance elements. Therefore the voltage amplification is a function of frequency, because X_L in Eqs. 42 and 43 is a function of frequency and R_L may be.

At a fixed frequency, the voltage amplification A depends on the components of the load impedance $\mathbf{Z}_L$ and the ratio $r_p/\mathbf{Z}_L$. For a pure resistance load,

$$\mathbf{Z}_L = R_L + j0, \tag{44}$$

and

$$A = \frac{\mu}{1 + \dfrac{r_p}{R_L}}, \tag{45}$$

and

$$\theta_A = 180°. \tag{46}$$

For a pure inductance load,

$$\mathbf{Z}_L = 0 + jX_L, \tag{47}$$

and

$$A = -\mu \, \frac{1}{1 + \dfrac{r_p}{jX_L}} \tag{48}$$

$$= +\mu \frac{1}{\sqrt{1 + \left(\dfrac{r_p}{X_L}\right)^2}} \left/ \underline{270° - \tan^{-1} \frac{X_L}{r_p}} \right. ; \tag{49}$$

thus

$$A = \mu \, \frac{1}{\sqrt{1 + \left(\dfrac{r_p}{X_L}\right)^2}}, \tag{50}$$

and

$$\theta_A = 270° - \tan^{-1} \frac{X_L}{r_p} = -90° - \tan^{-1} \frac{X_L}{r_p}. \tag{51}$$

Figure 13 shows the manner in which the voltage amplification A and the phase angle θ_A vary with Z_L/r_p for the limiting conditions of pure resistance and pure inductance loads. It is apparent that to produce the same voltage amplification, the impedance of the pure inductance load need not be so great in magnitude as that of the pure resistance load. An additional advantage, for operation at a given operating point on the plate characteristics—that is, for given quiescent-operating conditions—is that the plate-supply voltage E_{bb} for an amplifier with an inductive load may be smaller than for one with a resistive load, because of the steady component of voltage drop in the load resistor. The curve for a load comprising both resistance and reactance falls between the two curves in Fig. 13.

The *amplification factor* μ of the *tube* must be clearly distinguished from the *voltage amplification or gain A* of the amplifier *circuit*. The voltage amplification A approaches μ as a limit as the load impedance is increased; but, in practice, amplifiers with untuned loads are seldom designed to have a value of A greater than about 0.9μ, because of the expense associated with the large inductance or, if a resistance load is used, because of the large plate-supply voltage required to make A more nearly equal to μ.

The curves of Fig. 14 show the form of the variations of the voltage amplification with frequency for the pure resistance and the pure inductance loads. With the resistance load, the voltage amplification is independent of frequency, but, with the inductance load, it is zero for direct current and increases with the frequency. In practice, the curves of voltage amplification as a function of frequency for both types of loads differ from those shown, particularly at high frequencies, because of the effects of such factors as interelectrode capacitance, capacitance between the windings of inductors, and the effective resistance of coils and its change with frequency, which are discussed subsequently.

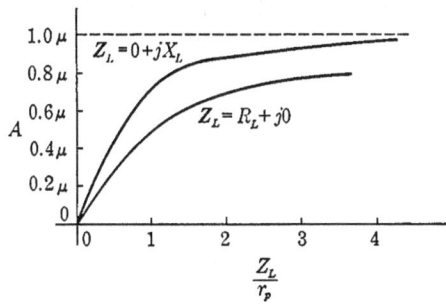

(a) Voltage amplification

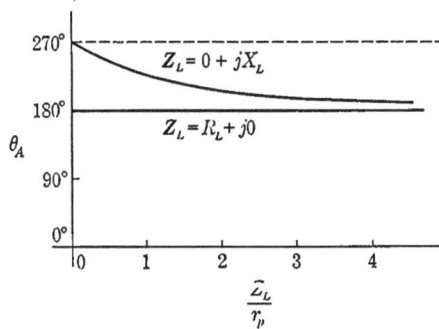

(b) Phase shift

Fig. 13. Amplifier characteristics as functions of the magnitude of the load impedance.

It is often convenient to derive the voltage amplification from the equivalent current source and shunt resistance shown in Fig. 9a. The complex voltage amplification then has the form

$$A = -g_m \frac{r_p Z_L}{r_p + Z_L},$$ ▶[52]

instead of that given in Eq. 40. The complex voltage amplification is thus equal to the mutual conductance multiplied by the impedance composed of the plate resistance and load impedance in parallel. This expression for A is particularly useful if the amplifier tube is a tetrode or pentode, because, as is stated in Arts. 8 and 9, Ch. IV, the plate resistance r_p for tetrodes and pentodes often has a high value—as much as one or two megohms for receiving tubes. A load impedance approximating this value is not, in general, economically practicable, even with tuned loads; in other words, r_p is generally large compared with Z_L in practice. Under these conditions, Eq. 52 reduces to the approximate expression

$$A \approx -g_m Z_L , \qquad \blacktriangleright [53]$$

which is equivalent to the statement that the approximate equivalent plate circuit of the vacuum tube is a constant-current generator of source current $g_m E_{g1}$ with the load impedance connected across its terminals. Equation 53 applies to the lower regions near the origin of the coordinates in Fig. 13a.

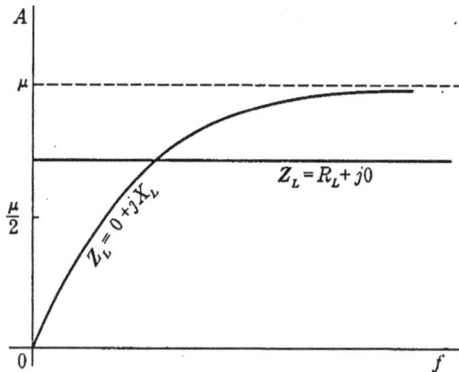

Fig. 14. Variation of the voltage amplification A as a function of frequency.

Because of the difficulty of realizing load impedances equal to or greater than r_p, the enormous voltage amplification that might be expected on the basis of amplification-factor values near or exceeding a thousand in tetrodes and pentodes cannot be realized in practice. Nevertheless, much higher stable values of voltage amplification are obtained from a single tube with these screen-grid tubes than with triodes.

7. VOLTAGE AMPLIFICATION WITH CATHODE IMPEDANCE

When a cathode-bias resistor is used to produce the grid-bias voltage in an amplifier, as in Fig. 5, a varying component e_s at the input causes operation about a quiescent point Q, which may be located on the plate characteristics in the manner described in Art. 3. For small increments in the linear range for the tube, analysis based on the equivalent circuit developed in Art. 5 is applicable. Figure 15a shows the over-all equivalent circuit for such incremental operation.

The tube is replaced by μe_g in series with r_p, and supplies the load R_L through the cathode circuit shown as C_k in parallel with R_k. Only the varying components of the currents and voltages in Fig. 5 are shown in Fig. 15. If the input voltage e_s is sinusoidal, all quantities may be represented by their corresponding complex values. Then, from the circuit,

$$I_p = \frac{\mu E_g}{r_p + Z_k + R_L}, \qquad [54]$$

where Z_k denotes the impedance of the parallel combination of R_k and

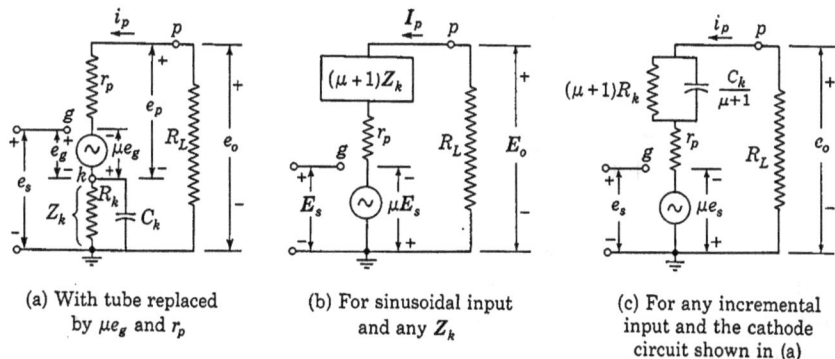

(a) With tube replaced by μe_g and r_p

(b) For sinusoidal input and any Z_k

(c) For any incremental input and the cathode circuit shown in (a)

Fig. 15.   Circuits equivalent for incremental quantities to the combination of the tube and the elements in its cathode circuit.

C_k or *any other impedance* that may be connected between the same points in the circuit. In the grid circuit

$$E_s = E_g + I_p Z_k, \qquad [55]$$

which, substituted in Eq. 54, gives

$$I_p = \frac{\mu E_s}{r_p + (\mu + 1)Z_k + R_L}. \qquad [56]$$

This relation shows that the current is the same as that given by a voltage generator μE_s having an internal impedance $r_p + (\mu + 1)Z_k$ supplying the load R_L, as is illustrated in Fig. 15b. Comparison with Fig. 9b indicates that the effect of a cathode impedance on the incremental quantities is the same as that of adding an impedance $(\mu + 1)Z_k$ to the plate resistance r_p in the equivalent circuit.

For the cathode impedance consisting of R_k and C_k in parallel,

$$(\mu + 1)Z_k = (\mu + 1) \frac{1}{\frac{1}{R_k} + j\omega C_k}, \qquad [57]$$

which may be rearranged as

$$(\mu + 1)Z_k = \cfrac{1}{\cfrac{1}{(\mu + 1)R_k} + j\omega \cfrac{C_k}{(\mu + 1)}} \cdot \qquad [58]$$

This relation expresses the impedance of a resistance $(\mu + 1)R_k$ in parallel with a capacitor $C_k/(\mu + 1)$; thus the equivalent circuit for sinusoidal increments is that shown in Fig. 15c. Since the operation is linear, and the circuit parameters are independent of frequency, this circuit is also applicable for increments having any other waveform.

The complex voltage amplification of an amplifier with a resistance load and a cathode-bias resistor shunted by a capacitor may be determined from Fig. 15c as

$$\frac{E_o}{E_s} = -\frac{\mu R_L}{r_p + R_L + (\mu + 1)R_k} \; \cfrac{1 + j\omega C_k R_k}{1 + j\omega C_k \cfrac{R_k(r_p + R_L)}{r_p + R_L + (\mu + 1)R_k}} \cdot \quad [59]$$

This expression has the form of a frequency-invariant term, which is the amplification for zero frequency or zero capacitance, multiplied by a dimensionless frequency-dependent factor. An alternative form of the expression is

$$\frac{E_o}{E_s} = -\frac{\mu R_L}{r_p + R_L} \; \cfrac{1 + j\omega C_k R_k}{1 + (\mu + 1)\cfrac{R_k}{r_p + R_L} + j\omega C_k R_k}, \qquad [60]$$

which is the complex voltage amplification that would be obtained with a constant grid-bias voltage multiplied by a dimensionless frequency-dependent factor. As the frequency or the capacitance is increased, the imaginary terms in the numerator and denominator of the frequency-dependent factor in Eq. 60 become large compared with the real terms, the factor approaches unity, and the complex voltage amplification approaches as a limit the value that would be obtained with a constant grid-bias voltage. This condition occurs when

$$\omega C_k \gg \frac{1}{R_k} + \frac{\mu + 1}{r_p + R_L}, \qquad [61]$$

in other words, when the reactance of C_k is small compared with the resistance of R_k and $(r_p + R_L)/(\mu + 1)$ in parallel. For most practical purposes, if the reactance of C_k is less than about one-third that resistance, the voltage amplification may be considered essentially equal to the value corresponding to constant grid-bias voltage, because the right-angle relation involved in the complex addition reduces the resulting discrepancy to less than five per cent. The capacitor C_k is called a *by-pass capacitor* because it by-passes the varying component of current around R_k and tends to prevent the development of an appreciable varying component in the voltage across R_k, which would counteract the input voltage.

8. INTERELECTRODE CAPACITANCES AND INPUT ADMITTANCE

In the preceding analysis of the operation of grid-controlled vacuum tubes with an incremental grid-signal voltage, no account has been taken of the effects of capacitances among the different elec-
trodes. Actually, these capaci-
tances may play an important role in the operation of the tube, particularly with rapidly changing pulse voltages, or at high fre-
quencies. For example, they may cause alternating-current power to be supplied from the tube into the *grid* circuit as well as into the plate circuit, thus causing instability and a tendency toward oscillation in an otherwise stable circuit.

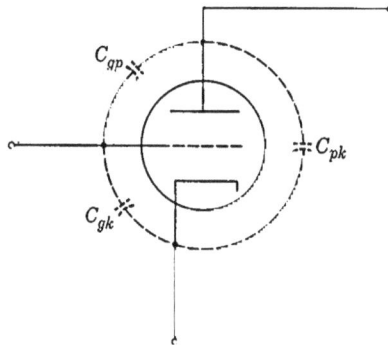

Fig. 16.   Interelectrode capacitances of a triode.

The interelectrode capacitances of a triode are shown in Fig. 16. These capacitances comprise not only the capacitances between the electrodes, but also the capacitances among the leads in the tube. In the actual circuit they are increased by the capacitances among all parts of the circuit connected to the electrodes, and these contributions may often predominate. The interelectrode capacitances are not readily measured individually because none can be isolated from the others. But measurement of the capacitance between each elec-
trode and the other two connected together gives values for the three *pairs* of interelectrode capacitances. Solution of the three simultaneous equations expressing the pair values as sums of individual values then gives the interelectrode capacitances. Because of the effects of space charge on interelectrode capacitances, the values measured with the cathode unheated differ somewhat from those effective under operating conditions.

The incremental equivalent circuit for the triode operating linearly, which is derived in Art. 5, becomes that in Fig. 17 when the interelectrode capacitances are included. As a result of the interelectrode

capacitances, the effect of the grid current often cannot be assumed negligible. Although the effect of the conduction grid current may be negligible, a displacement grid current of appreciable magnitude may occur. This displacement current must pass through the impedance of the source of grid-signal voltage and may, therefore, appreciably modify the magnitude and waveform of the grid-signal voltage when that impedance is large.

The relative importance of the three interelectrode capacitances may be understood from the following qualitative reasoning. The plate-to-

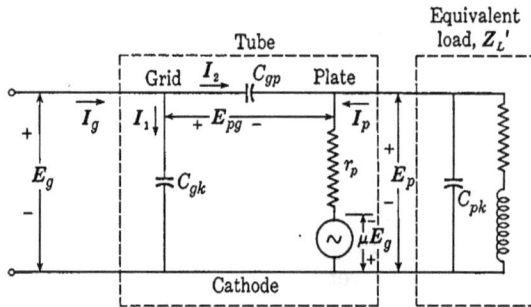

Fig. 17. Incremental equivalent circuit for a triode with the interelectrode capacitances included and C_{pk} considered as part of the effective load impedance Z_L'. For circuit analysis, sinusoidal increments are indicated.

cathode capacitance C_{pk} is directly across the output terminals, and hence acts as if it were a part of the load. The grid-to-cathode capacitance C_{gk} is directly across the input terminals, and hence acts as if it were a part of the source of grid-signal voltage. But the grid-to-plate capacitance C_{gp} is a mutual capacitance that interconnects the grid and plate circuits. In terms of subsequent discussion, it constitutes a feedback path from the plate to the grid, as well as a forward path from the grid to the plate. Because of it, the admittance or impedance effective at the input terminals is influenced by the load. Furthermore, its effect at the input terminals is much larger than would be caused by the same capacitance connected directly across them because the current through it results from the difference between the grid and plate voltages. For a resistance load discussed in Art. 4, an instantaneous increase of grid voltage is accompanied by a decrease of plate voltage—a decrease that is larger by the amount of the voltage amplification. The total voltage across the capacitor is thus larger than the input voltage. Accordingly, the effect of C_{gp} at the input is approximately that of a capacitor of the order of the voltage amplification times C_{gp} connected directly across the input terminals. These concepts underlie the analysis that follows.

For a sinusoidal grid-signal voltage, the current I_2 through the grid-plate capacitance C_{gp} in Fig. 17 is

$$I_2 = E_{pg} j\omega C_{gp},$$ [62]

where E_{pg} is the complex effective value of the voltage rise from the plate to the grid. This voltage is given by

$$E_{pg} = E_g - E_p = E_g - AE_g = E_g(1 - A),$$ [63]

where A is the complex voltage amplification for the amplifier. Consequently

$$I_2 = E_g j\omega C_{gp}(1 - A).$$ [64]

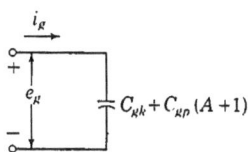

(a) For resistive plate load and low frequencies or small rates of change of voltage

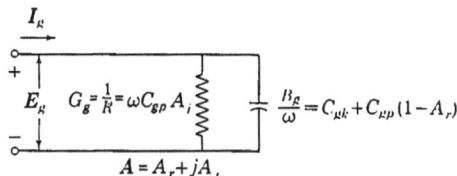

(b) For reactive plate load and any frequency (A_r and A_i are functions of frequency)

Fig. 18. Incremental equivalent input circuits for a triode showing the dependence of the circuit parameters on frequency and voltage amplification.

If the load is a pure resistance and the frequency is so low that the capacitive reactances of C_{gp} and C_{pk} are large compared with the load and plate resistances, these capacitances may be ignored in computing the voltage amplification, and A becomes $A\underline{/180°}$ or $-A$, as is discussed in Art. 6. Then

$$I_2 = E_g j\omega C_{gp}(A + 1).$$ [65]

Thus a capacitance $(A + 1)C_{gp}$ connected directly across the voltage E_g can account for the actual current I_2 through C_{gp} at the input. This apparent capacitance in parallel with the actual capacitance C_{gk} at the input therefore constitutes an incremental equivalent input circuit applicable for a resistance load and low frequencies as is shown in Fig. 18a—the total effective input capacitance is $C_{gk} + (A + 1)C_{gp}$. Since the equivalent circuit contains no frequency-dependent parameters, it also applies for any grid-signal waveform for which $-A$ expresses the voltage amplification. Thus it applies for instantaneous values of grid voltage as long as their rate of change is so small that displacement currents through C_{gp} and C_{pk} are negligible in comparison with the plate current. Accordingly, instantaneous values of grid voltage and current are indicated in Fig. 18a.

The fact that the grid-plate capacitance has the effect of a larger capacitance at the input terminals is known as the Miller effect.[8] In some applications, particularly pulse circuits, a capacitor is deliberately connected between the grid and the plate in order to simulate a much larger capacitance at the input terminals.

If the load contains inductance or capacitance, or if, with a resistance load, the frequency is so high that the capacitive reactances of C_{gp} and C_{pk} affect the magnitude or phase of the complex voltage amplification A, the fact that A is complex in Eq. 64 must be fully recognized. It may be expressed in terms of its real and imaginary parts A_r and A_i, respectively, as

$$A = A_r + jA_i . \tag{66}$$

Then Eq. 64 becomes

$$I_2 = E_g j\omega C_{gp}(1 - A_r - jA_i), \tag{67}$$

and, since

$$I_1 = E_g j\omega C_{gk} , \tag{68}$$

the input or effective grid-cathode admittance of the tube and load is

$$Y_g = G_g + jB_g = \frac{I_g}{E_g} = \frac{I_1 + I_2}{E_g} \tag{69}$$

$$= j\omega C_{gk} + j\omega C_{gp}(1 - A_r - jA_i). \tag{70}$$

Equating separately the real and imaginary parts in Eq. 70 gives for the input conductance

$$G_g = \omega C_{gp}A_i , \tag{71}$$

and for the input susceptance

$$B_g = \omega[C_{gk} + C_{gp}(1 - A_r)]. \tag{72}$$

These relations indicate that the incremental equivalent input circuit for any load and frequency is effectively a resistance equal to the reciprocal of G_g in Eq. 71 connected in parallel with a capacitance equal to the bracketed term in Eq. 72, as is shown in Fig. 18b. Note particularly that, because A_r and A_i are functions of frequency, the resistance and capacitance *are not constant* with frequency.

[8] J. M. Miller, "Dependence of the Input Impedance of a Three-Electrode Vacuum Tube upon the Load in the Plate Circuit," *Nat. Bur. Stand. Sci. Papers,* **15** (1919–1920), No. 351, 367–385.

Strictly, computation of the *voltage amplification* in the foregoing equations should include the effect of both C_{gp} and C_{pk}. Such a computation is lengthy, however, because the circuit contains two generators and several meshes. The resulting expressions for the components of the complex voltage amplification are often too unwieldy to be useful. Consequently, resort is frequently made to an approximate analysis involving combination of C_{pk} with the load, as in Fig. 17b, and neglect of the current through C_{gp} in comparison with the plate current. The complex voltage amplification is then

$$A = \frac{E_p}{E_g} = -\mu \frac{Z_L{}'}{r_p + Z_L{}'} = A_r + jA_i,$$ [73]

where

$$Z_L{}' = R_L{}' + jX_L{}'$$ [74]

is the equivalent load impedance including the effect of C_{pk}. From Eqs. 73 and 74,

$$A_r + jA_i = -\mu \frac{R_L{}' + jX_L}{r_p + R_L{}' + jX_L{}'},$$ [75]

which, when rationalized, yields

$$A_r + jA_i = -\mu \frac{R_L{}'(r_p + R_L{}') + (X_L{}')^2}{(r_p + R_L{}')^2 + (X_L{}')^2} - j\mu \frac{r_p X_L{}'}{(r_p + R_L{}')^2 + (X_L{}')^2}.$$ [76]

When Eq. 76 is substituted in Eqs. 71 and 72, the result for the input conductance is

$$G_g = -\mu\omega C_{gp} \frac{r_p X_L{}'}{(r_p + R_L{}')^2 + (X_L{}')^2},$$ ▶[77]

and for the input susceptance is

$$B_g = \omega \left[C_{gk} + C_{gp}\left(1 + \mu \frac{R_L{}'(r_p + R_L{}') + (X_L{}')^2}{(r_p + R_L{}')^2 + (X_L{}')^2} \right) \right].$$ ▶[78]

In general, the load may be a complicated network, and the reactance component $X_L{}'$ may thus be either positive or negative, depending on the frequency and the type of load. Equation 77 shows that the grid conductance is positive for a capacitive load but is negative for an inductive load. With the inductive load, then, power is supplied from the tube into both the grid and the plate circuits. When viewed from the grid-cathode terminals as well as from the plate-cathode terminals, the tube is not a passive network for alternating currents but is an active network. As such, it is capable of supporting sustained oscillations in the grid and plate circuits, as is discussed in Ch. XI. These results neglect the effect of the capacitance C_{gp} on the voltage amplification. When that effect is included, the results indicate that the input conductance may be negative for a range of inductive reactance, but is positive for values above and below that range.

To illustrate the order of magnitude of the quantities involved and the application of these relationships, the input admittance of a particular triode with a 10,000-cycle-per-second grid-signal voltage and a 30,000-ohm pure resistance or pure reactance load may be computed. The capacitances given by the manufacturer are for the tube alone, and they are here multiplied arbitrarily by a factor of two to account for the socket and wiring capacitances. The proper factor varies widely with the construction of the amplifier. If the equivalent load is a pure resistance—that is, if the actual load has sufficient inductive reactance at the operating frequency to compensate for the reactance of C_{pk} across it and thus to act as a noninductive equivalent load—the quantities involved are:

$$\mu = 20 \qquad\qquad R_L' = 30{,}000 \text{ ohms}$$
$$C_{gk} = 4 \times 2 = 8 \text{ micromicrofarads} \qquad r_p = 10{,}000 \text{ ohms}$$
$$C_{gp} = 1.8 \times 2 = 3.6 \text{ micromicrofarads} \qquad f = 10{,}000 \text{ cycles per}$$
$$C_{pk} = 13 \times 2 = 26 \text{ micromicrofarads} \qquad\qquad \text{second.}$$

Solution: Since the load is first assumed to be a pure resistance, X_L' is zero, and Eq. 77 gives

$$G_g = 0. \tag{79}$$

Equation 78 gives

$$B_g = 2\pi \times 10{,}000 \left[8 \times 10^{-12} + 3.6 \times 10^{-12} \left(1 + 20 \frac{30{,}000}{30{,}000 + 10{,}000} \right) \right] \tag{80}$$

$$= 62{,}800 \times 10^{-12} \, [8 + 3.6(1 + 15)] \tag{81}$$

$$= 62{,}800 \times 10^{-12} \, [65.6] = 4.12 \times 10^{-6} \text{ mho (capacitive susceptance).} \tag{82}$$

Thus the input impedance of the tube is $1/(4.12 \times 10^{-6})$, or 243,000 ohms, and is a capacitive reactance; the equivalent input circuit of Fig. 18a that applies for these particular conditions is a capacitor of 65.6 $\mu\mu$f with zero shunt conductance. Note that this capacitance is about 18 times C_{gp}, or 0.9 μC_{gp}.

On the other hand, if the equivalent load Z_L' is a pure inductive reactance, X_L' equals +30,000 ohms, and Eqs. 77 and 78 give

$$G_g = -20 \times 6.28 \times 10{,}000 \times 3.6 \times 10^{-12} \frac{10{,}000 \times 30{,}000}{(10{,}000)^2 + (30{,}000)^2} \tag{83}$$

$$= -1.36 \times 10^{-6} \text{ mho} \tag{84}$$

and

$$B_g = 2\pi \times 10{,}000 \bigg[8 \times 10^{-12}$$
$$+ 3.6 \times 10^{-12} \left(1 + 20 \frac{(30{,}000)^2}{(10{,}000)^2 + (30{,}000)^2} \right) \bigg] \tag{85}$$

$$= 2\pi \times 10{,}000 \, [76.4 \times 10^{-12}] \tag{86}$$

$$= 4.8 \times 10^{-6} \text{ mho.} \tag{87}$$

Therefore the equivalent input circuit given in Fig. 18b for this tube with a 30,000-ohm inductive load at the particular frequency of 10,000 cps is a *negative* resistance of

$$\frac{1}{1.36 \times 10^{-6}} = 735,000 \text{ ohms} \qquad [88]$$

in parallel with a 76.4 $\mu\mu$f capacitor having a *capacitive* reactance of

$$\frac{1}{4.8 \times 10^{-6}} = 208,000 \text{ ohms,} \qquad [89]$$

and the total input admittance is

$$\sqrt{(1.36)^2 + (4.8)^2} \times 10^{-6} = 4.98 \times 10^{-6} \text{ mho,} \qquad [90]$$

from which the shunt input impedance is

$$\frac{1}{4.98 \times 10^{-6}} = 201,000 \text{ ohms.} \qquad [91]$$

Apparent from the foregoing considerations is the desirability of a small value of C_{gp} in a tube if a high input impedance is to be realized; this is especially true for tubes with a high amplification factor. The negative input conductance that occurs with an inductive load is a source of trouble in certain types of amplifiers and places a limit on the amplification factor and capacitance of the tubes that may be used, because oscillations in the amplifier are likely to occur. Special circuits, called *neutralizing circuits*, and special tubes, the tetrode and pentode types, discussed in Art. 9 of this chapter and in Arts. 8 and 9, Ch. IV, have been developed primarily to avoid the troublesome effects of grid-plate capacitance.

9. INTERELECTRODE CAPACITANCES IN TETRODE AND PENTODE AMPLIFIERS

The capacitance between the control grid and the plate of a grid-controlled vacuum tube may be reduced by a factor of about 1,000—for example, from 10 micromicrofarads to 0.01 micromicrofarad—by the inclusion of a screen grid between the grid and the plate, as described in Art. 8, Ch. IV. This screen grid is so constructed that it almost entirely encloses the plate and reduces to a minimum the electrostatic coupling between the grid and the plate, but the mesh of the screen grid is sufficiently coarse for most of the electrons that approach it to pass through and on to the plate.

The basic circuit connections for a screen-grid pentode amplifier are shown in Fig. 19. The suppressor grid g_3 is connected to the cathode, and the screen grid g_2 is held at a constant positive potential with respect to the cathode by a direct screen-grid voltage supply E_{cc2}. The capacitor C is a by-pass capacitor included to insure a low-impedance path between the screen grid and the cathode for varying components. Its capacitance should be large enough to keep the screen-grid–to–cathode voltage essentially constant despite the changes in screen-grid current that accompany plate current changes. Otherwise the characteristic curves of which those shown in Fig. 19, Ch. IV, are typical are not applicable, and the large values of amplification factor and plate resistance that correspond to them do not occur. The series resistance R_S in Fig. 19 represents the internal impedance of the screen-grid voltage supply plus additional resistance that is often

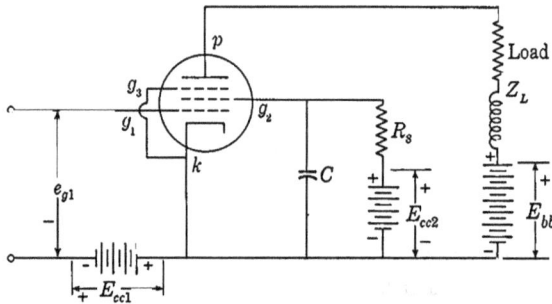

Fig. 19.   Schematic circuit diagram for a pentode amplifier.

included and adjusted so that the voltage drop in it caused by the average value of the screen-grid current corresponding to the desired operating conditions in the grid and plate circuits brings the screen-grid–to–cathode voltage to the required value. To permit supply of all electrode voltages from one source, E_{cc2} is frequently made identical with E_{bb}, and the grid-bias voltage E_{cc1} is obtained by a cathode-bias resistor and by-pass capacitor as is explained in Art. 3.

A partial equivalent circuit for the incremental components in Fig. 19 and linear Class A_1 operation is shown in Fig. 20a. With the introduction of the screen and suppressor grids, the number of interelectrode capacitances among all five electrodes becomes ten. Nine are included on the diagram. The tenth, C_{g3k}, does not appear because the suppressor grid is connected to the cathode. The by-pass capacitance C and series resistance R_S appear in parallel with the screen-grid–to–cathode capacitance, C_{g2k}, and the screen-grid–to–suppressor-grid capacitance C_{g2g3}. Actually, in such a multi-electrode tube there are transconductances, mu-factors, and electrode resistances corresponding to g_m, μ, and r_p, respectively, for each pair of electrodes. Except for μ and r_p these are neglected in Fig. 20a. When the capacitance C of the by-pass capacitor is so large that the incremental voltage developed across it is made negligible, the screen grid is effectively short-circuited to the cathode for incremental components, and the equivalent circuit becomes that shown in Fig. 20b. Since the grid g_1 is at a negative potential, and the incremental voltages between grid g_2 and g_3 and the cathode are zero for the conditions of Fig. 20b, all the transconductances and mu-factors except g_m and μ are zero, and all the electrode resistances except r_p are infinite. This equivalent circuit also applies for the tetrode or beam-power tube as an amplifier if the

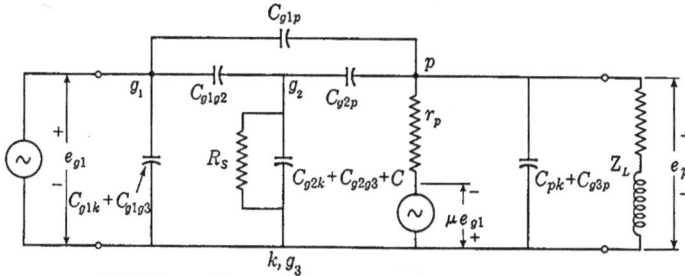

(a) With small by-pass capacitance C in Fig. 19

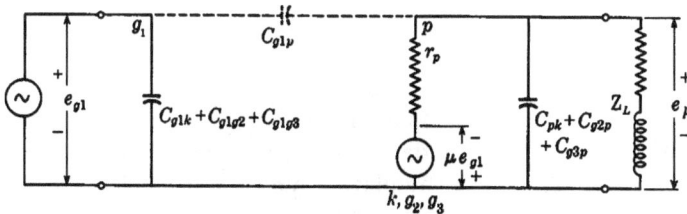

(b) With large by-pass capacitance C in Fig. 19

Fig. 20. Incremental equivalent circuits for a pentode amplifier. The transconductances involving the screen grid are neglected in (a).

interelectrode capacitances between the suppressor grid g_3 and the other electrodes are omitted from the diagram.

A comparison of the equivalent circuits for the triode in Fig. 17 and for the screen-grid tube in Fig. 20b shows a considerable degree of similarity. But an essential difference is the order of magnitude of the control-grid–to–plate capacitance C_{g1p}. This capacitance is reduced to a very small value in the tetrode or pentode because the screen grid surrounds the control grid and shields it from the plate; consequently it is shown dotted in Fig. 20b. Through this reduction, the screen grid mitigates the undesirable reactions mentioned previously in Art. 8. It should be emphasized, however, that the mere substitution of a screen-grid tube for a triode in an amplifier will not alone, in all probability, effect a great difference in reactions, because capacitances among connecting wires and the several circuit elements constitute a source of trouble almost equal in severity to that caused by the control-grid–to–plate capacitance. As this capacitance is made smaller, the interwiring capacitances grow in importance. Thus the correct use of screen-grid tubes presupposes electrostatic shielding between the entire control-grid and plate circuits. Tubes with all the electrode leads brought out through one end, called "single-ended" tubes, are superior to those in which the control-grid lead is brought out through the end opposite to the other leads, because the wiring can be made shorter and hence can be better shielded.

Although introduction of the screen grid decreases the control-

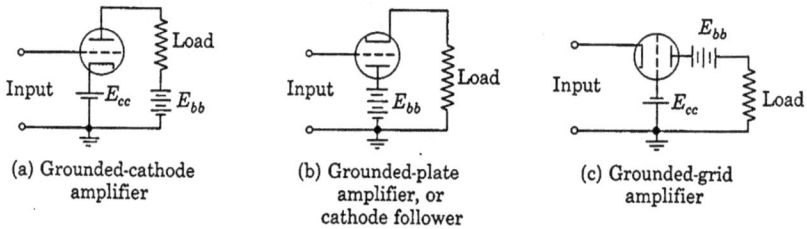

(a) Grounded-cathode amplifier

(b) Grounded-plate amplifier, or cathode follower

(c) Grounded-grid amplifier

Fig. 21.   Basic amplifier connections.

grid–to–plate capacitance, it tends to increase the capacitance at the input because of the capacitance between the control grid and the screen and suppressor grids. With the screen grid by-passed to the cathode, the three interelectrode capacitances from the control grid to all other electrodes except the plate appear in parallel at the input terminals as in Fig. 20b, and constitute the so-called *input capacitance* frequently given as data by tube manufacturers. Likewise the sum of the interelectrode capacitance from the plate to all other electrodes except the control grid constitutes the *output capacitance*, which appears across the load. Strictly, the effect of C_{g1p} in accordance with the considerations of Art. 8 must be added to $C_{g1k} + C_{g1g2} + C_{g1g3}$ to determine the effective input admittance. Practically, however, even with the large voltage amplification that can be realized with a screen-grid tube, the control-grid–to–plate capacitance C_{g1p} is so small that its effect at the input is often negligible, and the amplifier behavior may be determined from the equivalent circuit in Fig. 20b with that capacitance neglected.

10. CATHODE FOLLOWER

Thus far the discussion has concerned an amplifier circuit in which the cathode terminal is common to both the input and the output circuits. Such a connection, illustrated in Fig. 21a, is sometimes called the "grounded-cathode amplifier" because, for convenience in shielding parts of the circuit from one another and in supplying a number of tubes from one voltage source, the common cathode terminal is often connected to ground (through a bias voltage, which plays no part in the incremental behavior). Two other possibilities exist for amplifier connections in which the grid is connected to the input and the plate to the output. They are the "grounded-plate amplifier" in Fig. 21b, otherwise known as the *cathode follower*, and the "grounded-grid

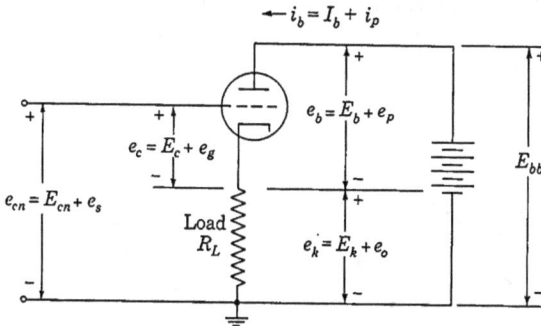

Fig. 22.　Basic cathode-follower amplifier circuit.

amplifier" in Fig. 21c. Each has special advantages, which are discussed in this article and the next. Three other possibilities for connection of an amplifier tube exist. They result from interchange of the input and the output in the three connections of Fig. 21, and are of little practical importance at present.

Redrawn with the tube in the conventional orientation, the cathode-follower amplifier appears as in Fig. 22. An increase of the grid-to-ground voltage e_{cn} tends to increase the plate current i_b and, consequently, to increase the cathode-to-ground voltage e_k. Thus, the direction of the cathode potential change *follows* that of the grid— actually, in normal operation, the cathode potential remains above that of the grid.

To find a relation between the plate current i_b and the input voltage e_{cn}, a graphical construction on the plate characteristics as shown in Fig. 23a is applicable. Although the load R_L appears in the cathode circuit, it is in series with the tube and plate-supply voltage E_{bb} just as in an ordinary amplifier. Hence,

$$e_b = E_{bb} - i_b R_L , \qquad [92]$$

and the path of operation on the plate characteristics is a straight line that may be constructed in the same manner as is discussed for the grounded-cathode amplifier in Art. 2. The current that corresponds to any particular value of input voltage e_{cn} is not explicitly shown by this diagram—a scale of e_{cn} along the load line is lacking. But the input voltage e_{cn} that corresponds to any particular value of current i_b or output voltage e_k is readily found because

$$e_{cn} = e_c + i_b R_L = e_c + e_k , \qquad [93]$$

and values of e_c and i_b appear on the diagram. To determine points on a graphical relation between i_b, or e_k, and e_{cn}, therefore, a suitable procedure is to select a value of e_c, read from the diagram the corresponding values of i_b and $i_b R_L$, or e_k, and compute e_{cn} from Eq. 93.

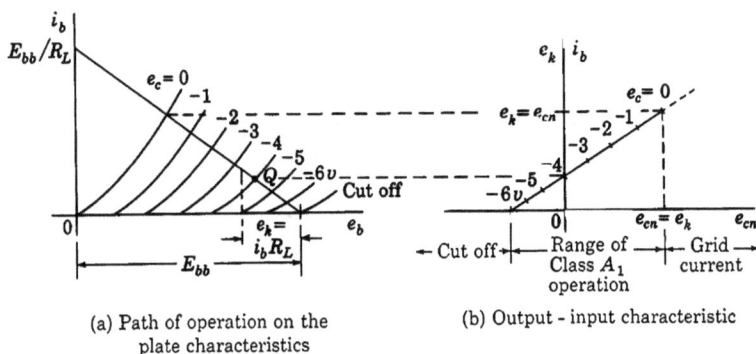

(a) Path of operation on the   (b) Output - input characteristic
plate characteristics

Fig. 23.   Graphical analysis of cathode-follower operation.

The graphical relation that results is shown in Fig. 23b. Transfer of values from the load line in Fig. 23a provides a scale of e_c along the curve. The diagram shows that as e_{cn} increases from zero, the grid-to-cathode voltage e_c approaches zero. When it reaches zero, the output voltage equals the input voltage in accordance with Eq. 93. As e_{cn} decreases from zero, the current and output voltage decrease and approach zero. They reach zero when e_c equals the cut-off grid-bias voltage, and e_{cn} then likewise equals this cut-off voltage. For Class A_1 operation, e_{cn} must remain between those limits, as is indicated in Fig. 23b.

For applications of the cathode follower that involve only increments from the values of current and voltages that exist prior to introduction of a signal voltage at the input, an analysis of the incremental behavior of the circuit is useful. To this end, each total voltage and the total plate current on Fig. 22 is expressed as the sum of a quiescent value, denoted by a capital letter, and an incremental component. Thus, e_s is the increment in the input voltage, or the *input signal voltage*, and e_o and i_p are the increments in the output voltage and the current caused by it. Equation 19 in Art. 5,

$$i_p = g_m e_g + \frac{e_p}{r_p},$$ [94]

gives the relation among the incremental components of the current and voltages at the tube terminals. But an expression relating the current increment i_p to the voltage increments e_s and e_o at the input

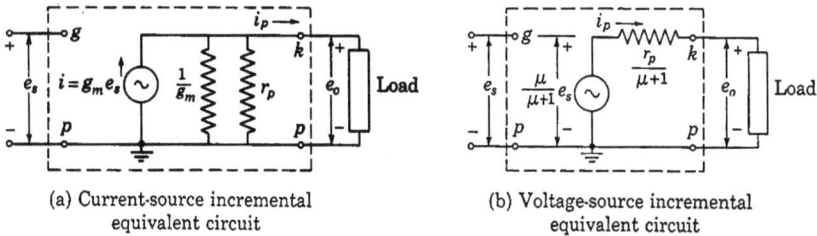

(a) Current-source incremental
equivalent circuit

(b) Voltage-source incremental
equivalent circuit

Fig. 24.   Incremental equivalent circuits for linear Class A_1
operation of a cathode follower.

and output terminals, respectively, is needed. To find it, e_g and e_p may be expressed in terms of e_s and e_o. Thus, in the grid circuit,

$$e_g = e_s - e_o,$$ [95]

and in the plate circuit,

$$e_p = -e_o.$$ [96]

Substitution of Eqs. 95 and 96 in Eq. 94 gives the desired relation,

$$i_p = g_m e_s - g_m e_o - \frac{e_o}{r_p}.$$ [97]

Synthesis of a linear circuit that satisfies this expression results in the circuit in the dotted rectangle of Fig. 24a, which is therefore the current-source equivalent circuit for linear incremental operation of the cathode-follower amplifier.

An alternative form of the equivalent circuit is found through multiplying all terms in Eq. 97 by r_p, which gives

$$i_p r_p = g_m r_p e_s - g_m r_p e_o - e_o.$$ [98]

Since $g_m r_p$ equals μ,

$$i_p r_p = \mu e_s - \mu e_o - e_o,$$ [99]

or

$$e_o = \frac{\mu}{\mu + 1} e_s - \frac{r_p}{\mu + 1} i_p.$$ [100]

This expression suggests the alternative voltage-source form of the equivalent circuit shown in Fig. 24b. The parallel combination of resistances $1/g_m$ and r_p in Fig. 24a may also be replaced by the equivalent value, $r_p/(\mu + 1)$, if desired.

Note that, for a resistance load, the output voltage e_o in Figs. 24a and 24b increases when the input voltage e_s increases, which agrees with the physical explanation above that the cathode follows the grid in potential changes. The reference directions of the sources and currents in Fig. 24 correspond to this in-phase operation and are opposite to those for the grounded-cathode amplifier circuit shown in Fig. 9.

Two important features of the cathode follower are apparent from the equivalent circuit. First, its voltage amplification, e_o/e_s, approaches $\mu/(\mu + 1)$ as the load impedance becomes very large. Thus, the maximum voltage amplification obtainable is somewhat less than unity. The cathode follower is, therefore, not useful as a voltage amplifier. Second, its internal output impedance, which the load faces, is $r_p/(\mu + 1)$ and approaches $1/g_m$ as μ becomes large compared with unity. This small output impedance makes the cathode follower useful for supplying a low-impedance load, because it permits closer matching of the load and internal impedances, a requirement for maximum power sensitivity explained in Art. 15. The cathode follower hence finds extensive application as an impedance transformer between a source having high internal impedance and a load having low impedance. Even though it lacks voltage amplification, the cathode follower can provide power amplification because the ratio of its input impedance to its output impedance is very large.

Suddenly applied increments of voltage at the input of a cathode follower can cause appreciable nonlinear operation even though they be small enough to result in linear operation if they were applied slowly. Such nonlinear behavior arises chiefly from the stray interelectrode capacitance or other circuit capacitance across the load. If the input voltage changes suddenly, the output voltage e_k across this capacitance tends to remain constant momentarily because a capacitor cannot discharge instantly—infinite current would be required. Thus, the operating point moves vertically up or down from its quiescent position on the load line in Fig. 23a. The tube plate current is, therefore, driven to zero by even a relatively small negative step of input voltage. Thereafter the output voltage e_k decays toward zero exponentially with time with a time constant determined only by the capacitance and the load resistance. As e_k decreases, conduction begins in the tube again, and the decay continues asymptotically toward a final value on the original load line, but with a new time constant determined by the capacitance and the parallel combination of the load resistance and $r_p/(\mu + 1)$. Actually, when the grid-to-cathode capacitance in the tube is comparable with the capacitance across the load, an additional component in the output voltage results from the effect of the two acting as a capacitive voltage divider.

A relatively small positive step of input voltage, e_{cn}, on the other hand, increases the grid voltage e_c to positive values, so that a large plate current and an appreciable grid current occur. The grid current produces a voltage drop in the internal impedance of the source of the

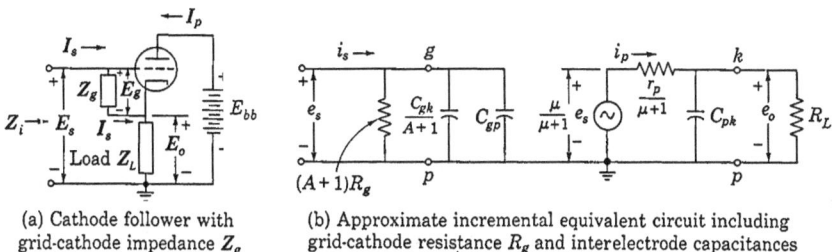

(a) Cathode follower with grid-cathode impedance Z_g

(b) Approximate incremental equivalent circuit including grid-cathode resistance R_g and interelectrode capacitances

Fig. 25.   Incremental input and output equivalent circuits of cathode follower.

input voltage that reduces the actual grid-to-ground voltage—clipping of the input wave occurs. For this reason, and because of the practical limit on the cathode current by the electron emission of the cathode, the load current does not increase linearly with the input voltage. The rate of rise of the output voltage can be no larger than the rate at which this maximum available cathode current can charge the capacitance that is in parallel with the load. Consequently, the behavior of the cathode follower is often unsymmetrical in response to input steps that are equal in magnitude, but opposite in sign.

The input impedance of the cathode follower, which is the impedance the source of input voltage faces, may be determined from a consideration of Fig. 25a. In this diagram, the impedance shown between the grid and the cathode, Z_g, may represent not only an actual circuit element connected between those points, such as a grid resistor, but also the reactance of the grid-to-cathode interelectrode capacitance C_{gk}, or the grid-to-cathode conductance associated with a conduction grid current, or any combination of the three.

In terms of the complex values of the incremental components of the voltages and currents for linear operation and a sinusoidal input signal, the input impedance is

$$Z_i = E_s/I_s .$$ [101]

But

$$I_s = E_g/Z_g ,$$ [102]

so that

$$Z_i = Z_g \frac{E_s}{E_g} = Z_g \frac{E_o + E_g}{E_g} \qquad [103]$$

or

$$Z_i = Z_g \left(\frac{E_o}{E_g} + 1\right). \qquad [104]$$

If the circuit parameters and the frequency are such that the magnitude of the input current I_s is small compared with that of I_p, the current in the load Z_L is then only I_p, and E_o/E_g may be interpreted as the voltage amplification of the tube and load that would occur if they were connected as an ordinary grounded-cathode amplifier. That is,

$$\frac{E_o}{E_g} = A = \frac{\mu Z_L}{r_p + Z_L}, \qquad I_s \ll I_p . \qquad [105]$$

Then, from substitution of Eq. 105 in Eq. 104,

$$Z_i = Z_g(A + 1). \qquad [106]$$

It should be emphasized that here A is *not* the voltage amplification of the cathode follower itself, but is the much larger value for the ordinary amplifier connection. Thus, Eq. 106 shows that the input impedance of the cathode follower is generally much larger than the impedance connected from grid to cathode. This is an important feature, for it permits use of the cathode follower when the impedance of the source is so high that the relatively low input impedance of a grounded-cathode amplifier would cause excessive voltage drop in the internal impedance of the source and thereby render such an amplifier useless.

When Z_g represents the grid-to-cathode interelectrode capacitance,

$$Z_g = \frac{1}{j\omega C_{gk}}, \qquad [107]$$

and

$$Z_i = \frac{1}{j\omega C_{gk}} (A + 1) = \frac{1}{j\omega \dfrac{C_{gk}}{A + 1}}. \qquad [108]$$

This relation shows that when the load can be considered a pure resistance so that A has no imaginary component, the effect of C_{gk} is the same as that of a new and smaller capacitance $C_{gk}/(A + 1)$ connected directly across the input terminals. Similarly, a resistance R_g connected between the grid and cathode is equivalent to $R_g(A + 1)$

across the input terminals. A complete equivalent circuit for the cathode follower including interelectrode capacitance and a grid resistor is then as shown in Fig. 25b. Note, however, that because of C_{pk} across the load, this representation applies strictly only for frequencies or rates of change of voltage so small that the currents through C_{pk} and C_{gk} are both small compared with the plate current. Comparison of Fig. 25b with Fig. 18a indicates that the effective input capacitance of the cathode follower is less than that of an ordinary amplifier by a factor $A + 1$ when each has the same resistance load.

A pentode is seldom used as such in a cathode follower, because the desired pentode behavior is not obtained unless the screen-grid potential remains fixed with respect to the *cathode*. If the screen-grid voltage is supplied by a voltage divider across the voltage E_{bb}, the screen-grid potential remains fixed with respect to *ground* and the *plate* in the cathode follower. Hence, the tube then behaves essentially as a triode-connected pentode. On the other hand, addition of a large by-pass capacitance between the screen grid and the cathode will maintain the voltage between them essentially constant during rapid changes of the input voltage as discussed in Art. 9. The tube then behaves as a pentode, but the resistance of the network supplying the screen grid appears across the load resistance of the incremental equivalent circuit because the by-pass capacitance is essentially a short circuit for incremental quantities.

11. GROUNDED-GRID AMPLIFIER

The grounded-grid amplifier, illustrated in Fig. 21c, has the advantage over the other two connections that the grounded grid acts as an electrostatic screen between the input and the output circuits just as the screen grid does in a tetrode or pentode amplifier. It thereby helps to prevent unwanted oscillations or instability. For this reason and the fact that triodes are found to introduce less random noise than pentodes, the grounded-grid amplifier is often used in high-frequency applications. A further useful feature of the circuit apparent from the considerations that follow is that the grounded-grid amplifier does not cause a phase reversal between the input and output voltages with respect to their common terminal as does the grounded-cathode amplifier.

Since the path for the plate current in a grounded-grid amplifier is through the source of input voltage as well as through the load, the internal impedance of the input source plays an important part in the circuit operation and should be included in analysis of it. Figure 26a shows a source having an electromotive force or open-circuit voltage

e_s and internal impedance R_s supplying the incremental input voltage e_g for such an amplifier. The basic incremental equivalent circuit obtained through replacing the tube by μe_g in series with r_p and

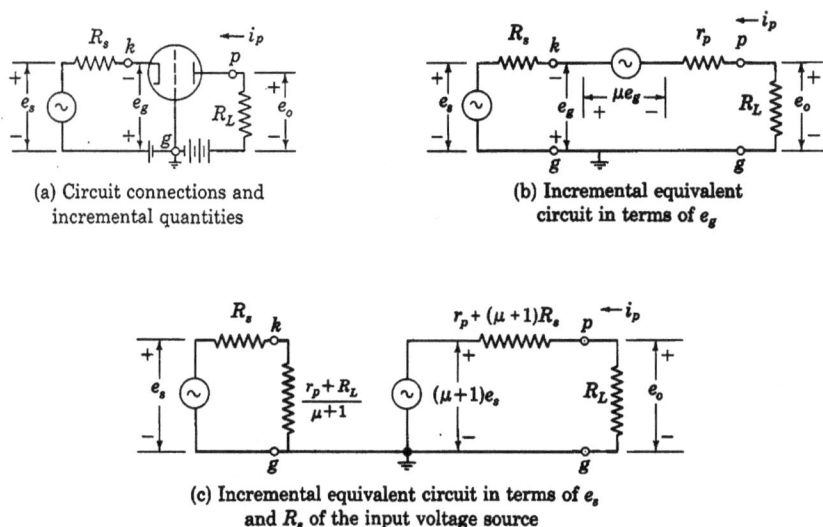

(a) Circuit connections and
incremental quantities

(b) Incremental equivalent
circuit in terms of e_g

(c) Incremental equivalent circuit in terms of e_s
and R_s of the input voltage source

Fig. 26.   Grounded-grid amplifier driven by voltage source having electromotive force e_s and internal impedance R_s

replacing all direct voltages by short circuits is given in Fig. 26b. The incremental plate current in this circuit is

$$i_p = \frac{\mu e_g - e_s}{r_p + R_L + R_s}. \qquad [109]$$

But in the grid circuit,

$$e_g = -e_s - i_p R_s . \qquad [110]$$

Substitution of Eq. 110 in Eq. 109 to eliminate e_g gives

$$i_p = -\frac{(\mu + 1)e_s}{r_p + (\mu + 1)R_s + R_L}. \qquad [111]$$

This relation suggests that the load is effectively supplied by a voltage source having an electromotive force $(\mu + 1)e_s$ and internal output impedance $r_p + (\mu + 1)R_s$ as shown in the right half of Fig. 26c. The voltage amplification in this circuit approaches $\mu + 1$ as the load resistance approaches infinity, and a load resistance equal to $r_p + (\mu + 1)R_s$ therefore provides maximum power sensitivity.

From Eq. 111, the impedance faced by the electromotive force e_s is given by

$$\frac{e_s}{-i_p} = R_s + \frac{r_p + R_L}{\mu + 1}. \qquad [112]$$

Thus the input impedance presented to the input source at the grid-cathode terminals is $(r_p + R_L)/(\mu + 1)$, as is shown at the left half of Fig. 26c.

12. WAVEFORM DISTORTION IN AMPLIFIERS

Although, as is stated earlier, the purpose of an amplifier is to produce an enlarged *reproduction* of the input wave, the reproduction in a practical amplifier is never exact. Imperfections in the amplifier always cause some differences between the waveforms of the output and the input—differences that are termed *distortion*. Distortion arises generally from causes associated either with the electronic phenomena within the tube, or with the characteristics of the networks inherent in the amplifier. An example of distortion resulting from the electronic phenomena is that caused by nonlinearity between the instantaneous values of the input wave and the output current. This type of distortion, which is discussed in Art. 13, is called *nonlinear distortion*. Other examples are the distortion resulting from the effects of nonlinearity in the relation between the grid current and the grid voltage, and from the dependence of the electron transit time on frequency.

For many applications, the distortion in an amplifier is conveniently described in terms of the response of the amplifier to a periodic input wave. When the input waveform is periodic, both it and the output waveform may be represented by Fourier series. In order that the output waveform be the same as that of the input, the amplifier must preserve among the Fourier components in the output the same relative amplitudes and the same relative time-phase differences as exist among the corresponding components in the input. Distortion therefore results if either the transmission of the amplitudes or the time-phase shift through the amplifier is dependent on either the amplitude or the frequency of the input components. Hence, four kinds of distortion may be recognized. These are describable in terms of the magnitude A and phase shift θ_A of the complex voltage amplification in Eq. 39. If A varies with the amplitude of the input wave, the amplitudes of the various components in the output do not bear to one another the same ratios that the corresponding components in the input do, and *amplitude distortion* is said to occur. This type of distortion is a form of nonlinear distortion, and is always accompanied by the production of new Fourier components, as is discussed in Art. 13. It is also accompanied by *intermodulation distortion*, discussed in Art. 11, Ch. XII. If A varies with the frequency of the input wave, *attenuation distortion or frequency distortion* is said to occur.

The angle θ_A in Eq. 39 is the shift in phase for each Fourier component as it passes through the amplifier measured in angular units. It is the shift in phase measured in time units multiplied by the angular frequency ω of the component. Thus if θ_A/ω is the same for all components, the phase of each is changed by the same amount *in time*, and the relative phase relations among the components are preserved as they pass through the amplifier. The waveform is therefore unchanged. This condition, although sufficient, is not necessary for preservation of the waveform, however, because, if all the Fourier components are shifted in phase by an integral multiple of 180 degrees, the waveform is not changed; the polarity of the complete wave may be altered, but polarity changes are not regarded as distortion. Hence, a criterion for distortion-free amplification is that the curve of phase-shift angle θ_A as a function of ω, called the *phase characteristic*, be a straight line intersecting the axis of θ_A at some multiple of π radians. If the phase characteristic departs from either of these ideal conditions, so-called *phase distortion* results. One type of phase distortion, termed *delay distortion*, occurs if the slope $d\theta_A/d\omega$ is not constant; and another type, termed *intercept distortion*, occurs if the curve does not intersect the axis of θ_A at a multiple of π radians. It has been shown[9] that often the effects of intercept distortion may be neglected, and the condition that $d\theta_A/d\omega$ equal a constant is ordinarily taken as being sufficient to insure satisfactory phase relations among the components of the output wave. If the phase-shift angle θ_A varies with the amplitude of the input, the resulting distortion is a type of phase modulation, which is discussed in Ch. XII.

13. WAVEFORM DISTORTION DUE TO NONLINEARITY OF THE TUBE CHARACTERISTICS

Since the characteristic curves of a tube are never straight, no amplifier is linear in the strict sense of the word. In practice, however, it is usually possible to design an amplifier in which the effects of the nonlinearity in the tube are negligible for a particular application, and the adjective "linear" in this chapter is meant to imply that condition. The degree of nonlinearity permissible depends on the particular application. This article discusses methods of determining the

[9] E. A. Guillemin, *Communication Networks*, Vol. II (New York: John Wiley & Sons, Inc., 1935), 490–491.

effects of the nonlinearity when they are small. Larger effects of nonlinearity are discussed in subsequent chapters.

Figure 27 shows the type of waveform distortion that often occurs as a result of nonlinearity of the tube characteristics. For simplicity, a pure resistance load is assumed, and the grid-signal voltage is assumed to be sinusoidal and to be given by

$$e_g = \sqrt{2}E_g \cos \omega t. \qquad [113]$$

The effects of interelectrode and stray capacitance are neglected in this analysis. In Fig. 27, both the plate characteristics and the transfer

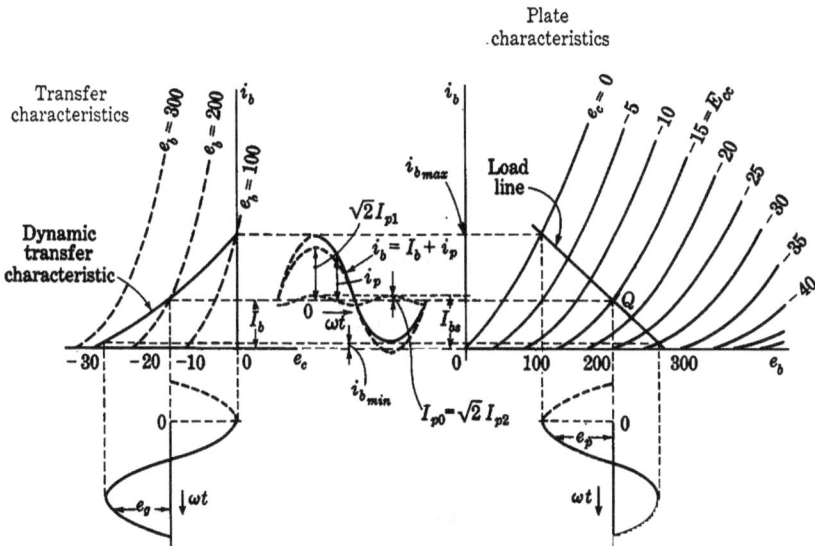

Fig. 27.   Distortion of waveform by nonlinearity of the tube characteristic curves.

characteristics are shown, the latter being obtained, by graphical construction, from the former. The path of the operating point across the transfer characteristics, which is the relation between i_p and e_g for the combined tube and load, is called the *dynamic transfer characteristic* and is also readily determined by the graphical method indicated. The term "dynamic" is included to indicate that the curve is the operating path for rapidly changing increments in the current and voltage. When the load is a pure resistance, as is assumed in Fig. 27, this dynamic path coincides with the locus for different steady values of the increments i_p and e_g. But when, as is considered later, the load is coupled to the tube through a large capacitance or a transformer, such coincidence does not occur. The dynamic transfer characteristic is then the operating path for rapidly changing increments only. This path is not straight, even for a resistance load, when the amplitude of the grid-signal voltage is large; the slope of the path increases for increasing grid voltages.

The dynamic transfer characteristic can be represented by an infinite power series, and, for a first approximation, sufficient accuracy may often be obtained from the first two terms of such a series,

$$i_p = ae_g + be_g{}^2 \cdots . \qquad [114]$$

Here a and b are constants determinable from points on the curve by the methods discussed in Art. 8, Ch. XII. Substitution of Eq. 113 into Eq. 114 gives

$$i_p = a\sqrt{2}E_g \cos \omega t + b(\sqrt{2}E_g \cos \omega t)^2 \qquad [115]$$

$$= a\sqrt{2}E_g \cos \omega t + 2bE_g{}^2 \tfrac{1}{2}(1 + \cos 2\omega t) \qquad [116]$$

$$= bE_g{}^2 + a\sqrt{2}E_g \cos \omega t + bE_g{}^2 \cos 2\omega t, \qquad [117]$$

or

$$i_p = I_{p0} + \sqrt{2}(I_{p1} \cos \omega t + I_{p2} \cos 2\omega t). \qquad [118]$$

Thus a Fourier series comprising a constant term, a fundamental, and a second harmonic describes the waveform when the dynamic transfer characteristic is representable by the quadratic expression, Eq. 114. Furthermore, the constant term is then equal to the amplitude of the second harmonic. Additional terms in the power series used to represent the dynamic characteristic give higher harmonic terms in the Fourier series for i_p.

The values of I_{p0}, I_{p1}, and I_{p2} can be determined from the expression for the dynamic transfer characteristic if the constants in it are available. However, the dynamic transfer characteristic is a function of the tube and circuit combined; hence it must be plotted point by point for each load resistance. It is therefore usually more convenient to determine the harmonic amplitudes from the plate characteristics upon which the path of operation is a readily constructed straight line. The three terms in the Fourier series are sketched in Fig. 27. The plate current goes through its maximum value when ωt equals zero, and its minimum value when ωt equals π. Thus, by substitution of these angular values in Eq. 118 and use of Eq. 12,

$$i_{b_{max}} = I_b + I_{p0} + \sqrt{2}I_{p1} + \sqrt{2}I_{p2} \qquad [119]$$

and

$$i_{b_{min}} = I_b + I_{p0} - \sqrt{2}I_{p1} + \sqrt{2}I_{p2} . \qquad [120]$$

Hence

$$\sqrt{2}I_{p1} = \frac{i_{b_{max}} - i_{b_{min}}}{2}, \qquad [121]$$

and, since Eq. 117 shows that $\sqrt{2}I_{p2}$ and I_{p0} are equal,

$$\sqrt{2}I_{p2} = I_{p0} = \tfrac{1}{2}\left[\frac{i_{bmax} + i_{bmin}}{2} - I_b\right]. \qquad [122]$$

The value of the steady component and the amplitudes of the fundamental and second-harmonic components of the plate current are given by Eqs. 121 and 122 in terms of quantities readily determined from the graphical construction using the plate characteristics shown in Fig. 27. A similar method is applicable to the plate-voltage wave, since, because of the resistance load, it is a reproduction of the current waveform. The average plate current as indicated by a direct-current ammeter increases by an amount I_{p0} when the signal voltage is applied, because of rectification in the plate circuit, and I_{p0} is known as the *rectified component of the plate current*. The average plate current in the presence of the signal is then

$$I_{bs} = I_b + I_{p0}, \qquad [123]$$

and the subscript s is similarly used to denote the "with-signal" average component of other electrode currents and voltages. In accordance with Eqs. 117, 122, and 123, the increase in the direct-current ammeter indication gives a measure of the amount of second-harmonic generation.

The per cent second harmonic in either the plate current or the plate voltage may be written as

$$\text{Per cent second harmonic} = \frac{\sqrt{2}I_{p2}}{\sqrt{2}I_{p1}} \times 100 \qquad [124]$$

and is therefore given by

$$\text{Per cent second harmonic} = \frac{\tfrac{1}{2}(i_{bmax} + i_{bmin}) - I_b}{(i_{bmax} - i_{bmin})} \times 100. \blacktriangleright[125]$$

The amount of harmonic generation that may be tolerated in amplifiers for sound reproduction depends on the sensitivity of the ear to waveform distortion. A value of 10 per cent total harmonic generation is commonly used as the permissible upper limit.[10]

The foregoing analysis was made on the assumption of a quadratic form of the dynamic characteristic, which is found to be equivalent to the assumption that no harmonics higher than the second exist in the plate current. Although this is generally a good approximation for a

[10] *Standards on Radio Receivers: Methods of Testing Frequency-Modulated Broadcast Receivers* (New York: The Institute of Radio Engineers, 1947), 2, 10–11; F. Massa, "Permissible Amplitude Distortion of Speech in an Audio Reproducing System," *I.R.E. Proc.*, *21* (1933), 682–689; M. G. Lloyd and P. G. Agnew, "Effect of Phase of Harmonics upon Acoustic Quality," *Nat. Bur. Stand. Sci. Paper No. 127*, *6* (1909), 255–263.

triode with a resistance load, it is not true for pentodes and some other tubes. A more general treatment of harmonic generation suitable for the determination of all the harmonics that may exist follows.

Since the plate current retraces the same waveform during each cycle, it is representable by a Fourier series that in general contains sine and cosine terms of all frequencies. However, if the origin for ωt is chosen at the point when the plate current is a maximum, if the

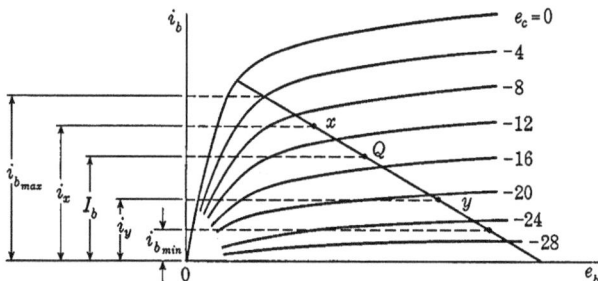

Fig. 28.   Points for determination of the harmonic generation in a pentode amplifier.

waveform of the grid-signal voltage e_g is sinusoidal, and if the load is pure resistance,

$$i_p(\omega t) = i_p(-\omega t);$$ [126]

that is, the curve is[11] an even function, symmetrical along the rising and falling portions, and is therefore representable by a Fourier series containing cosine terms only.[12] Thus

$$i_p = I_{p0} + \sqrt{2}(I_{p1} \cos \omega t + I_{p2} \cos 2\omega t + I_{p3} \cos 3\omega t + \cdots).$$ [127]

When no harmonics higher than the nth exist, the steady term and the harmonics up to the nth may be obtained by the Fischer-Hinnen or selected-ordinate method.[13] The half-cycle is divided into n equal intervals, and the $n + 1$ ordinates at the boundaries of these intervals, together with the corresponding values of ωt, are substituted in Eq. 127. The $n + 1$ simultaneous algebraic equations are then solved for the harmonic amplitudes in terms of the measured ordinates.

An example of considerable importance is the method of determining the third-harmonic generation in a pentode amplifier with a

---

[11] P. Franklin, *Differential Equations for Electrical Engineers* (New York: John Wiley & Sons, Inc., 1933), 48.

[12] See Franklin, reference 11, 64.

[13] J. Fischer-Hinnen, "Methode zur schnellen Bestimmung harmonischer Wellen," *E.T.Z.*, 22 (1901), 396–398; F. A. Laws, *Electrical Measurements* (2nd ed.; New York: McGraw-Hill Book Co., Inc., 1938), 687–695.

resistance load. The plate characteristics, together with the load line, are shown in Fig. 28, and the grid-signal voltage e_g is assumed to be representable by

$$e_g = \sqrt{2}E_g \cos \omega t. \qquad [128]$$

If the half-cycle is divided into three equal intervals, the angles at the limits of these intervals are 0, $\pi/3$, $2\pi/3$, and π. The corresponding values of the grid-signal voltage e_g are $\sqrt{2}E_g$, $+\frac{1}{2}\sqrt{2}E_g$, $-\frac{1}{2}\sqrt{2}E_g$, and $-\sqrt{2}E_g$, and the corresponding total plate currents are i_{bmax}, i_x, i_y, and i_{bmin}. These are indicated in Fig. 28. Substitution of these values in Eq. 127 and use of Eqs. 12 and 123 give, for

$$\omega t = 0, \qquad i_{bmax} = I_{bs} + \sqrt{2}(I_{p1} + I_{p2} + I_{p3}); \qquad [129]$$

$$\omega t = \pi/3, \qquad i_x = I_{bs} + \sqrt{2}(\tfrac{1}{2}I_{p1} - \tfrac{1}{2}I_{p2} - I_{p3}); \qquad [130]$$

$$\omega t = 2\pi/3, \qquad i_y = I_{bs} + \sqrt{2}(-\tfrac{1}{2}I_{p1} - \tfrac{1}{2}I_{p2} + I_{p3}); \qquad [131]$$

$$\omega t = \pi, \qquad i_{bmin} = I_{bs} + \sqrt{2}(-I_{p1} + I_{p2} - I_{p3}). \qquad [132]$$

Solution of Eqs. 129, 130, 131, and 132 for the harmonic and average components gives

$$I_{bs} = \frac{(i_{bmax} + i_{bmin}) + 2(i_x + i_y)}{6}, \qquad [133]$$

$$I_{p1} = \frac{(i_{bmax} - i_{bmin}) + (i_x - i_y)}{3\sqrt{2}}, \qquad [134]$$

$$I_{p2} = \frac{(i_{bmax} + i_{bmin}) - (i_x + i_y)}{3\sqrt{2}}, \qquad [135]$$

$$I_{p3} = \frac{(i_{bmax} - i_{bmin}) - 2(i_x - i_y)}{6\sqrt{2}}. \qquad [136]$$

The per cent second- and third-harmonic generations in the plate current or voltage waveform are then:

Per cent second harmonic $= 100\dfrac{I_{p2}}{I_{p1}} = \dfrac{(i_{bmax} + i_{bmin}) - (i_x + i_y)}{(i_{bmax} - i_{bmin}) + (i_x - i_y)} \times 100,$ ▶[137]

Per cent third harmonic $= 100\dfrac{I_{p3}}{I_{p1}} = \dfrac{(i_{bmax} - i_{bmin}) - 2(i_x - i_y)}{2[(i_{bmax} - i_{bmin}) + (i_x - i_y)]} \times 100.$ ▶[138]

The per cent total harmonic generation is given by

Per cent total harmonics $= \sqrt{(\text{Per cent 2nd})^2 + (\text{Per cent 3rd})^2}.$ [139]

The Fischer-Hinnen method of obtaining the harmonic components in the plate-circuit waveforms with a sinusoidal grid-signal voltage, which is illustrated by Eqs. 129 through 136, is readily extended to include higher harmonics than the third. It is applicable only for a pure resistance load, however, and it involves the assumption that there are no harmonics of higher order than the number of intervals into which the cycle is arbitrarily divided. It can be demonstrated also that this assumption is equivalent to the assumption that the power series similar to Eq. 114 that represents the dynamic transfer characteristic contains terms of the same order as the harmonics.

14. POWER OUTPUT AND EFFICIENCY

Power is supplied, in general, to the vacuum tube through three different circuits, the grid circuit, the plate circuit, and the cathode-heating circuit. The primary purpose of an amplifier tube is to inject suitably controlled power into the circuit and thereby to make the ratio of the output power to the grid-signal input power large. The average power supplied to the tube by the direct-current plate-power supply is direct-current power only (direct-current power being defined as the power given by the product of the average values of the voltage and current). The useful output power from the amplifier is often solely alternating-current power. Thus the function of the amplifier tube in many applications may be described as that of a converter of direct-current power to alternating-current power, under control of an external alternating-voltage source. In the amplifier tube, as in all power-conversion equipment, a factor of interest is the power-conversion efficiency—that is, the efficiency with which the device converts direct-current power into alternating-current power. This involves the various heat losses in the tube.

The impressed grid-signal voltage to be amplified sends some current through the input or grid circuit, thereby requiring some power to be supplied from the source of this voltage. In the Class B or Class C amplifiers usually used for high frequencies, the grid-input power may be appreciable, but in the linear Class A_1 amplifiers that are the subject of this chapter the grid voltage with respect to the cathode never becomes positive; hence the grid current is very small, and the grid-input power is often negligible compared with the other losses in the tube. The heating power for the cathode, $E_f I_f$, is constant and is readily included in a computation of the over-all efficiency of the circuit operation. The components of power in the plate circuit, however, are several in number, and an analysis of power relations in the plate circuit is of importance if the maximum efficiency is to be obtained with a given tube.

As a first approximation, it is assumed in the following that the tube operates as a linear Class A_1 amplifier and that the grid-signal voltage has no direct component. When the grid-signal voltage is zero, the plate current has its quiescent value I_b, and the total power input to the circuit from the plate-power supply is

$$P_{bb} = E_{bb}I_b .$$ [140]

This power divides between the tube and the load. The power absorbed in the load, called the *direct-current load power*, is

$$P_{dc} = I_b{}^2 R_L .$$ [141]

The power absorbed in the tube is the *direct-current power input* to the tube

$$P_b = E_b I_b .$$ [142]

It is transformed into heat at the plate, and is hence also the *quiescent plate dissipation*. Since, from Fig. 4,

$$E_{bb} = E_b + I_b R_L ,$$ [143]

$$P_{bb} = P_b + P_{dc} ,$$ [144]

as it must in accordance with the conservation of energy.

When the grid-signal voltage is applied to the circuit, the plate current varies, and the average value of the total power input is

$$P_{bb} = \frac{1}{2\pi} \int_0^{2\pi} E_{bb} i_b \, d(\omega t)$$ [145]

$$= E_{bb} \left[\frac{1}{2\pi} \int_0^{2\pi} i_b \, d(\omega t) \right].$$ [146]

The quantity within the brackets is the average value of the plate current; in general, it is denoted by I_{bs}, but when, as is here assumed, the operation of the circuit is linear and the grid-signal voltage has no direct component, it equals I_b. Thus the total average power input remains constant at its quiescent value given by Eq. 140, and is independent of the amplitude of the grid-signal voltage.

The power output to the load, however, becomes

$$P_L = \frac{1}{2\pi} \int_0^{2\pi} i_b{}^2 R_L \, d(\omega t)$$ [147]

$$= \frac{1}{2\pi} \int_0^{2\pi} (I_b + i_p)^2 R_L \, d(\omega t)$$ [148]

$$= \frac{1}{2\pi} \int_0^{2\pi} I_b{}^2 R_L \, d(\omega t) + \frac{1}{2\pi} \int_0^{2\pi} 2 I_b i_p R_L \, d(\omega t)$$

$$+ \frac{1}{2\pi} \int_0^{2\pi} i_p{}^2 R_L \, d(\omega t).$$ [149]

The first integral term is $I_b{}^2R_L$, the direct-current load power previously defined. The second integral is zero, since the current i_p has no direct component when the grid-signal voltage has none and the circuit operation is linear. The third integral term is the average power in the load caused by the varying components of the plate current, or the *alternating-current power output*, denoted here by P_{ac}. Hence,

$$P_L = P_{dc} + P_{ac} .   \blacktriangleright[150]$$

The power output thus increases with the grid-signal voltage, but the power input to the tube and load together does not. Consequently the power absorbed in the tube, or the plate dissipation, decreases by a like amount as the grid-signal voltage increases. The *plate dissipation in the presence of the grid-signal voltage*, denoted by P_p, is given by

$$P_p = \frac{1}{2\pi} \int_0^{2\pi} e_b i_b \, d(\omega t),   [151]$$

or, from the principle of conservation of energy and the previous expressions, by

$$P_p = P_{bb} - P_L .   [152]$$

By means of Eqs. 144 and 150, this can be expressed as

$$P_p = P_b - P_{ac} .   \blacktriangleright[153]$$

In this expression, the term P_b is the apparent direct-current power input to the tube that would be given by the product of the indications of d'Arsonval-type instruments used to measure the plate voltage and the plate current. For the conditions assumed, then, the plate dissipation P_p is the difference between the direct-current power input to the tube P_b and the useful power output P_{ac} caused by the varying component of the plate current. When the grid-signal voltage is sinusoidal, P_{ac} has the value

$$P_{ac} = E_p I_p \cos \theta,   \blacktriangleright[154]$$

where θ is the angle of the load impedance as defined in Eq. 24.

Since in the linear Class A_1 amplifier the total input power to the circuit remains constant, but the alternating-current power output increases as the grid-signal voltage increases, one interpretation of the circuit behavior might be that the tube acts as an alternating-current generator to furnish the alternating-current power output. Actually, however, the tube is not a generator; all the power comes from the direct-current plate-power supply, and the tube merely converts some of it into useful power output at the grid-signal frequency.

One way in which vacuum tubes are rated is on the basis of their allowable heat losses; that is, the plate dissipation that cannot be exceeded without raising the temperature of the plate excessively and thereby endangering the life of the tube. The plate dissipation P_p from Eq. 153, is the direct-current power supplied to the tube reduced by the amount of the alternating-current power converted by the tube and supplied to the plate-circuit load. Other things being constant, it follows that a given tube in a linear Class A_1 amplifier will operate at a lower temperature when it is delivering a large alternating-current output power than with a small output or none at all. Should the grid-signal voltage be accidentally removed from a tube that is operating at nearly its rated temperature, the temperature is likely to increase above the allowable limit and destroy the tube. The effect of an alternating-current output is actually to cool the tube.

The relations in Eq. 153 among the direct-current power input to the tube, the plate dissipation, and the alternating-current power output in the plate circuit of a linear Class A_1 amplifier with a resistance load may be illustrated graphically as in Fig. 29. The time-function wave-forms of the currents and voltages in Fig. 29 are similar to those in Fig. 8 but have larger amplitudes. In the lower part

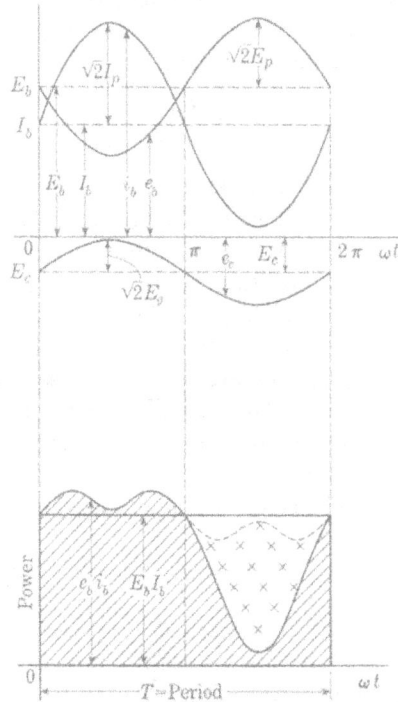

P_p = Plate dissipation = $\dfrac{\text{Shaded area}}{T}$

P_b = D-c power input to tube

$\quad = E_b I_b = \dfrac{\text{Area within heavy rectangle}}{T}$

P_{ac} = A-c power output

$\quad = \dfrac{\text{Area within heavy rectangle} - \text{shaded area}}{T}$

$\quad = \dfrac{\text{Area indicated by crosses}}{T}$

Fig. 29.   Power relations in a vacuum tube.

of the diagram, the instantaneous power supplied to the tube $e_b i_b$ is plotted as a solid curve. The average power dissipated in the tube, or the plate dissipation, is therefore equal to the shaded area divided by the period T. The direct-current power input to the tube, or the quiescent plate dissipation if the tube operation is linear, is given by the area within the heavy rectangle divided by the period T. In accordance with Eq. 153, the alternating-current power output to the load is given by the difference of the foregoing areas, which is the area shown with crosses, divided by the period T.

The efficiency with which the tube converts the direct-current power input to the tube into alternating-current power output in the load is called the *plate efficiency* η_p. For the linear Class A_1 amplifier considered here, its value is

$$\eta_p = \frac{P_{ac}}{P_b} = \frac{P_{ac}}{E_b I_b} = \frac{P_{ac}}{E_b I_b} \times 100 \text{ per cent.} \qquad \blacktriangleright[155]$$

The direct-current losses represented by $I_b{}^2 R_L$ contribute nothing to the useful output, because only the alternating components are generally useful in the plate circuit. The quiescent plate current I_b is therefore an undesirable quantity from the point of view of efficiency —it limits the plate efficiency obtainable in Class A amplifiers, as is evident in the following articles, and reduction or elimination of it contributes to the higher efficiencies obtainable with Class AB, Class B, and Class C amplifiers discussed in a subsequent chapter. In these amplifiers the operation is not linear, and for them, as well as for a Class A amplifier in which the excursions of voltage and current are so large that the operation is appreciably nonlinear, the plate efficiency must be expressed as

$$\eta_p = \frac{P_{ac}}{E_{bs} I_{bs}}, \qquad \blacktriangleright[156]$$

where the average values E_{bs} and I_{bs} replace the corresponding quiescent values in Eq. 155.

15. MAXIMUM POWER SENSITIVITY

The maximum value of the power output obtainable from a vacuum tube depends upon the choice of the quantities in the circuit that are allowed to vary, and of those that are maintained constant. For example, the maximum alternating-current power output that can be obtained from a given tube operating at a given point on the plate characteristics with a prescribed grid-signal voltage that is so small that the operation is restricted to the essentially linear region of the tube characteristics is one of two important and often-confused quantities that depend on the load impedance. The other quantity is the maximum power output with a prescribed amount of harmonic generation and a prescribed quiescent plate voltage; that is, the maximum power output that can be obtained from a given tube operating with a given quiescent plate voltage E_b but with a grid-signal voltage E_g limited only by the degree of waveform distortion that can be tolerated. This second quantity is discussed in a subsequent article.

If operation is restricted to the linear region of the tube characteristics, and the grid-signal voltage is prescribed, essentially linear operation occurs, with the tube acting as a fixed ideal voltage or current source and constant internal resistance, as is explained in Art. 5. Linear network theory is then applicable for finding the conditions for maximum power transfer from the tube to the load. The alternating-current power output is

$$P_{ac} = E_p I_p \cos 0 = I_p{}^2 R_L \, , \qquad [157]$$

where 0 is the impedance angle of the load and R_L is the real part of the complex expression for its impedance

$$Z_L = R_L + j X_L \, . \qquad [158]$$

By means of Eq. 36 the expression for the output power may be put into the form

$$P_{ac} = \mu^2 E_g{}^2 \frac{R_L}{(r_p + R_L)^2 + X_L{}^2} \, . \qquad [159]$$

Since the power output varies directly with the square of the grid-signal voltage, a significant measure of the usefulness of an amplifier is the *power sensitivity*, which, from Eq. 159, is given by

$$\text{Power sensitivity} = \frac{P_{ac}}{E_g{}^2} = \mu^2 \frac{R_L}{(r_p + R_L)^2 + X_L{}^2} \, . \qquad [160]$$

The particular values of R_L and X_L that give maximum power output or power sensitivity when the grid-signal voltage has a small and constant amplitude depend upon the choice of the independent variable. Three important cases are: first, R_L is the only variable; second, X_L is the only variable; and, third, 0 is held constant and Z_L is the only variable; that is, Q or X_L/R_L is held constant* and the magnitude of the load impedance is changed.

If R_L is the only variable, differentiating Eq. 159 with respect to R_L and equating the result to zero give

$$\frac{dP_{ac}}{dR_L} = \mu^2 E_g{}^2 \left[\frac{(r_p + R_L)^2 + X_L{}^2 - 2R_L(r_p + R_L)}{[(r_p + R_L)^2 + X_L{}^2]^2} \right] = 0 \quad [161]$$

or

$$(r_p + R_L)^2 + X_L{}^2 - 2R_L(r_p + R_L) = 0. \qquad [162]$$

* For the meaning and use of the quantity Q, see E. E. Staff, M.I.T., *Electric Circuits* (Cambridge, Massachusetts: The Technology Press of M.I.T.; New York: John Wiley & Sons, Inc., 1940), 319–325.

Thus the value of R_L for maximum alternating-current power output or power sensitivity when the reactance X_L is fixed and R_L is varied is

$$R_L = \sqrt{r_p{}^2 + X_L{}^2}. \qquad [163]$$

In other words, for maximum output, R_L is equal to the magnitude of the impedance formed by r_p and X_L in series.

If X_L is the only variable, P_{ac} is a maximum when X_L equals zero, as is evident by inspection of Eq. 159. Reactance added in series with a fixed load resistance reduces the power output and power sensitivity.

If the load impedance Z_L is to be a coil of fixed winding space, such as an inductor coil or a winding in a loud speaker, the Q of the coil can be shown to be approximately fixed at a given frequency independent of the value of Z_L. Thus Z_L may be varied through varying the number of turns and size of wire to fill the space, but θ, which equals $\tan^{-1}(\omega L/R_L)$, or $\tan^{-1}Q$, remains constant. Equation 159 can then be written as

$$P_{ac} = \mu^2 E_g{}^2 \frac{R_L}{(r_p + R_L)^2 + (QR_L)^2}. \qquad [164]$$

The value of Z_L for maximum power output or power sensitivity is then found by differentiation to be

$$Z_L = \sqrt{R_L{}^2 + X_L{}^2} = r_p, \qquad [165]$$

for Z_L variable but θ, or Q, constant.

When the load is pure resistance, Eq. **163 indicates that the power** output and the power sensitivity are maximum when R_L equals r_p. The maximum power output is then given by

$$P_{ac_{max}} = \frac{\mu^2 E_g{}^2}{4r_p} = \frac{E_g{}^2}{4}\mu g_m. \qquad \blacktriangleright[166]$$

This value is called the *available power*. It is the largest value of power output that can be obtained through adjustment of the load in any manner. The corresponding maximum power sensitivity is, from Eq. 166,

$$\text{Maximum power sensitivity} = \frac{P_{ac_{max}}}{E_g{}^2} = \frac{\mu g_m}{4}. \qquad [167]$$

Thus the factor μg_m, which depends only on the tube construction, is a valuable figure of merit for triodes. It is not equally important for tubes with additional grids, however, for generally with them the condition that R_L is equal to r_p is not economical or is precluded by considerations of distortion.

The power output as a function of the load resistance in the absence of reactance is given by

$$P_{ac} = \mu^2 E_g{}^2 \frac{R_L}{(r_p + R_L)^2} = \frac{\mu^2 E_g{}^2}{r_p} \cdot \frac{1}{\dfrac{r_p}{R_L} + 2 + \dfrac{R_L}{r_p}} . \qquad [168]$$

The symmetry in the denominator of Eq. 168 shows that the power output for a load resistance that is twice as great as the plate resistance is the same as that for a load resistance that is half as great as the plate

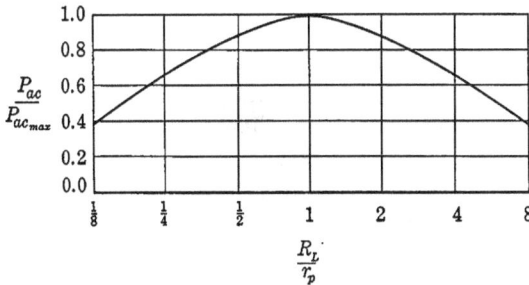

Fig. 30.　A-c power output as a function of load resistance.

resistance, and similarly for any other ratio. This symmetry is illustrated in the curves of Fig. 30. It should be noted that the power output is reduced only 11 per cent by a factor of two between the load and plate resistances; thus impedance matching in the plate circuit of a vacuum tube is not always so important as might be thought at first.

16. MAXIMUM POWER OUTPUT WITH A PRESCRIBED AMOUNT OF HARMONIC GENERATION AND A PRESCRIBED QUIESCENT PLATE VOLTAGE

In the previous article, the maximum power output obtainable from an amplifier tube with a fixed small grid-signal voltage is considered, but the accompanying distortion is disregarded. In general, the amount of harmonic generation increases with the power output as the grid-signal voltage is increased. Consequently, when an essentially unlimited grid-signal voltage is available, the permissible power output is limited by the degree of waveform distortion that can be tolerated, which depends on the service for which the amplifier is intended. For example, as previously stated, when the amplifier is a link in a sound-reproducing system, a maximum of 10 per cent total harmonic generation may be allowable. On the other hand, for measurements applications—such as in an amplifier used with an oscillograph—the allowable amount of harmonic generation may be smaller than one per cent. It has been conventional[14] to describe the maximum power output with 10 per cent total harmonic generation as the "maximum undistorted power output." Since the term is a misnomer, it is not used hereafter. The considerations of this article, however, deal with the maximum power output obtainable with a prescribed percentage harmonic generation. The power referred to is the power from the fundamental frequency components only, and the amount of waveform distortion is assumed to be small.

When distortion is considered, a value of power output of interest is the maximum power output that can be obtained with a prescribed percentage of harmonic generation from a given tube operating with a resistance load at a prescribed quiescent plate voltage. This value of power output is especially relevant when the load is coupled to the tube through a transformer, as is discussed in Art. 17. With the foregoing quantities fixed, the grid-signal voltage, load resistance, and grid-bias voltage become dependent quantities. The determination of the maximum power output and the accompanying values of grid-signal voltage, grid-bias voltage, load resistance, and quiescent plate current under the stated conditions can be made accurately only by a cut-and-try process—either graphically, using the tube characteristics, or electrically, using the tube itself and instruments for measurement of the power output and harmonic generation. If the determination is made graphically, the methods of Art. 13 may be employed. Usually when the tube is a triode, the results are a load resistance equal to about twice the plate resistance of the tube, and values of grid-bias and grid-signal voltages that depend on the amount of harmonic generation permitted and are chosen so that the instantaneous total grid voltage e_c reaches zero once during each cycle.

That the results as stated above should be true may be demonstrated by the following approximate analysis. If the harmonic generation is to be small, the operating region on the plate characteristics must have a lower boundary $i_{b_{min}}$ in Fig. 31, which is above the excessively curved lower portion of the curves. If the amplifier is to be operated as a Class A_1 amplifier, the upper boundary of the operating region must be the curve for zero grid voltage e_c. For maximum power output under the stated conditions, the load line terminates on these two boundaries; that is, P_1 lies on line $\overline{xy}$, and P_2 lies on line $\overline{yz}$. At the same time, if the quiescent plate voltage E_b is specified, and the distortion is to be small, the mid point of the operating line must lie at the abscissa E_b; thus $\overline{P_1Q}$ must equal $\overline{P_2Q}$. These conditions, together

[14] *Standards on Radio Receivers: Methods of Testing Frequency-Modulated Broadcast Receivers* (New York: The Institute of Radio Engineers, 1947), 2, 10–11.

with the value of the load resistance, determine the slope and position of the load line on the diagram.

For a particular value of load resistance R_L, the conditions are as shown in Fig. 31. Since the object of the analysis is to obtain an expression for the value of R_L that makes P_{ac} a maximum, a straightforward attack would be to obtain an expression for P_{ac} in terms of R_L, differentiate it with respect to R_L, set the derivative equal to zero,

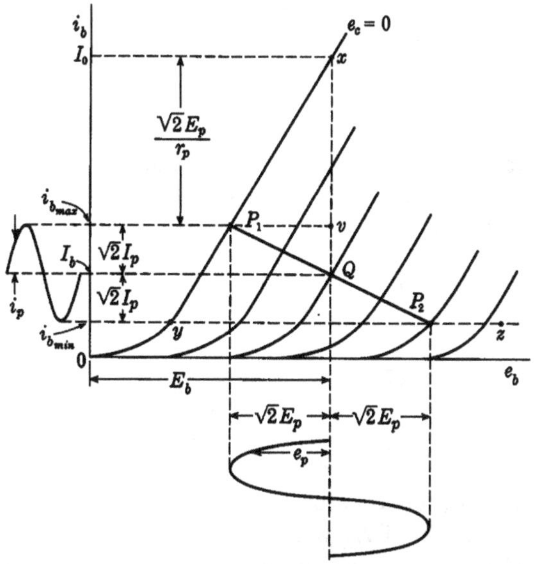

Fig. 31.   Limiting Class A_1 operating conditions in a triode for small harmonic generation and maximum power output with a prescribed quiescent plate voltage.

and solve for the desired value of R_L. Such a method, however, leads to rather complicated expressions when the constraints mentioned in the previous paragraph are involved. A simpler method, which is followed below, depends on recognition of the fact that when the plate characteristics in the region above $i_{b_{min}}$ are straight, parallel, and equally spaced lines for equal increments of grid voltage, the three quantities R_L, E_p, and I_p are related not only by the condition

$$R_L = E_p/I_p \qquad [169]$$

but also by another equation that results from the constraints stated above and is developed as a first step in the analysis that follows. Hence any one of the variables can equally well be chosen as the independent variable in the differentiation, since the other two are determined by any one. That is, the derivative of P_{ac} with respect to E_p or

I_p is zero for the same value of power as is the derivative of P_{ac} with respect to R_L.

The second relationship among the variables R_L, E_p, and I_p is obtained through expressing the current I_0, which is the plate current corresponding to the prescribed quiescent plate voltage and zero grid-signal voltage, as the sum of its components. Thus, from Fig. 31,

$$I_0 = i_{b_{min}} + 2\sqrt{2}I_p + \frac{\sqrt{2}E_p}{r_p} , \qquad [170]$$

where the last term is obtained through multiplying the length of the base of the right triangle P_1xv by the slope of the line $\overline{xy}$. The stated conditions under which the maximum power is to be found, then, require that as R_L is varied, E_p and I_p change in such a way that Eqs. 169 and 170 are simultaneously satisfied.

In accordance with the foregoing explanation, the maximum power occurs for

$$\frac{dP_{ac}}{dI_p} = \frac{d(E_p I_p)}{dI_p} = E_p + I_p \frac{dE_p}{dI_p} = 0. \qquad [171]$$

Thus, from Eqs. 169 and 171,

$$R_L = \frac{E_p}{I_p} = -\frac{dE_p}{dI_p}. \qquad [172]$$

The expression dE_p/dI_p remains to be evaluated. This derivative is not equal to R_L as it would be if R_L were fixed; rather it is determined by the constraints. It can therefore be found from Eq. 170. Since I_0 and $i_{b_{min}}$ do not vary with R_L, taking the differential of both sides of that equation gives the expression

$$0 = 0 + 2\sqrt{2}\, dI_p + \frac{\sqrt{2}}{r_p}\, dE_p; \qquad [173]$$

hence

$$\frac{dE_p}{dI_p} = -2r_p . \qquad [174]$$

Combination of Eqs. 172 and 174 gives

$$R_L = 2r_p , \qquad \blacktriangleright[175]$$

which, for the conditions of linearity of the plate characteristics above $i_{b_{min}}$, is the load resistance for maximum alternating-current power output from a particular tube with the quiescent plate voltage and allowable harmonic generation prescribed.

As is demonstrated in Art. 15, and shown by Fig. 30, the power output obtained with R_L/r_p equal to 2 is only 89 per cent of the power output obtained with R_L/r_p equal to 1 *for a given grid-signal voltage*. But when the limits of operation are set by *grid current, harmonic generation*, and a *prescribed* E_b, a larger grid-signal voltage can be used when R_L/r_p equals 2 than when R_L/r_p equals 1, and the maximum power output is larger when R_L/r_p equals 2.

The foregoing results apply only for the condition that the quiescent plate voltage is prescribed. If the plate-supply voltage E_{bb} or the plate

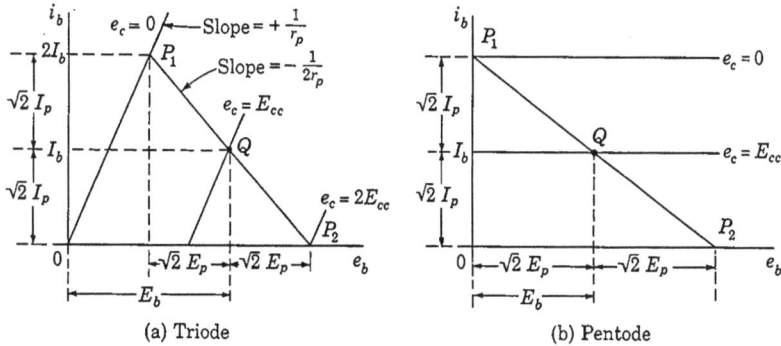

(a) Triode   (b) Pentode

Fig. 32. Maximum power-output conditions with idealized characteristics for Class A_1 operation and a prescribed quiescent plate voltage.

dissipation P_p is prescribed,[15] and the quiescent plate voltage is allowed to become a variable, the value of R_L equal to $2r_p$ is no longer optimum. A load resistance several times larger than $2r_p$ may be desirable, and the ratio of R_L to r_p depends upon the characteristics of the tube and the operating conditions specified.

A rough estimate of the theoretical plate efficiency for a triode operating under conditions of maximum power output as a Class A_1 amplifier with a prescribed quiescent plate voltage may be made if the plate characteristics are idealized as shown in Fig. 32. In Fig. 32a, the plate characteristics are represented as parallel straight lines, the load resistance R_L equals $2r_p$, and $i_{b_{min}}$ is assumed equal to zero.

From the geometry of the diagram,

$$\sqrt{2}E_p = E_b/2, \qquad [176]$$

and

$$\sqrt{2}I_p = I_b. \qquad [177]$$

Since the load is a pure resistance, cos θ in Eq. 154 is unity, and the

[15] W. B. Nottingham, "Optimum Conditions for Maximum Power in Class A Amplifiers," *I.R.E. Proc.*, *29* (1941), 620–623.

plate efficiency η_p is given by

$$\eta_p = \frac{E_p I_p \cos \theta}{E_b I_b} = \frac{\dfrac{E_b}{2\sqrt{2}} \times \dfrac{I_b}{\sqrt{2}}}{E_b I_b} \qquad [178]$$

$$= 25 \text{ per cent.} \qquad [179]$$

Thus the theoretical plate efficiency of a triode operating under conditions for maximum power output with Class A_1 operation and a prescribed quiescent plate voltage is 25 per cent. In practice, this figure is not attained with this type of amplifier because the curvature of the tube characteristics generally makes it undesirable to utilize the whole range of the load line, owing to distortion considerations. The Class A_1 triode amplifier is thus a very inefficient means of converting direct-current power into alternating-current power. Other types of amplifiers, Class B and Class C, make possible the realization of plate efficiencies up to 60 and 80 per cent, respectively, and are used when their limitations described in Ch. X are allowable in the particular application.

For maximum power output from a pentode operating in Class A_1 with a prescribed quiescent plate voltage, conditions in the plate circuit of a tube with idealized characteristics and a resistance load are as is shown in Fig. 32b. The plate characteristics are horizontal lines extending to the current axis. Since the operation is linear for any current between zero and the value for e_c equal to zero, the peak-to-peak current variation can be as large as this maximum current, and the peak-to-peak plate voltage can be as large as twice the quiescent value without excessive distortion, grid current, or cut-off. The maximum power output occurs when these optimum conditions are realized simultaneously. Thus the load resistance should be selected so that the load line intersects the current axis at P_1 and the voltage axis at P_2, where the plate voltage is twice the prescribed quiescent value. The required quiescent grid voltage is then the value for the middle curve of the family, for which e_c equals half the cut-off value. For the optimum load resistance, $\sqrt{2}E_p$ equals E_b, and $\sqrt{2}I_p$ equals I_b, as is shown in the diagram, so that the plate efficiency η_p as given by Eq. 156 is 50 per cent, an improvement by a factor of two over the theoretical maximum for a triode.

17. Output transformer for impedance matching

Advantages of adjustment of the load resistance to suit the tube are described in the preceding articles. In practice, however, an arbitrary choice of the load resistance to realize these advantages is usually not feasible because, for example, the load may be a device already avail-able, or one whose design involves inherent limitations of resistance. Hence, in power amplifiers, output transformers are generally used between the tube and the load. The characteristics of such trans-formers are described in *Magnetic Circuits and Transformers*. From the considerations of Art. 16, it is apparent that a value of load resistance equal to the plate resistance is not desirable when maximum power

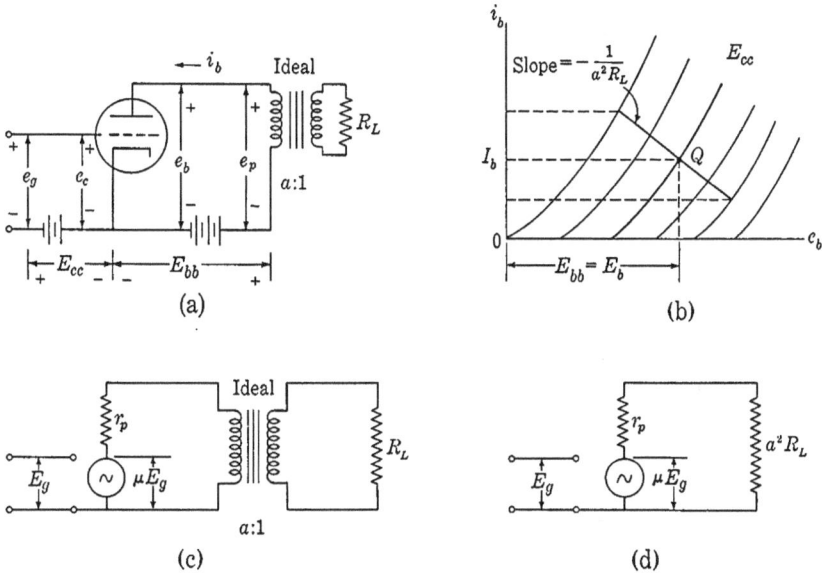

Fig. 33. Linear Class A_1 triode operation with an ideal output transformer.

with a prescribed amount of harmonic generation is wanted, and that a transformer ratio to cause the actual load resistance to have an apparent value in the plate circuit of about twice the plate resistance of the tube is needed when the tube is a triode and the quiescent plate voltage is specified.

The circuit diagram for an amplifier with an output transformer of turns ratio a and a resistance load is shown in Fig. 33a. As a first approximation, the transformer may be assumed to be ideal for alter-nating components—that is, it transforms all alternating voltages and currents by constant ratios a or $1/a$ regardless of frequency. It does not transmit steady or direct components. When the transformer is ideal, the path of operation on the plate characteristics is as shown in Fig. 33b. Because the ideal transformer has no losses, the windings have no resistance, and the quiescent operating point Q has an abscissa E_b that equals E_{bb}. The path of the operating point on Fig. 33b is along a load line having a slope $-1/(a^2 R_L)$, since the apparent impedance as viewed from the tube into the transformer is $a^2 R_L$. Thus the resistances used for the determination of the direct-current and alternating-current conditions on the plate characteristics are different.

If the path of operation remains in the linear region of the plate characteristics, the equivalent circuit for alternating components is that shown in Fig. 33c, which may be further simplified to that of Fig. 33d because of the impedance transformation property of the transformer. For maximum power output with a prescribed amount of harmonic generation and a prescribed quiescent plate voltage, the considerations of Art. 16 show that the turns ratio of the transformer should be

$$a = \sqrt{2r_p/R_L} \,. \qquad [180]$$

The output transformer serves a threefold purpose; namely, (a) it makes possible the realization of the conditions for maximum power output with almost any tube and load, since its turns ratio can be selected; (b) it eliminates $I_b{}^2 R_L$, the direct-current component of power in the load, because I_b is confined to the low-resistance primary winding of the transformer and does not exist in R_L; and (c) it serves as an electrical isolator between the tube and the load circuit, since with it the potential of any one point in the load circuit may be given any desired value without regard to potentials of points in the tube circuit.

18. Shift of dynamic load line with average plate current

A somewhat better approximation to actual transformer operation than that described in Art. 17 is obtained if the transformer is considered to have appreciable primary winding resistance R_{pri}, but is otherwise ideal in the sense described there. Direct or average components of the primary current then encounter a resistance R_{pri} at the primary terminals, but alternating components encounter $R_{pri} + a^2 R_L$—that is, the primary winding resistance plus the effect of the load resistance reflected into the primary circuit. Operating conditions are then as illustrated in Fig. 34.

The operating point for zero grid-signal voltage is point Q at the intersection of the plate characteristic corresponding to the grid-bias voltage and a load line that passes through E_{bb} on the e_b axis and has a slope corresponding to R_{pri}. This load line may be called the *d-c load line*, for it is the locus of operating points for steady values of the incremental grid-signal voltage—or for varying components that change so slowly that a practical transformer does not transmit them. Thus, if the incremental grid-signal voltage consists merely of a steady increment or contains an average component, the operating point corresponding to the average component lies at some point A on the d-c load line, and the average total plate current changes from I_b at Q to I_{bs} at A, as shown on the diagram.

If the incremental grid-signal voltage has alternating components, but no average component, and is so small that essentially linear operation occurs, the incremental plate current likewise has zero

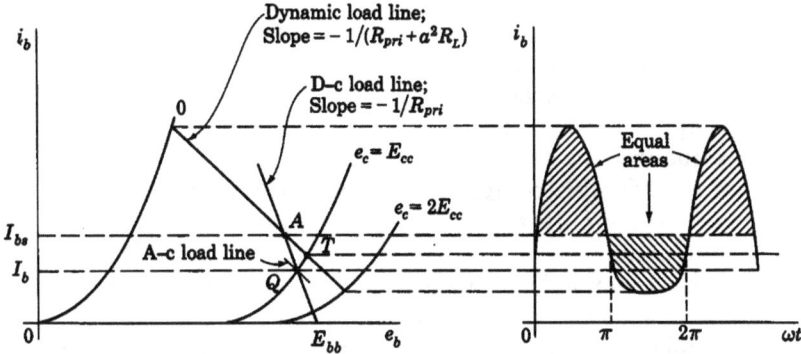

Fig. 34.   Shift of dynamic load line with average plate current.

average value, and the average values of e_b and i_b remain constant at E_b and I_b, respectively. The operating point corresponding to e_b and i_b then follows a path through Q with a slope corresponding to $R_{pri} + a^2 R_L$, which may be called the *a-c load line*, and is indicated by the shorter of the downward sloping lines through Q in Fig. 34.

When both average and alternating components are present in the incremental grid-signal voltage, and the operation is linear, their effects are linearly superposed on the quiescent condition in the plate circuit. Hence, the path of operation for the total plate current and plate voltage passes through a point such as A on the d-c load line, which corresponds to the sum of the quiescent and average-incremental components, and the path is a straight line having the slope corresponding to $R_{pri} + a^2 R_L$. This path is called the *dynamic load line*. The operating point on it that corresponds to zero instantaneous grid-signal voltage during a cycle lies at its intersection with the plate characteristic for the grid-bias voltage, as at T in Fig. 34.

When nonlinear distortion occurs, an average component of the incremental plate current is generally produced, as is explained in Art. 13, even though the grid-signal voltage does not contain an average component. Despite the nonlinearity in the tube, the principle of linear superposition may then be used to explain the path of operation, because the path is constrained to satisfy the current-voltage relations in the circuit external to the tube, and because that circuit is linear. Hence, the average component of incremental plate current produced by distortion may be considered to cause a shift of the dynamic operating path to a position through a point A on the d-c load line, just as it does when the grid-signal voltage contains an average component and the tube behavior is linear. When nonlinearity causes the average component of incremental plate current, however, its magnitude depends upon the position of the dynamic load line, and vice versa. Furthermore, the relation between them involves nonlinearity. Hence, a method of successive approximations is necessary for a determination of either. The dynamic load line shifts in position with the amplitude of the grid-signal voltage. Neglect of the shift is frequently justified in engineering work because consideration of it introduces considerable complexity of computation, though actually the shift often has only a small effect on the circuit operation for varying components.

Shift of the dynamic load line occurs only in vacuum-tube circuits in which the direct and alternating components of the plate current encounter different resistances at the terminals of the circuit external to the tube. Furthermore it occurs in such circuits only if the average component of the plate current with the signal applied, I_{bs}, differs from the current without the signal, I_b. A common example of such a circuit is one in which direct current is excluded from the load by a large coupling capacitor, as in the resistance-capacitance-coupled amplifier of Art. 5, Ch. IX.

19. PUSH-PULL CONNECTION; CLASS A_1

One method of obtaining greater power output than that obtainable from a single tube is the use of two tubes connected in parallel. A method of connecting two tubes that is often preferable to the parallel connection when increased power output is desired is shown in Fig. 35. The tubes are connected so that the plate current in one tube decreases when that in the other tube increases. This type of connection is commonly called the *push-pull* connection. The push-pull connection has numerous advantages over the parallel connection, one of the most important being the elimination of even-harmonic generation. As a result, the maximum power output with a prescribed amount of harmonic generation is greater than that from two tubes in parallel, and the push-pull circuit is extensively used not only for Class A operation but also for Class B and Class C operation.

It is assumed in the following analysis of the push-pull amplifier that: First, the operation is restricted to the negative-grid region of the tube characteristics and consequently the grid current is considered to be negligible; second, the transformers are ideal; third, the load is a pure resistance; and, fourth, the tubes have identical characteristics. Primed symbols are used to distinguish the quantities in one tube from those in the other, but the arbitrary assignments of direction and polarity are made symmetrically with respect to the common-cathode

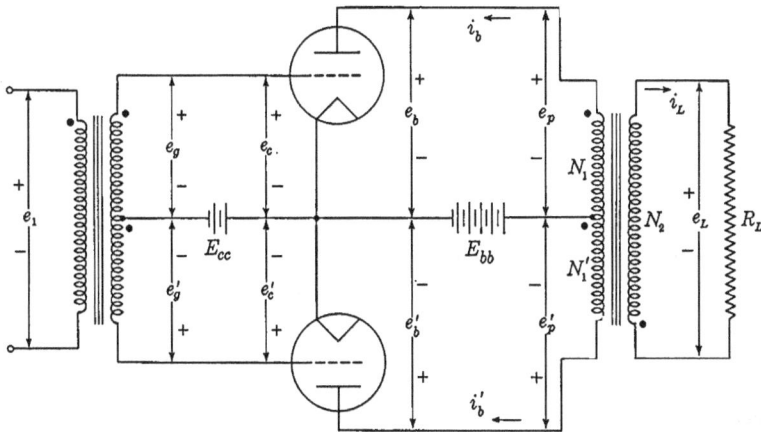

Fig. 35.   Push-pull connection of two triodes.

point on the diagram. The effects of nonlinearity on the alternating-current operation are neglected at first; later they are taken into account.

19a. *Determination of Quiescent Operating Point.* The determination of the quiescent operating point Q on the plate characteristics is made in the same manner as for a single tube in Art. 17, Fig. 33b. Thus when

$$e_1 = 0, \tag{181}$$

$$e_g = e_g{}' = 0, \tag{182}$$

and

$$e_c = e_c{}' = E_{cc} . \tag{183}$$

Also

$$e_p = e_p{}' = 0 \tag{184}$$

and

$$e_b = e_b{}' = E_b = E_{bb} . \tag{185}$$

The quiescent plate currents I_b and $I_b{}'$ are therefore equal in the two tubes and correspond to the point on the plate characteristics at the voltage co-ordinates given by Eqs. 183 and 185. The total quiescent plate current through the plate-power supply or battery is thus twice the current for one tube, and, as far as the quiescent operating conditions are concerned, the tubes operate *in parallel.*

The winding directions in the output transformer with its center-tapped primary winding are shown in Fig. 36. The dots used to indicate polarity mean that a rate of change of flux in the core that makes one of the dot-marked coil ends instantaneously positive with respect to the corresponding unmarked end also makes the other two dot-marked coil ends instantaneously positive with respect to their respective unmarked ends. For the zero-grid-signal condition

$$i_b = i_b' = I_b , \qquad [186]$$

the magnetomotive force in the upper winding of N_1 turns tends to send flux in a counterclockwise direction in the core, and that in the lower winding of N_1 turns tends to send flux in the clockwise direction.

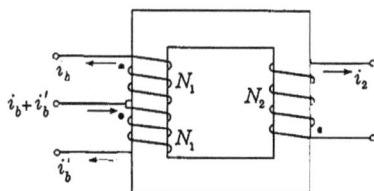

Fig. 36. Winding directions in the output transformer.

The net magnetization of the core resulting from the quiescent components of the plate currents is therefore zero. *This cancellation of magnetization is one of the principal advantages of the push-pull connection over the single-tube or parallel connection.* For a given power output and amount of harmonic generation caused by nonlinearity in the transformer, the transformer in a push-pull amplifier may be lighter in weight and less expensive than the one in a parallel-tube amplifier, because it is not necessary to provide a large core with an air gap to prevent the magnetic saturation and resulting waveform distortion caused by the average component of plate current. The effect of a direct current superposed on the alternating current in an iron-cored coil is discussed in *Magnetic Circuits and Transformers.*

19b. *Operation in a Linear Region.* For a small grid-signal voltage when the path of the operating point is *restricted to the linear region of the tube characteristics,* the operation for varying components of current and voltage may be obtained through use of the equivalent circuit for the tube. The total plate current and voltage are then obtainable through superposition of the quiescent and varying components. The equivalent circuit for varying components is that shown in Fig. 37.

If the input transformer is wound in a manner similar to the output transformer of Fig. 36, with equal numbers of turns on the two halves of the center-tapped secondary winding,

$$e_g' = -e_g . \qquad [187]$$

Because of the linearity and symmetry of the circuit, it follows that

$$i_p' = -i_p .$$ [188]

Thus the total current through the plate-power supply or battery is

$$i_b + i_b' = I_b + i_p + I_b - i_p$$ [189]

$$= 2I_b = \text{constant}$$ [190]

and contains no varying component. For this reason, it is not important that the plate-power supply have low internal impedance when
operation is restricted to the linear region of the curves. Also, if a

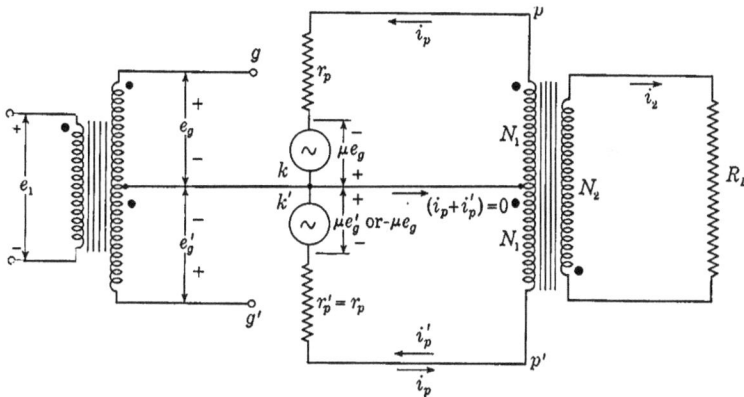

Fig. 37.   Incremental equivalent circuit for identical tubes with the
varying components of current and voltage restricted to the linear
region of the tube characteristics.

cathode-bias resistor is used (see Arts. 3 and 7), no by-pass capacitor is
required to prevent fluctuations in plate current from affecting the grid
voltage during strictly linear operation.

The ideal output transformer and load introduce an apparent resistance equal to $4(N_1/N_2)^2 R_L$ between points p and p' in the equivalent circuit of Fig. 37 because of the impedance-transformation
property of the transformer. The equivalent circuit including this
apparent resistance is thus that of Fig. 38a, where the center connection shown dotted is unnecessary, because no varying component
of current exists in it. From this circuit, the apparent load resistance
for each tube is seen to be $2(N_1/N_2)^2 R_L$, or half the *plate-to-plate
resistance* R_{pp}, where

$$R_{pp} = 4 (N_1/N_2)^2 R_L .$$ ▶[191]

If one tube were removed from its socket in the circuit of Fig. 35, the
second tube would then have an apparent load resistance $(N_1/N_2)^2 R_L$.

Re-insertion of the first tube would therefore double the effective load resistance of the second tube. This reaction of one circuit on the other would not occur if the output transformer were eliminated and a center-tapped load resistor were used. Thus it may be concluded that the effect is associated with the autotransformer action or coupling between the two plate circuits by the output transformer.

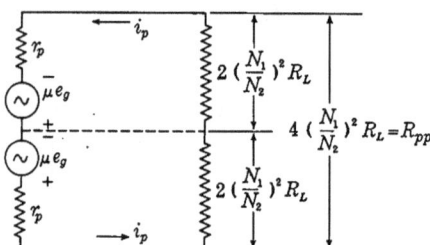

(a)

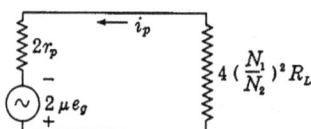

(b)

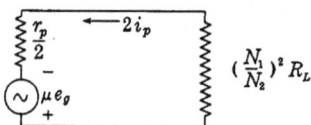

(c)

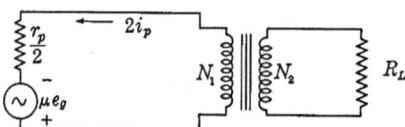

(d)

Fig. 38. Incremental equivalent circuits for the linear Class A_1 push-pull amplifier.

This analysis indicates that for small grid-signal voltages the load line, or the path of operation, of one tube on the plate characteristics in the linear region passes through the quiescent operating point, as shown in Fig. 33, but has a slope

$$-\frac{1}{2(N_1/N_2)^2 R_L}$$

instead of $-\dfrac{1}{(N_1/N_2)^2 R_L}$.

However, since the analysis here is restricted to operation in the linear region of the tube characteristics, it must not be inferred that the maximum power output as well as the corresponding optimum load resistance and harmonic generation can be obtained from this load line by the methods of Arts. 13 and 16. The reaction of the second tube through the transformer affects the path of operation in the first, and it is shown subsequently that over an extended region the path of operation is *not* a straight line on the plate characteristics of either tube, even though R_L is a pure resistance and the output transformer is ideal.

The circuit of Fig. 38a reduces to that of Fig. 38b. This diagram indicates that as far as varying components are concerned the two tubes are in series but, as was previously stated, they are in parallel as far as quiescent components are concerned.

However, the foregoing is not the only possible point of view. The circuit of Fig. 38c is equivalent to Figs. 38a and 38b for power considerations only; it is not equivalent for the voltage and current at the two plate terminals of the transformer primary winding. Furthermore, Fig. 38c is the equivalent circuit including the apparent resistance offered by the transformer and load in Fig. 38d, which is the equivalent circuit for two tubes connected in parallel and operating into one-half the transformer primary winding. Thus it may be stated that the push-pull connection is equivalent, as far as varying components and power considerations are concerned, to two such parallel-connected tubes. Either of the foregoing alternative concepts may be useful, but they are applicable only to operation over the linear region of the tube's characteristic curves.

19c. *Operation over a Range Extending beyond the Linear Region.* When the path of operation extends beyond the linear region of the plate characteristics, harmonics are generated in the plate current that are not present in the grid-signal voltage. For example, if the grid-signal voltage is sinusoidal and expressible as

$$e_g = \sqrt{2} E_g \cos \omega t, \qquad [192]$$

the plate current under conditions of no harmonic generation is

$$i_b = I_b + \sqrt{2} I_p \cos \omega t; \qquad [193]$$

but with harmonic generation it is expressible as the Fourier series,

$$i_b = I_b + I_{p0} + \sqrt{2}(I_{p1} \cos \omega t + I_{p2} \cos 2\omega t + I_{p3} \cos 3\omega t + \cdots), \quad [194]$$

as is demonstrated in Art. 13. Because of the symmetry in the push-pull circuit, the plate current in the second tube is similar to Eq. 194, but with ωt replaced by $\omega t + 180°$. Thus

$$i_b' = I_b + I_{p0} + \sqrt{2}[I_{p1} \cos (\omega t + 180°) + I_{p2} \cos (2\omega t + 360°)$$
$$+ I_{p3} \cos (3\omega t + 540°) + \cdots] \qquad [195]$$

$$= I_b + I_{p0} + \sqrt{2}[-I_{p1} \cos \omega t + I_{p2} \cos 2\omega t - I_{p3} \cos 3\omega t + \cdots]. [196]$$

One feature assumed in an ideal transformer is that the exciting current is negligible, which is equivalent to the statement that the magnetomotive force required to magnetize the core is zero. Consequently, the sum of the magnetomotive forces caused by currents in the windings is zero in a given direction around the core; and in a two-winding ideal transformer the current ratio is the inverse of the turns ratio. In the ideal push-pull output transformer, the sum of the magnetomotive forces in a given direction around the core caused by currents in the three windings is zero, just as in the two-winding transformer. Thus, for the transformer in Fig. 36,

$$i_b N_1 - i_b' N_1 - i_2 N_2 = 0, \qquad [197]$$

or

$$i_2 = \frac{N_1}{N_2}(i_b - i_b'). \qquad [198]$$

A rigorous analysis of the push-pull amplifier requires a consideration[16] of the finite magnetizing impedance and also of the leakage reactances among the three windings of the output transformer, but their effects are neglected here.

Substitution of Eqs. 194 and 196 in Eq. 198 gives the current in the load as

$$i_2 = 2\frac{N_1}{N_2}\sqrt{2}(I_{p1}\cos \omega t + I_{p3}\cos 3\omega t + \cdots). \qquad [199]$$

Thus the effect of the symmetrical arrangement is to cause a cancellation of the average components, the second harmonics, and all other even harmonics generated within the tube. However, if the *grid-signal voltage is nonsinusoidal*, all frequencies present in it, including even harmonics, are amplified as usual. The absence from the output signal of components resulting from even-harmonic generation is one of the advantages of the push-pull connection over the parallel connection of tubes.

The current through the plate-power supply or battery is the sum of the currents through the two tubes. From Eqs. 194 and 196, this is

Current through plate-power
supply $= i_b + i_b'$

$$= 2I_b + 2I_{p0} + \sqrt{2}(2I_{p2}\cos 2\omega t + 2I_{p4}\cos 4\omega t + \cdots). \quad [200]$$

The average value of this plate-power supply current as indicated by a direct-current ammeter increases by the amount $2I_{p0}$ when the grid-signal voltage is increased from zero to E_g. A change in the ammeter indication when the grid-signal voltage is applied therefore indicates waveform distortion caused by the generation of harmonics. Whereas only the odd harmonic-generation components exist in the output current and voltage, the current through the plate-power supply contains only the even harmonic-generation components.

Waveforms of the grid-signal voltages, plate currents, and output current, illustrating the operation when the path of the operating

---

[16] A. P-T. Sah, "Quasi Transients in Class B Audio-Frequency Push-Pull Amplifiers," *I.R.E. Proc.*, *24* (1936), 1522–1541.

point extends into the nonlinear region of the tube characteristics, are shown in Fig. 39. The grid-signal voltages e_g and $e_g{}'$ are sinusoidal and 180 degrees out of phase. The plate-current waveforms i_b and $i_b{}'$ are flattened at the bottom because of nonlinearity of the tube characteristics, but each is a replica of the other displaced by 180 degrees. The output current i_2 hence is flattened near both its crests, and, since the diagram shows that

$$i_2(\omega t) = -i_2(\omega t + \pi), \quad [201]$$

i_2 contains only odd harmonics,[17] as was deduced analytically in Eq. 199.

A graphical analysis for the path of operation on the plate characteristics over an extended range is not readily made for an individual tube, because of the coupling between the plate circuits of the two tubes through the output transformer. However, as is shown subsequently, the operation of the circuit can be represented graphically by construction of the plate characteristics for a *composite tube*, the com-

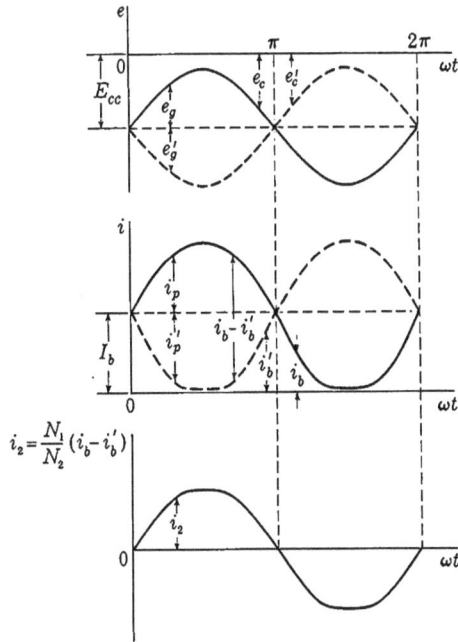

Fig. 39.   Waveforms in a push-pull amplifier when the operation extends into the nonlinear region of the tube characteristics.

posite tube being defined as one that, operating into one-half the output transformer primary winding with the other half open-circuited, gives the same current and power in the load as the two tubes in push-pull. The path of operation on these *composite characteristics* is a straight line, and the methods of finding the power output and harmonic generation given in Art. 13 are applicable. It is assumed again in this analysis that the output transformer is ideal, thus having no resistance, leakage reactance, exciting current, or losses, and that the load is a pure resistance. The circuit diagram is again that of Fig. 35, and Eq. 198 applies to the plate circuit.

Equation 198 shows that the output current i_2 is the same as the

[17] P. Franklin, *Differential Equations for Electrical Engineers* (New York: John Wiley & Sons, Inc., 1933), 65.

output current that would exist if an equivalent current

$$i_d = i_b - i_b'$$ [202]

existed in one-half the transformer primary winding. Thus the operation of the circuit in Fig. 35 is the same as that in Fig. 40, where the

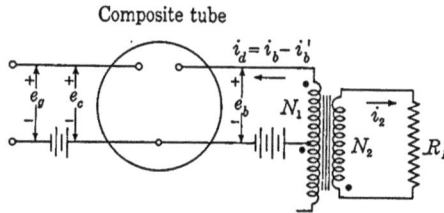

Fig. 40. Composite tube and circuit equivalent to that of Fig. 35.

composite tube has a plate current i_d and the relationships among i_d, e_b, and e_c are yet to be found.

In functional notation, Eq. 202 becomes

$$i_d(e_c, e_b) = i_b(e_c, e_b) - i_b'(e_c', e_b'),$$ [203]

where the two terms on the right-hand side of the equation represent the characteristics of the individual tubes. These can be combined in accordance with the circuit restrictions as follows: Since the input transformer is ideal,

$$e_g = -e_g';$$ [204]

thus

$$e_c' = E_{cc} + e_g' = E_{cc} - e_g = e_c - 2e_g .$$ [205]

Also, since the ideal output transformer acts as an autotransformer between the two plate circuits, it makes

$$e_p = -e_p'.$$ ▶[206]

With two separate load resistors substituted for the transformer and load, Eq. 206 would not be correct, and *the entire analysis that follows is therefore true only when the transformer is used.* Equation 188, which was obtained in the linear analysis, does not apply to the nonlinear operation of the tube.

Since the resistance of the transformer is negligible,

$$E_b = E_b' = E_{bb};$$ [207]

thus

$$e_b' = E_{bb} + e_p' = E_{bb} - e_p$$ [208]

and

$$e_b = E_{bb} + e_p ,$$ [209]

whence

$$e_b' = 2E_{bb} - e_b. \qquad [210]$$

Since the tubes are assumed to be identical, the function i_b has the same form as the function i_b', but they are functions of different variables; thus the prime may be dropped if the variables are indicated, and substitution of Eqs. 205 and 210 in Eq. 203 gives

$$i_d(E_{cc} + e_g, e_b) = i_b(E_{cc} + e_g, e_b) - i_b(E_{cc} - e_g, 2E_{bb} - e_b). \quad [211]$$

In this way, once the power-supply voltages E_{bb} and E_{cc} are selected, the independent variables are reduced to two, namely, e_g and e_b, and the characteristics may be graphically constructed as shown in Fig. 41,

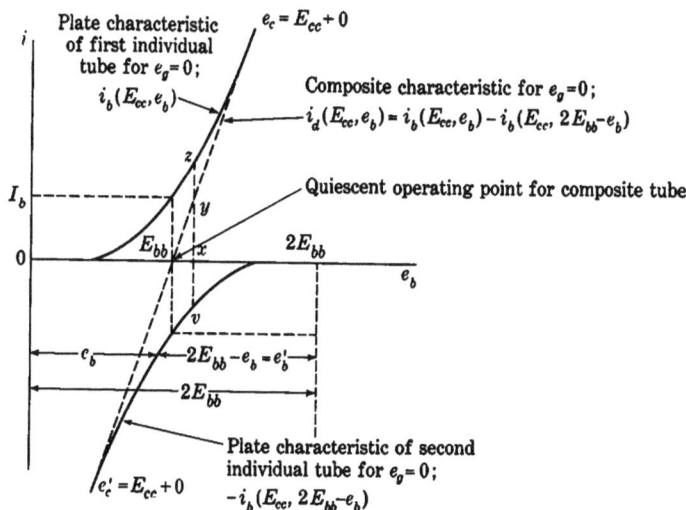

Fig. 41. Construction of a plate characteristic curve for the composite tube with zero grid-signal voltage.

where the curves for zero grid-signal voltage e_g are drawn. In accordance with Eq. 211, the plate-current characteristic curve for the composite tube is obtained through rotating the plate characteristic for an individual tube through 180 degrees about the origin, then displacing the curve along the axis of abscissas by an amount $2E_{bb}$ so that the point corresponding to the old origin falls at the point $2E_{bb}$ on the plate-voltage scale, and, finally, adding algebraically the ordinates of the two curves. Thus the length of the ordinate $\overline{xy}$ equals $\overline{xz}$ minus $\overline{xv}$ in Fig. 41. Several features of the composite characteristics are at once apparent; namely, (a) the quiescent plate current in the composite tube is zero; (b) the composite plate characteristic is much straighter than either of the individual tube characteristics, although it may have more curvature than that shown in Fig. 41; (c) the plate resistance of the composite tube, $\partial e_b / \partial i_d$, is one-half the plate resistance of either of the individual tubes at the quiescent operating point; and (d) the plate resistance for the composite tube is essentially constant over the range shown, though the plate resistances of the individual tubes vary considerably.

The construction of the plate characteristics of the composite tube corresponding to three particular values of grid-signal voltage e_g is shown in Fig. 42. One particular value of the grid-signal voltage is

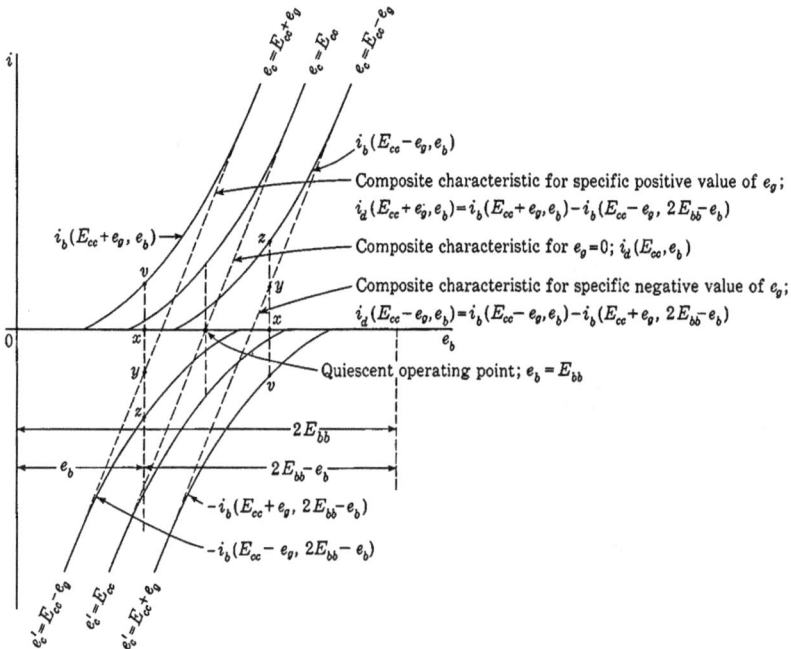

Fig. 42. Construction of the plate characteristics of the composite tube for three specific values of grid-signal voltage.

zero, another is positive, and the third is equal in magnitude to the positive one but is negative. The curves for zero grid-signal voltage lie between the others in the figure and are similar to those in Fig. 41. To obtain the composite characteristic curve for the particular positive value of grid-signal voltage, the curve for an equal negative grid-signal voltage, denoted by $i_b(E_{cc} - e_g, e_b)$ on the diagram, is rotated and displaced as previously explained, whereupon it becomes the curve denoted by $-i_b(E_{cc} - e_g, 2E_{bb} - e_b)$. The ordinates of this latter curve are then added algebraically to those of the curve for the particular positive value of grid-signal voltage, denoted by $i_b(E_{cc} + e_g, e_b)$, giving the left-hand dotted line with positive slope. Again, on this line, the length of the ordinate $\overline{xv}$ is subtracted from that of the ordinate $\overline{xz}$ to give the length $\overline{xy}$, and the resulting curve is the characteristic of the composite tube for the particular positive value of e_g. The construction of the right-hand dotted line, which is the characteristic of the composite tube for the particular negative value of e_g, is done in a similar manner. Note that composite characteristics for equal positive and negative grid-signal-voltage increments are images of each other about the quiescent operating point. This symmetry is the reason for several important operating features of the push-pull circuit with an output transformer.

Figure 43 shows a family of composite characteristics for two triodes in push-pull with a plate-supply voltage of 240 volts. The characteristics therefore spread over a range of 480 volts. The grid-bias voltage is −50 volts, and the heavy dotted lines with positive slopes are the two individual tube characteristics for zero grid-signal voltage. When the ordinates of these curves are added algebraically, the result is the heavy solid line with positive slope, which is the composite characteristic for zero grid-signal voltage. The light solid lines with positive slope are the composite characteristics for 10-volt increments of grid voltage constructed by the method shown in Fig. 42. The solid and dotted lines with negative slope are discussed subsequently.

Not only the characteristic corresponding to the zero grid-signal-voltage condition but all the composite characteristics over a wide range of grid voltage are essentially straight lines. The grid-bias voltage E_{cc}, chosen in Fig. 43 as −50 volts, is one for Class A operation. In later discussions of Class AB and Class B push-pull operation, it is shown that, for those operating conditions, the composite characteristics are not always straight. Note that the characteristics of the composite tube as defined here are dependent upon the values of E_{cc} and E_{bb} and thus depend on quantities external to the tube. The composite tube differs from an ordinary tube in this respect.

The composite plate characteristics from Fig. 43 are reproduced in Fig. 44. As far as the current, voltage, and power in the load resistor are concerned, the operation of the push-pull circuit is equivalent to that of Fig. 40, where the characteristics of the composite tube are given by Fig. 44. The quiescent operating point is on the abscissa axis at e_b equals E_{bb}, where i_d equals zero, and a load line having a slope

$$- \frac{1}{(N_1/N_2)^2 R_L}$$

gives the path of operation on the characteristics, as is shown by the solid line of negative slope in Figs. 43 and 44. The waveform distortion that occurs because of harmonic generation in the tubes may be obtained from this load line by the methods of Art. 13, if i_b is replaced by i_d, and negative values of i_d are recognized. Because of the symmetry of the composite characteristics mentioned previously,

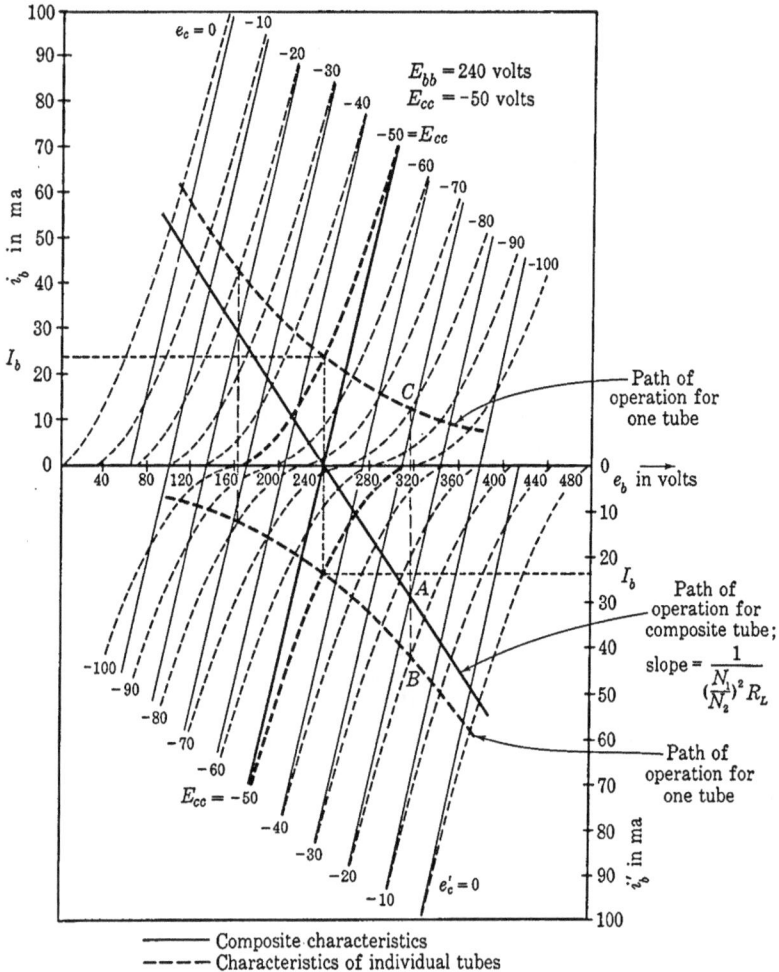

Fig. 43. Composite characteristics for two triodes in a push-pull circuit at specific values of grid-bias voltage and quiescent plate voltage.

no steady or even-harmonic components appear in i_d or in the load current.

Although the plate currents in the individual tubes may decrease to zero and remain there for an appreciable fraction of a half-cycle, such behavior is not apparent on the composite plate characteristics, since the operation along the load line is entirely symmetrical about the quiescent operating point for positive or negative values of grid-signal voltage. However, the paths of operation for the individual tubes can be found readily by a process that is the reverse of the one by which the composite characteristics are constructed. Thus, in Fig. 43, a vertical line through the intersection at A of the path of operation

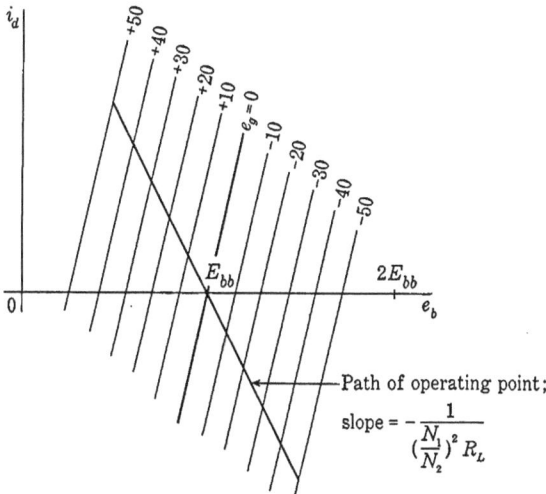

Fig. 44.   Plate characteristics of the composite tube with load line superposed.

and a particular composite characteristic intersects at B and C the two individual tube characteristics from which the composite characteristic is constructed, thereby disclosing the individual tube plate currents for a particular value of grid-signal voltage. By this method, the paths of operation for the individual tubes shown by the dotted curves with negative slopes are constructed. They are curved, even though the load is purely resistive and the output transformer is ideal.

For the particular conditions illustrated in Fig. 43, the individual tube plate currents do not fall to zero when the range of operation is limited by the two curves corresponding to zero grid voltage at the two tubes; thus the operation is Class A_1. However, if the grid-bias voltage is chosen somewhat larger, the individual tube currents may be zero for an appreciable fraction of the cycle, and operation changes to Class AB_1, which is discussed in more detail in Ch. X. Figure 45 shows a limiting example for Class A_1 operation, since the individual plate currents in it just reach zero when the grid voltage of the opposite tube reaches zero.

The considerations that govern the maximum power output with a prescribed amount of harmonic generation from a push-pull amplifier are quite different from those for a single-tube amplifier given in Art. 16. Since the even harmonics generated in the tubes are canceled in the

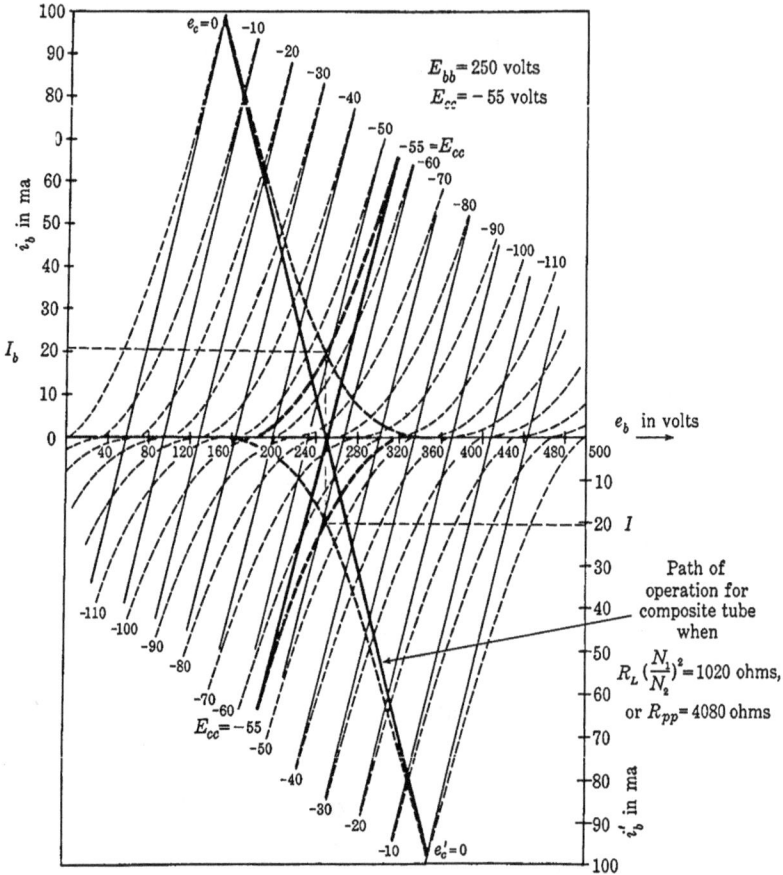

Fig. 45.   Composite characteristics for the two triodes of Fig. 43 in a push-pull circuit for the limiting condition of Class A_1 operation.*

output transformer, the total generation of harmonics in the amplifier is smaller in the push-pull amplifier than in the single-tube amplifier when the tubes have the same operating voltages and deliver the same power output individually. Consequently it follows that, for the same total harmonic generation in the amplifier, the maximum power output from each tube is larger in the push-pull amplifier than in the

* This diagram is adapted from B. J. Thompson, "Graphical Determination of Performance of Push-pull Audio Amplifiers," *I.R.E. Proc., 21* (1933), Fig. 8, p. 595, with permission.

single-tube amplifier. The increase may be as much as 50 per cent. This increased power output is made possible by the fact that changes both in the operating voltages and in the load resistance effective in the plate circuits of the individual tubes may be made under the specified conditions. Since the even harmonics are canceled, the path of operation may be extended farther into the lower region of the tube characteristics in a push-pull amplifier than is indicated in Fig. 31, Art. 16, for a single-tube amplifier when a prescribed amount of harmonic generation is not to be exceeded. Accordingly, for the same amount of harmonic generation, a larger magnitude of grid-bias voltage and a larger grid-signal voltage amplitude may be used in the push-pull amplifier than in the single-tube amplifier, and the power output from each tube is therefore larger.

The change of voltages described in the preceding paragraph is one factor that contributes to the increased value of maximum power output from the push-pull amplifier. Another factor is the change that may be made in the effective load resistance for each tube. Since the composite characteristics are symmetrical about the curve for zero grid-signal voltage, and are practically straight and parallel for Class A_1 operation, the amount of harmonic generation is essentially independent of the plate-to-plate load resistance for a particular value of grid-signal-voltage amplitude. Consequently, the considerations of Art. 16 and the result that the effective load resistance for a tube must be approximately equal to twice the plate resistance of the tube for maximum power output with a prescribed amount of harmonic generation and a prescribed quiescent plate voltage do *not* apply to the composite tube. Instead, the considerations of Art. 15 apply, and the slope of the path of operation on the composite characteristics for maximum power output or maximum power sensitivity is equal to the negative of the slope of the composite characteristics. The slope of the path of operation on the composite characteristics corresponds to $(N_1/N_2)^2 R_L$, which is one-fourth the plate-to-plate load resistance given by $4(N_1/N_2)^2 R_L$. Thus the plate-to-plate load resistance should equal four times the plate resistance of the composite tube. The plate resistance of the composite tube is, however, approximately one-half the plate resistance of the individual tubes at the quiescent operating point. The optimum plate-to-plate load resistance for the push-pull amplifier therefore is twice the plate resistance of the individual tubes, and the optimum value of the load resistance effective in the plate circuit of each tube hence is equal to the plate resistance of the tube. In accordance with Eq. 168, this condition results in an increase of power output from each tube, or of power sensitivity, over the value obtained when an effective load resistance equal to twice the plate resistance of the tube is used.

The discussion in this article applies to Class A_1 operation of triodes used with an output transformer for delivering power to a resistance load. The use of the push-pull connection as a balanced voltage amplifier without an output transformer is discussed in Ch. IX and Class AB, Class B, and Class C operation is discussed in Ch. X.

20. Symbols for vacuum-tube circuit analysis

In the preceding articles of this chapter a number of special symbols are introduced and defined. A large number of symbols are needed in analysis of vacuum-tube circuits, because the operation is complicated by the superposition of direct quantities and alternating quantities having several harmonic components. Confusion is likely to result if a consistent set of symbols is not defined and adhered to through all the analysis. The methods of circuit analysis given in *Electric Circuits* and elsewhere are directly applicable to vacuum-tube circuits and may be used with arbitrarily assigned positive reference directions of currents and voltages. The consistent set of definitions adopted in this volume is merely one of the innumerable possible sets. It is adopted because it is in substantial agreement with the latest standards[18] available; thus it is the set most likely to be encountered by the reader in other publications in the future. To eliminate one additional source of confusion, the definitions here are extended to include the assigned positive reference *direction* of the quantities. If symbols defined for the directions opposite to those chosen here are needed, other symbols can be used.

Table I gives the definitions and symbols of the current and voltage components that are fundamental to the operation of triode circuits. In Table I, the first four rows contain symbols pertaining to the total quantities, and the middle four rows, symbols that pertain to the varying components and are useful in circuit analysis when harmonic generation is neglected. The last four rows contain symbols useful in representing nonsinusoidal varying components as a Fourier series. Complex quantities are indicated by boldface italic type.

Tubes with more than one grid require additional symbols that are supplied in accordance with the principles outlined in Art. 8, Ch. IV. It should be noted that one possible source of confusion lies in the fact that some of the symbols in the last three rows of Table I are also used with a different meaning for multigrid tubes. However, this duplication does not lead to difficulty in any of the problems treated in this book.

---

[18] *Standards on Abbreviations, Graphical Symbols, Letter Symbols, and Mathematical Signs, 1948* (New York: The Institute of Radio Engineers, 1948), 2–3.

TABLE I. SYMBOLS FOR VACUUM-TUBE CIRCUITS

Component	Name and Assigned Positive Reference Direction* of Quantity			
	Voltage rise from cathode to grid	Voltage rise from cathode to plate	Current through the external circuit toward the grid	Current through the external circuit toward the plate
Instantaneous total value	e_c	e_b	i_c	i_b
Quiescent value; steady value when varying component of grid voltage is zero	E_c	E_b	I_c	I_b
Average value of the total quantity	E_{cs}	E_{bs}	I_{cs}	I_{bs}
Instantaneous maximum of the total quantity	E_{cm}	E_{bm}	I_{cm}	I_{bm}
Instantaneous value of the varying component	e_g	e_p	i_g	i_p
Complex effective value of the varying component	E_g	E_p	I_g	I_p
Effective value of the varying component	E_g	E_p	I_g	I_p
Amplitude of the varying component	E_{gm}	E_{pm}	I_{gm}	I_{pm}
Average value of the varying component	E_{g0}	E_{p0}	I_{g0}	I_{p0}
Instantaneous value of the harmonic components	$e_{g1}, e_{g2}, \cdots$	$e_{p1}, e_{p2}, \cdots$	$i_{g1}, i_{g2}, \cdots$	$i_{p1}, i_{p2}, \cdots$
Effective value of the harmonic components	$E_{g1}, E_{g2}, \cdots$	$E_{p1}, E_{p2}, \cdots$	$I_{g1}, I_{g2}, \cdots$	$I_{p1}, I_{p2}, \cdots$
Amplitude of the harmonic components	$E_{g1m}, E_{g2m}, \cdots$	$E_{p1m}, E_{p2m}, \cdots$	$I_{g1m}, I_{g2m}, \cdots$	$I_{p1m}, I_{p2m}, \cdots$

* Amplitudes and ordinary effective values, not complex, in the table have only magnitude. They do not have sense, or direction.

PROBLEMS

1. A triode having the plate characteristics of Fig. 7, Ch. IV, is used in the circuit of Fig. 3 with a plate-supply voltage E_{bb} of 400 volts, and a load resistance R_L of 100,000 ohms. For grid-bias voltages E_{cc} of -2, -3, and -4 volts, what are the values of the quiescent plate current I_b and quiescent plate voltage E_b?

2. A relay having a resistance of 1,000 ohms is to be operated by the plate current of a vacuum triode. If the available direct grid-signal voltage is 5 volts and the relay closes at 30 ma and opens at 20 ma, which of the triodes whose plate characteristics appear in some one manufacturer's literature should be satisfactory? For each triode selected, specify the plate-supply and grid-bias voltages that must be used.

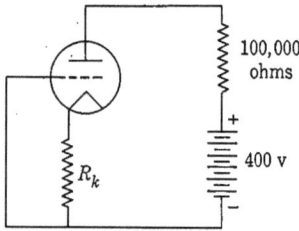

100,000 ohms

400 v

R_k

Fig. 46.   Triode circuit for Prob. 3.

3. A triode has the plate characteristics given in Fig. 7, Ch. IV, except that the grid-voltage scale is to be multiplied by ten—that is, the increment in grid voltage between adjacent curves is 10 volts instead of 1 volt. The tube is connected as shown in Fig. 46 with a 400-volt battery as a plate-power supply and a plate-load resistance of 100,000 ohms.

(a) First, the resistor R_k is so adjusted that there is a voltage of 50 volts between the grid and the cathode. Find the quiescent plate current I_b, the quiescent plate voltage E_b, and the required value of R_k.

(b) Second, the cathode resistor R_k is changed to 25,000 ohms. What are the new values of the quiescent plate current, plate voltage, and grid voltage?

4. The plate current of a particular triode is closely represented by the expression

$$i_b = 17 \times 10^{-5}\left(e_c + \frac{e_b}{8}\right)^{1.7} \text{amp,}$$

where e_c and e_b are in volts.

(a) Determine the plate current i_b corresponding to a grid voltage e_c of -15 volts and a plate voltage e_b of 200 volts.

(b) Find the dynamic plate resistance r_p and the mutual conductance g_m of the tube at the operating point specified in (a).

(c) If the tube is used as a Class A_1 voltage amplifier with a load resistance of 10,000 ohms and a grid-bias voltage E_{cc} of -15 volts, what plate-supply voltage E_{bb} is required to produce a quiescent plate current equal to that determined in (a)?

(d) Determine the voltage amplification of the amplifier for the conditions in (c).

5. A linear Class A_1 amplifier is designed with a vacuum triode having an amplification factor μ of 8 and a plate resistance r_p of 10,000 ohms. The load in the plate circuit is a resistor of 20,000 ohms and a capacitor of 0.01732 μf connected in parallel. A sinusoidal grid-signal voltage of 3 volts rms at 796 cps is applied.

(a) What is the complex voltage amplification?

(b) Sketch the vector diagram approximately to scale, making sure that it is consistent with your assigned positive reference directions of currents and voltages.

(c) What are the minimum allowable quiescent plate current I_b and the minimum allowable magnitude of the negative grid-bias voltage E_{cc} for operation in Class A_1?

6. A triode is used as a linear Class A_1 amplifier. The load resistance is 15,000 ohms, E_b is 250 volts, I_b is 9 ma, E_{cc} is -8 volts, g_m is 2,600 micromhos, μ is 20, and the frequency of the sinusoidal grid-signal voltage is 800 cps.

(a) What is the voltage amplification?
(b) What plate-supply voltage E_{bb} is needed?
(c) What inductance in henries substituted for the load resistance used above will give the same magnitude of voltage amplification?
(d) What new value of plate-supply voltage E_{bb} is needed in (c) to produce the same quiescent plate current as exists with the resistance load?

7. A triode is used in the circuit of Figs. 5a and 15 as a linear Class A_1 amplifier with a resistance load and a cathode-bias circuit. The resistance R_k is adjusted so that when the plate-supply voltage E_{bb} is 300 volts, the quiescent plate current is 8 ma, and the quiescent grid-bias voltage is -8 volts. In the circuit, μ is 20, r_p is 10,000 ohms, C_k is 10 μf, and R_L is 20,000 ohms.

(a) What must the value of R_k be?
(b) What is the quiescent value of the plate voltage?
(c) Assuming that all tube and wiring capacitances are negligible, plot a curve of the voltage gain E_o/E_s as a function of frequency, including points for frequencies of 1, 10, 100, and 1,000 cps.
(d) At what frequency is the reactance of C_k one-third that of R_k and $(r_p + R_L)/(\mu + 1)$ in parallel?
(e) What is the percentage discrepancy between the voltage amplification at the frequency found in (d) and the voltage amplification at 1,000 cps and higher, as indicated by the curve in (c)?

8. A triode is to be used as a linear amplifier with a cathode-bias resistor as in Fig. 5. The quiescent plate voltage is to be 250 volts, the quiescent plate current 5 ma, and the quiescent grid voltage -13.5 volts.

(a) What values of R_k and E_{bb} should be used if the load is a 10-henry inductor having 1,000 ohms resistance?
(b) With the circuit parameters arranged as in (a), μ is 13.8, and r_p is 9,500 ohms for the tube. Find the voltage amplification of the amplifier with a direct input-signal voltage, and with sinusoidal input-signal voltages having frequencies of 20, 200, and 2,000 cps when C_k is omitted.
(c) How large must the capacitance C_k be to make the voltage amplification for the sinusoidal input-signal voltages in (b) substantially the same as it would be with a constant grid-bias voltage?

9. A Type 6AU6 pentode is used as a Class A amplifier in the circuit of Fig. 47, with a plate-supply voltage E_{bb} of 300 volts and a load resistance R_L of 50,000 ohms. The cathode-bias resistor R_k produces a grid-bias voltage of -1.5 volts and the quiescent screen-grid current through the screen-grid resistor R_s reduces the screen-grid voltage so that its quiescent value is 100 volts. Find:

(a) The quiescent plate current and plate voltage.
(b) The quiescent screen-grid current on the assumption that the ratio of the screen-grid current to the plate current is independent of the negative

grid-bias voltage for any plate voltage well above that at the knee of the curves.

(c) The required value of R_s.

(d) The required value of R_k.

(e) The values of C_k and C_s required to make the voltage amplification substantially independent of frequency for frequencies above 100 cps.

(f) The voltage amplification for frequencies above 100 cps when the values from (e) are used.

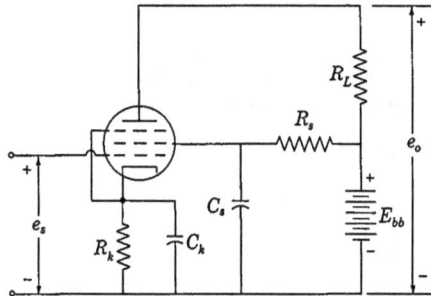

Fig. 47. Pentode amplifier connections for Prob. 9.

10. A high-mu triode is used as a Class A_1 amplifier at a frequency of 15,000 cps. For the conditions of operation, μ is 100, r_p is 65,000 ohms, C_{gp} is 2 $\mu\mu$f, C_{gk} is 2.2 $\mu\mu$f, and C_{pk} is 0.8 $\mu\mu$f. On the assumption that the effective interelectrode capacitances are increased by the wiring to three times those values, what are the effective input shunt conductance, shunt capacitance, and shunt admittance,

(a) for a resistance load of 500,000 ohms?

(b) for a pure inductance load of 500,000 ohms reactance?

11. A triode operated as a Class A_1 amplifier with a load resistance of 250,000 ohms has the coefficients and interelectrode capacitances stated in Prob. 10. Taking the interelectrode capacitances into account but neglecting other stray capacitances, determine for grid-signal-voltage frequencies of 100 cps and 10,000 cps:

(a) the voltage amplification of the stage,

(b) the ratio of the output and input voltages when a 1-megohm resistor is inserted in series with the source of input voltage so that the actual grid-signal voltage e_g is smaller than the input voltage by the amount of the voltage drop in the resistor.

12. A triode is used with a pure resistance load as a Class A_1 amplifier.

(a) Show that when interelectrode capacitances are taken into account as in Fig. 17, the incremental instantaneous grid voltage e_g and grid current i_g are related just as they are in the passive network with constant resistances and capacitances shown in Fig. 48. This network is thus shown to be an exact equivalent input circuit that does not involve the assumptions used in deriving Fig. 18.

(b) What special relation among the circuit parameters is required in order that the dotted connection can be made without altering the relation between i_g and e_g?

(c) With the condition in (b) satisfied, what additional condition is required in order that the equivalent input circuit reduce to that in Fig. 18a?

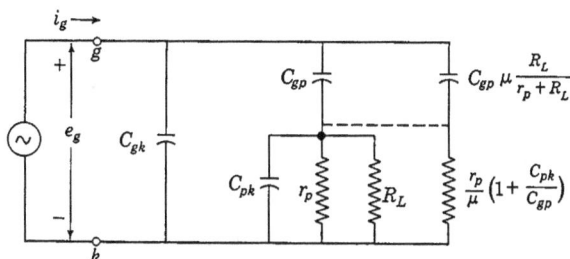

Fig. 48. Equivalent input circuit for a triode with a resistance load, for Prob. 12.

13. One unit of a Type 6SN7-GT twin triode is connected as a cathode-follower amplifier, as in Fig. 22, with a cathode-load resistance of 25,000 ohms and a plate-supply voltage of 300 volts.

(a) Using the manufacturer's plate characteristics, plot the output voltage as a function of the input voltage over the full range of Class A_1 operation.
(b) What positive input-signal voltage just causes the grid-to-cathode voltage to reach zero?
(c) What negative input-signal voltage just causes cut-off of the plate current?
(d) Determine from the curve from (a) the voltage amplification for small input-signal voltages and compare with the value computed from the incremental equivalent circuit.
(e) Using the manufacturer's values for the interelectrode capacitances, find the effective input capacitance of the circuit.

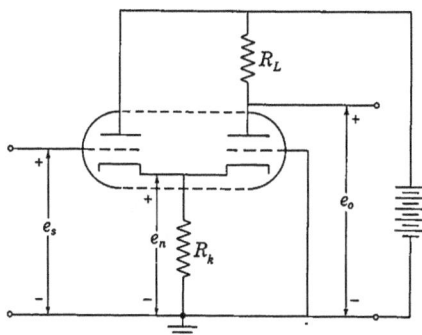

Fig. 49. Cathode follower driving a grounded-grid amplifier, for Prob. 14.

14. A twin triode connected as a cathode follower driving a grounded-grid amplifier is shown in Fig. 49. The electrode voltages indicated on the figure are the incremental values measured from ground. Assuming that the two triode units have identical values of μ and r_p, and using the incremental equivalent circuits of Figs. 24 and 26, determine in terms of μ, r_p, R_L, and R_k:

(a) The voltage amplification of the cathode follower, e_n/e_s,

(b) The voltage amplification of the grounded-grid amplifier, e_o/e_n,

(c) The over-all voltage amplification, e_o/e_s.

15. A Class A_1 amplifier stage comprises a Type 6J5 triode operating with a load resistance R_L of 100,000 ohms, a plate-supply voltage E_{bb} of 300 volts, and a grid-bias voltage E_{cc} of −8 volts. For a sinusoidal grid-signal voltage:

(a) What is the maximum possible peak value of the fundamental component of the output voltage for Class A_1 conditions?

(b) What is the percentage second-harmonic generation under the conditions of (a)?

(c) By what percentage does the indication of a d-c ammeter in the plate circuit change when the grid-signal voltage corresponding to (a) is applied?

16. In the diagrams of Figs. 6 and 7, E_{bb} is 400 volts, E_{cc} is −5 volts, R_L is 100,000 ohms, E_g is $5/\sqrt{2}$ volts, and ω is 377 radians per sec. The triode employed in this Class A_1 amplifier has the plate characteristics given in Fig. 7, Ch. IV.

(a) Determine the quiescent plate current and plate voltage for the operating point corresponding to the conditions specified.

(b) Determine the rms value of the fundamental output voltage developed across the load resistor.

(c) Determine the rms value of the second-harmonic voltage developed across the load, on the assumption that no higher harmonics are generated.

(d) Calculate the complex voltage amplification of the amplifier for the fundamental component.

(e) What is the indication of a d-c ammeter in the plate circuit when the grid-signal voltage is applied?

(f) What is the total rms value of the plate current when the grid-signal voltage is applied?

17. A triode is used in the circuit of Figs. 6 and 7 as a linear Class A_1 amplifier. In the circuit, E_{bb} is 350 volts, E_{cc} is −30 volts, E_g is $20/\sqrt{2}$ volts at an angular frequency ω of 377 radians per sec, and R_L is 2,500 ohms. At the quiescent operating point fixed by these conditions, I_b is 62 ma, μ is 4.2, and r_p is 800 ohms. Find:

(a) the magnitude of the voltage amplification,

(b) the rms value of the alternating component of the voltage developed across R_L,

(c) the power delivered by the plate-power supply,

(d) the plate dissipation of the tube under the quiescent operating conditions,

(e) the a-c power output of the amplifier,

(f) the plate dissipation of the tube with the grid-signal voltage applied,

(g) the plate efficiency η_p.

18. Consider a vacuum triode to be used as a Class A_1 amplifier.

(a) What features must the family of plate characteristics have in order that the amplification factor and the mutual conductance shall be constant over the operating region on the curves?

(b) When the grid-signal voltage is small, what relation must exist between the load line and these characteristics for maximum power transfer from the tube to the load, that is, for maximum power sensitivity?

19. A small sinusoidal alternating voltage is to be amplified by means of a Type 6B4-G or 2A3 triode and a transformer to drive a galvanometer-type oscillograph element. The triode operates with a quiescent plate voltage of 250 volts and a grid-bias voltage of -45 volts. For these operating conditions, the amplification factor μ is 4.2 and the plate resistance r_p is 800 ohms. The oscillograph element has a resistance of 1.2 ohms and requires a current of 0.005 amp per mm deflection by direct current. The transformer may be assumed ideal.

(a) What is the maximum sensitivity in millimeters per volt that may be obtained by this method?

(b) What turns ratio should the transformer have to secure this sensitivity?

(c) What maximum amplitude of current in the element can be obtained if the turns ratio of (b) is used and the operation is restricted to Class A_1? Is the corresponding deflection practicable in the usual oscillograph?

(d) What is the per cent second-harmonic generation with the sinusoidal input voltage that gives the maximum amplitude determined in (c)?

(e) What are the plate dissipation, plate efficiency, and a-c power output corresponding to the maximum amplitude determined in (c)?

20. A Type 6B4-G or 2A3 triode is used as a Class A_1 amplifier with an ideal output transformer to supply a resistance load of 10 ohms. The plate-supply voltage is 250 volts.

(a) If the grid-bias voltage is -45 volts, what transformer turns ratio should be used to supply maximum power to the load for small grid-signal voltages?

(b) Under these conditions, what is the maximum sinusoidal grid-signal voltage that can be used without producing more than 5 per cent second-harmonic generation?

(c) What are the a-c power output, plate dissipation, and plate efficiency corresponding to the grid-signal voltage in (b)?

The transformer turns ratio is now made $\sqrt{2}$ times that in (a), the plate-supply voltage remains the same, and a maximum of 5 per cent harmonic generation is prescribed.

(d) What grid-bias and grid-signal voltages should be used to obtain the maximum power output under these conditions if the operation is to remain in Class A_1?

(e) What are the a-c power output, plate dissipation, and plate efficiency for the conditions of (d)?

(f) Construct to scale the output-voltage waveform for the conditions of (d).

21. Two triodes are used in a direct-coupled push-pull Class A_1 amplifier stage as shown in Fig. 50. Assume that the amplification factor μ of the tubes is 3.5, that the plate resistance r_p is 1,700 ohms, and that these values are constant over the operating range. The quiescent plate current of each tube is 34 ma. When a sinusoidal voltage of 10 volts rms is applied at e_1, find the following:

(a) the rms value of the voltage e_2,

(b) the average value of the voltage e_2,

(c) the alternating component of the plate current in each tube,

(d) the alternating component of the current through the plate-power supply,

(e) the total a-c power output to the plate-load resistors under the foregoing conditions,

(f) the power rating in watts that the resistors must have for the output conditions of (e).

22. Two identical triodes are to be used in a Class A_1 push-pull amplifier having an output transformer. The amplification factor μ of each tube is 35, and the plate

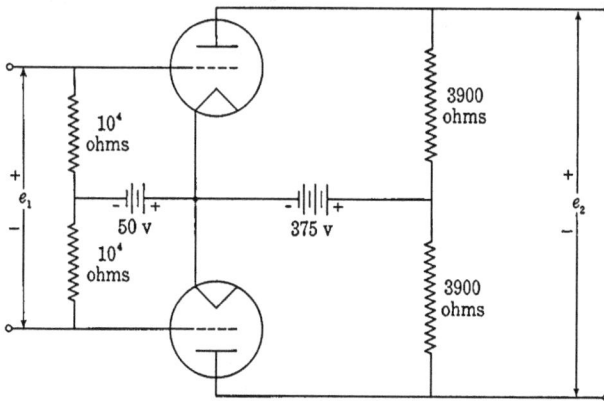

Fig. 50.   Push-pull amplifier circuit for Prob. 21.

resistance r_p of each is 22,000 ohms. The output transformer may be considered to be ideal. The circuit is to supply power to a pure resistance load of 8 ohms.

(a) If the power output is to be a maximum with small harmonic generation and a prescribed plate-supply voltage, approximately what should be the ratio of the total primary turns to secondary turns in the output transformer?

(b) If the grid-signal voltage is small, what is the proper transformer turns ratio?

23. Two identical triodes are used in a Class A_1 push-pull amplifier circuit with a transformer-coupled load. The transformer may be considered to be ideal except that its winding resistances must be considered. In the circuit,

μ = amplification factor of each tube,

r_p = plate resistance of each tube,

$2N_1$ = total number of transformer primary-winding turns,

N_2 = total number of transformer secondary-winding turns,

$2R_1$ = total transformer primary-winding resistance,

R_2 = total transformer secondary-winding resistance,

R_L = load resistance connected to transformer secondary.

The transformer may be represented as an ideal transformer having resistance R_1 in series with each primary winding and resistance R_2 in series with the secondary winding.

For small grid-signal voltages, what value should R_L have, in terms of the other given constants, in order that it shall absorb maximum power?

24. Two triodes are connected as a push-pull amplifier in the circuit of Fig. 35. For each tube, E_{bb} is 250 volts, E_{cc} is -35 volts, I_b is 22 ma, μ is 5.6, and r_p is 2,380 ohms.

(a) What turns ratio N_1/N_2 should the ideal output transformer have to deliver maximum power output to a 1,000-ohm load resistor for a sinusoidal grid-signal voltage E_g of 1 volt rms?

(b) What is the maximum effective value of a sinusoidal grid-signal voltage E_g that may be used without exceeding the limits of Class A_1 operation? Assume that the behavior of the tube remains linear over the full range of operation.

(c) What is the power output to the load under the conditions of (b)?

(d) What is the effective value of the voltage across the full primary winding of the transformer?

(e) What is the effective value of the alternating component of the plate current of an individual tube?

(f) What is the effective value of the alternating component of the current through the plate-power supply?

25. Two Type 6L6 beam power tubes operate as a Class A_1 amplifier with a prescribed maximum of 10 per cent total harmonic generation. The plate-supply and screen-grid-supply voltages are 250 volts, and the grid-bias voltage is obtained from a constant-voltage supply. The load is a pure resistance, and is coupled to the tubes by an output transformer that may be considered to be ideal. Compare the maximum power outputs available from the tubes and the corresponding values of percentage total harmonic generation when they are arranged in

(a) a parallel connection,
(b) a push-pull connection.

26. The plate resistance corresponding to the slope of the composite characteristics representing a pair of triodes operating in a push-pull Class A_1 amplifier circuit is 800 ohms.

(a) If the load is to be an 8-ohm loud-speaker voice coil, determine the turns ratio of the output transformer necessary to provide optimum power transfer to the load. The voice coil may be considered to be a constant pure resistance.

(b) On the assumption that departures of this transformer from the ideal do not affect the validity of Eq. 206 and the succeeding material based upon this equation in the text, specify values of minimum primary self-inductance and maximum total leakage inductance referred to the primary side that will provide an impedance match correct to within 10 per cent over a frequency range of 50 to 5,000 cps. Neglect winding resistances and all capacitance effects (see *Magnetic Circuits and Transformers*, Ch. XVIII, which discusses response characteristics of output transformers).

27. Assume that instead of the assigned positive reference directions of the symbols for vacuum-tube circuits summarized in Art. 20, new directions are assigned as follows:

e_b is the voltage drop from the cathode to the plate,
i_b is the current from the plate into the external circuit,
e_c is the voltage drop from the grid to the cathode.

Furthermore, assume that the definitions of μ, r_p, and g_m are changed so that they become:

$$g_m \equiv -\frac{\partial i_b}{\partial e_c} \qquad r_p \equiv \frac{\partial e_b}{\partial i_b} \qquad \mu \equiv -\frac{\partial e_b}{\partial e_c}.$$

(a) Draw the schematic circuit of the vacuum tube and show the assigned positive reference directions of e_b, i_b, and e_c that correspond to the foregoing definitions.

(b) Letting the subscripts p and g denote varying components of the quantities when they replace b and c, respectively, and using the same assigned positive reference directions for e_p, i_p, and e_g as for e_b, i_b, and e_c, respectively, draw the voltage-source equivalent circuit for varying components and linear operation, and show on it the assigned positive reference directions of all quantities including μe_g.

(c) Using the same symbols and assigned positive reference directions as in (b), draw the current-source equivalent circuit and show on it the assigned positive reference directions of all quantities including $g_m e_g$.

CHAPTER IX

Cascade Amplifiers; Class A_1

The amplifier circuits described in the previous chapter have as their ultimate theoretical limit of voltage gain a value equal to the amplification factor of the single tube, but the practical limit is considerably below this value. Since the voltage gain needed in engineering problems is often more than can be obtained with a single tube, recourse is frequently had to a *cascade amplifier* in which the first tube supplies the grid-signal voltage for the second, and so on. A coupling network is used to connect the plate circuit of each tube with the grid circuit of the succeeding tube. The choice of the type of coupling network depends on the frequency or range of frequencies that the amplifier must transmit, for the coupling network is the primary factor in determining this characteristic of the amplifier. The combination of a tube and its associated coupling networks is called a *stage* of amplification. Push-pull and parallel combinations of tubes are considered as single stages of amplification.

Since negligible power is required to supply the grid-signal voltage for a Class A_1 amplifier tube, all the stages except the last in a cascade Class A_1 amplifier are designed as voltage amplifiers—that is, designed to give the largest practicable amplification or output voltage with due regard to distortion but without regard to efficiency. The last or output stage, however, may have any one of three functions: (a) to supply power with appreciable voltage and current, (b) to supply voltage with negligible current (a very high-impedance load), or (c) to supply current with negligible voltage (a very low-impedance load). For example, power is of primary interest if the amplifier drives a loud speaker, voltage with negligible current is needed if it drives the deflecting plates of a cathode-ray oscilloscope tube, and current with negligible or small voltage is desired if it actuates an indicating device such as the galvanometer-type oscillograph element or a low-impedance cable and the use of an output transformer is not practical.

The particular application, then, governs the choice of coupling circuits throughout the amplifier. Accordingly this chapter deals with the types of coupling networks and the over-all amplifier characteristics that result from their use.

1. FREQUENCY-RANGE CLASSIFICATION OF AMPLIFIERS

Amplifiers find increasing application in measurement apparatus, control equipment, regulatory devices, and the like, but their principal use is in telegraph, telephone, radio, and television communication systems and radar systems; and the customary classification of them as to frequency range is based on their service in these last applications. In order to establish criteria for the fitness of an amplifier for a particular application, an understanding of the nature of the grid-signal voltage to be amplified and of the receiving mechanism is necessary. For example, the signal may be Morse code, voice, music, impulses from a television iconoscope, or a voltage to be measured; and the receiving mechanism may be a telegraph sounder, the human ear, a kinescope, or an indicating instrument. The following is an outline of several important aspects of the problems involved in a communications system.

The electrical counterpart of telegraphed intelligence consists of a voltage or current signal that alternates irregularly and abruptly between zero and a fixed value. Any single pulse can be expressed as a Fourier integral,* but, if the signal is idealized as a periodic series of flat-topped half-cycles, it may be resolved into a Fourier series and the spectrum of the signal may be considered directly. To transmit the signal without change of waveform, the amplifier and all other parts of the system must be able to amplify all voltages whose frequencies lie between the upper and lower limits of the spectrum of the signal; in other words, the amplifier must transmit a certain *band* of frequencies. Furthermore, it must preserve the phase relations among the different frequency components and reproduce them correctly in the output. The lower limit of the frequencies required for telegraph signals is zero, since there is a direct-current component in their waveform. The upper limit of the frequencies required seldom exceeds 100 cycles per second, and is given in Table I as 80 cycles per second along with the frequency limits of various other signals.

The electrical counterpart of speech, music, and other sound is generally vastly more complicated than that of the telegraph signal because the waveforms of these are more irregular. Again, the signal is essentially transient in nature; thus it is readily susceptible to frequency analysis by the Fourier integral method, and its components cover the complete frequency spectrum. However, it is ordinarily considered to have a range of about 16 to 20,000 cycles per second, which are the effective limits of audibility for a steady-state sinusoidal signal. An amplifier that transmits a band of frequencies within this range is called an *audio-frequency amplifier*. With some exceptions, current practice is to transmit a narrower range of frequencies in sound communication, as is indicated in Table I. Economic considerations

---

* See *The Mathematics of Circuit Analysis.*

are the reason for this practice, and inferior reproduction of the communicated intelligence is the result. The limits on the frequency band transmitted are often set by the cost of the transmission lines rather than by the cost of the amplifiers. The resulting waveform distortion is permissible only because of the insensitivity, tolerance, or adaptability of the human ear, for the ear is not only comparatively

TABLE I

FREQUENCY RANGES OF VARIOUS COMMUNICATION SERVICES*

Audio and Video Services

Type of service	Lower frequency limit	Upper frequency limit
Slow-speed telegraph	0	15 c
High-speed telegraph	0	80 c
Wire telephony (speech)	250 c	2,750 c
Wire telephony (music)	30 c	15,000 c
Sound motion pictures	40 c	10,000 c
Limits of audibility	16 c	20,000 c
Television (525 lines, 30 pictures per second)	30 c	4.5 Mc
Radar	500 c	8 Mc

Carrier Services

Type of service	Lower frequency limit	Upper frequency limit
Carrier telegraphy	425 c	10 kc
Carrier telephony	4,000 c	2 Mc
Radio, very-low-frequency	3 kc	30 kc
Radio, low-frequency	30 kc	300 kc
Radio, medium-frequency	300 kc	3 Mc
Radio, high-frequency	3 Mc	30 Mc
Radio, very-high-frequency	30 Mc	300 Mc
Radio, ultra-high-frequency	300 Mc	3,000 Mc
Radio, super-high-frequency	3,000 Mc	30,000 Mc
Radio, extremely-high-frequency	30,000 Mc	300,000 Mc
Radio, microwave	600 Mc	—
Television	54 Mc	890 Mc
Radar	100 Mc	30,000 Mc

c = cycles per sec; kc = kilocycles per sec; Mc = megacycles per sec.

* Parts of this table are adapted from D. G. Fink, *Engineering Electronics* (New York: McGraw-Hill Book Company, Inc., 1938), 272, with permission.

insensitive to moderate changes in the relative amplitudes of the harmonic components that make up an audible signal, but is almost completely insensitive to changes in their relative phase relations over a considerable range.

The signals from a television camera comprise an effective frequency range much greater than that in an audible signal. The frequency range in television corresponds to the amount of picture detail contained in the signal. United States practice is to divide the picture into approximately 525 strips, transmit signals in turn from each of them, and complete the transmission of one picture during an interval of one-thirtieth of a second. A new picture is transmitted during each succeeding interval of one-thirtieth second, and the appearance of the image is therefore similar to the ordinary "movie." If the maximum picture detail to be transmitted is considered to be that in a square picture in which the detail is equal along both dimensions and the light and dark patches form a "checkerboard" pattern, there are 525 times 525, or 275,625, light and dark patches to be transmitted in one-thirtieth second, and 275,625 times 30, or 8,268,750, light and dark patches per second. Each pair of light and dark patches makes up one cycle of the signal waveform. Thus the fundamental frequency in the waveform is approximately 4 megacycles per second. As a result, the frequency range required in the amplifiers used for television is very great—an upper frequency limit of several million cycles per second, with a lower limit about equal to that for an audio-frequency amplifier, is necessary. The term *video* is applied to the signals that carry the visible intelligence in a television transmission system, and an amplifier with an effective frequency range of about 30 to a few million cycles per second is called a *video-frequency* or *broad-band amplifier*.

Although power at audible frequencies is propagated through space, to utilize this propagation for communication is not practicable, because of the enormous size of the antennas required and the excessive power losses that occur. However, the transmission of intelligence by wave propagation through space is practicable at frequencies of the order of 100,000 cycles per second and higher, since the antennas then required are of reasonable size and the power losses encountered are small enough to be counteracted by amplifiers. For radio communication, it is therefore customary to use one of several modulation processes to translate the information contained in an audio-frequency signal into a band of frequencies of about the same range but of much higher mean frequency (see Ch. XII). After radiation by one antenna and reception by another, the information is then retranslated to its original frequency range by another modulation process, often called detection or demodulation, and is thereby made available for conversion to acoustic power and subsequent reception by the ear.

The bandwidth occupied by the information after the process of modulation is only a small fraction of the mean frequency in the resulting modulated radio-frequency wave. Amplifiers to transmit the modulated wave need therefore transmit a frequency band whose width is only a small fraction of the mean frequency—that is, they need have only a "narrow" transmission-band width—and in order to exclude similar signals having near-by mean frequencies at the receiver, they *should* have only a narrow bandwidth. Such performance may be obtained by the use of electrically tuned elements in the amplifier coupling networks; the amplifiers are called *radio-frequency* or *tuned* amplifiers.

2. METHODS OF REPRESENTATION OF AMPLIFIER CHARACTERISTICS

The way in which the ratio of the alternating components of the output and input voltages for steady-state operation depends upon the frequency is of primary importance as an indication of the performance of amplifiers. This ratio is defined in Art. 6, Ch. VIII, as the complex voltage gain A. Expressing it in polar form is convenient, for the amplifier characteristics may then be described by two curves, one giving the magnitude and the other giving the phase angle of the complex voltage gain—both as functions of frequency. The function that describes the variation of the magnitude of the voltage gain with frequency is often referred to as the *frequency characteristic* or *frequency response* of the amplifier, and the function that describes the variation of the phase angle between the input and output voltages with frequency is referred to as the *phase characteristic*. Curves of these functions are in common engineering use as a means of representing amplifier characteristics.

Although the ear senses absolute values as well as relative values of sound intensity or acoustic power, it compares or discriminates among these intensities more nearly by their ratios than by their absolute differences. The same is true for pitches, or frequencies. For example, an ear that can just detect a 1 per cent change in frequency at 100 cycles per second can, to a first approximation, also just detect a 1 per cent change in frequency at 10,000 cycles per second, although the absolute changes are 1 and 100 cycles per second, respectively. Since, on a logarithmic scale, equal ratios of the variable appear as equal increments of the logarithm everywhere on the scale, logarithmic scales are commonly used for both the voltage gain and the frequency in plotting the frequency characteristic, as shown in Fig. 1. The ideal frequency characteristic has, first, a constant voltage gain throughout the range or band of frequencies to be amplified—a "flat" frequency characteristic; and, second, a voltage gain that decreases rapidly

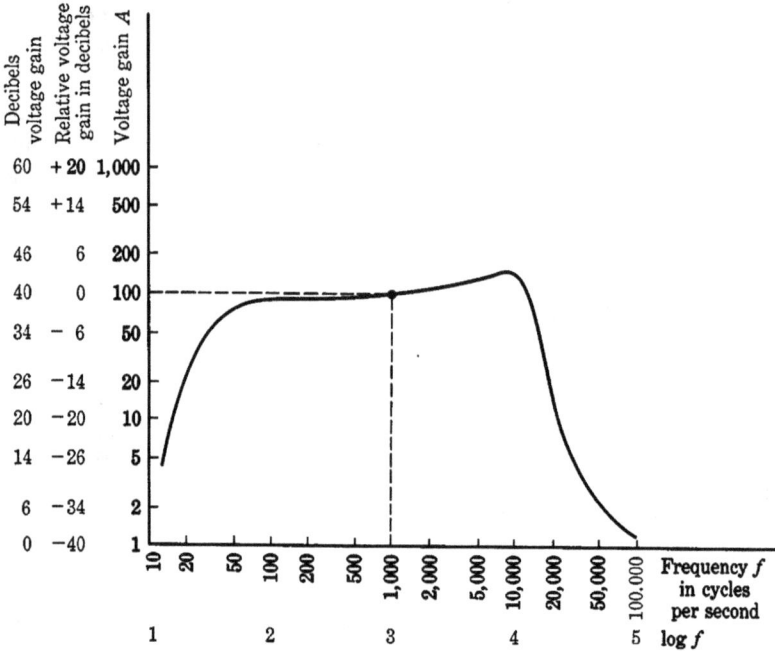

Fig. 1.   Frequency characteristic of an amplifier with equivalent scales.

toward zero outside the desired band. The first property assures a minimum of frequency distortion—variation of voltage gain with frequency; the second helps to minimize noise and interference through removing unwanted frequency components of voltage.

The *decibel*, abbreviated *db*, and used as a unit along the voltage-gain axis in Fig. 1, is fundamentally a measure of a *power ratio* and is defined as a unit such that

$$\left[\begin{array}{c}\text{Number of decibels by which} \\ P_2 \text{ exceeds } P_1\end{array}\right] = 10 \log \frac{P_2}{P_1}, \qquad \blacktriangleright[1]$$

where P_2 and P_1 are two amounts of power.

When two voltages E_1 and E_2 are applied successively to the *same* passive load or to *equal* load impedances, the amounts of power developed are proportional to the squares of E_1 and E_2; thus

$$\left[\begin{matrix}\text{Number of decibels by which} \\ E_2 \text{ exceeds } E_1\end{matrix}\right] = 10\log\left(\frac{E_2}{E_1}\right)^2 = 20\log\frac{E_2}{E_1}, \quad \blacktriangleright[2]$$

where E_2 and E_1 are the voltages across equal load impedances. Fundamentally, the ratio of two *voltages* cannot be expressed by a number of decibels given by Eq. 2 *except* when they are the voltages across *equal* load impedances, because the decibel denotes a *power* ratio.

If the output load of the amplifier is a pure resistance and thus remains fixed as the frequency is varied, the power output is proportional to the square of the output voltage; thus

$$\frac{P_2}{P_2'} = \left(\frac{E_2}{E_2'}\right)^2, \tag{3}$$

where the primed and unprimed quantities are at two different frequencies. If, furthermore, the input voltage at the two frequencies is the same, that is,

$$E_1 = E_1', \tag{4}$$

then, from Eqs. 3 and 4,

$$\frac{P_2}{P_2'} = \left(\frac{E_2/E_1}{E_2'/E_1'}\right)^2 = \left(\frac{A}{A'}\right)^2 \tag{5}$$

where A denotes voltage gain. From this expression,

$$\left[\begin{matrix}\text{Number of decibels by which} \\ A \text{ exceeds } A'\end{matrix}\right] = 20\log\frac{A}{A'}. \quad \blacktriangleright[6]$$

Therefore, on the assumption that the load is a pure resistance, the *relative* voltage gain in Fig. 1 may be expressed in decibels as given by the middle scale of ordinates. The scale refers all values of voltage gain to that at one frequency, the choice of which is arbitrary—in Fig. 1 it is chosen as the value at 1,000 cycles per second. When the voltage gain is larger than the reference value, the number of decibels is positive, and the voltage gain is said to be a certain number of decibels "above," or "up from," the reference value; when the voltage gain is smaller than the reference value, the number of decibels is negative, and the voltage gain is said to be a certain number of decibels "below," or "down from," the reference value.

The voltage gain of an amplifier is often a large number, and, when several stages of amplification are connected in cascade, the over-all voltage gain, which is the *product* of the voltage gains of the individual stages, becomes very large. For convenience, it has become common to express the voltage gain of an amplifier in decibels where the input and output voltages of the amplifier are used as E_1 and E_2, respectively, in Eq. 2. In this way the gain is expressed as a relatively small number, even though the gain is large, and the over-all voltage gain of a number of amplifiers in cascade becomes the *sum* of their voltage gains expressed in decibels. Such use is contrary to the fundamental definition of the decibel, however, because the power gain in an amplifier is not the square of the voltage gain but is often very much larger, since the input power to an amplifier is often negligible as a result of the very high input impedance. Conversely, the power gain in decibels of an ideal transformer is zero on the basis of Eq. 1, but the voltage gain may be very large if Eq. 2 is applied with E_2 and E_1 equal to the primary and secondary voltages, respectively. When Eq. 2 is used to express the ratio of input and output voltages in terms of the word decibel *without regard to the impedance levels*, the fact should be indicated by a careful statement, such as *decibels voltage gain*, to distinguish it from the defined use of the word. The names on the three vertical scales in Fig. 1 illustrate the distinction that should be made.

3. Noise in Amplifiers

As more and more stages of amplification are added to an amplifier it becomes sensitive to smaller and smaller signal voltages. But increasing the total amplification beyond a certain amount is of no avail, because noise is always present in an amplifier. Noise is a term used broadly to describe any extraneous electrical disturbance that causes an output from the amplifier when no signal voltage is impressed at the input. Regardless of where it originates in the amplifier, the noise is conveniently expressed as an equivalent noise voltage at the input that would cause the actual noise output—that is, the actual noise output voltage divided by the over-all voltage amplification of the amplifier. The noise voltage is amplified along with the signal, and tends to mask or cover up the amplified signal in the output. Input signal voltages smaller than the equivalent input noise voltage cannot be amplified to advantage, no matter how many amplifier stages are included. For a full understanding of this limitation, a knowledge of the characteristics of noise is needed. Since a complete treatment of the subject would be very extensive,[1] only a few of its elementary aspects are given here.

[1] G. E. Valley, Jr., and Henry Wallman, Editors, *Vacuum Tube Amplifiers*, Massachusetts Institute of Technology Radiation Laboratory Series, Vol. 18 (New York: McGraw-Hill Book Company, Inc., 1948), Chs. 12, 13, 14.

Noise occurs in amplifiers for many reasons. Some kinds of noise are subject to reduction or elimination through attention to the physical design of the amplifier, its components, and the tubes. For example, voltages induced by time-varying electric and magnetic fields, such as so-called hum voltages, may often be reduced through separation of elements and through shielding. Likewise, voltages called microphonic voltages, which develop because of mechanical vibration of the tubes, capacitor plates, or other circuit elements, may be reduced through cushioning and acoustic shielding. And voltages that result from current through poor contacts either at terminals or between the particles within such circuit elements as carbon resistors may be reduced through mechanical design and selection of the components. But when all these causes are reduced to negligible size by design, noise voltages are still present because of random fluctuation phenomena inherent in the circuit elements and the tubes. These voltages are caused by randomness in the motion of the discrete electrons that conduct the current.

A random fluctuation voltage exists between the terminals of every open-circuited resistance because the free electrons are in continuous motion as a result of their thermal energy.[2] On the average, the number of electrons flowing toward one end of the resistance equals the number flowing toward the other, and the average speeds of the two groups are equal. Hence the average net current is zero, and a direct-current voltmeter connected across the resistance indicates zero voltage. At any particular instant, however, the current in one direction may exceed that in the other because the electron motions are random in time and in space. Consequently, an instantaneous voltage develops between the terminals. This *thermal noise voltage* has a root-mean-square (rms) value given by

$$E_n = \sqrt{4kT\,R\,BW}\,,$$
[7]

where

 R is the resistance, in ohms,

 T is the temperature, in degrees Kelvin,

 k is Boltzmann's constant, 1.38×10^{-23} watt-second per degree Kelvin,

 BW is the effective frequency bandwidth of the system through which the voltage is measured, in cycles per second, and

 E_n is the rms thermal noise voltage, in volts.

The order of magnitude of the thermal noise voltage is ordinarily a

[2] W. Schottky, "Über spontane Stromschwankungen in verschiedenen Elektrizitäts-leitern," *Ann. d. Phys.*, *57* (1918), 541–567; J. B. Johnson, "Thermal Agitation of Electricity in Conductors," *Phys. Rev.*, *32* (1928), 97–109.

small fraction of a volt. For example, if the resistance is 1,000 ohms at room temperature, 300 degrees Kelvin, and the bandwidth is 1 megacycle per second, the rms noise voltage is 4 microvolts. When a resistance is connected in a circuit, the thermal noise voltage given by Eq. 7 may be considered to act in series with the otherwise noise-free resistance R.

It should be understood that this fluctuation noise has no amplitude or frequency in the ordinary sense. It is made up of an infinite number of components differing in frequency by infinitesimal amounts and having infinitesimal amplitudes and random phases. Nevertheless, because of its statistical nature, it has an instantaneous value, and over a period of time the amplitude of the highest peaks is about four times the root-mean-square voltage.[3]

When appreciable inductance or capacitance or both are associated with the resistance, the quantity R effective in Eq. 7 is the resistive component of the impedance between a pair of terminals, and is a function of frequency, $R(f)$. Often the transmission or voltage amplification of the system through which the noise voltage is measured or is effective is also a function of frequency, $A(f)$. In these circumstances, the total rms thermal noise voltage at the output caused by the resistance at the input is

$$E_n = \sqrt{4kT \int_0^\infty R(f)\, A^2(f)\, df}\,. \qquad [8]$$

Noise voltages originate in vacuum tubes because of several internal fluctuation phenomena.[4] Shot noise, partition noise, and induced grid noise are among the more important of these. *Shot noise* results from the fact that the time rate of emission of electrons from the cathode fluctuates in a random manner about a mean value. When the plate current is limited by the cathode temperature, the root-mean-square fluctuation in the current is

$$I_n = \sqrt{2Q_e I_b\, BW}\,, \qquad [9]$$

where

I_b is the average value of the current, in amperes,

Q_e is the charge on an electron, 1.6×10^{-19} coulombs,

BW is the effective frequency bandwidth of the system through which the current fluctuation is measured, in cycles per second, and

I_n is the rms shot-noise current, in amperes.

[3] V. D. Landon, "The Distribution of Amplitude with Time in Fluctuation Noise," *I.R.E. Proc.*, *29* (1941), 50–55.

[4] B. J. Thompson, D. O. North, and W. A. Harris, "Fluctuations in Space-Charge-Limited Currents at Moderately High Frequencies," *RCA Rev.*, *4* (1940), 269–285, 441–472; *5* (1941), 106–124, 244–260, 371–388, 505–524; *6* (1941), 114–124.

When the plate current is limited by space charge, however, as in an ordinary negative-grid vacuum tube, the fluctuations are smoothed by the retarding effect of the space charge, and are greatly reduced.

Vacuum diodes operating with the current limited by the cathode temperature so that Eq. 9 applies are frequently used as calibrated noise generators. The noise produced by the shot effect in such a tube is called *white noise* because the mean-square-current fluctuation per unit bandwidth, which is directly proportional to the noise power in that bandwidth delivered by the shot effect to a series-connected resistance load, is independent of the central frequency in the band just as the power per unit bandwidth in white light is uniform through the frequency spectrum.

Partition noise is the result of random fluctuations in the division of the cathode current among two or more positive electrodes. In the usual pentode, for example, the cathode current divides between the screen grid and the plate, and random fluctuations in the plate current and screen current occur. Partition noise does not occur in a triode.

Induced grid noise is the result of fluctuations in the grid voltage caused by irregularities in the space charge of electrons passing the grid on their way to the plate.[5] The fluctuating space charge induces currents in the grid circuit that develop noise voltages across the input impedance.

Although they are essentially independent of the temperature of the tube, the foregoing fluctuation noise voltages originating in vacuum tubes have the same type of randomness as thermal noise in resistors. Hence they are conveniently described in terms of an equivalent external grid resistance. It is the resistance required to increase the noise power output of the tube to twice the value obtained with the grid short-circuited to the cathode. This equivalent series resistance is found experimentally* to be

$$R_{eq} = \frac{2.5}{g_m}, \qquad [10]$$

for a triode, and

$$R_{eq} = \frac{I_b}{I_b + I_{c2}} \left(\frac{2.5}{g_m} + \frac{20 I_{c2}}{g_m{}^2} \right) \qquad [11]$$

for a pentode, where all quantities are expressed in mks or practical units. The total noise output for a vacuum tube may be computed as if it resulted from thermal noise alone in the sum of this equivalent resistance and the actual resistive component of the impedance connected at the input terminals.

[5] D. O. North and W. R. Ferris, "Fluctuations Induced in Vacuum-Tube Grids at High Frequencies," *I.R.E. Proc.*, *29* (1941), 49–50.

* See reference 4.

All resistances and tubes in an amplifier develop fluctuation noise voltages. But the voltage across the actual grid input impedance and the equivalent grid resistance of the first tube is amplified more than that developed elsewhere in the circuit. Hence its effect usually predominates and makes other fluctuation noise voltages negligible by comparison. To minimize the effect of fluctuation noise and thereby to maximize the signal-to-noise ratio at the output, the amplifier should be designed to have a bandwidth no larger than is necessary to accommodate the signal. Any wider bandwidth merely increases the noise output.

In addition to shot noise, partition noise, and induced grid noise, other types of noise are present in certain electron tubes. *Gas noise* occurs in gas tubes and in partially evacuated vacuum tubes in which the degree of vacuum is not adequate to make the effect of the gas ions negligible. Gas noise is caused by random fluctuations in the rate at which ions are produced in the gas.

Secondary emission noise results from random fluctuations in the number of secondary electrons produced by each primary electron at a surface. These fluctuations are the chief source of noise in multiplier phototubes.

Flicker noise, or the flicker effect, is a type of noise that results from fluctuations in the emission of electrons at the cathode of an electron tube. The mean-square voltage per unit bandwidth for this type of noise from an oxide-cathode tube is approximately directly proportional to the reciprocal of the frequency. In this respect flicker noise is similar to the noise that occurs in carbon-button telephone transmitters, carbon resistors, and the crystal diodes and point-contact transistors discussed in Ch. XIII when they carry current. Such noise differs from white noise because the power associated with it is concentrated at low frequencies instead of being spread uniformly throughout the frequency spectrum.

A quantity called the *noise-figure* serves as an index of the relative "noisiness" of amplifiers and other transmission systems, particularly radio receivers. The noise figure is defined somewhat differently by various authorities, but it is basically a measure of the ratio of the total noise power output to that part of the total that results from thermal noise power supplied by the internal impedance of the source of input voltage. A full understanding of the significance of the noise figure requires a knowledge of the special conditions for which it is intended, and under which it must be measured.[6] A noise-free amplifier

---

[6] H. T. Friis, "Noise Figures of Radio Receivers," *I.R.E. Proc.*, *32* (1944), 419–422, 729; H. Goldberg, "Some Notes on Noise Figures," *I.R.E. Proc.*, *36* (1948), 1205–1214; see also reference 1.

would have a noise figure of unity. Discussion of the noise figure for a specific application is given in Art. 9, Ch. XIII.

4. DIRECT-COUPLED AMPLIFIERS

Amplifier stages may be coupled in cascade in such a manner that either the steady, or direct, or sustained, increments in the plate voltage of one stage are transmitted by the coupling network to become the grid voltage for the succeeding stage, or that only time-varying increments are transmitted. The first method of coupling results in a *direct-coupled, or direct-current, amplifier*, and the second in an *alternating-current amplifier*. In a direct-coupled amplifier, a conductive path exists for the direct current from the plate of one stage to the grid of the next. In an alternating-current amplifier, as is explained in Art. 5, no such conductive path is present; each grid is electrically insulated from the preceding plate. Direct-coupled amplifiers are necessary whenever the time-average component in the incremental input-signal voltage or current must be amplified—for example, when the steady current or voltage supplied by a device, or the change in the steady value thereof, is too small to actuate directly a direct-current indicating instrument, cathode-ray tube, relay, or other mechanism. When only the time-varying component must be amplified, as in a sound-reproducing system, an alternating-current amplifier suffices.

A direct-coupled cascade amplifier circuit of basic design, together with the equivalent circuit for small changes in the voltages and currents, is shown in Fig. 2. Only two stages are shown, but the principles in the analysis are readily used for a number of stages. The resistor R_{L2} at the output may represent the resistance of the relay or other device that the amplifier is to actuate. In the incremental equivalent circuit, the incremental component of plate voltage in the first tube e_{p1} becomes the incremental component of grid voltage e_{g2} for the second tube because of the direct coupling.

From the equivalent circuit and the analysis given in Ch. VIII, the voltage gain for the first stage may be expressed as

$$\frac{e_{p1}}{e_{g1}} = -\frac{\mu_1}{1 + \dfrac{r_{p1}}{R_{L1}}}, \qquad [12]$$

and the voltage gain for the second as

$$\frac{e_{p2}}{e_{g2}} = -\frac{\mu_2}{1 + \dfrac{r_{p2}}{R_{L2}}}. \qquad [13]$$

Since the direct coupling makes

$$e_{g2} = e_{p1} \, , \qquad [14]$$

it follows that the over-all voltage gain is

$$\frac{e_{p2}}{e_{g1}} = \frac{\mu_1 \mu_2}{\left(1 + \dfrac{r_{p1}}{R_{L1}}\right)\left(1 + \dfrac{r_{p2}}{R_{L2}}\right)}. \qquad [15]$$

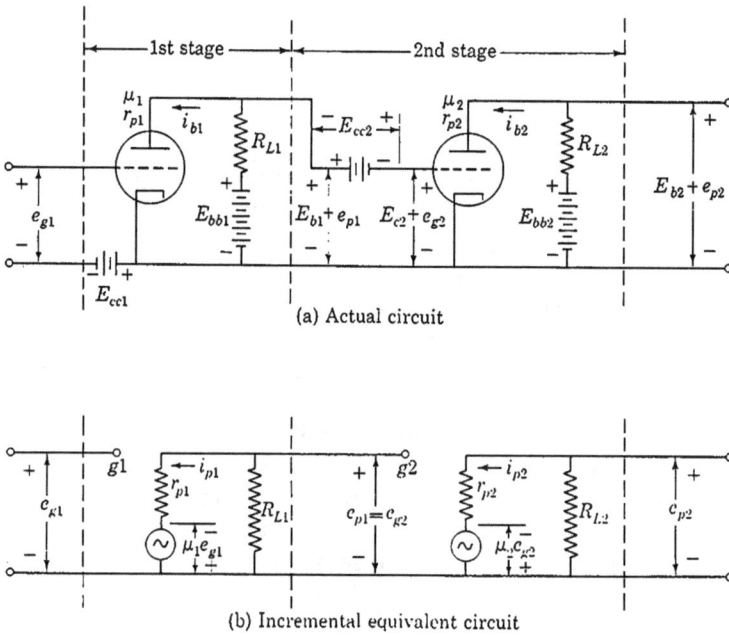

(a) Actual circuit

(b) Incremental equivalent circuit

Fig. 2.   Basic direct-coupled amplifier circuit.

The minus signs in Eqs. 12 and 13 must be interpreted in terms of the arbitrarily assigned positive reference directions of currents and voltages in Fig. 2. If the incremental component e_{g1} has a positive value, the potentials of the plate of the first tube and the grid of the second tube with respect to the cathodes decrease from their quiescent values, as is indicated by the minus sign in Eq. 12. Thus e_{p1} is negative, but the negative sign in Eq. 13, together with Eq. 14, indicates that the potential of the plate of the second tube with respect to the cathode increases under these conditions.

The direct-coupled amplifier circuit shown in Fig. 2a has the serious disadvantage that separate grid-bias batteries are required for each stage. Furthermore, since there is a steady component of voltage E_{b1} between the cathode and the plate of the first tube, the quiescent grid voltage of the second tube, denoted by E_{c2}, is not equal to E_{cc2}, but is given by

$$E_{c2} = E_{b1} + E_{cc2} .$$ [16]

If E_{c2} is to have the negative value required for Class A_1 operation, E_{cc2} must be negative and larger in magnitude than E_{b1}, or

$$|E_{cc2}| > (E_{bb1} - I_{b1}R_{L1}).$$ [17]

Thus the grid-bias battery voltage may be of the order of magnitude of one-half the plate-supply voltage for the previous stage. The battery

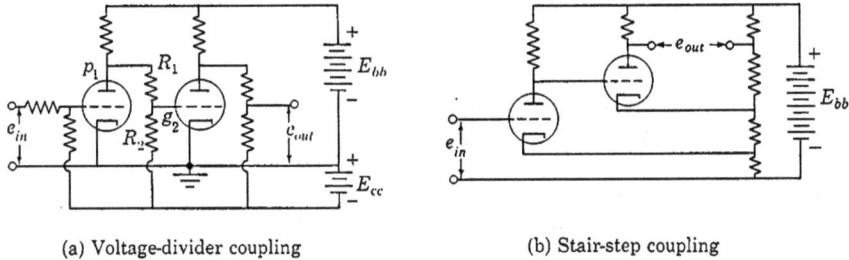

(a) Voltage-divider coupling   (b) Stair-step coupling

Fig. 3. Coupling methods that permit a common power supply.

for E_{cc2} is, therefore, expensive and cumbersome. Moreover, the stray capacitance from it to the cathode adds appreciably to C_{gk}, and, as is explained in Art. 5, reduces the amplification at high frequencies and the rate of rise of the output in response to a sudden incremental input.

Alternative cascade connections for direct-coupled amplifiers that avoid the requirement of separate grid-bias voltage supplies for each stage are shown in Fig. 3. The connection in Fig. 3a has the advantage that one terminal of the output and one terminal of the input are connected together and to one terminal of the common plate-voltage supply E_{bb}. The voltage supplies E_{bb} and E_{cc} may be furnished either from two separate sources as shown, or from one source with a tap at an appropriate point. A resistive voltage divider is connected from the first plate to the negative end of the supply voltage E_{cc}, so that the second grid g_2 may operate quiescently at a potential negative with respect to that of the cathodes. But disadvantages arise from the facts that this voltage divider acts also on the incremental signal components of the plate voltage, thereby reducing the over-all voltage amplification, and constitutes a part of the load for the first stage, thus further reducing the over-all amplification. To establish the necessary quiescent grid voltage for the second tube a considerable voltage drop must exist across R_1. Consequently, R_1/R_2 or E_{cc} must be relatively large. On the other hand, to avoid excessive loss of amplification in the voltage divider, R_2/R_1 should be large. In addition, a maximum permissible value for the parallel resistance of R_2 and R_1 is set by the grid-current tolerance limits on the second tube. In the design of the voltage divider, then, a compromise between power-supply voltage requirements and voltage amplification per stage must be made.

The connection[7] in Fig. 3b has the advantage of simplicity and large voltage amplification, for each grid is connected directly to the previous plate, but it is of limited usefulness because it lacks a common connection between the input and the output. With this coupling method, each cathode is elevated to a quiescent potential above that of the preceding plate. Ordinarily the impedance introduced in the plate circuit by use of the voltage divider must be considered in determining the over-all voltage amplification and the coupling between stages in this type of amplifier.

The grid-bias voltage for each stage can be obtained through use of a cathode-bias resistor, R_k, as in Fig. 5, Ch. VIII, but a by-pass capacitor C_k is of no avail because the input signal is a direct incremental voltage, and the loss of voltage amplification for zero frequency caused by the cathode resistor and indicated in Eq. 59, Ch. VIII, is often intolerable. Resort is therefore sometimes made[8] to by-passing the cathode resistor with the output of a cathode-follower amplifier, or, equivalently, the input of an unloaded grounded-grid amplifier stage. The by-pass path then has a relatively low resistance $r_p/(\mu + 1)$ for direct increments. Thus the increment in the cathode current is effectively by-passed around R_k, but the quiescent component is not.

The major disadvantage of direct-coupled amplifiers is the inherent instability associated with the direct coupling. This instability results primarily from the facts that the characteristics of vacuum tubes change slowly with time for a variety of reasons, and that battery voltages, or alternating line voltages if rectified power supplies are used, likewise change with time. These changes may be regarded as a type of noise comprising extremely low frequencies. The direct-coupled amplifier is particularly sensitive to such changes, because

[7] D. H. Loftin and S. Y. White, "Cascaded Direct-Coupled Tube Systems Operated from Alternating Current," *I.R.E. Proc.*, *18* (1930), 669–682; J. K. Clapp, "A-C-Operated Direct-Current Amplifier for Industrial Use," *Gen. Rad. Exp.*, *13* (February, 1939), 1–5.

[8] S. E. Miller, "Sensitive D-C Amplifier with A-C Operation," *Electronics*, *14* (November, 1941), 27–31, 105–109.

changes in the first stage are amplified by succeeding stages. Consequently, a small fractional change in the voltages or tube characteristics of the first stage may be sufficient to shift the quiescent operating point for the last stage to a region of high distortion or completely out of the linear range on the tube characteristics. After the amplifier is adjusted initially to have zero output for zero input, the output tends to change slowly with time even though the input remains zero—so-called "zero drift" occurs, and frequent resetting of the zero adjustment may be necessary. Such instability is perhaps the most important limitation of direct-coupled amplifiers, and makes special precautions necessary if many stages are to be used. These precautions are not necessary in an alternating-current amplifier because the coupling method that insulates one stage from another does not transmit the slowly changing increments. Balanced circuits[9] and modulation methods[10] are used to minimize this difficulty.

The cathode-coupled difference amplifier shown in Fig. 4 is a balanced amplifier circuit of great utility having several unique features useful in both direct-coupled and alternating-current amplifiers.[11] It is essentially a push-pull amplifier with an unby-passed cathode resistor. For simplicity in this diagram, the source of plate-supply voltage is not included. Here, and in subsequent diagrams, a battery, or rectified-power supply, connected with its positive terminal at the point designated $+E_{bb}$ and its negative terminal at the point

[9] C. E. Wynn-Williams, "The Application of a Valve Amplifier to the Measurement of X-Ray and Photoelectric Effects," *Phil. Mag., 6* (1928), 324–334; D. B. Penick, "Direct-Current Amplifier Circuits for Use with the Electrometer Tube," *R.S.I., 6* (1935), 115–120; H. Goldberg, "A High-Gain D-C Amplifier for Bioelectric Recording," *A.I.E.E. Trans., 59* (1940), 60–64; M. Artzt, "Survey of D-C Amplifiers," *Electronics, 18* (August, 1945), 112–118.

[10] T. R. Harrison, W. P. Wills, and F. W. Side, "A Self-Balancing Potentiometer," *Electronic Industries, 2* (May, 1943), 68–69, 153; M. D. Liston, C. E. Quinn, W. E. Sargeant, G. G. Scott, "A Contact Modulated Amplifier to Replace Sensitive Suspension Galvanometers," *R.S.I., 17* (1946), 194–198; A. J. Williams, Jr., R. E. Tarpley, and W. R. Clark, "D-C Amplifier Stabilized for Zero and Gain," *A.I.E.E. Trans., 67* (1948), 47–57; H. Palevsky, R. K. Swank, and R. Grenchik, "Design of Dynamic Condenser Electrometers," *R.S.I., 18* (1947), 298–314; S. A. Sherbatskoy, T. H. Gilmartin, and G. Swift, "The Capacitative Commutator," *R.S.I., 18* (1947), 415–421; E. A. Goldberg, "Stabilization of Wide-Band Direct-Current Amplifiers for Zero and Gain," *RCA Rev., 11* (1950), 296–300; F. R. Bradley and R. McCoy, "Driftless D-C Amplifier," *Electronics, 25* (April, 1952), 144–148.

[11] F. Offner, "Push-Pull Resistance-Coupled Amplifiers," *R.S.I., 8* (1937), 20–21; H. Goldberg, "A High Gain D-C Amplifier for Bio-Electric Recording," *A.I.E.E. Trans., 59* (1940), 60–64; E. Williams, "The Cathode-Coupled-Triode Stage," *Electronic Engineering, 16* (1944), 509–511; B. Chance, V. Hughes, E. F. MacNichol, D. Sayre, F. C. Williams, Editors, *Waveforms*, Massachusetts Institute of Technology Radiation Laboratory Series, Vol. 19 (New York: McGraw-Hill Book Company, Inc., 1949), 57–58, 340.

designated as ground is to be understood when no such power supply is actually shown. Ordinarily the resistance R_k is made large to achieve the advantages discussed subsequently. Hence, a source E_{cc} with its positive terminal grounded is usually needed in series with R_k as shown to provide satisfactory quiescent conditions. The letter designations of the voltages in Fig. 4 refer to the incremental components only, in accordance with the principles presented in Ch. VIII.

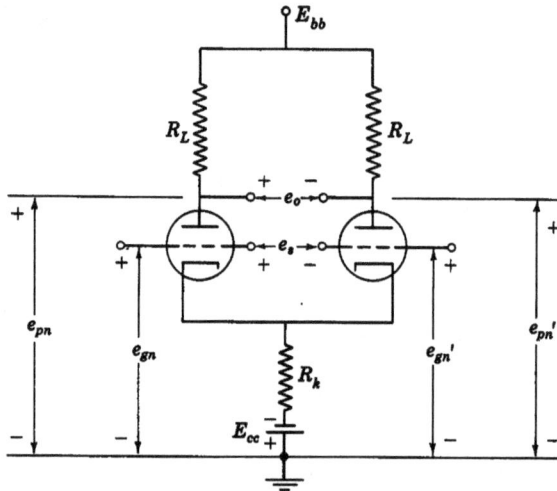

Fig. 4. Cathode-coupled difference amplifier. Incremental grid and plate voltages with respect to ground (not to cathode) are indicated.

When the circuit in Fig. 4 is symmetrical and the tubes are identical, any change in the plate-supply voltage affects both plate currents alike, and hence has no effect on the voltage between the plates. Similarly, a change in the heater-supply voltage, which effectively introduces equal incremental voltages in series with each cathode,[12] does not affect the plate-to-plate voltage. Thus, the output voltage between the plates caused by a direct signal voltage applied between the grids is, except for changes in the tube coefficients that may result, unaffected by changes in either supply voltage—an important consideration in any direct-coupled amplifier.

Since the plate-to-plate output voltage is independent of power-supply voltage fluctuations only when the circuit is symmetrical, an analysis of the circuit operation that takes advantage of and suits the

[12] S. E. Miller, "Sensitive D-C Amplifier with A-C Operation," *Electronics, 14* (November, 1941), 27–31, 105–109.

symmetry is appropriate. For such an analysis, each pair of incremental electrode voltages measured from ground and the plate currents may be expressed in terms of symmetrical components[13]—a *difference component*, sometimes called the signal or out-of-phase component, and a *sum component*, otherwise known as the common-mode, or mean, or average, or in-phase, component. The difference component in the incremental grid-to-ground voltages is

$$e_{gnd} = \frac{e_{gn} - e_{gn}'}{2},$$ [18]

and the sum component is

$$e_{gna} = \frac{e_{gn} + e_{gn}'}{2}.$$ [19]

Note again that e_{gn} and e_{gn}' are voltages measured with respect to ground, not with respect to the common cathode.

In terms of these symmetrical components, the incremental grid-to-ground voltages are

$$e_{gn} = e_{gnd} + e_{gna},$$ [20]

and

$$e_{gn}' = -e_{gnd} + e_{gna}.$$ [21]

The components e_{gnd} and e_{gna} may be computed for any pair of grid-to-ground voltages from Eqs. 18 and 19. From these relations, it is apparent that e_{gna} is the component that is identical in the two grid-to-ground voltages. It is their instantaneous average or mean value (*not* their time average). The component e_{gnd}, on the other hand, is the component that is equal and opposite in the two grid-to-ground voltages. Note that these are components of the instantaneous values and they need not be of the same frequency or waveform. The difference component, for example, might be a direct increment, and the sum component an extraneous 60-cycle-per-second voltage induced in the circuit.

The incremental plate-to-ground voltages may also be resolved into the difference component

$$e_{pnd} = \frac{e_{pn} - e_{pn}'}{2},$$ [22]

---

[13] F. Offner, "Balanced Amplifiers," *I.R.E. Proc.*, *33* (1945), 202. The analysis by components here is similar to that used for power systems: Edith Clarke, *Circuit Analysis of A-C Power Systems*, Vol. I (New York: John Wiley & Sons, Inc., 1943), 54–70, 286–288, 297. It is also an application of the theorem given by A. C. Bartlett, "An Extension of a Property of Artificial Lines," *Phil. Mag.*, Ser. 7, *4* (1927), 902–907; see also O. Brune, "Note on Bartlett's Bisection Theorem for 4-Terminal Electrical Networks," *Phil. Mag.*, Ser. 7, *14* (1932), 806–811.

and the sum component

$$e_{pna} = \frac{e_{pn} + e_{pn}'}{2},$$

[23]

in terms of which the incremental plate-to-ground voltages are

$$e_{pn} = e_{pnd} + e_{pna} ,$$

[24]

and

$$e_{pn}' = -e_{pnd} + e_{pna} ;$$

[25]

and so on for the incremental plate currents. Since the circuit behavior

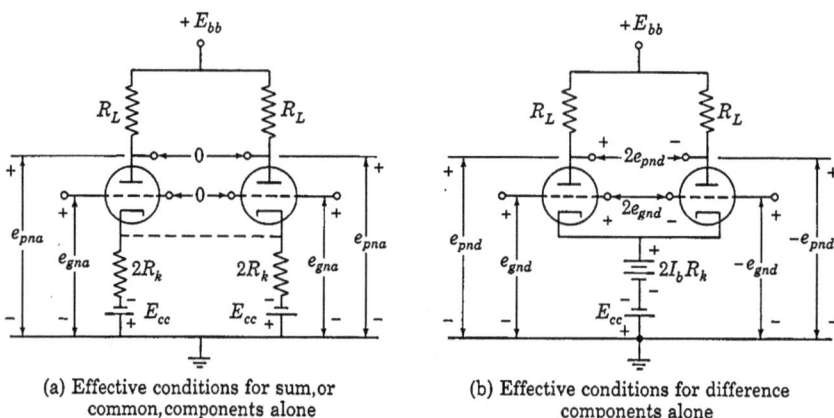

(a) Effective conditions for sum, or
common, components alone

(b) Effective conditions for difference
components alone

Fig. 5.   Circuit conditions for the sum and difference incremental components
applied separately.

is linear in the operating range, it may be analyzed through considering that each of the input components is applied separately, and that the results are superposed.

If the sum component is applied alone, the incremental grid-to-ground voltages are alike, and the voltage between the grids is zero, as is shown in Fig. 5a. The incremental plate currents are hence identical. The cathode resistor R_k in Fig. 4 is equivalent to two resistors $2R_k$ in parallel, as in Fig. 5a, with half the total cathode current in each. But since the incremental plate currents are equal, the individual cathode currents are also equal, and no current tends to flow in the connection shown dotted. Thus, the sum components behave as if the dotted connection were absent. The incremental plate-to-ground voltages are therefore equal, and, in accordance with Eqs. 24 and 25, each equals e_{pna}, as is indicated in Fig. 5a. The voltage gain for the sum components may therefore be computed from either half of the circuit, and, in accordance with the principles in Art. 7, Ch. VIII, is given by

$$\frac{e_{pna}}{e_{gna}} = -\frac{\mu R_L}{r_p + 2R_k(\mu + 1) + R_L}. \qquad [26]$$

If the difference component is applied alone, the incremental grid-to-ground voltages are equal in magnitude, but opposite in sign, as is illustrated in Fig. 5b. Thus, the circuit conditions are as described in the analysis of a push-pull amplifier, Art. 19, Ch. VIII. Since the circuit is linear for increments and is symmetrical, the incremental plate currents must also be equal in magnitude, but opposite in direction. The incremental current through R_k in Fig. 4, which is the sum of the incremental plate currents, is hence zero. Thus, the total current in R_k is constant and equals twice the quiescent plate current for either tube. The resistor R_k may, therefore, be replaced by a constant battery voltage $2I_bR_k$, as is shown in Fig. 5b, without altering the behavior of the difference components. The incremental plate-to-ground voltages produced by the plate currents in the equal load resistors are equal in magnitude and opposite in sign. Hence, they equal the difference component and its negative, as indicated, because, from Eq. 23, the sum component is then zero. The plate-to-plate voltage is therefore twice the difference component. (Equation 22 shows this relation to be true even when a sum component is present.) Since the cathode potential is constant, there is no coupling effective between the two halves of the circuit, and the voltage gain for the difference component can also be computed from either half. It is

$$\frac{e_{pnd}}{e_{gnd}} = -\frac{\mu R_L}{r_p + R_L}. \qquad [27]$$

To determine the plate-to-ground voltages, e_{pn} and $e_{pn}{}'$ for any pair of grid-to-ground voltages e_{gn} and $e_{gn}{}'$, then, Eqs. 18 and 19 are first used to find the difference and sum components of the grid voltages. Substitution of these in Eqs. 26 and 27, respectively, gives the components of the plate voltages, and a combination of these components, in accordance with Eqs. 24 and 25 gives the desired results. If needed, the grid-to-grid voltage as shown on Fig. 4 may then be found from

$$e_s = e_{gn} - e_{gn}{}', \qquad [28]$$

and the plate-to-plate voltage from

$$e_o = e_{pn} - e_{pn}{}'. \qquad [29]$$

Although the foregoing relationships are derived for a symmetrical circuit, their use is by no means restricted to operation with symmetrical input and output voltages. Equations 18, 22, and 27, for example, show that the voltage difference between the plates is directly proportional to the voltage difference between the grids, regardless of the presence of a sum or common component in the grid-to-ground voltages. Thus, for example, one volt applied between the grids gives the same voltage between the plates, regardless of whether one grid, the other grid, the midpoint between them, or any other common reference point for them is at ground potential. In this respect, the circuit amplifies only the difference voltage; it is hence called a *difference amplifier*.

Often an amplifier is desired that not only will amplify the difference voltage, but also will suppress or reject the sum or common component in the input voltages. For example, in a multistage amplifier consisting of a number of direct-coupled difference amplifiers in cascade[14] with the output plates connected to the deflection plates of a cathode-ray oscillograph tube, the difference component at the input results in deflection of the beam, and large voltage gain for this component is generally desired. The amplifier and cathode-ray tube supply voltages are usually arranged so that the anode of the cathode-ray tube is held at the quiescent potential of the amplifier output plates. Thus, the quiescent deflection-plate voltages are the same as that of the anode. Any common component in the input voltages to the amplifier drives the potentials of the deflection plates together above or below that of the anode and tends to defocus the cathode-ray beam. Furthermore, such a sum component may overload one of the later stages in the amplifier. Hence, essentially zero voltage gain for the sum components is desired. Comparison of Eqs. 26 and 27 shows that if $2R_k(\mu + 1)$ is made large compared with $r_p + R_L$, the voltage gain for the sum component becomes small compared with that for the difference component, and the desired performance is attained.

In some applications, only one grid is driven, or the output is taken from only one plate, or both. Such unbalanced operation is readily analyzed by the component method. If, for example, only e_{gn} is supplied, and e_{gn}' is zero, the difference and sum components from Eqs. 18 and 19 are equal. In general, however, they are amplified by different factors, as given by Eqs. 26 and 27, so that the corresponding components in the output voltage are not equal, and e_{pn}' is not then zero. In the limit, as $2R_k(\mu + 1)$ is made large compared with $r_p + R_L$, the sum component in the plate-to-ground voltages becomes negligible, as is discussed in the preceding paragraph. These voltages then contain

[14] H. Goldberg, "Bioelectric-Research Apparatus," *I.R.E. Proc.*, *32* (1944), 330–335; J. H. Reyner, "Direct-Coupled Oscilloscopes," *Electronics*, *21* (July, 1948), 102–106; P. O. Bishop and E. J. Harris, "A D.C. Amplifier for Biological Application," *R.S.I.*, *21* (1950), 366–377.

only a difference component; they are equal in magnitude and opposite in phase, and are hence suitable for driving a push-pull amplifier. The circuit is said to function as a phase inverter and to convert a single-ended input into a balanced-to-ground output.

Occasionally conversion from a push-pull or balanced-to-ground input to a single-ended output is desired. The input then contains only the difference component, and the output taken between either plate and ground equals that component amplified in accordance with Eq. 27.

When the input is applied between only one grid and ground, and the output is taken between the plate of the same tube and ground, the over-all voltage gain is half the sum of the voltage gains for the two components, Eqs. 26 and 27. When the output is taken between the plate of the other tube and ground, the over-all voltage gain is half the difference of the voltage gains for the two components.

A load connected between the plates has no effect on the voltage gain for the sum component because, as is indicated in Fig. 5a, the voltage between the plates is zero when only the sum components exist in the circuit. Such a load does affect the voltage gain for the difference components, however. When only the difference components exist in the circuit, as in Fig. 5b, the potential of the mid-point of the plate-to-plate load remains constant with respect to ground, because one plate increases and the other decreases in potential by like amounts. Hence, for the difference components, half that load may be considered to be in parallel with each of the resistors R_L, and the voltage gain for the difference components has the form of Eq. 27, with R_L replaced by the resistance of the parallel combination.

5. RESISTANCE–CAPACITANCE–COUPLED AMPLIFIERS

Probably the most widely used type of amplifier is the resistance-capacitance-coupled amplifier whose circuit is shown in Fig. 6a. It differs from the direct-coupled amplifier of Fig. 2a in the coupling circuit used between the stages. In Fig. 6a, a coupling capacitor C_c is employed whose capacitance is made large enough to give it a reactance small compared with the resistance of the grid resistor R_g. For that reason, practically all of the varying component of voltage e_{p1} across the load resistor R_L also appears across R_g and provides the grid-signal voltage for the second tube. However, no steady component of voltage exists across R_g, because the leakage resistance of the coupling capacitor C_c is made very large compared with R_g, and the direct current is thereby blocked. The network thus serves to couple the first and second tubes for the varying components of voltage and current but to block the steady components.

The grid-bias voltage is obtained in the circuit shown by means of the cathode resistor R_k as in Art. 3, Ch. VIII. Thus, a single source furnishes both the plate- and the grid-supply voltages. The by-pass capacitor C_k is considered to have a capacitance large enough to reduce the impedance of the parallel combination effectively to zero for the alternating components of the plate current, and to by-pass them

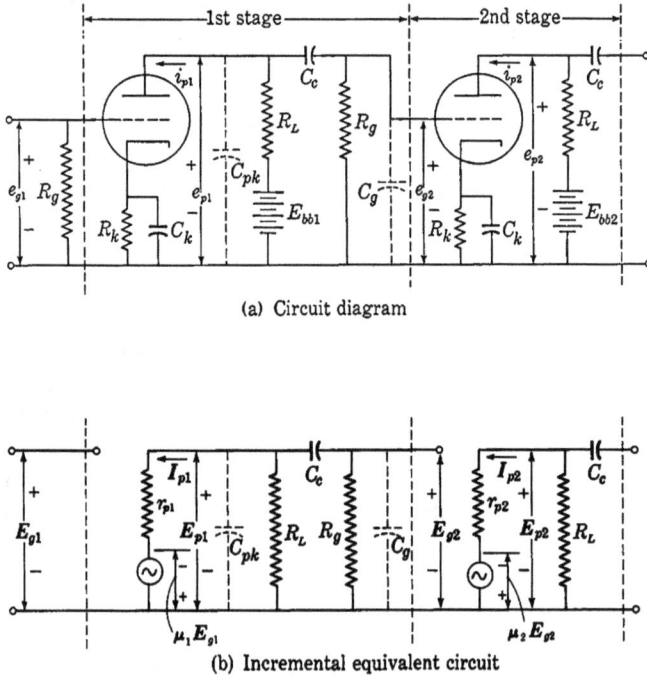

(a) Circuit diagram

(b) Incremental equivalent circuit

Fig. 6. Resistance-capacitance-coupled amplifier circuit. Only incremental components of the currents and voltages are indicated. Complex effective values representing sinusoidally varying increments are shown in (b).

around R_k; hence the impedance of the cathode circuit is neglected in the alternating-current incremental equivalent circuit.

The adoption of resistance-capacitance coupling eliminates many of the objectionable features of the direct-coupled amplifier. Large grid-bias batteries are not necessary, and slow drifts of quiescent operating conditions are not amplified by succeeding stages. However, the steady component of the input voltage also is not amplified. Sudden application of a sustained steady voltage at the input of the circuit in Fig. 6a results in a sudden change in the output current and voltage, but transients are set up in the coupling networks that finally disappear except for changes in the voltages across the coupling capacitor and load resistor of the first stage. The voltages and currents in the later stages return to their original quiescent values. Often a capacitor is inserted between the input terminal and the grid of the first tube so that the quiescent operation of the first stage also is not affected by a direct component of the input voltage.

The resistance-capacitance-coupled amplifier is stable, it can be built to have a good frequency characteristic, and it is relatively inexpensive because of the simplicity of its component elements. One precaution must be observed; namely, the coupling capacitors must have very large leakage resistance to prevent direct current supplied by the plate circuit of one tube from entering the grid circuit of the next. Furthermore, these capacitors must be capable of withstanding the whole of the plate-supply voltage E_{bb} when the cathodes are cold. Paper-dielectric capacitors are ordinarily used for the purpose. Electrolytic capacitors are not suitable, because of their large leakage current. The plate-power supplies may be common for all the stages in the resistance-capacitance-coupled amplifier, because these sources have one common connection, which is often connected to the ground. It is necessary, however, that the internal impedance of the common plate-power supply be small compared with the plate resistances of the tubes, in order that coupling between the various plate circuits through the common power supply may be avoided. A method of obtaining all voltages from a single source is shown in Fig. 24, Art. 10.

The frequency characteristic of a resistance-capacitance-coupled amplifier is determined primarily by the capacitances and the resistances in the circuit. In addition to coupling capacitor C_c, shunt capacitances C_{pk} and C_g exist. These two have a pronounced effect on the frequency characteristic at the high frequencies and therefore cannot be neglected. The capacitance C_{pk} includes the interelectrode capacitance between the plate and cathode of the first tube, and the stray capacitance in the wiring between the cathode, or ground, and that part of the network that connects the plate to the coupling capacitor. The capacitance C_g includes the effective input capacitance of the second tube, given by the analysis of Art. 8, Ch. VIII, and the wiring capacitance between the cathode, or ground, and that part of the network that connects the grid of the second tube to the coupling capacitor. Actually, the effective input admittance of the second tube is not a pure capacitive susceptance; in general, it comprises both conductance and susceptance whose magnitudes depend on *all* the plate and grid circuits associated with all the following tubes,[15] and is

[15] J. W. Sauber, "High-Frequency Characteristics of Resistance-Coupled Triode Amplifiers," *I.R.E. Proc., 40* (1952), 48–49.

a function of frequency. However, as a first approximation the conductance component is assumed in the following analysis to be negligible, and the input capacitance C_g is assumed to be independent of frequency and to have the value stated in the discussion of Eq. 65, Ch. VIII, plus some wiring capacitance. If this assumption is not justified in a particular amplifier, a modification of the method of analysis must be made. The grid-signal voltage is assumed to be sinusoidal, and the harmonic generation due to nonlinearity of the tube characteristics is considered negligible. The equivalent circuit of the amplifier for varying components of currents and voltages under the foregoing assumptions is that shown in Fig. 6b.

The determination of the frequency and phase characteristics of a single stage of amplification, such as that shown as the first stage in Fig. 6, is sufficient for the prediction of the performance of an amplifier of several resistance-capacitance-coupled stages, because the expressions involved are similar for all stages, although the values of the circuit parameters may be different. The straightforward application of the principles outlined in *Electric Circuits* leads to unwieldy expressions difficult to interpret, for various terms in the expressions become negligible at different frequencies. Instead of a complete analysis with engineering approximations introduced at the end, an equivalent method,[16] with physical approximations introduced in the circuit, is used here. It results in simplified expressions suitable only for certain ranges of frequencies. The method of analysis is based on linear operation and sinusoidal quantities, with the assumption that there is a range of frequencies, called the *middle range of frequencies*, in which the voltage gain is practically constant and is unaffected by changes in the reactances of the capacitances in the circuit. If the amplifier is not so designed that such a range of frequencies exists, a different[17] method of analysis must be used.

In the middle range of frequencies, the reactance of the coupling capacitor C_c is so small as to constitute a virtual short circuit for the varying components, while the reactances of C_{pk} and C_g are so large as to constitute virtual open circuits. At frequencies below the middle range, the susceptances of C_{pk} and C_g remain negligible, but the reactance of the coupling capacitor C_c increases, and E_{g2} therefore becomes a smaller and smaller fraction of E_{p1} as the frequency is decreased. Consequently, the voltage gain of the stage E_{g2}/E_{g1}

[16] F. E. Terman, *Radio Engineering* (3rd ed.; New York: McGraw-Hill Book Company, Inc., 1947), 230–243.

[17] D. G. C. Luck, "A Simplified General Method for Resistance-Capacity-Coupled-Amplifier Design," *I.R.E. Proc.*, *20* (1932), 1401–1406; E. A. Johnson, "Design of Tuned Resistance-Capacity-Coupled Amplifiers," *Physics*, *7* (1936), 130–132.

decreases as the frequency is decreased below the middle range. At frequencies higher than the middle range, the reactance of the coupling capacitor C_c remains a virtual short circuit, but the susceptances of C_{pk} and C_g increase and constitute an appreciable shunt susceptance across the load of the first tube. As the frequency is increased, the effect of the shunt susceptance becomes larger, the effective load impedance therefore becomes smaller, and the voltage gain of the stage decreases.

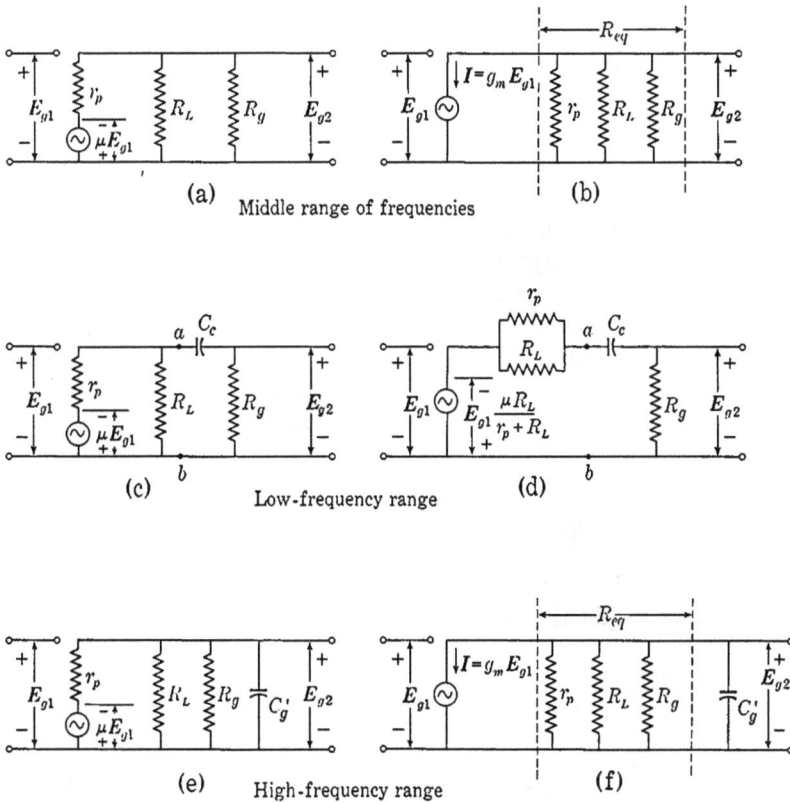

Fig. 7.   Approximate incremental equivalent circuits of the first stage of the resistance-capacitance-coupled amplifier of Fig. 6. Complex effective values of sinusoidally varying incremental currents and voltages are indicated.

Since, in the middle range of frequencies, the reactances of the capacitors have negligible effect on the voltage gain of the stage, they may be ignored in the circuit, and Fig. 6b reduces to the equivalent circuit shown in Fig. 7a for that frequency range. Complex effective values of voltages and currents are indicated in the equivalent circuits, since these values are involved in the determination of the magnitude and phase angle of the complex voltage gain. For analysis, it is convenient to replace the voltage source and series resistance r_p in Fig. 7a by the equivalent combination of a current source and a parallel resistance r_p, as shown in Fig. 7b. If the resistance of the parallel combination comprising r_p, R_L, and R_g in Fig. 7b is denoted by R_{eq}, A_{mid}, the complex voltage gain for the middle range of frequencies, becomes

$$A_{mid} = \frac{E_{g2}}{E_{g1}} = -g_m R_{eq} ,$$ ▶[30]

where

$$R_{eq} = \frac{r_p R_L R_g}{r_p R_L + r_p R_g + R_L R_g} .$$ ▶[31]

At frequencies below those in the middle range, the reactance of the coupling capacitor C_c is appreciable, but the susceptance of the shunt capacitors is negligible, and they may therefore be ignored. The equivalent circuit for low frequencies thus becomes that shown in Fig. 7c. By Thévenin's theorem, the circuit to the left of points a and b may be replaced by the equivalent source and internal resistance shown to the left of points a and b in Fig. 7d. If the resistance combination comprising R_g in series with R_L and r_p in parallel is denoted by R_{eq}', then

$$R_{eq}' = \frac{r_p R_L}{r_p + R_L} + R_g = \frac{r_p R_L + r_p R_g + R_g R_L}{r_p + R_L}.$$ ▶[32]

The complex voltage gain at the low frequencies, A_{low}, is then

$$A_{low} = \frac{E_{g2}}{E_{g1}} = -\mu \frac{R_L}{r_p + R_L} \frac{R_g}{R_{eq}' - jX_c} ,$$ [33]

where

$$X_c = \frac{1}{\omega C_c} .$$ ▶[34]

Substitution of Eq. 32 in Eq. 33 and simplification by means of Eq. 31 and the relation

$$g_m = \mu/r_p$$ [35]

give

$$A_{low} = -\frac{g_m R_{eq}}{1 - j\dfrac{X_c}{R_{eq}'}}$$ [36]

Thus the ratio of the complex voltage gains at low and middle frequencies is

$$\frac{A_{low}}{A_{mid}} = \frac{1}{1 - j\frac{X_c}{R_{eq}'}} \cdot \qquad \blacktriangleright[37]$$

An alternative form of the normalized low-frequency gain expressed by Eq. 37 is

$$\frac{A_{low}}{A_{mid}} = \frac{1}{1 - j\frac{f_1}{f}}, \qquad [38]$$

where

$$f_1 = \frac{1}{2\pi R_{eq}'C_c} \cdot \qquad [39]$$

The frequency f_1 is the value at which the reactance of C_c equals the resistance R_{eq}'. When f equals f_1 in Eq. 38 the magnitude of the voltage gain is $1/\sqrt{2}$, or 70.7 per cent, of its value in the middle range of frequencies. For a fixed amplitude of input voltage to the amplifier, the square of the output voltage is then one-half its value in the middle range of frequencies, and the power delivered to a fixed resistance load is also reduced by a factor of two, or three decibels. Hence f_1 is known as the *lower half-power frequency*.

At frequencies higher than those in the middle range, the reactance of the coupling capacitor C_c is negligible and may be ignored, but the shunt capacitances have an appreciable effect and must be included. Since the alternating voltage drop across the coupling capacitor C_c is practically zero, the voltages across the two capacitances C_{pk} and C_g in Fig. 6b are equal at the high frequencies, and for analysis they may be represented as one capacitance denoted by C_g' in the approximate equivalent circuit of Fig. 7e. Thus,

$$C_g' = C_{pk} + C_g, \qquad [40]$$

and the reactance of these lumped capacitances is given by

$$X_c' = \frac{1}{\omega C_g'} \cdot \qquad \blacktriangleright[41]$$

In Fig. 7f, the voltage source and series resistance r_p in the circuit of Fig. 7e are replaced by the equivalent current source and parallel resistance r_p, as is done in Fig. 7b. The voltage E_{o2} at the high frequencies is given by

$$E_{o2} = -IZ, \qquad [42]$$

where Z is the whole impedance supplied by the current I, that is, the impedance of R_{eq} in parallel with C_g'. In terms of these quantities, the impedance Z is given by

$$Z = \frac{-jR_{eq}X_c'}{R_{eq} - jX_c'} = \frac{R_{eq}}{1 + j\dfrac{R_{eq}}{X_c'}}, \qquad [43]$$

and the current from the current source is

$$I = g_m E_{g1}. \qquad [44]$$

Substitution of Eqs. 43 and 44 in Eq. 42 gives for the complex voltage gain at the high frequencies,

$$A_{high} = \frac{E_{g2}}{E_{g1}} = \frac{-g_m R_{eq}}{1 + j\dfrac{R_{eq}}{X_c'}}; \qquad [45]$$

whence the ratio of the complex voltage gains at the high and middle frequencies is

$$\frac{A_{high}}{A_{mid}} = \frac{1}{1 + j\dfrac{R_{eq}}{X_c'}}. \qquad \blacktriangleright[46]$$

This normalized high-frequency gain may also be expressed in terms of a half-power frequency, for an alternative form of Eq. 46 is

$$\frac{A_{high}}{A_{mid}} = \frac{1}{1 + j\dfrac{f}{f_2}}, \qquad [47]$$

where

$$f_2 = \frac{1}{2\pi R_{eq} C_g'}. \qquad [48]$$

This frequency f_2 is the value at which the reactance of C_g' equals the resistance R_{eq}. At the frequency f_2 the voltage gain is down three decibels from its value in the middle range of frequencies just as for the frequency f_1. Hence f_2 is known as the *upper half-power frequency*.

Vector, or phasor, diagrams based on Eqs. 30, 33, and 45 and the equivalent circuits in the three frequency ranges are shown in Fig. 8. The equivalent circuit diagrams are simplified somewhat in accordance with Eqs. 31 and 32. These vector diagrams are particularly instructive as a means of illustrating the phase relations between the voltages and currents.

The dependence of both the magnitude and the phase angle of the complex voltage gain of a single stage of a resistance-capacitance-coupled amplifier is completely described by Eqs. 30, 31, 32, 34, 37, 41, and 46, subject to the assumptions made at the beginning of the analysis. In order to show the magnitude and phase angle separately,

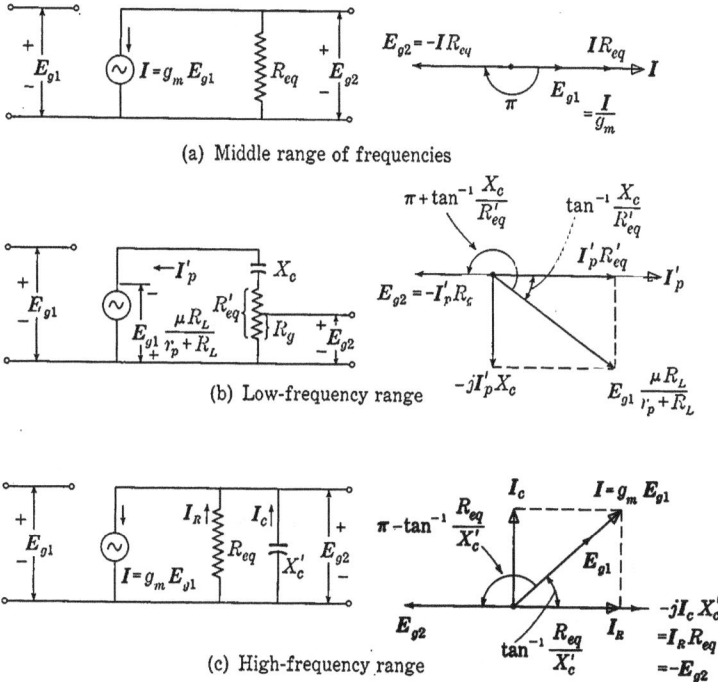

(a) Middle range of frequencies

(b) Low-frequency range

(c) High-frequency range

Fig. 8.   Vector, or phasor, relations in the resistance-capacitance-coupled amplifier.

the expressions may be put in polar form. Thus, from Eqs. 30, 37, and 46,

$$A_{mid} \equiv A_{mid}\underline{/\theta_{A\,mid}} = g_m R_{eq}\underline{/\pi}\,, \qquad [49]$$

$$\frac{A_{low}}{A_{mid}} = \frac{1}{\sqrt{1 + \left(\frac{f_1}{f}\right)^2}}\underline{/\tan^{-1}\frac{f_1}{f}}\,, \qquad [50]$$

$$\frac{A_{high}}{A_{mid}} = \frac{1}{\sqrt{1 + \left(\frac{f}{f_2}\right)^2}}\underline{/-\tan^{-1}\frac{f}{f_2}}\,. \qquad [51]$$

The basic relations describing the amplifier behavior, Eqs. 30, 38, and 47, may, for convenience, be combined into one,

$$\frac{A}{A_{mid}} = \frac{1}{1 - j \tan \theta}, \qquad -\frac{\pi}{2} \leqq \theta \leqq \frac{\pi}{2} \qquad [52]$$

or

$$\frac{A}{A_{mid}} = \cos \theta \; \epsilon^{j\theta}, \qquad -\frac{\pi}{2} \leqq \theta \leqq \frac{\pi}{2} \qquad [53]$$

where

$$\tan \theta = +\frac{f_1}{f} \text{ at low frequencies,}$$

$$\tan \theta \approx 0 \quad \text{ at middle frequencies,}$$

$$\tan \theta = -\frac{f}{f_2} \text{ at high frequencies.}$$

The terms in Eqs. 38 and 47 that express the magnitude of the voltage gain relative to its value in the middle range of frequencies,

$$\frac{A_{low}}{A_{mid}} = \frac{1}{\sqrt{1 + \left(\frac{f_1}{f}\right)^2}}, \qquad \blacktriangleright[54]$$

and

$$\frac{A_{high}}{A_{mid}} = \frac{1}{\sqrt{1 + \left(\frac{f}{f_2}\right)^2}}, \qquad \blacktriangleright[55]$$

show a form of symmetry between the high- and the low-frequency ends of the frequency characteristics. At a low frequency nf_1, the voltage gain has the same value as for a high frequency $(1/n)f_2$ where n has any value.

TABLE II

RELATIVE VOLTAGE GAIN AS A FUNCTION
OF FREQUENCY

The frequencies f_1 and f_2 are the half-power
frequencies defined by Eqs. 39 and 48.

f	$\dfrac{A}{A_{mid}}$	f
$10f_1$	0.995	$0.1f_2$
$5f_1$	0.98	$0.2f_2$
$2f_1$	0.895	$0.5f_2$
f_1	0.707	f_2
$0.5f_1$	0.447	$2f_2$
$0.2f_1$	0.196	$5f_2$
$0.1f_1$	0.100	$10f_2$

The *geometric-mean frequency*, otherwise known as the *midband frequency* and denoted here by f_0, is given by

$$f_0 = \sqrt{f_1 f_2} .$$ ▶[56]

It lies in the middle range of frequencies and is the frequency at which the voltage gain is a true maximum.

The results of computation of the voltage-gain ratios given by Eqs. 54 and 55 as functions of the ratio of the frequency to the half-power frequencies are given in Table II. These results are plotted in Fig. 9,

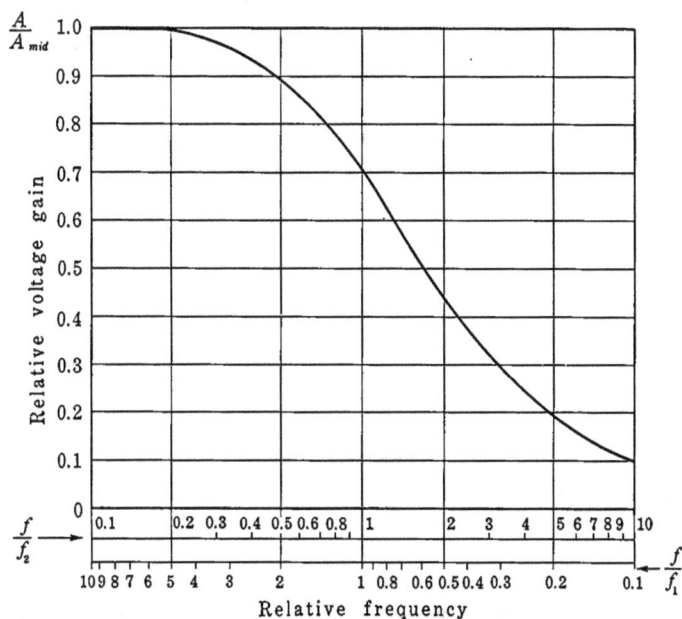

Fig. 9. Universal voltage-gain characteristic curve of a resistance-capacitance-coupled amplifier. This is the relative voltage gain as a function of the ratio of the actual frequency to the half-power frequencies f_1 and f_2 defined by Eqs. 39 and 48.

and the symmetry of the relations makes it possible to describe the frequency characteristic in terms of this nondimensional, or universal, or relative, curve, where the relative frequency with respect to the half-power frequencies is indicated on two logarithmic scales running in opposite directions with the unit points coinciding.

Figure 10 shows the voltage gain expressed by Eqs. 54 and 55 plotted on logarithmic scales of relative voltage gain and frequency. The scale for the relative voltage gain in decibels is thus uniform.

From Eq. 54, the relative voltage gain in decibels at low frequencies is

$$\frac{A_{low}}{A_{mid}}\bigg|_{db} = 20 \log_{10} \frac{A_{low}}{A_{mid}} = 20 \log_{10} \frac{1}{\sqrt{1 + \left(\frac{f_1}{f}\right)^2}} \qquad [57]$$

$$\approx 0, \qquad \frac{f}{f_1} \gg 1, \qquad\qquad [58]$$

$$\approx 20 \log_{10} \frac{f}{f_1}, \qquad \frac{f}{f_1} \ll 1. \qquad [59]$$

For frequencies large compared with f_1, Eq. 58 shows that the relative voltage gain approaches a constant. For frequencies small compared

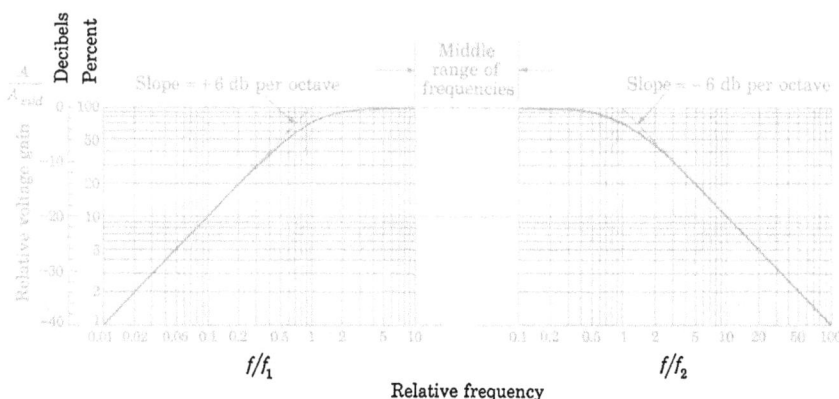

Fig. 10. Relative voltage gain as a function of frequency on logarithmic scales. The frequencies f_1 and f_2 are the half-power frequencies defined by Eqs. 39 and 48.

with f_1, Eq. 59 shows that the relative voltage gain increases 20 decibels for each decade increase in frequency, or approximately 6 decibels for each octave. For frequencies below the middle range of frequencies, therefore, the curve of relative voltage gain is asymptotic to two straight lines, as shown in the left half of Fig. 10. These lines intersect at the frequency f_1, in accordance with Eqs. 58 and 59. A similar analysis based on Eq. 55 shows that in the high-frequency range, the relative voltage gain is asymptotic to a straight line having a slope of -6 decibels per octave. The middle range of frequencies, wherein the voltage gain is constant within limits of about 0.5 per cent, extends from about $10f_1$ to about $0.1f_2$ when f_2/f_1 is so large that $0.1f_2$ exceeds $10f_1$.

As is previously stated, the foregoing analysis in terms of separate circuits for the low- and high-frequency ranges is based on the assumption that the series and shunt capacitances in the equivalent circuit are not simultaneously effective in influencing the voltage gain. Such an assumption is justified when the ratio f_2/f_1 is very large, say 1,000 or more. When the ratio is as small as, say, 10 or 100, a close approximation to the performance is obtained through assuming that the difference between the relative voltage gain and unity is the sum of the corresponding amounts computed from Eqs. 54 and 55. For example, if f_2/f_1 is 100, Eqs. 54 and 55 give the value 0.995, or $1 - 0.005$, for the relative voltage gain at the geometric mean frequency, $10f_1$ or $0.1f_2$. The true relative voltage gain with both capacitances considered simultaneously is nearly $1 - 0.005 - 0.005$ or 0.990. When the ratio f_2/f_1 is smaller than 10, this approximation is not valid, and the circuit performance is best determined through a straightforward analysis that accounts for the simultaneous presence of both capacitances.

In computing the frequency characteristic of a resistance-capacitance-coupled amplifier, the following are the steps required:

(a) Determine the circuit parameters, including C_g'.

(b) Compute R_{eq} and R_{eq}' by Eqs. 31 and 32.

(c) Compute f_1 and f_2 by Eqs. 39 and 48.

(d) Check to determine whether or not there is a true middle range of frequencies. The frequency f_2 should be 100 times f_1 if the error associated with the assumption that the amplifier has constant voltage gain for frequencies between $10f_1$ and $0.1f_2$ is to be less than 0.5 per cent; or f_2 should be 25 times f_1 if it is assumed that the voltage gain is constant from $5f_1$ to $0.2f_2$, and if the error is to be less than 2 per cent. If the middle range of frequencies does not exist, the method is not applicable.

(e) Compute the voltage gain in the middle range of frequencies by Eq. 30.

(f) Determine the voltage gain at frequencies outside the middle range by means of the foregoing results and the universal curves, Figs. 9 or 10, or Table II.

The effect of changes in the coupling capacitance C_c and the shunt capacitance C_g' while the resistances are held constant is to alter the values of the half-power frequencies f_1 and f_2, and hence to shift the falling portions of the frequency characteristic in the directions shown in Fig. 11.

The effect of changes in the load resistance R_L and the grid resistance R_g on the shape of the frequency characteristic is related to their orders of magnitude, which are governed not only by considerations based on the frequency characteristic but by other considerations as well. For a large value of voltage gain, R_{eq} in Eq. 30 should be large; and hence, from Eq. 31, R_L and R_g should both be large. However, the value of R_L is limited by various considerations such as harmonic generation and the plate-supply voltage available, as discussed in Ch. VIII. In general, manufacturers recommend values of load resistance that place the quiescent operating point in the region of low plate currents. The values recommended are a compromise between large voltage gain on the one hand, and small harmonic generation and

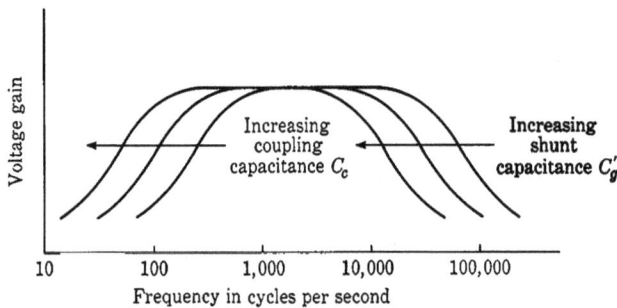

Fig. 11. Effect of capacitance variations on the shape of the frequency characteristic in a resistance-capacitance-coupled amplifier.

good high-frequency response on the other. A further consideration is that the mutual conductance g_m decreases as the quiescent operating point is moved to lower plate currents; consequently, the increase of voltage gain that might be expected from an increase of the load resistance is counteracted.

In contrast to the simple conditions discussed in Art. 4, Ch. VIII, the coupling network in the resistance-capacitance-coupled amplifier causes the load line to differ from a line drawn through the quiescent operating point and the point on the voltage axis corresponding to the plate-supply voltage. As is shown in Fig. 12, the quiescent operating point lies on a line with a slope corresponding to $-1/R_L$ through the point corresponding to E_{bb}. However, in the middle range of frequencies, in which the capacitances need not be considered, the resistance that determines the path of operation on the plate characteristics is the resistance of R_L and R_g in parallel, as is illustrated in Fig. 7a. Thus a line drawn through Q in Fig. 12, with a slope corresponding to $-(R_L + R_g)/(R_L R_g)$, is the path of the operating point. At frequencies outside the middle range, the equivalent circuits of Fig. 7c or 7e apply, and the path of operation becomes approximately an ellipse about the point Q. If the signal voltage is large enough to cause appreciable nonlinearity, the operating path shifts with amplitude, as is discussed in Art. 18, Ch. VIII.

The upper limit on the value of grid resistance R_g that can be used with a particular tube is governed primarily by the steady component of grid current that exists because of leakage, gaseous ionization, and other causes. In general, the upper limit is larger the smaller the tube, because of the decreased collecting area for ions in a small tube. If too large a grid resistor is used, the voltage drop produced in it by the grid

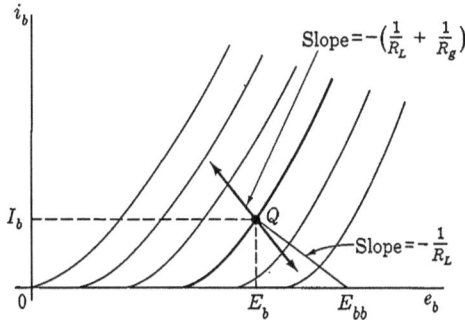

Fig. 12. Approximate path of the operating point in the middle range of frequencies.

current alters the net grid-bias voltage at the tube terminals and moves the operating point into the nonlinear region of the curves.

Within the limits set by the foregoing considerations, the values of the load resistance and the grid resistance may be altered to affect the frequency characteristic as desired. As shown by Eq. 32, variations in R_g with the other circuit parameters held constant affect R_{eq}' directly, since R_g is an additive term and the parallel combination of R_L and r_p often has a value considerably lower than R_g. However, variations in R_g have relatively little effect on R_{eq} given by Eq. 31, because r_p and R_L are usually lower and therefore predominate. As a result, variations in R_g have little effect upon the voltage gain in the middle range of frequencies given by Eq. 49, or upon the relative voltage gain at the high frequencies given by Eq. 51. However, variations in R_g have a pronounced effect upon the relative voltage gain at low frequencies, given by Eq. 50, and they therefore shift the frequency-characteristic curve at the low frequencies considerably, change the height at the middle frequencies only slightly, and have little effect on the shape of the high-frequency region, as is shown in Fig. 13a.

Variations of R_L have relatively little effect on R_{eq}', because R_g is the predominant term in Eq. 32, but their effect on R_{eq} in Eq. 31 is appreciable, as R_L is usually not more than a few times larger than r_p

for a triode amplifier and is generally much smaller than r_p for a tetrode or pentode amplifier. Variations in R_L therefore affect the voltage gain at the middle and high frequencies given by Eqs. 49 and 51 but have little effect on the relative voltage gain at the low frequencies; thus they affect the height of the frequency characteristic in the middle range of frequencies but influence the relative shape of the curve in the

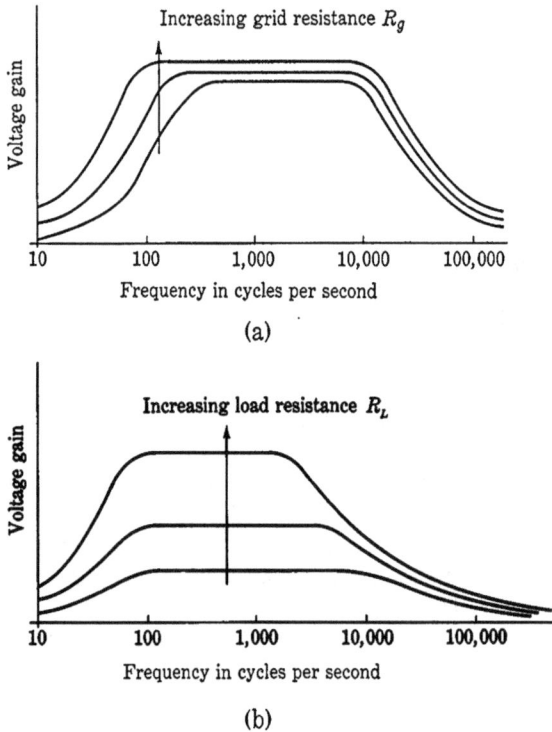

(a)

(b)

Fig. 13. Approximate effect of variations of the load and grid resistances on the shape of the frequency characteristic of a resistance-capacitance-coupled amplifier.

low-frequency region only slightly. At the high frequencies, as R_L is increased the half-power frequency in Eq. 48 is decreased and the frequency characteristic is raised in height, but the maximum-gain portion does not extend as far into the higher-frequency region. Consequently, with the other parameters fixed, a more nearly uniform voltage gain over a wider frequency range is obtained with a smaller load resistance, as is shown in Fig. 13b. This relationship is further clarified by a comparison of Eqs. 48 and 49. The first shows the upper half-power frequency to be inversely proportional to the equivalent shunt resistance R_{eq}, and the second shows the voltage amplification in the middle range of frequencies to be directly proportional to R_{eq}. The product,

$$A_{mid}f_2 = \frac{g_m}{2\pi C_g'} = \frac{g_m}{2\pi(C_{input} + C_{output})} \text{ cycles per second,} \quad [60]$$

where C_{input} and C_{output} include the stray and wiring capacitances, is therefore independent of the equivalent shunt resistance. An increase of midband gain can be obtained only at the expense of a reduction in the upper half-power frequency. Since in a broad-band amplifier the bandwidth defined by the difference between the half-power frequencies is approximately equal to the upper half-power frequency, one interpretation of Eq. 60 is that *the product of gain and bandwidth is constant* for a resistance-capacitance-coupled amplifier, and depends only on the g_m of the tube and the effective shunt capacitance in the circuit. That capacitance is the sum of the input capacitance of one tube and the output capacitance of the previous tube, both augmented by the wiring capacitance; hence, the ratio $g_m/(C_{input} + C_{output})$, where C_{input} and C_{output} include only the inter-electrode capacitances as discussed in Arts. 8 and 9, Ch. VIII, is a useful figure of merit for tubes to be used in broad-band amplifiers. When only the interelectrode capacitances are included, this ratio is the limiting value of the gain-bandwidth product in radians per second. It may be approached but not exceeded as the stray and wiring capacitances are reduced toward zero.

The *phase characteristic* of a resistance-capacitance-coupled amplifier —that is, the function giving the phase angle between the input and output voltages as a function of frequency—is described by the expressions for the phase angle in Eqs. 49, 50, and 51. The phase characteristic, like the frequency characteristic, is analyzed on the basis of three ranges of frequencies and the equivalent circuits of Fig. 7 for frequencies within those ranges. The phase angle in the amplifier depends on the assigned positive reference directions of the input and output voltages. From Eq. 49 and the vector diagram in Fig. 8a, it is evident that the phase angle between the input and output voltages is constant and equal to 180 degrees in the middle range of frequencies wherein the effect of the capacitances is neglected if the positive reference directions of the voltages are assigned as indicated. But the phase angle would be zero degrees if one or the other of the assigned reference directions were reversed.

At frequencies lower than those in the middle range, the phase angle in the amplifier is given by Eqs. 49 and 50 and by Fig. 8b. The vector diagram constructed with I_p' as a reference is shown, and the angle by which E_{g2} leads E_{g1} is

$$\measuredangle{}^{E_{g2}}_{E_{g1}} = \pi + \tan^{-1}\frac{X_c}{R_{eq}'} = \pi + \tan^{-1}\frac{1}{2\pi f C_c R_{eq}'}. \qquad \blacktriangleright[61]$$

At frequencies higher than those in the middle range, the phase angle in the amplifier is given by Eqs. 49 and 51 and by Fig. 8c. The vector diagram is constructed with the voltage across the output terminals as a reference, and at the high frequencies the angle by which E_{g2} leads E_{g1} is given by

$$\measuredangle{}^{E_{g2}}_{E_{g1}} = \pi - \tan^{-1}\frac{R_{eq}}{X_c'} = \pi - \tan^{-1} 2\pi f C_g' R_{eq}. \qquad \blacktriangleright[62]$$

Computations based on Eqs. 61 and 62 with Eqs. 39 and 48 give the results in Table III. Because of the symmetry of the relations, the departure of the phase angle from 180 degrees at a frequency nf_1 is the same in magnitude as the departure at a frequency $(1/n)f_2$, but is opposite in sign. From the data of Table III, the universal phase characteristic shown in Fig. 14 for a resistance-capacitance-coupled amplifier may be constructed. The approximation involved in dividing the frequency scale into three ranges is even more apparent in this figure than it is in Fig. 9, for the phase angle has a value of $180 + 5.8$ degrees at $10f_1$ and $180 - 5.8$ degrees at $0.1f_2$ instead of 180 degrees, as would be anticipated from the previous discussion.

TABLE III

UNIVERSAL PHASE CHARACTERISTIC OF A RESISTANCE-CAPACITANCE-COUPLED AMPLIFIER

The frequencies f_1 and f_2 are the half-power frequencies defined by Eqs. 39 and 48. Angles are *positive* if associated with the left-hand column of frequencies, *negative* if associated with the right-hand column.

f	$\left(\measuredangle{}^{E_{g2}}_{E_{g1}}\right) - 180°$	f
$10f_1$	5.8°	$0.1f_2$
$5f_1$	11.3°	$0.2f_2$
$2f_1$	26.6°	$0.5f_2$
$1f_1$	45.0°	$1f_2$
$0.5f_1$	63.5°	$2f_2$
$0.2f_1$	78.7°	$5f_2$
$0.1f_1$	84.3°	$10f_2$

The behavior of the resistance-capacitance-coupled amplifier is effectively summarized by a plot on the complex plane of the locus of the complex relative voltage gain expressed by Eqs. 38 and 47. Such a locus is constructed in Fig. 15. During increase of f from zero to about

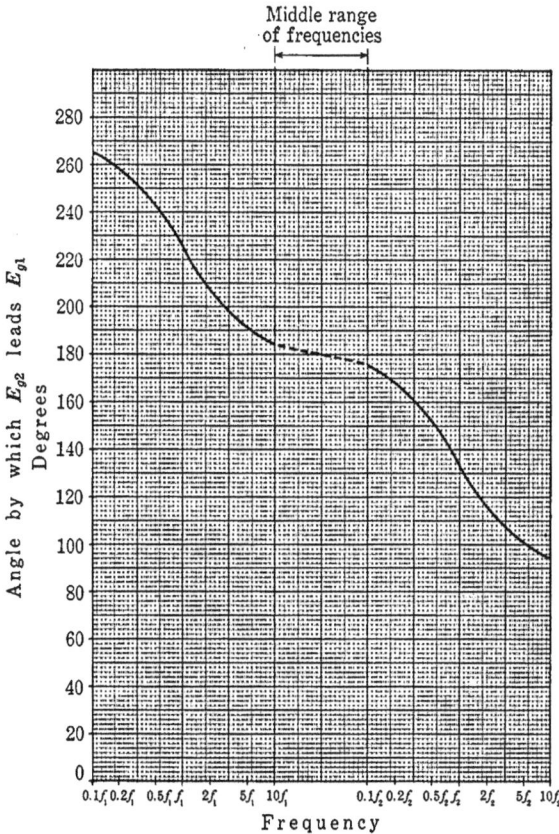

Fig. 14. Universal phase characteristic of a resistance-capacitance-coupled amplifier. Reference directions of E_{g1} and E_{g2} are defined in Fig. 6. The frequencies f_1 and f_2 are the half-power frequencies defined by Eqs. 39 and 48.

$10f_1$, Eq. 38 applies. The tip of the vector $1 - j(f_1/f)$ traces out a straight line segment from the point $1 - j\infty$ to the point $1 - j0.1$ as is shown in Fig. 15a. The tip of the reciprocal vector $1/[1 - j(f_1/f)]$ follows the dotted semicircle[18] in the clockwise direction from the origin to the point $0.99 + j0.099$ in Fig. 15a. This locus is reproduced in the upper right quadrant of Fig. 15c.

[18] E. E. Staff, M.I.T., *Electric Circuits* (Cambridge, Massachusetts: The Technology Press of M.I.T.; New York: John Wiley & Sons, Inc., 1940), 478–487.

When f lies between about $10f_1$ to $0.1f_2$, operation is in the middle range of frequencies, and the complex relative voltage gain A/A_{mid} is unity. The locus is represented by the single point $1 + j0$ in Fig. 15c.

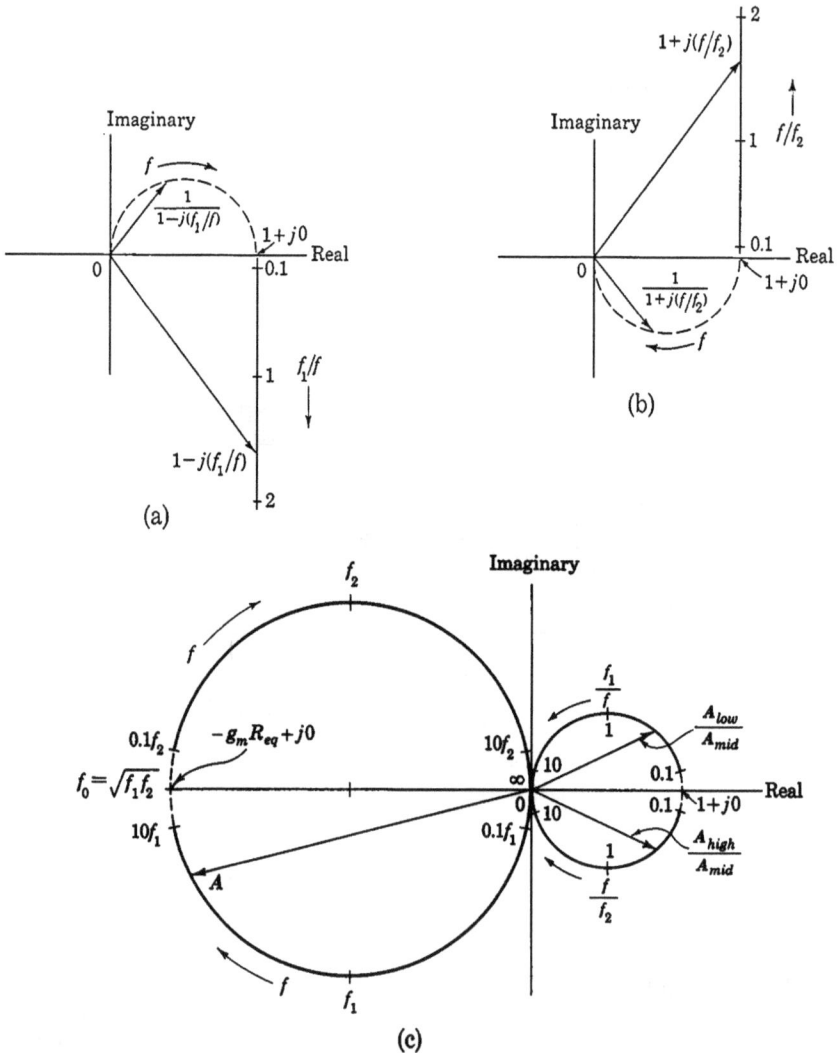

Fig. 15.   Locus of complex voltage gain A as a function of frequency.

During increase of f from $0.1f_2$ to infinity, Eq. 47 applies. The tip of the vector $1 + j(f/f_2)$ follows a straight line from the point $1 + j0.1$ to $1 + j\infty$ and that of the reciprocal vector the semicircle from the point $0.99 - j0.099$ to the origin. Thus the locus of A_{high}/A_{mid} is as shown in the lower right quadrant of Figs. 15b and 15c.

The locus of the complex voltage gain A is obtained from the loci for the relative values through multiplication by $-g_m R_{eq}$ in accordance with Eq. 30. Thus A follows the semicircular loci in the left half plane of Fig. 15c. Because of the approximations made in establishing the equivalent circuits for the three frequency ranges, an uncertainty and error exist in the frequency region covered by the dotted portion of the circle for A. This discrepancy is greater for the phase angle of A than for its magnitude. Construction of a complex locus based on the complete equivalent circuit in Fig. 6b avoids the discrepancy and gives accurate results.[19] The true locus for A is a circle, but one of somewhat smaller radius than that constructed in Fig. 15.

6. INDUCTANCE–CAPACITANCE–COUPLED AMPLIFIERS

To eliminate the voltage drop and power loss in the load resistance caused by the direct component of plate current I_b, and thus to permit operation with a lower plate-supply voltage and higher efficiency, an inductor is sometimes substituted for the resistor R_L of Fig. 6. The inductive reactance can be made very large over a considerable part of the frequency range, and, in accordance with the considerations of Art. 6, Ch. VIII, a large voltage gain results. Because of the decrease of inductive reactance with decrease of frequency, the frequency characteristic of an inductance-capacitance-coupled amplifier tends to slope downward more steeply as the frequency is decreased than does that of a resistance-capacitance-coupled amplifier. The extent of the middle-frequency range, in which the voltage gain is practically constant, is generally smaller than for a resistance-capacitance-coupled amplifier, but the magnitude of the voltage gain is greater.

In most amplifiers, inductance-capacitance coupling has no important advantages over transformer coupling and is economically inferior to resistance-capacitance coupling. Consequently, it is not often used at present for interstage coupling, although it was more common before high-quality transformers and high-mu tubes were developed.

The use of inductance and capacitance to couple a load to a tube is common, however, when the load absorbs considerable power. The inductance is connected in series with the plate-power supply, replacing R_L in Fig. 6, and the load replaces R_g and all the circuit to the right of R_g on the diagram. The inductance is selected to have a large

[19] A. C. Seletzky, "Amplification Loci of Resistance-Capacitance-Coupled Amplifiers," *A.I.E.E. Trans.*, 55 (1936), 1364–1371.

reactance and the capacitance to have a small reactance compared with the impedance of the load. Under these conditions, the effective impedance to alternating current, as viewed from the plate and cathode terminals, is equal to the load impedance that replaces R_g. The connection described is known as *shunt feed* of the plate circuit, and the alternative connection involving the load impedance in series with the plate and plate-power supply, discussed in Ch. VIII, is known as *series feed*. The advantages of the use of shunt feed over series feed are: First, the direct components of plate current and plate voltage are isolated from the load so that one terminal of the load may be practically at the cathode potential, which is often made that of ground; and, second, the undesirable effects of direct-current magnetic saturation are avoided when a device subject to such an effect is used as the load.

7. PHASE INVERTERS

The full advantages of the symmetry in a push-pull amplifier are realized only when it is supplied with input signal voltages that are equal in magnitude and opposite in direction with respect to the cathodes of the tubes. Such voltages are said to be balanced-to-ground because the cathodes of a push-pull amplifier, and those of most other amplifiers as well, are usually connected through a bias voltage to a metal plate, or chassis, which frequently is actually connected to ground. This connection is desirable because it permits use of a single power supply for several stages, facilitates interstage shielding, and serves to fix the stray capacitances in the circuit. Most sources of signal voltage do not and cannot furnish balanced-to-ground voltages. Although neither terminal of a two-terminal source is actually connected to ground, the stray leakage resistance and stray capacitance effective from one terminal to ground generally exceed those from the other. Hence, if an attempt is made to obtain balanced-to-ground voltages from a two-terminal source through the artifice of connecting a high-resistance voltage divider across the source with its center tap connected through a bias voltage to the cathodes of a push-pull amplifier, the stray effects tend to upset the voltage balance, especially at high frequencies. Some means for converting an unbalanced voltage into balanced-to-ground voltages is therefore needed. One means for accomplishing this conversion is the input transformer with a center-tapped secondary winding, shown in Art. 19, Ch. VIII. An internal grounded shield between the windings is frequently necessary to prevent unbalance due to the presence of interwinding capacitance. Another means is the *phase-inverter amplifier*.

In Art. 4 the cathode-coupled difference amplifier is shown to function as a phase inverter, for, when the cathode resistance R_k is made large compared with $(r_p + R_L)/(\mu + 1)$, the plate-to-ground voltages are almost balanced for any ratio and phase of the two input grid-to-ground voltages. Thus neither side of the input signal source need be grounded, although either side may be. The plates may be coupled to the grids of a succeeding push-pull amplifier stage either directly or through coupling capacitors combined with grid resistors as in Art. 5.

When connecting one terminal of the input signal source to the common or ground terminal of the amplifier is permissible, a phase

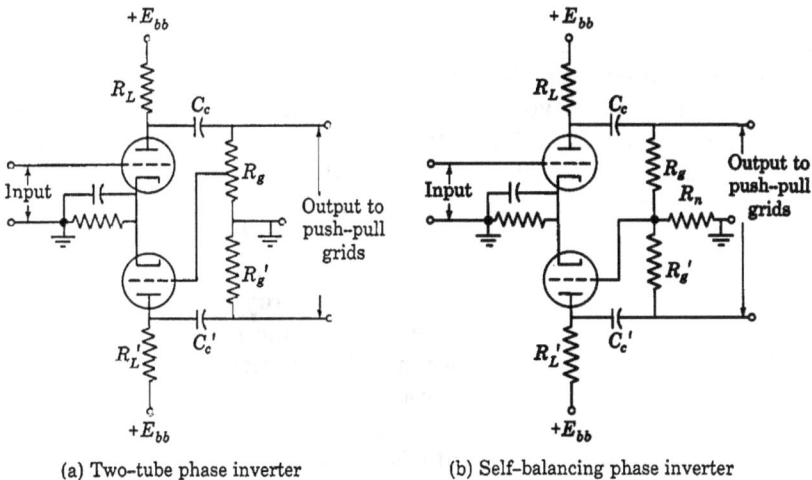

(a) Two-tube phase inverter   (b) Self-balancing phase inverter

Fig. 16.   Phase-inverter amplifiers for supplying a push-pull amplifier.

inverter of the types shown in Figs. 16 and 17 can be used. The two-tube phase inverter, Fig. 16a, is essentially two resistance-capacitance-coupled amplifier stages with the grid-signal voltage for one stage obtained from a tap on the output resistor of the other. The tap on the upper resistor R_g is adjusted to furnish to the lower tube a grid-signal voltage just sufficient to make it supply an output voltage equal in magnitude to that supplied by the upper tube. The output voltages across the two resistors R_g and R_g' are therefore equal in magnitude, but, because of the phase-inverting property of the lower tube, the potential of its plate is decreasing when the plate potential of the upper tube is increasing, and vice-versa—the voltages with respect to ground across the two resistors R_g and R_g' are 180 degrees out of phase.   The ratio of the total resistance R_g to the part of this resistance from which the lower tube receives its grid-signal voltage should be equal to the voltage amplification of the lower stage. The grids of the succeeding push-pull amplifier stage are connected as indicated, and the voltages supplied to them are 180 degrees out of phase in the range of frequencies in which the capacitive reactances in the circuit have no effect on the amplification. At frequencies above and below this range, however, the two output voltages are not 180 degrees out of phase, because the phase displacement in the lower stage is not then exactly 180 degrees, as is shown in Art. 5.

An alternative type of two-tube phase inverter is obtained if a resistance R_n is included in the ground connection from the two grid resistors R_g and R_g', and the lower grid is connected to the common point for the three, as is shown in Fig. 16b. The voltage across R_n supplied to the lower grid is then the difference between two components, one caused by plate voltage variations from the upper tube, and the other by plate voltage variations from the lower tube. The latter component results from plate-to-grid feedback around the lower tube; hence a full explanation of the circuit behavior is best based on the feedback-amplifier considerations of Art. 12. A qualitative understanding is obtained, however, if the voltage gain from the grid to the plate of the lower tube is considered to be so large that the incremental voltage across R_n required to produce any finite increment of plate voltage across the lower tube is negligible compared with the plate voltage output from the upper tube. The current through R_n then remains essentially zero during operation; hence, if R_g equals R_g', the two plate voltages must be equal in magnitude and opposite in sign—the feedback around the lower tube is a constraint that produces this desired equilibrium condition. In a practical amplifier exact balance of the plate voltages is never obtained because the voltage gain of the lower tube is finite, and some voltage must therefore exist across R_n. Analysis of the circuit operation may then follow that given for the feedback-amplifier circuit of Fig. 37b in Art. 12e, because the network R_g, R_g', and R_n and the lower tube constitute such a connection. The degree of balance is independent of the voltage gain of the upper tube and is not critically dependent on the coefficients μ, r_p, and g_m of the lower tube. Hence the circuit is said to be *self-balancing*. For least unbalance the resistance R_n is made as large as the grid-current requirements of the succeeding tubes will permit.

The single-tube phase inverter in Fig. 17 has the advantage of simplicity but the disadvantage that its voltage gain between the input and either of the two halves of its output is less than unity. It functions as a phase inverter because the same varying component of current exists in the resistors R_k and R_L, and the potential of point k with respect to the ground thus increases when that of p decreases. If the two resistors R_k and R_L are made equal in resistance, the alternating voltages across the two resistors R_g and R_g' are equal in magnitude and 180 degrees out of phase, as for the circuits in Fig. 16. The alternating component of voltage applied between the grid and cathode of the tube is the difference between the alternating components of the input voltage and the voltage across R_k. Consequently, the voltage across R_k cannot be greater than the input voltage, and the voltage available as a grid voltage for either of the succeeding push-pull amplifier tubes is less than the input voltage. An additional factor affecting the frequency characteristic of the circuit is that the product of the effective shunt capacitance across the resistor R_k and the resistance the capacitance faces is likely to be different from the corresponding product for the capacitance across R_L. Consequently the output voltages may differ appreciably at high frequencies. This phase inverter is a particular type of the degenerative amplifiers discussed in more detail in Art. 12.

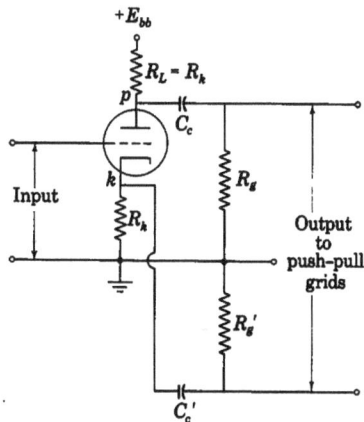

Fig. 17.   Single-tube phase inverter.

8. Amplifiers Coupled by Iron-Core Transformers

Transformers are often used in amplifiers to couple the source of input signal voltage to the first stage, the output of one stage to the input of the next stage, and the output of the last stage to the load. The interstage transformer serves to isolate the steady components of the voltages and currents in the different amplifier stages just as a coupling capacitor isolates those components, and it also provides means for stepping up or down voltage, current, and impedance. The distinguishing feature of the input and interstage transformers in Class A_1 amplifiers is that the impedance of their secondary load, which is the input to a grid-controlled vacuum tube, is usually so large that the effect of the stray and distributed capacitances in the transformer is important in determining the frequency characteristics of the amplifier. On the other hand, the output transformer and the transformers used in Class B and Class C amplifiers generally have a secondary load impedance so small that the stray and distributed capacitances have essentially negligible effect on the performance over the frequency range for which they are designed to operate. The characteristics of input, output, and interstage transformers used in amplifiers and the effects of the stray and distributed capacitances on their frequency characteristics are discussed in some detail in *Magnetic*

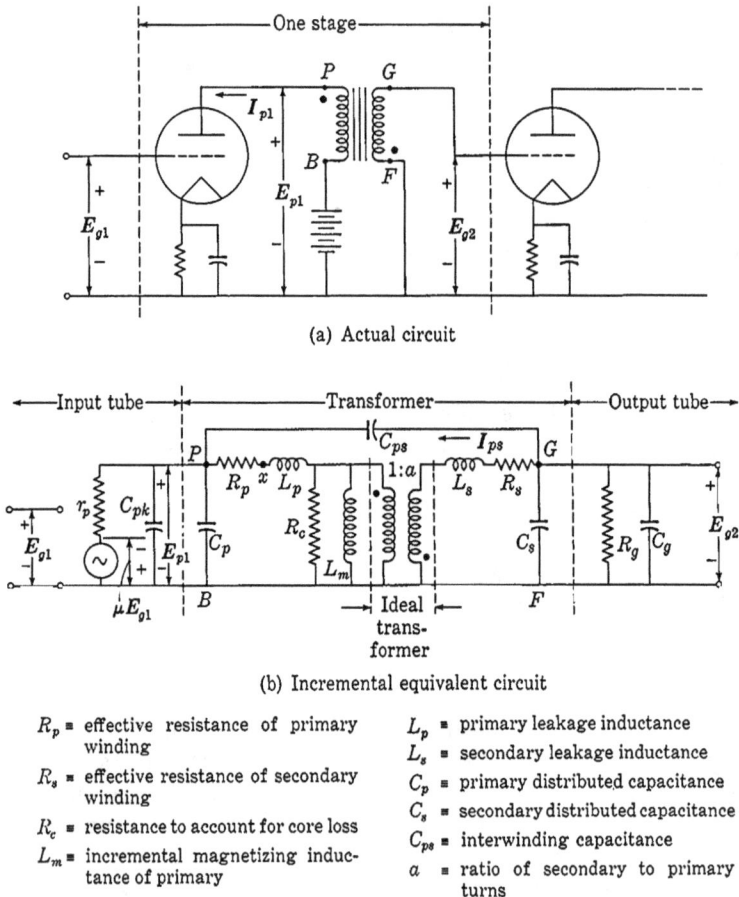

(a) Actual circuit

(b) Incremental equivalent circuit

R_p = effective resistance of primary winding		L_p = primary leakage inductance	
		L_s = secondary leakage inductance	
R_s = effective resistance of secondary winding		C_p = primary distributed capacitance	
		C_s = secondary distributed capacitance	
R_c = resistance to account for core loss		C_{ps} = interwinding capacitance	
L_m = incremental magnetizing inductance of primary		a = ratio of secondary to primary turns	

Fig. 18.   Transformer-coupled-amplifier circuit.

Circuits and Transformers. The discussion in this article is an approximate analysis confined to the characteristics of amplifiers in which interstage transformers are used, but appropriate also to applications of input transformers or to other applications in which the susceptance of the stray and distributed capacitances is appreciable.

The circuit of a transformer-coupled amplifier in which an interstage transformer is employed is shown in Fig. 18a. This type of amplifier is suitable for use at various frequencies; the type of transformer to be employed is determined by the frequency band the amplifier is required to transmit. When the band is several octaves wide, as in an audio-frequency amplifier, a transformer with a core of high-permeability material and a coupling coefficient near unity is generally used. When the frequency band covers only a small fraction of an octave, however, as in a radio-frequency amplifier, an air core is generally used, the coupling coefficient is generally appreciably less than unity, and tuning of the primary or secondary circuits, or both, is employed. In this article, amplifiers embodying ferromagnetic-core transformers with approximately unity coupling coefficient are discussed; amplifiers with air-core transformers are discussed subsequently in Art. 10.

The use of an iron-core transformer instead of a capacitor as a coupling element in an amplifier is advantageous for several reasons. It permits an increase in the voltage gain of an amplifier comprising tubes with a relatively low amplification factor, because a step-up turns ratio in the transformer can be employed. If the direct-current resistance of the primary winding is kept small, the required plate-supply voltage E_{bb} is only slightly greater than the desired quiescent plate voltage E_b. Thus, as compared with the resistance-capacitance-coupled amplifier, a transformer-coupled amplifier can produce a relatively large alternating voltage output in proportion to the plate-supply voltage when the limitation on amplitude is harmonic generation by the tube. Still another advantage of the transformer-coupled amplifier is that merely constructing the secondary winding with a center tap provides a source of grid-excitation voltage suitable for a push-pull stage.

The principal disadvantages of the use of a transformer for interstage coupling in an amplifier are economic ones. Subsequent discussion points out that the design and construction of the transformer involve certain difficulties, and the resulting high cost is a limitation on the use of transformer-coupled amplifiers. A second disadvantage, which also contributes to the cost of transformer-coupled amplifiers, is that magnetic coupling caused by the stray fields interlinking the transformers of different stages and by magnetic induction from extraneous sources often complicates the amplifier design.

For linear operation, the analysis of the performance of an amplifier stage with an iron-core transformer may be based on the equivalent circuits of the tubes and the transformer.[20] Figure 18b shows an equivalent circuit of the transformer and tubes. The equivalent circuit of the

20 G. Koehler, "The Design of Transformers for Audio-Frequency Amplifiers with Pre-assigned Characteristics," *I.R.E. Proc.*, *16* (1928), 1742–1770.

transformer is that of an ideal transformer with a ratio of secondary to primary turns equal to a, and with lumped series and shunt circuit elements to represent the effects of winding resistance, leakage inductance, core losses, primary coil inductance corresponding to the incremental permeability of the core (hereafter called the incremental magnetizing inductance; see *Magnetic Circuits and Transformers*), and winding capacitance.

The resistance R_p in Fig. 18 is the effective resistance of the primary winding to the alternating current, and L_p is the primary leakage inductance. Similarly, R_s and L_s are the effective resistance and leakage inductance, respectively, of the secondary winding. The inductance L_m is actually the difference between the incremental magnetizing inductance of the primary winding and the primary leakage inductance L_p; but in the usual iron-core transformer with a coupling coefficient near unity, L_m is approximately equal to the incremental magnetizing inductance. The resistance R_c accounts for the core losses.

In Fig. 18b, C_p and C_s are lumped capacitances that represent the effect of the distributed capacitances of the primary and secondary windings. Similarly, C_{ps} represents the distributed capacitance between the two windings. In general, at least six lumped capacitances connected between the various pairs of terminals of the transformer are needed to represent the effects of distributed capacitance. In the amplifier connection shown, however, one terminal of the primary winding and one terminal of the secondary winding have essentially zero impedance connected between them; hence one of the inter-terminal capacitances is short-circuited, and two others can be combined with C_p and C_s. The number of capacitances to be considered is thus reduced to three.

A complete solution for the voltage gain in the amplifier of Fig. 18 would hardly be worth while, because the complexity of the equivalent circuit is so great that the solution would undoubtedly be too unwieldy for effective engineering use. Also, the resistances and inductances shown in the circuit are often not constant; they change with frequency. Hence resort is made here to approximate methods somewhat similar in principle to those used previously to obtain a solution for the resistance-capacitance-coupled amplifier. The method is to simplify the equivalent circuit by reasonable physical approximations[21] before the mathematical analysis is attempted. An alternative method that accomplishes the same result is to simplify the mathematical expressions for the complex circuit, but a better physical

[21] F. E. Terman, *Radio Engineering* (2nd ed.; New York: McGraw-Hill Book Company, Inc., 1937), 188–202.

visualization of the approximations is perhaps obtained from the method used here.

Since the reactances of the distributed capacitance C_p and the interelectrode capacitance C_{pk} are generally large compared with the plate resistance of the tube, these capacitances have a negligible shunting

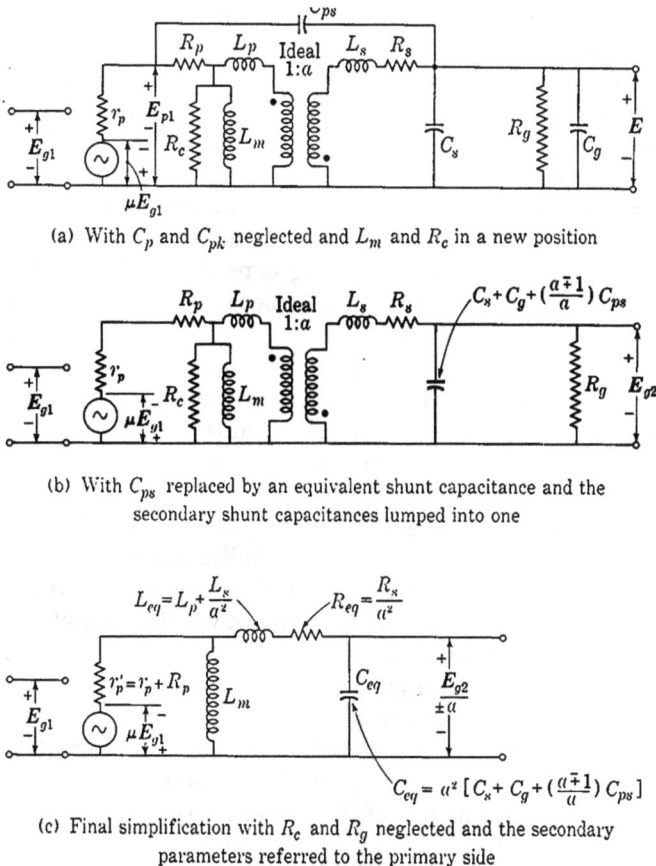

(a) With C_p and C_{pk} neglected and L_m and R_c in a new position

(b) With C_{ps} replaced by an equivalent shunt capacitance and the secondary shunt capacitances lumped into one

(c) Final simplification with R_c and R_g neglected and the secondary parameters referred to the primary side

Fig. 19. Simplification of the incremental equivalent circuit of the transformer-coupled amplifier.

effect and may be ignored. Also, the primary leakage inductance L_p is generally small compared with the primary incremental magnetizing inductance L_m if the coefficient of coupling in the transformer is near unity. Hence at frequencies at which the susceptance of the incremental inductance is appreciable, the voltage drop in the leakage inductance L_p is negligible, and the upper ends of the incremental magnetizing inductance L_m and the core-loss resistance R_c can be connected without appreciable error at point x instead of at the point shown in Fig. 18b. The circuit is then simplified by these approximations to that shown in Fig. 19a.

The analysis of the circuit would be further simplified if the interwinding capacitance C_{ps}, whose value ranges between 30 and 600 $\mu\mu$f for typical transformers,[22] could be replaced by an equivalent shunt capacitance in either the primary or secondary circuit. However, this replacement cannot be done rigorously. Since the interwinding capacitance interconnects the primary and secondary circuits, it causes equal components of current in both of them, regardless of the turns ratio a. Any shunt capacitance across the secondary terminals introduces components in the primary and secondary circuits having the ratio a, while any shunt capacitance across the primary terminals introduces no component of current at all in the secondary circuit. In general, however, the effect of the interwinding capacitance C_{ps} in the interstage transformer is important mainly because it tends to produce resonance with the leakage inductances L_s and L_p. Shunt capacitance across the secondary terminals also does so, and the analysis here assumes that the important effect of C_{ps} is approximately the same as that of a shunt capacitance across the secondary terminals when the shunt capacitance is adjusted to give the same component of secondary current. Conditions in the primary circuit are not truly represented by this approximation. Often the interwinding capacitance is not important in modern transformers, because a metallic shield is used between the windings. When the shield is connected to the negative terminal of the plate-power supply that is common to the various amplifier stages, the interwinding capacitance is reduced to a negligible value for the same reason that the control-grid-to-plate interelectrode capacitance is made negligible in a screen-grid pentode.

In general, a capacitance connected across the secondary terminals of the transformer can be adjusted to give a magnitude of secondary current equal to the magnitude of the current through the capacitance C_{ps}, but not in phase with this current, because the voltage across C_{ps} is not exactly in phase with the secondary terminal voltage E_{g2}. The voltage across C_{ps} is the difference of the terminal voltages E_{p1} and E_{g2} when the positive reference directions are assigned as shown. These terminal voltages are neither exactly in phase nor 180 degrees out of phase, because of the voltage drop in the leakage reactances and coil resistances. However, provided the departure in phase from these conditions is not great,

$$E_{g2} \approx \pm a E_{p1}, \qquad [63]$$

[22] G. Koehler, "The Design of Transformers for Audio-Frequency Amplifiers with Preassigned Characteristics," *I.R.E. Proc., 16* (1928), 1754, 1756.

where the negative sign applies for the two winding directions and the two assigned voltage reference directions shown in Fig. 18b, and the positive sign applies if an odd number of these directions are reversed. The current through the interwinding capacitance C_{ps}, shown as I_{ps} in Fig. 18b, is given by

$$I_{ps} = \frac{E_{g2} - E_{p1}}{1/(j\omega C_{ps})} = \frac{\left(1 - \frac{1}{\pm a}\right)E_{g2}}{1/(j\omega C_{ps})} \qquad [64]$$

$$= \frac{E_{g2}}{\dfrac{1}{j\omega C_{ps}\left(\dfrac{a \mp 1}{a}\right)}}. \qquad [65]$$

In Eq. 65, the positive sign applies to the conditions of Fig. 18b, and the negative sign applies for the opposite relative winding directions. Also, the value of C_{ps} depends on the winding directions. Thus a capacitance equal to $[(a \mp 1)/a]C_{ps}$ connected across the secondary terminals causes approximately the same component of current at the secondary terminals as that caused by C_{ps}. It should be recognized, however, that this approximation is inexact except at frequencies for which the voltage drops in the leakage reactances are small compared with the terminal voltages. Actually the approximation is used later in analysis of the operation at frequencies for which the voltage drops in the leakage reactances are comparable with the terminal voltages, although such use is not really justified. With C_{ps} represented by its approximately equivalent shunt capacitance, the secondary shunt capacitances may be combined into one, and the equivalent circuit is reduced to that shown in Fig. 19b.

Further simplification of the equivalent circuit may be accomplished by the use of Thévenin's theorem to replace the core-loss resistance R_c, the plate resistance r_p, the effective resistance of the primary winding R_p, and the electromotive force μE_{g1} in Fig. 19b by an equivalent generator and series resistance. The procedure is that used in Art. 5 to simplify Fig. 7c to the circuit in Fig. 7d. Since, in a well-designed transformer, R_c is frequently large compared with R_p of the transformer or r_p of a triode, R_c can often be ignored without undue error. The value of R_g depends on the load impedance and other circuit parameters in the stage of amplification that follows the one under consideration (see Art. 8, Ch. VIII). As a first approximation, R_g is ignored in the following analysis, but its effect is considered qualitatively in the subsequent discussion. Finally, the circuit is simplified further if the secondary-circuit parameters are referred to the primary side of the transformer, the turns ratio a, its reciprocal, its square, or the reciprocal of its square being used as a multiplying factor in the usual manner. The approximate equivalent circuit then has the form shown in Fig. 19c. In this diagram, the negative sign in $E_{g2}/(\pm a)$ applies for the winding directions shown.

Typical values of the parameters in Fig. 19c for an audio-frequency interstage coupling transformer are:[23]

$R_p = 800$ ohms

$R_s = 9,000$ ohms

$L_m = 20$ henries

$L_{eq} = 0.18$ henry

$C_{eq} = 627 \times 10^{-12}$ farad   (obtained from the series-resonant frequency of 15,000 cycles per second)

$a = 3$.

The ratio of L_m to L_{eq} is about 100 in the transformer described above, and it may be as great as 1,000 in high-quality transformers. Thus, in general, L_{eq}, the total leakage inductance referred to the primary side, is small compared with the primary incremental inductance L_m.

Since the two inductances in the equivalent circuit of Fig. 19c have different orders of magnitude, there are two essentially distinct frequencies of resonance in the circuit. One corresponds to parallel resonance between L_m and C_{eq}, which occurs at a frequency so low that the reactance of L_{eq} is negligible compared with that of C_{eq}. The other corresponds to series resonance of L_{eq} and C_{eq}, which occurs at a frequency so high that L_m is virtually an open circuit. From this reasoning it appears that in so far as resonance is a factor in governing the frequency response, the two inductances are not simultaneously important in determining the behavior of the circuit—one is important at high frequencies and the other at low frequencies.

At frequencies considerably below the resonant frequency of L_m and C_{eq}, the reactance of C_{eq} is so large as to constitute a virtual open circuit. The voltage drop in L_{eq} and C_{eq} is therefore negligible, and the behavior of the circuit for the relative winding directions shown is then that of the approximate equivalent circuit in Fig. 20a. From this circuit,

$$\frac{E_{g2}}{a} = \frac{\mu E_{g1} j\omega L_m}{r_p' + j\omega L_m}, \qquad [66]$$

[23] F. E. Terman, "Universal Amplification Charts," *Electronics*, *10* (June, 1937), 34–35.

or

$$A_{low} = \frac{E_{g2}}{E_{g1}} = \mu a \frac{j\omega L_m}{r_p{}' + j\omega L_m} ; \qquad \blacktriangleright[67]$$

thus

$$A_{low} = \mu a \frac{1}{\sqrt{1 + \left(\dfrac{r_p{}'}{\omega L_m}\right)^2}} \underline{/\theta_A} \qquad [68]$$

$$= \mu a \cos \theta_A \underline{/\theta_A} , \qquad [69]$$

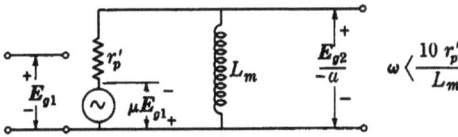

(a) Equivalent circuit for the low-frequency range

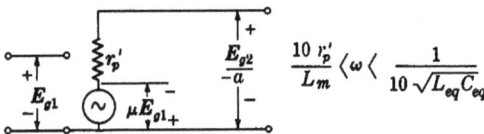

(b) Equivalent circuit for the middle range of frequencies

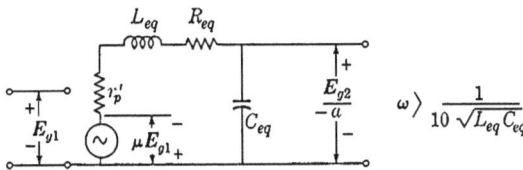

(c) Equivalent circuit for the high-frequency range

Fig. 20.   Incremental equivalent circuits for transformer-coupled amplifier for the various frequency ranges.

where A_{low} is the complex voltage gain of the amplifier stage at low frequencies and θ_A is the angle by which E_{g2} leads E_{g1}, namely,

$$\theta_A = \tan^{-1} \frac{r_p{}'}{\omega L_m} . \qquad [70]$$

Equation 68 shows that as the frequency is increased from a low value the voltage gain approaches μa because of the increase of the term ωL_m. However, when the frequency is larger than the minimum value for which ωL_m may be considered to be large compared with $r_p{}'$—say, when ωL_m is larger than $10r_p{}'$—the voltage gain is essentially constant and equal to μa. If the middle range of frequencies is defined as one in which the voltage gain is essentially constant, the preceding condition may be said to determine the boundary between the low and middle ranges. Thus the complex voltage gain in the low-frequency range is given by Eq. 69, but, when the frequency is so high that

$$\omega > 10 r_p'/L_m ,\qquad\qquad [71]$$

the voltage gain given by Eq. 69 is independent of the frequency. Hence the equivalent circuit for the middle range of frequencies is that of Fig. 20b; the complex voltage gain at these frequencies is

$$A_{mid} = \mu a\ \underline{/0^\circ}\ ,\qquad\qquad \blacktriangleright[72]$$

and the lower boundary of the middle range of frequencies is given by Eq. 71. Actually, somewhere in the middle range of frequencies, the capacitance C_{eq} resonates with the inductance L_m (the reactance of the inductor L_{eq} being negligible), thus increasing the impedance between the terminals of L_m. But this resonance does not appreciably affect the voltage gain, because it occurs in the frequency range where the reactance of L_m is already so large that it constitutes a virtual open circuit even when the increase of impedance between the terminals of L_m caused by resonance with C_{eq} is neglected.

Upon further increase of the frequency to values well above that corresponding to parallel resonance between L_m and C_{eq}, series resonance between L_{eq} and C_{eq} is approached. However, below a frequency of about one-tenth the resonant frequency of L_{eq} and C_{eq}, the effect of their reactances on the voltage gain is negligible, because that of C_{eq} is then more than one hundred times that of L_{eq} and is generally large compared with the resistance r_p'. Thus the high-frequency boundary of the middle range of frequencies is roughly one-tenth the series-resonant frequency of L_{eq} and C_{eq}, and Fig. 20b represents the behavior of the transformer in the middle range of frequencies for which

$$\frac{10 r_p'}{L_m} < \omega < \frac{1}{10\sqrt{L_{eq}C_{eq}}} .\qquad\qquad [73]$$

If the capacitance C_{eq} and the coefficient of coupling are relatively large, the susceptance of C_{eq} may become comparable with $1/r_p$ at frequencies between the parallel-resonant frequency of L_m and C_{eq} and the series-resonant frequency of L_{eq} and C_{eq}. The result is a "dip" in the frequency characteristic within that range of frequencies, but the analysis in this article does not account for it.

In the high-frequency range lying above that defined by Eq. 73, the susceptance of L_m is negligible, but L_{eq} and C_{eq} have an important influence on the circuit behavior; hence the equivalent circuit is that shown in Fig. 20c; whence,

$$A_{high} = \frac{E_{g2}}{E_{g1}} = \mu a \frac{1/(j\omega C_{eq})}{\left(r_p' + R_{eq}\right) + j\left(\omega L_{eq} - \frac{1}{\omega C_{eq}}\right)} \qquad [74]$$

where A_{high} is the complex voltage gain in the high-frequency range. For the purpose of converting Eq. 74 to a form suitable for application to any transformer, it is convenient to define

$$\omega_0 \equiv \frac{1}{\sqrt{L_{eq}C_{eq}}}, \text{ the series-resonant angular frequency} \qquad [75]$$
$$\text{of } L_{eq} \text{ and } C_{eq},$$

and

$$Q_0 \equiv \frac{\omega_0 L_{eq}}{r_p' + R_{eq}}, \text{ the effective } Q \text{ of the circuit}[24] \text{ at the} \qquad [76]$$
$$\text{resonant frequency } \omega_0.$$

When these quantities and Eq. 72 are substituted in Eq. 74, the result may be put in the rectangular-co-ordinate form

$$\frac{A_{high}}{A_{mid}} = \frac{1}{\left[1 - \left(\frac{\omega}{\omega_0}\right)^2\right] + j\left[\left(\frac{\omega}{\omega_0}\right)\frac{1}{Q_0}\right]}, \qquad \blacktriangleright[77]$$

or, in the polar-co-ordinate form,

$$\frac{A_{high}}{A_{mid}} = \frac{1}{\sqrt{\left[1 - \left(\frac{\omega}{\omega_0}\right)^2\right]^2 + \left[\left(\frac{\omega}{\omega_0}\right)\frac{1}{Q_0}\right]^2}} \bigg/ -\tan^{-1}\left[\frac{1/Q_0}{\left(\frac{\omega_0}{\omega}\right) - \left(\frac{\omega}{\omega_0}\right)}\right]. \qquad [78]$$

Although the inverse tangent expression for the phase angle in Eq. 78 is ambiguous as to the quadrant in which the angle lies, the expression when interpreted with reference to the circuit shows that the voltage E_{g2} lags E_{g1} by an angle that approaches zero as the frequency decreases from the series-resonant frequency, that becomes 90 degrees at the series-resonant frequency, and that approaches 180 degrees as the frequency increases above the series-resonant frequency. Furthermore, the symmetry in the denominator of the inverse tangent expression shows that for frequencies of, say, one-half and twice the series-resonant frequency, the changes in the phase angle from the series-resonant value of 90 degrees are equal and of opposite sign, and so on for any other corresponding pair of frequency ratios such as one-third and three. In other words, the curve of phase angle as a function

[24] E. E. Staff, M.I.T., *Electric Circuits* (Cambridge, Massachusetts: The Technology Press of M.I.T.; New York: John Wiley & Sons, Inc., 1940), 319–325.

of frequency on a logarithmic scale is symmetrical about the series-resonant frequency (see Art. 15, Ch. VIII, for an example of similar symmetry).

This symmetry is illustrated by the curves in Fig. 21, which show the voltage gain and phase angle in the amplifier stage. Obtained

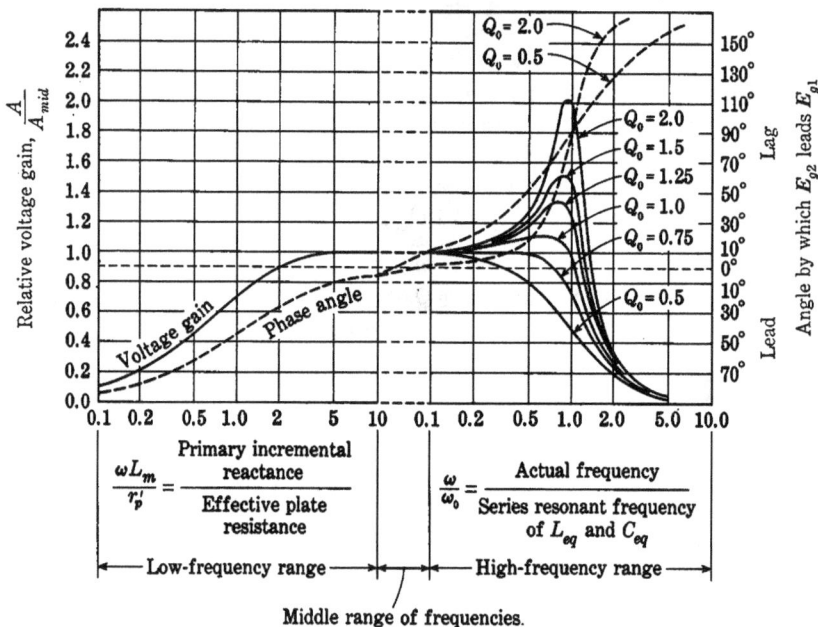

Fig. 21.   Dimensionless voltage-gain and phase-shift curves for a transformer-coupled amplifier.*

from Eqs. 69, 70, 72, and 78, these curves give the voltage gain relative to the value in the middle range of frequencies and the phase shift as functions of frequency in a dimensionless form, so that they are applicable to any tube and transformer when the circuit parameters in Fig. 20c are known. The assumptions regarding the frequency limits of the middle range of frequencies stated in Eq. 73 are seen to be reasonable so far as the voltage gain is concerned, for the relative voltage gain is very close to unity at both these frequencies. However, the phase angle is not zero at these limits; in fact, it is as much as 11.4 degrees for Q_0 equal to one-half; hence there is an appreciable error in the analysis concerning the phase angle.

The usefulness of these curves may be illustrated by a typical example as follows:

* This diagram is adapted from F. E. Terman, "Universal Amplification Charts," *Electronics, 10* (June, 1937), Fig. 2, p. 35, with permission.

Let it be desired to find the frequency characteristic of an amplifier stage in which the transformer has the circuit parameters given on page 540, and the tube has an amplification factor of 10 and a plate resistance of 10,000 ohms.

Solution: The voltage gain in the middle range of frequencies is

$$A_{mid} = 3 \times 10 = 30 \tag{79}$$

from Eq. 72. Also,

$$r_p' = 10,000 + 800 = 10,800 \text{ ohms,} \tag{80}$$

$$\omega_0 = \sqrt{\frac{1}{0.18 \times 627 \times 10^{-12}}} = 9.4 \times 10^4 \text{ radians per sec,} \tag{81}$$

$$\frac{\omega_0}{2\pi} = 15,000 \text{ cps}$$

$$= \text{series-resonant frequency of } L_{eq} \text{ and } C_{eq}, \tag{82}$$

and from Eq. 76,

$$Q_0 = \frac{9.4 \times 10^4 \times 0.18}{10,800 + \dfrac{9,000}{(3)^2}} = 1.43. \tag{83}$$

The frequency for which ωL_m equals r_p' is

$$f = \frac{r_p'}{2\pi L_m} = \frac{10,800}{2\pi \times 20} = 86 \text{ cps,} \tag{84}$$

and for this frequency the voltage gain given by the curve at $\omega L_m/r_p'$ equals 1 is 70.7 per cent of its value in the middle range of frequencies, or approximately 21. The voltage gain is approximately constant at the value 30 over the range from $\omega L_m/r_p'$ equals 10 to ω/ω_0 equals 0.1; that is, from 860 to 1,500 cps. In fact, it departs from constancy only slightly over a range of 430 to 3,000 cps. At the series-resonant frequency of the transformer (15,000 cps), ω/ω_0 equals 1; and by interpolation between the curves for Q_0 equals 1.25 and Q_0 equals 1.5, the voltage gain is found to be about

$$1.40 \times 30 = 42, \tag{85}$$

while its maximum value of

$$1.50 \times 30 = 45 \tag{86}$$

occurs at a slightly lower frequency. At 30,000 cps, ω/ω_0 equals 2, and the voltage gain is decreased to about

$$0.35 \times 30 = 10.5. \tag{87}$$

The voltage gain at other frequencies is readily determined from the curves in a similar manner, and the phase angle in the amplifier is given by the corresponding points on the phase-angle curve.

Several conclusions regarding the desirability of certain features in the design of the transformer may be drawn from the curves of Fig. 21. One conclusion is that the region of essentially constant voltage gain is extended to lower frequencies if (a) L_m, the incremental magnetizing inductance of the primary winding, is made larger; (b) R_p, the primary winding resistance, is made smaller; or (c) r_p, the plate resistance of the tube, is made smaller. A second conclusion is that the region of essentially constant voltage gain is extended to higher frequencies if ω_0 is made larger by a decrease in (a) the leakage inductance, (b) the distributed capacitance of the windings, or (c) the interwinding distributed capacitance. These two conclusions are incompatible, however, because an increase of primary magnetizing inductance requires an increase in the number of turns on the secondary winding if the same turns ratio is maintained and the same core is used, and because both the distributed capacitance and the leakage inductance increase with the number of turns. Correspondingly, increasing the turns ratio to increase the voltage gain is accompanied either by a decrease of the primary magnetizing inductance or by an increase in the secondary leakage inductance and distributed capacitance. Consequently, the frequency range over which the voltage gain is substantially constant usually decreases as the turns ratio is increased, and the practical maximum for the step-up turns ratio has been found to be about three for interstage transformers to be used to amplify the frequencies contained in speech or music; larger ratios cause excessive distortion.

A further deduction based on the curves of Fig. 21 is that to prevent an appreciable peak in the frequency response at the high frequencies near ω_0, the Q_0 of the circuit must not be more than about 0.8. The Q_0 may be reduced to this value through decreasing the leakage inductance or increasing the sum of the tube plate resistance and the transformer winding resistances. However, as has been seen, increasing the primary-winding resistance and tube plate resistance affects the low-frequency response of the amplifier adversely. On the other hand, the secondary-winding resistance does not affect the low-frequency response; hence some interstage transformers are deliberately wound with high-resistivity wire in the secondary winding to improve the high-frequency response.[25]

In the previous analysis, the effect of the load resistance R_g in Fig. 18b is neglected. Actually, the load tends to lower the voltage gain throughout the whole range of frequencies. Its greatest effect, however, occurs at frequencies near the resonant frequency ω_0. By introducing losses in the circuit, the load effectively reduces the Q_0 and thereby reduces the peak in the response at high frequencies. To improve the constancy of the voltage gain in an amplifier with a particular interstage transformer, it is therefore sometimes desirable to connect a

[25] G. Koehler, "The Design of Transformers for Audio-Frequency Amplifiers with Pre-assigned Characteristics," *I.R.E. Proc.*, *16* (1928), 1754–1755.

resistor across the secondary terminals of the transformer, although this practice results in some sacrifice of voltage gain at all frequencies.[26]

When pentode vacuum tubes are used with an interstage transformer in their plate circuits, the low-frequency response of the amplifier is likely to be inferior to that with a triode, because of the high plate resistance of the pentode vacuum tube. The low-frequency response may be improved by the connection of a resistor in shunt across the primary winding of the transformer, because, as may be demonstrated by application of Thévenin's theorem in a manner similar to that used to reduce Fig. 7c to Fig. 7d in Art. 5, one effect of such a resistor is similar to that of a decrease in the plate resistance. However, an additional effect is a decrease in the effective internal electromotive force of the equivalent generator; thus the voltage gain is reduced at the middle and high frequencies by the shunt resistor.

9. TUNED CIRCUITS IN AMPLIFIERS

When only a narrow band of frequencies is to be amplified, a coupling circuit or load impedance comprising one or more parallel-tuned or anti-resonant circuits is commonly employed. The first tubes in a radio receiver and the last ones in a radio transmitter have such circuits associated with them; the tuned circuits are adjusted so as to resonate at the frequencies to be amplified. The vacuum-tube oscillator, discussed in Ch. XI, often has one or more tuned circuits, and in this application tuned amplifiers are important not only at radio frequencies but also in the audio-frequency range.

The circuit diagram of a basic type of tuned amplifier is shown in Fig. 22a, where the load is represented by a capacitance C in parallel with a series combination of an inductance L and a resistance R. The dissipation of power that occurs in the tuned circuit is accounted for by the resistance R in series with the inductor. This resistance is the sum of two components: one a resistance that accounts for the series internal alternating-current resistance of the inductor and the losses in the capacitor, and the other a resistance that accounts for the load resistance in which the amplified alternating-current power is to be dissipated. In a practical application, the load resistor may be inductively coupled to the tuned circuit, but by the methods given in *Electric Circuits* the behavior of such a coupled circuit may often be expressed approximately in terms of the simple tuned circuit considered here. The resistance to direct current in the tuned circuit is

[26] P. W. Klipsch, "Design of Audio-Frequency Amplifier Circuits Using Transformers," *I.R.E. Proc.*, *24* (1936), 219–232.

generally so small that there is a negligible voltage drop in it from the average component of the plate current.

The impedance of the parallel-tuned circuit as a function of frequency is of interest, because this circuit is the external plate load of the amplifier and is the determining factor in the frequency characteristic. An analysis of the behavior of such a parallel resonant circuit

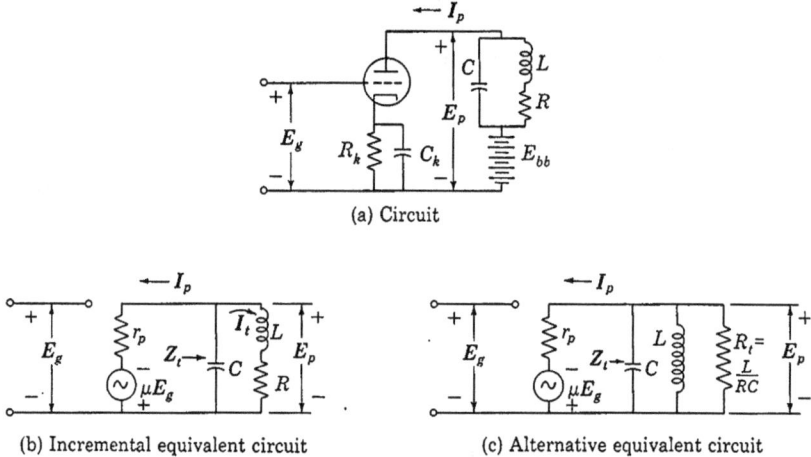

(a) Circuit

(b) Incremental equivalent circuit

(c) Alternative equivalent circuit

Fig. 22.   Basic tuned amplifier.

is given in most textbooks on electric circuits.[27] That analysis is summarized here, and extended to apply especially to tuned circuits used in amplifiers. In the equivalent circuit for the amplifier, Fig. 22b, the complex impedance of the tuned circuit is denoted by Z_t, and is given by

$$Z_t = \frac{(R + j\omega L)\dfrac{1}{j\omega C}}{R + j\omega L + \dfrac{1}{j\omega C}} \qquad [88]$$

$$= \frac{\dfrac{L}{RC} - j\dfrac{1}{\omega C}}{1 + j\left(\dfrac{\omega L}{R}\right)\left(1 - \dfrac{1}{\omega^2 LC}\right)} . \qquad [89]$$

Expressed in terms of

$$\omega_0 \equiv \frac{1}{\sqrt{LC}}, \qquad \text{the resonant angular frequency of the circuit with } R \text{ equal to } 0 \qquad [90]$$

---

[27] E. E. Staff, M.I.T., *Electric Circuits* (Cambridge, Massachusetts: The Technology Press of M.I.T.; New York: John Wiley & Sons, Inc., 1940), 329–332.

and

$$Q_0 \equiv \frac{\omega_0 L}{R},$$ 　the figure of merit, or Q, for the circuit at the resonant angular frequency ω_0, 　[91]

the impedance in Eq. 89 becomes

$$\mathbf{Z}_t = \frac{\dfrac{L}{RC} - j\dfrac{1}{\omega C}}{1 + jQ_0\left(\dfrac{\omega}{\omega_0} - \dfrac{\omega_0}{\omega}\right)} .$$ 　[92]

The numerator of the expression in Eq. 92 may be written as

$$\frac{1}{\omega C}\left(\frac{\omega L}{R} - j1\right) = \frac{1}{\omega C}\left(\frac{\omega}{\omega_0}Q_0 - j1\right),$$ 　[93]

and, under the assumption that ωL is large compared with R, which is true in most practical circuits used as plate impedances, the imaginary term may be neglected; hence

$$\mathbf{Z}_t \approx \frac{L}{RC}\frac{1}{1 + jQ_0\left(\dfrac{\omega}{\omega_0} - \dfrac{\omega_0}{\omega}\right)} , \text{ when } \frac{\omega L}{R} \gg 1.$$ 　▶[94]

　　If instead of being shown as in Fig. 22b, the tuned circuit is represented as in Fig. 22c—that is, as a parallel combination of L, C, and a resistance R_t having the value $L/(RC)$, where R is the resistance shown in Fig. 22b—Eq. 94 may be obtained directly without the assumption that ωL is large compared with R. Thus the neglect of the imaginary term in the numerator of Eq. 92 is equivalent to the assumption that the tuned circuit behaves as a parallel combination of those three elements rather than as the series-parallel circuit in Fig. 22b. Neither circuit is a strictly correct representation of an actual tuned circuit, since several causes contribute to the total power dissipation in the circuit; hence resistance at only one point in it is not sufficient to account for its behavior. However, *if the Q_0 of the tuned circuit is large compared with unity*, the variation of the impedance with frequency *in the vicinity of resonance* for the series-parallel circuit is indistinguishable from that of the parallel circuit, and each is a close approximation to the variation of the impedance with frequency for an actual tuned circuit. For these reasons, either the series-parallel circuit of Fig. 22b or the parallel circuit in Fig. 22c may be used to represent an actual tuned circuit, and Eq. 94 may be used to describe its behavior. In the subsequent discussion of tuned circuits in amplifiers, the series-parallel circuit is shown in the diagrams. The analysis, however, is based on Eq. 94, so that it applies equally well to the parallel circuit.

Frequencies in the vicinity of the resonant frequency $\omega_0/(2\pi)$ are mainly of interest when the circuit is sharply tuned. The term ω/ω_0 is then near unity, and the denominator involves the difference of two numbers that are almost equal; hence it is convenient to express the results in terms of

$$\delta = \frac{\omega - \omega_0}{\omega_0} = \frac{\omega}{\omega_0} - 1, \qquad [95]$$

which is the fractional departure of ω from the resonant angular frequency ω_0. In terms of δ, Eq. 94 may be written as

$$\mathbf{Z}_t \approx \frac{L}{RC} \frac{1}{1 + j2Q_0\delta \left(\frac{1 + \frac{\delta}{2}}{1 + \delta} \right)}, \text{ when } \frac{\omega L}{R} \gg 1. \qquad \blacktriangleright[96]$$

Equation 96 is an expression for the impedance of the tuned circuit in very useful form. It may be further simplified, if δ is small compared with unity, to

$$\mathbf{Z}_t \approx \frac{L}{RC} \frac{1}{1 + j2Q_0\delta}, \text{ when } \frac{\omega L}{R} \gg 1 \text{ and } \delta \ll 1. \qquad \blacktriangleright[97]$$

Resonance in a parallel-tuned circuit may be defined as either the condition of unity power factor or the condition of maximum impedance in Eq. 89. In general, these conditions may occur at different frequencies; when the Q_0 of the circuit is large, however, they occur at practically the same angular frequency ω_0, at which the imaginary term in the denominator of Eqs. 94 and 96 is zero. The impedance at resonance is thus a real quantity denoted by R_t and given by

$$R_t = \frac{L}{RC} = \text{impedance of parallel circuit at resonance,} \qquad [98]$$

or, in terms of other quantities,

$$R_t = \frac{L}{RC} = \frac{\omega_0 L}{R\omega_0 C} = \frac{X_{C0}{}^2}{R} = \frac{X_{L0}{}^2}{R} = Q_0{}^2 R = (\omega_0 L)Q_0 = Q_0\sqrt{\frac{L}{C}}, \quad \blacktriangleright[99]$$

where X_{C0} and X_{L0} are the reactances of the capacitance and inductance, respectively, at the resonant frequency, and $\omega_0 L/R$, or Q_0, is large compared with unity.

Normalized curves of the magnitude and phase angle of the impedance $\mathbf{Z}_t$ of a parallel tuned circuit as expressed by Eq. 96 are shown in Fig. 23 for values of 10 and infinity for Q_0. The abscissas on this diagram are the fractional angular frequency deviation from ω_0 relative to $1/Q_0$. Once values of the resonant impedance R_t, the resonant angular frequency ω_0, and the resonant value Q_0 are chosen or determined, scales of impedance and frequency become established for these curves and may be added to the diagram through simple arithmetic. Hence the curves serve universally for all parallel resonant circuits within the range of Q_0 indicated. The magnitude of the impedance decreases steeply on both sides of the resonant point, and the

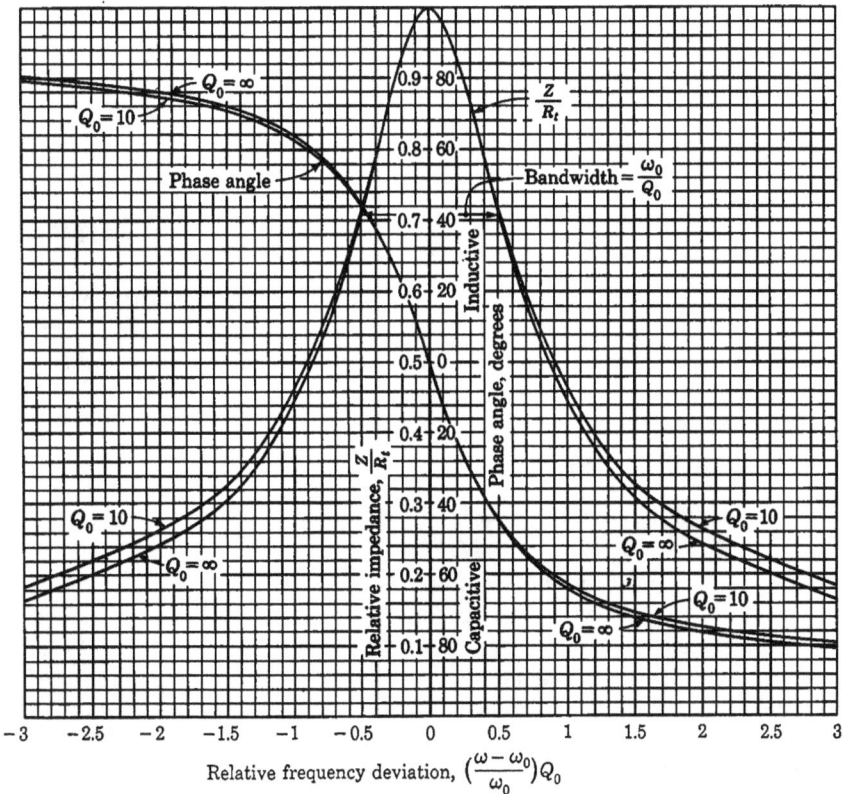

Fig. 23. Universal curves for the magnitude and phase angle of the impedance of a parallel resonant circuit in terms of ω_0, Q_0, and R_t to be selected or computed from Eqs. 90, 91, and 99.

phase angle is zero there. As the frequency deviates from the resonant value, the impedance approaches that of the coil and resistor at low frequencies and the reactance of the capacitor at high frequencies, while the phase angle approaches plus and minus 90 degrees for low and high frequencies, respectively, if Q_0 is large. (Actually, of course, the phase angle must reach zero at zero frequency, and the analysis here is applicable only for frequencies in the vicinity of resonance.) The curve for any value of Q_0 greater than 10 falls between the two shown. Since the curves differ so little in the vicinity of resonance, the curve for infinite Q_0 may be used with negligible error for most computations. Use of it is equivalent to the assumption that Eq. 97 is applicable.

When the imaginary term in the denominator of Eq. 96 or 97 has the value plus or minus unity, the impedance of the circuit is 70.7 per cent of its value at resonance. If the circuit is supplied from a constant-current source, the power dissipated in the circuit when the frequency is adjusted to produce either of these conditions is one-half its value at the resonant frequency. For this reason, the frequencies for these conditions are called the *half-power frequencies*. They are the frequencies for which

$$\delta = \pm \frac{1}{2Q_0}, \qquad \blacktriangleright[100]$$

and it is seen that, if Q_0 is large, the width of the curve of impedance as a function of frequency between these two frequencies is small; that is, the resonance curve is sharp. The frequency deviation between the two half-power frequencies, ω_0/Q_0 in radians per second, is commonly called the *bandwidth* of the resonance curve, as is indicated in Fig. 23. By means of Eqs. 90 and 99 it may be expressed as

$$\text{Bandwidth} \equiv \frac{\omega_0}{Q_0} = \frac{1}{R_t C} \text{ radians per second.} \qquad [101]$$

Thus it depends only on the resonant impedance R_t and the capacitance, and is independent of the resonant frequency.

In some tuned-amplifier circuits, particularly Class B and Class C tuned amplifiers, the plate current contains appreciable harmonic-frequency components. The relative impedance of the circuit at the harmonics of the resonant frequency is then of interest. At the frequencies $2\omega_0$, $3\omega_0$, $4\omega_0$, $\cdots$, δ has the values 1, 2, 3, $\cdots$. Since Q_0 is large compared with unity, the real term in the denominator of Eq. 96 may be neglected in comparison with the imaginary term at these high frequencies, and the impedance at the harmonic frequencies is approximately a capacitive reactance given by

$$X_2 = \frac{L}{RC} \frac{2}{3Q_0} \text{ for the second harmonic,} \qquad [102]$$

$$X_3 = \frac{L}{RC} \frac{3}{8Q_0} \text{ for the third harmonic,} \qquad [103]$$

$$X_4 = \frac{L}{RC} \frac{4}{15Q_0} \text{ for the fourth harmonic,} \tag{104}$$

$$X_5 = \frac{L}{RC} \frac{5}{24Q_0} \text{ for the fifth harmonic,} \tag{105}$$

and so on. If Q_0 has a representative value of 100, the impedance at the harmonic frequencies is a negligible fraction—less than one per cent—of the resonant impedance if R is considered constant. Actually, the ratio of the impedance at a harmonic frequency to the impedance at the resonant frequency is reduced further by the fact that skin and proximity effects increase the resistance of the coil at the harmonic frequency. By proper design, the resistance may be made to vary with frequency in such a way as to maintain the Q of the coil almost constant over a considerable range of frequencies. The values of Q depend on frequency, coil construction, and the coefficient of coupling to the load. At radio frequencies, values for Q of 100 to 400 or even higher may be obtained. The construction of suitable inductors is an important part of tuned-amplifier and oscillator design. Crystals that resonate mechanically and are coupled to the electrical circuit through the piezoelectric effect often serve as the tuned circuit. Also, at the higher radio frequencies, transmission lines and resonant cavities replace lumped-constant tuned circuits. Any such element that exhibits impedance or admittance in the form of Eq. 94 is said to have a resonant Q given by the coefficient occupying the position of Q_0 in that relation.

The current circulating in the tuned circuit at resonance may be expressed in terms of the plate current of the tube. The current I_t in the tuned circuit of Fig. 22b is given by

$$I_t = \frac{E_p}{\sqrt{R^2 + (\omega_0{}^2 L^2)}} \tag{106}$$

at the resonant frequency, and, since $\omega_0 L$ is large compared with R, this may be simplified to

$$I_t \approx \frac{I_p R_t}{\omega_0 L} = Q_0 I_p, \text{ when } \frac{\omega_0 L}{R} \gg 1. \tag{▶[107]}$$

This relation is sometimes very useful, for I_t, being many times larger than I_p, can be measured relatively easily, and then I_p, E_p, and other quantities can be found through dividing I_t by Q_0, and so forth. The power delivered to the tuned circuit by the source and dissipated in the resistance R at the resonant frequency may be computed by either

$$P = I_t{}^2 R, \tag{▶[108]}$$

or

$$P \approx I_p{}^2 R_t. \tag{▶[109]}$$

10. Tuned voltage amplifiers

The discussion of modulation in Ch. XII shows that the only components contained in the usual radio-frequency signal are those having frequencies close to the mean frequency. For example, the frequency components of a typical broadcast signal having a mean frequency of 1,000,000 cycles per second lie in the range of about 995,000 to 1,005,000 cycles per second, and the ratio of the bandwidth to the mean or carrier frequency is therefore about 1 to 100. An amplifier designed to transmit these signals need have a uniform voltage gain over only a relatively narrow frequency range. Voltages whose frequencies lie outside this range are undesired; they produce interference with the information to be transmitted. In order to reject unwanted signals and interference, the plate-load impedance or the transfer voltage ratio of the amplifier-coupling network should therefore fall steeply to low values outside the signal band, and thus reduce the voltage gain to negligibly low values. In some applications, it is desirable also that a simple adjustment shift the mean frequency to a new value. The preceding article indicates how a simple parallel-tuned circuit used as a plate-load impedance can partially accomplish this objective if a means of adjusting the resonant frequency is provided. Common practice is to provide adjustment by means of an adjustable capacitor, although in some applications the inductance is made adjustable.

10a. *Tuned-Plate Capacitance-Coupled Amplifier.* A simple type of cascade tuned voltage amplifier is shown in Fig. 24. It is essentially similar to the resistance-capacitance-coupled amplifier except that the plate-load impedances in series with the plate-power supply are parallel-tuned circuits.

Because the plate-load impedance is a tuned circuit, the tendency for the amplifier to oscillate owing to feedback through the grid-to-plate capacitance, discussed in Art. 9, Ch. VIII, and in Art. 7, Ch. XI, is particularly important. As the frequency is increased, this tendency becomes more pronounced with a given plate-load impedance, because the coupling susceptance through the tube capacitance is directly proportional to the frequency. Pentode vacuum tubes are used extensively in tuned radio-frequency amplifiers because their capacitance from control grid to plate is very small. In order to realize the full benefits that these tubes make possible, shielding of the grid and plate circuits is necessary, and is usually readily accomplished. Pentodes are therefore used for illustration in the circuit shown in Fig. 24.

All the electrode voltages in the amplifier of Fig. 24 are shown as being derived from a single source of direct voltage common to all stages. The grid-bias voltages are obtained from cathode-bias circuits as mentioned in Art. 3, Ch. VIII. Adjustment of the screen-grid and plate voltages E_{cc2} and E_{bb} is made by resistors in series with each electrode, and by-pass capacitors of low reactance at the signal

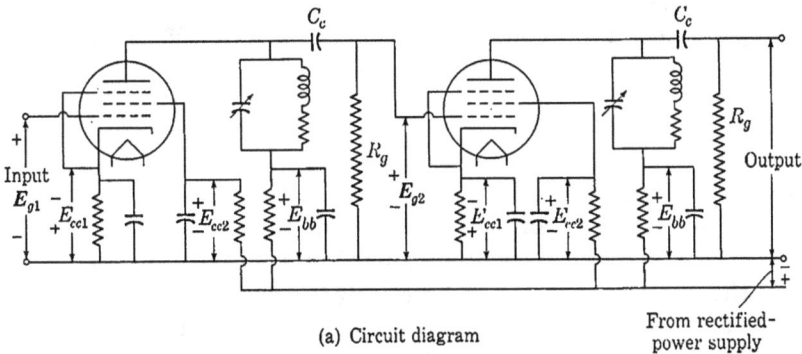

(a) Circuit diagram

From rectified-power supply

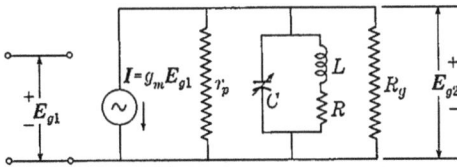

(b) Approximate incremental equivalent circuit for one stage

Fig. 24.   Capacitance-coupled tuned voltage amplifier.

frequencies are connected between each resistor and the common negative connection. These combinations of resistance and capacitance, called *decoupling circuits*, serve the dual purpose of providing the proper electrode voltages and of preventing variations in the current to one electrode from affecting the voltages of other electrodes in the same and other stages. When not made negligible by adequate means such as these decoupling circuits, coupling of that type through the common impedance of the power supply may cause unstable amplifier operation.

With suitable by-pass capacitors, the impedance between the cathode, suppressor grid, screen grid, or lower terminal of the tuned circuit and the common negative connection is very low. Thus the capacitance between the plate and these other electrodes in one stage of Fig. 24 is effectively in parallel with the tuning capacitor (shown adjustable), and these interelectrode capacitances and stray wiring capacitances affect only the adjustment of the tuning capacitor. The effective shunt input admittance of the second tube, which is predominantly a capacitive susceptance, is in parallel with the grid resistor R_g. The coupling capacitance C_c can, however, easily be made large enough to have a reactance that is small compared with the impedance of the parallel combination of R_g and the effective shunt input impedance of the second tube. Consequently, the effective shunt input admittance of the second tube also affects only the tuning adjustment.

In accordance with the foregoing considerations, the equivalent circuit for one stage of this amplifier when operation is restricted to the linear region of the tube characteristics is that shown in Fig. 24b, where the tube is represented as a constant-current source with r_p shunted across it. At the resonant angular frequency ω_0, which equals $1/\sqrt{LC}$, the impedance of the tuned circuit is a resistance R_t equal to $\omega_0 L Q_0$ as given by Eq. 99. The voltage amplification at resonance is then g_m times the impedance of r_p, R_g, and the tuned circuit in parallel; that is,

$$\frac{E_{g2}}{E_{g1}} = A_0 = g_m \frac{1}{\dfrac{1}{r_p} + \dfrac{1}{R_g} + \dfrac{1}{R_t}} \qquad \blacktriangleright[110]$$

$$= g_m \frac{\omega_0 L Q_0}{1 + \dfrac{\omega_0 L Q_0}{r_p} + \dfrac{\omega_0 L Q_0}{R_g}}. \qquad [111]$$

The equivalent circuit of Fig. 24b shows that the total impedance that determines the voltage gain of the amplifier is the combination of r_p, R_g, and the tuned circuit in parallel. On the assumption that the tuned circuit can be represented as a parallel combination of inductance, capacitance, and constant resistance R_t, rather than as shown in Fig. 24b, the curve of impedance as a function of frequency for the parallel combination including r_p and R_g is the same as though the resonant resistance of the tuned circuit were decreased from the value R_t to the resistance of the parallel combination of R_t, r_p, and R_g. Furthermore, the curve of voltage amplification as a function of frequency has this same shape, since the tube is considered as a constant-current generator supplying current to this impedance.

Since, from Eq. 99,

$$Q_0 = R_t \sqrt{\frac{C}{L}}, \qquad [112]$$

the effect of the resistance r_p and R_g in parallel with the tuned circuit is the same as though the Q_0 of the actual tuned circuit were reduced to the effective value

$$Q_0' = \frac{Q_0}{1 + \dfrac{R_t}{r_p} + \dfrac{R_t}{R_g}}, \qquad \blacktriangleright[113]$$

where

Q_0 is the resonant Q of the tuned circuit,

Q_0' is the effective resonant Q of the amplifier.

In terms of Q_0', the voltage amplification of the amplifier at resonance is, from Eqs. 111, 112, and 113,

$$A_0 = g_m Q_0' \sqrt{\frac{L}{C}}. \qquad [114]$$

The amplifier selectivity, which corresponds to Q_0', is always lower than the selectivity of the tuned circuit alone. From Eqs. 90 and 101, the bandwidth is

$$BW = \frac{f_0}{Q_0'} = \frac{1}{2\pi Q_0' \sqrt{LC}} \text{ cycles per second.} \qquad [115]$$

Thus the product of the voltage amplification at resonance and the bandwidth,

$$A_0 \times BW = \frac{g_m}{2\pi C} \text{ cycles per second,} \qquad [116]$$

depends only on the tube and the total capacitance, and is independent of the resonant frequency, the resonant Q_0, and the resonant impedance. This expression is seen to be essentially identical with Eq. 60 developed for the effective bandwidth of a resistance-capacitance-coupled amplifier. These conclusions hold only to the extent that the tuned circuit can be represented by a parallel combination of inductance, capacitance, and constant resistance. As is pointed out in the discussion of Eq. 94, however, if the Q_0 of the tuned circuit is large, this representation gives results indistinguishable from those of an actual circuit over a range of frequencies near the resonant value.

Equation 116 shows that when the capacitance is held constant, an increase in the voltage amplification can be obtained with a given tube only at the expense of bandwidth. The maximum value of the product of the voltage amplification and the bandwidth is determined by the minimum permissible capacitance in the tuned circuit. This minimum capacitance is usually set by the interelectrode and stray wiring capacitances inherent in the circuit. When the capacitance is larger than this minimum value, the resonant load impedance and, correspondingly, the voltage amplification may be increased through increasing either the L/C ratio or Q_0 as indicated by Eqs. 99 and 110. If the L/C ratio is increased for fixed values of Q_0 and ω_0, Eqs. 99 and 113 show that the effective Q_0 of the amplifier is decreased. Thus under these conditions an increase in voltage gain is indeed accomplished only at a sacrifice of frequency selectivity, or increase of bandwidth, in the amplifier, and a compromise between the two must be made. On the other hand, if the voltage gain of the amplifier is increased by an increase in the Q_0 of the tuned circuit at fixed values of ω_0 and L/C ratio, Eq. 113 shows that the effective Q_0' of the amplifier is increased and the frequency selectivity is also increased.

The voltage gain that can be obtained with triodes is limited because of the relatively low amplification factor of the tubes. Furthermore, special neutralizing methods are required in grounded-cathode triode amplifiers to prevent oscillations associated with the large values of grid-to-plate capacitance. Pentodes with their large values of amplification factor and plate resistance and their relative freedom from undesirable interelectrode capacitance make possible much higher voltage gain for the same sharpness of frequency characteristic and have therefore practically supplanted triodes in tuned amplifiers. When the relatively large noise voltage inherent in a pentode is not tolerable, however, a triode is often used as a grounded-grid amplifier.

10b. *Tuned-Secondary Transformer-Coupled Amplifier.* Instead of the capacitance coupling shown in Fig. 24, air-core transformer coupling is often used in radio-frequency amplifiers. A stage of tuned radio-frequency voltage amplification with an air-core transformer as a coupling element is shown in Fig. 25a. If all stray and interelectrode capacitances are neglected, the equivalent circuit for this amplifier is that shown in Fig. 25b. Such neglect is permissible if the primary inductance is low and the coupling coefficient is high, for then the capacitances affect principally the tuning capacitance.[28]

The expression for the voltage gain of the amplifier, which is the ratio of the output and input voltages, may be obtained as follows through simultaneous solution of the Kirchhoff's law equations for the two loops in Fig. 25b in a manner that eliminates the unwanted currents and voltages and leaves the input and output voltages and the circuit parameters in the final expression. The sum of the voltage rises taken clockwise around loop 1 and equated to zero is

$$-\mu E_{g1} + I_p(r_p + R_1 + j\omega L_1) - j\omega M I_2 = 0. \qquad [117]$$

[28] F. E. Terman, *Radio Engineering* (2nd ed.; New York: McGraw-Hill Book Company, Inc., 1937), 212–213.

Similarly, for the voltage rises taken counterclockwise around loop 2,

$$I_2 \left[R_2 + j \left(\omega L_2 - \frac{1}{\omega C_2} \right) \right] - j \omega M I_p = 0. \qquad [118]$$

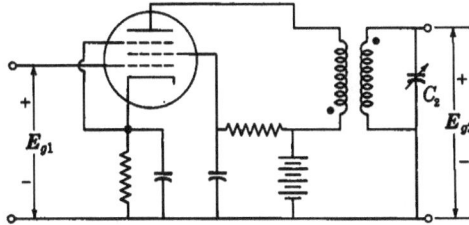

(a) Circuit diagram

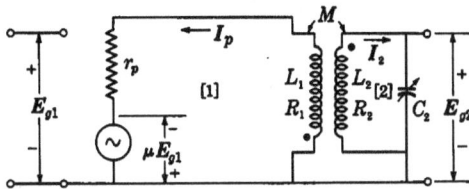

(b) Incremental equivalent circuit

Fig. 25. Tuned-secondary transformer-coupled voltage amplifier.

The complex voltage gain A is given by

$$A = \frac{E_{g2}}{E_{g1}} = \frac{I_2 \dfrac{1}{j\omega C_2}}{E_{g1}} \qquad [119]$$

Elimination of I_p from Eq. 117 by means of Eq. 118 and solution for the right-hand expression in Eq. 119 give

$$A = \mu \frac{M/C_2}{(r_p + R_1 + j\omega L_1) \left[R_2 + j \left(\omega L_2 - \dfrac{1}{\omega C_2} \right) \right] + \omega^2 M^2}. \qquad [120]$$

Generally, R_1 and ωL_1 are negligible compared with r_p, and Eq. 120 can be simplified to

$$A = \mu \frac{M/C_2}{r_p \left[R_2 + j \left(\omega L_2 - \dfrac{1}{\omega C_2} \right) \right] + \omega^2 M^2}. \qquad [121]$$

If the tuning capacitance C_2 is varied while the other parameters are fixed, both the numerator and denominator in Eq. 121 change. However, the numerator varies inversely with C_2, but the reactance term in the denominator is a difference term that approaches zero rapidly as C_2 is varied toward the value for which

$$\omega = \omega_0 = 1/\sqrt{L_2 C_2}.$$ [122]

Hence, in general, the voltage gain is a maximum at practically the resonant angular frequency defined by Eq. 122. At the resonant frequency, or with the optimum value of C_2 that makes the circuit resonate at any given frequency, the voltage gain is therefore

$$A = \mu \frac{M/C_2}{r_p R_2 + \omega_0{}^2 M^2}, \quad \text{for optimum } C_2, \text{ or } \omega = \omega_0.$$ [123]

This result may also be expressed in terms of Q_0, or $\omega_0 L_2/R_2$, by means of Eq. 122 as

$$A = g_m \frac{\omega_0 M Q_0}{1 + \dfrac{\omega_0{}^2 M^2}{r_p R_2}}, \quad \text{for optimum } C_2.$$ ▶[124]

The denominator of this equation is the ratio of the apparent series resistance[29] in the secondary circuit, $R_2 + (\omega_0{}^2 M^2/r_p)$, to the actual series resistance R_2, the apparent resistance being larger because of the reflected effect of r_p from the primary circuit. The over-all Q_0 of the amplifier circuit is for this reason lower than the Q_0 of the tuned circuit alone. The shape of the curve of voltage gain of the amplifier as a function of frequency, or its selectivity, corresponds to a reduced value of Q_0—the value obtained through dividing the Q_0 of the tuned circuit by the factor $1 + [\omega_0{}^2 M^2/(r_p R_2)]$.

It is explained subsequently that with tetrodes or pentodes, r_p is generally large compared with the values of $\omega_0{}^2 M^2/R_2$ that can be obtained practically in radio-frequency amplifiers; thus, when these tubes are used, the denominator in Eq. 124 is approximately unity, the effective Q_0 is nearly as large as the actual Q_0, and the peak in the frequency characteristic is nearly as narrow as the resonance curve of the tuned circuit itself. The voltage gain at resonance with pentodes or tetrodes is therefore given approximately by

$$A \approx g_m \omega M Q_0, \text{ for optimum } C_2 \text{ and } r_p \gg \frac{\omega_0{}^2 M^2}{R_2}.$$ ▶[125]

[29] E. E. Staff, M.I.T., *Electric Circuits* (Cambridge, Massachusetts: The Technology Press of M.I.T.; New York: John Wiley & Sons, Inc., 1940), 378–384. See also, in this book, the discussion of Eq. 134 in this chapter and Art. 5, Ch. X.

Equation 125 alone indicates that the voltage gain increases without limit as M is increased. However, this condition is not true, because the $\omega_0{}^2M^2$ term in the denominator of Eq. 124 becomes important with increasing M, and there is an optimum value of M for maximum voltage gain that may be derived through equating to zero the partial derivative of A with respect to M as follows. From Eq. 123,

$$\frac{\partial A}{\partial M} = \mu \, \frac{(r_pR_2 + \omega_0{}^2M^2)\dfrac{1}{C_2} - \dfrac{M}{C_2}(2\omega_0{}^2M)}{(r_pR_2 + \omega_0{}^2M^2)^2} = 0; \qquad [126]$$

whence, the optimum value of M is

$$M_{opt} = \frac{1}{\omega_0}\sqrt{r_pR_2}\,; \qquad [127]$$

or, by substitution of Eq. 122,

$$M_{opt} = \sqrt{r_pR_2L_2C_2}\,. \qquad \blacktriangleright[128]$$

The maximum voltage gain that can be obtained by adjustment of C_2 and M with the other parameters constant is, by substitution of Eq. 127 in Eq. 123,

$$A = \mu \, \frac{1}{2\omega_0C_2\sqrt{r_pR_2}}, \quad \text{for } M = M_{opt} \text{ and optimum } C_2. \quad [129]$$

By means of Eq. 122, this maximum voltage gain may be expressed as

$$A = \frac{\mu}{\sqrt{r_p}}\,\frac{\omega_0L_2}{2\sqrt{R_2}}; \qquad [130]$$

whence,

$$A = \frac{1}{2}\sqrt{\mu g_m \frac{L_2}{R_2C_2}}, \quad \text{for } M = M_{opt} \text{ and optimum } C_2, \quad [131]$$

or, in terms of Q_0 and the equivalent expressions in Eq. 99,

$$A = \frac{1}{2}\sqrt{\mu g_m \omega_0 L_2 Q_0} = \frac{Q_0}{2}\sqrt{\mu g_m R_2} \qquad [132]$$

$$= \frac{\mu Q_0}{2}\sqrt{\frac{R_2}{r_p}}, \quad \text{for } M = M_{opt} \text{ and optimum } C_2. \quad \blacktriangleright[133]$$

Equations 131 and 132 show that in order for the optimum voltage gain A to be large, the quantity μg_m, which depends only on the tube, must be large, as also must the parallel-tuned impedance of the secondary circuit of the transformer.

The effective plate-load impedance offered by the transformer and secondary load—that is, the impedance as viewed from the tube into the transformer primary-winding terminals—is:[30]

$$\text{Effective plate-load impedance} = Z_{11} - \frac{(Z_{12})^2}{Z_{22}}, \qquad [134]$$

where Z_{11}, Z_{22}, and Z_{12} are the self-impedances of the primary and secondary circuits and the mutual impedance between them, respectively. For the circuit of Fig. 25b, these are given by

$$Z_{11} = R_1 + j\omega L_1 , \qquad [135]$$

$$Z_{22} = R_2 + j\left(\omega L_2 - \frac{1}{\omega C_2}\right), \qquad [136]$$

$$Z_{12} = j\omega M. \qquad [137]$$

When the mutual inductance M and the capacitance C_2 have their optimum values given by Eqs. 122 and 127, Eq. 134 becomes

$$\text{Effective plate-load impedance} = R_1 + j\omega L_1 + r_p. \qquad [138]$$

Since, in general, $R_1 + j\omega L_1$ is small compared with r_p, the optimum conditions are essentially the impedance-matched conditions at which the tube delivers maximum power to its load.

Because of practical considerations, the optimum value of M is not of much importance in present-day amplifiers. With tetrodes or pentodes, the value of r_p is so high that it may be difficult to realize the value of M required in Eq. 127 and at the same time to have coils that behave like inductances at the operating frequencies—that is, coils whose self-resonant frequencies are above the operating frequencies. Furthermore, if the optimum value of M could be realized, difficulties with oscillations would be likely to occur in spite of the low control-grid–to–plate interelectrode capacitance and careful shielding, because the plate-load impedance would then be comparable to the reactance of the interelectrode capacitance. Consequently, the mutual inductance is usually chosen far below the optimum value in pentode vacuum-tube amplifiers, and Eq. 125 applies. In triode amplifiers, neutralizing methods to prevent oscillations are practicable, and these

[30] E. E. Staff, M.I.T., *Electric Circuits* (Cambridge, Massachusetts: The Technology Press of M.I.T.; New York: John Wiley & Sons, Inc., 1940), 378–379.

—together with the relatively low values of r_p—make feasible the utilization of the optimum value of M; but, even so, the values of voltage gain that can be realized are usually lower and the circuit required is considerably more complicated than with pentodes.

When a frequency characteristic with a flatter top than that obtained with a single resonant circuit is desired, double-tuned transformer coupling is often utilized.[31] This type of coupling is accomplished by the addition of a second tuning capacitor connected across the primary terminals of the coupling transformer, and is sometimes called *band-pass* coupling. By adjustment of the parameters in the circuit, the frequency characteristic may be made to have a relatively uniform voltage gain over a band of frequencies and sharply defined boundaries at the limits of the band. A still wider band-pass characteristic is often obtained through *stagger tuning*; that is, through cascading several tuned amplifier stages having different resonant frequencies. Such circuits are often used in the intermediate-frequency amplifiers of superheterodyne radio, radar, and television receivers, which amplify over a fixed frequency band.

11. COMPENSATED BROAD-BAND AMPLIFIERS

In a simple resistance-capacitance-coupled amplifier, as the discussion of Eq. 60 in Art. 5 points out, an increase of the upper half-power frequency can be obtained only at the expense of the voltage amplification in the middle range of frequencies. Various circuit artifices[32] are available for extending the region of relatively uniform voltage amplification to higher frequencies without sacrificing voltage amplification. One of the simpler methods is that of adding an inductor L in series with the plate-load resistor, as is shown in Fig. 26. The inductance is so small that it has a negligible effect on the voltage gain in the low or middle ranges of frequency, but at high frequencies the tendency for resonance between it and the shunt capacitance of the circuit increases the effective load impedance and therefore the voltage gain, as is indicated in Fig. 26d.

The equivalent circuit for this amplifier, shown in Fig. 26b, is similar to that shown in Fig. 6b except for the added inductance L. At the high frequencies, the coupling capacitor C_c has negligible reactance;

---

[31] W. L. Everitt, *Communication Engineering* (2nd ed.; New York: McGraw-Hill Book Company, Inc., 1937), 496–503; E. A. Guillemin, *Communication Networks*, Vol. I (New York: John Wiley & Sons, Inc., 1931), 323–339; L. B. Arguimbau, *Vacuum-Tube Circuits* (New York: John Wiley & Sons, Inc., 1948), 207–217.

[32] R. L. Freeman and J. D. Schantz, "Video-Amplifier Design," *Electronics, 10* (August, 1937), 22–25, 60, 62; D. E. Foster and J. A. Rankin, "Video Output Systems," *RCA Rev., 5* (1941), 428–438.

and, since R_L is ordinarily made small compared with the resistances r_p and R_g to increase the high-frequency limit of the frequency

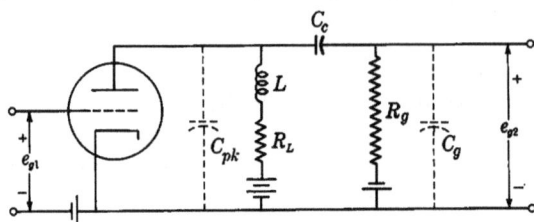

(a) Circuit diagram

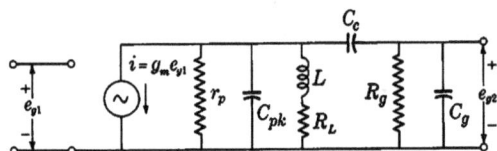

(b) Incremental equivalent circuit

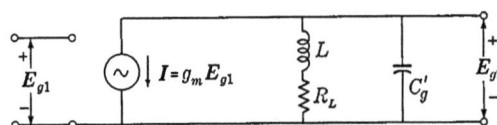

(c) Approximate incremental equivalent circuit

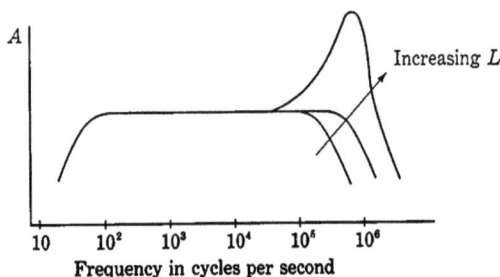

(d) Frequency characteristic

Fig. 26.   Shunt-compensated broad-band amplifier.

response, the effective load for the constant-current generator is that shown in Fig. 26c, where

$$C_g' = C_g + C_{pk}.$$ [139]

The complex voltage gain for the amplifier stage is, then,

$$A = \frac{E_{g2}}{E_{g1}} = -g_m \frac{1}{\dfrac{1}{R_L + j\omega L} + j\omega C_g'},$$ [140]

where the complex quantities E_{g2} and E_{g1} represent the corresponding sinusoidal voltages.

Rearranging Eq. 140 into a normalized or dimensionless form makes possible plotting the essential features of the relation in a concise graphical form that is universally applicable for any combination of values of the circuit parameters. For this purpose, ω may be expressed relative to ω_2, the half-power angular frequency in the absence of inductance, given by

$$\omega_2 = \frac{1}{R_L C_g'} \; ; \qquad [141]$$

and A may be expressed relative to $g_m R_L$, the magnitude of the voltage amplification in the middle range of frequencies when L is zero. Equation 140 then takes the form

$$\frac{A}{g_m R_L} = - \frac{1 + j\left(\dfrac{\omega}{\omega_2}\right) \dfrac{L}{R_L{}^2 C_g'}}{1 - \left(\dfrac{\omega}{\omega_2}\right)^2 \dfrac{L}{R_L{}^2 C_g'} + j\left(\dfrac{\omega}{\omega_2}\right)} \qquad [142]$$

But

$$\frac{L}{R_L{}^2 C_g'} = \frac{\omega_2 L}{R_L} = Q_2, \qquad [143]$$

where Q_2 is the Q of the parallel resonant circuit at the half-power angular frequency ω_2. Thus Eq. 142 becomes

$$\frac{A}{g_m R_L} = - \frac{1 + jQ_2\left(\dfrac{\omega}{\omega_2}\right)}{1 - Q_2\left(\dfrac{\omega}{\omega_2}\right)^2 + j\left(\dfrac{\omega}{\omega_2}\right)} , \qquad [144]$$

which expresses the characteristics of the circuit in the form of a relation between two dimensionless variables with a third, Q_2, as a parameter. The magnitude of the relative voltage amplification is given by the magnitude of Eq. 144,

$$\frac{A}{g_m R_L} = \sqrt{\frac{1 + \left(Q_2 \dfrac{\omega}{\omega_2}\right)^2}{\left[1 - Q_2\left(\dfrac{\omega}{\omega_2}\right)^2\right]^2 + \left(\dfrac{\omega}{\omega_2}\right)^2}} , \qquad [145]$$

and the phase shift is

$$\theta_A = \tan^{-1}\left\{ -\left(\frac{\omega}{\omega_2}\right)\left[1 - Q_2 + \left(Q_2 \frac{\omega}{\omega_2}\right)^2\right]\right\}. \qquad [146]$$

The inverse tangent expression for the phase angle in Eq. 146 is ambiguous by an additive multiple of 180 degrees. As ω and Q_2 decrease, however, the angle must approach the 180-degree value corresponding to that of a simple resistance-capacitance-coupled amplifier operating in the middle range of frequencies. Hence it is apparent that the angle lies between 90 and 180 degrees if Q_2 is small compared with one, as it always is for the small values of L used in broad-band amplifiers. The phase angle

$$\theta = \theta_A - 180 \text{ degrees} \qquad [147]$$

is the phase shift measured from the 180-degree value inherent because of the tube, and lies between 0 and -90 degrees.

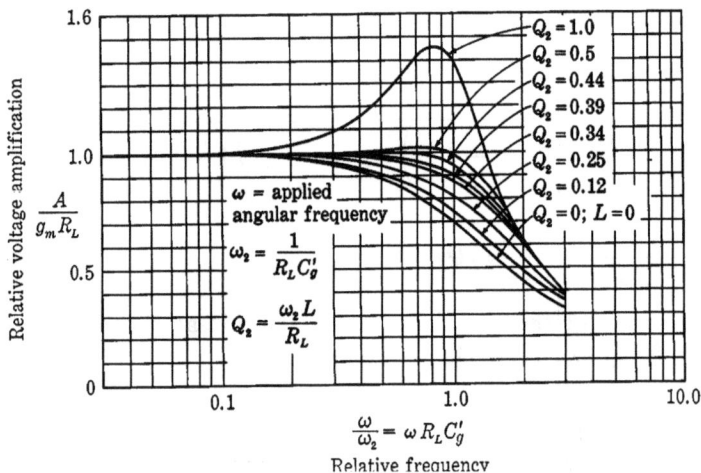

Fig. 27.   Universal voltage-gain curves for the shunt-compensated broad-band amplifier.*

A family of normalized curves relating the relative voltage amplification $A/(g_m R_L)$ to the relative frequency ω/ω_2 as expressed by Eq. 145 is shown in Fig. 27. From these curves the circuit parameters required to give a particular shape to the high-frequency end of the frequency characteristic may be obtained. For example, if Q_2 is chosen as 0.5, the amplifier has the same amplification at the angular frequency ω_2 or frequency f_2 as in the middle range of frequencies. For any value of C_g' inherent in the circuit and any chosen value of ω_2, the values of R_L and L required to produce the frequency characteristic

---

* This diagram is adapted from A. V. Bedford and G. L. Fredendall, "Transient Response of Multistage Video-Frequency Amplifiers," *I.R.E. Proc.*, 27 (1939), Fig. 3, p. 280, with permission.

corresponding to this value of Q_2 are found from

$$R_L = \frac{1}{\omega_2 C_g'},$$ [148]

and

$$\omega_2 L = \frac{R_L}{2} = \frac{1}{2\omega_2 C_g'}.$$ [149]

Thus, if the voltage amplification must be practically constant up to a particular high frequency and have the same value at the particular

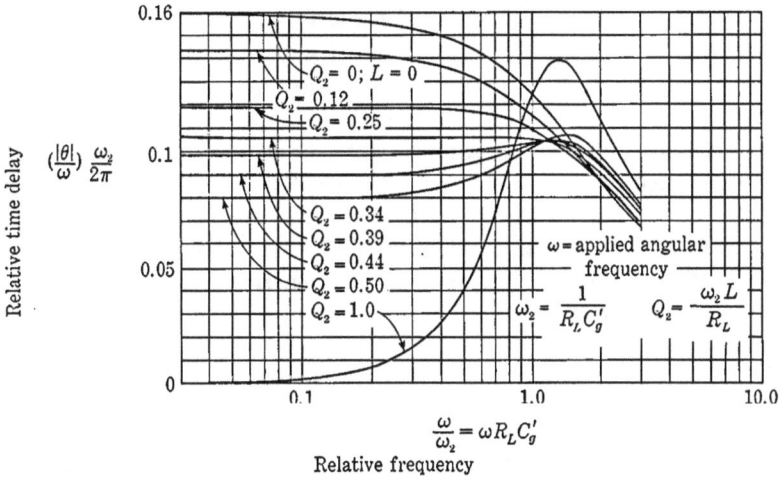

Fig. 28.   Universal curves of $|\theta|/\omega$ as a function of ω for the shunt-compensated broad-band amplifier.*

high frequency as in the middle range of frequencies, the load resistance should equal the reactance of the effective shunt capacitance and the inductive reactance should equal half the load resistance at that frequency.

It is pointed out in Art. 12, Ch. VIII, that when θ is directly proportional to ω, or, alternatively, θ/ω is independent of ω, the amplifier causes no phase distortion. To show the range of frequencies over which the time delay θ/ω is essentially constant and the phase distortion is hence zero, a family of curves that is universally applicable for any values of R_L, L, and C_g' is shown in Fig. 28. The ordinates of the curves are the time delay $|\theta|/\omega$ relative to the time of one cycle, $2\pi R_L C_g'$, at the reference angular frequency ω_2, where $|\theta|$ is the magnitude of the part of θ_A that varies with frequency, as given by

* This diagram is adapted from A. V. Bedford and G. L. Fredendall, "Transient Response of Multistage Video-Frequency Amplifiers," *I.R.E. Proc.*, **27** (1939), Fig. 4, p. 280, with permission.

Eq. 147. The abscissas of the curves are the relative frequency ω/ω_2.

With ω/ω_2 and Q_2 given, the phase shift at any frequency is readily determined from Fig. 28. For example, if L is zero, Q_2 is zero, and the amplifier is then the ordinary resistance-capacitance-coupled amplifier. At the angular frequency ω_2, the value read from the ordinate scale for the curve for Q_2 equals zero is

$$\frac{|\theta|}{\omega_2} \frac{1}{2\pi R_L C_g'} = 0.125. \qquad [150]$$

Thus

$$|\theta| = 2\pi \times 0.125 = 0.785 \text{ radian} = 45 \text{ degrees}. \qquad [151]$$

Since, as was previously explained, θ is a negative angle, this result means that E_{g2} leads E_{g1} by 180 — 45 degrees, or 135 degrees, which agrees with the result in Fig. 8c when the frequency is f_2.

In Fig. 28 the curve for Q_2 equal to 0.34 shows the least variation of $|\theta|/\omega$ and hence produces the least phase distortion. On the other hand, the curve for Q_2 equal to 0.44 shows the least variation of voltage gain in Fig. 27, and hence produces the least frequency distortion—that is, variation of amplitude with frequency. In general, therefore, the value of Q_2 should lie between 0.34 and 0.44, but distortion from one cause or the other cannot be avoided by the simple expedient of insertion of inductance in series with the plate-load resistor.

The response of a shunt-compensated amplifier to a suddenly impressed constant input voltage is shown by the curves in Fig. 29. These curves, which may be derived by ordinary methods,[33] show the instantaneous output voltage e_{g2} relative to its final value as a function of time relative to the time of one cycle at the angular frequency ω_2. For all values of Q_2 larger than 0.25, the curve overshoots the final value. This value of 0.25 for Q_2 thus corresponds to the condition of critical damping in the oscillatory RLC system. The curve for Q_2 equal to 0.25 reaches the final value sooner than any curve for a smaller value of Q_2. Hence, for best transient response Q_2 should equal 0.25, or a value somewhat larger if some overshoot is permissible. Both the value 0.44, which gives the flattest frequency characteristic, and the value 0.34, which gives the least phase distortion, exceed the value 0.25, which gives the maximum rate of rise of the output for no overshoot. No one value can give optimum results in all three respects.

Other methods of compensating for the shunting effect of the inherent capacitance at the high-frequency end of the frequency characteristic are the use of an inductor in series with the coupling

---

[33] E. E. Staff, M.I.T., *Electric Circuits* (Cambridge, Massachusetts: The Technology Press of M.I.T.; New York: John Wiley & Sons, Inc., 1940), 220–250.

capacitor and combinations of inductors in series with both the load resistor and the coupling capacitor.[34] These more elaborate four-terminal networks give some improvement in response over that obtainable through use of the simple series inductor illustrated in Fig. 26, but they have the serious disadvantage of requiring a much more complicated adjustment procedure.

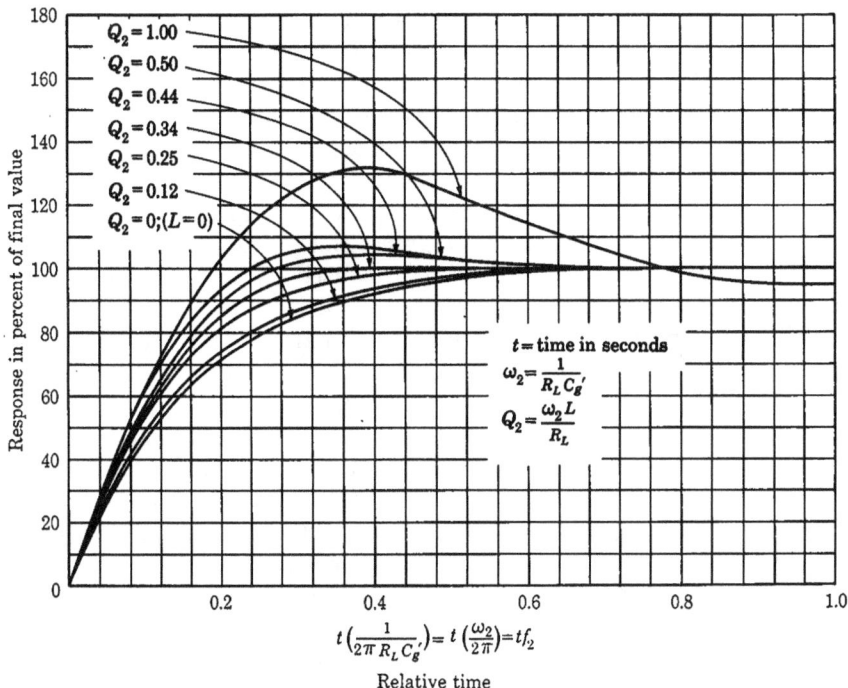

Fig. 29.   Transient response of shunt-compensated amplifier to a suddenly applied constant input voltage.*

Compensation to minimize the phase distortion at the low frequencies is also desirable and is accomplished by proper adjustment of the time constants in the resistance-capacitance circuits used to couple

[34] S. W. Seeley and C. N. Kimball, "Analysis and Design of Video Amplifiers," *RCA Rev.*, 2 (1937), 171–183; 3 (1939), 290–308; V. D. Landon, "Cascade Amplifiers with Maximal Flatness," *RCA Rev.*, 5 (1941), 347–362; G. E. Valley, Jr., and Henry Wallman, Editors, *Vacuum Tube Amplifiers*, Massachusetts Institute of Technology Radiation Laboratory Series, Vol. 18 (New York: McGraw-Hill Book Company, Inc., 1948), 75–77.

* This diagram is adapted from A. V. Bedford and G. L. Fredendall, "Transient Response of Multistage Video-Frequency Amplifiers," *I.R.E. Proc.*, 27 (1939), Fig. 5, p. 280, with permission.

the successive stages and to supply the grid-bias and plate-supply voltages to each stage.

12. FEEDBACK AMPLIFIERS[35]

Significant modifications of the performance of an amplifier occur if a voltage or current proportional to the amplifier's output voltage or current, or a combination of the two, is fed back and superposed on the input signal voltage or current. The voltage or current may be fed back so as either to increase or to decrease the amplification. If it increases the amplification, the method is called *positive, or regenerative, feedback*. If it decreases the amplification, the method is called *negative, inverse, or degenerative, feedback*. Both methods have advantages as well as disadvantages. Among the advantages of one or the other of the two methods are: (1) *Improved stability of gain*, that is, a reduction in the changes in over-all gain caused by variations of the electrode supply voltage and by variations of tube parameters with age. (2) *Reduction of nonlinear distortion* at any particular level of output signal voltage in an amplifier. (3) *Reduction of noise* relative to the signal for any particular level of output signal when the noise originates at a point other than the input of the amplifier. (4) *Provision of adjustable input and output* impedances so as best to suit the characteristics of the source of signal voltage and of the load. (5) *Provision of adjustable frequency response*. Feedback may permit a material reduction of frequency and phase distortion. On the other hand, it may be used to accentuate the frequency selectivity of an amplifier, and to convert an amplifier into an oscillator.

The chief disadvantages of negative feedback in amplifiers are: (1) a reduction in the gain per stage so that more stages are required to obtain the same over-all gain, and (2) an increased tendency toward oscillation at some frequency outside the range for which the amplifier is designed to be useful. The chief disadvantage of positive feedback is that it usually results in poor gain stability. These characteristics of feedback are mentioned here at the outset to emphasize its wide importance. They apply not only to vacuum-tube amplifiers, but also to other electrical amplifiers, such as transistor and magnetic amplifiers, and even to nonelectrical amplifiers as well. Further discussion of vacuum-tube amplifiers is given along with the analysis that follows.

12a. *General Theory.* Although a more complete treatment is necessary to analyze some feedback amplifiers, a simplified analysis

[35] I. Langmuir, United States Patent 1,273,627 (July 23, 1918); H. S. Black, United States Patent 2,102,671 (December 21, 1937); H. S. Black, "Stabilized Feedback Amplifiers," *E.E.*, *53* (1934), 114–120; *B.S.T.J.*, *13* (1934), 1–18.

that is applicable to one basic class of vacuum-tube amplifiers is useful as an introduction and for determining many of their important characteristics. Assumptions for such an analysis are: (1) The circuit parameters, including the tube coefficients over the operating range, are constant and positive. (2) The circuits operate in the steady state with sinusoidal excitation. (3) The voltage fed back is connected in series with the grid of the amplifier, and the input impedance of the amplifier is large compared with other associated impedances. (4) The amplifier is unilateral—it transmits in only one direction. In accordance with the first two assumptions, all voltages and currents in the circuit are sinusoidal and have the same frequency; hence they may be represented by complex quantities and vector diagrams. The third assumption permits neglect of any reaction the amplifier and source might have on the feedback network.

All methods of feedback in vacuum-tube amplifiers involve feedback of a *voltage* to the input terminals of some stage. If this voltage fed back is proportional to the output voltage of the same or a succeeding stage by a factor that is independent of the magnitude of the load impedance, the feedback is called *voltage feedback*. If the voltage fed back is proportional to the output current by a factor that is independent of the load impedance, the feedback is *current feedback*. These two methods produce different over-all behavior of the amplifier in many respects.

The connections for an elementary feedback amplifier of the voltage-feedback type are shown in Fig. 30a. A feedback network returns a portion E_{fb} of the output voltage E_o to the input voltage E_g of the internal amplifier. The combination of internal amplifier and feedback network makes up the over-all *feedback amplifier* contained in the dotted rectangle.

The complex voltage gain of the internal amplifier may be represented by the complex quantity

$$A = \frac{E_o}{E_g}.$$

[152]

Note that the total load on the internal amplifier, which enters into a determination of the gain A, is the parallel combination of the load Z_L and the impedance presented by the feedback network at its input terminals a–a. Also, the voltage E_g denotes the input voltage to the internal amplifier, which is usually, but not always, the grid-to-cathode voltage. The transfer voltage ratio of the feedback network may be represented by the complex expression

$$\beta = \frac{E_{fb}}{E_o}.$$

[153]

Hence, from Eqs. 152 and 153,

$$E_{fb} = A\beta E_g. \qquad [154]$$

The quantity $A\beta$ is called the *loop transmission, loop gain,* feedback

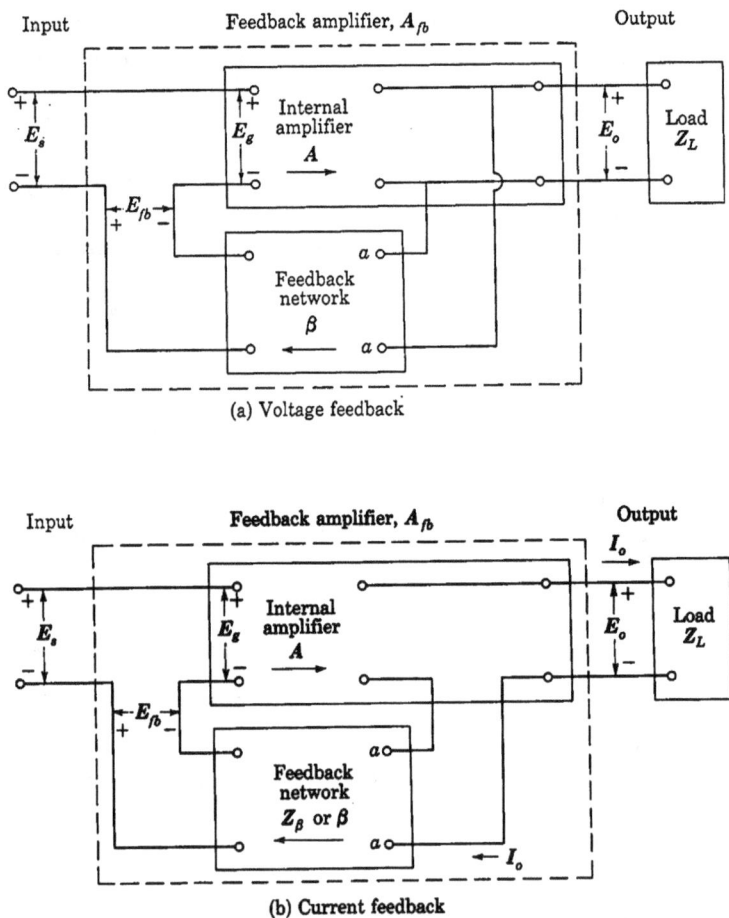

(a) Voltage feedback

(b) Current feedback

Fig. 30.   Elements in a feedback amplifier.   A voltage proportional to the output is returned through the feedback network to the input. For analysis, sinusoidal operation is assumed.

factor, or return ratio. It is the factor by which the input voltage E_g is multiplied in a traverse through the amplifier and back around the loop through the feedback network to the input.

For a determination of the behavior of the complete system, the relationship of the output voltage E_o to the input voltage E_s of the over-all feedback amplifier, rather than to E_g, is of interest. The quantity E_g may be eliminated through the use of the Kirchhoff's law relation at the input,

$$E_g = E_s + E_{fb}.$$

[155]

Combination of Eqs. 154 and 155 gives

$$E_g = \frac{E_s}{1 - A\beta},$$

[156]

and substitution of Eq. 156 in Eq. 152 gives the basic feedback relationship

$$A_{fb} = \frac{E_o}{E_s} = \frac{A}{1 - A\beta},$$

▶[157]

where A_{fb} is the over-all complex voltage gain of the complete feedback amplifier, which is enclosed in the dotted rectangle of Fig. 30a.

Connections for an elementary feedback amplifier utilizing current feedback instead of voltage feedback are shown in Fig. 30b. Here the output load current I_o passes through the feedback network and develops the feedback voltage E_{fb}. The feedback network may then be represented by a complex transfer impedance

$$Z_\beta = \frac{E_{fb}}{I_o}.$$

[158]

The analysis leading to Eq. 157 applies to this amplifier connection as well if the feedback voltage ratio β is recognized as

$$\beta = \frac{E_{fb}}{E_o} = \frac{E_{fb}}{I_o} \times \frac{I_o}{E_o} = \frac{Z_\beta}{Z_L},$$

[159]

where Z_L is the impedance of the load. For current feedback, the ratio β is thus a function of the load impedance, whereas for voltage feedback, it is not. This dependence of β on Z_L distinguishes current feedback from voltage feedback. From Eqs. 153 and 159 and Figs. 30a and 30b, it is evident that a practical test to determine the type of feedback in any particular amplifier is to examine the values of E_{fb} for an open-circuit and for a short-circuit at the load. The feedback is identified as voltage feedback if E_{fb} becomes zero when the load is short-circuited, because a short-circuit at the load reduces the voltage at terminals *a–a* in Fig. 30a to zero. It is identified as current feedback if E_{fb} becomes zero when the load is open-circuited, because such an open-circuit reduces the load current and the voltage at the terminals *a–a* in Fig. 30b to zero. Other effects of the two types of feedback on the over-all behavior of the amplifier are discussed subsequently.

Frequently, clear-cut identification of separate sections of a feed-back amplifier to be considered as the internal amplifier A and the feedback network β or Z_β is impossible. Moreover the feedback net-work is not always arranged so that its output voltage is directly in series with the input of an internal amplifier having infinite input impedance; consequently, assumption (3) stated earlier is not always valid. In general, the input signal voltage and voltage fed back may be effectively added by any six-terminal passive network to provide the input to the internal amplifier. Furthermore, many amplifiers have multiple-loop feedback—for example, a major loop around two stages each of which involves a minor loop. Analysis of these more compli-cated amplifiers requires an extension of the foregoing concepts and

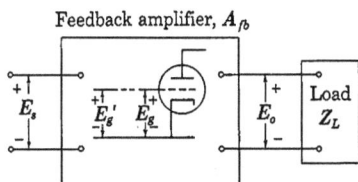

Feedback amplifier, A_{fb}

Fig. 31. Generalized feedback amplifier. One of the several vacuum tubes is shown with its grid lead broken.

further mathematical formaliza-tion.[36] For many particular amplifier circuits, straightforward application of Kirchhoff's laws and solution of the resulting simultaneous equations is the most direct method of analysis. These equations generally involve voltages or currents such as μE_g or $g_m E_g$, which are those of dependent sources that represent the active elements in the circuit. Equations expressing the voltage or current of these dependent sources in terms of the other variables in the circuit must be written and included in the set of simultaneous equations. For other amplifier circuits, a simpler method, and one that often gives a clearer insight into the circuit operation, is the application of the principle of superposition. By this method, quantities are found that, in the general case, are closely analogous to A and $A\beta$ and are identical with A and $A\beta$ for the series connection at the input in Fig. 30.

Analysis of a feedback amplifier through use of the principle of superposition may be based on the representation of the circuit shown in Fig. 31. Here the complete amplifier, including the load, is shown with input terminals at E_s, output terminals at E_o, and a grid voltage E_g existing at some one tube selected from the many that may be inside it. The grid lead to this tube is considered to be temporarily broken, so that its grid voltage E_g is an *independent* variable, just as the input voltage E_s is an independent variable. The voltage returned to the open grid connection when E_s and E_g are impressed is E_g'. It

[36] H. W. Bode, *Network Analysis and Feedback Amplifier Design* (New York: D. Van Nostrand Company, Inc., 1945); J. S. Brown and F. D. Bennett, "The Application of Matrices to Vacuum-Tube Circuits." *I.R.E. Proc.*, 36 (1948), 844–852.

depends on both E_s and E_g, as well as the circuit parameters. The break is considered to be made inside the interelectrode capacitances so that these capacitances become a part of the return circuit affecting E_g', and E_g faces infinite impedance. The returned grid voltage E_g' and the output voltage E_o are the *dependent* variables to be determined. Since the circuit is considered to be linear, each dependent variable is linearly related to all the independent variables, and a set of linear equations between the dependent and the independent variables is applicable. In other words, the response of the network is a super-position, or sum, of the responses to the individual sources. Thus,

$$E_o = A_{so}E_s + A_{go}E_g \, ,$$
[160]

and

$$E_g' = A_{sg}E_s + A_{gg}E_g \, ,$$
[161]

where the A's are constant coefficients with subscripts indicating the points between which they apply. Each coefficient is a transfer voltage ratio that corresponds to conditions in the circuit when one of the independent variables is made zero. The coefficients may thus be designated as

$$A_{so} = \frac{E_o}{E_s}\bigg|_{E_g = 0} = \text{direct transmission,}$$
[162]

$$A_{go} = \frac{E_o}{E_g}\bigg|_{E_s = 0} \, ,$$
[163]

$$A_{sg} = \frac{E_g'}{E_s}\bigg|_{E_g = 0} \, ,$$
[164]

$$A_{gg} = \frac{E_g'}{E_g}\bigg|_{E_s = 0} = \text{loop transmission.}$$
[165]

The coefficient A_{so} is called the *direct transmission* because it is the voltage ratio that would occur if E_g for the tube were zero. Alternatively, it is the voltage ratio that occurs if μ or g_m, but not r_p, for the tube becomes zero. It accounts, therefore, for all paths from the input to the output that by-pass the tube. The coefficient A_{gg}, called the *loop transmission*, is the voltage returned to the open grid connection per volt applied at the grid when the input terminals are short-circuited. Similarly, the coefficient A_{sg} is the transfer voltage ratio from the input to the open grid when E_g, μ, or g_m, but not r_p, is considered zero; and A_{go} is the transfer voltage ratio from the grid to the output when the input terminals are short-circuited.

When the open grid lead in Fig. 31 is reconnected to provide normal operation of the amplifier, E_g is no longer independent; it is constrained to equal E_g'. But since E_g faces infinite impedance, restoring the broken grid lead does not affect the impedances in the circuit; Eqs. 160 and 161 still apply, and the coefficients are unchanged. Solution of Eqs. 160 and 161 for the over-all voltage ratio subject to this constraint gives

$$A_{fb} = \frac{E_o}{E_s} = A_{so} + \frac{A_{sg}A_{go}}{1 - A_{gg}}. \qquad \blacktriangleright[166]$$

Application of this method to the voltage-feedback circuit of Fig. 30a gives 0, 1, A, and $A\beta$ for A_{so}, A_{sg}, A_{go}, and A_{gg}, respectively. Thus the product $A_{sg}A_{go}$ is analogous to A, and A_{gg} is analogous to $A\beta$ in Fig. 30a. The direct transmission A_{so} has no counterpart in this elementary feedback circuit with a series connection at the input, because there is no transmission through the circuit if μ of one of the tubes in the internal amplifier cascade becomes zero.

This analysis by superposition reduces the problem of determining the amplifier voltage gain to one of determining four coefficients or voltage ratios. Since each ratio is found for a simplifying condition of E_s or a μ equal to zero in the network, the individual coefficients are often readily determined. Furthermore, each has a physical significance that facilitates an understanding of the circuit behavior. The method may be employed in multiple-loop feedback circuits through successive application of the principle to each of the loops in turn and is also applicable to linear amplifying systems in which the amplification is accomplished by a device other than an electron tube, for example, by a transistor or by a mechanical amplifier. Note that this formulation of the superposition method depends on the concept that the circuit may be broken at an infinite-impedance point. Otherwise, restoring the broken connection would invalidate evaluation of the coefficients in Eqs. 160 and 161 by the method indicated in Eqs. 162 through 165. The coefficients applicable when the circuit is closed would then differ from those applicable when it is open.

12b. *Stability in Feedback Amplifiers.* For an understanding of the significance of the feedback equations, Eqs. 157 and 166, the A's and β must be recognized as being functions of frequency that may be either real or complex. Each equation thus expresses a phase as well as an amplitude relation.

In the relatively simple circumstance when both the A's and β are real and each may be either positive or negative, the resulting loop gain $A\beta$ or A_{gg} is also real and may be either positive or negative. When $A\beta$ or A_{gg} is real, positive, and less than unity, $1 - A\beta$ or $1 - A_{gg}$ is real, positive, and less than unity, so that the magnitude of the over-all voltage gain with the feedback, A_{fb}, is larger than A in Eq. 157, and the feedback is of the regenerative or positive type. Vectorially, the voltage fed back, E_{fb}, is in phase with the input-signal voltage, E_s, which is, therefore, reinforced so that E_g is made larger than E_s. If the reinforcement is increased so that $A\beta$ or A_{gg} approaches unity, the over-all gain A_{fb} approaches infinity. The amplifier is then capable of sustaining the output E_o even though the input E_s becomes zero; thus the amplifier becomes an oscillator, and can serve as a source of alternating current. Further discussion of the significance of this condition is given in Ch. XI.

When $A\beta$ or A_{gg} is real and negative, $1 - A\beta$ or $1 - A_{gg}$ is real, positive, and larger than unity; hence, A_{fb} is smaller than A in Eq. 157, and the feedback is degenerative or negative. Vectorially E_{fb} is then opposite in phase to the input signal E_s so that E_g is smaller in magnitude than E_s.

When the A's and β are complex, $A\beta$, $1 - A\beta$, and $1 - A_{gg}$ are also complex, and their behavior as functions of frequency may be represented by a locus of $A\beta$ or A_{gg} on the complex plane—that is, the locus of the curve traced by the terminus of the vector $A\beta$ or A_{gg} as the frequency varies. From this locus, the behavior of $1 - A\beta$ and $1 - A_{gg}$ may be inferred. The discussion that follows relates to $A\beta$, but is applicable to A_{gg} as well.

The relation between $A\beta$ and $1 - A\beta$ is illustrated in Fig. 32. In this diagram, P represents the point corresponding to one frequency on the locus of $A\beta$. A line drawn from the end of the vector $A\beta$, computed for any particular value of frequency, to the point $1 + j0$ then represents $1 - A\beta$. If the point P lies within the circle of unit radius about the point $1 + j0$, the magnitude of $1 - A\beta$ is evidently less than unity. Hence, from Eq. 157, the magnitude of the over-all voltage gain A_{fb} is larger than that of A, and the feedback is regenerative. Thus the whole area within the circle corresponds to regenerative or positive feedback. But if P lies outside the circle of unit radius, the magnitude of $1 - A\beta$ is larger than unity, and the feedback is degenerative or negative. If P coincides with the point $1 + j0$, A_{fb} becomes infinite, and oscillation occurs. Generally, the amount of feedback expressed in decibels is

$$\text{db of feedback} = 20 \log_{10} \frac{1}{|1 - A\beta|}, \text{ or } 20 \log_{10} \frac{1}{|1 - A_{gg}|}, \quad [167]$$

and the sign of the feedback, positive or negative, is the same as the sign of the logarithm in Eq. 167.

By an extension of the reasoning in the previous paragraph,

Nyquist[37] has deduced a more general criterion for stability—that is, freedom from oscillation—which is applicable to feedback amplifiers and to many other physical systems as well. This criterion may be stated as follows: If, when plotted for all frequencies from zero to infinity, the curve traced out by the terminus of $A\beta$ is closed and passes through or encloses the point $1 + j0$, the system is unstable—it will oscillate. If the curve does not pass through or enclose the point $1 + j0$, the system is stable. If the locus of $A\beta$ is not a closed curve, the

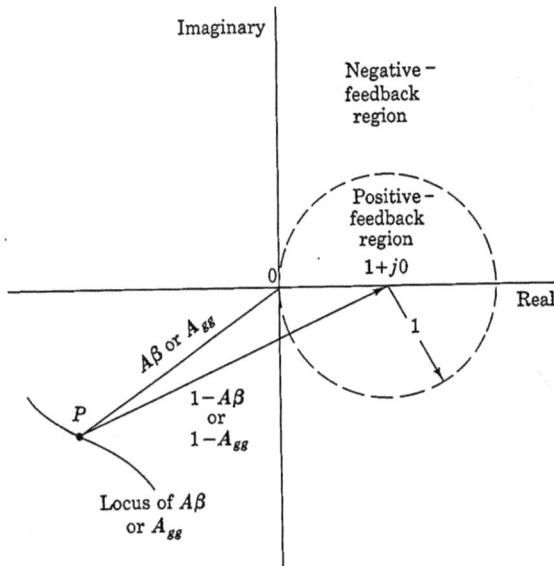

Fig. 32.   Representation of $A\beta$ and $1 - A\beta$, or A_{gg} and $1 - A_{gg}$
in the complex plane.

locus of the conjugate of $A\beta$, which is merely the mirror image of $A\beta$ in the real axis, must also be plotted. The two loci together then form a closed curve to which the criterion may be applied.

Examples of $A\beta$ loci are shown in Fig. 33. The locus in Fig. 33a represents the behavior of a particular transformer-coupled alternating-current amplifier with voltage feedback. The feedback is seen to be negative or degenerative for all values of ω below 50,000. The locus coincides with the origin at ω equals zero for this as well as other types of alternating-current amplifiers because the transformer or other means of interstage coupling does not pass direct current. The locus returns to the origin when ω equals infinity because shunt capacitance

[37] H. Nyquist, "Regeneration Theory," *B.S.T.J.*, *11* (1932), 126-147.

then reduces the voltage gain to zero. Since this particular locus is a closed curve and does not reach or enclose the point $1 + j0$, the Nyquist criterion indicates that the amplifier is stable. Figure 33b shows the $A\beta$ locus for a typical direct-coupled amplifier. This locus does not coincide with the origin when ω equals zero, because, for this type of amplifier, neither A nor β is zero for a steady input voltage.

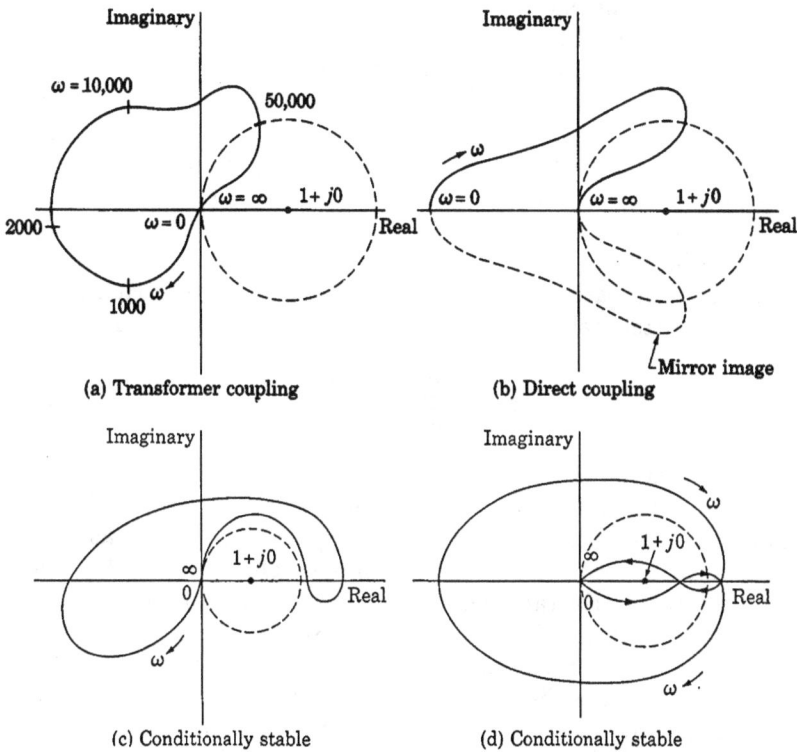

(a) Transformer coupling

(b) Direct coupling

(c) Conditionally stable

(d) Conditionally stable

Fig. 33.   Illustrative $A\beta$ or A_{gg} loci.

Shunt capacitance again brings the locus to the origin when ω equals infinity, however. Since the locus is not a closed curve, the locus of the conjugate of $A\beta$ is also shown. The locus and its mirror image form a closed curve that does not enclose the point $1 + j0$; hence, by the Nyquist criterion, the amplifier is stable.

Figure 33c represents the $A\beta$ locus of a hypothetical amplifier that is stable as it stands, would oscillate if A or β were reduced in magnitude for all frequencies, and would cease to oscillate if A or β were still further reduced. Such an amplifier is said to be *conditionally stable*. If the plate voltage is applied before the heater voltage in a conditionally stable amplifier, the $A\beta$ locus grows outward from the origin as the cathode temperature and emission increase toward their final values. Oscillation may thereby be initiated, and once started is likely to continue at the final cathode temperature.

Some question arises whether a locus such as that shown in Fig. 33d encloses the point $1 + j0$. This question may be resolved as follows: Draw a line from the point $1 + j0$ to the point on the locus for ω equals zero. Hold one end of the line fixed on the point $1 + j0$ while letting the other end traverse the locus in the direction of increasing ω. If the net angle through which the line turns as ω increases from zero to infinity is zero, the locus does not enclose the point $1 + j0$, and operation is stable. Application of this rule to Fig. 33d shows that the point $1 + j0$ is outside this locus. Consequently, it, like that in Fig. 33c, represents a conditionally stable type of amplifier.

12c. *Properties of Feedback Amplifiers.* As the magnitude of the loop gain is increased in a negative feedback amplifier, $A\beta$ or A_{gg} becomes large compared with unity, and $|1 - A\beta|$ approaches $A\beta$. Then the feedback relation for zero direct transmission, Eq. 157, becomes

$$A_{fb} = \frac{E_o}{E_s} \approx -\frac{1}{\beta}.$$ [168]

In this circumstance, Eq. 154 shows that E_{fb} becomes large compared with E_g so that E_{fb} approaches E_s. That is, the voltage fed back almost cancels the voltage E_s originally applied. One result as $A\beta$ is increased is, then, that the over-all complex amplification A_{fb} becomes independent, to an arbitrary degree, of the complex voltage gain A of the internal amplifier. Changes in A caused by changes in supply voltages, tube coefficients, and characteristic curves are counteracted by the feedback so that their effect on A_{fb} is made negligible. The over-all gain is thus stabilized against such changes. A second result is that the over-all frequency characteristic becomes independent of the amplifier in the region for which $A\beta$ is large, and depends only upon the frequency characteristic of the feedback network. It approaches an inverse of the frequency characteristic of the feedback network[38] and can, therefore, be adjusted by appropriate design of that network. These results are true, of course, only if the increase in $A\beta$ is accomplished without violation of the Nyquist stability criterion.

The degree of stabilization of the gain when the amount of feedback is finite may be determined through differentiation of Eq. 157.[39] If a change in A caused by variation of any parameter is designated by

[38] G. H. Fritzinger, "Frequency Discrimination by Inverse Feedback," *I.R.E. Proc.*, 26 (1938), 207–225.

[39] F. S. Woods, *Advanced Calculus* (New ed.; Boston: Ginn and Company, 1934), 345.

dA, the corresponding change in the over-all gain A_{fb} of the amplifier with feedback is

$$dA_{fb} = \frac{dA_{fb}}{dA} dA = \frac{1}{(1 - A\beta)^2} dA ,$$ [169]

β being assumed constant. It is convenient to express this change dA_{fb} as a fraction of the over-all gain A_{fb} that existed before the change occurred; thus

$$\frac{dA_{fb}}{A_{fb}} = \frac{1}{(1 - A\beta)^2} \frac{dA}{A_{fb}} ,$$ [170]

and substitution of Eq. 157 gives

$$\frac{dA_{fb}/A_{fb}}{dA/A} = \frac{1}{1 - A\beta} .$$ [171]

When compared with Eq. 157, Eq. 171 shows that the ratio of the fractional change in the over-all gain to the fractional change of any amplifier parameter whose variation causes a proportional change in gain is modified by the same factor as is the gain itself. Since this relation is developed in terms of complex variables, it is general for linear operation and for both positive and negative feedback. For negative feedback, it means that the addition of feedback stabilizes the gain against disturbances to the same degree that it reduces the gain. For example, if sufficient negative feedback is applied to reduce the over-all voltage gain to one-tenth that of the internal amplifier, a change in μ, g_m, or supply voltage that alters the internal-amplifier gain by 20 per cent changes the over-all gain by only 2 per cent.

Feedback may be used to reduce the effects of certain types of extraneous voltages that occur in amplifiers, but it has no effect on others. The term extraneous voltage is here intended to include many types of unwanted effects, such as the noise voltages discussed in Art. 3 and harmonic voltages resulting from nonlinearity in the characteristic curves of the tubes. The effect of feedback on such voltages depends greatly upon the point at which the voltage is introduced into the circuit, as may be deduced from an analysis based on Fig. 34. In this diagram, E_n represents the extraneous voltage, which is considered to be introduced at an arbitrary point in the amplifier chain. Thus it is preceded by a complex voltage gain A_1 and followed by a complex voltage gain A_2, the product A_1A_2 being the total internal-amplifier complex voltage gain A. In general, the extraneous voltage has a different frequency or frequency spectrum from that of the signal voltage, and the frequency dependence of the amplifier and feedback network must be considered in this type of analysis. For simplicity, it is assumed here, however, that the quantities A_1, A_2, and β have the same complex values for the extraneous voltage that they have for the signal voltage. This condition is true when the series reactances and shunt susceptances in the amplifier network are negligibly small so that, for increments, the network behaves as if it comprises only resistances.

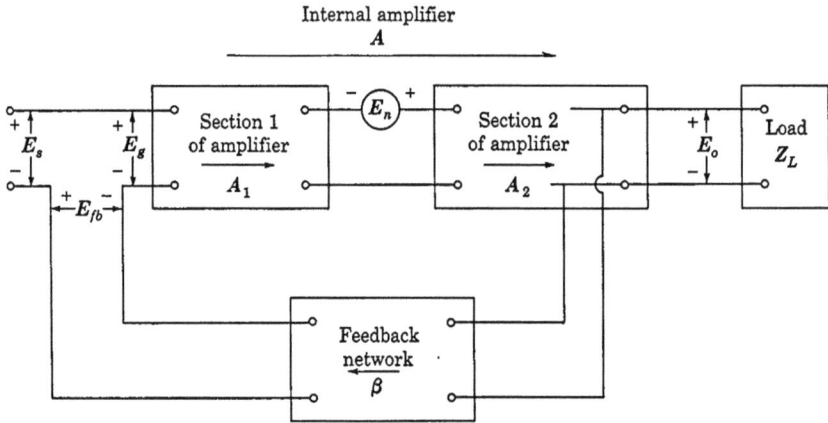

Fig. 34.   Extraneous voltage in a feedback amplifier.

In most amplifier applications, a design that increases the signal component at the output of the amplifier system relative to the output component corresponding to the extraneous voltage is desirable—in other words, increase of the output signal-to-noise ratio is desired. Consequently, an examination of the effect of feedback on this ratio is of interest.

In the absence of feedback—that is, for β and E_{fb} equal to zero—in Fig. 34, E_g equals E_s. The signal component in the output is then $A_1A_2E_s$, and the noise component is A_2E_n, so that

$$E_o = A_1A_2E_s + A_2E_n \ . \qquad [172]$$

The output signal-to-noise ratio in the absence of feedback, is, therefore,

$$\frac{\text{Output signal voltage}}{\text{Output noise voltage}} = \frac{|A_1A_2E_s|}{|A_2E_n|} = \frac{|A_1E_s|}{|E_n|} \ . \qquad [173]$$

This ratio could evidently be increased through increasing the magnitude of the signal voltage E_s relative to that of the noise voltage E_n, or through increasing A_1, the gain of the amplifier section *preceding* the point where the noise is introduced. The ratio would evidently not be affected through increasing A_2.

In the presence of feedback—that is, for β and E_{fb} not zero—

$$E_o = A_1 A_2 E_g + A_2 E_n \,, \qquad [174]$$

and, since

$$E_g = E_s + \beta E_o \,, \qquad [175]$$

$$E_o = A_1 A_2 (E_s + \beta E_o) + A_2 E_n \,, \qquad [176]$$

from which

$$E_o = \frac{A_1 A_2}{1 - A_1 A_2 \beta} E_s + \frac{A_2}{1 - A_1 A_2 \beta} E_n \,. \qquad [177]$$

From Eq. 177, the output signal-to-noise ratio in the presence of feedback is

$$\frac{\text{Output signal voltage with feedback}}{\text{Output noise voltage with feedback}} = \frac{|A_1 E_s|}{|E_n|} \,, \qquad [178]$$

which is identical with the result in Eq. 173 for no feedback. Thus the application of feedback alone has no direct effect on the output signal-to-noise ratio when the ratio of the signal voltage to the noise voltage at their points of entry and the complex voltage gain A_1 remain unchanged. The feedback changes the signal and the noise components by the same factor.

Feedback does have an indirect effect, however, that makes improvement of the output signal-to-noise ratio possible. Equations 173 and 178 show that this ratio may be increased through increase of $|A_1 E_s|$ relative to $|E_n|$. In the amplifier without feedback, such an increase is often precluded, because the resulting increased output-signal component would cause excessive distortion, or overloading, in the output stage. But, if negative feedback is applied at the same time that $|A_1 E_s|$ is increased, the output-signal component may be held to a moderate value while the noise component at the output is reduced by the factor $1/|1 - A_1 A_2 \beta|$. The output signal-to-noise ratio is then increased without occurrence of overloading. If, for example, negative feedback is applied, and the input signal voltage, E_s, or the gain preceding the noise, A_1, or both, are increased so that the output signal remains at the same value it had before the feedback was applied, the only net effect at the output is a decrease of the output noise component by the factor $1/|1 - A_1 A_2 \beta|$. Thus *for the same output-signal component*, negative feedback increases the output signal-to-noise ratio by the factor $|1 - A_1 A_2 \beta|$. In general, then, the output signal-to-noise ratio can be increased only through increasing either the input-signal voltage or the voltage gain preceding the point of entry of the noise, and feedback is helpful only insofar as it makes such an increase permissible. If neither additional input signal voltage nor additional noise-free amplification is available, feedback is of no avail in improving the output signal-to-noise ratio.

When the extraneous voltage originates at the input—that is, at the same point that the signal voltage enters—the gain preceding this voltage, A_1, is fixed at the value unity. The signal and the noise add directly and are amplified by the same amount. The signal-to-noise ratio at the output is then the same as that ratio at the input, regardless of the amount of feedback. Thus when the extraneous voltage results from such causes as thermal agitation of the electrons in an input resistor, slow changes of the grid-to-cathode contact potential in the input tube of a direct-coupled amplifier, or induction of hum voltage in an input transformer, feedback cannot increase the corresponding output signal-to-noise ratio.

When the extraneous voltage originates at the output, A_2 is fixed at unity, and A_1 is the same as the whole gain A of the internal amplifier. Equation 177 then becomes

$$E_o = \frac{A}{1 - A\beta} E_s + \frac{1}{1 - A\beta} E_n , \qquad [179]$$

and improvement of the output signal-to-noise ratio is possible by simultaneous increase of the internal-amplifier voltage gain and application of negative feedback, as discussed above. One example of extraneous voltage originating at the output is ripple in the plate power-supply voltage of the output stage of an amplifier. The effect of such a ripple voltage may be made negligible, and the necessity for large ripple-filter components in the power supply obviated, by simultaneous increase of A and β.

Another example of extraneous voltage that originates at the output is the voltage component produced by nonlinear distortion in the output stage of an amplifier. The preceding analysis for the effect of feedback is applicable, but one additional fact must be recognized; namely, the amplitude of the distortion voltage is a function of that of the output-signal component, as is discussed in Art. 13, Ch. VIII. Thus, the extraneous voltage E_n effectively introduced by the distortion is a function of the output-signal component $AE_s/(1 - A\beta)$ in Eq. 179. But, if $|AE_s|$ is increased when negative feedback is applied so that the same output-signal component is maintained, the distortion voltage E_n also remains the same with as without feedback. The only net effect at the output is then a decrease of the distortion component by the factor $1/|1 - A\beta|$, just as for noise, as discussed previously. Hence, *for the same amplitude of output-signal component*, addition of negative feedback reduces the distortion component in the output by the factor $1/|1 - A\beta|$. Thus it reduces the percentage harmonic distortion in the output by the factor $1/|1 - A\beta|$ at all output signal-voltage levels. Evidently, the application of negative feedback together with additional distortionless voltage gain ahead of a particular amplifier stage can reduce the nonlinear distortion from that stage to any desired value for a particular output signal-voltage level.

In qualitative terms, the negative feedback supplies a voltage $A\beta E_n$ that combines with E_n to reduce the net distortion component in the output while the increased voltage gain A counteracts the effect of the feedback and maintains the output signal at the original level. This analysis neglects the fact that non-linearity in the output stage acts on the distortion component $A\beta E_n$ returned through the input by the feedback—that is, it neglects the distortion of the distortion—but this is ordinarily only a small second-order effect.

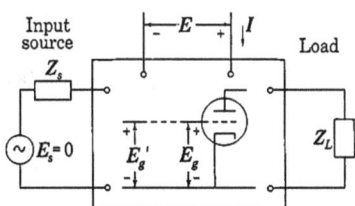

Fig. 35. Driving-point impedance at an arbitrarily selected terminal pair is E/I.

12d. *Input and Output Impedances of Feedback Amplifiers.* One of the more important effects of feedback is its influence on the impedance that the amplifier presents to any external circuit connected to it—for example, its effect on the input impedance faced by the source of signal voltage, or the output impedance faced by the load. These effective impedances are of primary concern because they affect such quantities as the efficiency of power transfer from the source to the amplifier, and from the amplifier to the load, the reflections on a transmission line connected to the input or the output, and the like.

To determine the impedance presented at any arbitrarily selected pair of terminals, called a *driving-point impedance*, the amplifier may be represented as in Fig. 35 for linear operation. In general, the terminals for E and I may be selected anywhere in the circuit, but other parts of the circuit are considered to have their normal operating values. Thus, the load is connected; likewise, the source of input signal voltage is also connected so that its internal impedance is a part of the circuit, but its internal electromotive force is zero. The driving-point impedance to be determined is the ratio E/I. It may be expressed in terms of coefficients in a set of linear algebraic equations, since the circuit behavior is considered to be linear, just as the complex voltage gain is so expressed in Eq. 166. Thus, if I at the chosen terminal and the E_g at the grid of any tube considered to have a broken grid lead are selected as independent variables, the terminal voltage E and the voltage E_g' returned to the break in the grid connection are dependent variables that are given by

$$E = ZI + A_{g1}E_g ,$$ [180]

and

$$E_g' = Z_{1g}I + A_{gg_{oc}}E_g ,$$ [181]

where

$$Z = \left. \frac{E}{I} \right|_{E_g = 0}$$ [182]

is the impedance encountered by E and I when E_g is zero or when the grid lead is reconnected, but μ and g_m, but not r_p, of the tube are equal to zero—that is, Z is the impedance with the feedback ineffective;

$$Z_{1g} = \left. \frac{E_g'}{I} \right|_{E_g = 0}$$ [183]

is the transfer impedance from the selected terminals to the grid of the tube for E_g or μ and g_m equal to zero as above;

$$A_{g1} = \left. \frac{E}{E_g} \right|_{I = 0}$$ [184]

is the transfer voltage ratio from the grid to the selected terminals when the terminals are open-circuited; and

$$A_{gg_{oc}} = \left. \frac{E_g'}{E_g} \right|_{I = 0}$$ [185]

is the loop gain when the selected terminals are open-circuited.

If the broken grid lead is reconnected, E_g is constrained to equal E_g'. Solution of Eqs. 180 and 181, subject to this constraint, gives for the desired impedance,

$$Z_{fb} = \frac{E}{I} = Z \frac{1 - A_{gg_{oc}} + \dfrac{A_{g1}Z_{1g}}{Z}}{1 - A_{gg_{oc}}}.$$ [186]

This expression may be put in more convenient form for computation through recognition of the fact that, for E equal to zero, direct substitution from Eqs. 180 and 181 gives

$$A_{gg_{oc}} - \frac{A_{g1}Z_{1g}}{Z} = \left. \frac{E_g'}{E_g} \right|_{E = 0}$$ [187]

For E equal to zero, the right-hand side of this relation is evidently the loop gain, E_g'/E_g—in other words, it is the loop gain when the selected terminals are short-circuited. But, by definition,

$$\frac{E_g'}{E_g}\bigg|_{E=0} = A_{gg_{sc}} .$$

[188]

In other words, the terms on the left-hand side of Eq. 187 represent the loop gain when the selected terminals are short-circuited. Thus, Eq. 186 may be written

$$Z_{fb} = Z\frac{1 - A_{gg_{sc}}}{1 - A_{gg_{oc}}}.$$

▶[189]

This relation applies to feedback generally and is of fundamental importance. For the types of feedback for which A and β are readily evaluated, its equivalent form,

$$Z_{fb} = Z\frac{1 - (A\beta)_{sc}}{1 - (A\beta)_{oc}},$$

▶[190]

is useful. In this relation also, the short-circuit and open-circuit designations refer to conditions at the terminal pair at which Z_{fb} and Z are measured.

As an example, Eq. 190 may be applied to computation of the output impedance of the cathode-follower amplifier shown in Fig. 22, Ch. VIII. In this circuit, the complex output voltage corresponding to the instantaneous voltage e_o in the diagram is E_o, and the complex output current is I_p. For μ or g_m, but not r_p, equal to zero, the output impedance faced by the load is

$$Z = \frac{E_o}{-I_p}\bigg|_{\mu=0} = r_p .$$

[191]

The loop gain when the output terminals are short-circuited, $(A\beta)_{sc}$ or $A_{gg_{sc}}$, is zero, because a short circuit at the load reduces the voltage fed back in series with the input to zero. In accordance with the definitions given earlier, the feedback in this circuit is, therefore, of the voltage-feedback type. The loop gain when the output terminals are open-circuited, $(A\beta)_{oc}$ or $A_{gg_{oc}}$, is $-\mu$ for the reference directions of voltage chosen in the figure. Thus,

$$Z_{fb} = Z\frac{1 - (A\beta)_{sc}}{1 - (A\beta)_{oc}} = \frac{r_p}{\mu + 1} ,$$

[192]

which is in accord with the output impedance shown in the incremental equivalent circuit in Fig. 24, Ch. VIII, and computed in Eq. 100, Ch. VIII, by other means.

The reduction in output impedance apparent in Eq. 192 is explained in qualitative terms by the fact that when the load impedance is decreased in a circuit with voltage inverse feedback, the output voltage decreases; hence, the magnitude of the voltage fed back also decreases. But this decrease in the voltage fed back tends to increase the output voltage and thereby to maintain it nearly constant. Negative voltage feedback thus tends to stabilize the output voltage against changes in load. In other words, it reduces the output impedance so that the amplifier becomes more nearly a constant-voltage source at the output for any particular value of input signal voltage.

Determination of the output impedance of the amplifier with cathode bias shown in Fig. 15, Ch. VIII, serves as a second example of the application of Eq. 190. When μ or g_m, but not r_p, is zero in that circuit, the value of Z is $r_p + Z_k$. For an open circuit at the output, the current through Z_k is zero, and the voltage fed back is hence zero; thus the open-circuit loop gain, $(A\beta)_{oc}$ or $A_{g g_{oc}}$, is zero. In accordance with the definitions given earlier, the feedback is, therefore, of the current-feedback type. If the output terminals are short-circuited, the loop gain is given by

$$(A\beta)_{sc} \text{ or } A_{g g_{sc}} = \frac{-I_p Z_k}{E_g} = -\mu \frac{Z_k}{r_p + Z_k}, \qquad [193]$$

where the minus sign is a result of the choice of reference directions for the voltages in Figs. 30a and 31, and Fig. 15, Ch. VIII. Substitution of Eq. 193 in Eqs. 189 or 190 for the output impedance faced by the load gives

$$Z_{fb} = (r_p + Z_k)\left(1 + \mu \frac{Z_k}{r_p + Z_k}\right) \qquad [194]$$

$$= r_p + (\mu + 1)Z_k, \qquad [195]$$

which agrees with the value shown in Fig. 15b, Ch. VIII, and computed in Eq. 56, Ch. VIII, by other means.

Again, the increase in output impedance represented by Eq. 195 is explained in qualitative terms by the fact that when the load impedance is decreased in a circuit with current inverse feedback, the output current increases; hence the magnitude of the voltage fed back increases. But this increase in the voltage fed back tends to reduce the output current and thereby to maintain it nearly constant. Negative current feedback thus tends to stabilize the output current against changes in load. In other words, it increases the output impedance so that the amplifier becomes more nearly a constant-current source at the output for any particular value of input signal voltage.

The increase in the output impedance produced by the current feedback and indicated in Eq. 195 is illustrated graphically by the cathode-bias curve in Fig. 5, Ch. VIII. This line represents i_b as a function of e_b when e_c is determined by the feedback impedance in the cathode circuit. The reciprocal of the slope of the curve is, therefore, the output impedance at the tube terminals, which is the output impedance $r_p + (\mu + 1)Z_k$ effective at the load terminals minus the actual Z_k in series with the tube, giving the net value $r_p + \mu Z_k$.

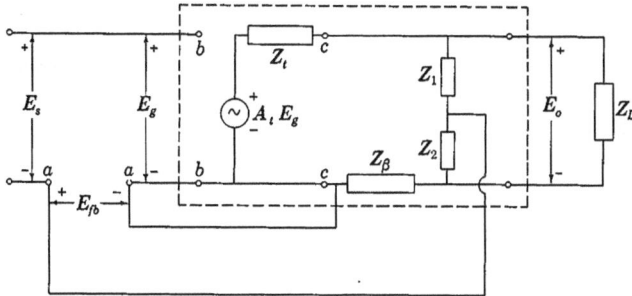

Fig. 36. Bridge-feedback circuit for simultaneous voltage and current feedback.

12e. *Shunt and Bridge Feedback.* The advantageous features of both voltage and current feedback may be combined in a single amplifier in a number of ways, for example, as is shown in Fig. 36. In this diagram, $A_t E_g$ and Z_t are the Thévenin's equivalent of the internal amplifier circuit between the terminals b–b and c–c; Z_β is the series- or current-feedback impedance, and Z_1 and Z_2 form a voltage divider across the load for shunt or voltage feedback.

The output impedance Z faced by the load for μ and, correspondingly, A_t equal to zero is the parallel combination

$$Z = \frac{(Z_t + Z_\beta)\,(Z_1 + Z_2)}{Z_t + Z_\beta + Z_1 + Z_2}.$$ [196]

For application of Eq. 190, the loop gain with the output short-circuited, $(A\beta)_{sc}$, is given by

$$(A\beta)_{sc} = A_t \frac{Z_\beta}{Z_t + Z_\beta},$$ [197]

and the loop gain with the output open-circuited is given by

$$(A\beta)_{oc} = A_t \frac{Z_\beta + Z_2}{Z_t + Z_\beta + Z_1 + Z_2}.$$ [198]

Substitution of Eqs. 197 and 198 in Eq. 190 gives for the output impedance

$$Z_{fb} = \frac{(Z_t + Z_\beta)(Z_1 + Z_2)}{Z_t + Z_\beta + Z_1 + Z_2} \; \frac{1 - A_t \dfrac{Z_\beta}{Z_t + Z_\beta}}{1 - A_t \dfrac{Z_\beta + Z_2}{Z_t + Z_\beta + Z_1 + Z_2}}. \quad [199]$$

In general, the series and shunt feedback impedances in Fig. 36 may be adjusted so as either to increase or to decrease the output impedance. The output impedance is obviously unchanged by the feedback if $(A\beta)_{oc}$ equals $(A\beta)_{sc}$ in Eq. 190. Equating these values from Eqs. 197 and 198 gives

$$\frac{Z_1}{Z_\xi} = \frac{Z_t}{Z_\beta} \quad\quad [200]$$

as the relationship required if the output impedance is to be unchanged by the feedback. Since this adjustment makes the loop gain the same for zero as for infinite load impedance, and the circuit is linear, it makes the loop gain independent of the load impedance, a condition that is not true for either current or voltage feedback alone. It should be noted that in accordance with the basic definition in Fig. 30b and Eq. 158, Z_β in Fig. 36 does not give true current feedback because it conducts a current component in addition to that through the load.

The feedback arrangement in Fig. 36 is frequently called *bridge feedback* because the impedances Z_1, Z_2, Z_t, and Z_β form the four arms of a Wheatstone bridge for which the load terminals and the terminals at E_{fb} may be considered to constitute the generator and detector terminals. Equation 200 is the condition for balance in such a bridge.

Thus far, the voltage E_{fb} fed back is considered to be connected in series with the grid-cathode terminals of a tube having infinite input impedance. Frequently, however, a shunt or bridge arrangement is used at the input as well as the output. Figure 37a illustrates such a circuit in general form. The impedances Z_1, Z_2, and Z_3 are included as an equivalent-π representation of any actual network that may exist between the two terminal pairs they connect. Since the input impedance of the remainder of the circuit is not infinite, the impedance of the source of input voltage, Z_s, plays a part in the circuit behavior and must be considered.

An important example of a feedback amplifier having shunt connections at both the input and the output is shown in Fig. 37b. This arrangement is the residue if certain obvious impedances in Fig. 37a become zero and others are combined in parallel. A complete analysis of the circuit may be made through application of the superposition method, but the partial solution that follows is useful for an understanding in this case. Kirchhoff's current law requires that the sum of the currents through paths leading to the upper input terminal of the internal amplifier, which is assumed to have infinite input impedance,

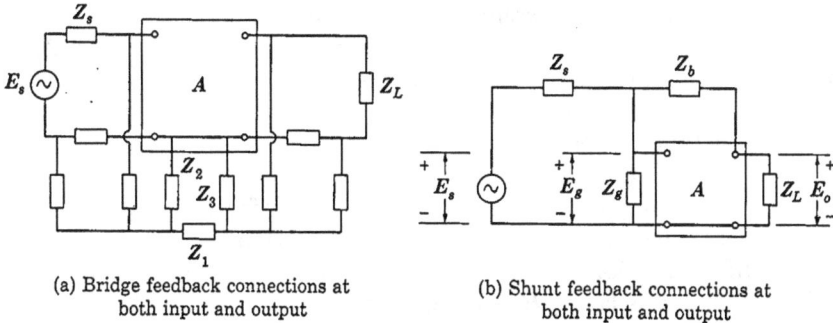

(a) Bridge feedback connections at both input and output

(b) Shunt feedback connections at both input and output

Fig. 37.　Circuits with similar feedback connections at both the input and the output.

must be zero. Expressing these currents as voltage drops divided by impedances gives

$$\frac{E_s - E_g}{Z_s} - \frac{E_g}{Z_g} + \frac{E_o - E_g}{Z_b} = 0. \qquad [201]$$

But

$$E_o = AE_g , \qquad [202]$$

where A is the complex voltage gain from the grid to the output *with the impedance* Z_b *connected.* Substitution of Eq. 202 in Eq. 201 gives for the over-all complex voltage gain

$$A_{fb} = \frac{E_o}{E_s} = -\frac{Z_b}{Z_s} \frac{1}{1 - \frac{1}{A}\left[1 + \frac{Z_b}{Z_s} + \frac{Z_b}{Z_g}\right]}, \qquad [203]$$

which is arranged in the form of an impedance ratio Z_b/Z_s multiplied by a correction factor that is dependent on the effective internal voltage gain A. This expression shows that, if

$$|A| \gg \left|1 + \frac{Z_b}{Z_s} + \frac{Z_b}{Z_g}\right|, \qquad [204]$$

the over-all complex voltage gain becomes

$$A_{fb} \approx -\frac{Z_b}{Z_s}. \qquad \blacktriangleright[205]$$

The negative sign in this complex expression indicates a 180-degree phase shift. A qualitative explanation for the important result in Eq. 205 is that for a large value of A, E_g remains negligibly small for any finite value of E_o that can occur. Consequently, the current through Z_g is negligible, and currents through Z_s and Z_b are equal. The voltages E_s and E_o are then constrained to equal the voltages across Z_s and Z_b, respectively, caused by the same current in each, and Eq. 205 results. The circuit behaves as an electrical see-saw pivoted at E_g, and Z_g could as well be omitted, except that it is sometimes necessary to provide a path for the grid current.

Use of resistance, capacitance, inductance, or logarithmic elements for Z_s and Z_b in Fig. 37 leads to a large number of circuits[40] suitable for phase inversion, for shaping waveforms, and for performing the mathematical operations of addition, subtraction, multiplication, division, differentiation, and integration. If, for example, Z_b is made equal to Z_s and A is large, E_o and E_s become equal in magnitude and opposite in phase. The circuit then serves as a phase inverter, as is explained in Art. 7, and is sometimes known as a *gain-of-minus-one amplifier*. When capacitance is used for Z_b, the circuit is often called a "Miller" circuit because it utilizes an effect first described by Miller,[41] which is discussed in Art. 8, Ch. VIII.

12f. *Feedback Around Resistance-Capacitance-Coupled Amplifiers.* Feedback is frequently applied around one or more stages of resistance-capacitance-coupled amplification to alter the frequency characteristics and the output impedance and to reduce the nonlinear distortion. The effectiveness of the feedback depends on the number of stages and the details of the circuit connections. For example, straightforward application of voltage feedback to a single stage may be accomplished through use of a resistance voltage divider across the output, as is shown in Fig. 38a. This circuit is of little practical importance because it does not have a common connection between the input and the output. Stray capacitance and leakage between the input and the output circuit elements is effectively in parallel with the resistance βR_g, as is discussed in Art. 7, and may greatly alter the feedback. Nevertheless, since it applies in part to all types of voltage feedback around one

[40] B. Chance, V. Hughes, E. F. MacNichol, D. P. Sayre, and F. C. Williams, Editors, *Waveforms*, Massachusetts Institute of Technology Radiation Laboratory Series, Vol. 19 (New York: McGraw-Hill Book Company, Inc., 1948), 27–37, 643–674; I. A. Greenwood, Jr., J. V. Holdam, Jr., and V. MacRae, Jr., Editors, *Electronic Instruments*, Massachusetts Institute of Technology Radiation Laboratory Series, Vol. 21 (New York: McGraw-Hill Book Company Inc., 1948), 17, 33, 64–74, 78–83.

[41] J. M. Miller, "Dependence of the Input Impedance of a Three-Electrode Vacuum Tube upon the Load in the Plate Circuit," *Nat. Bur. Standards Sci. Papers*, 15 (1919–1920), No. 351, 367–385.

stage that are essentially resistive, an analysis of this amplifier that neglects such stray effects is instructive.

The relation between the complex voltage gain A_{fb} of the amplifier with feedback to the gain A without feedback—that is, with the tap

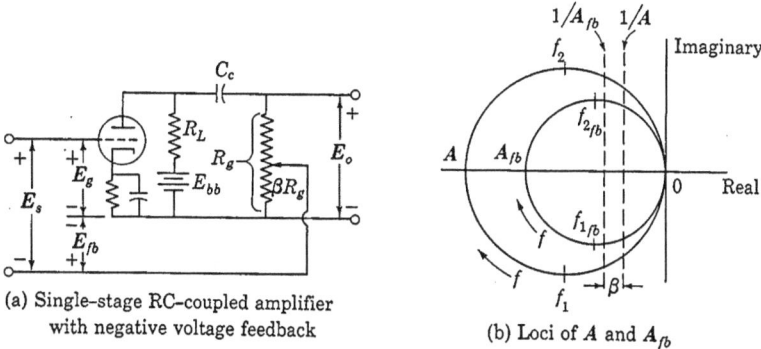

(a) Single-stage RC-coupled amplifier with negative voltage feedback

(b) Loci of A and A_{fb}

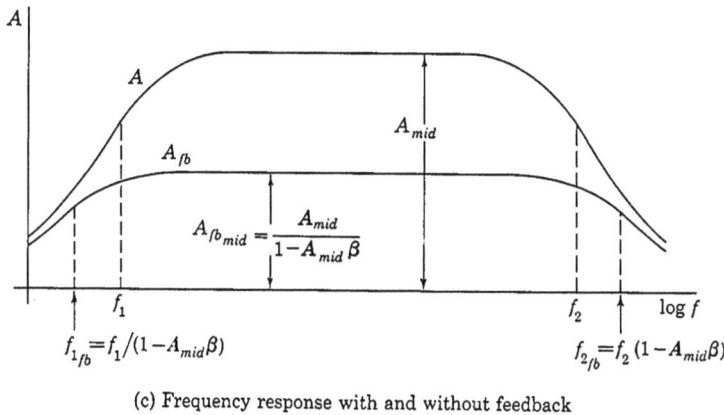

(c) Frequency response with and without feedback

$$f_{1fb} = f_1/(1 - A_{mid}\beta)$$

$$f_{2fb} = f_2(1 - A_{mid}\beta)$$

Fig. 38. Effect of negative voltage feedback in single-stage RC-coupled amplifier on the complex voltage amplification as a function of frequency.

on R_g at the bottom of the resistor—may be represented graphically on the complex plane as in Fig. 38b. The locus of A is a circle, as derived previously in Art. 5. To determine the locus of A_{fb}: First, express it as

$$A_{fb} = \frac{A}{1 - A\beta} = \frac{1}{\dfrac{1}{A} - \beta}. \qquad [206]$$

In accordance with this expression, take the reciprocal $1/A$ of the circle A. Just as the reciprocal relation converts the vertical lines of Figs. 15a and 15b into the right-hand circle in Fig. 15c, so here the reciprocal relation applied to the circular locus A yields the straight vertical line $1/A$ in Fig. 38b. Add $-\beta$ to $1/A$. Since, in the circuit of Fig. 38a, β defined by Eq. 153 is a positive constant, performing this addition yields a second straight line on the left of the first as shown. Take the reciprocal $1/[(1/A) - \beta]$. This operation produces a second and smaller circle as the locus of A_{fb}. The shape of the A_{fb} locus indicates that the magnitude and phase responses of the over-all amplifier with feedback as functions of frequency are similar in shape to the responses without feedback. They differ only by scale factors.

The similarity of behavior of this amplifier with and without feedback leads to a relatively simple relation between the effect of this particular type of feedback on the midband gain and its effect on the half-power frequencies.[42] Since Eq. 157 applies for any frequency, in the low range of frequencies for which Eq. 38 is applicable, the complex voltage gain is given by

$$A_{fb_{low}} = \frac{A_{low}}{1 - A_{low}\beta} . \qquad [207]$$

Substitution of Eq. 38 in Eq. 207 gives

$$A_{fb_{low}} = \frac{\dfrac{A_{mid}}{1 - j(f_1/f)}}{1 - \dfrac{A_{mid}\beta}{1 - j(f_1/f)}} = \frac{\dfrac{A_{mid}}{1 - A_{mid}\beta}}{1 - j\dfrac{f_1}{f}\dfrac{1 - A_{mid}\beta}{f}} . \qquad [208]$$

Comparison with Eq. 38 indicates that Eq. 208 has a similar form,

$$\frac{A_{fb_{low}}}{A_{fb_{mid}}} = \frac{1}{1 - j\dfrac{f_{1fb}}{f}} , \qquad [209]$$

where

$$A_{fb_{mid}} = \frac{A_{mid}}{1 - A_{mid}\beta} , \qquad [210]$$

and

$$f_{1_{fb}} = \frac{f_1}{1 - A_{mid}\beta} . \qquad [211]$$

[42] F. E. Terman and W. T. Pan, "Frequency Response Characteristics of Amplifiers Employing Negative Feedback," *Communications*, 19 (March, 1939), 5–7, 42–45.

Similarly, from Eqs. 47 and 157,

$$A_{fb_{high}} = \frac{A_{high}}{1 - A_{high}\beta} = \frac{\dfrac{A_{mid}}{1 + j(f/f_2)}}{1 - \dfrac{A_{mid}\beta}{1 + j(f/f_2)}}, \qquad [212]$$

or

$$A_{fb_{high}} = \frac{\dfrac{A_{mid}}{1 - A_{mid}\beta}}{1 + j\dfrac{f}{f_2(1 - A_{mid}\beta)}} = \frac{A_{fb_{mid}}}{1 + j(f/f_{2_{fb}})}, \qquad [213]$$

where

$$f_{2_{fb}} = f_2(1 - A_{mid}\beta). \qquad [214]$$

Note that appearance of complex quantities in the right-hand sides of Eqs. 211 and 214 does not imply that the values of $f_{1_{fb}}$ and $f_{2_{fb}}$ are complex, because A_{mid} and β are both real for the conditions analyzed here.

Equations 210, 211, and 214 for the complex midband voltage gain and the half-power frequencies with feedback show that negative feedback shifts each of the half-power frequencies outward from the middle range by the same factor that it reduces the gain in the middle range. The frequency response with and without feedback thus appears as in Fig. 38c. The gain stability and the linearity of the amplifier are also improved, and the output impedance is reduced in accordance with the considerations given earlier. This analysis ignores the feedback through the grid-to-plate capacitance of the tube, and may hence be appreciably in error at high frequencies.

Increase of the upper half-power frequency and reduction of the output impedance can also be effected simply through reduction of the load resistance R_L. But such a change would also reduce the permissible amplitude of the signal component of the plate voltage for Class A_1 operation. The advantages of the use of feedback are that it does not reduce this amplitude, and that it does provide increased stability of gain.

Use of a transformer in the feedback network or at the input or output of an amplifier permits connecting one terminal of the input to one of the output, with its attendant advantages. However, the leakage inductance and interwinding capacitances of the transformer may then materially affect the frequency characteristics and stability of the amplifier. A transformerless circuit having a common input and output terminal and negative feedback around one stage is shown in Fig. 39. The left-hand tube is a buffer stage. Its output impedance serves as a source impedance, so that the feedback around the second stage is similar to the basic type illustrated in Fig. 37b.

If feedback were applied across two stages of an RC-coupled amplifier by the method illustrated in Fig. 38 for a single stage, a regenerative or an oscillatory effect rather than a degenerative effect would be obtained. In the middle range of frequencies for a two-stage amplifier, A is a positive real number. Hence, to effect degeneration over the operating range of frequencies, a phase shift of 180 degrees is needed in the feedback network. A transformer used as in Fig. 40 illustrates a method of obtaining this phase shift, though, from a practical standpoint, the frequency characteristics of the transformer would have a pronounced effect on the behavior of the circuit. A transformerless amplifier having essentially this type of feedback is shown in the Wien-bridge oscillator circuit, Fig. 9, Ch. XI.

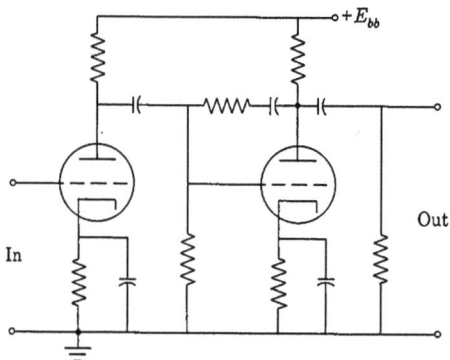

Fig. 39. Circuit having a common connection between input and output and feedback around one RC-coupled stage.

Analysis of the circuit in Fig. 40 on the assumption that the transformer is ideal makes evident several important characteristics of feedback around two stages. The feedback ratio β is negative for the transformer winding directions shown in Fig. 40, so that

$$\beta = -\frac{N_1}{N_2} = \frac{N_1}{N_2} \epsilon^{j\pi}. \qquad [215]$$

If A_1 denotes the complex voltage gain of one RC-coupled amplifier stage, it may be expressed in terms of Eqs. 30 and 53 as

$$A_1 = A_{1_{mid}} \cos \theta \ \epsilon^{j(\pi + \theta)}, \qquad [216]$$

where

$\theta = \tan^{-1} (f_1/f)$ at low frequencies,

$\theta \approx 0$ at middle frequencies,

$\theta = \tan^{-1} (-f/f_2)$ at high frequencies.

When plotted, this expression for A_1 yields a circle similar to that shown dotted in Fig. 40b. For two identical stages in tandem, the total internal complex voltage gain is

$$A = A_1 \times A_1 = A_{1_{mid}}^2 (\cos^2\theta)\epsilon^{j(2\pi + 2\theta)} \qquad [217]$$

$$= \frac{A_{1_{mid}}^2}{2} (1 + \cos 2\theta)\epsilon^{j2\theta} \qquad [218]$$

$$= \frac{A_{1_{mid}}^2}{2} (1 + \cos \phi)\epsilon^{j\phi}, \qquad [219]$$

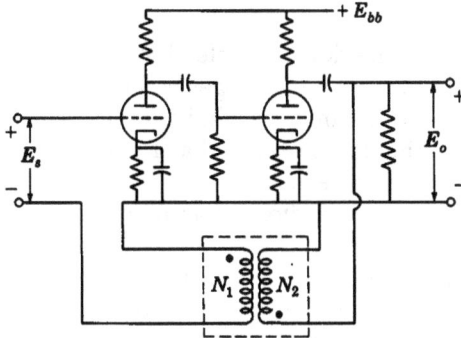

(a) Circuit with negative feedback across two stages

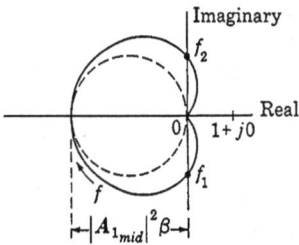

(b) $A\beta$ loci for idealized circuit of (a)

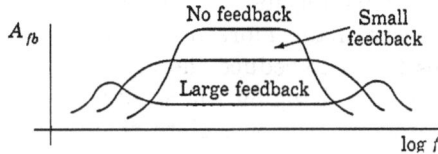

(c) Effect of amount of feedback in idealized circuit of (a)

Fig. 40.   Effect of negative voltage feedback across two stages of an RC-coupled amplifier on the complex voltage amplification as a function of frequency.

where ϕ equals 2θ. From Eqs. 215 and 219,

$$A\beta = \frac{1}{2} \frac{N_1}{N_2} A_{1_{mid}}^2 (1 + \cos \phi)\epsilon^{j(\pi + \phi)}, \qquad [220]$$

which will be recognized as the polar equation for a cardioid as drawn in Fig. 40b. The half-power frequencies for the *individual* stages lie on the vertical axis as shown.

If one stage is replaced by an ideal stage having a complex voltage gain $A_{1_{mid}}\epsilon^{j\pi}$ independent of frequency, the internal complex voltage gain becomes

$$A = A_{1_{mid}}^2 (\cos \theta)\epsilon^{j(2\pi + \theta)} \qquad [221]$$

$$= A_{1_{mid}}^2 (\cos \theta)\epsilon^{j\theta}, \qquad [222]$$

and

$$A\beta = \frac{N_1}{N_2} A_{1_{mid}}^2 (\cos \theta)\epsilon^{j(\pi + \theta)}, \qquad [223]$$

which is the equation of the dotted circle shown in Fig. 40b.

In any practical amplifier, the two stages will not be mathematically identical, nor will either of them have ideal characteristics. Hence, the cardioid and the circle should be regarded as limits between which the actual $A\beta$ locus lies. The more nearly alike the two stages are, the nearer the $A\beta$ locus is to the cardioid. The more unlike they are, the nearer the locus is to the circle. The amplifier is free from unwanted oscillations because the locus does not surround the point $1 + j0$.

The effect of the feedback on the frequency response of the amplifier may be qualitatively inferred from Fig. 40b. At low and high frequencies, the gain is increased by the feedback, because any $A\beta$ locus lving between the circle and the cardioid traverses a part of the region lying within the unit circle with center at $1 + j0$. The tendency toward this regeneration is increased as β is increased and as the stages are made more nearly alike. Conversely, dissimilarity of the stages tends to reduce the regeneration. The effect on the frequency response of changing the amount of feedback through changing β is indicated in Fig. 40c. Note that here the curves cross, whereas for feedback around one stage as illustrated in Fig. 38c they do not.

The $A\beta$ locus for any practical circuit will differ from that shown in Fig. 40b because of the frequency characteristics of a practical transformer. Since, as is explained in Art. 8, the transformer introduces an additional phase shift tending toward 90 degrees at the low and high frequencies,[43] the locus may surround the point $1 + j0$ for large amounts of feedback. The amount of feedback must be limited, therefore, in order to prevent oscillations.

Negative voltage feedback across three RC-coupled amplifier stages may be accomplished as shown in Fig. 41a. The corresponding $A\beta$ locus is the solid curve in Fig. 41b when the stages are identical. This circuit has the practical disadvantage that the input and output

---

[43] E. E. Staff, M.I.T., *Magnetic Circuits and Transformers* (Cambridge, Massachusetts: The Technology Press of M.I.T.; New York: John Wiley & Sons, Inc., 1943), 477.

terminals have no point in common that may be grounded. The disadvantage may be overcome by adoption of the basic scheme involving an extra driving stage at the input shown in Fig. 39. A more serious limitation, however, is the tendency of the circuit to oscillate.

As the frequency of the input voltage E_s is increased from zero, the phase shift introduced by each stage decreases from 270 degrees toward 90 degrees; and the angle by which E_o leads E_s decreases from 810 degrees toward 270 degrees. At the frequencies for which E_o leads E_s by 720 degrees or 360 degrees, oscillations occur if $|A\beta|$ exceeds

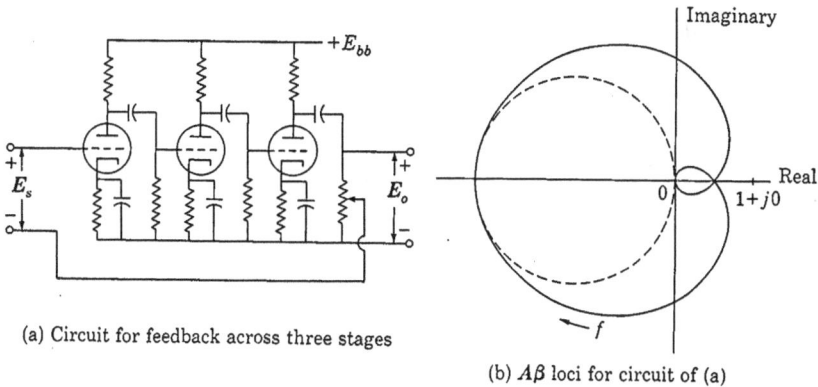

(a) Circuit for feedback across three stages

(b) $A\beta$ loci for circuit of (a)

Fig. 41. Negative voltage feedback across three RC-coupled stages.

unity. At these frequencies, the phase shift in each stage is 240 degrees or 120 degrees, respectively, if the stages are alike and the phase shift in β is assumed zero. Each stage then operates with a phase shift 60 degrees more or less than the middle-frequency phase shift, which is 180 degrees. The voltage gain of each stage is $A_{1mid}/2$, where A_{1mid} is the voltage gain of a single stage in the middle range of frequencies. The total internal voltage gain of the amplifier at either frequency for which E_o and E_s are in phase is

$$A = \left[\frac{A_{1mid}}{2}\right]^3 \tag{224}$$

Hence, if $\dfrac{A^3_{1mid}}{8}\beta$ is greater than unity, oscillations occur. Alternatively, $A^3_{1mid}\beta$ must be less than 8 to prevent oscillations. This severe limitation on the amount of feedback means that feedback around three identical stages can hardly be considered a practical arrangement.

The tendency to oscillate may be reduced and the allowable feedback increased through designing one or two of the stages to have a much broader middle range of frequencies than the remainder. To

insure that the $A\beta$ locus does not enclose the point $1 + j0$, it is neces-
sary that $|A\beta|$ remain less than unity until the frequency increases
above the value for which the total phase shift is 720 degrees and that
it decrease to less than unity before the frequency decreases below the
value for which the total phase shift is 360 degrees. If one broad-
band stage is used, the $A\beta$ locus tends to become a cardioid as in Fig.
40b; if two such stages are used, the locus approaches a circle as in Fig.
38b. Either tendency fulfills the necessary requirement for avoiding
oscillations.

Although, in general, inverse feedback has one important disadvantage—it decreases the over-all voltage gain of the amplifier—
the advantages often outweigh the disadvantages because the voltage
gain can usually be brought back to the desired level by the addition
of more stages of amplification.

The advantages to be gained by inverse feedback are numerous, and
the preceding discussion merely gives a basis for an understanding of
the possibilities. The fundamental principle of inverse feedback is used
not only in vacuum-tube amplifiers but also in other types of electrical
and mechanical systems and in all control systems in which the result
of a control operation tends to neutralize or decrease the original
control signal. Servomechanisms, hydraulic governors, automatic
pilots for ships and aircraft, closed-cycle process-control devices, and
many others employ the basic principle here called inverse feedback,
and are susceptible to the same type of analysis that is applicable to
feedback amplifiers.[44]

PROBLEMS

1. A direct-coupled amplifier comprises two Type 6J5 triodes in the circuit of
Fig. 3a. The plate-supply voltage E_{bb} is 300 volts and E_{cc} is -100 volts. For each
tube, the zero-signal value of the plate voltage is to be 120 volts, of the grid
voltage, -2 volts, and of the plate current, 8 ma. An approximate value for μ is
then 20 and for r_p is 6,700 ohms. Each of the three resistances connected to the
negative terminal of the grid-bias supply is 1 megohm. The impedance of the
source of input voltage e_{in} may be considered zero. Find:

    (a) the required values of all resistances,
    (b) the over-all voltage amplification,
    (c) the maximum permissible positive and negative values of e_{in} if the grid
        voltages for both tubes are to remain negative and larger than the cut-off
        value.

[44] L. A. MacColl, *Fundamental Theory of Servomechanisms* (New York: D. Van Nos-
trand Company, 1945); H. M. James, N. B. Nichols, R. S. Philips, Editors, *Theory of
Servomechanisms*, Massachusetts Institute of Technology Radiation Laboratory Series,
Vol. 25 (New York: McGraw-Hill Book Company, Inc., 1947); G. S. Brown and D. P.
Campbell, *Principles of Servomechanisms* (New York: John Wiley & Sons, Inc., 1948).

2. A cathode-coupled difference amplifier consists of two triodes in the circuit of Fig. 4. The grid-to-grid voltage to be amplified, e_g, is 50 millivolts at 1 cps. There exists also an extraneous sum component, or common mode, in the grid-to-ground voltages of 10 millivolts at 60 cps. The plate-supply voltage E_{bb} is 200 volts, E_{cc} is 100 volts with the polarity shown, and each tube is to operate so that E_b is 120 volts, E_c is -6 volts, I_b is 0.5 ma, μ is 20, and r_p is 10,000 ohms. Find:

(a) the load resistance R_L and the cathode resistance R_k,

(b) the plate-to-plate output voltage, e_o,

(c) the sum component in the plate-to-ground voltages, e_{pna}.

3. A triode with an amplification factor of 20 and a plate resistance of 10,000 ohms is used as a linear Class A_1 amplifier with resistance-capacitance coupling.

If the grid resistance R_g for the succeeding stage is limited to 10^6 ohms, what values of load resistance R_L and coupling capacitance C_c will give a voltage gain of 15 in the middle range of frequencies and a lower half-power frequency of 30 cps?

4. A Type 6J5 triode is used in one stage of the resistance-capacitance-coupled Class A_1 amplifier circuit of Fig. 6a with a battery substituted for the cathode-bias arrangement of R_k and C_k. The plate-supply voltage E_{bb} is 300 volts and the grid-bias voltage E_{cc} is -6 volts. The load resistance R_L is 10^5 ohms, the coupling capacitance C_c is 0.01 μf, the grid resistance R_g is 10^6 ohms, and the total shunt capacitance C_g' is 50 $\mu\mu$f.

Determine the voltage gain of the stage in the middle range of frequencies and the upper and lower half-power frequencies.

5. A Type 6J5 triode is used in one stage of the resistance-capacitance-coupled amplifier of Fig. 6a. The load resistance R_L is 10^4 ohms, the coupling capacitance C_c is 0.01 μf, the grid resistance R_g is 10^5 ohms, and the total shunt capacitance C_g' is 150 $\mu\mu$f.

(a) Specify suitable plate-supply and grid-bias voltages for linear Class A_1 operation of the stage.

(b) Determine the value of the cathode resistance R_k that must be used to provide the grid-bias voltage specified in (a).

(c) On the assumption that the reactance of C_k is essentially zero at all frequencies of interest, plot the voltage gain of the stage as a function of frequency.

6. One stage of a resistance-capacitance-coupled Class A_1 amplifier employing a triode is arranged as shown in Fig. 6a. The quiescent operating point of the tube is such that μ is 60 and r_p is 15,000 ohms. The load resistance R_L is 100,000 ohms, the grid resistance R_g is 500,000 ohms, the total shunt capacitance C_g' is 20 $\mu\mu$f, and C_k may be assumed to be so large that its reactance is essentially zero at all frequencies of interest.

(a) Calculate and plot the voltage gain of the stage at a frequency of 50 cps as a function of C_c for C_c equal to 0.001, 0.005, 0.01, 0.02, 0.05, and 0.1 μf.

(b) Plot the frequency-response characteristic of the stage for C_c equal to 0.05 μf.

7. The first stage of a resistance-capacitance-coupled Class A_1 amplifier consists of a triode in conjunction with a load resistor R_L of 100,000 ohms, a coupling capacitor C_c of 0.01 μf, and a grid resistor R_g of 250,000 ohms. The total shunt capacitance C_g', which includes the stray capacitance of the leads, the output capacitance of the first tube, and the input capacitance of the next tube, is 150 $\mu\mu$f. The plate resistance and the amplification factor of the tube may be

assumed to be 10,000 ohms and 20 respectively. The capacitance C_k may be assumed to be so large that its reactance is essentially zero at all frequencies of interest.

(a) Find the voltage gain of this amplifier in the middle range of frequencies, and sketch the frequency characteristic on logarithmic scales of gain and frequency.

(b) It is desired to change the coupling capacitor so that the over-all voltage gain at a frequency of 60 cps will be 90 per cent of the maximum possible value. How large should the capacitance of the coupling capacitor be?

(c) What new value should the load resistance have in order to increase the upper half-power frequency to twice its value in (a)?

(d) What is the ratio of the voltage gain in the middle range of frequencies obtained in (c) to that in (a)?

8. The first stage of a resistance-capacitance-coupled Class A_1 amplifier utilizes a Type 6SJ7 pentode with a plate-supply voltage of 400 volts, a constant screen-grid voltage of 100 volts, and a control-grid-bias voltage of -3 volts. The suppressor grid is maintained at cathode potential. The load resistance R_L is 10^5 ohms, the coupling capacitance C_c is 0.05 μf, and the grid resistor R_g of the following stage is 10^5 ohms.

The second stage of the amplifier employs a Type 6B4-G triode as a power amplifier with a resistance load of 8 ohms coupled through an essentially ideal output transformer having a turns ratio of 10 to 1. The plate-supply voltage is 250 volts, and the grid-bias voltage is -45 volts. Neglect the wiring capacitances.

(a) Compute the over-all frequency-response characteristic of the amplifier over the range between the half-power frequencies.

(b) As the input voltage is increased, which tube first exceeds the limits of Class A_1 operation in the middle range of frequencies?

9. Design a single-stage resistance-capacitance-coupled Class A_1 amplifier for energizing the deflecting plates of a cathode-ray oscilloscope that are shunted by a 5-megohm resistor. Requirements call for a voltage gain of 50 uniform within ± 5 per cent over a frequency range extending from 10 to 20,000 cps with a sinusoidal peak input of one volt. One half of a Type 6SL7-GT twin triode is to be used, and second-harmonic distortion is to be limited to 5 per cent. Specify suitable plate-supply and grid-bias voltages, load resistance, and coupling capacitance, and determine the plate current I_b and plate voltage E_b corresponding to the resulting quiescent operating point. Give reasons for your choice of design constants.

10. A two-stage resistance-capacitance-coupled Class A_1 amplifier is to be designed, using a twin triode. The voltage gain as a function of frequency is to vary by not more than 6 db between 30 and 70,000 cps. The output of the amplifier is to supply a vacuum-tube voltmeter that has an effective input shunt capacitance C_g and grid resistance R_g equal to those of each of the two amplifier stages, which are to be identical. Assume for analysis that C_{gp} is 3.3 $\mu\mu$f, C_{gk} is 3.2 $\mu\mu$f, and C_{pk} is 3.5 $\mu\mu$f for each triode unit, that the interstage wiring capacitance is 15 $\mu\mu$f, and that the effective input shunt conductance for each amplifier stage is zero. The tubes are to operate under the following conditions:

$$\text{Quiescent plate current} = 2.3 \text{ ma}$$
$$\text{Quiescent plate voltage} = 250 \text{ volts}$$
$$\text{Grid-bias voltage} = -2 \text{ volts}$$
$$\text{Amplification factor} = 70$$
$$\text{Plate resistance} = 44{,}000 \text{ ohms.}$$

The grid resistance R_g in each grid circuit should not exceed 1 megohm. A plate-voltage supply of 400 volts and a 2-volt cell for grid bias are available.

(a) Design the amplifier, specifying the ratings of all resistors and capacitors.

(b) What is the maximum voltage gain of the amplifier in decibels?

11. The circuit shown in Fig. 42 is one stage of a pseudo-tuned Class A_1 amplifier for use at 60 cps. At frequencies well above 60 cps, the reactance of C_1 becomes small and reduces the gain of the stage, and at frequencies well below 60 cps the reactance of C_2 becomes large relative to the resistance of R_g and the gain is again reduced. Although the discrimination of one such stage against unwanted frequencies is relatively small, several stages are effective.

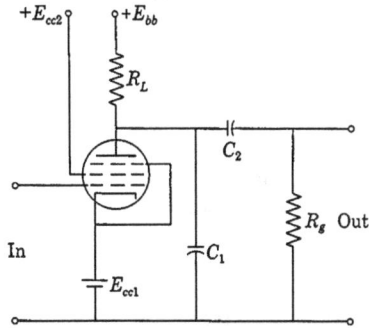

Fig. 42. Pseudo-tuned amplifier stage for Prob. 11.

(a) **Draw an equivalent circuit for one stage of this amplifier.**

(b) **Show that at an angular frequency ω the complex voltage gain is given by**

$$A = -\frac{\mu R_g}{r_p} \frac{1}{1 + \dfrac{C_1}{C_2} + \dfrac{R_g}{R_L} + \dfrac{R_g}{r_p} + j\left(\omega R_g C_1 - \dfrac{r_p + R_L}{\omega R_L r_p C_2}\right)}.$$

(c) Plot the locus of A in the complex plane as a function of ω.

(d) Find the value of C_2 required to make the voltage gain a maximum at 60 cps when g_m is 1,575 μmhos, r_p is 0.7 megohm, R_L is 0.25 megohm, R_g is 0.25 megohm, and C_1 is 0.01 $\mu\mu$f.

(e) Determine the magnitude of A at 30 and 120 cps relative to its magnitude at 60 cps when the circuit parameters are as in (d).

(f) Determine for four such stages in cascade the magnitude of the over-all voltage gain at 120 cps relative to its magnitude at 60 cps.

12. Two triodes are to be used in a push-pull Class A_1 voltage amplifier similar to that shown in Fig. 43. The plate-supply voltage E_{bb} is 348 volts and the grid-bias voltage is to be -8 volts. In the circuit, R_L is 10,000 ohms, R_g is 100,000

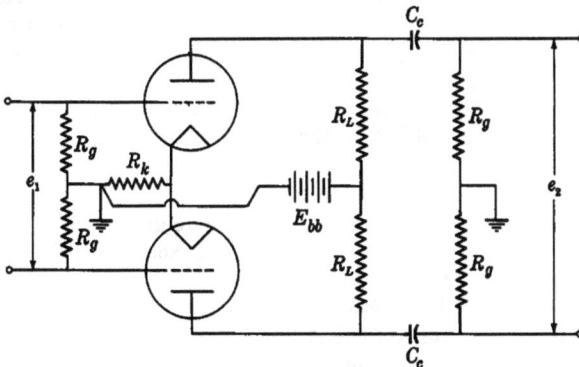

Fig. 43. Push-pull voltage amplifier stage for Prob. 12.

ohms, and C_c is 0.02 μf. The quiescent plate current at the operating point fixed by the supply voltages and circuit constants is 9 ma per tube. At this operating point, the amplification factor μ is 20 and the plate resistance r_p is 7,700 ohms for each tube.

(a) Determine the value of the cathode-bias resistance R_k necessary to provide the specified grid-bias voltage.

(b) Determine the over-all voltage gain of the stage in the middle range of frequencies.

(c) What is the quiescent plate dissipation in one tube?

(d) Determine the low frequency at which the power developed in the output resistors R_g is one-half the constant value developed in the middle range of frequencies.

13. A public-address system is to be so designed as to supply an audio-frequency power output of 10 watts when energized by a crystal microphone that supplies an input of 0.01 volt rms. Show in block diagram form a complete amplifier that would produce the desired results. Indicate what tubes should be used, give the approximate voltage gain per stage, and show what output voltage or power could be expected from each stage. Indicate where a volume control would be inserted in the circuit.

14. Consider the first stage of the Class A_1 amplifier circuit of Fig. 6a, with the cathode-bias combination R_k and C_k adjusted so that the voltage across it is constant for all pertinent frequencies. The following data apply:

$$\mu = 20 \qquad\qquad C_c = 0.01 \ \mu f$$
$$r_p = 7{,}700 \text{ ohms} \qquad\qquad C_g' = 100 \ \mu\mu f$$
$$R_g = 10^6 \text{ ohms}.$$

The value of the capacitance C_g' given above accounts for all the shunt capacitance in the tubes and wiring. If the value of R_L is chosen so as to make the voltage gain in the middle range of frequencies equal to 15, what are the values of the following:

(a) the resistance of R_L,

(b) the lower half-power frequency,

(c) the upper half-power frequency?

Next, consider that a tuned circuit consisting of a parallel combination of a capacitor C with a coil of inductance L and series resistance R is substituted for the resistor R_L in the circuit. The data that now apply to the circuit are:

μ, C_c, R_g, C_g', and r_p remain the same; the parameters of the tuned circuit are such that if an additional capacitance equal to C_g' were connected across it, the resonant frequency would be 10^6 cps, and the Q at that frequency would be 200. The voltage gain of the amplifier at the resonant frequency is 15 as before. What are the values of

(d) the inductance of the tuned circuit?

(e) the series resistance of the coil in the tuned circuit?

(f) the required capacitance of the tuned circuit, exclusive of C_g'?

(g) the lower half-power frequency?

(h) the upper half-power frequency?

15. A triode is used as a tuned Class A_1 amplifier with a plate-supply voltage E_{bb} of 300 volts and a grid-bias voltage E_{cc} of −7.5 volts. The tube has a μ of 20 and a plate resistance r_p of 10,000 ohms, and the tuned circuit is composed of a

capacitor of 100 $\mu\mu$f capacitance, including the stray and interelectrode capacitances, in parallel with a coil of 0.1 mh inductance. The Q of the coil is 100 at the resonant frequency of the tuned circuit.

Determine the voltage gain at the resonant frequency and the frequencies at which the voltage gain is 0.707 times its value at the resonant frequency.

16. A Class A_1 radio-frequency amplifier uses a pentode for which g_m is 5,000 μmho and r_p is 0.8 megohm. The tuned load comprises a 1-mh coil in parallel with an effective total capacitance of 100 $\mu\mu$f. The resonant Q_0 of the load is 200.

(a) Determine the voltage gain of the stage at the resonant frequency of the tuned load.
(b) Determine the voltage gain of the stage at 10,000 cps above and below resonance.
(c) Find the half-power frequencies by the use of Eqs. 100 and 113.

17. A certain pentode operates as a tuned radio-frequency Class A_1 amplifier with a tuned load circuit comprising a coil of 280 μh inductance and a Q_0 of 80 at the resonant frequency of 10^6 cps, which is the frequency of the grid-signal voltage. The load circuit is tuned to resonance at this frequency by means of a suitable shunt capacitance. The tube has an amplification factor μ of 620 and a plate resistance r_n of 10^6 ohms.

(a) What is the voltage gain of the stage at the resonant frequency?
(b) What is the voltage gain for an interfering signal 50 kc per sec higher than the resonant frequency?

18. A pentode with an amplification factor μ of 630 and a plate resistance r_p of 4×10^5 ohms operates as a linear Class A_1 amplifier with a tuned load circuit comprising a 1-mh coil in series with the primary of an ideal transformer. The series combination is tuned to the grid-signal frequency of 10^6 cps by an adjustable shunt capacitance. The transformer is used to couple a 250,000-ohm load resistance into the tuned circuit.

On the assumption that the coil has negligible dissipation, what transformer ratio will provide maximum power transfer to the 250,000-ohm load for small grid-signal voltages?

19. A one-stage capacitance-coupled tuned-radio-frequency voltage amplifier of the type shown in Fig. 24a employs a load circuit that is tuned to the grid-signal frequency of $10^6/(2\pi)$ cps. The tuned circuit comprises a coil with a Q of 100 in parallel with a capacitive reactance of 150 ohms at the resonant frequency. The value of the grid-resistance R_g is 10^6 ohms, and the by-pass and coupling capacitors offer negligible reactance.

Assuming that the bandwidth equals the difference between the half-power frequencies, find the ratio of the maximum voltage gain to the bandwidth for the amplifier using:

(a) a triode with an amplification factor μ of 20 and a plate resistance r_p of 10,000 ohms,
(b) a pentode with an amplification factor μ of 1,000 and a plate resistance r_p of 500,000 ohms.

20. A broad-band amplifier of the type shown in Fig. 26 is to be so designed as to have a substantially constant voltage gain over a frequency range from 100 to 150,000 cps. The effective shunt capacitance of the stage when connected to the succeeding stage is 200 $\mu\mu$f. A high-mu triode is employed for which g_m is 1,500 μmhos and r_p is 66,000 ohms.

(a) Determine the values of all resistances, capacitances, and inductances in the circuit.

(b) Plot the voltage gain as a function of frequency.

(c) If the inductor L of Fig. 26 were omitted, show what change would have to be made in the design to satisfy the frequency requirements, and compare the voltage gain in the middle range of frequencies given by the new design with that of (b).

21. Determine for several representative high-frequency pentodes, the limiting value of the gain-bandwidth product in Eq. 60 that is approached as all stray and wiring capacitances approach zero. Express the results in cycles per second and also in cycles per second per dollar of tube cost.

22. A triode with an amplification factor μ of 100 and a plate resistance r_p of 48,000 ohms is used in a resistance-capacitance-coupled voltage-amplifier stage as shown in Fig. 44. The values of the load resistance R_L and the grid resistances R_g are each 10^5 ohms and that of the coupling capacitor C_c is 0.01 μf.

The rectified-power supply for the amplifier delivers a direct voltage of 300 volts with a ripple factor of 0.01. It is connected as shown in the figure so as to supply both the plate and grid-bias voltages.

(a) What is the rms value of the ripple voltage appearing at the amplifier output terminals?

(b) What would the answer to (a) be if the grid-bias voltage were obtained from a separate battery instead of from the rectified-power supply?

23. Assuming the cathode-bias voltage is constant at all frequencies, find as functions of frequency the voltage gains from the input voltage to each of the two output voltages with respect to ground in the self-balancing phase inverter of Fig. 16b. Can the two voltage gains be equal at any frequency? At all frequencies?

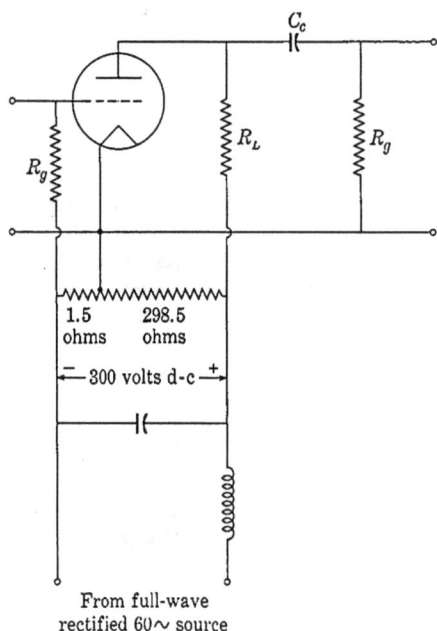

Fig. 44. Circuit for Prob. 22 showing power-supply connections.

24. An amplifier supplied from a rectified-power source has a voltage gain of 10,000 when the a-c supply voltage has the normal value, 120 volts, and a voltage gain of 9,000 when the a-c supply voltage decreases to 108 volts. Negative feedback sufficient to reduce the voltage gain by a factor of 100 when the a-c supply voltage has the normal value is now applied.

(a) What are the new values of the voltage gain for the two values of a-c supply voltage?

(b) By what factor is the percentage variation of the voltage gain with a-c supply voltage reduced?

(c) What is the amount of the feedback in decibels?

25. A resistance-capacitance-coupled Class A_1 amplifier having the circuit shown in Fig. 38a except that a direct-voltage source is substituted for the cathode-bias arrangement has a voltage gain of 20 in the middle range of frequencies and lower and upper half-power frequencies of 50 and 15,000 cps, respectively, when the feedback ratio β is made zero.

(a) Plot to scale the polar diagram of $A\beta$ when β is 10 per cent and indicate a scale of frequency on the diagram.

(b) Find the new half-power frequencies under the conditions of (a).

26. A cascade resistance-capacitance-coupled Class A_1 amplifier composed of two identical stages having fixed grid-bias voltage and a midband voltage gain of -25 each employs 40 db of negative feedback in the middle range of frequencies. To provide that amount of feedback, an appropriate fraction of the output voltage of the second stage is fed back to the input of the first stage through an essentially resistive feedback network.

(a) Sketch the polar plot of the product $A\beta$.

(b) Indicate the points on the polar plot that correspond to the frequencies at which the voltage gain of the internal amplifier, exclusive of the effect of the feedback, is one-half of its maximum value.

(c) Discuss the possibilities of oscillation in this feedback amplifier.

(d) Is the voltage gain of the amplifier reduced by the feedback at all frequencies?

27. The circuit of Fig. 45 is that of an output stage for a direct-coupled amplifier. The supply voltages and resistors are so adjusted that for an input voltage e_{g1} of

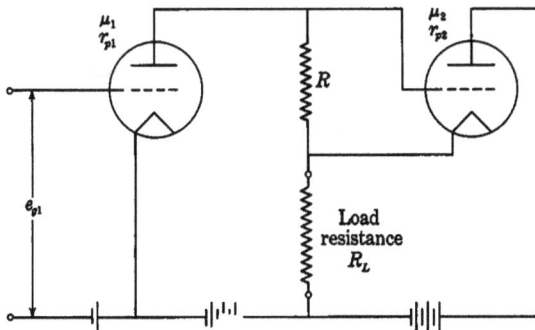

Fig. 45. Circuit diagram for output stage of direct-coupled amplifier of Prob. 27.

zero the current through the load R_L is zero. The tubes operate in the linear regions of their characteristic curves, so that for small input voltages they may be characterized by constant values of amplification factors and plate resistances μ_1, μ_2, r_{p1}, and r_{p2}.

(a) Develop expressions for the Thévenin's-theorem equivalent of the amplifier —that is, the ideal voltage source and internal series resistance that are equivalent to the amplifier as viewed from the load terminals. Note that the quantity R_L should not appear in the expressions for the ideal source voltage and internal series resistance.

(b) What relationship must exist among the parameters in the circuit if the two tubes are to contribute components of equal magnitude to the total load current when an input voltage e_{g1} is applied?

(c) With the conditions of (b) established, what is the over-all transconductance of the amplifier?

28. Show that regardless of the waveform of the input voltage, when the element labeled Z_s is a resistance and that labeled Z_b is a capacitance in Fig. 37b, the instantaneous output voltage approaches the time integral of the instantaneous input voltage as the voltage gain A approaches infinity.

Show that the same is true when the element labeled Z_s is an inductance and that labeled Z_b is a resistance.

29. Show that regardless of the waveform of the input voltage, when the element labeled Z_s is a capacitance and that labeled Z_b is a resistance in Fig. 37b, the instantaneous output voltage approaches the time derivative of the instantaneous input voltage as the voltage gain A approaches infinity.

Show that the same is true when the element labeled Z_s is a resistance and that labeled Z_b is an inductance.

30. Devise a two-stage amplifier circuit comprising positive current feedback and negative voltage feedback combined so as to make the output impedance of the amplifier zero and the voltage gain of the amplifier the same as it would be with both types of feedback eliminated.

CHAPTER X

Amplifiers with Operation Extending beyond the Linear Range of the Tube Characteristic Curves; Class AB, Class B, and Class C Amplifiers

In the amplifiers discussed in previous chapters, the excursion of the operating point is restricted to the approximately linear region of the tube characteristics, and the waveform of the plate current closely approximates that of the grid-signal voltage. The amplitudes of the harmonic-frequency components in the plate current generated by nonlinearity of the tube characteristics are of a lower order of magnitude than is the fundamental-frequency component. Also, in order to maintain the grid current at a value that is negligible in most applications, the grid-bias voltage is adjusted and the grid-signal voltage limited to values that at all times maintain the grid at a negative potential with respect to the cathode. If this precaution is not taken, and if the effective impedance in series with the grid is appreciable, the resulting nonlinear resistance between the grid and cathode may cause waveform distortion of the grid-signal voltage. Because of these limitations, Class A operation suffers the disadvantage that only small plate efficiencies can be realized, as is shown in Art. 16, Ch. VIII.

This chapter considers amplifiers in which the excursion of the operating point extends beyond the linear range of the tube characteristics. The primary objective of such operation is to reduce the ratio of the average value of the plate current to the value of the varying component and thereby to increase the plate efficiency, discussed in Art. 14, Ch. VIII. However, the increase in plate efficiency is accomplished at the expense of a restriction on the applicability, for the harmonic components in the plate current are then large, and the method is useful only in applications in which the circuit connections tend to reduce or eliminate the harmonic components in the output current and voltage.

In general, the two types of circuits in which the distortion components in the output are substantially smaller than those in the tube plate current are those utilizing the push-pull connection and those utilizing the parallel-tuned load. The push-pull connection is applicable over a relatively wide band of frequencies in the low- or audio-frequency range; the tuned-load connection is applicable over a relatively narrow band of frequencies and is used mainly at radio frequencies. A combination of the two types of circuit is sometimes used at radio frequencies. The discussion in this chapter is devoted to these two types.

1. Push-pull Class AB audio-frequency power amplifiers

The essential distinguishing feature in the classification of amplifiers given in Art. 4, Ch. VIII, is the fraction of a cycle during which the plate current of the tube is zero. In the Class AB amplifier, the tube plate current is greater than zero for appreciably more than a half-cycle but is zero for an appreciable fraction of a cycle. Under such conditions of operation, considerable harmonic generation takes place within the tube, and the plate-current waveform is not a close approximation to the grid-signal-voltage waveform. Consequently, single-tube Class AB amplifiers are never used for sound reproduction. Satisfactory reproduction of audio-frequency signals for reception by the ear requires the use of the push-pull connection to eliminate the even harmonics and thereby to reduce the total harmonic generation when Class AB operating conditions are used. The same considerations apply to the audio-frequency Class B amplifiers discussed in the next article.

As an introduction to the operation of push-pull Class AB amplifiers, it is of interest to examine the changes in the shapes of the characteristics of the composite tube and in the paths of operation that take place when the grid-bias voltage is changed in a push-pull amplifier comprising an ideal output transformer and a resistance load, in which (a) a specified plate-supply voltage is maintained and (b) the grid-signal-voltage amplitude is kept equal to the magnitude of the negative grid-bias voltage. The trend of the shapes of the family of composite plate characteristics as the operation is varied from Class A through Class AB to Class B for the same triode tube is shown in Figs. 43 and 45 of Ch. VIII and Figs. 1 and 2 of this chapter. Figures 43 and 45 of Ch. VIII show the change that takes place when the grid-bias voltage is altered from −50 volts in the first figure to −55 volts in the second. The effect of the change in the plate-supply voltage from 240 volts to 250 volts is negligible by comparison. A further change in grid-bias voltage by an equal increment to the value −60 volts has the effect shown in Fig. 1 of this article. Figure 2 in the next article shows the effect of an even larger negative grid-bias voltage.

The grid-bias voltage in Fig. 43, Ch. VIII, is well within the range for Class A_1 operation in accordance with the definition in Art. 4, Ch. VIII, since the instantaneous plate current in an individual tube, shown by the dotted locus, does not become zero for any grid-signal-voltage amplitude smaller than the magnitude of the grid-bias voltage. A somewhat larger grid-signal voltage results in Class A_2 operation. The grid-bias voltage for Fig. 45, Ch. VIII, is the limiting value for

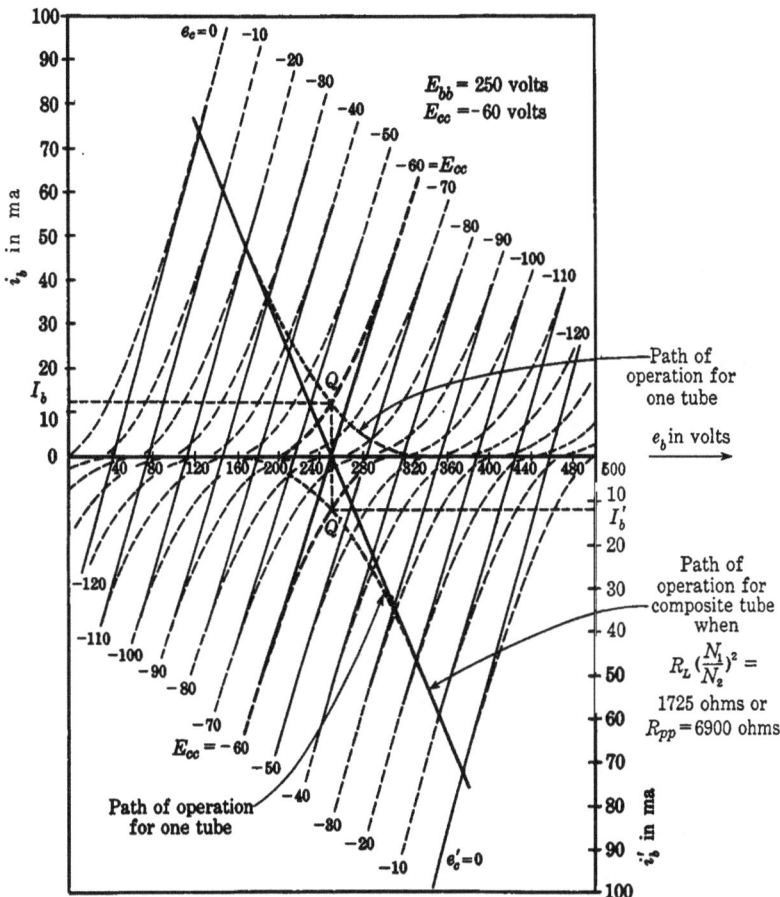

Fig. 1.   Composite characteristics of the two triodes of Fig. 43, Ch. VIII, in a push-pull circuit with a grid-bias voltage giving Class AB operation.

Class A_1 operation, since the instantaneous plate current of an individual tube approaches zero at one time in the cycle as the amplitude of the grid-signal voltage approaches the magnitude of the grid-bias voltage.

For a grid-signal-voltage amplitude equal to the magnitude of the grid-bias voltage, the instantaneous plate currents shown by the dotted loci in Fig. 1 of this article reach zero and remain zero for an appreciable part of the cycle. Hence, by the definitions in Art. 4, Ch. VIII, the operation in Fig. 1 is Class A_1 for small amplitudes such that the plate currents do not become zero; Class AB_1 for large amplitudes up to the value that equals the magnitude of the grid-bias voltage; and Class AB_2 for still larger amplitudes, since the instantaneous grid voltages e_c and e_c' then become positive during the cycle, and the grid current is therefore appreciable. In this discussion the grid current is assumed to be negligible for negative grid voltages but appreciable for positive grid voltages, an assumption that is approximately true in most applications of grid-controlled vacuum tubes.

Although the slope of the composite characteristics changes with the grid-bias voltage, the curves remain practically straight, parallel, and equally spaced for equal increments of the grid voltage, even for push-pull Class AB operation. Therefore the harmonic generation resulting from nonlinearity in the tube is small in the output of a push-pull Class AB amplifier with an output transformer having a coupling coefficient near unity, despite the fact that the individual plate currents in the tubes have a "flat-bottomed" wave containing appreciable harmonics. The push-pull connection and the close coupling between the two sections of the transformer primary winding make the circuit useful for linear amplification even though the excursion of the operating points extends beyond the linear range of the individual characteristics. The values of the harmonic-frequency components in the output current or voltage can be obtained by the methods given in Art. 13, Ch. VIII. However, the higher-order harmonics may be appreciable; hence the more nearly accurate method of plotting waveforms from the operating points on the composite characteristics and analyzing the results by a schedule method, or the method of direct electrical measurement in the circuit, may often be preferable.

In Class AB operation, the quiescent value of the plate current I_b is smaller for a particular quiescent plate voltage than in Class A operation because of the increased magnitude of the negative grid-bias voltage; consequently the quiescent plate voltage for a given zero-grid-signal plate dissipation may be increased. However, when a grid-signal voltage is impressed, the average value of the plate current increases to the value

$$I_{bs} = I_b + I_{p0}, \qquad [1]$$

where I_{p0} is the average value of the component of plate current caused by the grid-signal voltage; that is, I_{p0} is the rectified component of the plate current. The power supplied to the tube by the plate-power supply therefore increases with the grid-signal voltage at the same time that the alternating-current power output to the load increases, whereas in the linear Class A amplifier the power supplied to the tube is constant. The plate dissipation is the difference between the foregoing two amounts of power, and it may either increase or decrease, depending upon the relative rates at which they change. In general, for large negative grid-bias voltages in Class AB operation, the plate dissipation increases with the grid-signal voltage; and, for small negative grid-bias voltages, it decreases as for Class A operation. The plate efficiency for Class AB_1 operation at full output is higher than for Class A_1 operation; thus the maximum power output for a given plate dissipation or given tube is higher for Class AB_1 operation. This condition is perhaps best understood after a study of the analysis of the push-pull Class B amplifier in the next article, because Class AB operation is in many respects intermediate between Class A and Class B. The plate-to-plate load resistance for maximum power output is also higher for Class AB than for Class A operation, as is evident from Fig. 1.

Since the average plate current increases with the grid-signal voltage in a Class AB amplifier, the magnitude of the negative grid-bias voltage also increases if a cathode-bias circuit is used. This effect tends to increase the harmonic generation, because the excursion of the operating point extends farther into the nonlinear region of the plate characteristics. Consequently, the bias resistor must be adjusted to give a grid-bias voltage near the limiting Class A_1 value for small grid-signal voltages in order to prevent the bias voltage from becoming too large for large grid-signal voltages. Also, since the sum of the two plate currents is not constant in the push-pull Class AB amplifier, a by-pass capacitor must be used around the cathode-bias resistor to prevent degeneration, although it is unnecessary in a push-pull linear Class A amplifier.

In order to obtain greater power output without an increase in plate-supply voltage, the Class AB amplifier with a grid-signal-voltage amplitude sufficient to extend the operation to positive values of the grid voltage e_c is sometimes used. The grid current in this Class AB_2 amplifier is then appreciable, and the preceding amplifier stage must furnish appreciable power. In order to prevent harmonic generation caused by the nonlinear relation between the grid current and grid voltage, the preceding stage usually has a step-down output transformer to reduce its apparent internal impedance as a source of grid-signal voltage for the Class AB stage.

The problem of determining the harmonic generation and optimum operating conditions in the Class AB amplifier is not readily solved by analytical means, and graphical or direct electrical-measurement methods are necessary for an accurate solution.

2. PUSH-PULL CLASS B AUDIO-FREQUENCY POWER AMPLIFIERS

When the magnitude of the negative grid-bias voltage is increased to a value that reduces the quiescent plate currents in the individual

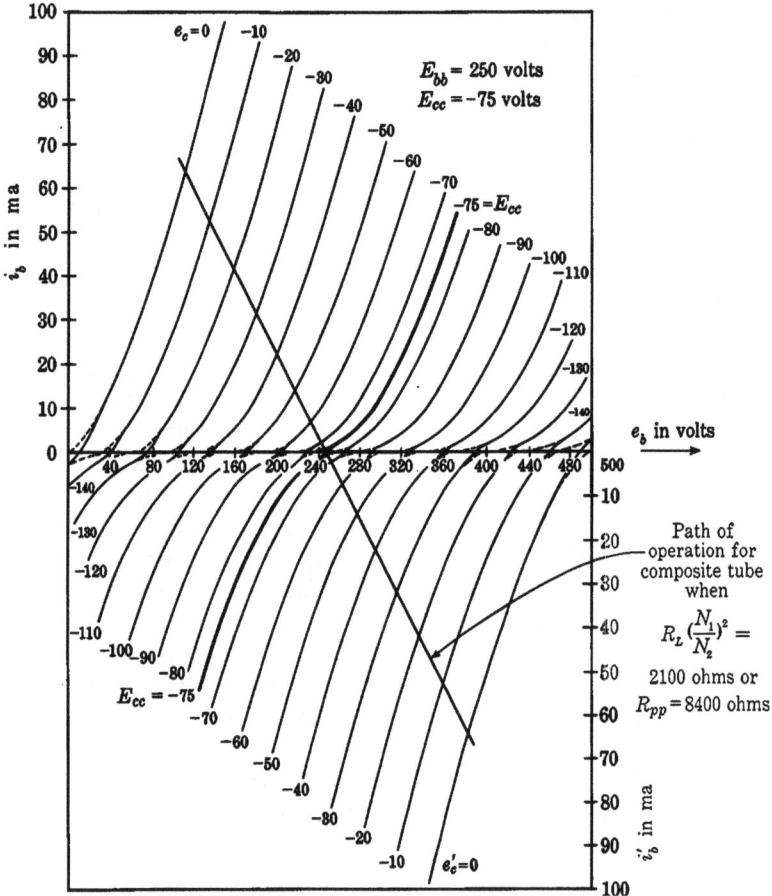

Fig. 2. Composite characteristics of the two triodes of Fig. 43, Ch. VIII, in a push-pull circuit with a grid-bias voltage giving Class B operation.

tubes practically to zero—that is, to the cut-off value—the composite plate characteristics for the circuits in Figs. 35 and 40 of Ch. VIII have the form shown in Fig. 2 of this chapter. The composite characteristics follow one or the other of the individual tube characteristic curves over essentially their entire range and are appreciably curved in the region of small grid-signal-voltage amplitudes. Their average slope is roughly half that of the curves for Class A operation shown in

Fig. 45, Ch. VIII, and is smaller than that of the curves for Class AB operation shown in Fig. 1. Hence the average apparent plate resistance of the composite tube supplying one-half the primary winding of the output transformer (see Art. 19c, Ch. VIII) is larger than for Class A or Class AB operation.

The locus of the operating point for an individual tube is along the straight path of operation for the composite tube in Fig. 2 for positive values of grid-signal voltage (which tend to increase one individual plate current) and along the e_b axis for negative values of grid-signal voltage, since the individual plate currents are zero at quiescent conditions. (See Figs. 43 and 45, Ch. VIII, and Fig. 1 of this chapter for the trend of the shape of the paths of operation for individual tubes as the magnitude of the negative grid-bias voltage is increased.) In the push-pull circuit with the grid-bias voltage adjusted to the cut-off value, the plate current in the first tube is zero while the second is conducting, and vice versa. With a sinusoidal input voltage e_1, therefore, the individual plate currents are approximately a series of half-sinusoids with zero current in the alternate half-cycles, and the operation is Class B as defined in Art. 4, Ch. VIII. A similarity to the full-wave rectifier is evident in the output-transformer currents, but power flows in the direction opposite to that in a rectifier transformer; that is, from the direct-current supply toward the alternating-current load, rather than from an alternating-current supply to a direct-current load.

The percentage harmonic generation caused by nonlinearity is greatest for small grid-signal amplitudes, since the characteristic curves have their greatest curvature in the region of operation for these amplitudes. The determination of the percentage harmonic generation from the graphical characteristic curves is best made through plotting the waveforms and making a Fourier analysis by a schedule method, instead of using the approximate methods given in Art. 13, Ch. VIII, since the high-order harmonics are appreciable.

An approximate analysis of the amplifier power output, plate dissipation, and plate efficiency can be made as follows on the assumptions that the input voltage is sinusoidal, the output transformer is ideal, and the plate characteristics of the individual tube are a linear family of straight parallel lines equally spaced for equal increments of grid voltage. The composite characteristics then have the form shown in Fig. 44, Ch. VIII. The composite-tube current i_d is sinusoidal, and the individual tube plate currents are a series of half-sinusoids with zero current in the alternate half-cycles. During a cycle,

$$I_{dm} = I_{bm}, \qquad [2]$$

where I_{dm} and I_{bm} are the maximum values of the current in the composite tube and an individual tube, respectively. Since the composite-tube current is sinusoidal, and the load as viewed from the tubes into the transformer primary winding is a resistance, the power output of the *two* tubes is

$$P_{ac} = \left(\frac{I_{dm}}{\sqrt{2}}\right)\left(\frac{E_{pm}}{\sqrt{2}}\right) = \tfrac{1}{2}I_{bm}E_{pm}, \qquad \blacktriangleright[3]$$

where E_{pm} is the amplitude of the varying component of the plate voltage of one of the tubes—that is, the voltage across one-half of the transformer winding, as shown in Fig. 40, Ch. VIII.

The average value of the plate current of an individual tube for the half-sinusoid waveform is

$$I_{bs} = \frac{1}{2\pi}\int_0^{2\pi} i_b\, d(\omega t) = \frac{1}{2\pi}\int_0^{\pi} I_{bm}\sin \omega t\, d(\omega t) \qquad [4]$$

$$= \frac{I_{bm}}{\pi}, \qquad \blacktriangleright[5]$$

which is identical with the expression in an earlier chapter for the half-wave rectifier. Since the ideal transformer has zero resistance in its windings, the considerations of Art. 17, Ch. VIII, are applicable, and the direct-current power input to the tubes, P_{bs}, is the product of the plate-supply voltage and the average value of the current through it, the latter being the sum of the individual tube currents, $2I_{bs}$; thus

$$P_{bs} = 2E_{bb}I_{bs} = \frac{2}{\pi}E_{bb}I_{bm}, \qquad \blacktriangleright[6]$$

where P_{bs} is the power supplied to the *two* tubes; the plate efficiency is therefore

$$\eta_p = \frac{P_{ac}}{P_{bs}} = \frac{\tfrac{1}{2}E_{pm}I_{bm}}{\frac{2}{\pi}E_{bb}I_{bm}} = \frac{E_{pm}}{E_{bb}} \times 78.5 \text{ per cent.} \qquad \blacktriangleright[7]$$

The plate efficiency in Eq. 7 increases with E_{pm} and therefore with the grid-signal voltage. The amplitude E_{pm} is equal to half the projection of the excursion of the operating point on the e_b axis in Fig. 2. This quantity has an ultimate limit equal to the plate-supply voltage E_{bb}, since the plate voltage e_b cannot become negative; hence the maximum plate efficiency for the Class B amplifier is 78.5 per cent. However, in order that E_{pm} equal E_{bb}, either the path of the operating point must approach the e_b axis or the grid voltage for the individual tubes must become highly positive during the cycle. The first condition implies a very high load resistance and a low power output; the second implies a relatively large amount of power to supply the grid-signal voltage. Neither condition represents practical operation. In practice, the grid voltage of Class B amplifiers is ordinarily driven somewhat positive during the cycle in order to realize a high efficiency, but because of the magnitude of the grid-driving power required the plate efficiency of Class B audio-frequency power amplifiers is seldom made to exceed about 60 per cent. Even though it is not the theoretical maximum, this represents a considerable increase in available power output from a particular tube over Class A operation, for, since the plate efficiency of any amplifier is

$$\eta_p = \frac{P_{ac}}{P_{bs}} = \frac{P_{ac}}{P_{ac} + P_p},$$ [8]

the alternating-current power output is expressible in the form

$$P_{ac} = P_p \frac{\eta_p}{1 - \eta_p},$$ ▶[9]

where P_p is the plate dissipation defined by Eq. 151, Ch. VIII. There-fore, with Class A operation at, say, 20 per cent plate efficiency, the available power output from a pair of tubes is one-fourth the sum of their allowable plate dissipations; whereas with Class B operation at, say, 60 per cent plate efficiency, it is one and one-half times that sum —an increase in power output by a factor of six.

The direct current through the plate-voltage supply, $2I_{bs}$, increases directly with the output current and grid-signal voltage. The power output may be expressed in terms of $2I_{bs}$ and the plate-to-plate resistance R_{pp} by the following approximate analysis. The alternating-current power output is

$$P_{ac} = I_2{}^2 R_L,$$ [10]

where I_2 is the effective value of the current and R_L is the load resistance on the load side of the output transformer. Since in the ideal transformer,

$$i_d = \left(\frac{N_2}{N_1}\right) i_2,$$ [11]

$$I_2 = \left(\frac{N_1}{N_2}\right) \frac{I_{dm}}{\sqrt{2}} = \left(\frac{N_1}{N_2}\right) \frac{I_{bm}}{\sqrt{2}},$$ [12]

from Eq. 2. Substitution of Eq. 12 in Eq. 10 gives

$$P_{ac} = \left(\frac{I_{bm}}{\sqrt{2}}\right)^2 \left(\frac{N_1}{N_2}\right)^2 R_L,$$ [13]

and substitution of Eq. 191, Ch. VIII, gives

$$P_{ac} = \frac{I_{bm}^2}{2} \frac{R_{pp}}{4}.$$ [14]

By means of Eq. 5, Eq. 14 is expressible as

$$P_{ac} = \frac{\pi^2}{32} (2I_{bs})^2 R_{pp} = 0.308(2I_{bs})^2 R_{pp}.$$ [15]

Thus the alternating-current power output is approximately three-tenths of the direct-current power that would be developed by the plate-voltage-supply current, as indicated by a direct-current ammeter, if its path were the plate-to-plate load resistance.

Since the direct component of current through the plate-voltage supply increases from zero with increasing grid-signal voltage, the method previously described of obtaining the grid-bias voltage from a cathode-bias resistor and by-pass capacitor is not applicable to the Class B audio-frequency power amplifier. A fixed grid-bias voltage from a rectifier or battery is necessary. A plate-voltage supply with especially small voltage regulation is also necessary to prevent fluctuation of the plate-supply voltage as the grid-signal voltage changes.

Pentode vacuum tubes may be used in push-pull Class B amplifiers. The method of graphically constructing their composite characteristics is identical with that for triodes. However, the resulting composite characteristic curves have knees similar to those of the characteristics of the individual tubes, and, if the harmonic components in the output generated by nonlinearity in the tube are to be small, it is essential that the path of operation be confined to the approximately linear region.

For maximum power output from triodes, the slope of the load line for the composite tube should correspond roughly to the plate resistance of a single tube, as is illustrated in Fig. 2. The plate-to-plate resistance (which is four times the composite-tube plate resistance) should therefore be roughly four times the plate resistance of one tube. However, for pentode vacuum tubes, the amount of harmonic generation increases to excessive values as the plate-to-plate resistance is made larger than the value for which the path of the operating points intersects the curves near the knee; hence, when they are used, the plate-to-plate resistance should not be adjusted arbitrarily to equal four times the plate resistance of one tube.

The selection of the best plate-to-plate resistance must be made with due regard to harmonic generation, power output, plate dissipation, and grid-driving power (among other quantities). The grid-driving power increases with the grid current and the amplitude of the grid-signal voltage. To prevent excessive harmonic generation due to non-linearity in the *grid* circuit, the impedance as viewed from the grid-to-cathode terminals of the Class B stage back toward the previous amplifier stage (called the *driver stage*) must be small compared with the effective impedance as viewed from the same terminals into the Class B stage. Hence the grid-signal voltage of the Class B audio-frequency power amplifier is usually supplied from the previous amplifier stage through a step-down transformer. Increased grid-driving power makes necessary a greater step-down ratio in the input trans-former, which results in a lower effective voltage gain in the previous stage. Thus, in a deter-mination of the optimum plate-to-plate load resistance, the per-formance of the previous stage should be considered, and graphical methods for finding the optimum load resistance under specified limiting con-ditions of grid-driving power and plate dissipation are ex-pedient.[1]

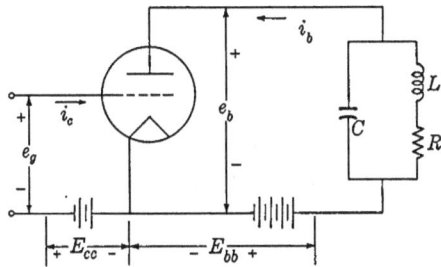

Fig. 3.   Basic circuit diagram of a tuned amplifier.

In general, the Class B audio-frequency power amplifier gives a large power output relative to the plate dissipation. A plate-voltage supply with good regulation is necessary, however, and the tubes must have essentially identical characteristics in order to prevent excessive harmonic generation. Since the grid-driving power with Class AB_2 operation is generally smaller than with Class B operation, and since cathode-bias methods are practical, Class AB_2 is sometimes considered to be preferable to Class B operation despite its smaller power output and plate efficiency.

3. Tuned Class B Power Amplifiers

The basic circuit diagram of an amplifier with a parallel-tuned load is shown in Fig. 3. The circuit is the same for any class of amplifier; the essential difference in Class A, B, or C operation is in the value of the grid-bias voltage relative to the plate-supply voltage. For Class B and Class C operation, larger grid-bias voltages are employed; hence the harmonics generated in the plate current are comparable in amplitude

---

[1] H. J. Reich, *Theory and Applications of Electron Tubes* (2nd ed.; New York: McGraw-Hill Book Company, Inc., 1944), 269–278.

to the fundamental component. Nevertheless, if the load circuit is tuned to the frequency of the grid-signal voltage and has a large Q_0—say 10 or more—the harmonic components in the voltage across the load are small compared with the fundamental component, because the impedance of the load at each harmonic frequency is small compared with the resonant impedance, as is shown in Art. 9, Ch. IX. Thus the effect of the harmonic components of the tube plate current is suppressed by the characteristics of the load, and the usefulness of the amplifier *depends* on the fact that the load has a large Q_0.

For Class B operation, the grid-bias supply voltage E_{cc} is made negative by an amount sufficient to reduce the plate current to zero for zero grid-signal voltage e_g. Consequently, for a periodic grid-signal voltage, the plate-current waveform is a series of pulses during the positive half-cycles of grid-signal voltage alternating with current-free periods during the negative half-cycles. A quantitative analysis of the performance of this amplifier is difficult unless several simplifying assumptions are made. The semi-graphical analysis given subsequently in Art. 4a for Class C amplifiers is applicable also to the Class B amplifier and gives good results over a wide range of operation. The semi-graphical analysis, however, requires at the start a selection of certain operating conditions in order to give a solution that includes such desired results as the power output, plate dissipation, and plate efficiency. It is therefore suitable only as a guide in a cut-and-try process of solving for the best conditions of operation for a given tube.

If further simplifying assumptions are made in addition to those required for the semi-graphical solution, an approximate but useful analytical solution[2] for the performance of the Class B tuned amplifier is readily obtained. It has the advantage over the semi-graphical analysis that in the results such desired quantities as the power output, plate current, and plate efficiency are expressed in terms of quantities known, limited, or subject to adjustment, so that the possibilities for application of a particular tube are quickly investigated. The approximate analytical solution is given in this article, and the semi-graphical method is given in Art. 4a.

The object of the following analysis is to derive explicit expressions for the power output and plate efficiency in terms of the usual rated quantities for the tube, under the assumption that the tube is utilized to best advantage as a Class B amplifier with a high-Q_0 tuned load. Since the analysis is somewhat long and involved, it is perhaps desirable to outline briefly the steps taken, before discussing them in detail.

[2] W. L. Everitt, *Communication Engineering* (2nd ed.; New York: McGraw-Hill Book Company, Inc., 1937), 582–590.

The first step in the analysis is the development of a relatively simple approximate analytical expression for the tube characteristics in order that the remaining steps may be carried out analytically rather than graphically. The second step consists of the formulation of a relationship between the varying components of the tube currents and voltages by means of, first, a relationship between the tube supply voltages inherent in the definition of the Class B amplifier, and, second, an assumption regarding the behavior of the tuned load. The third step is the development of a linear equivalent circuit for the fundamental components of the currents and voltages in the circuit; this is derived from a deduction as to the phase relation of these components. The fourth step is the utilization of the equivalent circuit to arrive at expressions for the voltage gain, power output, plate efficiency, plate dissipation, and other related quantities in terms of the voltages and parameters in the circuit. At this point the straightforward circuit analysis is complete, but the engineering limitations imposed by the tube ratings are yet to be considered.

The fifth step is based on a conclusion as to the operating conditions that utilize the tube to best advantage; it consists of a solution for the resonant impedance of the tuned circuit that is required to satisfy those operating conditions. The final step is the substitution of this value of resonant impedance in the expressions previously obtained, whereupon the desired expressions for the power output, plate efficiency, and related quantities are obtained in terms of the tube coefficients and the rated values of the plate dissipation and the plate-supply voltage.

In order that the need for graphical methods may be eliminated, it is necessary as a first step in the analysis to express the tube characteristics in analytical form. As is shown in Art. 7, Ch. IV, the relationship between the total instantaneous values of the plate current and the grid and plate voltages of a triode may be expressed approximately as

$$i_b = g_m \left(e_c + \frac{e_b}{\mu} \right), \qquad \text{for} \left(e_c + \frac{e_b}{\mu} \right) \geqq 0. \qquad \blacktriangleright [16]$$

In Fig. 4, i_b is plotted as a function of the voltage $e_c + (e_b/\mu)$, called the *control voltage* (see Art. 3, Ch. XI, and Art. 9, Ch. XII).

The second step in the analysis depends on the fact, shown in Art. 9, Ch. IX, that the impedance of a parallel-tuned circuit having a large Q_0 is negligible at harmonics of the resonant frequency in comparison with its impedance at the resonant frequency. Since the plate current in the Class B amplifier is a pulsating wave, it contains appreciable harmonics, and the voltage drop in the plate load may be

expressed as a Fourier series, each term of which is the product of a
current harmonic and the impedance of the load at the corresponding
harmonic frequency. However, this analysis assumes that the parallel-
tuned load has a Q_0 sufficiently large to make the impedance at the
harmonic frequencies negligible compared with its value at the
resonant frequency; consequently, if the grid-signal voltage is
sinusoidal and the load is tuned to the frequency of the grid-signal
voltage, the voltage drop across the load is also practically sinusoidal
and of the same frequency.* The voltage drop in the plate load caused
by the steady component of plate current is considered negligible
because the direct-current resistance of
the inductor is ordinarily very small.

When the grid-signal voltage is zero,
the value of $e_c + (e_b/\mu)$ is

$$e_c + \frac{e_b}{\mu} = E_{cc} + \frac{E_{bb}}{\mu}. \qquad [17]$$

For Class B operation, the grid-bias
supply voltage E_{cc} is adjusted to the
cut-off value

$$E_{cc} = -\frac{E_{bb}}{\mu}, \qquad \blacktriangleright[18]$$

Fig. 4. Approximate representation of triode characteristics.

so that the control voltage $e_c + (e_b/\mu)$ in Eqs. 16 and 17 is zero for zero
grid-signal voltage. Hence the plate current, expressed by Eq. 16, is
zero for zero grid-signal voltage.

In general, then, for any grid-signal voltage e_g, with the grid-bias
voltage adjusted to the Class B value given in Eq. 18,

$$e_c + \frac{e_b}{\mu} = (E_{cc} + e_g) + \left(\frac{E_{bb} + e_p}{\mu}\right) = e_g + \frac{e_p}{\mu}; \qquad [19]$$

and, from Eqs. 16 and 19,

$$i_b = g_m \left(e_g + \frac{e_p}{\mu}\right), \qquad \text{for} \left(e_g + \frac{e_p}{\mu}\right) \geqq 0. \qquad \blacktriangleright[20]$$

The third step in the analysis follows from a qualitative consideration of the phase relations in the circuit. The plate voltage e_p is
sinusoidal because of the tuned circuit; and, if e_g is sinusoidal at the
resonant frequency of the tuned circuit, the sum $e_g + (e_p/\mu)$ is also

---

* If the load is tuned to a particular harmonic of the grid-signal-voltage frequency, the
only component of appreciable amplitude in the voltage drop across the load is the one
having that particular harmonic frequency, and the circuit becomes a frequency *doubler*,
tripler, or other *multiplier*.

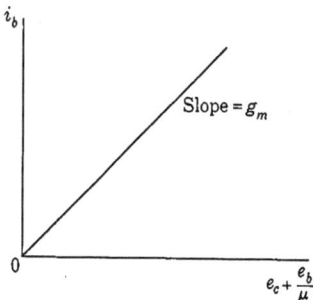

sinusoidal regardless of the phase relation between the grid and plate voltages. The waveform of the current i_b given by Eq. 20 consists therefore of a series of alternate half-sinusoid and zero-current half-cycles, as shown in Fig. 5. If the origin for time is chosen at the maximum point of one of the current pulses, the current waveform is an even function[3] and is representable by a Fourier series comprising

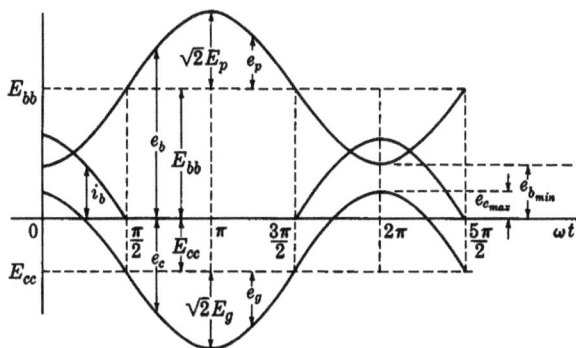

(a)

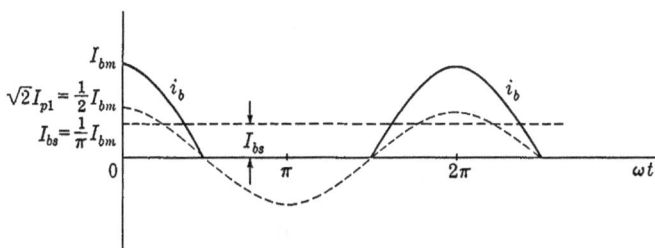

(b)

Fig. 5. Waveforms in the Class B tuned amplifier.

only cosine terms. Therefore the fundamental component of the plate current reaches its maximum value at the above selected reference point of time. It is assumed that the circuit is so tuned that its impedance at the fundamental frequency is a pure resistance R_t. Hence, if I_{p1} denotes the fundamental component of the plate current,

$$E_p = -I_{p1}R_t,$$ [21]

[3] P. Franklin, *Differential Equations for Electrical Engineers* (New York: John Wiley & Sons, Inc., 1933), 64.

in accordance with the selected reference directions in the circuit, where R_t is the impedance of the tuned circuit at resonance given by Eq. 98, Ch. IX. Thus e_p reaches its negative maximum value at the particular instant when i_b reaches its positive maximum value. In order that this may be true in Eq. 20, e_g must reach its positive maximum value at the same instant. The grid-signal voltage e_g is therefore 180 degrees out of phase with the plate voltage e_p, and the phase relations and waveforms in the circuit are those shown in Fig. 5a.

The plate current i_b may be expressed in terms of its maximum value I_{bm} as

$$i_b = I_{bm} \cos \omega t, \quad \text{for } (2n - \tfrac{1}{2})\pi \leq \omega t \leq (2n + \tfrac{1}{2})\pi, \qquad [22]$$

$$i_b = 0, \qquad \text{for } (2n - \tfrac{3}{2})\pi \leq \omega t \leq (2n - \tfrac{1}{2})\pi, \qquad [23]$$

where n is zero or any integer. The average value of the plate current I_{bs} is

$$I_{bs} = \frac{1}{2\pi} \int_0^{2\pi} i_b \, d(\omega t) = \frac{1}{\pi} \int_0^{\pi/2} I_{bm} \cos \omega t \, d(\omega t); \qquad [24]$$

thus,

$$I_{bs} = I_{bm}/\pi. \qquad \blacktriangleright[25]$$

The value of I_{bs} is the same as that obtained in Art. 2, since the same current waveform is encountered. By Fourier analysis, the amplitude of the fundamental component of the plate current, $\sqrt{2}I_{p1}$, is

$$\sqrt{2}I_{p1} = \frac{1}{\pi} \int_0^{2\pi} i_b \cos \omega t \, d(\omega t) = \frac{2}{\pi} \int_0^{\pi/2} I_{bm} \cos^2 \omega t \, d(\omega t); \qquad [26]$$

hence

$$\sqrt{2}I_{p1} = \tfrac{1}{2} I_{bm}, \qquad \blacktriangleright[27]$$

where I_{p1} is the effective value of the fundamental component of the plate current. These components of i_b are plotted in Fig. 5b and are similar to those in the half-wave rectifier.

The expressions for the grid-signal voltage e_g and the plate voltage e_p are

$$e_g = \sqrt{2}E_g \cos \omega t \qquad [28]$$

and

$$e_p = -\sqrt{2}E_p \cos \omega t, \qquad [29]$$

where E_g and E_p are the respective effective values. Substitution of Eqs. 22, 28, and 29 in Eq. 20 gives

$$I_{bm} = \sqrt{2}g_m \left(E_g - \frac{E_p}{\mu} \right). \qquad [30]$$

From Eq. 21,

$$I_{p1} = E_p/R_t ,$$ [31]

and substitution of Eqs. 27 and 31 in Eq. 30 gives

$$I_{p1} = \frac{\mu E_g}{2r_p + R_t} .$$ ▶[32]

From Eqs. 25, 27, and 32,

$$I_{bs} = \frac{2\sqrt{2}}{\pi} \frac{\mu E_g}{2r_p + R_t} .$$ ▶[33]

By the foregoing method, the two components of the plate current are expressed in terms of the grid-signal voltage and the tube and circuit parameters. In the Class B tuned amplifier, the average value of the plate current in Eq. 33 is directly proportional to the amplitude of the grid-signal voltage. The effective value of the fundamental component of plate current in Eq. 32 is also

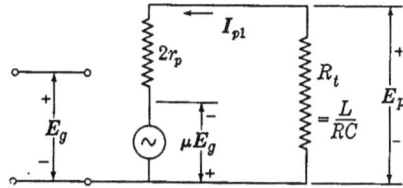

Fig. 6.   Equivalent circuit of the Class B tuned amplifier for current and voltage components at the resonant frequency.

directly proportional to the grid-signal voltage, and this fact implies that the circuit is linear for the fundamental components. Equation 32 shows that as far as the fundamental components of currents and voltages are concerned, the circuit is equivalent to that shown in Fig. 6, where the assigned positive reference directions are the same as those for the Class A amplifier. The equivalent circuit for the Class B tuned amplifier differs from that for the Class A tuned amplifier only in the fact that the internal resistance of the tube is doubled if fundamental component quantities only are considered.

As a fourth step in the analysis, certain expressions may be derived to correspond with the equivalent circuit. The alternating plate voltage and the voltage gain at the resonant frequency are

$$E_p = \mu E_g \frac{R_t}{2r_p + R_t},$$ [34]

from which

$$A = \mu \frac{R_t}{2r_p + R_t} .$$ ▶[35]

The fact that the voltage gain is constant for resonant-frequency components, regardless of their amplitude, is important in the application of Class B tuned amplifiers, because it means that a sinusoidal voltage of that frequency whose amplitude is slowly varying—that is, a modulated sinusoidal voltage—can be amplified without serious distortion. In practice, deviations from this linear relation may occur, but, by proper adjustment of the amplifier, a constant voltage gain over a wide range of grid-signal voltage can be obtained. Amplifiers so adjusted are frequently called *linear Class B amplifiers.*

The direct-current power input to the plate P_{bs}, the alternating-current power output to the load P_{ac}, the plate dissipation P_p, and the plate efficiency η_p are given in terms of the fundamental component of plate current I_{p1} by the following relations:

$$P_{bs} = E_{bb}I_{bs} = \frac{2\sqrt{2}}{\pi} E_{bb}I_{p1} ,$$

▶[36]

$$P_{ac} = I_{p1}{}^2 R_t ,$$

▶[37]

$$P_p = P_{bs} - P_{ac} = \frac{2\sqrt{2}}{\pi} E_{bb}I_{p1} - I_{p1}{}^2 R_t ,$$

▶[38]

$$\eta_p = \frac{P_{ac}}{P_{bs}} = \frac{\pi}{2\sqrt{2}} \frac{I_{p1}R_t}{E_{bb}} = \frac{\pi}{2\sqrt{2}} \frac{E_p}{E_{bb}}$$

▶[39]

$$= \frac{\pi}{4} \frac{\sqrt{2}E_p}{E_{bb}} = 78.5 \frac{\sqrt{2}E_p}{E_{bb}} \text{ per cent.}$$

▶[40]

Hence the Class B tuned amplifier, like the Class B push-pull amplifier (see Eq. 7), has a plate efficiency that increases directly with the grid-signal voltage and approaches 78.5 per cent as the amplitude of the plate-voltage variation $\sqrt{2}E_p$ approaches the plate-supply voltage. However, the other limitations discussed subsequently make it impractical to realize the full theoretical efficiency.

The fifth step in the analysis follows from a consideration of the range of operating conditions within which the approximate representation of the tube characteristics in Fig. 4 is satisfactory. As the grid-signal-voltage amplitude is increased, the alternating component of plate voltage E_p increases, and a condition is finally reached at which

$$e_{c_{max}} = e_{b_{min}}$$

[41]

in the waveforms shown in Fig. 5a. Further increase of grid-signal-voltage amplitude results in the grid's becoming *positive* with respect to the *plate* during a portion of the cycle. Under these conditions, the linear relation in Eq. 20 no longer holds,[4] an increasing fraction of the

[4] W. G. Dow, *Fundamentals of Engineering Electronics* (2nd ed.; New York: John Wiley & Sons, Inc., 1952), 76–78.

cathode current is diverted to the grid, secondary electrons are attracted from the plate to the grid, and, instead of increasing, the plate current decreases with increasing grid-signal voltage. The foregoing expressions for the circuit performance then no longer hold. The power output ceases to increase with the square of the grid-signal voltage, and the grid current and grid-driving power increase rapidly. The power amplification tends to decrease, the plate-current waveform tends toward a double-peaked curve, and the percentage of harmonics in the plate-current waveform increases. This so-called overexcited condition, which perhaps can be understood more readily from the semi-graphical analysis of the next article, is a condition of diminishing returns and constitutes a limit on the power output and plate efficiency of the amplifier. In practice, therefore, the condition expressed in Eq. 41 is rarely exceeded, and often the ratio of $e_{b_{min}}$ to $e_{c_{max}}$ is maintained greater than 2 or 3.

For the condition expressed by Eq. 41,

$$E_{cc} + \sqrt{2}E_g = E_{bb} - \sqrt{2}E_p .$$ [42]

Elimination of E_{cc} and E_p in Eq. 42 by means of Eqs. 18 and 34 gives

$$E_g = \frac{E_{bb}}{\sqrt{2}}\left(1 + \frac{1}{\mu}\right)\frac{2r_p + R_t}{2r_p + (\mu + 1)R_t}.$$ ▶[43]

For optimum power output at a given plate-supply voltage E_{bb} and resonant-circuit resistance R_t, then, the grid-signal voltage should have the value given by Eq. 43. If this value is substituted in Eqs. 32 and 33, the results are:

$$I_{bs} = \frac{2}{\pi}\frac{(\mu + 1)E_{bb}}{2r_p + (\mu + 1)R_t},$$ ▶[44]

$$I_{p1} = \frac{1}{\sqrt{2}}\frac{(\mu + 1)E_{bb}}{2r_p + (\mu + 1)R_t}.$$ ▶[45]

When the value from Eq. 45 is substituted in Eqs. 36 through 39, the important operating conditions are determined.

Certain of the quantities in the expressions above—such as μ and r_p (or g_m)—are limited or determined by the tube. The maximum value of the plate voltage e_b is approximately twice the plate-supply voltage E_{bb} under the optimum conditions; thus the allowable maximum value of E_{bb} is limited to about one-half the limiting value of e_b set by insulation or gaseous-ionization and cathode-bombardment considerations. Also the plate dissipation P_p is limited to the value of power that can be radiated or conducted from the plate without exceeding a temperature that will cause damage or shorten the life of the tube. It is therefore desirable to express in terms of μ, r_p, P_p, and E_{bb} for the tube the results that describe the circuit operation.

The one additional quantity upon which all the other results depend is R_t. Thus, if R_t for the optimum conditions is expressed in terms of μ, r_p, P_p, and E_{bb}, the tube performance is then readily expressed in terms of them. From Eqs. 38 and 45, the plate dissipation is

$$P_p = \frac{2}{\pi} \frac{(\mu+1)E_{bb}{}^2}{2r_p + (\mu+1)R_t} - \frac{1}{2} \frac{(\mu+1)^2 E_{bb}{}^2 R_t}{[2r_p + (\mu+1)R_t]^2}. \qquad [46]$$

When this expression is rearranged, the following quadratic equation is obtained:

$$R_t{}^2 + \left[\frac{4r_p}{\mu+1} - \frac{E_{bb}{}^2}{P_p}\left(\frac{2}{\pi} - \frac{1}{2}\right)\right]R_t + \left[\frac{4r_p{}^2}{(\mu+1)^2} - \frac{E_{bb}{}^2}{P_p}\frac{4r_p}{\pi(\mu+1)}\right] = 0. \quad \blacktriangleright[47]$$

With μ, r_p, E_{bb}, and P_p as known data, the solution of Eq. 47 gives a value of R_t, which, when operation is subject to the condition that $e_{c_{max}}$ equals $e_{b_{min}}$, *must* be provided by the tuned plate-load circuit in order to insure maximum power output for the prescribed values of plate-supply voltage and plate dissipation. With the tuned circuit designed for this value of R_t, the value of I_{p1} from Eq. 45, substituted in Eqs. 36, 37, and 39, gives the power output, power input, and plate efficiency; and the final step in the analysis is then complete.

Substitution of Eq. 45 in Eq. 39 gives the plate efficiency in terms of the tube and resonant load resistances.

$$\eta_p = 78.5 \frac{R_t}{\dfrac{2r_p}{\mu+1} + R_t} \quad \text{per cent.} \qquad [48]$$

This equation indicates that, when a tube is operated with a grid-signal voltage satisfying the optimum condition expressed by Eq. 41, its plate efficiency depends only on the value of the resonant load resistance R_t and the tube coefficients. As R_t is made large compared with $2r_p/(\mu+1)$, which is approximately $2/g_m$, the plate efficiency approaches 78.5 per cent. The power output is then relatively small, but it may be increased through increasing the plate-supply voltage. In general, to utilize the tube to full advantage, the resonant load resistance given by Eq. 47 is required, and the plate efficiency may be considerably lower than 78.5 per cent.

The foregoing analysis is valuable as a means of determining approximately the operation of a Class B tuned amplifier. Because of the assumption regarding the linearity of the tube characteristics, the results may be appreciably in error, especially when operation is near the overexcited condition. As a check on the results, it is best to apply the semi-graphical analysis of the following article, which does not involve the assumption of linear tube characteristics.

4. TUNED CLASS C POWER AMPLIFIERS

If the magnitude of the negative grid-bias voltage is changed from the value for Class A operation to the larger value required for Class B operation, the plate efficiency and power output for a given tube increase, as is shown in the previous articles of this chapter. When the

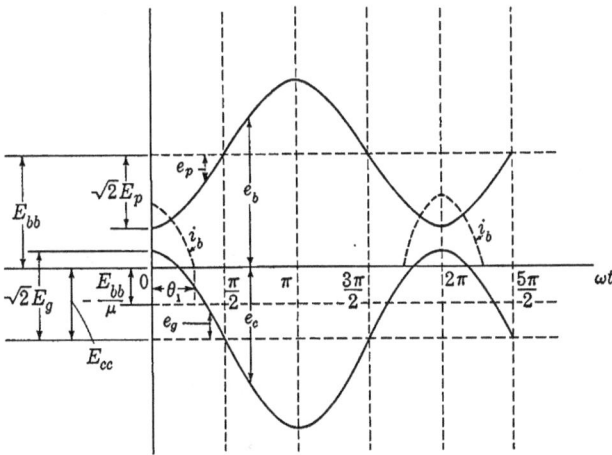

Fig. 7.   Waveforms in a Class C amplifier.

magnitude of the negative grid-bias voltage is increased beyond the cut-off value used in Class B operation, further increases in plate efficiency and power output are achieved, and the resulting operation is useful for certain applications. The waveforms in the tuned amplifier of Fig. 3 for the increased negative grid-bias voltage are sketched in Fig. 7. A sinusoidal grid-signal voltage e_g and a load circuit having a high Q_0 and tuned to resonate at the grid-signal frequency are assumed; hence, as is explained in Art. 3, the varying component of plate voltage e_p is sinusoidal and 180 degrees out of phase with the grid-signal voltage. Since the grid-bias voltage has a negative value such that

$$|E_{cc}| > E_{bb}/\mu, \tag{49}$$

the plate current is zero for zero or negative grid-signal voltages e_g, and is also zero for positive values of e_g so small that e_c remains less than $-e_b/\mu$. The waveform of the plate current then consists of pulses during intervals of less than 180 degrees of the cycle. The operation, by the definitions of Art. 4, Ch. VIII, is therefore Class C amplification. There is no output for small grid-signal voltages, because the plate current is zero; thus the amplitude of the output voltage is not proportional to that of the input voltage. The Class C amplifier is therefore not suitable for amplifying a grid-signal voltage of varying amplitude and is ordinarily used only for amplifying a signal of fixed amplitude.

An important use of the Class C amplifier is in radio communication as a plate-modulated amplifier, as discussed in Ch. XII. In this application, the grid-signal voltage E_g is fixed, and the desired characteristic is that E_p be directly proportional to E_{bb}. As is pointed out in Ch. XII, under the condition of 100 per cent modulation, the effective plate voltage of the Class C amplifier varies from zero to twice E_{bb} at the frequency of the modulating signal. Since the modulation frequency is usually low compared with the grid-signal frequency, a number of plate-current pulses may flow while the plate voltage is temporarily equal to twice E_{bb}. In order to maintain high plate efficiency, it is desirable to limit the period of the plate current pulses to one-half of a cycle or less, as is discussed in Art. 4b. Consequently, the magnitude of the negative grid-bias voltage should be made at least large enough to equal the cut-off value for a plate voltage of twice E_{bb}. For that reason it is common practice to specify, for a Class C amplifier that is to be modulated, a magnitude of grid-bias voltage $|E_{cc}|$ equal to or greater than $2E_{bb}/\mu$, which is twice the cut-off value for the normal plate voltage E_{bb}. Frequently the magnitude of the grid-bias voltage is made three or more times the cut-off value in order to permit an even higher plate efficiency.

An analysis of the operation of the Class C tuned amplifier can be made on the basis of the assumption of a linear tube characteristic, as is made in Art. 3 for the Class B amplifier.[5] However, one new parameter, the grid-bias supply voltage E_{cc}, is introduced. The analysis is greatly complicated thereby, and resort to design charts is finally made. Although the analysis is only approximate and is somewhat lengthy, it has the advantage over others of giving an explicit solution for the optimum operating conditions and is therefore directly useful in design. Other approximate analyses are sometimes used[6] when the data for the tube characteristics are incomplete or results are desired with a minimum of computation.

[5] W. L. Everitt, *Communication Engineering* (2nd ed.; New York: McGraw-Hill Book Company, Inc., 1937), 565–594.

[6] W. G. Wagener, "Simplified Methods for Computing Performance of Transmitting Tubes," *I.R.E. Proc.*, 25 (1937), 47–77; F. E. Terman, *Radio Engineering* (3rd ed.; New York: McGraw-Hill Book Company, Inc., 1947), 374-393.

4a. *Semi-graphical Analysis Suitable for Class B and Class C Amplifiers.* The analysis given in this article makes use of the actual static characteristics of the tube.[7] The first objective is to obtain the waveform of the plate current and, from it, values for the average and the fundamental components of the plate current. These values are then used to give the power output, plate efficiency, and other results similar to those obtained in Art. 3 for the Class B amplifier.

Since the varying components of the grid and plate voltages are sinusoidal, but that of the plate current is not, the path of the operating point is not a straight line on either the $i_b(e_c)$ or the $i_b(e_b)$ tube-characteristic curves. However, the rectangular–co-ordinate graph of one variable as a function of another is readily shown to be a straight line if both variables are sinusoidal, have the same frequency, and differ in phase by an integral multiple of π radians.[8] Thus, if instantaneous values of e_c and e_b are used as the two variables on a rectangular–co-ordinate graph, the locus of the values that correspond to the same instant of time is a straight line. The path of operation on the $e_c(e_b)$, or constant-current, characteristic curves (see Fig. 8, Ch. IV) is therefore a straight line for an amplifier with a parallel-tuned load, and it is advantageous to use those characteristics in analyzing that type of amplifier.

The position of one-half of the straight path of operation in the constant-current curves is shown in Fig. 8. The waveforms of the grid and plate voltages from Fig. 7 are drawn as sinusoidal variations about the steady values E_{cc} and E_{bb}, respectively. At ωt equal to zero, the point P of the line is determined by $e_{c_{max}}$ and $e_{b_{min}}$. At ωt equal to $\pi/2$, or 90 degrees, the point Q of the line is determined by E_{cc} and E_{bb}. A locus of constant ratio $E_{cc}/$(cut-off bias), which is equal to $E_{cc}/(-E_{bb}/\mu)$, may be drawn from the origin through point Q, and a locus of constant $e_{b_{min}}/e_{c_{max}}$ may be drawn from the origin through point P, as is shown by dotted lines. The geometry of the figure indicates that the co-ordinates of P and Q are determined by any two pairs of *independent* quantities in the following list of eight: E_{bb}, E_{cc}, E_p, E_g, $e_{b_{min}}/e_{c_{max}}$, $E_{cc}/$(cut-off bias), $e_{c_{max}}$, $e_{b_{min}}$.

The path of operation shown in Fig. 8 is for the first and last quarters of the cycle only. During the second and third quarter-cycles, it is a straight line of equal length and is a projection of PQ through Q. However, during the second and third quarter-cycles the grid-signal voltage is negative and the plate current is zero for a Class B or Class

[7] I. E. Mouromtseff and H. N. Kozanowski, "Analysis of the Operation of Vacuum Tubes as Class C Amplifiers," *I.R.E. Proc.*, 23 (1935), 752–778.

[8] E. E. Staff, M.I.T., *Electric Circuits* (Cambridge, Massachusetts: The Technology Press of M.I.T.; New York: John Wiley & Sons, Inc., 1940), 333–336.

C amplifier; thus this half of the path is not of interest and is not drawn. The position of the operating point at various phase angles in the first quarter-cycle is indicated along the line. The distance along the line between the operating point and point Q varies as the cosine of the phase angle in the cycle.

If the path of operation in Fig. 8 is superposed on the constant-current curves for a tube, as is shown in Fig. 9, the plate current and grid

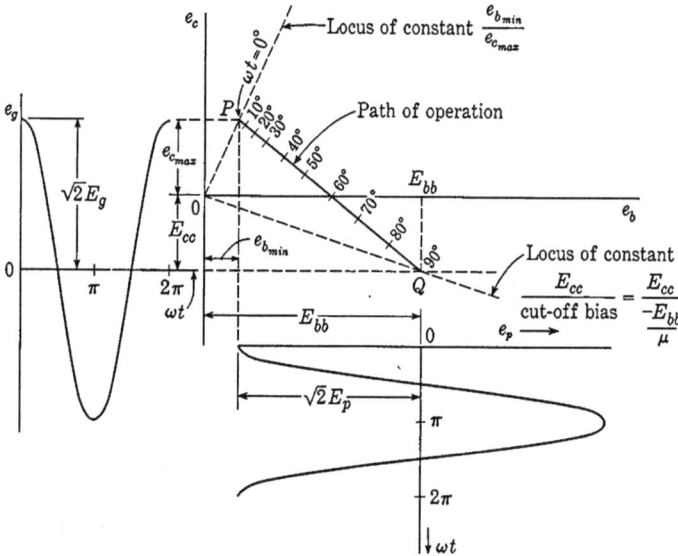

Fig. 8.   Path of operation on the constant-current characteristic curves.

current at particular instants in the cycle may be determined by interpolation between the curves. The maximum plate current $i_{b_{max}}$ occurs at ωt equal to zero in the figure, and its value provides an additional co-ordinate for determining the P terminus of the line; it may therefore be added to the list in the previous paragraph. The particular set of values selected to determine the line PQ in Fig. 9 is:

$$\text{Point } Q \left\{ \begin{array}{l} E_{bb} = 500 \text{ volts} \\[2mm] \dfrac{E_{cc}}{\text{cut-off bias}} = 2 \end{array} \right.$$

$$\text{Point } P \left\{ \begin{array}{l} i_{b_{max}} = 300 \text{ ma} \\[2mm] \dfrac{e_{b_{min}}}{e_{c_{max}}} = 1. \end{array} \right.$$

A means of systematizing the analysis of the amplifier performance as obtained from a diagram such as Fig. 9 is given in Table I, and the following is a description of the steps to be taken. The nine quantities at the top of the table are determined by the geometry of the figure when any independent four of them are specified. The length l of the line is measured and recorded, the half-cycle of operation is divided into n equal parts, and values of cos $(k\pi/n)$ where k is an integer are entered in the third row of the table. The particular value of n chosen here is 18 in order to give 10-degree increments in phase angle.

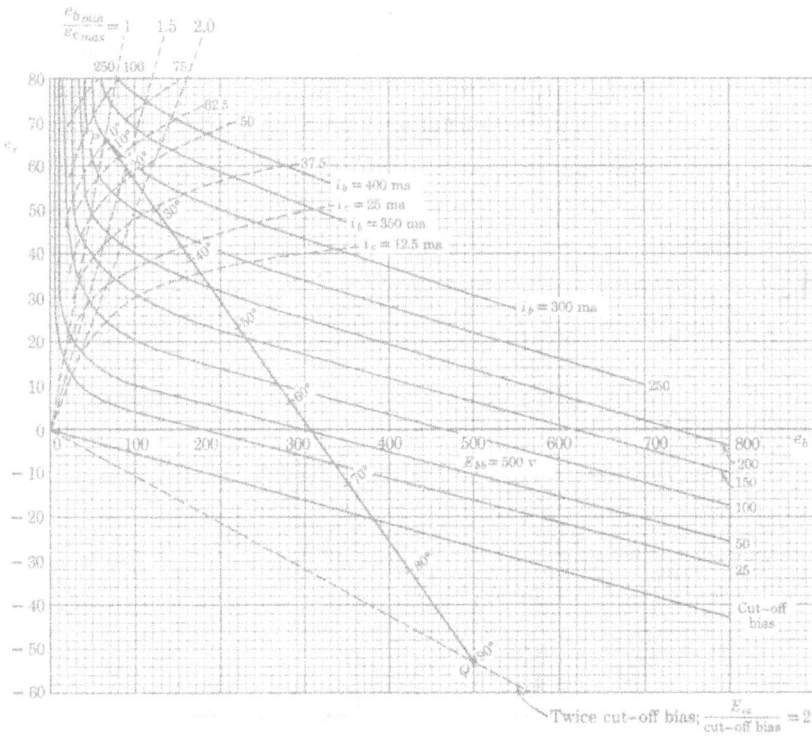

Fig. 9. Path of operation on the constant-current characteristic curves of a particular tube.

Eighteen is about the minimum value for acceptable accuracy, and any convenient larger value would improve the accuracy. The measured length l of the line PQ may be multiplied by cos $(k\pi/n)$ and the fourth row of the table filled in except for the last column. These distances are then laid off along the line from Q, determining the points on the line that correspond to the particular angles in the first quarter-cycle. The values of plate current and grid current at the angles $k\pi/n$,

TABLE I. ANALYSIS OF CLASS B AND CLASS C TUBE OPERATION

Tube:

$$E_{bb} = \qquad E_{cc}/\text{cut-off bias} = \qquad e_{bmin} =$$
$$E_g = \qquad\qquad E_{co} = \qquad\qquad E_p =$$
$$i_{bmax} = \qquad\qquad e_{cmax} = \qquad\qquad \frac{e_{bmin}}{e_{cmax}} =$$

l = length of line from Q to P; $n = \pi/\Delta\omega t = 18$; k = an integer; θ_1 = cut-off angle.

1	k	0	1	2	3	4	5	6	7	8	9	
2	$\dfrac{k\pi}{n} \times \dfrac{360°}{2\pi}$	0°	10°	20°	30°	40°	50°	60°	70°	80°	90°	θ_1
3	$\cos\left(\dfrac{k\pi}{n}\right)$	1.0	0.985	0.940	0.866	0.766	0.643	0.500	0.342	0.174	0.000	
4	$l\cos\left(\dfrac{k\pi}{n}\right)$										0	
5	$i_b\left(\dfrac{k\pi}{n}\right)$										0	0
6	$i_c\left(\dfrac{k\pi}{n}\right)$										0	0
7	$i_b\left(\dfrac{k\pi}{n}\right)\cos\left(\dfrac{k\pi}{n}\right)$										0	0
8	$e_g\left(\dfrac{k\pi}{n}\right)$										0	
9	$e_c\left(\dfrac{k\pi}{n}\right) = e_g\left(\dfrac{k\pi}{n}\right) + E_{cc}$											
10	$i_c\left(\dfrac{k\pi}{n}\right)e_c\left(\dfrac{k\pi}{n}\right)$										0	0

If

$$i_b \equiv I_{bs} + \sqrt{2}I_{p1}\cos\omega t + \sqrt{2}I_{p2}\cos 2\omega t + \cdots,$$

then

$$I_{bs} = \frac{1}{n}\left[\frac{i_b(0)}{2} + \sum_{k=1,2,3,\ldots} i_b\left(\frac{k\pi}{n}\right)\right] =$$

$$I_{p1} = \frac{\sqrt{2}}{n}\left[\frac{i_b(0)\cos(0)}{2} + \sum_{k=1,2,3,\ldots} i_b\left(\frac{k\pi}{n}\right)\cos\left(\frac{k\pi}{n}\right)\right] =$$

$$I_{cs} = \frac{1}{n}\left[\frac{i_c(0)}{2} + \sum_{k=1,2,3,\ldots} i_c\left(\frac{k\pi}{n}\right)\right] =$$

$$P_{cs} = \frac{1}{n}\left[\frac{i_c(0)e_c(0)}{2} + \sum_{k=1,2,3\ldots} i_c\left(\frac{k\pi}{n}\right)e_c\left(\frac{k\pi}{n}\right)\right] =$$

indicated in the table by the functional notation $i_b(k\pi/n)$ and $i_c(k\pi/n)$ in rows 5 and 6, are read by interpolation between the corresponding constant-current curves. The remainder of the table is for the purpose of obtaining the average components of the plate and grid currents, the fundamental harmonic component of the plate current, and the average power required by the tube in the grid circuit.

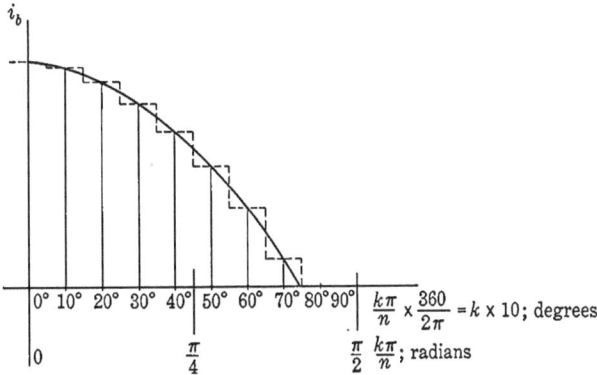

Fig. 10.   Current waveform from Fig. 9 and its approximate representation. Plate current is plotted; the grid-current curve can be approximated similarly.

The waveforms of the plate current and grid current are given by the data in the table and may be plotted as shown in Fig. 10, except for the abscissa when i_b equals zero. This abscissa is θ_1, the cut-off angle, shown in Fig. 7, and appears in the last column of Table I. It is the angle whose cosine is the distance from Q to the point at which the cut-off-bias line crosses the line PQ divided by the length l, and is found through filling in rows 4, 3, and 2 of the last column of the table in that order. Figure 7 shows that θ_1 is equal to one-half the angle of conduction in the tube.

In general, it is not necessary to plot the grid- and plate-current waveforms if the following methods of analysis are used to find their Fourier components.

The plate current may be expressed as the Fourier series

$$i_b = I_{bs} + \sqrt{2}(I_{p1}\cos \omega t + I_{p2}\cos 2\omega t + \cdots) \qquad [50]$$

in accordance with the considerations of Art. 13, Ch. VIII. The average value of the plate current I_{bs} is given by

$$I_{bs} = \frac{1}{2\pi}\int_0^{2\pi} i_b \, d(\omega t), \qquad [51]$$

but, since the wave is symmetrical about the ordinate axis at the origin,

$$I_{bs} = \frac{1}{\pi} \int_0^{\pi} i_b \, d(\omega t). \qquad [52]$$

The integral in Eq. 52 is the area under the curve of Fig. 10. This is approximately the area under the dotted step-function curve, which is the sum of the areas under the individual steps. The width of the first step is $\frac{1}{2}(\pi/n)$ radians, and its height is $i_b(0)$; thus its area is $(\pi/2n)i_b(0)$. The width of the other steps, except the last, is π/n radians, and their heights are $i_b(k\pi/n)$; thus their areas are $(\pi/n)i_b(k\pi/n)$, where k is a positive integer greater than zero. The width of the last step is uncertain, but this step is only a small contribution to the total area, and its width may be considered to be approximately equal to π/n radians. The average plate current I_{bs}, from Eq. 52, is therefore given approximately by

$$I_{bs} = \frac{1}{\pi} \left[\frac{\pi}{2n} i_b(0) + \sum_{k=1,2,3,\ldots} \frac{\pi}{n} i_b \left(\frac{k\pi}{n} \right) \right], \qquad [53]$$

or

$$I_{bs} = \frac{1}{n} \left[\frac{i_b(0)}{2} + \sum_{k=1,2,3,\ldots} i_b \left(\frac{k\pi}{n} \right) \right]. \qquad [54]$$

The average value of the plate current I_{bs} in Eq. 54 is readily evaluated from the data in row 5 of the table. The average value of the grid current is obtained from the data in row 6 of the table in the same manner—the subscript c takes the place of the subscript b.

The amplitude of the fundamental harmonic component of the plate current in Eq. 50 is $\sqrt{2}I_{p1}$. This coefficient in the Fourier series is given[9] by

$$\sqrt{2}I_{p1} = \frac{1}{\pi} \int_0^{2\pi} i_b(\omega t) \cos \omega t \, d(\omega t), \qquad [55]$$

and, since both $i_b(\omega t)$ and $\cos \omega t$ are symmetrical about the ordinate axis at the origin, it is also given by

$$\sqrt{2}I_{p1} = \frac{2}{\pi} \int_0^{\pi} i_b(\omega t) \cos \omega t \, d(\omega t). \qquad [56]$$

The integral in Eq. 56 may be expressed as a summation by the approximate method used before. Thus

$$\sqrt{2}I_{p1} = \frac{2}{\pi} \left[\frac{\pi}{2n} i_b(0) \cos(0) + \sum_{k=1,2,3,\ldots} \frac{\pi}{n} i_b \left(\frac{k\pi}{n} \right) \cos \left(\frac{k\pi}{n} \right) \right]; \qquad [57]$$

[9] See, for example, P. Franklin, *Differential Equations for Electrical Engineers* (New York: John Wiley & Sons, Inc., 1933), 69.

whence

$$I_{p1} = \frac{\sqrt{2}}{n} \left[\frac{i_b(0) \cos (0)}{2} + \sum_{k=1,2,3,\,\cdots} i_b \left(\frac{k\pi}{n} \right) \cos \left(\frac{k\pi}{n} \right) \right].\quad \blacktriangleright[58]$$

Row 7 in the table is the product of the quantities in rows 3 and 5 of a given column. The effective value of the fundamental harmonic component of the plate current I_{p1} is determined from the data in row 7 by means of Eq. 58.

The average grid-driving power required by the tube at the grid and cathode terminals is

$$P_{cs} = \frac{1}{2\pi} \int_0^{2\pi} i_c(\omega t) e_c(\omega t)\, d(\omega t),\qquad [59]$$

and, again, because of the symmetry of the functions,

$$P_{cs} = \frac{1}{\pi} \int_0^{\pi} i_c(\omega t) e_c(\omega t)\, d(\omega t).\qquad [60]$$

The values of $i_c(\omega t)$ required for the evaluation of Eq. 60 by the summation process previously used are available in row 6 of the table. The values of $e_c(\omega t)$ may be read from the ordinate scale in Fig. 9 and entered in row 9, or they may be computed by means of the values in row 8, which are the products of the amplitude of the grid-signal voltage and the corresponding values in row 3. The average grid-driving power expressed by the summation method is then

$$P_{cs} = \frac{1}{n} \left[\frac{i_c(0) e_c(0)}{2} + \sum_{k=1,2,3,\,\cdots} i_c \left(\frac{k\pi}{n} \right) e_c \left(\frac{k\pi}{n} \right) \right],\qquad \blacktriangleright[61]$$

and this summation is evaluated from the data in row 10, which are the products of corresponding quantities in rows 6 and 9.

From the results of the foregoing analysis, the alternating-current power output to the tuned load,

$$P_{ac} = E_p I_{p1},\qquad [62]$$

is readily determined, and the value of the input resistance of the tuned load required to satisfy the assumed operating conditions is the ratio of E_p to I_{p1}. Methods of coupling an actual load to the tube in such a manner as to satisfy this requirement are discussed in Art. 5. Also the direct-current power input to the plate, which is equal to the average power furnished by the plate-voltage supply when the direct-current power dissipated in the plate-load resistance is negligible, is given by

$$P_{bs} = E_{bb} I_{bs}.\qquad [63]$$

The plate efficiency is the ratio of Eq. 62 to Eq. 63, or

$$\eta_p = \frac{P_{ac}}{P_{bs}} = \frac{E_p I_{p1}}{E_{bb} I_{bs}},$$ ▶[64]

and the plate dissipation is the difference of Eqs. 63 and 62; that is,

$$P_p = P_{bs} - P_{ac} = E_{bb} I_{bs} - E_p I_{p1}.$$ ▶[65]

4b. *Design Considerations for Class C Amplifiers.* The analysis of the Class C amplifier presented above is based on the assumption that the locus of the operating point on the tube characteristics is known. However, an important matter of engineering design is the selection of the operating conditions that govern the locus to give high plate efficiency and other desirable results. It is instructive in connection with such a selection to consider qualitatively the similarities between the Class C amplifier having a tuned circuit and the clock having an escapement and pendulum mechanism.

When the pendulum swings, the stored energy is in potential form at the limits of the swing where the motion stops and the bob is highest in the gravitational field. The stored energy is all in the kinetic form at the center of the swing where the speed of motion is the greatest and the position of the bob the lowest. During a cycle, the stored energy is transferred from the kinetic form to the potential form and back again. If the pendulum swings in a viscous fluid, power is dissipated by friction; and a source of power to drive the pendulum is necessary if the amplitude of the motion is to remain constant.

Similarly, the stored energy in the parallel-tuned circuit of an amplifier is all in the capacitor when the current is zero, and is all in the inductor when the current is a maximum and the rate of change of current with time is momentarily zero. During the cycle, the stored energy is transferred back and forth between the capacitance and the inductance. Power is dissipated in the resistance of the circuit, and a source of power is necessary if the amplitude of the electric oscillations is to be maintained.

In the clock, power to maintain the pendulum oscillation is supplied by a spring or weight through the escapement mechanism. The escapement serves to deliver a power impulse to the pendulum at the end of its swing. Despite the fact that the power is delivered as an impulse over one short interval in the cycle, the motion of the pendulum is closely sinusoidal for small amplitudes if the bob is heavy and the pendulum is long, and if the viscous medium (air) provides only a small power dissipation. These conditions are equivalent to a large ratio of stored energy to energy dissipated during a cycle.

In the Class C amplifier, the tube acts like an escapement to control the flow of energy from the plate-power supply to the load, and thus to maintain the electric oscillations. Power is delivered to the load during the short interval when the plate-current pulse exists. If the ratio of energy stored to energy dissipated during a cycle in the tuned load circuit is large, the oscillations are essentially sinusoidal, since the operation is analogous to that of the pendulum and escapement discussed above.

The circuit conditions corresponding to a large ratio of energy stored to energy dissipated during a cycle may be expressed by the following approximate analysis. When a parallel-tuned circuit is excited by a source of sinusoidal voltage or current at its resonant frequency, the circulating current in the circuit is also sinusoidal. If the series resistance in the tuned circuit is small compared with the inductive and capacitive reactances (which are equal), the energy stored is essentially constant, since, in the limit, as the resistance approaches zero the energy input and the energy dissipated during a cycle both approach zero, and the total energy stored must therefore be constant. The total energy stored is then equal to the maximum cyclic value of the energy stored in either the inductance or the capacitance; that is, it is equal to LI_t^2 or CE^2, where I_t and E are the effective values of the current in, and the voltage across, the tuned circuit.

The amount of energy dissipated during a cycle is $I_t^2 R/f_0$, where f_0 is the resonant frequency and R is the series resistance in the circuit. When this amount of energy is so small that the amount stored can be regarded as constant, the ratio of the amount stored to the amount dissipated per cycle is large and is given by

$$\frac{\text{Energy stored}}{\text{Energy dissipated per cycle}} = f_0 \frac{LI_t^2}{RI_t^2} = \frac{\omega_0 LI_t^2}{2\pi RI_t^2} \qquad [66]$$

$$- \frac{\text{Reactive volt-amperes}}{2\pi \text{ watts}} \qquad \blacktriangleright[67]$$

$$= \frac{\omega_0 L}{2\pi R} = \frac{Q_0}{2\pi}. \qquad \blacktriangleright[68]$$

It follows that a circuit having a large ratio of energy stored to energy dissipated during a cycle has essentially constant energy storage and a value of Q_0 that is large compared with 2π.

Equation 68 is based on the assumption of sinusoidal excitation for the circuit, whereas, in the Class C amplifier, the excitation is the synchronized unidirectional pulsating wave of the tube plate current. The solution for the circuit behavior with the pulsating excitation is not so simple as that leading to Eq. 68. It involves the decay of the oscillations between pulses, and the oscillations are not strictly sinusoidal. However, the analogy with the pendulum and escapement indicates that, if the circuit has a large ratio of energy stored to energy dissipated during a cycle for sinusoidal excitation, it will also have essentially sinusoidal oscillations even for a synchronized unidirectional pulsating excitation. Equation 68 is therefore applicable and the criterion for essentially sinusoidal current and voltage in the load is that the Q_0 of the load circuit be large compared with 2π.

The degree by which the oscillations depart from strictly sinusoidal operation may be determined approximately through an extension of the analysis outlined in Table I. If the departure from a sinusoidal waveform of voltage across the load circuit is small, the path of operation in Fig. 9 and the current waveform that results are not affected appreciably. The harmonic components in the plate current are then obtained by a summation method similar to that indicated by Eq. 58 but $\cos m(k\pi/n)$ is substituted for $\cos (k\pi/n)$, where m is the order of the harmonic component desired. The impedance of the load circuit is determined for the corresponding harmonic frequencies by the methods outlined in Art. 9, Ch. IX, and the harmonic components in the output voltage across the load circuit are the products of the harmonic components of the plate current and the corresponding values of the load impedance. Additional considerations regarding the harmonic content in the output waveform are discussed in the following article.

Although the function of the tube in the Class C amplifier is merely that of an escapement or valve, one must recognize that the tube has losses. These decrease the efficiency with which power is transferred from the plate-power supply to the load—that is, the plate efficiency. The instantaneous power loss in the tube is the product of the instantaneous values of the plate current and voltage. The smallest tube losses and highest plate efficiency are obtained, therefore, if the plate-current pulse takes place when the plate voltage e_b has its smallest value in the cycle, and if the amplitude of the alternating component of the plate voltage is sufficiently large to reduce $e_{b_{min}}$ to a small value.

The wave shape of the plate-current pulse also has considerable effect on the plate efficiency. Since the expression for I_{p1}, Eq. 56, involves the factor $\cos \omega t$, the center portion of the plate-current pulse—near ωt equal to zero—contributes the greater part to the integral, and the current at the sides of the pulse where ωt is large has little effect on I_{p1} or on the alternating-current power output in Eq. 62. On the other hand, the expression for I_{bs} in Eq. 52 does not contain the cos ωt factor, and the sides of the plate-current pulse do have an appreciable effect on I_{bs} and the power input in Eq. 63. It follows therefore that for high plate efficiency, a sharp current pulse with steeply falling sides is desirable. Such a waveform is obtained if the magnitude of the negative grid-bias voltage E_{cc} is made large compared with the cut-off-bias voltage, and high plate efficiencies—up to 80 or 90 per cent—may be obtained. It does not follow that a large magnitude of E_{cc} is always the most desirable operating condition, however, for the following considerations show that the grid-driving power is often increased by such operation.

The value of P_{cs} obtained by the method indicated in Eq. 61 is the power required by the tube at the grid-cathode terminals. In Fig. 3 the direct component of grid current I_{cs} passes through the grid-bias-voltage supply in such a direction that power is absorbed rather than delivered by the voltage source—if a battery is used it is charged rather than discharged. The power absorbed comes from the source of grid-signal voltage; hence the total power required of the grid-signal-voltage supply is

$$P_g = P_{cs} + |E_{cc}| I_{cs}. \qquad \blacktriangleright[69]$$

Thus, for a constant $e_{c_{max}}$, an increase of $|E_{cc}|$ tends to increase the power required in the grid circuit of a Class C amplifier at the same time as it provides an increase in plate efficiency. The over-all efficiency of the amplifier circuit, which is the ratio of the output to the total input power is

$$\eta = \frac{E_p I_{p1}}{E_{bb} I_{bs} + P_{cs} + |E_{cc}| I_{cs} + E_f I_f}. \qquad [70]$$

For given values of $e_{c_{max}}$, $e_{b_{min}}$, and E_{bb}, an increase in the amount of negative grid-bias voltage results in decreased values of I_{bs} and I_{cs}, but the presence of $|E_{cc}|$ in the denominator of Eq. 70 indicates that a particular value of $|E_{cc}|$ gives the highest over-all efficiency.

The total power required of the grid-signal-voltage supply, P_g, is given by

$$P_g = \frac{1}{2\pi} \int_0^{2\pi} e_g i_c \, d(\omega t). \qquad [71]$$

This integral may be evaluated by means of Fig. 9 or in the terms given by Eq. 69. However, since the waveform of the grid current i_c usually consists of a pulse of short time duration near the instant when the grid-signal voltage has its maximum value $\sqrt{2}E_g$, the e_g term may be considered[10] a constant of that value, and Eq. 71 leads to the

---

[10] H. P. Thomas, "The Determination of Grid-Driving Power in Radio-Frequency Power Amplifiers," *I.R.E. Proc.*, 21 (1933), 1134–1141.

approximate result

$$P_g \approx \sqrt{2}E_g I_{cs}.\qquad\qquad\blacktriangleright[72]$$

Thus a direct-current ammeter to indicate I_{cs} and an alternating-current voltmeter to indicate E_g are sufficient for an approximate determination of P_g.

4c. *Methods of Obtaining Grid-Bias Voltage.* Since the Class C amplifier usually operates with a fixed amplitude of grid-signal voltage, the cathode-bias method of obtaining the grid-bias voltage described in Arts. 3 and 7, Ch. VIII, is adaptable to it. However, the

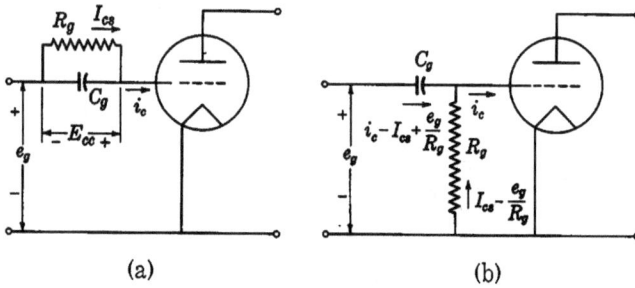

(a)   (b)

Fig. 11.   Grid-leak-and-capacitor methods of obtaining the grid-bias voltage.

power loss in the bias resistor is large in a high-power amplifier, because both $|E_{cc}|$ and I_{bs} are large. It is hence desirable to use the grid-leak-and-capacitor method shown in Fig. 11, because in it the plate current is not utilized. Instead, the direct component of grid current in the grid resistor provides the grid-bias voltage. Two alternative arrangements are shown in Fig. 11. In both of them the magnitude of the grid-bias voltage is equal to $I_{cs}R_g$, and a by-pass capacitor C_g having a reactance small compared with the resistance R_g at the operating frequency is used to minimize the voltage drop caused by the varying component of grid current.

The circuit in Fig. 11b has the advantage over the one in Fig. 11a that the direct component of i_c does not exist in the source of grid-signal voltage. However, the circuit in Fig. 11a has the advantage that the power loss in the grid leak is merely $I_{cs}{}^2 R_g$, whereas in Fig. 11b there is an additional component of loss due to the alternating component of grid voltage, and the power loss in R_g is $I_{cs}{}^2 R_g + (E_g{}^2/R_g)$. This disadvantage may be obviated by the insertion of an inductor in series with R_g to reduce the alternating component of current. Either grid-leak-and-capacitor method has the disadvantage that, if the grid-signal voltage is removed, the grid-bias voltage also disappears, and the plate-current may rise to a destructively high value. On the other hand, either method has the advantage that the value of $e_{c_{max}}$ obtained is not critically dependent on the value of R_g or E_g. The grid current is a steeply rising function of the grid voltage at $e_{c_{max}}$ in the usual Class C amplifier, and an increase of E_g results in a relatively large increase of I_{cs}, which increases the magnitude of the grid-bias voltage and reduces the change in $e_{c_{max}}$ to a small amount. Similarly, if R_g is decreased, I_{cs} increases and maintains E_{cc} and $e_{c_{max}}$ nearly constant. Because of its simplicity and economy, the grid-leak-and-capacitor method of obtaining the grid-bias voltage is generally used with Class C amplifiers, but it is sometimes combined with a direct voltage from a battery, power supply, or generator to improve the amplifier's modulation characteristic, as is discussed in Ch. XII.

4d. *Vector Diagram for the Tuned Amplifier.* The vector diagram is useful for an understanding of the operation of the tuned amplifier. It is assumed that the tuned circuit is of high Q_0 and is adjusted to resonate at the grid-signal frequency. Since the harmonics in the output are essentially eliminated by the tuned circuit for operation over the full range of grid-bias voltage from the Class A to the Class C value, fundamental-frequency quantities only are of interest, and the vector diagram is the same in form for the whole range of grid-bias voltage. The circuit diagram with complex notation for the fundamental-frequency components of the currents and voltages is shown in Fig. 12a. Note that in this diagram I_c denotes capacitor current, not grid current. The phase relations among the grid-signal voltage, the plate voltage, and the plate current are discussed in Art. 3. Since the parallel circuit is assumed to be so tuned that its impedance at the fundamental frequency is a pure resistance, I_{p1} is in phase with $-E_p$, and E_g, E_p, and I_{p1} are all in phase or 180 degrees out of phase with each other. The vector diagram for the Class C amplifier is therefore that shown in Fig. 12b. In this diagram, the length of the vector for the current I_{p1} is greatly exaggerated with respect to the lengths of the other current vectors. This vector diagram is useful for comparison later with the vector diagram of the circuit when operating as a tuned-plate oscillator.

The vector μE_g in Fig. 12b is a fictitious one analogous to the quantity existing in a Class A amplifier, for μ is indeterminate during the zero-plate-current intervals in the Class C amplifier and may vary considerably during the intervals when the current is not zero. Nevertheless, when the vector μE_g proportional to E_g is included, it follows that the tube in the Class C amplifier behaves, as far as quantities of fundamental frequency are concerned, as if it were a generator of internal source voltage μE_g and series resistance $\bar{r}_p$, since all the vectors μE_g, E_g, and E_p are in phase. The resistance $\bar{r}_p$, however, is a fictitious or average plate resistance accounting for the action of the tube but not determinable at any one point on the tube characteristics. Further discussion of the significance of these fictitious quantities occurs in Arts. 3 and 5, Ch. XI.

5. DESIGN CONSIDERATIONS FOR TUBE-TO-LOAD COUPLING NETWORKS

The analysis of the operation of the tuned power amplifiers in the preceding two articles is based on the assumption that the tube feeds

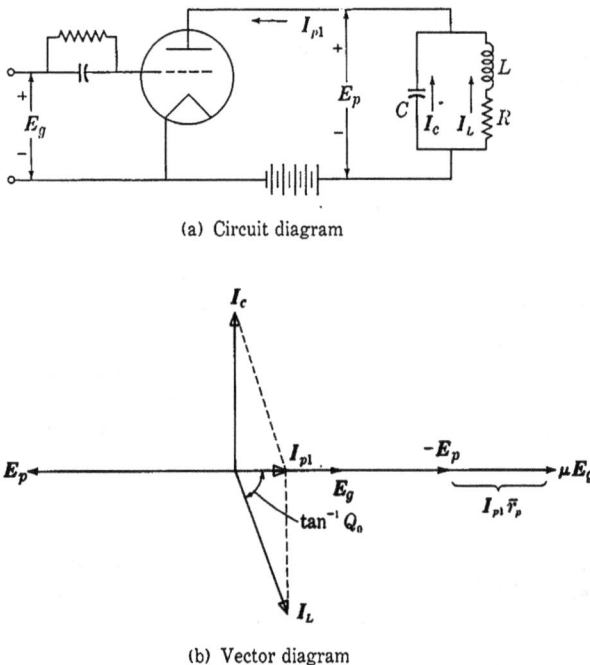

(a) Circuit diagram

(b) Vector diagram

Fig. 12.   Vector diagram for the tuned amplifier.

power into a tuned circuit whose input impedance is purely resistive at the grid-signal frequency and is negligible at harmonics of this frequency. The plate efficiency of the tube then indicates the effectiveness with which the tube transforms the direct-current power from the plate-power supply into alternating-current power delivered to the input terminals of the tuned circuit. Actually, the tuned circuit connected in the plate circuit of the tube is almost never a simple parallel-tuned circuit such as is shown in Fig. 3, because the required impedance at the tube terminals is fixed, and the actual load impedance is generally also fixed by conditions external to the tube; hence a more complicated circuit is needed to perform the function of impedance transformation. However, for the purpose of computing the effective impedance in the plate circuit of the tube at the grid-signal frequency, the coupling network and load can usually be reduced to an equivalent simple parallel-tuned circuit at that one frequency. The actual load element is commonly connected to the tube through a tuned coupling network in which the mutual impedance between the meshes may comprise resistance, capacitance, self-inductance, and mutual inductance in a variety of combinations to suit the particular requirements.

Since the elements in the coupling network introduce appreciable resistance, there is a power loss in the coupling network. As an index of this power loss, the efficiency of the coupling network may be defined as the power delivered to the actual load divided by the power delivered from the tube to the input terminals of the coupling network. The over-all efficiency of power conversion from the direct-current power input to the tube into the alternating-current power output at the actual load terminals is then the product of the plate efficiency and the efficiency of the coupling network. As the analysis in the previous two articles indicates, the plate efficiency is determined by the tube operating conditions, which are selected to give a specified power output at a specified plate dissipation and plate-supply voltage. On the other hand, the design of the coupling network for highly efficient power transfer is independent of the tube characteristics and involves only linear circuit analysis.

Several considerations that may govern the design of the coupling network can be summarized as follows:

First, the input, or driving-point, impedance of the coupling network with the load connected should have the value at the grid-signal frequency that meets the tube requirements at the specified operating conditions, as discussed in the previous article.

Second, the coupling-network efficiency should be as large as is practicable.

Third, the input impedance of the loaded coupling network at the harmonic frequencies should be small compared with its impedance at the fundamental frequency, in order that the harmonic generation in the amplifier (or the power output from the tube at the harmonic frequencies) shall be negligible.

Fourth, the selectivity of the coupling network should be adequate to suppress the unwanted frequency components and prevent their reaching the load; in other words, the transfer impedance of the coupling network at the harmonic frequencies should be small compared with the transfer impedance at the resonant frequency.

Fifth, the selectivity of the coupling network should not be too large
if the amplifier is to deliver power over a band of frequencies near
the resonant frequency, as, for example, when a Class B amplifier
is used with a modulated input signal, or a Class C amplifier is
plate-modulated (see Ch. XII).

The relative importance and stringency of these individual requirements differ for different applications, and the final design of the
coupling network varies accordingly.[11]

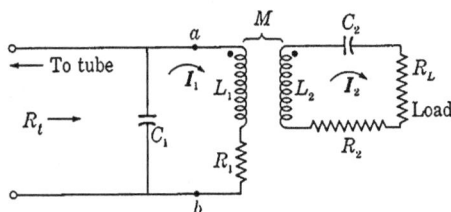

(a) Actual circuit

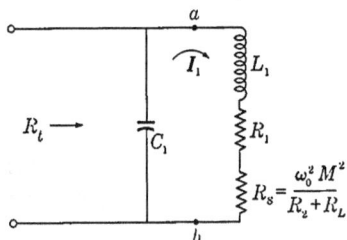

(b) Simple parallel-tuned circuit equivalent to (a)
at the resonant frequency

Fig. 13.   A loaded coupling network and its equivalent circuit.

A general treatment of the effect of the foregoing requirements on
the design of all types of tuned coupling networks is not practicable at
this point. Consequently, only one typical network is considered here,
although the analysis of others may be carried out along similar lines.

A typical coupling network is that shown in Fig. 13a, in which the
load R_L is coupled to the tube through mutual inductance only. To
simplify the analysis, the tuning of the two loops in the circuit by

[11] W. L. Everitt, "Output Networks for Radio-Frequency Amplifiers," *I.R.E. Proc.,*
19 (1931), 725–737; P. H. Osborn, "A Study of Class B and C Amplifier Tank Circuits,"
I.R.E. Proc., 20 (1932), 813–834; K. S. Kunz, "Bilinear Transformations Applied to the
Tuning of the Output Network of a Transmitter," *I.R.E. Proc., 37* (1949), 1211–1217;
R. A. Martin, "Input Admittance Characteristics of a Tuned Coupled Circuit," *I.R.E.
Proc., 40* (1952), 57–61; W. B. Bruene, "How to Design R-F Coupling Circuits," *Elec-
tronics, 25* (May, 1952), 134–139.

means of the capacitors C_1 and C_2 is assumed to be so done that two conditions are fulfilled: First, the capacitor C_2 is adjusted so that $1/\sqrt{L_2 C_2}$ equals the operating frequency; and, second, the capacitor C_1 is then adjusted so that the input impedance across C_1 is a pure resistance R_t at the operating frequency. It is not to be implied, however, that such tuning necessarily gives the best operating conditions. The resistances R_1 and R_2 are the actual coil resistances in the primary and secondary circuits, respectively, and the capacitors are considered to be dissipationless.

In order to determine the input impedance of the loaded coupling network at the grid-signal frequency, which is of interest in connection with the first of the foregoing five listed requirements, it is convenient to replace the circuit to the right of points a and b by an equivalent (for that frequency) series combination of circuit elements. The equivalent series circuit can be found by means of the relations developed in books on electric circuits,[12] where it is shown that the input impedance of a two-loop network is

$$Z_{1a} = Z_{11} - \frac{Z_{12}{}^2}{Z_{22}}, \qquad [73]$$

in which Z_{1a} is the input impedance, Z_{11} is the self-impedance of the first mesh, Z_{22} is the self-impedance of the second mesh, and Z_{12} is the mutual impedance between the meshes. Since the second mesh is assumed to be tuned to resonance at the grid-signal frequency, the self-impedance Z_{22} at that one frequency is merely a resistance; and, in terms of the circuit notation in Fig. 13a, the relation for the input impedance Z_{ab} of the circuit to the right of points ab becomes

$$Z_{ab} = R_1 + j\omega_0 L_1 + \frac{\omega_0{}^2 M^2}{R_2 + R_L}, \qquad [74]$$

where ω_0 is the resonant angular frequency. Thus, at the resonant frequency, the impedance Z_{ab} is the same as if the secondary circuit were removed and a resistance

$$R_s = \frac{\omega_0{}^2 M^2}{R_2 + R_L}, \qquad [75]$$

called the *reflected resistance*, were placed in series with the primary coil, as is shown in Fig. 13b. It follows that the circuit in Fig. 13b is equivalent to the circuit in Fig. 13a *at the resonant frequency*, and the input impedance of the whole circuit in Fig. 13a can be computed on the basis of Fig. 13b for that frequency.

[12] E. E. Staff, M.I.T., *Electric Circuits* (Cambridge, Massachusetts: The Technology Press of M.I.T.; New York: John Wiley & Sons, Inc., 1940), 378–379.

The value of the input impedance of the loaded coupling network required to meet the tube operating conditions is

$$R_t = E_p/I_{p1} , \qquad [76]$$

where E_p and I_{p1} are the fundamental alternating components of the tube plate voltage and current determined by the methods outlined in Art. 4a. This value of input impedance can be obtained through application of the relations given in Eq. 99, Ch. IX, to the equivalent parallel-tuned circuit in Fig. 13b. It is shown in that equation that

$$R_t = Q_0 \sqrt{L/C} , \qquad [77]$$

where Q_0 is defined as the ratio of the reactance to the total resistance in the circuit at the resonant frequency. For the coupled circuit, Q_0 therefore includes the effect of the reflected resistance R_s and may be called the *loaded* Q_0. Equation 77 shows that the value of the input impedance required by the tube, as stated in Eq. 76, can be obtained by simultaneous selection of the loaded Q_0 and the L/C ratio of the equivalent circuit. This selection is subject to the practical limitation that the value of C cannot be smaller than the capacitance inherent in the tube and wiring. Also it is shown subsequently that the additional requirements previously listed may determine the desired value of Q_0 within close limits, and the L/C ratio therefore must often be selected to satisfy the tube requirements.

The coupling-network efficiency, mentioned in the second requirement listed above, obviously is related to the losses in the resistances R_1 and R_2 as compared with the power dissipated in R_L. Since the Q of a coil is a valuable index of its quality, it is convenient to express the coupling-network efficiency in terms of the Q's in the circuit as follows: In Fig. 13b, the power dissipated in R_1 is power used in heating the primary coil, while the power dissipated in the reflected resistance R_s actually is power transferred to the secondary circuit and includes both the power dissipated in the coil resistance R_2 and the load resistance R_L. Since the current in the primary coil I_1 is the current in both R_1 and R_s, the ratio of the power delivered to the secondary circuit to the total power input, which may be defined as the efficiency of the primary circuit, is

$$\eta_1 = \frac{I_1{}^2 R_s}{I_1{}^2 (R_1 + R_s)} = \frac{R_s}{R_1 + R_s}. \qquad [78]$$

The Q of the primary circuit at the resonant frequency, with the secondary load removed, is the Q of the primary coil alone; thus

$$Q_{01} \equiv \frac{\omega_0 L_1}{R_1} \text{ is the } Q_0 \text{ of the primary } coil \text{ at resonance (unloaded).} \qquad [79]$$

The effective Q of the equivalent simple parallel-tuned circuit at the resonant frequency includes the reflected resistance. Hence

$$Q_{0_L} \equiv \frac{\omega_0 L_1}{R_1 + R_s} \text{ is the } Q_0 \text{ of the loaded primary } circuit \text{ at resonance.} \quad [80]$$

This value is to be used in Eq. 77. When these values of Q_0 are substituted in Eq. 78, the efficiency becomes

$$\eta_1 = 1 - \frac{Q_{0_L}}{Q_{01}}. \qquad \blacktriangleright[81]$$

Similarly, in the secondary circuit, the ratio of the power dissipated in the load resistance R_L to the power input to the secondary circuit, defined as the efficiency of the secondary circuit, is

$$\eta_2 = \frac{I_2{}^2 R_L}{I_2{}^2 (R_2 + R_L)} = \frac{R_L}{R_2 + R_L} = 1 - \frac{Q_{02_L}}{Q_{02}}, \qquad \blacktriangleright[82]$$

where

$$Q_{02} \equiv \frac{\omega_0 L_2}{R_2} \text{ is the } Q_0 \text{ of the unloaded secondary } coil \text{ at resonance,} \quad [83]$$

and

$$Q_{02_L} \equiv \frac{\omega_0 L_2}{R_2 + R_L} \text{ is the } Q_0 \text{ of the loaded secondary } circuit \text{ at resonance.} \quad [84]$$

The over-all efficiency of the coupling network—that is, the ratio of the power delivered to the load to the power input to the coupling network—is the product of the primary and secondary efficiencies $\eta_1 \eta_2$.

From the foregoing expressions it is evident that for the highest efficiency in the coupling network, the unloaded Q_0's of the coils must be as large as practical and the loaded Q_0's as small as practical.

Once the values of the four Q_0's defined above are selected, the values of the unknown parameters in the networks in Fig. 13a are uniquely determined and can be found in the following manner. At the start, the known quantities are the load resistance, the grid-signal frequency, and the required input resistance of the coupling network at that frequency. The R's, L's, and C's, and the M in the circuit are unknown. Equation 77 gives the L_1/C_1 ratio for the primary circuit in terms of the selected value of Q_{0_L} and the known value of R_t. The known value of the grid-signal frequency establishes a value of the $L_1 C_1$ product, and the required values of L_1 and C_1 can be found from the resulting values of their ratio and their product. Next, the resistance R_1 of the primary coil can be found from the selected value of

Q_{01}, the grid-signal frequency, and the value of L_1. Then the value of R_s can be determined by means of the selected values of Q_{0_L} and Q_{01} and Eqs. 78 and 81.

Attention is next directed to the secondary circuit. The ratio of the load resistance to the secondary-coil resistance can be found by means of the selected values of Q_{02} and Q_{02_L} and Eq. 82. Since the load resistance is a known value, this ratio leads to a value of R_2. Then, by means of the values selected for Q_{02} and the grid-signal frequency, the values of L_2 and C_2 for the secondary circuit are found by the use of Eq. 83 and the $L_2 C_2$ product corresponding to the specified resonant frequency. The one remaining parameter to be found is the value of M. This quantity is determinable by Eq. 75, since the values of ω_0 and R_L are known and the values of R_s and R_2 are included in the previous results.

It is evident from the preceding considerations that the problem of the coupling-network design can often be resolved to one of selection of the loaded and unloaded Q_0's of the circuits at the resonant frequency. The maximum practical values of the unloaded Q_0's of the coils are usually limited by space, cost, and other considerations. Typical maximum practical values for coils used at broadcast frequencies are 100 to 200 for coils having dimensions of a few inches as used in relatively low-power amplifiers, up to values of 500 to 800 for coils having dimensions of a few feet as used in amplifiers of high power—in the range of 50 to 500 kilowatts.

The minimum practical values of the loaded Q_0's of the circuits are not so clearly defined. Only a qualitative discussion of a few of the factors and limitations that affect their selection is given here.

The loaded Q_0 of the first mesh or primary circuit Q_{0_L} influences primarily the input impedance of the circuit, although, as discussed subsequently, its influence on the frequency dependence of the input impedance is not simple in the general case. One or more of several requirements may determine the lowest allowable value of Q_{0_L}. First, as stated previously, the minimum practical value of the tuning capacitance across the tube terminals C_1 places a lower limit on Q_{0_L} in accordance with Eq. 77. Some capacitance in addition to that inherent in the tube and wiring is generally considered desirable in order to provide a means of tuning and to decrease the effect on the tuning of any changes in the capacitance of the wiring and tube. Second, if the loaded Q_0 is reduced below a value of about 10, the tuning adjustments for unity power factor and for maximum impedance at the tube output terminals are no longer the same when the capacitance is adjusted. As a result, it is no longer sufficient to accomplish the tuning by the usual simple method of adjusting the capacitance until the indication of the plate-current ammeter is a minimum. Bridge measurements or their equivalent are necessary, and, as these are often impractical with the equipment available, a lower limit of about 10 for Q_{0_L} is commonly specified.[13] Third, when the loaded Q_0 of the primary circuit is reduced, the percentage of harmonics in the plate voltage of the tube is generally increased, and the third requirement in the list of five given at the beginning of this article may not be satisfied. The presence of appreciable harmonics in the plate voltage results in a loss of efficiency in the tube, but no data on the magnitude of the effect are published at present. Values of two to three for the loaded Q_0 of the primary circuit are used in the amplifiers of broadcast transmitters. The assumption of a sinusoidal waveform for the plate-voltage variation used in the analysis of Art. 4 is then not justified and the analysis is made more complicated.

The loaded Q_0's of the meshes other than the first, in the coupling network, influence the transfer impedance primarily. Consequently, their main effect is on the selectivity of the coupling network and on the amount of power output to the actual load at frequencies near the resonant frequency and at harmonic frequencies. Thus the fourth and fifth requirements in the list of five given at the beginning of this article are determining factors in the selection of these loaded Q_0's. For example, in the circuit of Fig. 13a, if the loaded Q_0 of the secondary circuit is made too low, the required L_2 is decreased and the coupling coefficient is increased. With an increase of coupling coefficient, the selectivity in this type of coupling network generally decreases, and furthermore there is a practical limit to the coupling coefficient that can be obtained. No minimum values for the loaded Q_0's can be specified to cover all amplifiers. Each application requires a separate study of the effect of the loaded Q_0's on the requirements stated.

In the particular but unusual application in which the load and coupling network form a simple parallel-tuned circuit, as shown in Fig. 3, the relations between the requirements listed previously and the Q_0's of the circuit are established by the analysis in the preceding article and the derived expressions in Art. 9, Ch. IX. It is stated in Art. 4 that, for this circuit, the Q_0 must be large if the harmonic components in the output voltage are to be small. The magnitude of the harmonic components in the output voltage from this circuit can be estimated as follows. Since the plate-current waveform in the Class C amplifier is

[13] F. E. Terman, *Radio Engineering* (3rd ed.; New York: McGraw-Hill Book Company, Inc., 1947), 383; R. S. Glasgow, *Principles of Radio Engineering* (New York: McGraw-Hill Book Company, Inc., 1936), 272; D. C. Prince, "Vacuum Tubes as Power Oscillators, Part I," *I.R.E. Proc.*, *11* (1923), 301; A. B. Newhouse, "Power Amplifier Plate Tank Circuits," *Electronics*, *14* (November, 1941), 32–35.

somewhat similar in shape to that of a half-wave rectifier but has sharper peaks of shorter duration, the percentage of second harmonic in the current may be assumed to be somewhat larger than in a half-wave rectifier. In the ideal half-wave rectifier with a resistance load, the amplitude of the second-harmonic component is approximately 42 per cent of the fundamental; thus, in a Class C amplifier, it may be estimated as, say, 100 per cent as a pessimistic value. Under these conditions, the percentage second harmonic in the plate voltage is 100 times the ratio of the impedance of the tuned circuit at the second-harmonic frequency to the impedance at the fundamental frequency. This ratio may be derived from Eqs. 98 and 102, Ch. IX. When the Q_0 of the circuit is 13.3 and the parameters in the circuit are independent of frequency, the value of the ratio is 5 per cent. It may also be shown that, under these conditions, the power output at the second-harmonic frequency is less than 0.1 per cent of the power output at the fundamental frequency. The reason that the power ratio is so much smaller than the impedance ratio is that the power output is related to the real part of the impedance, and the impedance at the second-harmonic frequency has a very low power factor.

It may be concluded, then, that if the power output at the second-harmonic frequency is to be less than 0.1 per cent of the fundamental power output, the loaded Q_0 of the simple parallel-tuned circuit should be not less than about 13. Note, however, that this result must not be applied to a coupled circuit such as that of Fig. 13a. The loaded Q_0 of the primary coil of that circuit is the ratio of the reactance to the resistance in the equivalent circuit of Fig. 13b, but Fig. 13b is an equivalent circuit only at the resonant frequency. The reflected impedance is a function of frequency; thus, at any frequency other than the resonant frequency, the parameters in the circuit are different, and the simple relations regarding the impedance, phase angle, and selectivity given in Art. 9, Ch. IX, do not apply. It follows that the loaded Q_0 of the equivalent circuit has no simple, generally applicable relation to the resonance curve of the input impedance of a complicated tuned network, and conclusions regarding the input impedance at harmonic frequencies or the selectivity cannot be drawn from the loaded Q_0 of the primary circuit for coupling networks in general.

PROBLEMS

1. A Type 6N7 twin triode is used in a push-pull power-amplifier circuit similar to that of Fig. 35, Ch. VIII. The tubes are biased to cut-off (-7.5 volts) so that the operation is in Class B. The plate-supply voltage is 250 volts, and the input grid-signal voltage from one grid to the center tap of the transformer is given by

$$e_g = 22.5 \sin \omega t \text{ volts.}$$

Assume that the output transformer is ideal.

(a) What output-transformer turns ratio N_1/N_2 gives the maximum power transfer to the load if R_L is 600 ohms?

(b) Determine the power output, plate dissipation, and plate efficiency corresponding to the grid-signal voltage given above.

2. Two triodes are operated in a push-pull power-amplifier circuit similar to that of Fig. 35, Ch. VIII, as an audio-frequency Class B_1 amplifier. A plate-supply voltage of 300 volts and a fixed grid-bias voltage of -30 volts are used. The tubes are coupled to a resistance load of 250 ohms by means of a transformer whose characteristics may be assumed to be ideal. In order to simplify the analysis, the plate characteristics of the tubes may be assumed to be straight, parallel, equally spaced lines, corresponding to an amplification factor μ of 10 and plate resistance r_p of 5,000 ohms for each tube. A sinusoidal grid-signal voltage having the maximum allowable amplitude for Class B_1 operation is used.

(a) What should the turns ratio of the transformer be to give maximum power transfer to the load?

(b) Find the Fourier series for the individual plate currents when the transformer turns ratio found in (a) is used. Give numerical values of the coefficients up to that of the 4th harmonic.

(c) Repeat (b) for the load current and the plate-supply current.

(d) What are the total plate dissipation and the plate efficiency for the two tubes?

3. A Class B audio-frequency push-pull power amplifier utilizes two triode-connected Type 6K6-GT pentodes to supply power to a resistance load of 60 ohms through a 10/1 output transformer whose characteristics may be assumed to be ideal. The plate-supply voltage is 300 volts.

(a) Compare the power output and plate efficiency for maximum Class B_1 conditions with those for Class B_2 conditions when the grid-signal-voltage amplitude under the latter conditions is limited to the value that corresponds to a maximum positive instantaneous grid voltage of 30 volts. Neglect harmonic generation in the stage.

(b) Plot the plate dissipation of the stage as a function of the grid-signal voltage. Show from this plot that the plate dissipation is not a limiting factor in determining the maximum allowable power output of this arrangement.

4. A current source delivers, to the circuit of Fig. 14, a flat-topped wave that can be represented by the Fourier series

$$i = 0.5 + \frac{4}{\pi}[\sin 377t + \tfrac{1}{3}\sin(3 \times 377t) + \cdots],$$

where i is in amperes. Measurements of the parameters of the tuned load show that L is 0.5 henry, C is 3.52 μf, and R is 10.0 ohms.

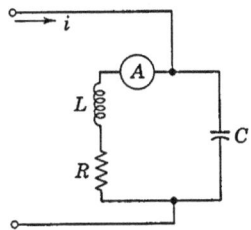

Fig. 14. Circuit for Prob. 4.

(a) Find the rms value of the current indicated by the ammeter A.

(b) Find the rms value of the voltage developed across the capacitor C.

5. A triode is used in the tuned amplifier circuit shown in Fig. 3. The plate characteristics of the tube are given by

$$i_b = 2 \times 10^{-4}(8e_c + e_b) \text{ amp}, \qquad \text{for } (8e_c + e_b) > 0,$$
$$i_b = 0, \qquad \text{for } (8e_c + e_b) < 0.$$

The plate-supply voltage E_{bb} is 300 volts, the grid-bias voltage E_{cc} is -37.5 volts, and the grid-signal voltage is given by

$$e_g = 50 \cos (2\pi \times 10^6)t \text{ volts.}$$

The resonant circuit is tuned to the grid-signal frequency, and at that frequency has an impedance of 10,000 ohms and a Q_0 of 50. Determine:

  (a) the class in which the amplifier operates,
  (b) the average value I_{bs} of the plate current,
  (c) the effective values of the fundamental and second-harmonic components of the voltage across the tuned circuit,
  (d) the a-c power output,
  (e) the plate dissipation.

6. A triode having an amplification factor μ of 10 and a plate resistance r_p of 3,000 ohms is used with a plate-supply voltage of 1,000 volts as a Class B tuned amplifier. The grid-signal voltage is adjusted so that the circuit operates with a plate efficiency of 60 per cent and a value of $e_{b_{min}}$ equal to that of $e_{c_{max}}$.
What are the tuned load impedance, the rms grid-signal voltage, and the a-c power output?

7. A transmitting triode has the following ratings as a Class B r-f amplifier:

$$\text{Amplification factor} = 25$$
$$\text{Plate resistance} = 6{,}300 \text{ ohms}$$
$$\text{Maximum direct plate voltage} = 2{,}500 \text{ volts}$$
$$\text{Maximum direct plate current} = 0.225 \text{ amp}$$
$$\text{Maximum plate dissipation} = 250 \text{ watts.}$$

This tube is to be used as a Class B tuned amplifier with a parallel-tuned load circuit having an effective loaded Q_0 of 10. The grid-signal voltage is sinusoidal at a frequency of 500,000 cps. It is desired to secure the maximum output possible without exceeding any of the rated values for the tube. A d-c generator having a substantially constant terminal voltage of 2,000 volts is available as a power source.

Specify the operating conditions for this amplifier, including the following:

  (a) Impedance of the tuned circuit at resonance, R_t,
  (b) Direct grid-bias voltage, E_{cc},
  (c) Direct plate current, I_{bs},
  (d) Fundamental component of the alternating plate current,
  (e) Effective value of the voltage across the tuned load circuit,
  (f) Effective value of the alternating component of grid voltage,
  (g) Effective value of the circulating current in the tuned load circuit,
  (h) A-c power output,
  (i) Power supplied by the direct plate-voltage source,
  (j) Plate dissipation,
  (k) Plate efficiency,
  (l) Inductance in the tuned circuit,
  (m) Capacitance in the tuned circuit.

8. A transmitting triode has the following ratings for use as a Class B r-f amplifier:

$$\mu = 12.6,$$
$$r_p = 4{,}000 \text{ ohms,}$$
$$\text{Maximum direct plate voltage} = 3{,}000 \text{ volts,}$$
$$\text{Maximum plate dissipation} = 150 \text{ watts.}$$

If this tube is used with a parallel-tuned load circuit having a coil with an effective loaded Q_0 of 10, specify the following for optimum operating conditions at a frequency of 10^6 cps:

(a) Impedance of the tuned circuit at resonance, R_t,
(b) Direct grid-bias voltage, E_{cc},
(c) Direct plate current, I_{bs},
(d) Effective value of the circulating current in the tuned load circuit,
(e) Effective value of the alternating component of grid voltage,
(f) Effective value of the voltage across the tuned load circuit,
(g) A-c power output,
(h) Power furnished by the plate-power supply,
(i) Plate efficiency,
(j) Inductance in the tuned circuit,
(k) Capacitance in the tuned circuit.

9. A transmitting triode having the constant-current characteristics shown in Fig. 9 is used as a Class C amplifier with a plate-supply voltage of 550 volts, a grid-bias voltage of twice the cut-off value, a peak plate current of 350 ma, and a value of $e_{b_{min}}/e_{c_{max}}$ equal to 1.5. Determine:

(a) A-c power output,
(b) Power furnished by the plate-power supply,
(c) Plate efficiency,
(d) Plate dissipation,
(e) Grid-driving power,
(f) The values of L, R, and C required in the tuned load circuit if the loaded Q_0 is 12 and the frequency of the grid-signal source is 10^6 cps.

10. A transmitting triode having the constant-current characteristics shown in Fig. 9 is used as a Class C amplifier with a tuned plate load at a frequency of 100,000 cps. If the plate-supply voltage is 750 volts, the grid-bias voltage is -60 volts, the peak plate current is 200 ma, and a sinusoidal grid-signal voltage of 75.5 volts rms is applied, what are:

(a) The rms value of the alternating component of the voltage across the tuned circuit?
(b) The portion of each cycle during which plate current flows?
(c) The power supplied by the plate-power supply?
(d) The a-c power output to the tuned circuit?
(e) The plate efficiency?

11. The transmitting triode that has the constant-current characteristics shown in Fig. 9 has the following ratings for operation as a Class C amplifier at all frequencies less than 60 megacycles per sec:

$$\text{Maximum direct plate voltage} = 750 \text{ volts}$$
$$\text{Maximum direct plate current} = 105 \text{ ma}$$
$$\text{Maximum direct grid current} = 35 \text{ ma}$$
$$\text{Maximum plate dissipation} = 25 \text{ watts.}$$

This tube is used as a Class C amplifier with a grid-signal frequency of 1.595 megacycles per sec, a plate-supply voltage of 750 volts, a grid-bias voltage of three times the cut-off value, a peak plate current of 350 ma, and a value of $e_{b_{min}}/e_{c_{max}}$ equal to 1.5.

(a) Find:
    (1) A-c power output,
    (2) Power supplied by the plate-power supply,
    (3) Plate efficiency,
    (4) Plate dissipation,
    (5) Grid-driving power,
    (6) Impedance of the tuned circuit at resonance,
    (7) The values of L and C in the tuned circuit if the loaded resonant Q_0 is 12.

(b) Which, if any, of the rated values of the tube are exceeded?

(c) What changes in the operating conditions would you recommend, and why?

CHAPTER XI

Vacuum-Tube Oscillators

In amplifiers, direct-current power from the plate-power supply is converted by the electron tube into alternating-current power at the load, the process being controlled by a separate, or external, source connected in the grid circuit. Though this controlling grid voltage is necessary for the control of the power, the output power is obtained primarily from the direct-current source, not from the source of grid voltage.

The electron tube is also used to convert direct-current power to alternating-current power in circuits called *vacuum-tube oscillators;** in these, however, the conversion is effected without the aid of an external source of controlling voltage. The frequency and waveform of the sustained oscillations that they generate consequently depend only upon the circuit elements and the way in which they are connected.

In general, any device that amplifies or that possesses a negative dynamic resistance can be made to form an oscillator. For example, if, in place of the external source, part of the output power from an electronic amplifier is used to supply the controlling grid voltage and associated grid power, the remainder of the output power is available to the load. Analogous conditions may exist in electric circuits that do not contain vacuum tubes, but do contain other amplifiers such as transistors and magnetic amplifiers. Furthermore, examples of self-excited oscillations are to be found in nonelectrical systems such as the spring- or weight-driven clock, the steam-driven reciprocating feed-water pump, the pneumatic drill, the pneumatic windshield wiper, and the reciprocating steam engine. In all of them, a steady force results in a steady-state periodic motion.

The vacuum-tube oscillator can produce, from a direct-current supply, an alternating current having a constancy of frequency and a purity of waveform far in excess of that generated by any other means at present available. It can operate at frequencies from a few cycles per hour up to several hundred million cycles per second with almost equal ease, and produce power outputs up to hundreds of kilowatts. The oscillator is hence one of the more important

---

* A rectifier converts alternating-current power into direct-current power. By contrast, devices for converting direct-current power into alternating-current power by means of gas tubes are often called *inverters,* but the oscillator and the self-excited inverter involve essentially the same type of circuit.

applications of vacuum tubes and finds a wide variety of uses. Modern electrical communication systems employ vacuum-tube oscillators, in power ratings from a fraction of a watt up to hundreds of kilowatts, for many different functions. Radio and television broadcasting and other developments of the communication art have been made possible by this means of generating very high frequencies. For measurements and induction heating, and generally as auxiliary equipment, the oscillator finds extensive application in all branches of electrical engineering.

1. TYPES OF OSCILLATORS

Of the great variety of oscillator circuits that have been developed, all contain some device for converting the direct-current power into the desired alternating-current output power. This device cannot be an ordinary linear resistance, capacitance, self-inductance, or mutual inductance, for a basic principle of the theory of circuits containing only these elements is that, in the steady state, all currents and voltages caused by a single source of direct current or sinusoidal alternating current must be of the same frequency as the source. Oscillators hence are conveniently classified according to the particular device that makes oscillations possible.

One class comprises the *negative-resistance oscillators*, which are those that contain a circuit element having a current-voltage characteristic curve of negative slope within some range of operation, that is, having a negative dynamic resistance within this range. In contrast to the fact that power is absorbed by a positive resistance, power is delivered by a negative resistance. Since the direct-current supply cannot deliver alternating-current power, the negative resistance in these oscillators—produced by the direct-current power supply, tube, and circuit—serves as a source of alternating-current power; it converts the direct-current power into alternating-current power at a frequency determined by the circuit.

A second class contains *feedback oscillators*. In these, the particular device that makes the oscillations possible is an amplifier, and a circuit connection is made to return part of the amplifier output to the input circuit. If this connection is so made as to supply alternating components of input voltage and power of proper magnitude and phase to sustain the alternating components of the output voltage and power, the amplifier supplies its own excitation and produces sustained oscillations. The division between feedback and negative-resistance oscillators is not sharply defined, for many oscillators may be considered as belonging to either class in accordance with the point of view taken in the analysis of them. However, in most oscillators, either a negative-resistance element or an amplifier with a feedback connection is prominent as the means for producing the oscillations, and the oscillator is considered to belong to the corresponding class.

A second convenient method of classification of oscillators depends upon the form of the output wave. Those that produce a nearly sinusoidal waveform may be called *sinusoidal oscillators*. Those for which the output waveform is markedly nonsinusoidal may be called *relaxation oscillators*, since their oscillations are characterized by a sudden change, or relaxation, from one state of unstable equilibrium to another. The name "relaxation oscillator" was first applied to those oscillators having a circuit in which the oscillation is produced through the periodic charge and discharge of a capacitor in series with a resistor. Accordingly, relaxation oscillators have often been defined as those in which combinations of resistance and capacitance determine the frequency.[1] Since waves of relaxation form may also be produced by oscillators in which inductances are important in the determination of the frequency, and since sinusoidal waves are produced by some oscillators containing resistance and capacitance, limitation of the term "relaxation" to oscillators containing only resistance and capacitance is not desirable, and definition of the terms on the basis of the waveform is to be preferred. Classification of oscillators on the basis of the two types of waveform produced is, furthermore, in accordance with the fact that different methods of analysis are suited to oscillators having the two types of waveform. Again, the line of demarcation between the classes is not sharp, for in many oscillators the output wave may be made to vary continuously from sinusoidal to relaxation form through the change of a circuit parameter.

Oscillators of any of these classes may take many physical forms. Connection of tuned circuits or other linear networks to the vacuum tube in different ways produces a wide variety of oscillators. Some oscillators, such as *piezoelectric* (or *crystal-controlled*) *oscillators* and *magnetrostriction oscillators*, contain mechanically resonant elements coupled to the electric circuit. Others contain special tubes having negative-resistance characteristics; examples are the *dynatron oscillator* (see Ch. IV), the *magnetron oscillator* (see Ch. I), the *Barkhausen-Kurtz oscillator*, the *Gill-Morrell oscillator*, and the *gas-diode relaxation oscillator*.

Because there are so many oscillator circuits, an attempt to analyze

---

[1] H. A. Thomas, *Theory and Design of Valve Oscillators* (London: Chapman & Hall, Ltd., 2nd ed.; 1951), 210, 215; W. A. Edson, *Vacuum-Tube Oscillators* (New York: John Wiley & Sons, Inc., 1953), 59-62, 265-310.

all of them here is not practicable. Instead, a discussion of some of the fundamental principles of feedback oscillators is given, and an illustrative application of these principles to the tuned-plate oscillator is made. The amplifier or negative-resistance element in an oscillator cannot be completely linear because, as is discussed in Art. 3, some nonlinearity is always needed to limit the amplitude of the oscillations. Hence no analysis based on a linear approximation can be complete; in particular, accurate conclusions as to amplitude and frequency stability are not obtainable by such methods. For this reason, adequate analysis of oscillators[2] is often a difficult problem, and graphical or experimental methods of design and adjustment are sometimes expedient. Certain approximate and relatively simple analyses, such as those in the remainder of this chapter, however, yield valuable information.

2. CONDITIONS FOR SELF-EXCITATION IN FEEDBACK OSCILLATORS

All the many types of feedback oscillators utilize the principle that any device having a periodic output with an output power greater than the required controlling input power can be made self-exciting if a definite part of the output is fed back in the proper magnitude and phase to the input. There are many analyses that give the approximate conditions for self-excitation. One simple one consists of a determination of the conditions under which the damping term becomes zero in the differential equation for the circuit.[3] Another is based on the Kirchhoff's law equations for the steady-state conditions in the circuit. Since the circuit contains no alternating-voltage source, the determinant of the coefficients in the simultaneous equations for the currents must vanish in order that an alternating current greater than zero may exist in the circuit. The analysis chosen here differs from both the foregoing analyses and is similar in some respects to that of Barkhausen.[4] It emphasizes the feedback feature of the circuit and follows closely the analysis of inverse-feedback amplifiers given in Art. 12, Ch. IX, where the possibility of oscillations is mentioned. Still

[2] P. le Corbeiller, "The Nonlinear Theory of the Maintenance of Oscillations," *I.E.E.J.*, *79* (1936), 361–378; B. Van der Pol, "The Nonlinear Theory of Electric Oscillations," *I.R.E. Proc.*, *22* (1934), 1051–1086; J. G. Brainerd and C. N. Weygandt, "Unsymmetrical Self-Excited Oscillations in Certain Simple Nonlinear Systems," *I.R.E. Proc.*, *24* (1936), 914–922.

[3] H. J. Van der Bijl, *The Thermionic Vacuum Tube and Its Applications* (New York: McGraw-Hill Book Company, Inc., 1920), 266–285.

[4] H. Barkhausen, *Lehrbuch der Elektronen-Röhren*, *Band III* (3rd and 4th ed.; Leipzig: Verlag S. Hirzel, 1935).

another approach, which is sometimes useful, views the combination of the tube and feedback connection as a negative-resistance element.

The usual single-tube feedback-oscillator circuit can generally be represented as shown in Fig. 1a, where a triode and a coupling network are employed. Initially, in this analysis, the tube is assumed to operate as a linear Class A_1 amplifier with a sinusoidal input voltage; hence sinusoidal quantities only are considered, and the currents and voltages are representable by complex or vector symbols with arbitrarily assigned positive reference directions. In a practical oscillator

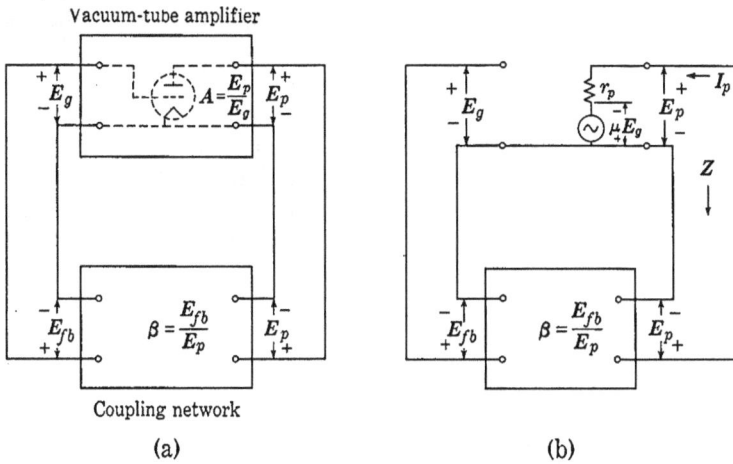

Fig. 1. Feedback-oscillator circuit.

circuit, the operation of the tube may be Class B or Class C for the sake of improved efficiency, and a resonant circuit is then usually employed to minimize the harmonics in the currents and voltages, as discussed in Ch. X.

The complex voltage amplification of the amplifier is, as previously defined,

$$A \equiv E_p/E_g \,. \qquad \blacktriangleright[1]$$

Similarly, the complex voltage ratio β in the feedback coupling network may be defined as

$$\beta \equiv E_{fb}/E_p \,. \qquad \blacktriangleright[2]$$

Suppose that initially E_g is supplied from an external source not shown in the diagram. If the coupling network has such characteristics that the voltage E_{fb} equals the voltage E_g in both magnitude and phase, the voltage E_{fb} may be substituted for the externally supplied E_g without observable alteration in the voltages or currents in the circuit. The tube then furnishes its own excitation, and it will continue to sustain the alternating current without the help of a separate alternating-current source; it generates self-excited oscillations. The question of the stability of the amplitude of the oscillations is reserved for consideration later in connection with a specific example.

When E_{fb} in Eq. 2 is equated to E_g in Eq. 1, the resulting relation

$$\beta = 1/A \qquad \blacktriangleright[3]$$

is the condition for oscillation, also derived in Art. 12, Ch. IX. Generally, the magnitude of A is larger than unity; hence that of β must be less than unity, and the larger the voltage gain of the amplifier the smaller is the value of the feedback voltage ratio required to give steady oscillations.

When the amplifier comprises a single tube operating under linear Class A_1 conditions, the grid current is negligible, and the voltage gain can be evaluated from the equivalent circuit shown in Fig. 1b. If Z denotes the impedance as viewed from the plate side of the tube into the coupling network,

$$E_p = -\mu E_g \frac{Z}{r_p + Z} ; \qquad [4]$$

whence

$$A = -\mu \frac{Z}{r_p + Z} , \qquad [5]$$

as derived in Art. 6, Ch. VIII. From this relation,

$$\frac{1}{A} = -\left(\frac{1}{\mu} + \frac{r_p}{\mu Z}\right) = -\left(\frac{1}{\mu} + \frac{1}{g_m Z}\right). \qquad [6]$$

The substitution of Eq. 6 into Eq. 3 gives the *Barkhausen criterion* for self-excitation of the single-tube oscillator, namely,

$$\beta = -\left(\frac{1}{\mu} + \frac{1}{g_m Z}\right) \quad \text{(for sustained oscillations).} \qquad \blacktriangleright[7]$$

The negative sign is equivalent to a 180-degree phase shift in the coupling network. This complex or vector relationship contains two factors, μ and g_m, that depend on the tube alone; and two others, β and Z, that are determined solely by the circuit. The values of the latter are substantially independent of the current and voltage magnitudes when the circuit external to the tube is linear.

The significance of the complex form of the self-excitation criterion, Eq. 7, should be understood. The quantities Z and β are both complex and are functions of the frequency. If the polar form is used, the definitions

$$\beta \equiv \beta \epsilon^{j\phi} \qquad [8]$$

and

$$\frac{1}{\mu} + \frac{1}{g_m Z} \equiv D \epsilon^{j\theta} \qquad [9]$$

may be made. The self-excitation criterion then becomes

$$D \epsilon^{j\theta} = \beta \epsilon^{j(\phi + \pi)}, \qquad [10]$$

where the angle π takes the place of the minus sign in Eq. 7. This equality can be satisfied only if the magnitudes of the two complex quantities and their angles are separately equal. Thus Eq. 10 is equivalent to the two following equations, which give the magnitude and the angle equilibriums separately:

Magnitude equilibrium: $D = \beta,$ [11]

Angle equilibrium: $\theta = \phi + \pi.$ [12]

From Eq. 9,

$$D = \frac{1}{\mu} \left| 1 + \frac{\mu}{g_m Z} \right| = \frac{1}{\mu} \left| 1 + \frac{r_p}{Z} \right|. \qquad [13]$$

When substituted into Eq. 11, the relation for magnitude equilibrium, Eq. 13, leads to the relation

$$\beta = \frac{1}{\mu} \left| 1 + \frac{r_p}{Z} \right| \quad \text{(for sustained oscillations).} \qquad [14]$$

This relation shows that the required magnitude of the feedback ratio is reduced if the amplification factor of the tube is large compared with unity, and Z, the magnitude of the plate-circuit impedance, is large compared with the plate resistance of the tube. Often a parallel-resonant circuit is included in the network, and, as is evident in Fig. 23, Ch. IX, it is then easier to excite oscillations in the neighborhood of the resonant frequency. At the resonant frequency, Z may be a pure resistance of magnitude R_t; hence θ may be zero, and the relation giving the angle equilibrium, Eq. 12, requires that the coupling network in Fig. 1 transform E_p to E_g with a phase shift of 180 degrees. This fact means qualitatively that the instantaneous grid potential must increase when the instantaneous plate potential decreases. If the angle of the feedback ratio β differs from 180 degrees, the impedance Z must also have an angle different from zero, and the frequency of the oscillations must depart slightly from the resonant frequency of the tuned circuit. Further discussion of this question is given in the next article in connection with a particular oscillator circuit.

3. CONDITIONS FOR OSCILLATION IN THE TUNED-PLATE OSCILLATOR

As an example of the use of the Barkhausen self-excitation criterion, Eq. 7 may be applied to the commonly used oscillator circuit shown in

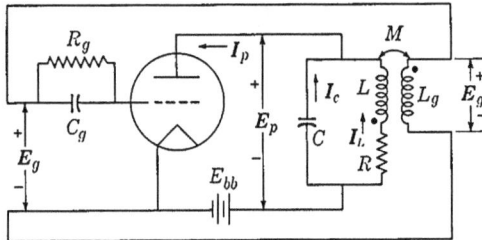

(a) Circuit diagram

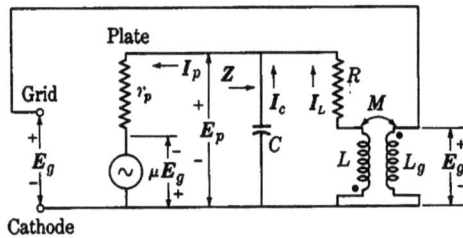

(b) Incremental equivalent circuit
for Class A_1 operation

Fig. 2.   The tuned-plate oscillator circuit.

Fig. 2. A parallel-resonant circuit is located directly in series with the plate of the tube; hence the oscillator is called the *tuned-plate oscillator*. The self-excitation voltage is supplied by induction from the current I_L through one coil of the mutual inductance M. A grid leak and capacitor are usually included as shown in the circuit, to supply a grid-bias voltage.

Ordinarily, nonlinearity is deliberately employed in this circuit—that is, the tube does not operate as a linear Class A_1 amplifier—for two reasons. In the first place, nonlinearity must be present in the grid circuit to cause rectification if the grid leak and capacitor are to supply the grid-bias voltage, and appreciable grid current must exist for at least part of the cycle, as is discussed in Art. 4c, Ch. X. A desirable result of the use of the grid leak and capacitor, explained in a subsequent discussion of the transient behavior of the oscillator given in Art. 4, is that they have an important stabilizing effect on the steady-state behavior. In the second place, in order that the plate efficiency of the circuit may be high, the grid-bias voltage is ordinarily adjusted so that the operation of the tube is in Class C. The plate current hence flows in short pulses, and nonlinearity occurs in the plate circuit.

Because of the nonlinearity, both the grid and the plate currents contain appreciable harmonics. The Barkhausen criterion is not directly applicable to such circuit behavior, because it is developed on the assumption that the currents and voltages are sinusoidal, and it therefore cannot include the effects of nonlinearity. A first approximation to the behavior of the circuit shown in Fig. 2a can, however, be found through application of the Barkhausen criterion if, as is done below, the nonlinearity of the grid current is eliminated from consideration by the assumption that the grid current is zero at all times, and the nonlinearity in the plate circuit is eliminated by the assumption that the operation is linear and in Class A. Linear Class A_1 operation is therefore assumed. Some oscillators, but not those having the circuit of Fig. 2a, actually do operate under these conditions.

The equivalent circuit for the oscillator, based on the assumption of linear Class A_1 operation of the tube, is shown in Fig. 2b. The quantities to be found from the circuit for substitution in the Barkhausen criterion are β and $\mathbf{Z}$. The expression for β is

$$\beta = \frac{\mathbf{E}_g}{\mathbf{E}_p} = \frac{j\omega M \mathbf{I}_L}{-(R + j\omega L)\mathbf{I}_L} = -\frac{j\omega M}{R + j\omega L}, \qquad [15]$$

where the minus sign results from the assigned positive reference directions of the currents and voltages in the figure. The impedance $\mathbf{Z}$ as viewed from the tube is

$$\mathbf{Z} = \frac{(R + j\omega L)\dfrac{1}{j\omega C}}{R + j\left(\omega L - \dfrac{1}{\omega C}\right)}; \qquad [16]$$

that is, $\mathbf{Z}$ is the impedance of the inductor and the capacitor in parallel, since no energy is assumed to flow through the mutual inductance to the grid circuit. When these values of β and $\mathbf{Z}$ are substituted in the Barkhausen self-excitation criterion, Eq. 7, and the real and imaginary parts are separated, the result may be put in the form

$$\left[\frac{L}{\mu C} + \frac{R}{g_m} - \frac{M}{C}\right] + j\left[\frac{1}{g_m}\left(\omega L - \frac{1}{\omega C}\right) - \frac{R}{\mu \omega C}\right] = 0. \qquad [17]$$

The circuit must therefore satisfy Eq. 17 if oscillations are to be sustained. This relation is equivalent to the two relations obtained through equating to zero the real and the imaginary parts separately, because a complex expression vanishes only if both its real and imaginary parts vanish. The equivalent relations are

$$\frac{L}{\mu C} + \frac{R}{g_m} - \frac{M}{C} = 0 \qquad [18]$$

and

$$\frac{1}{g_m}\left(\omega L - \frac{1}{\omega C}\right) - \frac{R}{\mu\omega C} = 0, \qquad [19]$$

both of which must be satisfied for self-excited oscillations to occur.

In this particular example, Eq. 18 for the real parts does not contain the frequency as a variable. The value

$$g_m = \frac{\mu RC}{\mu M - L}, \qquad \blacktriangleright[20]$$

obtained from it, is the value the mutual conductance of the tube *must* have for sustained oscillations with a specified set of values of μ, R, C, L, and M. From Eq. 19 for the imaginary parts, the square of the angular frequency is found to be

$$\omega^2 = \frac{1}{LC}\left(1 + \frac{g_m R}{\mu}\right); \qquad [21]$$

and, by substitution of Eq. 20, the mutual conductance is eliminated, giving

$$\omega^2 = \frac{1}{LC}\left(1 + \frac{R^2 C}{\mu M - L}\right). \qquad [22]$$

Equation 22 is an explicit statement of the square of the angular frequency in terms of the constants in the circuit and the amplification factor of the tube. A more useful form of the expression for the angular frequency is obtained if g_m is left in the equation and ω_0 is substituted. where ω_0 is the resonant angular frequency of the tuned circuit with its resistance neglected; that is,

$$\omega_0 = 1/\sqrt{LC}. \qquad [23]$$

Then

$$\omega^2 = \omega_0^2\left(1 + \frac{g_m R}{\mu}\right), \qquad [24]$$

or

$$\omega = \omega_0\sqrt{1 + \frac{R}{r_p}} \approx \omega_0\left(1 + \frac{R}{2r_p}\right), \qquad R \ll r_p. \qquad \blacktriangleright[25]$$

Equation 25 appears to indicate that the frequency of oscillation is slightly higher than the resonant frequency of the tuned circuit. The amount of the increase is normally small, however, since R is small when the Q_0 of the tuned circuit is large. Furthermore, the effect indicated by Eq. 25 is usually small compared with the effects that result from changes of tube capacitances, the effect of harmonics in the waveform, and so on. An important limitation of the applicability of Eq. 25 is that it results from the assumption that the effect of the loss in the tuned circuit is correctly accounted for by a resistor in series with the inductor. This conclusion is often not true, and the increase of frequency brought about by the loss in the circuit may be expected to be less than that indicated by the equation. If a resistor having a resistance comparable with the resonant reactance of the inductor is inserted in series with the inductor, the effect indicated by the equation may be expected to occur.

Since the external circuit essentially determines the frequency of their oscillations, vacuum-tube oscillators can be made to oscillate over a great range of frequency. It must be recognized, however, that the foregoing analysis is based on lumped-parameter circuit theory and negligible transit time of the electrons in the tube. These approximations fail to represent accurately the conditions in ultra-high-frequency oscillators. In general, when the wavelength becomes comparable with the dimensions of the circuit, or the transit time becomes comparable with the period, other methods of analysis are required.

Although this analysis is valid only when the operation is restricted to the linear region of the tube characteristics, it may be extended to apply approximately to the nonlinear region as well. The amplification factor μ may, with reasonable justification, be assumed to be constant, but g_m, which is the slope of the curves of i_b as a function of e_c, cannot be assumed to be constant for large amplitudes. However, it is shown in Art. 5, Ch. IV, that, when μ is constant, all these curves have the same shape and are displaced from one another along the e_c axis by amounts proportional to the differences in plate voltage among them. Thus the family of curves is representable by the functional relationship

$$i_b = f\left(e_c + \frac{e_b}{\mu}\right), \qquad [26]$$

and may be plotted as a single curve, as is shown in Fig. 3, where

$$e_0 = e_c + \frac{e_b}{\mu}, \qquad [27]$$

and is called the control voltage (see Art. 3, Ch. X, and Art. 9, Ch.

XII). The slope of the $i_b(e_0)$ characteristics is also g_m. For the purposes of this analysis, it is convenient to define the average mutual conductance $\bar{g}_m$ as the slope of the line connecting the two extreme points on the characteristic reached by the alternation, as is depicted in Fig. 3; that is,

$$\bar{g}_m = \frac{\Delta i_b}{\Delta e_0}. \qquad \blacktriangleright [28]$$

In the oscillator circuit, the voltage fed back to the grid circuit is proportional to g_m if the other parameters in the circuit are fixed. If g_m is smaller than the value given by Eq. 20, oscillations must die away, because the voltage fed back is smaller than the grid voltage required to sustain them. If g_m has a value greater than that given by Eq. 20, the voltage fed back exceeds the required grid voltage, and the oscillations tend to increase in amplitude. Hence, on the assumption that the average value of mutual conductance given by Eq. 28 is effective in the circuit, three conditions are possible as the amplitude

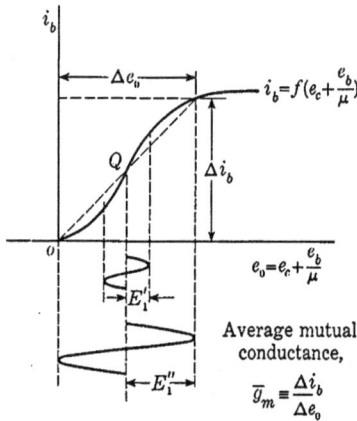

Fig. 3. Relation between the average mutual conductance and the characteristic curve.

and $\bar{g}_m$ vary together, namely,

$$\bar{g}_m \begin{cases} < \dfrac{\omega RC}{\mu M - L} & \text{(decaying oscillations)} \\[2ex] = \dfrac{\mu RC}{\mu M - L} & \text{(sustained oscillations)} \qquad \blacktriangleright [29] \\[2ex] > \dfrac{\mu RC}{\mu M - L} & \text{(growing oscillations).} \end{cases}$$

Suppose that an oscillator with a plate-supply voltage and grid-bias-supply voltage corresponding to the operating point Q in Fig. 3 is so adjusted that initially $\bar{g}_m$ is greater than $\mu RC/(\mu M - L)$. The closing of a switch in the plate circuit, the thermal agitation of the electrons in the conductors, or the fluctuations in the plate current that are always present because of the motion of the discrete electrons are sufficient to start oscillations of very small amplitude. Under the foregoing prescribed conditions, these oscillations build up in amplitude until the value E_1' in Fig. 3 is reached; after which, $\bar{g}_m$ grows smaller, and the building up continues at a decreasing rate until $\bar{g}_m$ just equals $\mu RC/(\mu M - L)$—for example, at E_1'' in Fig. 3. At this amplitude, the oscillations are sustained in the steady state.

Although the nonlinearity of the tube plate characteristics causes harmonics, they are ordinarily minimized in the input and output voltages by the resonant circuit. However, the *nonlinearity of the characteristics is responsible for the stable operation of the oscillator.* This conclusion follows from a consideration of the conditions for

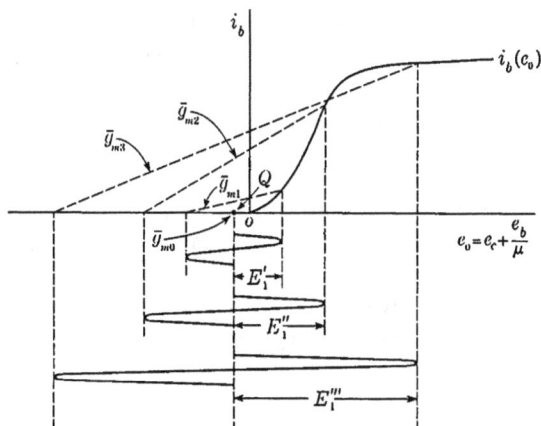

Fig. 4.   Variation of the average mutual conductance with the oscillation amplitude for a grid-bias voltage beyond the cut-off value.

oscillations in a strictly linear circuit. In practice, to initiate oscillations in any oscillator circuit, the condition for build-up given in Eq. 29 must be established. But if the tube characteristics were linear and g_m were a constant, there would be no limiting value to the amplitude once the build-up conditions were established, and a stable state could not be reached. Actually the nonlinearity of the grid-current characteristics also tends to limit the amplitude, as is explained in Art. 4, and it is often used to regulate and stabilize the amplitude of the oscillations.

If the direct-voltage supplies are such as to establish the quiescent operating condition beyond the cut-off value, as at Q in Fig. 4, the initial value of $\bar{g}_m$, denoted by $\bar{g}_{m0}$, is zero, and Eq. 29 shows that oscillations do not tend to start. However, if oscillations are started by some auxiliary means, the value of $\bar{g}_m$ increases as their amplitude increases from zero to E_1' and on to E_1'' in the figure; but further increase of amplitude to E_1''' results in a decrease of $\bar{g}_m$. If the value

g_{m1} that corresponds to $E_1{}'$ is slightly smaller than the value for build-up or maintenance of oscillations in Eq. 29, oscillations will not start of themselves or even with the application of an external grid voltage sufficient to cause the amplitude $E_1{}'$. However, if an amplitude larger than $E_1{}'$, say $E_1{}''$, is established in the circuit by some cause such as the transient voltage produced by closure of the plate-supply switch, and if the corresponding value of average mutual conductance $\bar{g}_{m2}$, which is greater than $\bar{g}_{m1}$, is sufficiently large to produce the build-up condition in Eq. 29, the oscillations tend to increase in amplitude. After increasing somewhat, the alternations become so large that $\bar{g}_m$ is again reduced in magnitude, as is brought out in the figure; and, finally, when the average mutual conductance becomes $\bar{g}_{m3}$, which is assumed to satisfy exactly the condition of $\bar{g}_m$ equal to

$$\mu RC/(\mu M - L),$$

the stable condition is reached.

From this discussion the conclusion can be drawn that, when an oscillator has a fixed grid-bias voltage equal to or greater in magnitude than the cut-off value, it is not self-starting. Such a value of grid-bias voltage is the one most often desired in oscillators, because otherwise the plate efficiency is very low, as is explained in Ch. X. Since, as is explained in the next article, the self-starting feature may be obtained if the grid-leak-and-capacitor combination described in Art. 4c, Ch. X, is used to provide the grid-bias voltage, this combination is frequently employed.

4. BUILD-UP OF GRID-BIAS VOLTAGE IN THE GRID-LEAK-AND-CAPACITOR CIRCUIT

The self-starting feature and a grid-bias voltage of magnitude equal to or greater than the cut-off value are both realized through the use of the grid-leak-and-capacitor combination shown in Fig. 5. As explained in Art. 4c, Ch. X, this method is an economical one for obtaining the grid-bias voltage and is frequently used in oscillator circuits as well as in grid-circuit detectors and Class C amplifiers.

The operation is illustrated by the curves in Fig. 5 and may be described as follows: When the circuit is first put into operation by the closing of the plate-supply switch, the grid-bias voltage E_{cs} is zero, and the quiescent operating point is high on the $i_b(e_c)$ characteristic, where the value of $\bar{g}_m$ is large and the build-up condition is well established. Consequently oscillations start and begin to build up. During the positive half-cycle of grid voltage, a pulse of grid current occurs, and the capacitor C_g receives a charge. The action is similar to that in a half-wave rectifier circuit with a resistance load and a smoothing capacitor. The capacitance of the capacitor should be large enough to have negligible reactance at the operating frequency in comparison with the grid-leak resistance R_g. Nevertheless, C_g can be too large and cause intermittent operation (see Art. 5). Under the proper conditions, then,

$$1/(\omega C_g) \ll R_g; \qquad [30]$$

hence the time constant of the biasing circuit must satisfy the relation

$$R_g C_g \gg 1/(2\pi f). \qquad [31]$$

Thus

$$R_g C_g \gg T/(2\pi), \qquad [32]$$

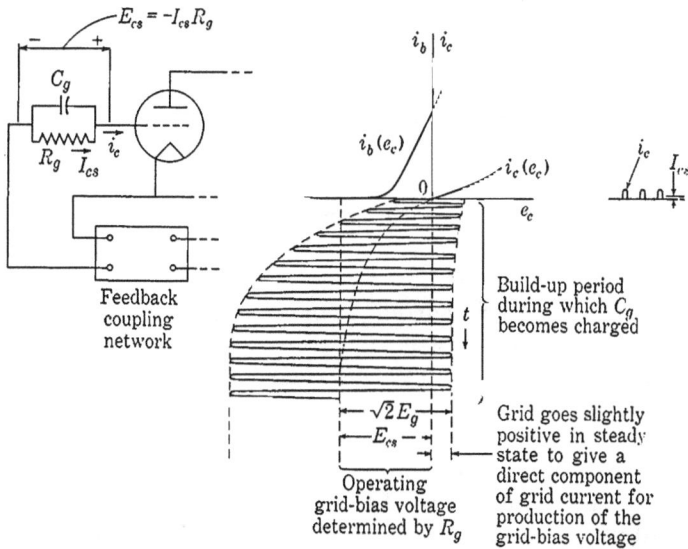

Fig. 5.   Build-up of the grid-bias voltage in the grid-leak-and-capacitor circuit.

where T is the period of the oscillation. During the part of the cycle in which e_c is negative, the charge leaks off the capacitor through the grid leak; but, under the conditions of Eq. 32, the rate of leakage is small, and on the average some charge is accumulated on C_g during the cycle. This process is cumulative, and the charge on the capacitor builds up, the negative grid-bias voltage increases, and the grid-current pulses decrease until, in the steady state, the total charge received by the capacitor during a cycle equals the total charge lost by it through R_g. There can then be no steady component of grid current I_{cs} through the capacitor; the path of the steady component is only through R_g, and that of the alternating components is through C_g.

Through a variation of the resistance R_g, the grid-bias voltage can be regulated to any value between zero and almost $\sqrt{2}E_g$. The value of R_g lies in the range between 10^3 and 10^6 ohms for most types and sizes of tubes. The correct values for R_g and C_g are not critical, and they are often determined by experiment. The general effect of the shift of the operating point to the left that accompanies an increase of R_g or E_g may be considered as a decrease in $\bar{g}_m$. On this basis, together with Eq. 29, the limitation of the amplitude of oscillation by the grid-leak-and-capacitor circuit is explained as a result of the reduction of the average mutual conductance by the automatic grid-bias-voltage regulation characteristic of the circuit. This arrangement also stabilizes the amplitude against other changes in the circuit. For example, if the coupling-network voltage ratio β is increased, E_g also tends to increase, but a small increase in it causes a sharp increase in the magnitude of the grid-bias voltage E_{cs} (which equals $-I_{cs}R_g$) because of the steepness of the $i_c(e_c)$ characteristic, thereby reducing $\bar{g}_m$ and keeping the amplitude of oscillation substantially unchanged.

According to the foregoing discussion, the amplitude of oscillation is always just large enough to allow a small grid current. Variations in the direct component of grid current as measured with a direct-current ammeter are used in practice to indicate the presence of oscillations and to determine roughly their relative amplitude. The chief objectionable effect of this grid current is that it introduces an unwanted damping in the coupling network and thereby reduces the value of β and lowers the amplitude of oscillation. In addition, when the oscillator is overexcited, the reduction in the plate current caused by the grid current's subtracting from the cathode current lowers the $i_b(e_c)$ characteristic, thereby reducing $\bar{g}_m$ and the amplitude of the oscillations.

5. DESIGN OF THE TUNED-PLATE OSCILLATOR

Although the analysis in the previous articles gives several significant conclusions regarding the operation of the circuit, that analysis is often not the most practical method to use in the actual design of the circuit. Since the tube operating conditions in most highly efficient oscillators are similar to those in a Class C amplifier, a method capable of giving more precise results for the design of many power oscillator circuits is one that utilizes an analysis of the Class C amplifier such as that given in Art. 4, Ch. X, with an extension to include the effect of the feedback circuit. For example, the circuit of Fig. 2a is identical with that of the Class C amplifier in Fig. 12a, Ch. X, except for the addition of a coil L_g, which is inductively coupled to the coil L in the parallel-resonant circuit and is used to supply the grid-excitation voltage E_g. The determination of the equilibrium operating conditions of the tuned-plate oscillator may therefore be based on the design of the Class C amplifier, but the stability of the equilibrium conditions should be investigated and verified, as discussed subsequently.

The equivalent circuit in Fig. 2b is based on linear Class A_1 operating conditions for the tube. When Class C operating conditions are involved and the resonant circuit has a large Q_0, only the fundamental component of the plate current is of interest, and I_{p1} may be substituted for I_p in Fig. 2b. Also, the plate resistance r_p is changed to a fictitious plate resistance $\bar{r}_p$, where $\bar{r}_p$ equals $\mu/\bar{g}_m$ and $\bar{g}_m$ has the significance discussed in Art. 3.

On the assumption that the effects of the load and of the losses in the tuned circuit are such as to be correctly representable by the effect of a resistance R as shown in series with the inductor, and with the foregoing modifications made in the equivalent circuit of Fig. 2b, the vector diagram for the oscillator can be constructed. The voltage E_g supplied to the grid circuit in Fig. 2b is $j\omega M I_L$ if grid current is neglected, and inspection of the vector diagram in Fig. 12b, Ch. X, shows that the phase of a vector $j\omega M I_L$ is not the same as that of the vector E_g in a Class C amplifier. In the oscillator circuit, however, these two vectors are constrained to equality in both amplitude and phase. Hence the vector diagram of the oscillator must differ from that of the Class C amplifier. The difference lies in the phase relations, as may be seen upon construction of a diagram, such as Fig. 6, that satisfies the constraint. The connection of the mutual inductance must be made to satisfy the constraint, as is indicated by the polarity dots in Fig. 2. The order of constructing the vectors, which is entirely arbitrary, and the equations used are:

(a) I_L

(b) $I_L R$

(c) $I_L j\omega L$

(d) $-E_p = I_L(R + j\omega L)$

(e) $I_c = (-E_p)j\omega C$

(f) $I_{p1} = I_L + I_c$

(g) $I_{p1}\bar{r}_p$

(h) $\mu E_g = -E_p + I_{p1}\bar{r}_p$

(i) $E_g = \dfrac{\mu E_g}{\mu}$

(j) $E_g = j\omega M I_L$ (check)

The resulting vector diagram for the tuned-plate oscillator is seen to be similar to that of the Class C amplifier except for a phase shift among the vectors. Since the vector I_{p1} leads the vector $-E_p$, the input impedance of the load circuit is somewhat capacitive, which means that the frequency at which the circuit operates is slightly higher than the resonant frequency. This result is in agreement with that given in Eq. 25.

This vector diagram, like Eq. 25 for the angular frequency of oscillation, applies only to an oscillator for which the losses in the tuned circuit may be represented by a resistance in series with the inductance. Frequently, a representation of the losses by a resistance

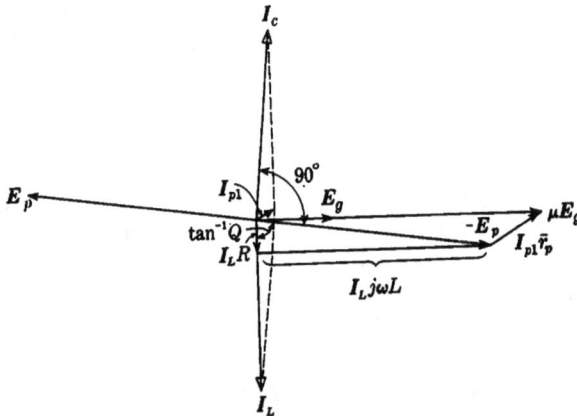

Fig. 6.   Vector diagram for the tuned-plate oscillator of Fig. 2.

in parallel with the inductor and capacitor is more nearly correct. Then E_g and E_p are exactly 180 degrees out of phase, and the frequency of oscillation is exactly the resonant frequency of the tuned circuit, except for additional effects not included in this analysis— changes of tube capacitances, harmonics in the waveform, grid current, and so on. Even if the series-resistor representation of the loss is appropriate, the magnitude of the phase shift of the vectors in the tuned-plate oscillator is generally much smaller than that indicated in Fig. 6. In this figure, the vectors are not drawn to a common scale, and the angles are exaggerated for clarity of the diagram. The actual magnitude of the phase shift may be estimated from a knowledge of the loaded Q_0 of the primary circuit in the tube-to-load coupling network of the oscillator. The practical lower limit of this loaded Q_0 is not as low as that in the Class C amplifier, because there must be sufficient energy storage in the oscillatory circuit to sustain the grid-signal voltage during the intervals when the plate current is zero. A practical lower limit of about 4π for the loaded Q_0 is commonly specified,[5] because lower values tend to give erratic operation. As a result of this limit, the phase relations in a well-designed oscillator differ only slightly from those in a Class C amplifier, and the difference may usually be neglected in design computations.

The design of the tuned-plate oscillator, then, follows closely that of the Class C amplifier. A selection of the tube operating conditions is first made, and the amplitudes of the components of the various currents and voltages are determined. However, in designing the resonant load circuit, it must be recognized that not all the power output from the tube is available for the load, because some of it is used as grid-driving power. The grid-driving power is part of the load for the tube plate circuit, and the additional useful load that may be coupled into the plate circuit is the value that absorbs the difference between the computed power output of the tube and its grid-driving power. The power available for the load at the input of the tube-to-load coupling network is therefore

$$P = E_p I_{p1} - P_{cs} - I_{cs}{}^2 R_g , \qquad\qquad \blacktriangleright[33]$$

and this must be multiplied by the per unit efficiency of the coupling network to give the useful power output. With these additional considerations, the design of the coupling network may be made as for the Class C amplifier. The required value of mutual inductance is then computed from

$$M = \frac{E_g}{\omega_0 I_L} \approx \frac{E_g}{\omega_0 Q_0 I_{p1}} . \qquad\qquad \blacktriangleright[34]$$

In practice, the final adjustment of the mutual inductance is usually made experimentally. Note that this analysis is approximate in several respects, one being that grid current is neglected in computation of the value of M but is considered in determination of the available power output.

Although the assumed operating conditions give results that correspond to an equilibrium in the circuit, it is by no means certain that the equilibrium is stable—in other words, that the amplitudes of the alternating components of the currents and voltages in the circuit will remain constant. The equilibrium may be unstable, with a tendency for the amplitudes to increase or decrease. Hence the stability of the equilibrium should be examined, for example, through a consideration of the effects of a variation of the amplitude of an alternating component of voltage in the circuit from its equilibrium value. In the following discussion, an examination is made of the influence of such a

[5] D. C. Prince, "Vacuum Tubes as Power Oscillators," *I.R.E. Proc.*, *11* (1923), 309–313.

variation on the available power from the tube plate circuit and the power required by the load and the grid circuit.

The method of analysis used here depends on the fact that for non-equilibrium conditions the power output from the plate circuit of the tube is not equal to the sum of the grid-driving power and the load power. For small departures from the equilibrium amplitudes, there is an excess or deficiency of power output from the tube, which is available as an excess or deficiency of grid-driving power and can cause a tendency for the amplitudes to increase or decrease. In particular, the method involves superposed plots of the power required by the load at the input terminals of the coupling network and of the net power output from the tube as functions of the amplitude of the grid-voltage or plate-voltage variations in the range above and below their equilibrium values. The net power output from the tube in these plots is the power output from the plate circuit minus the grid-driving power. If, for an incremental increase in the amplitude, the load power exceeds the net power output, there is a deficiency of available grid-driving power, with a resulting tendency for the amplitude to return to its original value, and the equilibrium is stable. If not, there is an excess of grid-driving power, with a resulting tendency for the amplitude to increase further, and the equilibrium is unstable. Similarly, for an incremental decrease in amplitude, if the load power is less than the net power output, there is an excess of grid-driving power, with a resulting tendency for the amplitude to increase, and the equilibrium is stable. If not, there is a deficiency of grid-driving power, with a resulting tendency for the amplitude to decrease further, and the equilibrium is unstable. A criterion for the stability of the equilibrium operating conditions may therefore be stated as follows: If the curve of load power as a function of amplitude is steeper than the curve of net power output at the selected equilibrium operating conditions, the equilibrium is stable; if not, it is unstable.

In order to apply this criterion to an oscillator circuit, several approximations must ordinarily be made. In the first place, for small variations from the equilibrium values, the ratio of the amplitudes of the alternating components of the plate and grid voltages may be assumed to be fixed by the circuit. Actually some change in this ratio will occur. In the second place, if a grid-leak-and-capacitor circuit is used, an assumption must be made as to the manner in which the grid-bias voltage changes when the amplitude of oscillation changes. With this circuit, the grid-bias voltage is proportional to the direct component of grid current and varies with the amplitude of the alternating component of grid voltage in a complicated manner not readily expressed analytically. This circuit tends to hold a constant value of grid-bias voltage for sudden changes in amplitude, however, since the voltage across the capacitor cannot change suddenly. As a first approximation, therefore, the grid-bias voltage may be assumed to be constant, whereupon the power output, grid-driving power, and load power may be determined by the methods of Art. 4a, Ch. X, for amplitudes slightly larger and slightly smaller than the value that corresponds to the selected operating conditions. The stability criterion may then be applied to the resulting data.

The assumption of constant grid-bias voltage may not always give the correct indication of stability, because of the sluggishness of the resonant circuit in allowing changes of the amplitude of the oscillations. When the resonant circuit has a large loaded Q_0, the ratio of the stored energy to the energy dissipated during a cycle is large. If the power input and output are changed slightly, several cycles may elapse before the increased or decreased value of energy dissipated during a cycle can result in a change in the stored energy and a corresponding change in the amplitude. Consequently, if the time constant of the grid-leak-and-capacitor circuit is only a few times larger than the period of the oscillations, and if the loaded Q_0 of the resonant circuit is large, the grid-bias voltage may readjust itself as rapidly as is necessary to maintain essentially the value determined by the grid leak and the direct component of grid current despite a continuous change in the amplitude. When the stability criterion for these conditions is applied, the net power output should be plotted as a function of amplitude for a varying grid-bias voltage determined by the grid resistor and direct component of grid current. In general, this method requires that a family of these curves be plotted for different fixed grid-bias voltages in the neighborhood of the selected equilibrium value.[6] If the assumption of constant grid-bias voltage shows the equilibrium to be stable, the more complex analysis involving the varying grid-bias voltage also gives the same result, because the grid-leak-and-capacitor circuit tends to increase the grid-bias voltage as the amplitude of the grid-voltage variation increases. This circuit thereby tends to maintain a constant peak grid current and the use of it results in a less rapidly increasing grid-driving power with increasing amplitude than does the use of constant grid-bias voltage. However, an equilibrium condition that appears to be unstable on the basis of the constant grid-bias assumption may be found to be stable when the varying grid-bias voltage corresponding to the grid resistor and direct component of grid current is taken into account. The operating conditions are then stable for a small grid capacitor and a large loaded Q_0

[6] D. C. Prince, "Vacuum Tubes as Power Oscillators," *I.R.E. Proc.*, 11 (1923), 309–313.

in the resonant circuit, but may be unstable with intermittent operation for the opposite conditions.* The oscillations start and stop intermittently as the grid capacitor charges and discharges—so-called *squegging* occurs. If the circuit is designed so that sufficient grid-bias voltage develops during the first cycle, the oscillation terminates before the cycle is complete, and the circuit is called a *blocking oscillator*.

The oscillator with Class C operating conditions is even more closely analogous to the pendulum clock than is the Class C amplifier. As mentioned in Art. 4b, Ch. X, the pendulum is analogous to the resonant circuit. In the oscillator, the feedback of voltage to the grid circuit is analogous to the coupling mechanism that enables the pendulum to drive the escapement, which in turn converts the direct-acting force of the spring or weight into a synchronized pulsating force. Similarly the tube converts the steady source of voltage in the plate-power supply into a synchronized pulsating source to drive the resonant circuit.

6. OTHER FEEDBACK-OSCILLATOR CIRCUITS

Among the large number of feedback-oscillator circuits, a few are basic and are sometimes given distinguishing names for convenience. Four of the more common of these are shown in Fig. 7. Each contains one or more parallel-resonant circuits and provision for coupling between the grid and plate circuits.

One of the simpler circuits is the Hartley circuit shown in Fig. 7a. The inductance is divided into two parts, L_1 and L_2, and their common point is connected to the cathode. There may be mutual inductance between L_1 and L_2, but it is not necessary. The circulating current in the resonant circuit passes through both portions of the inductance and develops the voltage for the grid circuit. The connection of the plate-voltage supply shown is known as *shunt feed* of the plate (see Art. 6, Ch. IX). The direct component of the plate current is furnished by the plate-voltage supply through an inductor, or choke, which has a sufficiently large inductance to reduce to a negligible value the alternating component of current through it. The path of the alternating component is through the blocking capacitor, which has a small reactance compared with the load impedance, and on through the load. The method of shunt feed is applicable to many oscillator and amplifier circuits. It has the advantages that: First, the direct component of the plate current is eliminated from the load, which prevents magnetic saturation and its attendant deleterious effects in circuits

* See footnote 6 on page 677.

with iron-cored coils; second, the load circuit is maintained at the direct potential of the cathode; and, third,[7] a means is provided for adjustment of the phase angle between the grid and the plate voltages, with a resulting increase of efficiency.

The Colpitts oscillator circuit shown in Fig. 7b has a parallel-resonant circuit in which the capacitor is divided into two parts, C_1 and

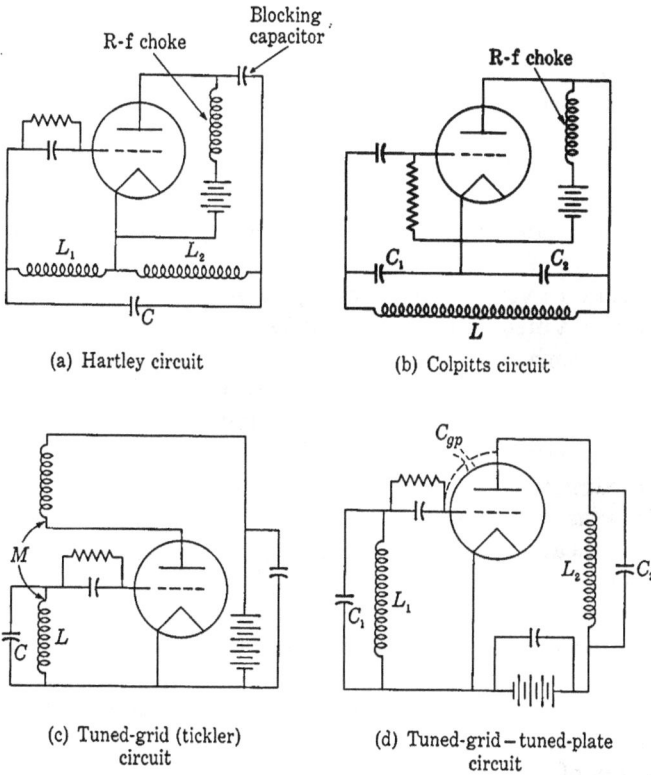

(a) Hartley circuit   (b) Colpitts circuit

(c) Tuned-grid (tickler)   (d) Tuned-grid – tuned-plate
circuit   circuit

Fig. 7.   Typical oscillator circuits.

C_2. The division of capacitance is so made as to give the required ratio of grid to plate voltage—the β of the Barkhausen analysis in Art. 2. In the tuned-grid oscillator circuit, Fig. 7c, the parallel-resonant circuit is placed in the grid circuit, and the feedback is provided by the mutual inductance from a coil in the plate circuit. The tuned-grid–tuned-plate circuit shown in Fig. 7d differs from the others in that coupling between the plate and grid circuits other than that through the tube is not purposely employed. Separate parallel-resonant circuits are placed

[7] D. C. Prince, "Vacuum Tubes as Oscillation Generators, Part IV," *G.E. Rev.*, *21* (1928), 147–149.

in the grid and plate circuits and are tuned to approximately the same frequency. As shown by the curves of Fig. 23, Ch. IX, the plate load is inductive at a frequency below the resonant frequency, and the considerations of Art. 8, Ch. VIII, show that, under these conditions, the grid-to-plate interelectrode capacitance results in a negative value of the effective conductance between the grid and the cathode. This negative conductance may be interpreted as supplying power to the parallel-resonant circuit in series with the grid and thereby making possible the sustained oscillations.

Many special oscillator circuits have been designed to meet the requirement of essentially constant frequency over long periods of time. Oscillators have been constructed whose frequency is constant to an almost unbelievable degree. Among the more satisfactory constant-frequency oscillators are those whose frequency is controlled by piezoelectric crystals; with them, frequency variations of only a few parts in a million over a period of days can be secured. The piezoelectric crystal is a mechanically resonant system that may be made to oscillate at its resonant frequency if it is placed in an electric field of that frequency, and that then produces an alternating voltage between its surfaces because of its oscillations. Since the effective Q_0 of the crystal is very high—of the order of tens of thousands—and since, as compared with that of an ordinary tuned circuit, its frequency of oscillation is relatively independent of connected electrical apparatus, it is capable of controlling the frequency of an oscillator within very close limits when it is used in place of a resonant circuit in the oscillator. In the design of these oscillators, the change of frequency of the crystal with temperature is of importance. To produce a high degree of frequency stability, the temperature of the crystal is held as nearly constant as is practicable, and, through use of appropriate directions in cutting the crystal from its natural form, the temperature coefficient of the crystal is minimized. A typical circuit using a piezoelectric crystal is similar to that of Fig. 7d, except that the crystal is substituted for the tuned circuit connected to the grid. Since the crystal does not conduct direct current, the grid-leak resistor must be connected between grid and cathode, and the grid capacitor may be omitted.

The tuned-plate oscillator was chosen in Art. 3 as an example for the application of the Barkhausen self-excitation criterion because of the simplicity of the resulting analysis. This simplicity occurs, as previously mentioned, because there is no power flow through the mutual inductance into the grid circuit under the assumption of negligible grid current. The expressions for β and Z are therefore relatively simple. The conditions in other oscillators are not the same, and the expressions obtained from the Barkhausen self-excitation criterion are often complicated and unwieldy even though grid current is neglected. In general, neither the real nor the imaginary part of the expression is independent of the frequency, and the difficulty of solving for the amplitude and frequency is often so great as to render this method of analysis impractical. In these circumstances, one of the other methods of attack mentioned in Art. 2 may be preferable.[8]

7. OSCILLATION IN AMPLIFIERS

The tendency for self-excited oscillations to develop is one of the limitations in amplifier circuits. Figure 8 shows a typical circuit in

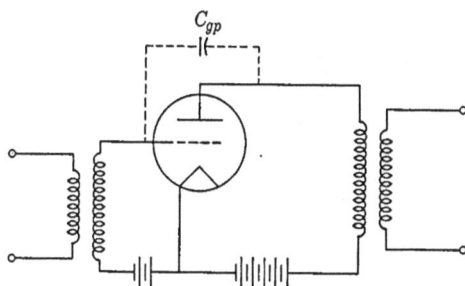

Fig. 8.   Amplifier circuit subject to oscillations.

which this phenomenon may occur. Inductance is assumed to be used in both the grid and the plate circuits either as a coupling impedance or as an internal parameter in a transformer. As may be seen by inspection, these inductances, together with the grid-to-plate interelectrode capacitance, form a parallel-resonant circuit for oscillations exactly as in the Hartley circuit of Fig. 7a. Thus oscillations may occur at a frequency set by the circuit constants, which is usually very high. Often, such oscillations in an audio-frequency amplifier remain undetected because, owing to their high frequency, they do not affect the instruments used. The oscillations disturb the desired operating conditions in the amplifier and normally cause distortion or increased losses in the tube. They can sometimes be eliminated by neutralizing methods, mention of which is made in Art. 4, Ch. XII, or by decrease of the capacitance C_{gp}. The latter is accomplished in the screen-grid tube. One explanation of the effect of the screen grid is that the reduction of C_{gp} increases the natural oscillation frequency of the resonant circuit to such a large value that the resulting increased loss

[8] C. K. Jen, "A New Treatment of Electron-Tube Oscillators with Feedback Coupling," *I.R.E. Proc., 19* (1931), 2109–2144.

in the inductors is large enough to damp out the oscillations. Sometimes a small resistance inserted in the grid circuit, next to the grid, is
effective in preventing these oscillations by increasing the loss in the
resonant circuit associated with them.

8. RESISTANCE-CAPACITANCE OSCILLATORS

Although inductance and capacitance are included in the feedback
network of all the sine-wave feedback oscillators described previously,
the presence of both is not necessary in such an oscillator. The essential

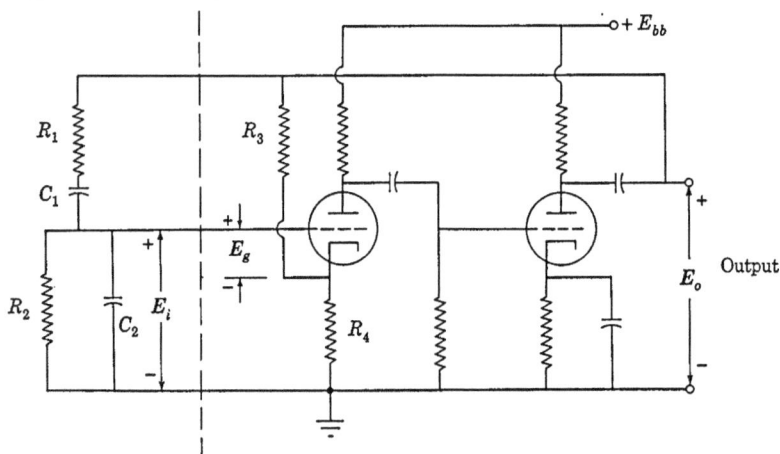

Fig. 9.   Wien-bridge resistance-capacitance sine-wave oscillator circuit.

requirement of the feedback network is that it provide the magnitude
and phase of feedback required to maintain a voltage at the desired
frequency, and a markedly different magnitude and different phase
for harmonics of that frequency, so that the harmonics shall not be
maintained. Resistances with either capacitances or inductances in
the feedback network can satisfy this requirement. Several frequency-
sensitive feedback networks that comprise only resistance and capacitance and have been applied in oscillators are the Wien bridge,[9] the
parallel-T network,[10] and the resistance-capacitance iterated, or ladder,
network.[11]   Figure 9 shows the basic circuit diagram of a so-called

[9] F. E. Terman, R. R. Buss, W. R. Hewlett, and F. C. Cahill, "Some Applications of
Negative Feedback with Particular Reference to Laboratory Equipment," *I.R.E. Proc.*,
27 (1939), 649–655.

[10] H. H. Scott, "A New Type of Selective Circuit and Some Applications," *I.R.E.
Proc.*, *26* (1938), 226–235.

[11] E. L. Ginzton and L. M. Hollingsworth, "Phase-Shift Oscillators," *I.R.E. Proc.*, *29*
(1941), 43–49.

Wien-bridge oscillator. For analysis, linear sinusoidal operation is assumed—the assumption is justified later. The two frequency-sensitive branches comprising the series combination R_1 and C_1 and the parallel combination R_2 and C_2 may be considered to constitute two arms of a Wien bridge, and the resistances R_3 and R_4 the other two. The output of the bridge is E_g, which is amplified by a two-stage resistance-capacitance-coupled amplifier and is fed back so that it becomes the bridge supply voltage E_o. The oscillator does not operate with the bridge balanced, however, because the voltage E_g would then be zero. Rather, it operates with the bridge unbalanced and sufficient voltage amplification in the amplifier to establish a net gain of unity around the feedback loop.

Any complete analysis of the circuit must take account of two facts: First, the alternating current in R_4 is the sum of the alternating component of the plate current in the first tube and the current through R_3 and R_4 due to E_o; and, second, the load for the second tube includes the resistance-capacitance circuit on the left of the dotted line. Such a solution is not simple, but approximate results for the frequency of oscillation and the required voltage amplification are readily obtained if, first, the circuit on the right of the vertical dotted line in Fig. 9 is assumed to be a resistance-capacitance-coupled amplifier having essentially zero total phase shift and negligible output impedance within the operating range of frequencies; and second, the circuit on the left of the dotted line is considered to be a frequency-sensitive voltage divider that feeds a fraction E_i of the output voltage E_o back to the input of the amplifier. At the voltage divider,

$$\frac{E_i}{E_o} = \frac{R_2/(j\omega C_2)}{R_2 + \dfrac{1}{j\omega C_2}} \quad \frac{1}{R_1 + \dfrac{1}{j\omega C_1} + \dfrac{R_2/(j\omega C_2)}{R_2 + \dfrac{1}{j\omega C_2}}}, \qquad [35]$$

which simplifies to the form

$$\frac{E_i}{E_o} = \frac{1}{\left(1 + \dfrac{R_1}{R_2} + \dfrac{C_2}{C_1}\right) + j\left(\omega R_1 C_2 - \dfrac{1}{\omega R_2 C_1}\right)}. \qquad [36]$$

If, as is assumed, the amplifier has zero phase shift, it constrains E_o to remain in phase with E_i. Hence the imaginary term in Eq. 36 must be zero, and the angular frequency of operation will be

$$\omega = 1/\sqrt{R_1 R_2 C_1 C_2}. \qquad [37]$$

This relation indicates that if R_1 and R_2 or C_1 and C_2 are ganged so that one dial rotation varies either pair simultaneously, a frequency variation of as much as ten to one is feasible. With Eq. 37 satisfied, the voltage amplification that the amplifier must have to maintain the oscillations is, from Eq. 36,

$$\text{Required voltage amplification} = 1 + \frac{R_1}{R_2} + \frac{C_2}{C_1}. \qquad [38]$$

This voltage amplification is not large—for example, if R_1 equals R_2 and C_1 equals C_2, it is only three. Such a low value is readily obtained if the ratio of the voltage divider composed of R_3 and R_4 is adjusted so that a substantial and sufficient fraction of the output voltage is fed back in series with the first cathode. At the same time, this large amount of voltage feedback reduces the output impedance of the amplifier so that the resistance-capacitance voltage divider, which constitutes a load at the output terminals, has only a small effect on the voltage amplification, and the assumption that the voltage amplification is independent of frequency is nearly realized.[12]

The fact that the only frequency satisfying the requirement that the total phase shift through the amplifier and back through the Wien bridge be zero or an integral multiple of 2π radians is the frequency given by Eq. 37 justifies the original assumption that the operation is sinusoidal when the operation is restricted to the linear region of the tube characteristics.

In this oscillator, as in the others discussed previously, the amplitude of the oscillations tends to build up until, on the average, the voltage amplification during a cycle is reduced to the value stated in Eq. 38. If the reduction is accomplished through occurrence of plate-current cut-off or positive grid voltages, substantial harmonic distortion in the output voltage may be produced. To cause limiting of the voltage amplification without appreciable distortion, a temperature-sensitive resistor, such as a tungsten-filament lamp that has a positive temperature coefficient and a thermal time constant long compared with a cycle of the lowest frequency to be produced, is often included as the resistance R_4. Then, as the amplitude builds up, R_4 increases and the resulting increased feedback decreases the voltage amplification to the value corresponding to Eq. 38, but R_4 acts as a linear resistance *within each cycle*, and the distortion is maintained small. Alternatively, a resistance having a negative temperature coefficient, such as a thermistor, can be used as R_3 to fulfill the same purpose.

This type of oscillator is widely used for production of frequencies

---

[12] J. A. B. Davidson, "Variable Frequency Resistance–Capacity Oscillators," *Electronic Engineering*, *16* (January, 1944), 316–319, (February, 1944), 361–364.

in and near the audio-frequency range chiefly because it contains no bulky and expensive inductors, tunes over a relatively wide frequency range with rotation of one dial, has low distortion, has good frequency stability, and delivers a relatively constant output voltage over a wide frequency range.

9. NEGATIVE-TRANSCONDUCTANCE OSCILLATOR

The circuit in Fig. 10 is one form of a *negative-transconductance oscillator*. The production of oscillations depends on the fact that the potential of the suppressor grid determines the ratio in which the current of electrons divides between the plate and the screen grid. An *increase* of the potential of the suppressor grid permits a larger fraction of the electrons to reach the plate, and therefore causes a *decrease* in the rate of arrival of electrons at the screen grid.[13] The suppressor-to-screen transconductance is hence negative, and there is no inherent volt-age polarity inversion when the tube is used as an amplifier with the input applied between the suppressor grid and the cathode, and the load connected between the screen grid and the cathode. In the circuit the resistances and capacitances constitute a frequency-sensitive feedback network around such an amplifier, and oscillations occur at the frequency for which the total phase shift around the feedback loop through the tube and this network is zero.

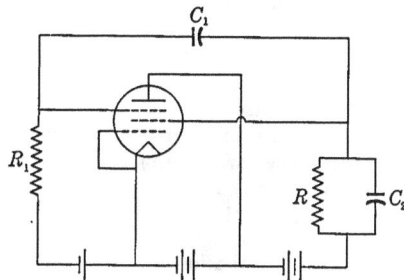

Fig. 10. Circuit of negative-transconductance oscillator.

PROBLEMS

1. In the tuned-plate oscillator circuit of Fig. 2, consider that the tuned circuit is a parallel combination of inductance L, capacitance C, and resistance R_t, rather than the series-parallel combination shown in the figure. Assume that the tube operates as a Class A_1 amplifier with an amplification factor μ and an average mutual conductance $\bar{g}_m$.

(a) Apply the Barkhausen criterion to determine, in terms of the given parameters, the value of $\bar{g}_m$ necessary for sustained oscillations, and the frequency of oscillation.

(b) Draw a vector diagram for this oscillator.

(c) For what reasons might the frequency of oscillation of an actual tuned-plate oscillator differ from that found in (a)?

[13] *The Use of the 57 or 6C6 to Obtain Negative Transconductance and Negative Resistance* —*Application Note No. 45* (Harrison, N. J.: RCA Manufacturing Company, Inc., 1935).

2. A point of view that is sometimes useful in the analysis of feedback oscillators is that the circuit external to the tube controls the grid voltage, and hence the plate current, in such a way that the plate current increases when the plate voltage decreases. Thus the plate circuit of the tube may be replaced by a negative resistance (sometimes in combination with other circuit elements), from which a-c power is supplied to the positive resistances in other parts of the circuit. An equivalent of the Barkhausen criterion is obtained through requiring that in the steady state the current supplied by this equivalent of the plate circuit of the tube be equal to that required by the load. To illustrate the method, apply it to the tuned-plate oscillator described in Prob. 1, using the circuit parameters given there and, in addition, the average plate resistance $\bar{r}_p$.

(a) Draw an equivalent circuit for the oscillator in which the tube is represented by a current source in parallel with a resistance.

(b) Using the relation between the grid and plate voltages imposed by the external circuit, show that the current source in this circuit may be replaced by a negative resistance. Draw the resulting equivalent circuit.

(c) Determine in terms of the parameters of this circuit the value of $\bar{g}_m$ necessary for sustained oscillations, and the frequency of oscillation.

(d) Show that the relations obtained in (c) are equivalent to those obtained in (a) of Prob. 1.

3. Two identical triodes connected as shown in Fig. 11 operate in Class A_1 and have the constant coefficients μ, r_p, and g_m. For analysis of the circuit operation,

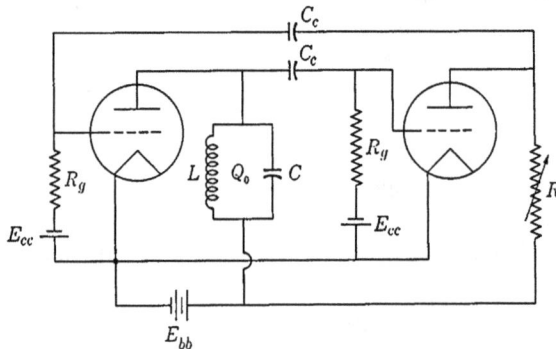

Fig. 11. Two-stage tuned-plate oscillator circuit for Prob. 3.

neglect the voltages across the coupling capacitors C_c, the currents through the grids and the grid resistors R_g, and any effect that the direct voltage drop across R might have upon the parameters of the second triode because of shift of the quiescent operating point. Assume that the power absorbed in the tuned circuit, which determines its Q_0, may be represented as that absorbed in a resistance in parallel with the inductance and capacitance.

Determine in terms of the circuit parameters:

(a) the expression for the critical value of the resistance R at which oscillations will just start,

(b) the frequency of oscillation.

4. In an oscillator circuit arranged as shown in Fig. 12, Z_1 equals $j\omega L_1$, Z_2 equals $1/j\omega C$, and Z_3 equals $j\omega L_2$. Assume that there is no mutual flux coupling L_1 and L_2, that except for the tube the circuit elements are all dissipationless, and that the operation of the tube is in Class A_1.

(a) Determine for the oscillatory condition the required relation of the average amplification factor $\bar{\mu}$ to the circuit parameters and the frequency of oscillation.
(b) Draw a vector diagram for this oscillator.
(c) How are these results altered if mutual flux between L_1 and L_2 is present so that the coefficient of coupling between them is k?
(d) By what name is this oscillator circuit commonly known?

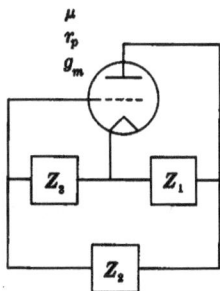

Fig. 12. Generalized oscillator circuit for Probs. 4 and 7. Supply voltages are omitted from the figure for simplicity.

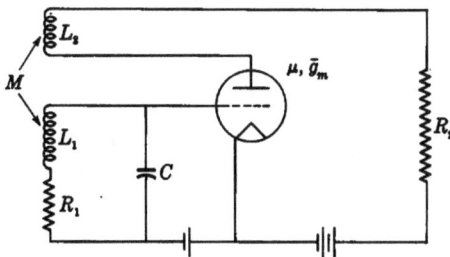

Fig. 13. Tuned-grid oscillator circuit for Prob. 5.

5. The circuit of a tuned-grid oscillator with a resistance load R_2 is shown in Fig. 13. For analysis, the grid current may be neglected.

(a) Draw an equivalent circuit for the oscillator, in which the tube appears as a voltage source in series with a resistance.
(b) Obtain in terms of the circuit parameters indicated in the figure an expression for the frequency of oscillation.
(c) Obtain an expression for the average mutual conductance necessary for the production of sustained oscillations.
(d) Show the relative winding directions of L_1 and L_2 that must be used in order that oscillations may occur.
(e) Draw a vector diagram for this oscillator.

6. To simplify the analysis of the tuned-grid–tuned-plate oscillator circuit shown in Fig. 7d, neglect dissipation in the elements of the tuned circuits, neglect all components of grid current except that through the grid-to-plate capacitance C_{gp}, and assume that both tuned circuits have the same resonant frequency.

(a) Show that sustained oscillations may be produced by this circuit at a frequency slightly below the resonant frequency of the tuned circuits, and determine in terms of the given circuit parameters the value of the average amplification factor $\bar{\mu}$ necessary for oscillation.
(b) Determine the frequency of oscillation.

  (c) Since dissipation in the tuned circuits is neglected, oscillations in the tuned circuit connected to the grid should be sustained if the tube parameters and the impedance in the plate circuit are so adjusted that the input conductance of the tube is zero at the natural oscillation frequency of the grid circuit. Why, then, cannot this point of view and Eqs. 77 and 78, Ch. VIII, be used to determine the results found above?

  (d) Might oscillations still be produced if the resonant frequencies of the two tuned circuits were not the same? Justify your answer.

7. Figure 12 shows a generalized circuit that may be made to include several common oscillator circuits through proper choice of the impedances Z_1, Z_2, and Z_3. To simplify the analysis of the circuit, grid current may be neglected and the operation of the tube may be assumed to be in Class A_1.

  (a) Show that the Barkhausen criterion takes the form

$$r_p(Z_1 + Z_2 + Z_3) + Z_1[Z_2 + (\mu + 1)Z_3] = 0.$$

  (b) If the impedances Z_1, Z_2, and Z_3 are pure reactances jX_1, jX_2, and jX_3, respectively, determine the relations among X_1, X_2, and X_3 necessary to satisfy this criterion.

  (c) If each of the impedances Z_1, Z_2, and Z_3 is to be a single inductor or capacitor, show that oscillations are possible for only two of the conceivable arrangements. Sketch and name the circuits that correspond to these arrangements.

  (d) Obtain the critical condition for oscillation and the frequency of oscillation for the oscillator of Prob. 4 as a special case of the criterion obtained in (a).

8. A tuned-plate oscillator having the circuit of Fig. 2a comprises a transmitting triode having the constant-current characteristics given in Fig. 9, Ch. X. The operation of the tube is in Class C with a plate-supply voltage of 550 volts, a grid-bias voltage of twice the cut-off value, a maximum instantaneous plate current of 350 ma, and a minimum instantaneous plate voltage 1.5 times the maximum instantaneous grid voltage.

  (a) If the oscillation frequency is 10^6 cps, and the loaded Q_0 of the tuned circuit is 12, state the values of L and C to be used in this circuit.

  (b) Assuming that the efficiency of the tuned circuit as a plate-to-load-and-grid coupling network is 100 per cent, determine the necessary mutual inductance and grid-leak resistance, the power loss in the grid leak, the available power output, and the proper order of magnitude of the capacitance of the grid capacitor.

9. Assuming that the change of screen-grid current is a linear function of the changes in the suppressor-grid and screen-grid voltages, and the suppressor-grid current is negligible in the negative-transconductance oscillator of Fig. 10:

  (a) Draw an equivalent circuit for the oscillator, in which the screen-grid circuit appears as a current source shunted by a resistance. Define carefully, with due regard to signs, the parameters used to represent the tube.

  (b) Determine in terms of the parameters of the circuit the value of the suppressor-to-screen transconductance necessary to sustain oscillations. What would result if this transconductance had a greater value? A smaller value?

  (c) Determine the frequency of oscillation for the conditions assumed.

CHAPTER XII

Modulation and Demodulation, or Detection

A periodic electrical wave—that is, a periodic electrical "disturbance which is a function of time or space or both,"[1] may serve to transmit energy from one point to another, but it cannot transmit information either for interpretation by a human being or for use in a control process because each new cycle is exactly like all the previous ones. Information can be transmitted, however, if the otherwise periodic wave is varied. In electrical engineering the term *modulation* is used* to describe this "process by which some characteristic of [one wave called] a *carrier* is varied in accordance with [some characteristic of a second wave called] a *modulating wave*." Generally the carrier wave to be modulated is a current or voltage that is direct, sinusoidal, or a recurrent series of pulses. The modulating wave, on the other hand, may have a nonperiodic, or transient, waveform.

1. THE MODULATION PROCESS

Perhaps the most widely used type of modulation occurs in the carbon-granule telephone transmitter in telephony. In this application, the direct current supplied by a battery is varied (modulated) in accordance with the sound pressure on the transmitter diaphragm, which causes the resistance of the carbon granules to vary with time. Another example of modulation is the action of the common amplifier tube. In preceding chapters, the amplifier tube is represented by an alternating current or voltage source with an internal resistance. An alternative point of view, however, is that the amplifier tube is merely a resistance whose magnitude is controlled by the grid-signal voltage. The grid-signal voltage produces a change of resistance and thereby varies (modulates) the current from the direct-current plate-power supply. Broadly speaking, then, modulation includes the behavior of an amplifier tube and certain processes involving conversion of power from one form to another—for instance, conversion from acoustical power to electrical power. It may be accomplished by a time variation of a parameter in a circuit or, as is evident from subsequent analysis, by a nonlinear parameter.

When the modulated wave is an alternating current, additional

---

[1] *Standards on Antennas, Modulation Systems, Transmitters—Definitions of Terms, 1948* (New York: The Institute of Radio Engineers, 1948), 9.

* See page 8 of reference 1.

features of modulation are useful. The usefulness of one aspect results from the fact that often a wave that is to be transmitted electrically—such as the wave derived from the voice, from a telegraph key, from an optical image to be televised, or from another nonelectrical quantity—has a frequency or band of frequencies, transmission of which to a distant point is difficult or not economical. Under these conditions it is expedient to translate the frequency of the signal wave to some new value by means of modulation, to transmit the altered wave, and finally to translate it back to its original frequency. Although it is not apparent from the definition, subsequent discussion shows that frequency translation is often present in the process of modulation; in fact, it is the basic reason for modulation in many instances.

An important application in which modulation is useful is radio transmission. In it, the frequency-translating feature of modulation is utilized. The frequency associated with the power used in radio must be suitable for radiation from an antenna. Direct-current power, for example, is not radiated and hence cannot be transmitted through space. Power at audible frequencies is radiated from the circuits that carry currents of these frequencies, but only a small portion of the total power involved is radiated electrically unless the dimensions of the circuit are comparable with the wavelength of the alternations. Higher frequencies, however, have a shorter wavelength. Hence it is feasible to build circuits that have dimensions comparable with the wavelength at high frequencies and permit efficient radiation. To transmit audible frequencies, therefore, the process that has evolved is first to modulate a high frequency (say 100,000 or more cycles per second) with an audible frequency. The resulting modulated high-frequency wave can then be efficiently transmitted through space. In amplitude-modulated radio transmission, for example, a high-frequency wave is radiated from an antenna, and the circuit is arranged so that the amplitude of the high-frequency wave may be varied at a rate corresponding to the audio-frequency signal. For instance, the amplitude of a 1,000,000-cycle-per-second wave may be varied at the rate of, say, 1,000 cycles per second. This is comparable to changing the amplitude of a 60-cycle-per-second wave at a rate of 0.06 cycle per second, or 16.7 seconds per cycle—a slow rate compared to the frequency of the 60-cycle-per-second source. One thousand cycles of the high frequency occur during one cycle of the low frequency in this example. At the receiving end of the radio system, *detection* is performed,* which is "the process by which a wave

---

* See page 7 of reference 1.

corresponding to the modulating wave is obtained in response to a modulated wave." Detection is often called *demodulation.*

The principles of modulation and demodulation utilized in radio transmission are also utilized in carrier-current telephony. Without modulation a pair of wires used as a transmission line cannot carry more than one voice conversation at once without interference between the conversations; but, if additional voice conversations are translated to successively higher frequency ranges, all of them can be transmitted at once without mutual interference, and separate demodulators tuned to receive distinct ranges of frequency can be used to separate and recover the conversations at the receiving end. For example, speech transmission requires a frequency range up to about 3,000 cycles per second. One conversation in carrier-current telephony can therefore use the range from about 0 to 3,000 cycles per second; another, the range from, say, 4 to 7 kilocycles per second; a third, the range from 8 to 11 kilocycles per second, and so on for additional conversations. Modulation thereby increases the information-carrying capacity of the transmission line in wire telephony, and in the same way it permits the simultaneous transmission of a multitude of messages or television signals through the air without wires by making possible the separation and recovery of them by tuned circuits at the receiving station.

2. Types of modulation

There are several methods of modulating an alternating-current wave. Usually the wave to be modulated is sinusoidal and can therefore be represented by

$$i = A \cos (\omega t + \theta). \qquad [1]$$

In this expression, A is the amplitude, and the whole term in parentheses is an angle; thus the expression is basically

$$i = A \cos \phi(t). \qquad [2]$$

The angle $\phi(t)$ is here indicated as being a function of time; and Eq. 1 shows that, for the sinusoidal wave, the angle increases linearly with time.

The frequency of the wave is the number of times the wave repeats itself per second, and it is proportional to the time rate of change of the angle, being given by

$$f = \frac{1}{2\pi} \frac{d\phi(t)}{dt}, \qquad [3]$$

since

$$\omega = \frac{d\phi(t)}{dt}, \qquad \blacktriangleright[4]$$

where ω is called the *angular frequency*. In terms of the usual vector representation of the sinusoidal wave, ω is the angular velocity of the vector.

In general, there are two basic methods for varying the sine wave to effect the process of modulation. One is to vary the amplitude in accordance with the instantaneous value of the modulating wave; the other is to vary the way the angle changes with time. These two methods are called *amplitude modulation* and *angular modulation*, respectively.[2] Angular modulation is commonly subdivided into two types, *frequency modulation* and *phase modulation*. Subsequent discussion shows that they are closely interrelated and perhaps should not be very sharply distinguished, since both are often utilized in a single modulation system. Still another method for varying a wave is called *pulse modulation*. This method involves either variation of a periodic wave in accordance with modulating pulses, or variation of some characteristic of a sequence of pulses in accordance with a modulating wave. In its broadest sense, the term pulse modulation describes a system that may embrace both amplitude and angular modulation.

Amplitude modulation is produced through varying the amplitude of the wave to be modulated, the deviation of the amplitude from the unmodulated value being directly proportional to the instantaneous value of the modulating wave but independent of its frequency.* For simplicity, the modulating wave may be considered to be sinusoidal, having a waveform given by

$$A_m \cos \omega_m t. \qquad [5]$$

In practice, ω_m is small compared with the angular frequency of the wave to be modulated—hereafter denoted by ω_c.† In accordance with the foregoing definition, the amplitude of the modulated wave for sinusoidal modulation is a function of time given by

$$A(t) = A_c + k_a A_m \cos \omega_m t, \qquad [6]$$

where k_a is a constant of proportionality and A_c is the amplitude of the wave before it is modulated. The second term on the right-hand side of Eq. 6 is the deviation in amplitude that is superposed on the constant unmodulated value of the amplitude, A_c. The complete expression for the wave with sinusoidal amplitude modulation, obtained by substitution of Eq. 6 for the amplitude and $\omega_c t$ for the phase angle in Eq. 2, is

$$i = (A_c + k_a A_m \cos \omega_m t) \cos \omega_c t. \qquad \blacktriangleright[7]$$

---

[2] W. L. Everitt, "Frequency Modulation," *A.I.E.E. Trans.*, *59* (1940), 613–624.

* In the sense used here, a deviation means a change from an otherwise constant value.

† An exception to this statement is the modulation of a direct current.

A vector representation for the wave with sinusoidal amplitude modulation is shown in Fig. 1. The total length of the vector is composed of the constant component A_c and the sinusoidally varying component $k_a A_m \cos \omega_m t$. The vector rotates at a constant angular velocity ω_c, and the projection on the real axis is the instantaneous value of the amplitude-modulated wave. The tip of the vector describes a spiral as

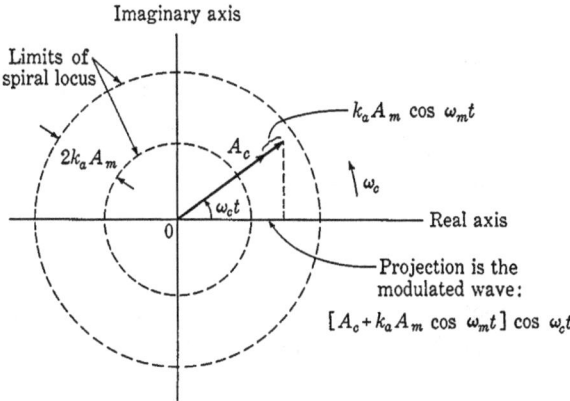

Fig. 1. A vector representation of an amplitude-modulated wave. The locus of the tip of the rotating vector is a spiral that remains between the dotted circles.

the length of the vector varies between the limits of $A_c \pm k_a A_m$. Such a wave might be produced through superposing a sinusoidal variation on the direct field current of an alternator driven at constant angular speed. Further discussion of amplitude modulation follows in Art. 3.

Frequency modulation is produced through varying the instantaneous frequency of the wave to be modulated, the instantaneous deviation of the frequency from the unmodulated value being directly proportional to the instantaneous value of the modulating wave but independent of its frequency. Accordingly, the instantaneous angular frequency of the frequency-modulated wave with sinusoidal modulation may be expressed as the function of time

$$\omega(t) = \omega_c + k_f A_m \cos \omega_m t, \qquad [8]$$

where k_f is a constant of proportionality. In this equation, the second term on the right-hand side is the deviation in angular frequency that is superposed on the constant unmodulated angular frequency, ω_c. Whereas in amplitude modulation, the phase angle of the modulated wave increases linearly with time, in frequency modulation a deviation or modulation component is superposed on the linearly increasing component. This conclusion follows from Eq. 4, from which it is evident that the phase angle is the time integral of the instantaneous angular frequency; thus, from Eq. 4,

$$\phi(t) = \int_0^t \omega(t)\, dt, \qquad [9]$$

and substitution from Eq. 8 gives

$$\phi(t) = \omega_c t + \frac{k_f A_m}{\omega_m} \sin \omega_m t, \qquad [10]$$

where t is measured from a point at which ϕ is zero. In Eq. 10, the last term is the deviation in phase angle superposed on the linearly increasing term that corresponds to the unmodulated wave. The complete expression for the frequency-modulated wave, given by the substitution of the constant amplitude A_c and Eq. 10 in Eq. 2, is

$$i = A_c \cos\left(\omega_c t + \frac{k_f A_m}{\omega_m} \sin \omega_m t\right). \qquad \blacktriangleright [11]$$

Note that the direct substitution of $\omega(t)$ from Eq. 8 for ω in Eq. 1 to obtain the equation of the frequency-modulated wave is *incorrect*. The product ωt is equal to the instantaneous phase angle of the wave *only* when ω is constant. If ω is a variable, the angle is related to it only by the integral in Eq. 9 and is not given by $\omega(t)$ times t.

The physical significance of frequency modulation may be visualized through reference to an alternator in which the stator is mounted on bearings concentric with the shaft so that the stator as well as the rotor may be rotated. The rotor is assumed to be driven at constant angular speed, which accounts for the term $\omega_c t$ in Eq. 11. If the stator is then rocked with sinusoidal angular motion about its midposition, with an instantaneous *angular velocity* directly proportional to the instantaneous value of the modulating wave, the resulting output wave of the alternator is frequency modulated. Alternatively, the motion of the stator may be described as a sinusoidal angular motion with an amplitude of angular variation directly proportional to the quotient of the amplitude divided by the frequency of the modulating wave—in mathematical terms, $k_f A_m / \omega_m$. Actually, the motion of the stator would change the amplitude of the wave as well as its frequency and phase, but this effect is small and is to be neglected in this analogy. In the 60-cycle analogy, then, the stator is turned very slowly but through a relatively large number of degrees or revolutions.

Phase modulation is produced through varying the instantaneous phase angle of the wave to be modulated, the instantaneous deviation of the phase angle from the unmodulated value being directly proportional to the instantaneous value of the modulating wave but independent of its frequency. The deviation in phase angle in this type of modulation is not superposed on a constant value but is superposed on the linearly increasing term $\omega_c t$ that corresponds to the constant unmodulated angular frequency. Thus the instantaneous phase angle $\phi(t)$ of the phase-modulated wave with sinusoidal modulation may be expressed as

$$\phi(t) = \omega_c t + k_p A_m \cos \omega_m t. \qquad [12]$$

Through substitution of A_c for the constant amplitude of the unmodulated wave, and of $\phi(t)$ from Eq. 12, Eq. 2 becomes the complete expression for the phase-modulated wave with sinusoidal modulation,

$$i = A_c \cos (\omega_c t + k_p A_m \cos \omega_m t). \qquad \blacktriangleright[13]$$

The instantaneous frequency also undergoes a deviation with time in the phase-modulated wave, in accordance with Eq. 3. Thus

$$\omega(t) = \frac{d\phi(t)}{dt} = \omega_c - k_p \omega_m A_m \sin \omega_m t \qquad [14]$$

in the phase-modulated wave.

The physical significance of phase modulation may be visualized by reference to the same alternator discussed above in connection with frequency modulation. If the stator is rocked with a sinusoidal angular motion about its midposition with an instantaneous *angular deviation* directly proportional to the instantaneous value of the modulating wave, the resulting output wave of the alternator is phase-modulated.

From the foregoing discussion and from Eqs. 8, 10, 12, and 14, it is evident that frequency and phase modulation are closely related. In each, the deviation in both the frequency and the phase is sinusoidal for a sinusoidal modulating wave. However, the distinguishing feature in frequency modulation is that the amplitude of the frequency deviation is independent of the modulation frequency, but the amplitude of the phase-angle deviation is inversely proportional to the modulation frequency. Correspondingly, in phase modulation, the amplitude of the phase-angle deviation is independent of the modulation frequency, but the amplitude of the frequency deviation is directly proportional to the modulation frequency. As is discussed in Art. 18, pure phase modulation is almost never used in practice, and some so-called frequency-modulation systems utilize phase modulation over a part of the range of modulation frequencies that they are designed to accommodate.

The reason for the apparent similarity between frequency modulation and phase modulation in the foregoing analysis is that both the time integral and the time derivative of the sinusoidal modulating wave chosen for illustration are also sinusoidal in waveform. A clearer distinction between the two types of modulation is gained from a consideration of the effects of a square wave as the modulating wave.[3] These effects are illustrated in Fig. 2. With a square modulating wave and frequency modulation, the deviation in phase angle superposed on the linear increase given by $\omega_c t$ is the integral of the variation in ω. The

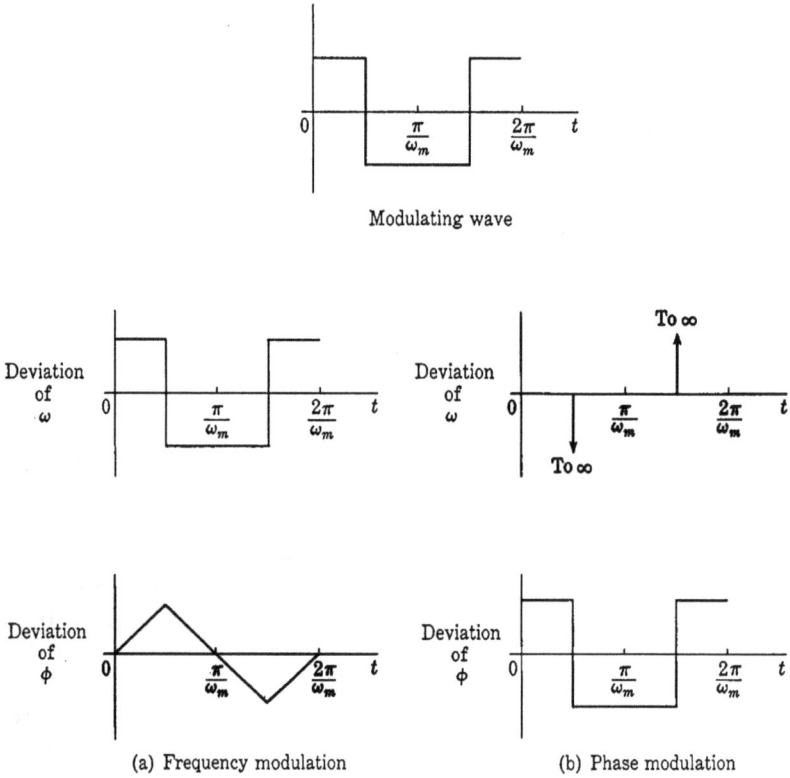

Modulating wave

(a) Frequency modulation   (b) Phase modulation

Fig. 2.   Comparison of frequency and phase modulation.

phase-angle deviation has a saw-toothed waveform and does not resemble the modulating wave. In contrast, with a square modulating wave and phase modulation, the deviation in frequency superposed on ω_c is the derivative of the deviation in phase angle. The frequency deviation is then a series of pulses having infinite magnitude but a finite time integral. In general, then, modulation of the frequency or of the phase angle always involves modulation of the other quantity. In frequency modulation, the waveform of the deviation in *frequency* resembles the waveform of the modulating voltage; in phase

[3] G. W. O. Howe, "Frequency or Phase Modulation?" *Wireless Engr.*, *16* (1939), 547.

modulation, the waveform of the deviation of the *phase angle* resembles that of the modulating voltage.[4]

Amplitude modulation is often accompanied by the other two types of modulation. For example, one method sometimes used to produce an amplitude-modulated wave is to vary the plate-supply voltage of a vacuum-tube oscillator in accordance with the modulating waveform. With proper adjustment of the circuit parameters, the amplitude of the high-frequency wave from the oscillator varies in accordance with the instantaneous value of the modulating wave. However, a deviation of frequency also occurs in the usual oscillator circuit, and the result is not pure amplitude modulation. Although the amplitude variation predominates, the frequency deviation is sufficiently great to render the method impractical for many purposes. Other methods described in subsequent articles reduce the frequency deviation. When the modulation is pure amplitude modulation and there is no angular modulation, all the intervals between the points at which the wave has values of zero are equal. When these intervals are not equal, angular modulation is present.

Pulse modulation[5] is widely used in telegraphy, television, radar, multichannel microwave transmission, and other communication and control systems. The early telegraph utilized a form of pulse modulation—a direct-current carrier was modulated by pulses produced by a manually operated switch, or key, in accordance with the Morse code. A second and later example of pulse modulation familiar to many is a high-frequency oscillator started or stopped by a similarly operated key. The output is a series of short and long bursts of high-frequency power separated by intervals of zero output, which may be used for wireless telegraphy. From these elementary types of pulse modulation, advance has been made to a variety of more complex modulation systems involving pulses, aimed chiefly at bettering the quality of long-distance transmission of information or at reducing the required bandwidth. One of the more highly developed forms is a system dependent on the binary arithmetic that is the basis for the operation of many high-speed electronic digital computers.

As ordinarily used in connection with these improved systems of modulation, the term pulse modulation refers to a two-step process:

[4] D. G. C. Luck and H. Roder, "Discussion on 'Amplitude, Phase, and Frequency Modulation,' " *I.R.E. Proc.*, *20* (1932), 884–887.

[5] V. D. Landon, "Theoretical Analysis of Various Systems of Multiplex Transmission," *RCA Rev.*, *9* (1948), 287–351, 433–482; E. M. Deloraine, "Pulse Modulation," *I.R.E. Proc.*, *37* (1949), 702–705; "Standards on Pulses: Definitions of Terms," *I.R.E. Proc.*, *39* (1951), 624–626; *40* (1952), 552–554; *Standards on Antennas, Modulation Systems, Transmitters—Definitions of Terms*, 1948 (New York: The Institute of Radio Engineers, 1948), 9.

First, the wave to be transmitted, such as a speech wave, is applied as a modulating wave to vary some characteristic of a *pulse train*—a sequence of pulses—which is used here as a *pulse carrier*; and, second, the resulting modulated pulse carrier is applied as a new modulating wave to modulate a high-frequency sinusoidal carrier wave. The pulse carrier is hence said to serve as a *subcarrier*. The modulation of the pulse carrier may involve varying either the amplitude or the timing of the pulses. Accordingly, *pulse-amplitude modulation* or *pulse-time modulation* occurs. Since there are many time relationships among the characteristics of a train of pulses that can be altered, there are several types of pulse-time modulation. *Pulse-duration modulation* results if the duration, or length, or width, of the pulses is varied. The modulating wave may vary the time of the leading edge, the trailing edge, or both edges of the pulses in the recurrent series. *Pulse-position modulation* occurs if the position in time, or the phase, of the pulses is varied. And *pulse-frequency modulation* results if the frequency, or the repetition rate, of the pulses is varied.

Only instantaneous *samples* of the modulating wave are transmitted in some types of pulse modulation. But if the pulse-repetition rate, or sampling rate, is large enough, sufficient information is transmitted to permit reproduction of a continuous modulating wave with reasonable accuracy at the receiving point. According to Shannon,[6] "if a function $f(t)$ contains no frequencies higher than W cycles per second, it is completely determined by giving its ordinates at a series of points spaced $1/(2W)$ seconds apart." Exact values of the samples of the amplitude of the modulating wave are not always transmitted. In one class of pulse modulation, the range of amplitudes is divided into a number of finite intervals and a value is assigned to each. At each sampling instant, the amplitude is approximated by the value for the interval within which it falls, and that value is transmitted. Thus only a discrete number of so-called *quantized* amplitude values is transmitted. Often these values are transformed into a code, such as the binary code, before transmission, and the modulation process is then termed *pulse-code modulation.*[7]

3. AMPLITUDE MODULATION

The mathematical expression for the amplitude-modulated wave with sinusoidal modulation, Eq. 7, may be rearranged in the form

$$i = A_c(1 + m \cos \omega_m t) \cos \omega_c t, \qquad \blacktriangleright[15]$$

[6] C. E. Shannon, "Communication in the Presence of Noise," *I.R.E. Proc.*, *37* (1949), 10–21.

[7] B. M. Oliver, J. R. Pierce, and C. E. Shannon, "The Philosophy of PCM," *I.R.E. Proc.*, *36* (1948), 1324–1331.

where the quantity m is called the *modulation factor*. The *modulation frequency* $\omega_m/(2\pi)$ is smaller than the modulated frequency $\omega_c/(2\pi)$, which is called the *carrier frequency*. In radio transmission, the ratio of ω_c to ω_m may be very large—1,000 or more.

The waveform of the sinusoidally modulated wave is shown in Fig. 3. Instantaneous values of i/A_c are plotted to illustrate the fractional variation. The term in parentheses, $(1 + m \cos \omega_m t)$, which varies

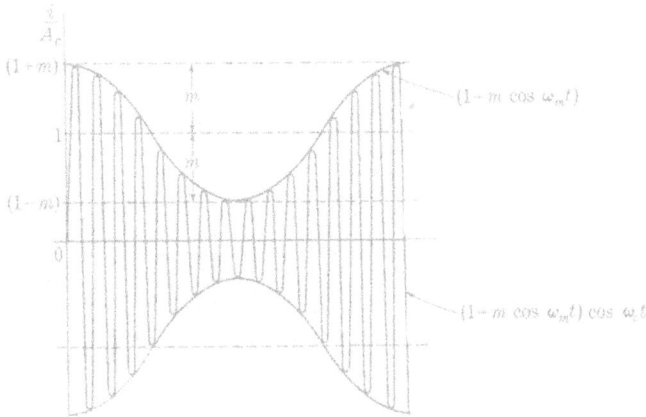

Fig. 3.   Amplitude-modulated wave with sinusoidal modulation.

slowly compared with $\cos \omega_c t$, may be considered to be approximately the amplitude of the high-frequency alternations; it is called the *envelope* of the wave. The modulation factor m denotes the fractional extent to which the modulation varies the amplitude, or the per unit modulation. The relation

$$\text{Per cent modulation} = m \times 100 \qquad [16]$$

therefore follows. Since the amplitude cannot become smaller than zero, 100 per cent modulation is the maximum that can be obtained by means of the simple conventional modulating devices. Percentage modulations smaller than 100 are common. An attempt to obtain a higher percentage modulation usually results in an interval of zero amplitude.

The signal encountered in voice transmission is of an irregular shape such as that shown in Fig. 4a. The envelope and the amplitude-modulated high-frequency wave then have the form shown in Fig. 4b.

Since the amplitude term in an amplitude-modulated wave is a function of time, the high-frequency wave is not strictly sinusoidal even during a single cycle; the over-all expression for it is a much more complex function of time than that for the pure sinusoid. The amplitude-modulated wave is nonsinusoidal and contains components with frequencies different from the carrier frequency. The presence of these components and important concepts of the process of amplitude modulation are brought out by use of the trigonometric identity:

$$\cos \theta \cos \phi = \tfrac{1}{2} \cos (\theta + \phi) + \tfrac{1}{2} \cos (\theta - \phi). \qquad [17]$$

By means of this expression, Eq. 15 may be put in the equivalent form

$$i = A_c \left[\cos \omega_c t + \frac{m}{2} \cos (\omega_c + \omega_m)t + \frac{m}{2} \cos (\omega_c - \omega_m)t \right]. \quad \blacktriangleright[18]$$

The sinusoidally modulated wave is thus shown to consist of the sum of

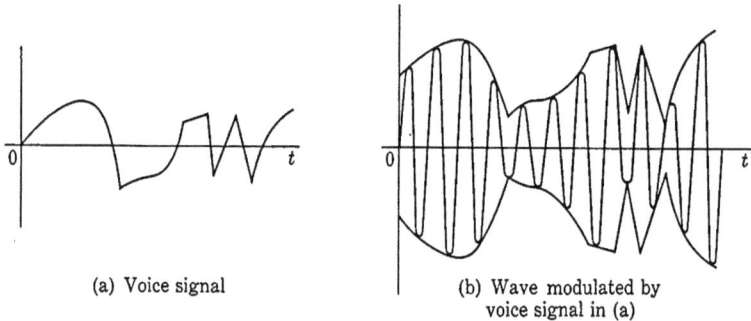

(a) Voice signal

(b) Wave modulated by
voice signal in (a)

Fig. 4.   Amplitude modulation with an irregular modulating wave.

three sinusoidal components of different frequencies, named the *carrier*, *upper side frequency*, and *lower side frequency*, which are composed as follows:

	Angular Frequency	Relative Amplitude	Relative Power
Carrier	ω_c	1	1
Upper side frequency	$\omega_c + \omega_m$	$m/2$	$m^2/4$
Lower side frequency	$\omega_c - \omega_m$	$m/2$	$m^2/4$

On the basis of Eq. 18, a vector representation of the carrier and side frequencies can be constructed as shown in Fig. 5, which is an extension of Fig. 1. The lower side-frequency vector rotates more slowly than the carrier vector, and the upper side-frequency vector rotates more rapidly. The projection of the sum of the three vectors on the real axis is the amplitude-modulated wave. The relative motion of the three vectors may be understood through considering that the carrier vector

A_c is held stationary; in other words, that ω_c is made zero. The two side-frequency vectors then rotate in opposite directions with respect to the carrier vector—one at an angular velocity $+\omega_m$, and the other at the angular velocity $-\omega_m$. All components normal to A_c cancel; hence, the sum of the carrier and side-frequency vectors is a vector along the direction of A_c and varying in length between the limits

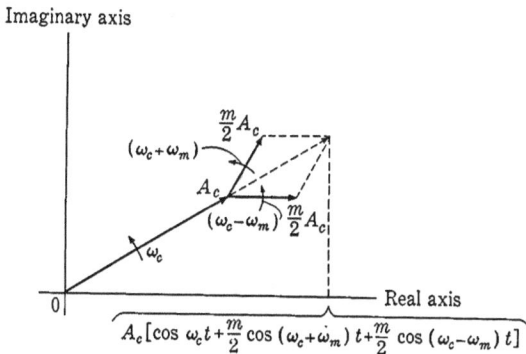

Fig. 5.   Vector representation of the carrier and side frequencies in an amplitude-modulated wave.

$A_c(1 + m)$ and $A_c(1 - m)$. Because of this cancellation of the normal components, the resultant vector rotates at constant angular velocity. In particular, the intervals between the points at which the instantaneous amplitude is zero are constant—a fact indicating that no frequency or phase modulation is present in such a wave. This result does not hold if, as is discussed later, one side frequency is suppressed.

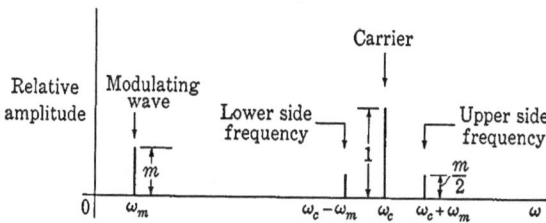

Fig. 6.   Frequency spectrum of a sinusoidally modulated wave.

Figure 6 shows graphically the distribution of the components of the sinusoidally modulated wave in the frequency spectrum. Here the frequencies are plotted as abscissas and the amplitudes as ordinates. A sinusoidal component in the wave is therefore represented as a vertical line at the particular frequency with a height corresponding to the amplitude.

In general, the modulating wave is not a pure sinusoid but consists of a number of components of different frequencies and amplitudes and is representable by a Fourier series for a periodic function or a Fourier integral for a nonperiodic function. The frequency spectrum can then be represented[8] by such a curve as $G(\omega)$ in Fig. 7, where height of the vertical lines represents the amplitudes, and position represents the frequencies of the separate components. Each frequency in the modulating wave then produces a pair of side frequencies in the modulated wave, and the modulated wave has upper and lower *side bands*

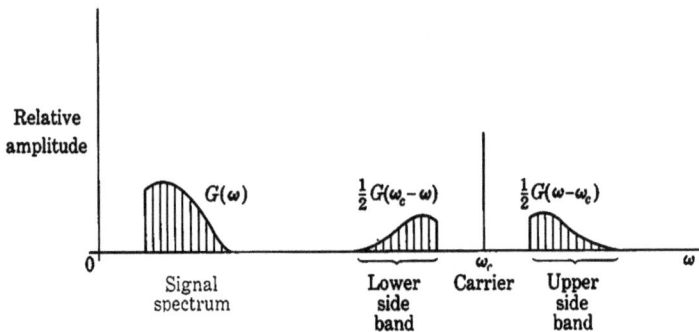

Fig. 7.   Frequency spectrum of a wave with nonsinusoidal modulation.

$\frac{1}{2}G(\omega - \omega_c)$ and $\frac{1}{2}G(\omega_c - \omega)$, respectively, disposed symmetrically about the carrier, each of which contains all the essential characteristics of the modulating wave. For example, if an audio spectrum from 0 to 5,000 cycles per second is to be transmitted on a 1,000,000-cycle-per-second carrier frequency, the modulated wave covers a 10,000-cycle-per-second band centered about 1,000,000 cycles per second. The modulated wave then is said to require a *channel* 10,000 cycles per second wide.

These considerations illustrate the frequency translation mentioned in Art. 1 as being inherent in the modulation process. The signal $G(\omega)$, originally located near the origin, can be translated to any other part of the spectrum by appropriate choice of the carrier frequency. The economic importance of this process cannot be overemphasized, for, without it, no radio-telephonic communication would be possible and long-distance wire communication would be very much more expensive and limited than it is. Radiation of electromagnetic waves for communication at audio frequencies with any antenna of a size that might conceivably be constructed on this planet is so inefficient as to be

---

[8] R. V. L. Hartley, "Relations of Carrier and Side Bands in Radio Transmission," *B.S.T.J.*, *2* (1923), 90–112.

unthinkable; but, by being shifted to a band of sufficiently high frequencies, the information may be effectively radiated and transmitted over great distances—even around the world. Carrier-current telephony, whereby one pair of wires is used for transmitting different conversations simultaneously, also depends on the modulation process for shifting each audio-frequency signal to its appointed private channel, or frequency band, for transmission over wires.

The frequency-translating effect of modulation is by no means limited to shifting an audio signal to a higher location in the spectrum. A high-frequency signal may be shifted to a still higher frequency location in the spectrum—say from 50,000 cycles per second to 10,000,000 cycles per second. It may also be employed to shift a frequency down to a lower value—say from 30,000,000 cycles per second to 456,000 cycles per second, as is often done in superheterodyne radio receivers in order that the signal may be amplified effectively.

Although frequency translation occurs with amplitude modulation, the effect is different from mere multiplication of the frequencies. If all the frequencies in a wave were multiplied by the same factor, the ratio of the bandwidth to the mean frequency would remain constant. In the process of modulation, however, the bandwidth itself remains constant, and the channel required is of constant width regardless of the carrier frequency employed.

The amplitude of each side frequency for a sinusoidally modulated wave is $m/2$, which has a maximum value of $\frac{1}{2}$ for 100 per cent modulation. The relative amounts of power dissipated in a resistive load by the carrier and the side frequencies are therefore:

$$P_{carrier} = 1,\qquad\qquad [19]$$

$$P_{lower\ s.f.} = \frac{m^2}{4},\qquad\qquad [20]$$

$$P_{upper\ s.f.} = \frac{m^2}{4},\qquad\qquad [21]$$

as is stated in a previous table. Hence

$$P_{total\ s.f.} = \frac{m^2}{2}\,P_{carrier};\qquad\qquad \blacktriangleright[22]$$

or, with 100 per cent sinusoidal modulation, the total power in the side frequencies is one-half that in the carrier wave. However, if the modulation is 100 per cent, and the modulating wave is a square wave such as might result from a telegraph signal, the amplitude of the wave is double the carrier value for half the time and zero for the other half.

Since the power varies as the square of the amplitude, the total power in the wave is therefore twice the carrier power, and the total power in the side bands is equal to that in the carrier wave.

With sinusoidal modulation, the total power in the side frequencies varies with the square of the modulation factor. Thus, though it is one-half the carrier power for 100 per cent modulation, it is only one two-hundredth of the carrier power for 10 per cent modulation. The carrier wave does not in itself contain the information to be transmitted, and it therefore represents a waste of power. In carrier-current wire telephony and in transoceanic radio telephony, the carrier is suppressed and not transmitted, but the added complications and cost of the suppressed-carrier systems make their use in most applications uneconomical. Since one side band alone is adequate to convey the complete information, perfect communication can be effected if, with suitable apparatus, both the carrier and the other side band are suppressed. Radio transmission, however, with the exception noted above, employs both side bands and the carrier. Suppressed-carrier modulation is widely used in many measurement and automatic-control systems.

The foregoing description of sinusoidal modulation is based on the assumption that the amplitude of the high-frequency modulated wave is directly proportional to the instantaneous value of the low-frequency modulating wave. Actually, in practical modulating methods, this condition of direct proportionality is never exactly obtained. There is always some nonlinearity in the relation between the two quantities. Consequently the variation of the amplitude of the modulated wave is not exactly symmetrical above and below the carrier value. The same dissymmetry occurs when the modulating wave is nonsinusoidal. With these conditions, the analysis becomes more complex, and the percentage modulation must be defined in terms of separate values for the percentage increase and decrease of the amplitude of the modulated wave during the cycle.[9]

In the following articles, practical methods of modulation and demodulation are discussed. So-called *linear* methods are dealt with first, and the so-called *square-law* devices are treated in the later articles. The behavior of the square-law methods lends itself to analysis by means of a power-series representation of the electrical characteristics of the circuit, but for the linear devices such a method of attack is precluded by the fact that the current is zero a considerable portion

[9] L. F. Gaudernack, "Some Notes on the Practical Measurement of the Degree of Amplitude Modulation," *I.R.E. Proc.*, 22 (1934), 819–846; *Standards on Antennas, Modulation Systems, Transmitters—Definitions of Terms, 1948* (New York: The Institute of Radio Engineers, 1948), 17.

of the time. An approximate treatment that does not involve the power-series representation is used for the linear devices. No attempt is made to discuss all the numerous methods of modulation; only a few of fundamental importance are included.

4. PLATE MODULATION OF CLASS C AMPLIFIERS

The discussion in the previous article indicates that one way of representing the characteristics of a device suitable for effecting modulation is that shown in Fig. 8. If the curve is a straight line, the amplitude of the modulated wave is directly proportional to the instantaneous value of the modulating wave; no harmonics are generated in the process of modulation, and the modulation is distortionless.

In practice, the straight-line characteristic curve in Fig. 8 can be closely approached. A common and useful point of view in the analysis of modulation devices is the assumption that the amplitude of the carrier-frequency output of the device varies with the instantaneous

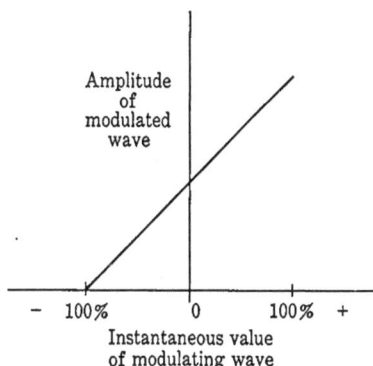

Fig. 8. Characteristic curve of an ideal modulator.

value of the modulating wave in the same way that it varies with changes in a steady supply voltage of the circuit. This assumption is based on the concept that, since the modulating frequency is small compared with a carrier frequency, the modulating wave can therefore be considered as having a frequency of practically zero in comparison with that of the modulated wave. In other words, the modulating wave behaves in the circuit as a direct voltage whose value is changed slowly during a modulation cycle and imperceptibly during a carrier-frequency cycle. This point of view can be summarized by the statement that the dynamic and the static behaviors of the circuit are assumed to be the same.

Such a point of view is subject to at least two limitations. In the first place, owing to blocking capacitors, inductors or chokes, internal impedance of the power supply, and the variation of other impedances in the circuit with frequency, the circuit does not behave the same for the varying voltage as it does for the steady supply voltage. In the second place, sharply tuned portions of the circuit may not permit rapid changes in the amplitude of the carrier-frequency components in the currents or voltages, as has been mentioned in Art. 5, Ch. XI. This condition exists when the tuned portions of the circuit have appreciably different impedances at the carrier and at the side-band frequencies. Nevertheless, the point of view is often a useful approximation in the analysis of modulation circuits, and the curve of the carrier-frequency output as a function of a steady supply voltage is commonly called the *modulation characteristic* of the circuit. In general,

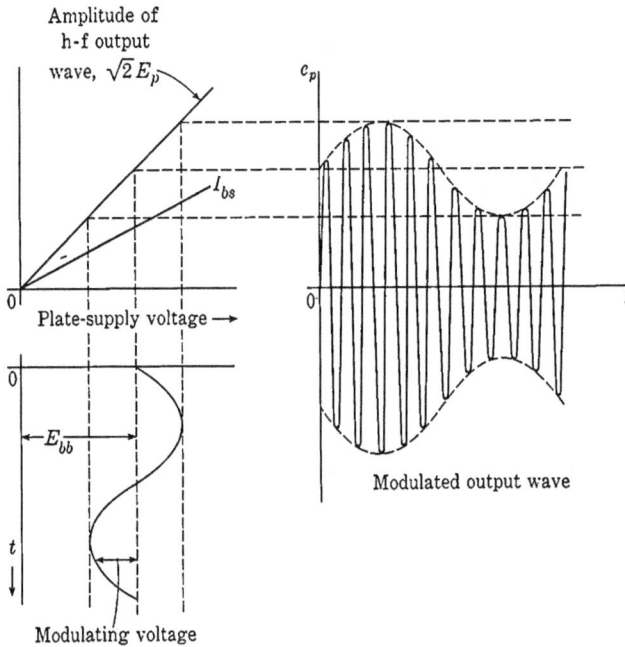

Fig. 9.   Idealized conditions during plate modulation of a Class C amplifier.

the supply voltage of any one of the tube electrodes may be varied to produce the modulated wave, and names that correspond to the particular electrode chosen are given to the various methods of modulation.

With proper design and adjustment, a Class C amplifier can be made to have approximately the modulation characteristics shown in Fig. 9. The amplitude of the high-frequency components of the plate voltage and current, to which the output circuit is tuned, vary practically linearly with the plate-supply voltage, and the average value of the plate current is also linearly related to the plate-supply voltage. To obtain these characteristics, it is generally necessary to employ a combination of fixed grid-bias and grid-leak-and-capacitor grid-bias voltages and to adjust the impedance of the tuned load circuit to a favorable value.[10] The final adjustments are often made experimentally because of the lengthy methods required for an analytical solution.

In one use of a Class C amplifier for the production of a modulated wave, termed *plate modulation*, the carrier frequency is introduced into the grid circuit and the modulating voltage is connected in series with the direct plate-supply voltage E_{bb}, as is shown schematically in Fig. 10. When the modulating voltage is a sine wave, the circuit behaves as illustrated in Fig. 9, where the total plate-supply voltage comprises the sum of a direct and a sinusoidal component, and where the envelope of the output wave also contains a sinusoidal component.

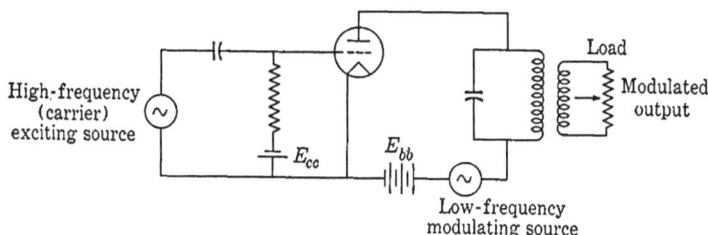

Fig. 10.   Simplified basic circuit of a plate-modulated Class C amplifier.

In actual practice, the modulating wave is usually transmitted to the modulated Class C amplifier through an amplifier, and various means are used to couple the modulating amplifier to the modulated amplifier. In early work, the modulating wave was introduced in the plate circuit of an oscillator by the method called Heising modulation.[11] One practical circuit diagram of a modulated Class C amplifier and its modulating amplifier, which involves principles similar to those underlying Heising modulation, is shown in Fig. 11. The Class C amplifier has a combination of fixed grid-bias and grid-leak-and-capacitor grid-bias supply, and neutralization for preventing oscillations is accomplished by means of a neutralizing capacitor and the tapped arrangement of the coil in the tuned circuit. The modulating amplifier, frequently called the modulator, is of the push-pull type; to realize high efficiency, it is often operated in Class B.[12] The output voltage of

[10] I. E. Mouromtseff and H. N. Kozanowski, "Analysis of the Operation of Vacuum Tubes as Class C Amplifiers," *I.R.E. Proc.*, *23* (1935), 752–778; W. L. Everitt, *Communication Engineering* (2nd ed.; New York: McGraw-Hill Book Company, Inc., 1937), 530–545.

[11] E. B. Craft and E. H. Colpitts, "Radio Telephony," *A.I.E.E. Trans.*, *38* (1919), 328.

[12] J. A. Hutcheson, "Application of Transformer-Coupled Modulators," *I.R.E. Proc.*, *21* (1933), 944–957.

the modulating amplifier $\sqrt{2}E_m \cos \omega_m t$ is connected in series with the direct plate-supply voltage E_{bb}, so that the total effective plate-supply voltage for the Class C amplifier is $E_{bb} + \sqrt{2}E_m \cos \omega_m t$. A high-frequency choke and a high-frequency by-pass capacitor are used in combination to isolate the high-frequency components of the plate-current and voltage in the modulated amplifier and to exclude them from the modulating amplifier and direct-current plate-power supply.

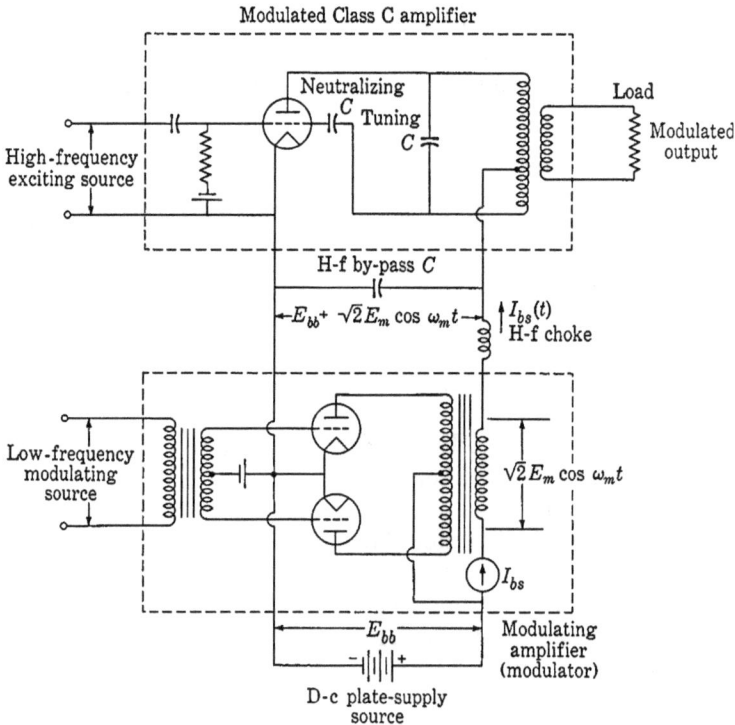

Fig. 11.   Plate-modulated Class C amplifier with a Class B transformer-coupled modulating stage.

The direct-current plate power for the push-pull modulating amplifier may be obtained from the same source as is that for the modulated amplifier, but this common source is not essential, and the requirements of the tubes may make it undesirable.

Among the requirements of interest for the design of amplifiers used in plate modulation are the plate dissipation of the tubes and the turns ratio of the output transformer of the push-pull stage. These are determined by the following considerations. The total effective plate-supply voltage $E_{bb} + \sqrt{2}E_m \cos \omega_m t$ may be expressed as

Total effective plate-supply voltage $= E_{bb}(1 + m \cos \omega_m t)$,   [23]

since, for 100 per cent modulation, Fig. 9 shows that the amplitude of the modulating voltage $\sqrt{2}E_m$ is equal to the direct plate-supply voltage E_{bb}. The total plate current in the Class C modulated stage comprises carrier-frequency and modulation-frequency components with harmonics, together with other frequencies introduced by non-linearities of the tube characteristic, and a steady or average component. However, for the purpose of computing the average power input to the modulated amplifier, only the components having the frequencies contained in the total effective plate-supply voltage are important, because in any circuit the average power comprises contributions from the products of current and voltage terms of like frequency only. The average plate current I_{bs} in Fig. 9, which varies linearly with the plate-supply voltage, has the usual significance when the plate-supply voltage is constant. But when the plate-supply voltage varies at a rate slow compared with the carrier-frequency—that is, at the modulation frequency—the average plate current also varies, and the average plate current then takes on a new significance. It is an average over the carrier-frequency cycle but a variable during the modulation-frequency cycle, and may therefore be denoted by $I_{bs}(t)$. The expression for the modulation-frequency and steady components of the current furnished by the combined modulating source and direct-current plate-supply voltage is therefore

$$I_{bs}(t) = I_{bs}(1 + m \cos \omega_m t),\qquad [24]$$

where I_{bs} is the average current as indicated by an ordinary direct-current ammeter. In fact, the physical significance of $I_{bs}(t)$ can also be interpreted in terms of the response of a measuring instrument. It is the current indicated by the deflection of an oscillograph such as the bifilar type that, because of its inertia, does not respond to the high carrier-frequency components but does respond to the low modulation-frequency components. Finally, the total plate current i_b, which includes the carrier-frequency components, is the current indicated by an instrument such as the cathode-ray oscilloscope, which responds to all components, including those of very high frequency.

In accordance with the foregoing concepts, the average power supplied to the modulated amplifier is

$$P_{bs} = \frac{1}{2\pi}\int_0^{2\pi} E_{bb}(1 + m \cos \omega_m t)I_{bs}(1 + m \cos \omega_m t)\, d(\omega_m t),\quad [25]$$

or

$$P_{bs} = E_{bb}I_{bs} + \frac{m^2}{2}E_{bb}I_{bs}.$$ ▶[26]

Also the power supplied by the direct-current plate-power supply is

$$P_{bb} = \frac{1}{2\pi}\int_0^{2\pi} E_{bb}I_{bs}(1 + m \cos \omega_m t)\, d(\omega_m t),$$ [27]

or

$$P_{bb} = E_{bb}I_{bs} .$$ ▶[28]

Thus the power supplied by the modulating amplifier is

$$P_m = \frac{m^2}{2}E_{bb}I_{bs} .$$ ▶[29]

Consequently, for 100 per cent sinusoidal linear modulation, the modulating amplifier must deliver half as much power as the direct-current plate-power supply.

If the plate efficiency of the modulated amplifier is constant over the modulation-frequency cycle (as is approximately true in practice), the power output in the unmodulated carrier wave is

$$P_h = \eta_c P_{bb} = \eta_c E_{bb}I_{bs},$$ [30]

where η_c is the plate efficiency of the modulated Class C amplifier. The power output with sinusoidal modulation is

$$P_{ac} = \eta_c P_{bs} = \eta_c E_{bb}I_{bs}\left(1 + \frac{m^2}{2}\right).$$ [31]

Equation 31 contains terms proportional to P_{bb} and P_m in Eqs. 28 and 29; hence it may be concluded that the direct-current plate-power supply furnishes the power to produce the carrier wave, and the modulating amplifier furnishes the power to produce the side frequencies or side bands.

Since the peak amplitude of the modulated output wave is twice that of the carrier wave for 100 per cent modulation, the peak power output of the modulated amplifier is four times the unmodulated value. The plate dissipation also increases with modulation, and the amplifier must *not* be designed for the maximum allowable plate dissipation with an unmodulated output. The average plate dissipation of the tubes of the modulated amplifier is

$$P_{p_c} = P_{bs} - P_{ac} = P_{bs}(1 - \eta_c),$$ [32]

or, from Eqs. 26 and 28,

$$P_{p_c} = P_{bb}\left(1 + \frac{m^2}{2}\right)(1 - \eta_c).$$ [33]

The factor $m^2/2$ shows that there is an increase in plate dissipation of 50 per cent caused by 100 per cent sinusoidal modulation. Consequently, if the modulated amplifier is to be capable of delivering a sustained 100 per cent sinusoidally modulated output, it should be designed so that the plate dissipation in the absence of modulation is two-thirds of the maximum allowable value. For reproduction of sound, 100 per cent modulation is almost never sustained over an appreciable period of time; hence the factor of two-thirds is generally increased somewhat for this service.

The plate dissipation of the tubes in the modulating amplifier is

$$P_{p_m} = P_m \left(\frac{1}{\eta_m} - 1 \right), \qquad [34]$$

where η_m is the plate efficiency of the modulating amplifier. From Eqs. 28 and 29, this becomes

$$P_{p_m} = \frac{m^2}{2} P_{bb} \left(\frac{1}{\eta_m} - 1 \right). \qquad [35]$$

Thus the ratio of the plate dissipation of the tubes in the modulating amplifier and the modulated amplifier is

$$\frac{P_{p_m}}{P_{p_c}} = \frac{\dfrac{m^2}{2} P_{bb} \left(\dfrac{1}{\eta_m} - 1 \right)}{\left(1 + \dfrac{m^2}{2} \right) P_{bb}(1 - \eta_c)} \qquad [36]$$

$$= \left(\frac{m^2}{2 + m^2} \right) \left(\frac{1 - \eta_m}{1 - \eta_c} \right) \frac{1}{\eta_m}. \qquad \blacktriangleright[37]$$

This ratio is considerably larger than unity under certain operating conditions, with the result that the cost of the tubes in the modulating amplifier is sometimes greater than that of those in the modulated amplifier. As an example, it may be assumed first that the modulating amplifier is operated in Class A$_1$ with a plate efficiency η_m of 0.2. The modulated Class C amplifier may have a plate efficiency η_c of, say, 0.7. The ratio of the plate dissipations for 100 per cent continuous sinusoidal modulation is then

$$\frac{P_{p_m}}{P_{p_c}} = \left(\frac{1}{2 + 1} \right) \left(\frac{1 - 0.2}{1 - 0.7} \right) \frac{1}{0.2} = 4.44. \qquad [38]$$

With these operating conditions, the tubes in the modulating amplifier therefore must have a total allowable plate dissipation of about $4\frac{1}{2}$ times that of the modulated-amplifier tubes.

As an alternative example, it may be assumed that the modulating amplifier is operated in Class B with a plate efficiency η_m of 0.5 and that η_c equals 0.7 as before. The ratio of the plate dissipations is then

$$\frac{P_{p_m}}{P_{p_c}} = \left(\frac{1}{2+1}\right)\left(\frac{1-0.5}{1-0.7}\right)\frac{1}{0.5} = 1.11. \qquad [39]$$

Thus, with the Class B modulating amplifier, the plate dissipation of the modulating tubes must be approximately equal to that of the modulated amplifier tubes. The ratio of the total number of tubes of a specified plate dissipation needed in the two cases is

$$\frac{1+4.44}{1+1.11} = 2.58. \qquad [40]$$

The total plate-dissipation rating of the tubes needed to produce a modulated wave with a Class B modulating amplifier is therefore only about 40 per cent of that needed with a Class A_1 amplifier. An example of the use of a Class B modulating amplifier is a radio transmitter having a carrier power of 500 kilowatts in which eight 100-kilowatt tubes are used in the modulating amplifier, which operates at approximately 50 per cent plate efficiency at its full output.[13] The Class C amplifier comprises twelve 100-kilowatt tubes.

Actually, the efficiency of the transformer or other circuit element used to couple the modulating amplifier to the modulated amplifier is a factor in the determination of the relative plate dissipations required. Dividing the right-hand side of Eqs. 34 through 37 by this efficiency brings it properly into the analysis.

A knowledge of the effective impedance across the secondary terminals of the output transformer of the modulating amplifier is needed in order that the turns ratio of the transformer may be chosen to have the value that reflects the optimum value of impedance into the plate circuits of the modulating tubes. The transfer of power at the modulation frequency is of interest; hence the effective impedance as viewed from the secondary terminals toward the modulated amplifier is the ratio of the modulation-frequency alternating secondary voltage and the modulation-frequency component of the secondary current. From Fig. 9 and the preceding discussion, this ratio is

$$R_b = E_{bb}/I_{bs}. \qquad \blacktriangleright[41]$$

Thus the effective impedance across the output terminals of the

[13] J. A. Chambers, L. F. Jones, G. W. Fyler, R. H. Williamson, E. A. Leach, and J. A. Hutcheson, "The WLW 500-kw Broadcast Transmitter," *I.R.E. Proc.*, 22 (1934), 1154.

modulating transformer is a resistance R_b given by E_{bb}/I_{bs}, and the turns ratio of the transformer should be selected to correspond to this load.

The plate-modulated amplifier is widely used in radio transmitters. It has the advantage that the linear characteristics shown in Fig. 9 can be approached in practice by reasonably simple methods, whereas other methods of modulation require more complicated adjustments and often result in inferior linearity. The main disadvantage of the plate-modulated amplifier is that a comparatively large amount of

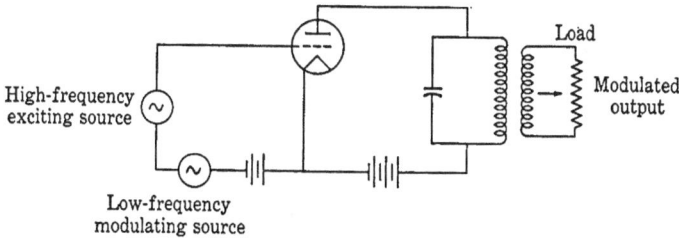

Fig. 12.   Basic circuit for grid-bias modulation.

power at the modulation frequency is required, and the resulting cost of the equipment is sometimes greater than that of other methods.

5. ADDITIONAL METHODS FOR LINEAR MODULATION OF CLASS C AMPLIFIERS

Modulation may be accomplished through connecting the low-frequency modulating source in series with one of the grid-voltage sources instead of in series with the plate-power supply as is done in the plate-modulated amplifier discussed in the previous article. The basic simplified circuit for this method of modulation is shown in Fig. 12. In effect, the grid-bias voltage is varied by the low-frequency modulating source, and the amplitude of the high-frequency output is thereby varied. The method is therefore called *grid-bias modulation*, although this name is sometimes shortened to *grid modulation*. The distinction resulting from the use of the longer term should be made, however, because there are other methods of modulation that also involve the introduction of the modulating voltage into the grid circuit but operate on an essentially different principle. For example, the Van der Bijl modulator discussed in Art. 12 also employs a series connection of the low- and high-frequency sources in the grid circuit. However, the Van der Bijl modulator operates with Class A conditions over the nonlinear region of the tube characteristics, whereas grid-bias modulation involves Class C operating conditions.

Grid-bias modulation has the advantage that only a small amount of low-frequency modulating power is required, but this is not to be interpreted as meaning that the total over-all efficiency, which includes the effects of the losses at both the plate and the grid, is greater than that of plate modulation, because the average plate efficiency during the modulation cycle is much lower. The circuit adjustments for a linear modulation characteristic are more difficult with grid-bias modulation and they are sensitive to variations in the high-frequency exciting voltage, the plate-supply voltage, and the magnitude of the tuned load impedance.

A modulated output wave is produced if a modulating voltage is introduced in series with the cathode of a Class C amplifier tube. This method is called *cathode modulation*.[14] Since the modulating voltage is effectively introduced into both the plate and the grid circuits, the characteristics of the method are a combination of those of grid-bias and of plate modulation. The values of the plate efficiency of the modulated amplifier and the amount of modulating power required are intermediate between the corresponding values for grid-bias and for plate modulation.

Tetrodes and pentodes are often used in Class C amplifiers, because the addition of a screen grid makes it unnecessary to provide for neutralization of the tendency of the amplifier to oscillate, and hence simplifies the circuit. The introduction of additional grids also increases the possibilities for modulating the amplifier, and several methods are in use.

Plate modulation of tetrode amplifiers is accompanied by distortion or harmonic generation unless the screen-grid voltage is also varied in accordance with the modulating voltage, because secondary emission causes pronounced dips and curvature in the modulation characteristics for a fixed screen-grid supply voltage. If the modulating voltage is introduced simultaneously into the screen-grid and plate circuits, characteristics having satisfactory linearity may be obtained.[15]

Grid-bias modulation of tetrode or pentode amplifiers is accomplished in the same manner as for triodes.

The suppressor grid of the pentode amplifier may be used as an additional control grid for the purpose of modulation.[16] The modulating

[14] E. E. Spitzer, A. G. Neukut, and L. C. Waller, "Cathode Modulation," *RCA Ham Tips*, *3* (January-February, 1940), 1; F. C. Jones and F. W. Edmonds, "Cathode Modulation," *QST*, *23* (November, 1939), 23–25, 102; F. W. Edmonds, "More on Cathode Modulation," *QST*, *23* (December, 1939), 52–54, 57; F. C. Jones, "Cathode Modulation," *Radio*, No. 242 (October, 1939), 14–18.

[15] H. A. Robinson, "An Experimental Study of the Tetrode as a Modulated Radio-Frequency Amplifier," *I.R.E. Proc.*, *20* (1932), 131–160.

[16] C. J. de Lussanet de la Sabloniere, "The New Transmitting Pentode P.C. 1.5/100," *Philips Transmitting News*, *1* (December, 1934); C. B. Green, "Suppressor-Grid Modulation," *Bell Lab. Rec.*, *17* (1938), 41–44.

voltage is connected in series with the suppressor grid and a suitable direct bias voltage. The results are somewhat similar to those of grid-bias modulation of a triode, but the circuit adjustments are simpler, because in the pentode amplifier they are made in two different circuits whose effects are relatively independent, whereas in the triode amplifier the modulating voltage, high-frequency exciting voltage, and direct grid-bias voltage must be mutually suitable.

In all the foregoing methods of modulation, the modulating voltage is introduced into the circuit in series with one of the electrodes of the high-frequency amplifier tube. All the methods have the common characteristics of relatively low over-all efficiency and considerable nonlinearity with resultant harmonic generation. Several methods have been devised to improve the efficiency, some of which make use of the impedance-inverting property of a quarter-wave transmission line and require careful adjustment.[17]

In general, the amount of nonlinearity in the modulation characteristic increases with the per cent modulation. Negative feedback has been effectively employed to overcome the effects of this nonlinearity. The modulated output of the amplifier is fed to a linear demodulator, which recovers the modulation-frequency components of the modulated wave. These components are then fed back to the input of the modulating amplifier in such a manner as to decrease the output, and the over-all harmonic generation in the complete modulating system is thus decreased, as is explained in Art. 12, Ch. IX.

6. LINEAR DETECTION

The process of detection or demodulation is used to recover the original frequency components in the modulating wave from the carrier and side-band components when the modulated wave arrives at the point in the transmission system where it is desired to reproduce the original modulating wave. The modulation frequencies which were translated to a higher part of the frequency spectrum by modulation are translated back to their original part of the spectrum by demodulation. The demodulation is generally accomplished by a process of rectification using some form of nonlinear circuit element. The common methods may be separated into two classes—so-called *linear*

[17] H. Cheirex, "High-Power Outphasing Modulation," *I.R.E. Proc.*, *23* (1935), 1370–1392; W. H. Doherty, "A New High-Efficiency Power Amplifier for Modulated Waves," *I.R.E. Proc.*, *24* (1936), 1163–1182; F. E. Terman and J. R. Woodyard, "A High-Efficiency Grid-Modulated Amplifier," *I.R.E. Proc.*, *26* (1938), 929–945; R. B. Dome, "High-Efficiency Modulation System," *I.R.E. Proc.*, *26* (1938), 963–982; L. F. Gaudernack, "A Phase-Opposition System of Amplitude Modulation," *I.R.E. Proc.*, *26* (1938), 983–1008.

detection and *square-law detection*. These classes are distinguished by the fact that, in linear detection, the current through the detecting circuit element is zero part of the time and flows in pulses, but in the circuit element used in square-law detection the current is unidirectional and greater than zero at all times. When the current is continuous, it is convenient to analyze the behavior of the device by the use of a power-series representation of the tube characteristics, as

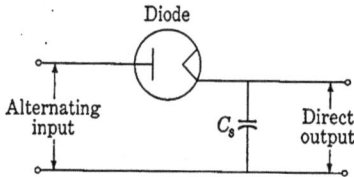

Fig. 13.   Basic linear-detection circuit.

is done in the subsequent articles of this chapter. The phenomena in linear detection, which are discussed in this article, do not lend themselves to analysis by the power-series method, however, and the method of analysis generally used is an extension of that used in the previous chapters on rectifiers.

The essential principle of linear detection is that, in a diode rectifier circuit with a capacitance load such as is shown in Fig. 13, the direct voltage developed across the capacitor C_s is equal to the crest value[18] of the alternating voltage at the input terminals, as is explained in the discussion of rectifiers, Art. 9, Ch. VI. If, however, a modulated input voltage is used, the capacitor retains the charge acquired from the greatest amplitude of the input voltage during the modulation cycle, and its voltage remains fixed at that value, because no output current

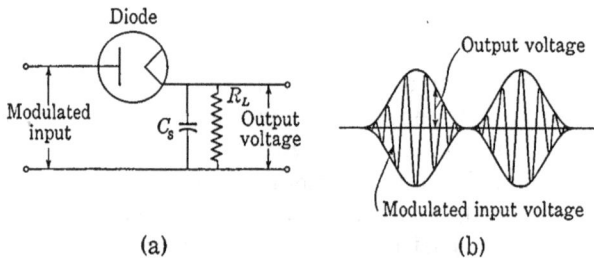

Fig. 14.   Detection in an idealized linear detector.

is drawn from the capacitor and its voltage therefore cannot decrease. Consequently, in order that the output voltage be able to decrease with the amplitude of the input voltage, a load such as R_L in Fig. 14a is necessary. The ideal desired in the operation of the circuit is that the capacitor should charge to the peak of the input wave during each carrier-frequency cycle, but that there should be sufficient current

---

[18] C. H. Sharp and E. D. Doyle, "Crest Voltmeters," *A.I.E.E. Trans.*, *35* (1916), 99–107; C. H. Sharp, United States Patent 1,209,766, December 26, 1916.

drain through R_L to cause the output voltage to decrease as rapidly as does the envelope of the modulated input voltage, as shown in Fig. 14b. In these circumstances, the output voltage has the waveform of the envelope of the modulated wave, which comprises a steady component and a component having the waveform of the original modulating voltage used to produce the modulated input wave. Since the steady component can be isolated by the use of a transformer or a resistance-

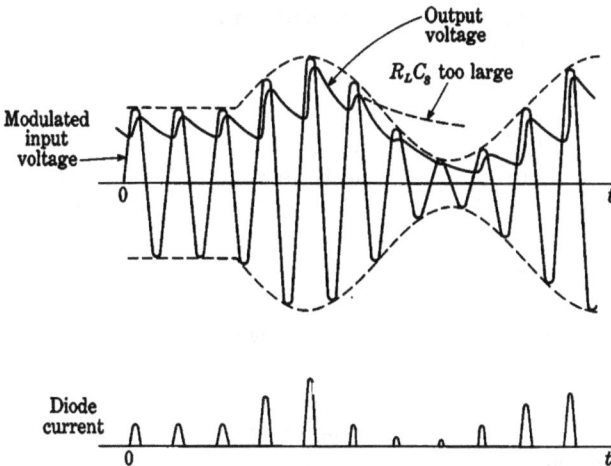

Fig. 15. Waveforms in the circuit of Fig. 14. The irregularities in the output waveform are greatly exaggerated, since ordinarily during a cycle of the envelope there would be many more cycles of the high frequency than are shown.

capacitance coupling circuit, the essentials of detection are accomplished by the diode used in this way.

Actually, the previously mentioned objectives, that (a) the capacitor charge to the peak of the input wave during each cycle and (b) the resistance load drain the charge fast enough to enable the output voltage to decrease with the envelope, are incompatible. The circuit of Fig. 14a will be recognized as the same as that discussed in Art. 9, Ch. VI, where a capacitor is used for smoothing the rectified output voltage. The one additional consideration peculiar to the detector circuit is that the alternating input voltage varies in amplitude. The waveforms of the input voltage, output voltage, and tube current that result in the circuit of Fig. 14a are roughly those shown in Fig. 15. Near the top of each positive peak of input voltage, the tube passes a pulse of current that charges the capacitor to a value near the envelope of the input voltage, but the voltage drop in the tube during the conduction period prevents the output voltage from actually reaching the peak value. When the input voltage drops below the capacitor voltage, the tube ceases to conduct, and the capacitor discharges slowly through the load resistance until the next positive peak of input voltage occurs and recharges it. The resulting waveform of the output voltage[19] is therefore somewhat "jagged." The diagram of the waveforms in Fig. 15 exaggerates this effect, however, because the carrier frequency is ordinarily much greater than the modulation frequency, so that the pulses of current and voltage are very much closer together and the jaggedness is less pronounced. Essentially, the output voltage of the detector follows the envelope of the modulated input wave and has the waveform of the modulating voltage.

If a small amplitude of the high-frequency variations in the output voltage is to be secured, $R_L C_s$, the time constant of the load circuit, must be large compared with the period of a carrier-frequency cycle. There is a maximum permissible value for this time constant, however, because it must not be so large that the output voltage cannot decay as rapidly as the envelope decreases. If the time constant were larger than this value, the output voltage would follow a curve such as the dotted one in Fig. 15, labeled $R_L C_s$ *too large*, and would not be a reproduction of the envelope. The negative peaks of the modulating wave would be "clipped" off. The greater the frequency of the modulating wave, the greater the restriction on the time constant, since, for a particular percentage modulation, the maximum rate of decrease of the envelope is proportional to the highest modulating frequency.

The maximum permissible value for the time constant may be determined approximately as follows.[20] Under the most unfavorable conditions, the envelope waveform may be represented by

$$e = E(1 + m \cos \omega_m t), \qquad [42]$$

where ω_m is the largest angular modulating frequency the detector is designed to handle. At any particular time t_0, the slope of the envelope is

$$\frac{de}{dt}\bigg|_{t_0} = -mE\omega_m \sin \omega_m t_0, \qquad [43]$$

and the particular value of the envelope at that time is

$$e_0 = E(1 + m \cos \omega_m t_0). \qquad [44]$$

---

[19] A more nearly complete analysis is given by W. B. Lewis, "The Detector," *Wireless Engr.*, *9* (1932), 487–499.

[20] F. E. Terman and N. R. Morgan, "Some Properties of Grid-leak Power Detection," *I.R.E. Proc.*, *18* (1930), 2160–2175.

If the capacitor voltage equals e_0 at the time t_0, the equation of its decay thereafter is

$$e_c = e_0 \epsilon^{-\frac{t-t_0}{R_L C_s}},$$ [45]

and its initial rate of change is

$$\frac{de_c}{d(t-t_0)}\bigg|_{t_0} = -\frac{e_0}{R_L C_s}.$$ [46]

If the capacitor voltage is to remain smaller than the envelope voltage after the time t_0, then at the time t_0 the slope of the curve of capacitor voltage, given by Eq. 46, must be smaller than (a negative number of greater magnitude than) the slope of the envelope of the modulated wave, given by Eq. 43. Thus

$$-\frac{e_0}{R_L C_s} \leqq -mE\omega_m \sin \omega_m t_0,$$ [47]

or

$$\frac{e_0}{R_L C_s} \geqq mE\omega_m \sin \omega_m t_0.$$ [48]

Substitution of Eq. 44 for e_0 and rearrangement of the expression give

$$\frac{1}{R_L C_s} \geqq \omega_m \left[\frac{m \sin \omega_m t_0}{1 + m \cos \omega_m t_0}\right].$$ [49]

Hence, in order that the initial rate of decay of the capacitor voltage be greater than the rate of decay of the envelope voltage at any time of equality of the voltages t_0, the time constant of the load circuit $R_L C_s$ must satisfy the relation

$$R_L C_s \leqq \frac{1}{\omega_m \left[\dfrac{m \sin \omega_m t_0}{1 + m \cos \omega_m t_0}\right]}.$$ [50]

Since this relation involves the assumption of equality of the capacitor and envelope voltages at the time t_0, the bracketed term in the denominator of the right-hand member of Eq. 50 is a function of t_0. By the usual process of maximization, the value of t_0 for which the bracketed term is a maximum and the required time constant hence is a minimum is found to be that for which

$$\cos \omega_m t_0 = -m,$$ [51]

or

$$\sin \omega_m t_0 = \sqrt{1 - m^2}.$$ [52]

Substitution of Eqs. 51 and 52 in Eq. 50 gives

$$R_L C_s \leqq \frac{1}{\omega_m} \frac{\sqrt{1 - m^2}}{m}.$$ ▶[53]

A time constant that satisfies this relation insures that the capacitor voltage will have a greater rate of decay than the envelope voltage at the time of equality of the voltages t_0, regardless of the point in the cycle at which this time occurs. Since the charging pulses of current supplied through the detector tube repeatedly restore the capacitor voltage to practical equality with the envelope voltage during the modulation cycle, the time of equality t_0 takes on essentially all values of time during the cycle. Hence Eq. 53 must be satisfied if the output voltage is to follow the waveform of the envelope. According to this equation, as the modulation approaches 100 per cent the required time constant approaches zero. In other words, the capacitance should be zero if the modulation is to be 100 per cent. Under these conditions, the output voltage would contain the carrier as well as the modulating frequency, and the output would not follow the envelope of the input voltage. Actually, additional factors, such as the impedance of the source supplying the modulated voltage, enter into the behavior of the detector,[21] and it has been shown[22] experimentally that the amount of harmonic generation, or distortion, in the detector is not excessive for reproduction of sound if

$$R_L C_s \leqq 1/(\omega_m m). \qquad \blacktriangleright[54]$$

Equation 53 or 54 governs the value of the product of R_L and C_s to be used in the circuit, but their individual values are subject to additional design considerations. For the current through the tube and the voltage drop in it to be small, R_L should be large compared with the effective resistance of the tube. On the other hand, the capacitance C_s should be large compared with the capacitance of the diode, in order that the displacement current through the two capacitances in series may not develop an appreciable voltage component of the carrier frequency in the output. Generally a compromise adjustment is made corresponding to a value of the load resistance given by Eq. 53 or 54 and a value of C_s five to ten times the plate-to-cathode capacitance of the diode.

In addition to the desired component of waveform corresponding to that of the original modulating voltage, the output of the circuit in Fig. 14a also contains a steady component that corresponds to the amplitude of the carrier voltage. If the detector is to supply the input voltage for a succeeding stage of amplification, elimination of this steady component is often desirable, and a blocking capacitor with

[21] S. Bennon, "Note on Large Signal-Diode Detection," *I.R.E. Proc.*, *25* (1937), 1565–1573.

[22] F. E. Terman and J. R. Nelson, "Discussion of 'Some Notes on Grid-Circuit and Diode Rectification,'" *I.R.E. Proc.*, *20* (1932), 1971–1974.

resistance-capacitance coupling to the succeeding amplifier stage is frequently used for this purpose, as is shown in Fig. 16. The circuit then has many of the characteristics of the resistance-capacitance-coupled amplifier discussed in Art. 5, Ch. IX.

To analyze its behavior, the *rectification characteristics* of the detector tube, which are measured in the circuit[23] of Fig. 17, are useful. In this circuit, an alternating voltage and a direct voltage are connected in series across the diode, and the steady component of the plate current is measured as the dependent variable. A capacitance C_s is connected across the direct-voltage supply in order to prevent varia-

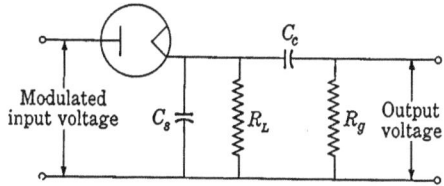

Fig. 16.　Diode-detector circuit with a blocking capacitor to eliminate the steady component in the output.

tions in the voltage at the tube caused by the pulses of current flowing through the slide-wire rheostat used to adjust the voltage. If the alternating-current supply voltage is held constant at various values as a parameter, and the rectified current is measured as a function of the direct voltage, curves similar to those of Fig. 18 result. The negative sign for the direct voltage along the axis of abscissas is a result of the choice made in the assigning of the positive reference

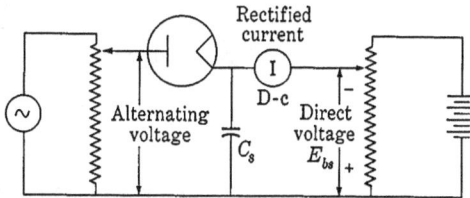

Fig. 17.　Circuit for measurement of the rectification characteristics of a diode. The capacitance C_s is assumed to be large enough to reduce the ripple voltage across it to a negligible value.

direction for voltage in Fig. 17. These curves are characteristics of the tube alone, it being assumed that a capacitor C_s sufficiently large to reduce the carrier-frequency component in the output voltage to a negligible value is included in the circuit. Each curve is a direct-current volt-ampere characteristic of the tube for a fixed value of the parameter (the alternating input voltage), and the curves are similar in that sense to the plate characteristics of a triode. When a resistance

---

[23] C. E. Kilgour and J. M. Glessner, "Diode-Detection Analysis," *I.R.E. Proc.*, **21** (1933), 930–943.

load is connected in the plate circuit of the tube, a solution for the direct current and corresponding output voltage may be obtained by the same graphical procedure as is outlined for the triode in Art. 2, Ch. VIII. The straight-line volt-ampere characteristic of the load is super-posed on the rectification characteristics, and the direct current and direct voltage developed in the load are given by the intersection of a particular rectification characteristic and the load line. Since there is no direct-voltage supply in series with the tube, the load line passes through the origin for the circuits of Figs. 14a and 16. Several load lines for different values of load resistance R_L are drawn in Fig. 18.

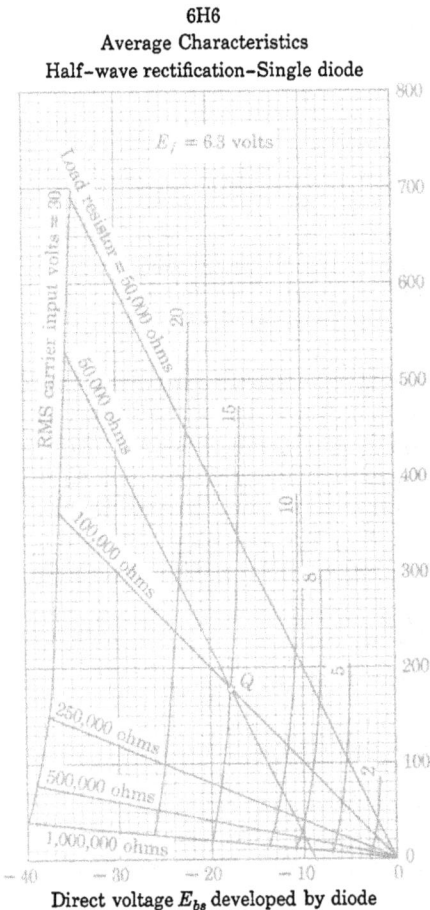

6H6
Average Characteristics
Half-wave rectification–Single diode

Fig. 18. Rectification characteristics of a typical diode-detector tube.*

To illustrate operation with the circuit of Fig. 14a, assume that the load resistance is 100,000 ohms, and the carrier voltage of the modulated input wave is 15 volts. Figure 18 shows that at the quiescent point Q, which corresponds to zero modulation, the steady component of the output current is then 178 microamperes and the steady component of the output voltage is −17.8 volts. If the modulation is 100 per cent, the input voltage varies between the limits of 0 and 30 volts, and the corresponding intersections on the 100,000-ohm load line show that the output voltage then varies between 0 and −36.0 volts.

The fact that the rectification characteristics in Fig. 18 are somewhat

* This diagram is adapted from *RCA Receiving Tube Manual*, Technical Series RC-16 (Harrison, N. J.; RCA Manufacturing Company, Inc., 1950), Fig. 92CM-4446T, p. 146, with permission.

curved does not imply that the detection is therefore nonlinear and that excessive harmonic generation will occur. Even if the volt-ampere characteristic of the diode itself is strictly linear, an analysis similar to that of Art. 9, Ch. VI, shows that the rectification characteristics are curved.[24] The methods of Art. 13, Ch. VIII, are applicable for a determination of the amount of harmonic generation in the detector circuit; and a study of the curves shows that, along any one load line, the direct-current output is directly proportional within close limits to the parameter (amplitude or root-mean-square value of the input voltage). Thus the harmonic generation is generally negligible. For example, for strictly linear detection, the direct output voltage at the peak of the modulation cycle would be -17.8×2, or -35.6 volts, instead of the measured -36.0 volts, a negligible difference. In practice, linear detectors are made that introduce much less than one per cent of harmonic generation.

When the detector is resistance-capacitance coupled to the succeeding amplifier stage, as is done in Fig. 16, the path of the operating point does not follow the load line through the origin, because the impedance of the combined load to the steady component of current corresponding to the carrier input is different from its impedance to the modulation-frequency component of current. The conditions are analogous to those described for the resistance-capacitance-coupled amplifier (see Fig. 12, Ch. IX). A line drawn through the origin with a slope corresponding to R_L intersects the curve corresponding to the carrier voltage at the point giving the output current and voltage for an unmodulated input. For a modulated input, operation takes place along a line through Q with a slope corresponding to $R_L R_g/(R_L + R_g)$, where R_g is the resistance coupled to the diode through the capacitance C_c. The reactance of C_c at the modulation frequency is assumed to be negligible compared with R_g. If, for example, R_L and R_g in the circuit of Fig. 16 are 100,000 ohms each, and the effective value of the carrier voltage is 15 volts, the operating point for the unmodulated input is the same as in the example above. For a modulated input, however, the operation takes place about Q along a line through Q with a slope corresponding to 50,000 ohms.

The figure shows that, as a result of this new operating path, the output current decreases to zero not when the envelope of the modulated wave decreases to zero but when it decreases to an effective value of about 6.0 volts (approximately midway between the curves for 8 and

---

[24] F. M. Colebrook, "The Theory of the Straight-Line Rectifier," *Exp. Wireless and Wireless Eng.*, 7 (1930), 595–603; C. E. Kilgour and J. M. Glessner, "Diode-Detection Analysis," *I.R.E. Proc.*, 21 (1933), 936.

5 volts). The maximum percentage modulation that the detector can accommodate without exceeding the limits of linear operation is therefore

$$\frac{15 - 6.0}{15} 100 = 60.0 \text{ per cent.} \qquad [55]$$

For higher degrees of modulation, the envelope decreases below 6.0 volts, the peaks are clipped, and appreciable harmonic generation takes place; hence the output wave contains harmonics of the original modulation frequency. To minimize this undesirable effect, the grid resistor R_g should be made large compared with R_L. However, the grid-current requirements of the amplifier tube that follows the detector generally place a practical upper limit on the value of R_g.

As previously stated, a basic assumption underlying the foregoing analysis is that the capacitance C_s is large enough to suppress the carrier-frequency components in the output of the detector, but small enough to allow the output to follow the envelope of the modulated wave. A second assumption, previously stated, is that the reactance of the coupling capacitance C_c is small compared with the resistance R_g. When these assumptions are not fulfilled, the graphical analysis based on the straight load lines through the point Q is not applicable. If the capacitance C_c is not large, the analysis must be extended to take into account the resulting elliptical operating paths.[25] The shift of the operating point caused by harmonic generation and discussed in Art. 18, Ch. VIII, is a further consideration.

This analysis should be recognized as approximate; it does not include many aspects of diode detection. One additional consideration is that the combination of the diode detector and load absorbs power from the circuit supplying the modulated input voltage. As a result, the diode circuit can be represented as having a particular input impedance that loads the input circuit and may affect its characteristics —for instance, its selectivity when the input circuit is tuned. Another result of the absorption of power by the detector and load is that the effective internal impedance of the source of modulated voltage has an appreciable effect on the harmonic generation taking place in the detector circuit.[26] A precise analysis of the diode detector circuit must include the effect of the initial velocities of emission of the electrons from the cathode. They can be represented as establishing an effective direct bias voltage in the plate circuit of the diode.[26] One complete

[25] W. P. N. Court, "Diode Operating Conditions," *Wireless Engr.*, *16* (1939), 548–555.

[26] H. A. Wheeler, "Design Formulas for Diode Detectors," *I.R.E. Proc.*, *26* (1938), 745–780.

analysis includes the representation of the curved lower part of the tube characteristic by an equation of exponential form.[27]

The linear diode detector is widely used in radio receivers because it has the advantage over other types of detectors of causing less harmonic generation, particularly for large percentages of modulation. Its one disadvantage is that it requires appreciable input power for its operation, but this is offset to a considerable extent by the fact that it delivers a larger output voltage for a given input voltage than do some other types of detectors.

7. POWER-SERIES REPRESENTATION OF NONLINEAR FUNCTIONS

At this point in the discussion of modulation and demodulation, a digression is made to develop and illustrate some of the methods of representation of the characteristics of vacuum tubes by analytical expressions in the form of infinite series. These methods are applicable to the analysis of the behavior of nonlinear elements in general; they are by no means restricted to vacuum-tube applications. In fact, semiconductor rectifier elements are widely used as modulators.[28]

In a power-series form, the $i_b(e_b)$ functional relation for a nonlinear element such as a diode becomes

$$i_b = a_0 + a_1 e_b + a_2 e_b{}^2 + a_3 e_b{}^3 + \cdots \qquad [56]$$

Such a power-series representation has limited utility, first, because of the relatively large number of terms often required for representing accurately a given function; and, second, because of the often unwieldy manipulations that the use of the power series involves. In certain applications, nevertheless, such representation is very helpful.

Reference is made in Art. 13, Ch. VIII, to the fact that the dynamic characteristic of a triode with a resistance load can be represented by a power series, and the relation between the generated harmonics and the presence of terms in the series is pointed out. In this article, a method is presented for finding the power series representing a given function, such as $i_b(e_b)$, defined by a set of numerical experimental data. The procedure involves, first, the evaluation by arithmetical and graphical means of the derivatives of this numerically defined function; and, second, the calculation of the coefficients a_0, a_1, a_2, $\cdots$ from the values of these derivatives.

The relation between the coefficients and the derivatives may be

[27] C. B. Aiken, "Theory of the Diode Voltmeter," *I.R.E. Proc.*, *26* (1938), 859–876.

[28] R. S. Caruthers, "Copper-Oxide Modulators in Carrier Telephone Systems," *E.E.*, *58* (1939), 253–260.

determined by successive differentiation of Eq. 56, which yields the following expressions:

$$\frac{di_b}{de_b} = a_1 + 2a_2 e_b + 3a_3 e_b{}^2 + 4a_4 e_b{}^3 + \cdots, \tag{57}$$

$$\frac{d^2 i_b}{de_b{}^2} = 2a_2 + 6a_3 e_b + 12a_4 e_b{}^2 + \cdots, \tag{58}$$

$$\frac{d^3 i_b}{de_b{}^3} = 6a_3 + 24a_4 e_b + \cdots. \tag{59}$$

Suppose that, in a given example, the numerical differentiation when performed as indicated below shows that the third and higher derivatives are either negligible or so poorly defined as to be useless. It can then be reasonably concluded, from Eq. 59 and expressions for the higher derivatives, that a_3, a_4, and higher-order coefficients are substantially zero or that the terms containing them will have no significance. The term $a_2 e_b{}^2$ in Eq. 56 is therefore the last one that is useful in that example; or, if the third derivative is well defined and constant, the fourth and higher derivatives are zero, and the $a_3 e_b{}^3$ term can close the series. Under these conditions, a_4, a_5, and higher-order coefficients are zero; and by substitution of known values of the first three derivatives and of e_b for any convenient point on the curve into Eqs. 57 to 59, the coefficients a_3, a_2, and a_1 can be computed in turn. The first step in the process, then, is the method of computation of the derivatives of a numerically defined function, which is a modification of the method of finite differences.

This method can be understood from a numerical example for which the data in the first two columns of Table I describe the characteristics of a particular diode or other nonlinear device. The general procedure is to take differences Δi_b between successive values of i_b for equally spaced values of e_b. These so-called first-order differences $\Delta_1 i_b$ are then plotted as a function of the average e_b in the interval, and a smooth curve is drawn, from which the column of "smoothed" $\Delta_1 i_b$'s is obtained. Second-order differences $\Delta(\Delta i_b)$, or alternatively $\Delta_2 i_b$, of the smoothed $\Delta_1 i_b$'s are obtained by subtraction and plotted. From a smooth curve determined by these points, smoothed values are again read and tabulated for uniformly spaced increments of e_b. Third-order differences $\Delta(\Delta_2 i_b)$ or $\Delta_3 i_b$ are computed, and the process is repeated until the differences either become substantially constant or so erratic as to be meaningless.

The results of the application of the above processes to the tube data are shown in Table I. The smoothed curves are shown in Fig. 19. The third-order differences are so near to being constant, as indicated by a linearly varying second-order difference, that the fourth and higher derivatives may be assumed to be negligible.

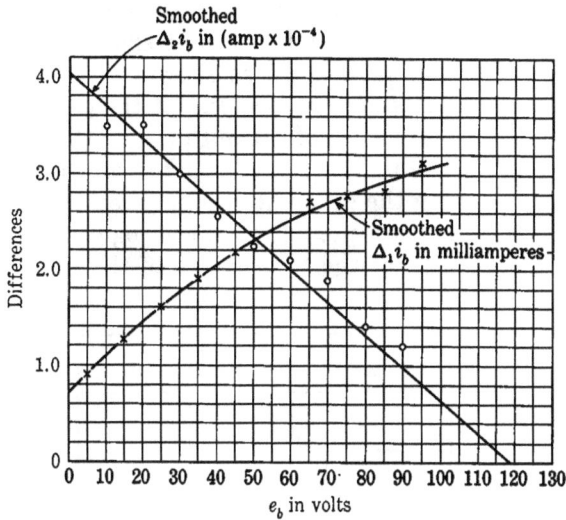

Fig. 19.   Determination of the smoothed differences.

TABLE I

e_b volts	i_b ma	$\Delta_1 i_b$ ma	Smoothed $\Delta_1 i_b$ ma	$\Delta_2 i_b$ ma	Smoothed $\Delta_2 i_b$ ma	$\Delta_3 i_b$ ma
0	0					
		0.9	0.9			
10	0.9			0.35	0.37	
		1.25	1.25			−0.034
20	2.15			0.35	0.336	
		1.60	1.60			−0.034
30	3.75			0.30	0.302	
		1.90	1.90			−0.034
40	5.65			0.255	0.268	
		2.15	2.155			−0.034
50	7.80			0.225	0.234	
		2.40	2.38			−0.034
60	10.20			0.21	0.200	
		2.70	2.59			−0.034
70	12.90			0.19	0.166	
		2.78	2.78			−0.034
80	15.68			0.14	0.132	
		2.82	2.92			−0.034
90	18.50			0.12	0.098	
		3.10	3.04			
100	21.60					

The smoothing of the differences can be justified by the following consideration. Suppose that in the example all the i_b's were precisely on a third-degree curve except one, which was too large by an amount δ. If differences are taken numerically without smoothing, this one small error spreads in size and extent, as shown by the tabulation below, from which all entries are omitted except the errors. A single irregularity δ in the original values of i_b can thus give the appearance of erratic behavior to the differences of an otherwise perfect third-degree function. Smoothing the differences largely reduces the effects of small chance errors that are present to some degree in all experimental data, and hence makes the true nature of the function more apparent.

i_b	$\Delta_1 i_b$	$\Delta_2 i_b$	$\Delta_3 i_b$
—			
—	—		
—		—	$+\delta$
		$+\delta$	
	$+\delta$		-3δ
δ		-2δ	
	$-\delta$		$+3\delta$
—		$+\delta$	
	—		$-\delta$
—	—		
—			

A test of the accuracy of this process of taking differences is the accuracy with which a power series determined by this process agrees with the original data. To obtain the coefficients of the power series, it is necessary to have the derivatives at some value of voltage e_b. These are obtained approximately as follows:

$$\frac{di_b}{de_b} \approx \frac{\Delta_1 i_b}{\Delta e_b}, \tag{60}$$

$$\frac{d^2 i_b}{de_b{}^2} \approx \frac{\Delta\left(\dfrac{\Delta_1 i_b}{\Delta e_b}\right)}{\Delta e_b} = \frac{\Delta_2 i_b}{(\Delta e_b)^2}, \tag{61}$$

$$\frac{d^3 i_b}{de_b{}^3} \approx \frac{\Delta\left[\dfrac{\Delta_2 i_b}{(\Delta e_b)^2}\right]}{\Delta e_b} = \frac{\Delta_3 i_b}{(\Delta e_b)^3}. \tag{62}$$

In the example used here, Δe_b is 10 volts throughout. The coefficient a_3 is computed from Eq. 59 and Table I as

$$a_3 = \frac{1}{6}\frac{d^3 i_b}{de_b{}^3} \approx \frac{1}{6}\frac{\Delta_3 i_b}{(\Delta e_b)^3} = \frac{-0.0340}{6 \times 10^3} \qquad [63]$$

$$= -0.567 \times 10^{-5} \text{ milliampere per volt cubed.} \quad [64]$$

The values of the other coefficients depend slightly on the value of e_b at which they are evaluated. By trial, approximately the best agreement between the power series and the original $i_b(e_b)$ data is found to be obtained when the coefficients are evaluated at the point where e_b is 30 volts. At this point, the coefficients are computed as follows from Eqs. 56 through 59 and the curves of Fig. 19:

$$a_2 = \frac{1}{2}\left[\frac{d^2 i_b}{de_b{}^2} - 6a_3 e_b\right] \approx \frac{1}{2}\left[\frac{\Delta_2 i_b}{(\Delta e_b)^2} - 6a_3 e_b\right] \qquad [65]$$

$$= \frac{1}{2}\left[\frac{0.302}{10^2} - 6(-0.567 \times 10^{-5})30\right] \qquad [66]$$

$$= 0.00202 \text{ milliampere per volt squared.} \qquad [67]$$

$$a_1 = \frac{d i_b}{de_b} - 2a_2 e_b - 3a_3 e_b{}^2 \approx \frac{\Delta_1 i_b}{\Delta e_b} - 2a_2 e_b - 3a_3 e_b{}^2 \qquad [68]$$

$$= \frac{1.76}{10} - 2(0.00202)30 - 3(-0.567 \times 10^{-5})30^2 \qquad [69]$$

$$= 0.0701 \text{ milliampere per volt.} \qquad [70]$$

$$a_0 = i_b - a_1 e_b - a_2 e_b{}^2 - a_3 e_b{}^3 \qquad [71]$$

$$\approx 3.75 - (0.0701 \times 30) - (0.00202 \times 30^2) + (0.567 \times 10^{-5} \times 30^3) \quad [72]$$

$$= -0.018 \text{ milliampere.} \qquad [73]$$

With the coefficients evaluated, the power series becomes

$$i_b = -0.018 + 0.0701 e_b + 0.00202 e_b{}^2 - 0.567 \times 10^{-5} e_b{}^3, \quad [74]$$

where i_b is in milliamperes and e_b is in volts.

The following tabulation gives the current i_b as determined from the series, together with the error of the series:

e_b	0	10	20	30	40	50	60	70	80	90	100
i_b from series	−0.02	0.88	2.15	3.75	5.66	7.83	10.24	12.94	15.62	18.53	21.52
error	−0.02	−0.02	0.00	0.00	0.01	0.03	0.04	0.04	−0.06	0.03	−0.08

Evaluation of the coefficients at other values of e_b gives slightly greater errors, but the maximum deviations are about 0.2 milliampere, and the average is much smaller. Thus it is shown that these particular data, which were not in any way selected to suit the purpose of illustration, can be represented within the usual experimental error of their measurement by a power series ending with the cube term. Such a good agreement or "fit" cannot generally be obtained with so few terms.

As may be concluded from the foregoing illustration, the evaluation of the coefficients of a power series is usually a tedious process, and it is seldom carried through in practice. Nevertheless the concepts involved in the power-series treatment of many circuits comprising nonlinear elements are of the utmost importance for an understanding of the behavior of the circuits. For example, it may often be more important to know that the even-order harmonics can be balanced out in a push-pull amplifier, or that the amplitude of the second harmonic increases with the square of the input voltage, than to know, for instance, that the amplitude of a particular harmonic for 10 volts applied is 60 microamperes.

8. TAYLOR-SERIES REPRESENTATION OF NONLINEAR FUNCTIONS

In this article, the method of representation of the volt-ampere characteristic of a device by means of a Taylor series is developed. The Taylor series is a type of power series and is useful when operation is of interest in the neighborhood of some point on the curve at which no discontinuity of slope exists.[29] It is not a practicable mathematical tool when in a diode the excursion of the operating point extends on both sides of the origin, as in the operation of the linear detector discussed in Art. 6. The utility of the Taylor series is contingent upon the characteristic's being of such simple form, over the range actually used, that relatively few terms of the series are required.

As a starting point, the fairly simple problem of the operation of the circuit of Fig. 20 may be analyzed. In this circuit, a direct-voltage supply E_{bb} provides a bias voltage in the plate circuit of a diode or other nonlinear device, and a varying voltage e_p connected in series with the device causes a varying component of current i_p. The object of the analysis is to determine a power-series relation between the varying or incremental components of the plate current and plate voltage. The total plate voltage e_b applied to the tube may be considered to

[29] F. S. Woods and F. H. Bailey, *Elementary Calculus* (revised ed.; Boston: Ginn and Company, 1928), 205–207.

consist of two parts—one being E_b, which may have any predetermined value, and the other being an increment of voltage e_p such that e_b is equal to $E_b + e_p$. When e_p is zero, e_b equals E_b, and the current has a value I_b that can be determined from the volt-ampere characteristic of the nonlinear element. If e_b is then changed by an incremental amount e_p, the current changes from I_b to $I_b + i_p$, where i_p is an

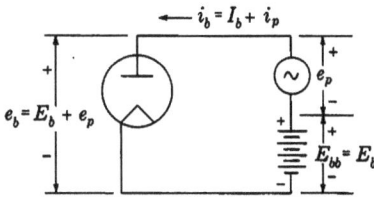

Fig. 20. Diode circuit with direct-current and incremental sources in series.

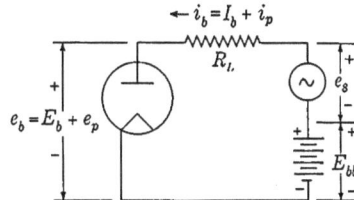

Fig. 21. Diode in series with a resistance load and direct-current and incremental sources.

increment in plate current. The operating point on the nonlinear characteristic curve is (E_b, I_b), and the Taylor-series expansion about this point is

$$i_b = I_b + i_p$$

$$= I_b + \frac{di_b}{de_b}\Big|_{E_b} e_p + \frac{1}{2!}\frac{d^2i_b}{de_b{}^2}\Big|_{E_b} e_p{}^2 + \frac{1}{3!}\frac{d^3i_b}{de_b{}^3}\Big|_{E_b} e_p{}^3 + \cdots, \qquad [75]$$

in which, for example, $\dfrac{d^2i_b}{de_b{}^2}\Big|_{E_b}$ means that the second derivative of the functional relation $i_b(e_b)$ is to be evaluated at the point where e_b equals E_b. By subtraction of I_b from both sides, Eq. 75 can be condensed to

$$i_p = b_1e_p + b_2e_p{}^2 + b_3e_p{}^3 + \cdots, \qquad [76]$$

where

$$b_1 \equiv \frac{di_b}{de_b}\Big|_{E_b}, \quad b_2 \equiv \frac{1}{2!}\frac{d^2i_b}{de_b{}^2}\Big|_{E_b}, \quad b_3 \equiv \frac{1}{3!}\frac{d^3i_b}{de_b{}^3}\Big|_{E_b} \qquad [77]$$

The coefficients (b's) can be evaluated approximately from smoothed finite differences taken in the vicinity of E_b, or, as in the example involving an illustrative vacuum-tube characteristic curve in Art. 7, they can be obtained from the power series, Eq. 74, that represents the entire characteristic. For example, at E_b equal to 50 volts, the coefficients are given by substitution of Eqs. 64, 67, and 70 in Eqs. 57,

58, and 59 as follows:

$$b_1 = \frac{di_b}{de_b}\bigg|_{50v}$$

$$= [0.0701 + (2 \times 0.00202 \times 50) - (3 \times 5.67 \times 10^{-6} \times 50^2)] \times 10^{-3} \quad [78]$$

$$= 2.3 \times 10^{-4} \text{ ampere per volt,} \quad [79]$$

$$b_2 = \frac{1}{2!}\frac{d^2i_b}{de_b^2}\bigg|_{50v}$$

$$= \tfrac{1}{2}[2 \times 0.00202 - 3 \times 2 \times 5.67 \times 10^{-6} \times 50] \times 10^{-3} \quad [80]$$

$$= 1.17 \times 10^{-6} \text{ ampere per volt squared,} \quad [81]$$

$$b_3 = \frac{1}{3!}\frac{d^3i_b}{de_b^3}\bigg|_{50v} = \tfrac{1}{6}[-3 \times 2 \times 5.67 \times 10^{-6}] \times 10^{-3} \quad [82]$$

$$= -0.567 \times 10^{-8} \text{ ampere per volt cubed.} \quad [83]$$

The coefficient b_1 is a quantity for the diode analogous to that defined as $1/r_p$ in Art. 6, Ch. IV, for the triode. Thus the coefficients may be replaced by their equivalents in terms of the dynamic resistance r_p of the diode, which are

$$b_1 = 1/r_p, \quad [84]$$

$$b_2 = \frac{1}{2!}\frac{d^2i_b}{de_b^2}\bigg|_{E_b} = \frac{1}{2!}\frac{d(1/r_p)}{de_b}\bigg|_{E_b} = -\frac{1}{2r_p^2}\frac{dr_p}{de_b}\bigg|_{E_b}, \quad [85]$$

$$b_3 = \frac{1}{3!}\frac{d^3i_b}{de_b^3}\bigg|_{E_b} = \frac{1}{3!}\frac{d^2(1/r_p)}{de_b^2}\bigg|_{E_b} = \frac{2}{3!r_p^3}\left[\frac{dr_p}{de_b}\bigg|_{E_b}\right]^2 - \frac{1}{3!r_p^2}\frac{d^2r_p}{de_b^2}\bigg|_{E_b}, \quad [86]$$

$$b_4 = \cdots, \quad [87]$$

so that the power series becomes

$$i_p = \frac{e_p}{r_p} - \frac{e_p^2}{2r_p^2}\frac{dr_p}{de_b}\bigg|_{E_b} + \frac{2e_p^3}{3!r_p^3}\left[\frac{dr_p}{de_b}\bigg|_{E_b}\right]^2 - \frac{e_p^3}{3!r_p^2}\frac{d^2r_p}{de_b^2}\bigg|_{E_b} + \cdots. \quad [88]$$

Thus far, the power series represents the behavior of the nonlinear device alone, which is of no utility until some type of load is connected in series with it. Accordingly, the analysis may be continued by a consideration of the circuit of Fig. 21, which includes a resistance load. Here the incremental voltage source is denoted by e_s in order that it may be distinguished from the incremental tube voltage drop e_p, and similarly the direct supply voltage E_{bb} is distinguished from the corresponding component of voltage across the tube, E_b. When e_s is zero, the current has a value I_b, the voltage e_b has a value E_b, and Kirchhoff's law gives the relation

$$E_b = E_{bb} - I_b R_L. \quad [89]$$

When the voltage e_s is not zero, the voltage e_b is changed to $e_p + E_b$, and the current to $i_p + I_b$, and Kirchhoff's law then gives

$$e_b = (E_{bb} + e_s) - i_b R_L ,$$ [90]

or

$$e_p + E_b = (E_{bb} + e_s) - (i_p + I_b)R_L .$$ [91]

Subtraction of Eq. 89 from Eq. 91 gives

$$e_p = e_s - i_p R_L ,$$ [92]

a relation between the incremental quantities alone. In order to ascertain the current through the circuit in terms of the incremental input voltage e_s, Eqs. 76 and 92 may be combined to eliminate e_p, giving

$$i_p = b_1(e_s - i_p R_L) + b_2(e_s - i_p R_L)^2 + b_3(e_s - i_p R_L)^3 + \cdots .$$ [93]

This equation could be solved to give an explicit expression for i_p in terms of e_s, but only at the expense of considerable effort. An alternative and useful method of obtaining the result is the method of *successive approximations*.[30] In this method it is assumed—and this is an important restriction—that the b_2 term and all succeeding terms, while not negligible, are relatively small, so that to a first approximation the current i_p is given by a solution using the first term only. If this so-called first-order approximation is denoted by i_{1p},

$$i_{1p} = b_1(e_s - i_{1p} R_L)$$ [94]

or

$$i_{1p} = b_1 \frac{e_s}{1 + b_1 R_L}.$$ (First-order approximation)   ▶[95]

The object of the next step is to determine a second approximation function i_{2p} which is nearer the correct solution than the first. To determine this "second-order approximation," it may be recognized that, since the b_2 term is small, the substitution in it of an approximately correct i_p, such as the first approximation, makes the resulting two-term equation

$$i_p = b_1(e_s - i_p R_L) + b_2(e_s - i_{1p} R_L)^2$$ [96]

a source of a much better approximation for i_p. Thus this second-order approximation is given by

$$i_{2p} = b_1(e_s - i_{2p} R_L) + b_2 \left(e_s - b_1 \frac{e_s R_L}{1 + b_1 R_L}\right)^2$$ [97]

$$= b_1(e_s - i_{2p} R_L) + b_2 \left(\frac{e_s}{1 + b_1 R_L}\right)^2 ;$$ [98]

[30] J. R. Carson, "A Theoretical Study of the Three-Element Vacuum Tube," *I.R.E. Proc.*, 7 (1919), 187–200.

whence,

$$i_{2p} = \frac{b_1 e_s}{1 + b_1 R_L} + \frac{b_2 e_s{}^2}{(1 + b_1 R_L)^3} \cdot \qquad \text{(Second-order approximation)} \qquad \blacktriangleright [99]$$

Further approximations can be obtained through a continuation of the same process. Thus, if terms in Eq. 93 up to the cube are considered, substitution of the first approximation Eq. 95 into the cube term, and of the second-order approximation into the square term, followed by solution of the resulting equation for i_p, yields the third-order approximation, and so on for higher-order approximations. For most purposes where the power-series representation proves useful, the second-order approximation is sufficient. When it is not sufficient, the continuation of this method becomes so cumbersome that the method loses its utility altogether and is usually displaced by the method described in Art. 7 or by a similar one.

If Eq. 99 is sufficient, the subscript 2 in i_{2p} can be dropped and the equation further abbreviated through defining the coefficients

$$a_1 \equiv \frac{b_1}{1 + b_1 R_L}, \qquad a_2 \equiv \frac{b_2}{(1 + b_1 R_L)^3} \qquad [100]$$

so that Eq. 99 becomes

$$i_p = a_1 e_s + a_2 e_s{}^2 . \qquad [101]$$

Equation 101 is a general relation between the incremental current and the incremental applied voltage (which may be functions of time) in the circuit of Fig. 21. It is based on two assumptions: first, that two terms of a Taylor series are sufficient to describe the nonlinear characteristic for increments; and, second, that the size of the terms of Eq. 93 are such that the method of successive approximations used to obtain Eq. 97 is justified.

When the load is not a pure resistance but is an impedance that varies with the frequency, the Taylor series can also be developed by the method of successive approximations. However, the form of the series is greatly complicated by the presence of the variable impedance.[31] Note that it is not correct merely to substitute the function $Z(\omega)$ for R_L in the foregoing method.

9. TAYLOR-SERIES REPRESENTATION OF TRIODE CHARACTERISTICS

The Taylor series of Eq. 99 for the diode can be extended to apply to the triode as well, by recognition of the fact that, to the extent that μ

[31] J. R. Carson, "A Theoretical Study of the Three-Element Vacuum Tube," *I.R.E. Proc.*, 7 (1919), 187–200.

is a constant, the functional relation between the plate current and the electrode voltage is

$$i_b = f\left(e_c + \frac{e_b}{\mu}\right) = f(e_0),   \qquad \blacktriangleright[102]$$

where e_0 is called the control voltage (see Art. 3, Ch. X, and Art. 3, Ch. XI). In the triode, the relations

$$i_b = I_b + i_p ,   \qquad [103]$$

$$e_c = E_c + e_g ,   \qquad [104]$$

$$e_b = E_b + e_p   \qquad [105]$$

apply where the capital letters denote steady components corresponding to the quiescent conditions for e_g equal to zero, and the quantities i_p, e_g, and e_p are the incremental variations from these quiescent values.

A general form for the Taylor-series expansion of $f(x)$ about the point x equals a is[32]

$$f(x) = f(a) + (x - a)\frac{df}{dx}\bigg|_{x=a} + \frac{(x-a)^2}{2!}\frac{d^2 f}{dx^2}\bigg|_{x=a} + \cdots . \quad [106]$$

In the notation used here,

$$f(x) = i_b ,   \qquad [107]$$

$$f(a) = I_b ,   \qquad [108]$$

$$a = E_c + \frac{E_b}{\mu} = E_0 ,   \qquad [109]$$

$$x = e_0 = E_c + e_g + \frac{E_b + e_p}{\mu},   \qquad [110]$$

$$x - a = e_g + \frac{e_p}{\mu}.   \qquad [111]$$

Thus the Taylor series becomes

$$i_b = I_b + i_p$$

$$= I_b + \left(e_g + \frac{e_p}{\mu}\right)\frac{di_b}{de_0}\bigg|_{E_0} + \frac{\left(e_g + \frac{e_p}{\mu}\right)^2}{2!}\frac{d^2 i_b}{de_0^2}\bigg|_{E_0} + \cdots , \qquad [112]$$

where the subscript E_0 denotes that the derivative is to be evaluated at the quiescent operating point. Since in general either the grid

---

[32] R. G. Hudson, *The Engineer's Manual* (2nd ed.; New York: John Wiley & Sons, Inc., 1939), 34.

voltage or the plate voltage may be an independent variable, it is convenient to convert the derivatives in Eq. 112 to their equivalents in terms of one or the other of these variables. It can be shown[33] that

$$g_m \equiv \frac{\partial i_b}{\partial e_c} = \frac{di_b}{de_0}\frac{\partial e_0}{\partial e_c} = \frac{di_b}{de_0} \qquad [113]$$

and

$$g_m' \equiv \frac{\partial g_m}{\partial e_c} \equiv \frac{\partial^2 i_b}{\partial e_c{}^2} = \frac{d^2 i_b}{de_0{}^2}\left(\frac{\partial e_0}{\partial e_c}\right)^2 + \frac{di_b}{de_0}\frac{\partial^2 e_0}{\partial e_c{}^2} = \frac{d^2 i_b}{de_0{}^2}. \qquad [114]$$

Thus, after subtraction of I_b from both sides, and substitution of Eqs. 113 and 114, the Taylor series for the triode, Eq. 112, becomes

$$i_p = g_m\left(e_g + \frac{e_p}{\mu}\right) + \frac{1}{2!}g_m'\left(e_g + \frac{e_p}{\mu}\right)^2 + \frac{1}{3!}g_m''\left(e_g + \frac{e_p}{\mu}\right)^3 + \qquad [115]$$

where the third-order term has been included for completeness, and g_m and its derivatives are to be evaluated at the quiescent operating point. This power series involves the assumption, implicit in Eq. 102, that μ is a constant. When μ is not constant, a more general Taylor series of two variables applies.[34]

When a load impedance is included in the plate circuit, and the only source of incremental voltage is in the grid circuit, the incremental plate voltage e_p becomes equal to the voltage across the load (with a negative sign in accordance with the assigned reference directions for the voltages and currents introduced for triodes in Ch. VIII). The solution for the current i_p in terms of e_g can then be made by a successive-approximation method similar to that used to derive Eq. 99 in Art. 8. The first two terms of the series when the load is a resistance are

$$i_p = \frac{g_m e_g}{1 + \dfrac{g_m}{\mu}R_L} + \frac{1}{2!}g_m'\frac{e_g{}^2}{\left(1 + \dfrac{g_m}{\mu}R_L\right)^3}, \qquad [116]$$

which may be written

$$i_p = g_m\frac{e_g}{1 + \dfrac{R_L}{r_p}} + \frac{1}{2!}g_m'\frac{e_g{}^2}{\left(1 + \dfrac{R_L}{r_p}\right)^3}. \qquad \blacktriangleright[117]$$

[33] F. S. Woods, *Advanced Calculus* (new ed.; Boston: Ginn and Company, 1934), 71.
[34] F. B. Llewellyn, "Operation of Thermionic-Vacuum-Tube Circuits," *B.S.T.J.*, 5 (1926), 433–462.

When the load impedance is a function of frequency, the successive-approximation method is considerably more cumbersome, as is stated in the previous article, and there is no simple approach to the problem.

The first term of Eq. 117 is a different form of Eq. 26, Ch. VIII, for a pure resistance load and linear operation. When g_m is a constant, the second term in Eq. 117 is zero. Thus the Taylor-series representation of the tube behavior includes linear operation as the first term and agrees with the relations derived in Ch. VIII for the operation of the tube over the linear range of the characteristic curves in which g_m is constant. The second-order and higher-order terms account for the nonlinearity in a tube. For example, Eq. 117 is an expression for the dynamic transfer characteristic discussed in Art. 13, Ch. VIII, and it is shown there that the second-harmonic generation in the tube is a direct result of the square term in the power series (see Eq. 114, Ch. VIII). Also, a rectified component is created in the plate current—the alternating-current input is partially converted into a direct current. These effects also occur in the two-element device discussed in the previous article, since the power series has the same form for both. The magnitude of the rectified component is proportional to the coefficient in the second-order term and therefore proportional to the second derivative, or rate of change of slope at the quiescent point on the dynamic transfer characteristic of the triode, or diode, and load. A change of slope in the operating range is sufficient and necessary to cause rectification and harmonic generation.

The dependence of the harmonic generation on the coefficients of the power series leads to one convenient method used in practice for the evaluation of the coefficients. A pure sinusoidal voltage is impressed on the nonlinear device, and the amplitude and frequency of the harmonics generated are measured.[35] The coefficients are then found from these experimental results by a process that is essentially the reverse of that given in Art. 13, Ch. VIII. If the input amplitude is varied, the approximation involved in neglecting high-order terms can be investigated. A knowledge of the coefficients obtained by this experimental approach is of value in an interpretation of the more complicated modulation processes in which two or more input voltages are superposed.

Subsequent articles show that modulation and demodulation can be accomplished by a nonlinear device whose power-series representation contains a second-order term. Thus, while the second-order term is a source of unwanted harmonic generation in an amplifier, it is of great practical utility for the purpose of modulation.

[35] L. B. Arguimbau, "The New Wave Analyzer—Some of Its Features," *Gen. Rad. Exp.*, *13* (December, 1938), 1–5; L. B. Arguimbau, "Wave Analysis," *Gen. Rad. Exp., 8* (June–July, 1933), 12–14.

The limitations of the power-series representation are worthy of further emphasis. In the first place, it is useful only over that part of a curve where there are no discontinuities in the function or its derivatives—the curve must be smooth without sudden changes in slope. Actually, if the function has discontinuities only in the first or higher derivatives (not in the function itself), it is frequently possible to obtain a sufficiently good approximation to the function by means of a power series. However, this limitation often makes the use of the Taylor series less desirable than alternative methods, and ineffective for the purpose of representing the behavior of a circuit in which the voltage across the device reverses in direction and in which the device is a rectifier that passes current in one direction but not in the other.

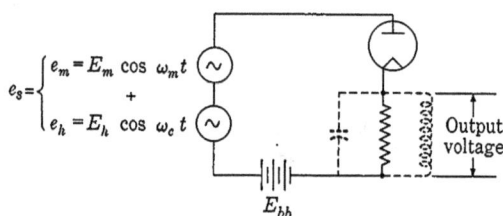

$$e_s = \begin{cases} e_m = E_m \cos \omega_m t \\ \quad + \\ e_h = E_h \cos \omega_c t \end{cases}$$

E_{bb}

Fig. 22. Modulation by means of a square-law diode.

The usefulness of the power series is much greater for operation over a range of currents in the conducting direction in a rectifier. The same restrictions apply to the operation of triodes; the power (or Taylor) series is therefore not applicable for the analysis of Class B and Class C amplifiers. In the second place, the power series obtained by successive approximations from a Taylor series and containing only two or three terms is a good approximation only when the second term is of a lower order of magnitude than the first, and the third is of a lower order of magnitude than the second, and so forth. In other words, it is of value only when the magnitude of the terms decreases rapidly as their order increases.

10. Square-law diode modulation

Although it is not often used as a modulator, the diode circuit of Fig. 22 offers perhaps the simplest illustration of the modulating action of a device having a square-law characteristic. This is the same circuit as that of Fig. 21, with the total input voltage e_s made up of the sum of two components—one a high-frequency carrier voltage source

$$e_h = E_h \cos \omega_c t, \qquad [118]$$

and the other a low-frequency modulating voltage source

$$e_m = E_m \cos \omega_m t. \tag{119}$$

Thus

$$e_s = E_h \cos \omega_c t + E_m \cos \omega_m t. \tag{120}$$

It is assumed that the over-all volt-ampere characteristic curve of the series combination of the diode and resistance load is describable by the Taylor series

$$i_p = a_1 e_s + a_2 e_s{}^2, \tag{121}$$

whose coefficients can be derived as described in Arts. 7 and 8.

When Eq. 120 is substituted in Eq. 121,

$$i_p = a_1 E_h \cos \omega_c t + a_1 E_m \cos \omega_m t + a_2 E_h{}^2 \cos^2 \omega_c t$$
$$+ 2a_2 E_h E_m \cos \omega_m t \cos \omega_c t + a_2 E_m{}^2 \cos^2 \omega_m t. \tag{122}$$

By means of the trigonometric identities

$$\cos^2 \alpha = \frac{1 + \cos 2\alpha}{2} \tag{123}$$

and Eq. 17, Eq. 122 may be rearranged in the form:

Component	*Identification*
$i_p = \dfrac{a_2}{2} [E_h{}^2 + E_m{}^2]$	(a) Direct-current component; rectified
$+ a_1 E_m \cos \omega_m t$	(b) Modulation frequency
$+ \dfrac{a_2}{2} E_m{}^2 \cos 2\omega_m t$	(c) Second harmonic of modulation frequency
$+ a_2 E_h E_m \cos (\omega_c - \omega_m)t$	(d) Lower side frequency ▶[124]
$+ a_1 E_h \cos \omega_c t$	(e) Carrier
$+ a_2 E_h E_m \cos (\omega_c + \omega_m)t$	(f) Upper side frequency
$+ \dfrac{a_2}{2} E_h{}^2 \cos 2\omega_c t$	(g) Second harmonic of carrier frequency.

Thus there are terms of seven different frequencies in the output of the square-law modulation circuit, these terms being identified as shown. The linear term in the power series produces in the output the carrier frequency shown as term (e) in Eq. 124, but the square-law term produces the side frequencies. Terms (d), (e), and (f) constitute the amplitude-modulated component in the current; thus

Modulated current component $= a_1 E_h \cos \omega_c t$

$$+ a_2 E_h E_m \cos (\omega_c + \omega_m)t + a_2 E_h E_m \cos (\omega_c - \omega_m)t, \quad [125]$$

and by means of Eq. 17 this may be put in the form

Modulated current component

$$= a_1 E_h \left[1 + \frac{2a_2 E_m}{a_1} \cos \omega_m t \right] \cos \omega_c t, \qquad \blacktriangleright[126]$$

from which it is seen that the degree of modulation is given by

$$\text{Modulation factor} = 2 \frac{a_2}{a_1} E_m. \qquad \blacktriangleright[127]$$

It therefore follows that the larger the ratio a_2/a_1, the larger the degree of modulation. However, the ratio a_2/a_1 is generally less than unity in vacuum tubes, and it must be small if the successive-approximation method is to be used for a determination of the power series.

From this discussion it may be said that in general the square term in the power series produces components of three categories: first, components having the sum and difference of all the impressed frequencies; second, components having double the frequency of all the impressed frequencies; and, third, direct-current components. Or, in summary, *the square term produces double-frequency, sum-and-difference-frequency, and zero-frequency components.* An alternative statement is that the square-law term produces components having the sum and difference frequencies of all frequencies in the input voltage, including each frequency taken with itself. This last step results in the production of the double- and zero-frequency components.

If a sinusoidally modulated wave is *impressed* on the circuit of Fig. 22, the input voltage e_s contains the carrier frequency and two side frequencies. The output contains frequencies that are the sums and differences of those of these three components and therefore contains a term having the original modulation frequency. The original modulating waveform is thereby recovered, and the process of detection takes place. *The square-law device serves equally well as a modulator or as a demodulator.*

Both the a_1 and the a_2 coefficients of the power series involve the load resistance R_L (see Eq. 100). The ratio a_2/a_1 decreases as the load resistance increases; hence, from Eq. 127, the modulated power output varies with the load resistance. It has been shown[36] that the modulated power output is a maximum when the load resistance is one-fifth of the dynamic plate resistance.

The behavior of the two-terminal square-law modulator may be

[36] J. R. Carson, "The Equivalent Circuit of the Vacuum-Tube Modulator," *I.R.E. Proc., 9* (1921), 243–249.

illustrated in a qualitative manner by Fig. 23. Because of the non-linearity in the volt-ampere characteristic of the combined tube and load, the amplitude of the high-frequency component in the current waveform is greater during the positive half-cycle of the modulating voltage than during the negative half-cycle. The average value of the

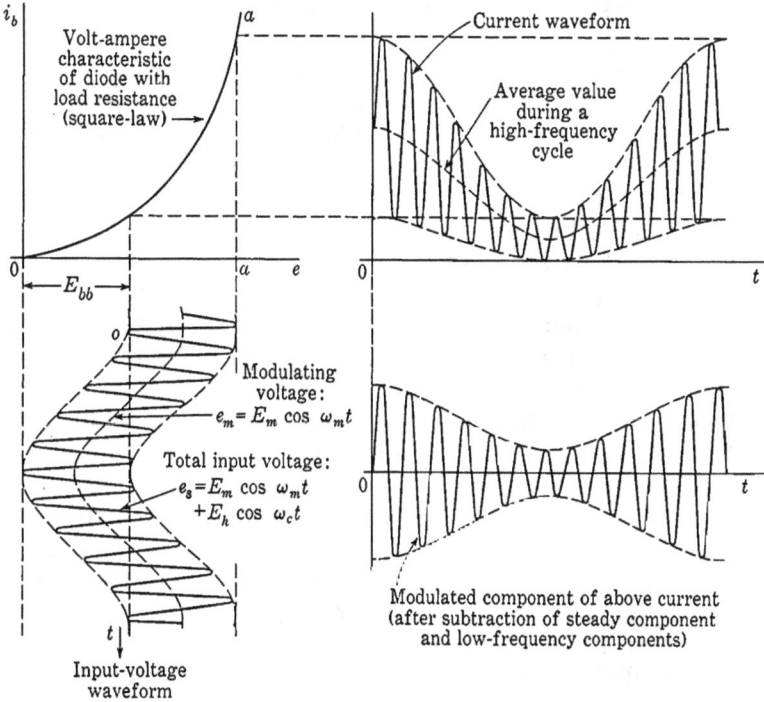

Fig. 23. Waveforms in a square-law diode modulator with resistance load. In this example, both E_m and E_h are assumed equal to $E_{bb}/2$.

current over a high-frequency cycle varies with time and contains the low or modulation frequency. After this average value and the steady component have been subtracted, the remainder has the form of an amplitude-modulated wave, as is shown in the figure, but the presence of the second harmonics of the modulation and carrier frequencies that is revealed by the mathematical analysis is not apparent without careful graphical construction. In a sense, the low-frequency modulating component in the input voltage can be thought of as causing a variation of the operating point and thereby of the average slope over which the carrier-frequency variation occurs. Thus the carrier-frequency component in the output varies in amplitude in accordance with the modulating wave.

The modulating voltage may also be considered to cause a time variation of the effective internal dynamic resistance r_p of the tube. Since this dynamic resistance is in series with the input carrier voltage and the load resistance, its variation effects modulation in a manner similar to that which occurs in the superheterodyne mixer tube described in Art. 17.

In both the modulator and the detector with a resistance load, unwanted frequency components exist in the output voltage. In the modulator, only the terms in the output voltage that correspond to the fourth, fifth, and sixth terms of the current in Eq. 124 are desired. To eliminate the others effectively, it is expedient to substitute a parallel-resonant circuit for the resistance load, as is indicated by the dotted coil and capacitor in Fig. 22. If the carrier frequency is sufficiently large compared with the modulation frequency, the tuned circuit can be so designed that its impedance is effectively a resistance for the range of angular frequencies from $\omega_c - \omega_m$ to $\omega_c + \omega_m$ (which includes the three desired frequency components) but is effectively zero for the frequency components in the current that exist outside that range, thus eliminating the unwanted components from the output voltage. In the detector, not only the desired modulation-frequency component exists in the output but also, as is discussed in more detail in Art. 13, terms that include the second harmonic of the modulation frequency terms are present. A capacitance is ordinarily connected across the load to reduce the load impedance at the high frequencies and thus effectively to eliminate from the output voltage the unwanted components having frequencies comparable to the carrier frequency.

Note that when the load impedance is changed from a pure resistance to a parallel-resonant circuit or a parallel combination of resistance and capacitance, as indicated above, the analysis given in this article for the resistance load *does not apply*, despite the fact that the load is effectively a pure resistance at the frequencies of the desired components. The same frequency components are present, and no new ones are created, but their amplitudes are changed. The analysis for the amplitudes requires the use of the more complicated series expansion mentioned in Art. 8, and it develops that a considerable increase in the modulated power output occurs[37] with the substitution of the resonant load.

11. POSSIBILITIES FOR SQUARE-LAW MODULATION WITH TRIODES

The triode, having two circuits into which the carrier and modulating voltages may be impressed, presents several possibilities for

[37] J. R. Carson, "The Equivalent Circuit of the Vacuum-Tube Modulator," *I.R.E. Proc., 9* (1921), 243–249.

square-law modulation. These may be understood by reference to the circuit of Fig. 24. The circuit contains a resistance load, and alternating-voltage sources e_g and e_2 are included in the grid and plate circuits, respectively. The tube is assumed to be biased to a point of considerable nonlinearity on its dynamic transfer characteristic, so that the square term in the power series that describes the variations about the operating point on the dynamic transfer characteristic, Eq. 115, has appreciable magnitude. In the circuit,

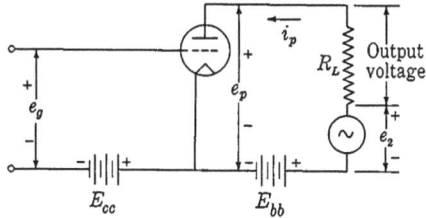

Fig. 24.   Basic circuit for modulation by a triode.

$$e_p = -i_p R_L + e_2, \quad [128]$$

and substitution of this relation in Eq. 115 gives

$$i_p = g_m \left[\left(e_g + \frac{e_2}{\mu} \right) - \frac{i_p R_L}{\mu} \right]$$

$$+ \frac{1}{2!} g_m' \left[\left(e_g + \frac{e_2}{\mu} \right) - \frac{i_p R_L}{\mu} \right]^2 + \cdots. \quad [129]$$

If the successive-approximation method is applied for the solution for the current in Eq. 129, it may be seen by comparison with Eqs. 115, 116, and 117 that, since the term $e_g + (e_2/\mu)$ in Eq. 129 replaces the term e_g in Eq. 115, the result for the second-order approximation is

$$i_p = g_m \frac{e_g + \dfrac{e_2}{\mu}}{1 + \dfrac{R_L}{r_p}} + \frac{1}{2!} g_m' \frac{\left(e_g + \dfrac{e_2}{\mu} \right)^2}{\left(1 + \dfrac{R_L}{r_p} \right)^3}, \qquad \blacktriangleright [130]$$

which may be written as

$$i_p = d_1 \left(e_g + \frac{e_2}{\mu} \right) + d_2 \left(e_g + \frac{e_2}{\mu} \right)^2, \qquad [131]$$

where d_1 and d_2 are constants that depend on the tube and circuit. Expansion of the terms in Eq. 131 gives

$$i_p = d_1 e_g + d_1 \frac{e_2}{\mu} + d_2 e_g^2 + d_2 \left(\frac{e_2}{\mu} \right)^2 + d_2 \left(2 e_g \frac{e_2}{\mu} \right). \qquad [132]$$

$$(1) \qquad\qquad (2) \qquad (3) \qquad\qquad (4) \qquad\qquad (5)$$

If a carrier voltage e_h equal to $E_h \cos \omega_c t$ and a modulating voltage e_m equal to $E_m \cos \omega_m t$ are utilized, it follows from Eq. 132 that four possibilities for modulation exist, which may be summarized as follows:

Method	e_g	e_2	Term in Eq. 132 that produces the carrier	Term in Eq. 132 that produces the side frequencies
I	$e_h + e_m$	0	(1)	(3)
II	0	$e_h + e_m$	(2)	(4)
III	e_m	e_h	(2)	(5)
IV	e_h	e_m	(1)	(5)

Method I is known as *Van der Bijl modulation*. In practice, resonant or selective load circuits are substituted for the resistance load, and the relative amplitudes and phases of the various components are hence changed.

The type of modulation indicated as Method I also occurs in every Class A amplifier since nonlinearity of the dynamic transfer characteristic is always present to some extent. When more than one frequency is present in the grid-signal voltage, the square term (3) in Eq. 132 causes not only the second harmonic distortion discussed in Art. 13, Ch. VIII, but also a cross-product term as in Eq. 122 that gives rise to the sum and difference frequencies in the output. The components having these frequencies are the *intermodulation distortion components*. They are particularly objectionable in sound reproduction because they are not harmonics of any of the signal components, and may therefore be dissonant to the ear.

12. Van der Bijl modulation

The circuit for the use of a triode over the square-law portion of its dynamic characteristic known as the *Van der Bijl modulator*[38] is shown in Fig. 25. Here a tuned plate circuit is employed, and the selectivity of the tuned circuit is so chosen as practically to eliminate the low-frequency or modulation-frequency components from the output but to pass the carrier and side frequencies. In this article, only an approximate analysis of the circuit behavior is given, for it is assumed that the load for the tube behaves as a pure resistance at all frequencies. An accurate analysis would require the use of the complex power series, because the impedance in the plate circuit actually varies with frequency (see Art. 8).

[38] H. J. Van der Bijl, *The Thermionic Vacuum Tube and Its Applications* (New York: McGraw-Hill Book Company, Inc., 1920), 318–322.

In so far as the load can be considered a pure resistance, the behavior of the circuit can be illustrated graphically by a diagram similar to Fig. 23. The only change required is to substitute the dynamic characteristic of the triode and load in the Van der Bijl modulator for the combined volt-ampere characteristic of the diode

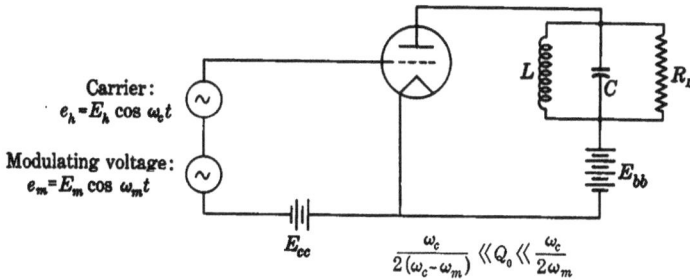

Fig. 25. Van der Bijl modulation circuit.

and load in the diode modulator. This change entails a shift of the origin of the co-ordinates to the right, say to the line aa in Fig. 23, so that operation takes place over the negative-grid-voltage region of the curve.

Since the grid current in the Van der Bijl modulator may be made very small by the use of the negative grid-bias voltage, little power is required from the sources of carrier and modulating voltage; this is the important advantage of the Van der Bijl device over the diode modulator. The efficiency of the Van der Bijl modulator nevertheless is low because of the undesired direct component of its plate current, which is relatively large in comparison with the desired carrier and side-frequency components. Consequently it is used only in low-power applications such as carrier-current telephony and measurement or control applications where the efficiency is unimportant.

The plate current in the Van der Bijl modulator with a resistance load, having a dynamic characteristic given by Eq. 117, is describable by Eq. 124 if the coefficients a_1 and a_2 are replaced by

$$\frac{g_m}{1 + \dfrac{R_L}{r_p}} \quad \text{and} \quad \frac{1}{2}\frac{g_m{}'}{\left(1 + \dfrac{R_L}{r_p}\right)^3},$$

respectively.* The modulated output component is, from Eq. 126,

---

* When the values of L and C in Fig. 25 are so chosen that the load impedance as viewed from the tube is essentially a constant resistance R_L for the carrier and side frequencies, but is essentially zero for all higher or lower frequencies, the term a_2 is changed to the value $\dfrac{1}{2}\dfrac{g_m{}'}{[1 + (R_L/r_p)]^2}$, and Eqs. 133 and 134 are modified accordingly.

Modulated current component

$$= \frac{E_h g_m}{1 + \dfrac{R_L}{r_p}} \left[1 + \frac{g_m{}'}{g_m} \frac{E_m}{\left(1 + \dfrac{R_L}{r_p}\right)^2} \cos \omega_m t \right] \cos \omega_c t. \qquad \blacktriangleright [133]$$

The modulation factor is therefore

$$\text{Modulation factor} = \frac{g_m{}'}{g_m} \frac{E_m}{\left(1 + \dfrac{R_L}{r_p}\right)^2}. \qquad \blacktriangleright [134]$$

From this relation it is apparent that the modulation factor increases with the ratio $g_m{}'/g_m$ and with a decrease of the load resistance.

In addition to the low-frequency terms corresponding to (b) and (c) in Eq. 124, which may be suppressed in the output voltage by the tuned circuit of Fig. 25 or in a subsequent part of the system, the output also contains a term of double the carrier frequency corresponding to (g) in Eq. 124. This high-frequency term can be suppressed in a similar manner. Third-, fourth-, and higher-power terms in the power series for the dynamic transfer characteristic, Eq. 117, contribute to the output additional unwanted components that constitute distortion in the modulation. Further analysis shows that the ratio of the amplitudes of the unwanted side frequencies to the desired side frequencies increases with the modulation factor. This system therefore is applicable only where modulation of small percentage is required at small outputs.

13. SQUARE-LAW DETECTION WITH A TRIODE

As is mentioned in Art. 10, a square-law device is applicable for demodulation. The triode with a square-law dynamic transfer characteristic can hence be used as a detector to translate the information or modulating voltage from an inaudible frequency region, to which it was shifted for economical transmission, back to its original position in the audio region of the frequency spectrum.

The fundamental principle of square-law detection with a triode can be illustrated by the circuit shown in Fig. 26a, where the amplitude-modulated wave

$$e_a = A_c[1 + G(f)] \cos \omega_c t \qquad [135]$$

is impressed on the grid of the vacuum tube. The term $G(f)$ represents the modulating voltage, which may be an audio-frequency signal in a Fourier-series or Fourier-integral form. The detection process must give a component of current in the output of the form $G(f)$.

The plate-load resistance R_L in the detector circuit is generally shunted by a by-pass capacitor C in order to provide a path for high-frequency components of current and to exclude them from the load, while an inductance L may also be used to enhance this effect. Under these conditions, the combined load impedance for the tube is a function of frequency, and an accurate analysis of the circuit behavior

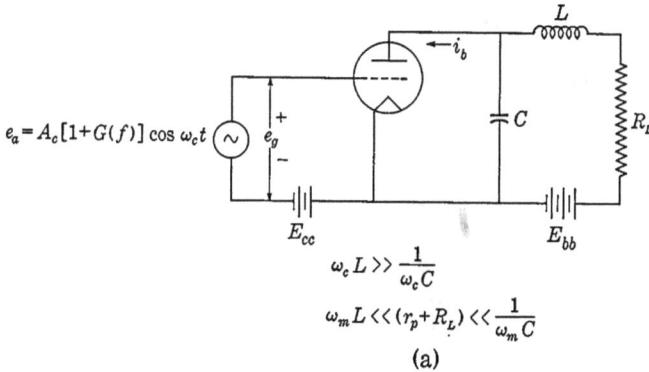

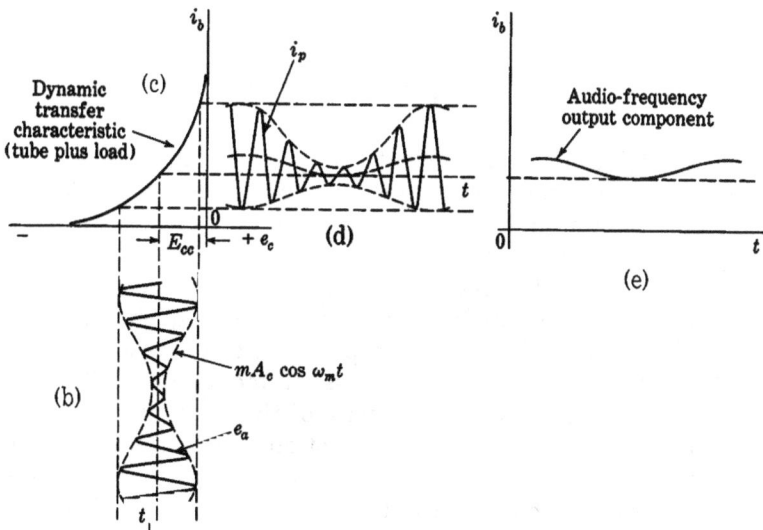

$$e_a = A_c[1+G(f)] \cos \omega_c t$$

$$\omega_c L \gg \frac{1}{\omega_c C}$$

$$\omega_m L \ll (r_p + R_L) \ll \frac{1}{\omega_m C}$$

(a)

$mA_c \cos \omega_m t$

(b)

Dynamic transfer characteristic (tube plus load)

(c)

(d)

Audio-frequency output component

(e)

Fig. 26.   Square-law detection with a triode.

requires the use of the complex power series (see Art. 8). The analysis given here is approximate and involves the assumption that the capacitor and inductor are not used—the load is assumed to be a pure resistance at all the frequencies involved. In the complete circuit of

Fig. 26a, the frequencies of the current components in the load can be shown to be the same as if a pure resistance load were used, but their amplitudes are different.

When the load is a pure resistance, the behavior of the circuit as a detector can be illustrated as in Fig. 26. The tube is assumed to be biased to a point of large curvature on its dynamic transfer characteristic, so that the varying component of its plate current i_p can be represented approximately by the power series

$$i_p = a_1 e_g + a_2 e_g{}^2 + \cdots, \qquad [136]$$

where the coefficients a_1 and a_2 are defined* in Eq. 117. Substitution of e_a from Eq. 135 for e_g in Eq. 136 gives

$$i_p = a_1 e_a + \tfrac{1}{2} a_2 A_c{}^2 [1 + G(f)]^2 \cos 2\omega_c t + \tfrac{1}{2} a_2 A_c{}^2 [1 + G(f)]^2. \quad [137]$$

The first two terms in Eq. 137 are of no value in the process of detection, because they represent inaudible high-frequency currents that are generally by-passed through the capacitor. The last term may be written

Direct and audio-frequency components of i_p

$$= \underbrace{\tfrac{1}{2} a_2 A_c{}^2}_{\substack{\text{Direct-}\\\text{current}\\\text{component}}} + \underbrace{a_2 A_c{}^2 G(f)}_{\substack{\text{Facsimile}\\\text{of signal}}} + \underbrace{\tfrac{1}{2} a_2 A_c{}^2 G^2(f)}_{\substack{\text{Distortion}\\\text{of signal}}}. \qquad [138]$$

The term $\tfrac{1}{2} a_2 A_c{}^2$ is a direct-current component and may be disregarded, for it is generally eliminated by the coupling network used between the detector and the succeeding stage. The second term $a_2 A_c{}^2 G(f)$ is the desired audio-frequency facsimile of the original signal $G(f)$ that modulated the carrier before transmission. Thus, in effect, the side bands are translated from their high-frequency position down to their correct audio-frequency position; this translation may be considered to be the inverse process of modulation. The proportionality to $A_c{}^2$—that is, to the square of the carrier amplitude— does not disturb the facsimile reproduction, since the carrier is generally unaltered by the modulation. The term $\tfrac{1}{2} a_2 A_c{}^2 G^2(f)$ represents frequencies other than those of the signal; it is hence a distortion term in the output.

---

* When the values of L and C in Fig. 26 are so chosen that the load impedance as viewed from the tube is essentially a constant resistance R_L for the audio frequencies and their harmonics, but is essentially zero for the carrier and all higher frequencies, the term a_2 appearing in Eqs. 136 through 140 is changed from its value corresponding to Eq. 117 to the new value $\dfrac{1}{2} \dfrac{g_m{}'}{[1 + (R_L/r_p)]^2}$.

As a means of further analysis, the conditions when the modulation is sinusoidal, or

$$G(f) = m \cos \omega_m t,$$ [139]

may be considered. The components of audio frequency in the output current are, from Eq. 138,

Audio-frequency components of i_p

$$= a_2 A_c{}^2 [m \cos \omega_m t + \frac{m^2}{4} \cos 2\omega_m t];$$ ▶[140]

thus the square-law detection process produces in the output a second-harmonic audio-frequency component proportional to the square of the modulation factor. If the second-harmonic amplitude is to be kept smaller than, say, one-tenth of the fundamental, it is necessary that

$$\frac{m^2}{4} < \frac{1}{10} m$$ [141]

or

$$m < \frac{4}{10}.$$ [142]

To maintain the second-harmonic distortion at less than 10 per cent, the percentage modulation must thus be less than 40 per cent for sinusoidal modulation.

If the dynamic characteristic, Eq. 136, contains third-, fourth-, and higher-degree terms, the audio-frequency output contains third-, fourth-, and higher-order harmonic currents proportional to m^3, m^4, and correspondingly higher powers of m. These components are usually negligible when m is small. When $G(f)$ contains more than one sinusoidal Fourier component, as it does in actual voice transmission, additional new frequencies produced by terms involving the cross products of the sinusoids appear in the output.

In the graphical analysis of the square-law detector shown in Fig. 26, the modulated input wave is shown in Fig. 26b. By projection of this wave on the dynamic transfer characteristic of Fig. 26c, the output current wave of Fig. 26d is found. The average value of the output wave over a high-frequency cycle has the form dotted in Fig. 26d; thus the audio-frequency component in the output is as shown in Fig. 26e. The operation of the circuit depends upon the nonlinearity of the curve of plate current as a function of grid voltage; hence this method is sometimes called *plate* or *anode-bend detection*.

An alternative method of square-law detection with a triode makes use of the nonlinearity of the grid-current grid-voltage characteristic.[39]

[39] F. B. Llewellyn, "Operation of Thermionic Vacuum-Tube Circuits," *B.S.T.J.*, *5* (1926), 433–462; F. E. Terman and N. R. Morgan, "Some Properties of Grid-Leak Power Detection," *I.R.E. Proc.*, *18* (1930), 2160–2175.

This method has improved voltage sensitivity, because the amplification property of the tube becomes effective; but it has the disadvantage that the presence of appreciable grid current requires that the source of grid current furnish appreciable power. In the early days of the development of radio communication, this method of detection was customarily used, and a parallel combination of grid leak and capacitor of suitable size was employed in series with the grid, the tube being biased to a point of large curvature on its $i_c(e_c)$ characteristic.

The square-law triode plate detector is inferior to the linear-diode detector in that it generates harmonics of the modulation frequencies. It does, however, have the advantage over the linear diode that it can be so biased as to require only negligible power from the source of modulated voltage when a triode is used.

The analysis given here is directly applicable to the square-law diode detection mentioned in Art. 10. Since the initial velocities of emission of the electrons from the cathode of a diode have a Maxwellian distribution that includes all values, the current approaches zero gradually, and the combined diode-and-load volt-ampere characteristic has appreciable curvature near the zero of current regardless of the load resistance. Consequently even the so-called linear diode detector reverts to a square-law detector for small amplitudes of the modulated input voltage.

As is mentioned in Art. 6, the modulation-frequency output of a linear-diode detector may be considered approximately as a slowly fluctuating direct current that varies, in some cases, in the same way with the dynamically varied carrier envelope as the direct-current output would vary with the amplitude of the applied carrier voltage. Similarly the variation of the direct current with the input carrier amplitude in a square-law detector may be measured experimentally and used to estimate, by a graphical analysis, the modulation component in the output with a dynamically varying carrier input. Such an approach often gives results more readily than does a power-series treatment.

14. THE BALANCED MODULATOR

A modulator circuit of considerable importance is the so-called *balanced modulator*[40] shown in Fig. 27. Its importance lies in the fact that in it certain frequency components are suppressed and do not appear in the output. In many respects it is similar to the push-pull amplifier, and the frequency suppression in the modulator is related

[40] E. Peterson and C. R. Keith, "Grid-Current Modulation," *B.S.T.J.*, 7 (1928), 131.

to the fact, developed in Art. 19c, Ch. VIII, that in the push-pull amplifier only the odd harmonics generated in the tubes appear in the output current, while the even harmonics appear in the plate-power-supply current.

The circuit as shown has two sets of terminals at e_1 and e_2 for control of the grid voltages of the tubes, and two other sets of terminals at e_3 and e_4 from which output power or voltage can be taken from the

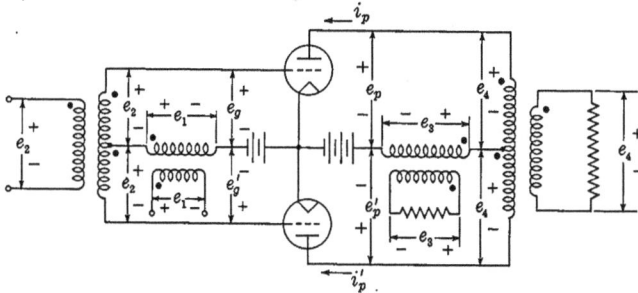

Fig. 27.　Balanced-modulator circuit.

plate circuits. It is assumed that the transformers are ideal and that a resistance load is connected across the terminals at e_3 and e_4. The transformer windings have the polarities indicated by dots, and it is convenient though not essential to assume that their turns ratios are unity, as indicated.

The grid-bias and plate-supply voltages are adjusted so that operation takes place over the portion of the dynamic transfer characteristics that has considerable curvature. The considerations of Art. 19c, Ch. VIII, show that the coupling in the output transformer at e_4 has an effect on this curvature, and that the resulting dynamic transfer characteristic for a single tube is not the same as that for a single tube with a pure resistance load. Only varying components of the electrode voltages and currents need be considered, since the direct components are not of interest. It is assumed that the dynamic transfer characteristics can be represented to a sufficient degree of approximation by the power series

$$i_p = a_1 e_g + a_2 e_g{}^2 + \cdots \tag{143}$$

for one tube, with a similar expression with primed letters for voltage and current for the other tube, the two tubes being identical. In accordance with the assigned positive reference directions of the voltages in the circuit,

$$e_g = e_1 + e_2, \tag{144}$$

$$e_g{}' = e_1 - e_2. \tag{145}$$

Thus

$$i_p = a_1(e_1 + e_2) + a_2(e_1 + e_2)^2 \qquad [146]$$

and

$$i_p' = a_1(e_1 - e_2) + a_2(e_1 - e_2)^2. \qquad [147]$$

Because the transformers are ideal and the load is a pure resistance, the voltage e_3 is proportional to the sum of the currents i_p and i_p', and the voltage e_4 is proportional to their difference. Thus

$$e_3 = K_3(i_p + i_p') \qquad [148]$$

and

$$e_4 = K_4(i_p - i_p'), \qquad [149]$$

where K_3 and K_4 are constants. The results of the substitution of Eqs. 146 and 147 in Eqs. 148 and 149 are

$$e_3 = 2K_3[a_1e_1 + a_2(e_1^2 + e_2^2)] \qquad [150]$$

and

$$e_4 = 2K_4[a_1e_2 + 2a_2e_1e_2]. \qquad [151]$$

As is stated in Art. 10, the squared terms in Eq. 150 result in sum-frequency, difference-frequency, and double-frequency components of all the frequencies contained in the squared voltage, while consideration of the trigonometric transformation, Eq. 17, shows that the cross product e_1e_2 in Eq. 151 produces the sums and differences of all the frequency components in e_1 and e_2. From these considerations, the results in Table II may be derived.

TABLE II

FREQUENCIES IN THE BALANCED MODULATOR

	Input frequencies at		Output frequencies at	
	e_1	e_2	e_3	e_4
1	ω_c	ω_m	$2\omega_m,\ \omega_c,\ 2\omega_c$	$\omega_m,\ (\omega_c \pm \omega_m)$
2	ω_m	ω_c	$\omega_m,\ 2\omega_m,\ 2\omega_c$	$\omega_c,\ (\omega_c \pm \omega_m)$
3	$\omega_m,\ \omega_c$	0	$\omega_m,\ 2\omega_m,\ \omega_c,\ (\omega_c \pm \omega_m),\ 2\omega_c$	0
4	0	$\omega_m,\ \omega_c$	$2\omega_m,\ (\omega_c \pm \omega_m),\ 2\omega_c$	$\omega_m,\ \omega_c$

There are four possible combinations for impressing the carrier angular frequency ω_c and the modulating angular frequency ω_m at the two positions, as is shown in the table. The side frequencies, as indicated by $\omega_c \pm \omega_m$, appear at one or the other of the two output positions e_3 and e_4. In combinations 2 and 3, the carrier frequency appears with the side frequencies, so that a modulated output is obtained. Combination 2 is a type of push-pull grid-bias modulation. Combination 3 is merely operation of the two tubes in parallel as Van der Bijl modulators. In combinations 1 and 4, however, the carrier frequency does not appear at the same output position as the side frequencies but is diverted to the other output position. In this way, the carrier frequency is suppressed, and the side frequencies alone may be obtained through the filtering out of unwanted frequency components, which are remote from the side frequencies. As is stated in Art. 3, suppression of the carrier is desirable to save power, and the balanced modulator is commonly used in carrier-current telephony. One side band is usually filtered out to reduce the width of the channel required for the transmission, and the remaining side band is transmitted to the receiving station where a local oscillator supplies a carrier frequency to be combined with the side band before detection by a square-law demodulator takes place. Although the balanced modulator finds its greatest use in carrier-current telephony, it is also widely used in measurement apparatus[41] and in control systems.

15. RADIO COMMUNICATION

One important application of modulation and demodulation is radio communication. As previously mentioned, communication without wires by electromagnetic waves could not be realized without means for shifting the audio-frequency signal *up* to higher frequencies at the sending end and *down* to the original frequencies at the receiving end. Modulation serves a double purpose—it makes possible efficient radiation of the signal, and it also makes possible the separation of one signal from another when they are received simultaneously at the same station.

Most of the essential components of a radio-telephone system are discussed in this chapter and the previous ones, and a discussion of the simple type of radio system shown in Fig. 28 can now be made. The microphone translates the sound wave into electrical currents of substantially the same waveform. These currents are small, but they are amplified by cascaded vacuum-tube amplifiers to a power level of several watts in the small stations to 100 kilowatts and more in the

[41] C. R. Moore and A. S. Curtis, "An Analyzer for the Voice-Frequency Range," *B.S.T.J.*, *6* (1927), 217–229; A. G. Landeen, "Analyzer for Complex Electric Waves," *B.S.T.J.*, *6* (1927), 230–247; L. B. Arguimbau, "The New Wave Analyzer—Some of Its Features," *Gen. Rad. Exp.*, *13* (December, 1938), 1–5; L. B. Arguimbau, "Wave Analysis," *Gen. Rad. Exp.*, *8* (June–July, 1933), 12–14.

larger stations now operating. A carefully regulated *master* vacuum-tube oscillator (usually with a temperature-controlled piezoelectric crystal controlling the frequency) supplies a carrier voltage of remarkably constant frequency (a government regulation forbids a frequency deviation of more than 20 cycles per second, or 2 parts in 100,000 at a 1-megacycle carrier frequency) to a cascade of high-efficiency tuned radio-frequency amplifiers. These successively increase the power to approximately the same magnitude as the audio-frequency power output from the signal amplifiers. The signal and the carrier frequencies meet in the modulated vacuum-tube amplifier, where the audio signal modulates the carrier to form the two side bands. The carrier and the

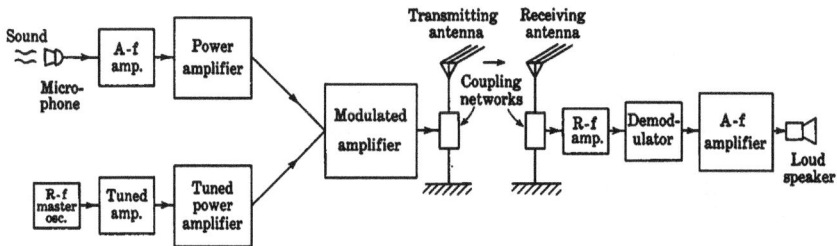

Fig. 28.   Radio-telephone communication system.

side bands are transferred to the antenna system by a coupling network, and energy is radiated into the surrounding space by the antenna, from which it travels to the receiving point or points. There the electromagnetic or radio wave induces an electromotive force in another antenna. These minute voltages, of orders of magnitude of 0.0001 volt, are amplified before being demodulated by a vacuum-tube detector, which shifts the information from the high-frequency location to its original audible position. Subsequent audio-frequency amplifiers increase the power to a value lying between 0.5 watt and 20 watts or more before it is delivered to a loud speaker for conversion into sound waves.

In another and now generally used type of radio receiving circuit, the *superheterodyne* receiver, the signal is demodulated in two steps. This type of receiver is discussed in Art. 17.

16. Beat-frequency, or heterodyne, oscillator

A second important application of modulation is the beat-frequency or heterodyne oscillator. This type of oscillator is one of the most convenient for the production of audio-frequency currents for measurement and for other purposes requiring a continuous and rapid variation of the frequency over a wide range, which it allows with a single adjustment. The schematic diagram of this type of oscillator is shown in Fig. 29. Two high-frequency oscillators having almost the same frequencies ω and $\omega + \Delta\omega$, ω being fixed and $\Delta\omega$ being variable, induce electromotive forces in the two coils, as shown. The two voltages e_x and e_y act in series on the demodulator, whose function is to produce a current of audio frequency $\Delta\omega/(2\pi)$. Suppose that the

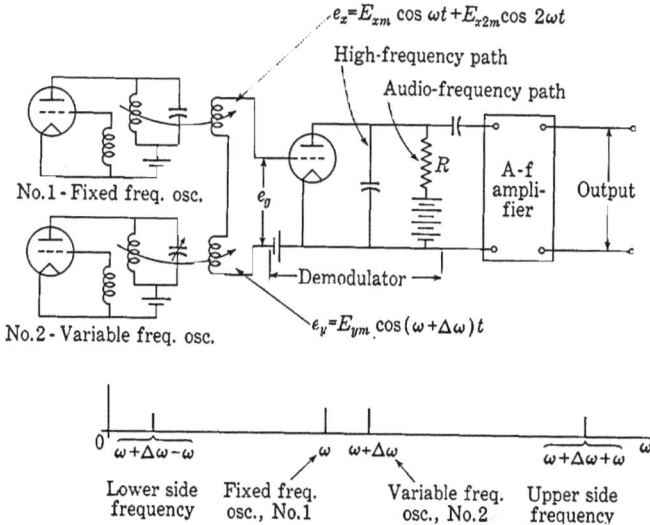

Fig. 29. Simplified circuit of a beat-frequency oscillator, and the spectrum of the frequencies in it.

variable-frequency oscillator has a substantially sinusoidal output, so that

$$e_y = E_{ym} \cos (\omega + \Delta\omega)t, \qquad [152]$$

but the fixed-frequency oscillator has an appreciable second harmonic and is representable by

$$e_x = E_{xm} \cos \omega t + E_{x2m} \cos 2\omega t. \qquad [153]$$

In the previous analysis of modulation, the quadratic term and higher terms are found to be the only ones giving components of new frequencies. Hence, if the demodulator is of the square-law type, and if its load is assumed to behave as a pure resistance, the term of the output current that contains the new frequencies can be written as

$$i = a_2 e_g^2 = a_2(e_x + e_y)^2 \qquad [154]$$

$$= a_2[E_{xm}^2 \cos^2 \omega t + E_{x2m}^2 \cos^2 2\omega t + E_{ym}^2 \cos^2 (\omega + \Delta\omega)t$$

$$+ 2E_{xm}E_{x2m} \cos \omega t \cos 2\omega t + 2E_{xm}E_{ym} \cos \omega t \cos (\omega + \Delta\omega)t$$

$$+ E_{x2m}E_{ym} \cos 2\omega t \cos (\omega + \Delta\omega)t]. \qquad [155]$$

Study of the expanded terms of this expression shows that the only audio frequency comes from the fifth term, all other components being of higher frequency. After expansion, the fifth term becomes

$$2a_2E_{xm}E_{ym}[\tfrac{1}{2}\cos(\omega + \Delta\omega + \omega)t + \tfrac{1}{2}\cos(\omega + \Delta\omega - \omega)t],$$

and therefore the audio-frequency current is

$$i_{audio} = a_2E_{xm}E_{ym}\cos\Delta\omega t. \qquad \blacktriangleright[156]$$

The voltage produced across the resistance R by this current is amplified to the desired level by the following amplifier. A pure sinusoidal output is produced, provided one of the oscillators has a simple-harmonic waveform, even though the other oscillator output contains harmonics. The use of an appropriately high ω and a small variable capacitor of the type used in radio apparatus can produce a beat-frequency oscillator that covers continuously the entire audio-frequency spectrum with a single adjustment.

When two waves of almost the same frequency, such as e_x and e_y above, beat together, their sum has an appearance very similar to that of a modulated wave. However, between the beat wave and a modulated wave there is a distinct difference that may be illustrated as follows. Let

$$e_1 = E_1 \cos \omega_1 t \qquad [157]$$

and

$$e_2 = E_2 \cos \omega_2 t \qquad [158]$$

be the two waves whose sum is e. Then

$$e = e_1 + e_2 = E_1 \cos \omega_1 t + E_2 \cos \omega_2 t \qquad [159]$$

$$= E_1 \cos \omega_1 t + E_2 \cos \omega_2 t + [E_2 \cos \omega_1 t - E_2 \cos \omega_1 t] \qquad [160]$$

$$= E_2[\cos \omega_1 t + \cos \omega_2 t] + [E_1 - E_2] \cos \omega_1 t \qquad [161]$$

$$= \left[2E_2 \cos\left(\frac{\omega_1 - \omega_2}{2}\right)t\right] \cos\left(\frac{\omega_1 + \omega_2}{2}\right)t + [E_1 - E_2] \cos \omega_1 t. \qquad [162]$$

For simplicity, let

$$\frac{\omega_1 + \omega_2}{2} = \omega \qquad [163]$$

and

$$E_1 = E_2. \qquad [164]$$

Then
$$e = E(t) \cos \omega t, \qquad [165]$$
where

$$E(t) = 2E_2 \cos \left(\frac{\omega_1 - \omega_2}{2}\right)t. \qquad [166]$$

The waveform of the beat wave, Eq. 165, is drawn in Fig. 30, which shows that the term $E(t)$ may be visualized as the varying amplitude or envelope of a sinusoidal wave of frequency ω. In the beat wave, however, the sign of the apparent amplitude reverses during the beat cycle. In other words, the phase of the oscillation reverses once during each beat cycle. There is actually no oscillation of simple-harmonic

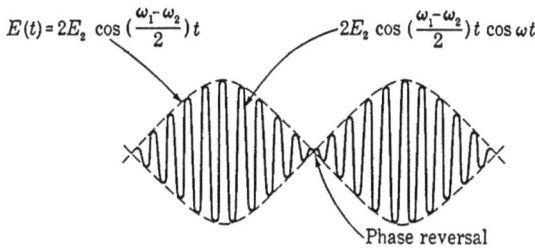

Fig. 30.   Beats between two sinusoidal waves of equal amplitude.

form of angular frequency ω, and no power can be supplied at this angular frequency to a sharply resonant circuit tuned to ω. The wave does not contain an additive component of the beat frequency $(1/2\pi)(\omega_1 - \omega_2)/2$ for the mere addition of two sinusoidal waves of different frequency cannot create a new frequency. Such a component can be obtained only through distortion of the waveform by some non-linear process such as that involved in the demodulator of the beat-frequency oscillator. (See Art. 18d for further discussion of superposed waves.)

17. SUPERHETERODYNE RECEIVERS, MIXERS, AND CONVERTERS

The heterodyne principle is very generally used in the super-heterodyne type of radio receiver. The modulated voltage from the antenna is combined with the voltage from a local oscillator and demodulated in a *frequency-converter tube* or device, sometimes called a first detector, which effectively changes the carrier frequency of the modulated wave to the resonant frequency of a tuned amplifier called the *intermediate-frequency amplifier*. The tuning of the intermediate-frequency amplifier is fixed, and selection of the signal to be amplified is made by adjustment of the local-oscillator frequency.

After amplification in the intermediate-frequency amplifier, the signal is again demodulated by a detector, which translates the signal to its original place in the audio-frequency spectrum. The output of this detector is then amplified and delivered to a loud speaker or telephone receiver.

There are numerous tube and circuit combinations for producing the local-oscillator voltage, combining it with the incoming modulated wave, and performing the desired change of frequency.[42] A single

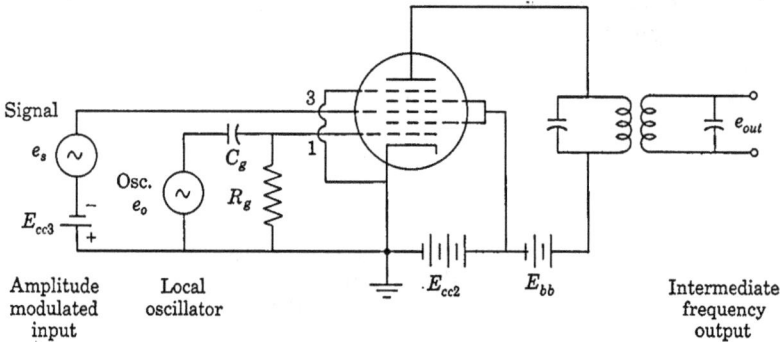

Fig. 31. Pentagrid-mixer circuit.

device that performs all these functions is called a *converter*. One which requires a separate local oscillator is called a *mixer*. A simple diode or crystal rectifier may be used as a mixer and has advantages at very high carrier frequencies. Triodes, tetrodes, and pentodes are also used as mixers. Tubes having more than three grids are particularly useful as mixers at moderate carrier frequencies (see Art. 13, Ch. IV). In the use of one such tube called a *pentagrid mixer*, the incoming amplitude-modulated voltage is impressed on one control grid, as at e_s in Fig. 31, and the local-oscillator voltage is impressed on another control grid, as at e_o. A screen grid is included in the tube to minimize interaction between the local oscillator and the input-signal circuits, as is a suppressor grid to eliminate the effects of secondary emission. The load is connected to the plate circuit through a transformer, and is tuned to the desired intermediate frequency.

The local-oscillator voltage e_o, which has an angular frequency ω_0, is

[42] M. J. O. Strutt, "On Conversion Detectors," *I.R.E. Proc.*, *22* (1934), 981–1008 (contains a bibliography of 89 references); E. W. Herold, "The Operation of Frequency Converters and Mixers for Superheterodyne Reception," *I.R.E. Proc.*, *30* (1942), 84–103; E. W. Herold and L. Malter, "Some Aspects of Radio Reception at Ultra-High Frequencies, Part V. Frequency Mixing in Diodes," *I.R.E. Proc.*, *31* (1943), 575–582; E. W. Herold, "Superheterodyne Frequency Conversion Using Phase-Reversal Modulation," *I.R.E. Proc.*, *34* (1946), 184–198.

supplied through a grid leak and capacitor, R_g and C_g, to grid 1 of the mixer tube in Fig. 31. This voltage, together with the bias voltage it produces by grid rectification, is sufficient to reduce the plate current to small values or even to cut it off for an appreciable part of the cycle. The oscillator voltage thus effectively controls the transconductance g_m between grid 3 and the plate in the manner shown in Fig. 32. The transconductance from the signal electrode to the plate is hence a time-varying quantity that varies periodically at the oscillator frequency.

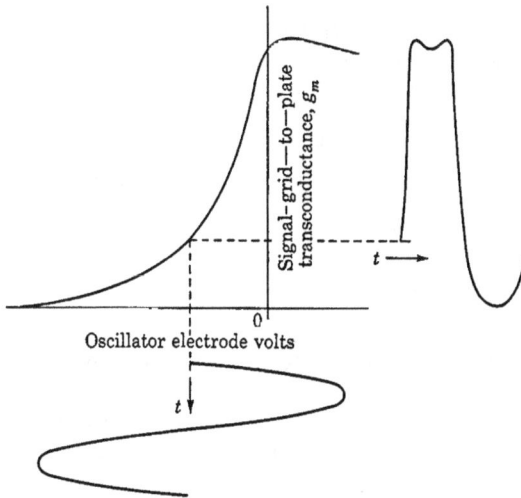

Fig. 32.   Time-varying transconductance between the signal grid and the plate in a mixer tube.*

The amplitude-modulated signal voltage e_s is ordinarily much smaller than the oscillator voltage; hence it has an inappreciable effect on the instantaneous value of g_m. Also, the dynamic plate resistance is very large compared with the load impedance because the screen and suppressor grids have effects similar to those in a pentode. Hence, the varying component of the plate current associated with the simultaneous presence of the signal voltage and the oscillator voltage is given approximately by

$$i_p = g_m e_s,$$ [167]

where both e_s and g_m are *functions of time*. Other varying components

* This diagram is adapted from E. W. Herold, "The Operation of Frequency Converters and Mixers for Superheterodyne Reception," *I.R.E. Proc.*, *30* (1942), Fig. 1, p. 87, with permission.

of the plate current exist, but they are not of interest in the frequency-conversion application.

If the oscillator voltage is represented as a cosine wave in Fig. 32, the transconductance is an even periodic function, and may therefore be represented by the Fourier series

$$g_m = a_0 + a_1 \cos \omega_0 t + a_2 \cos 2\omega_0 t + \cdots. \qquad [168]$$

The amplitude-modulated signal voltage may be represented as

$$e_s = A(t) \cos \omega_c t, \qquad [169]$$

where $A(t)$ is the modulated amplitude and ω_c is the carrier angular frequency. The varying component of the plate current given by substitution of Eqs. 168 and 169 in Eq. 167 is then

$$i_p = a_0 A(t) \cos \omega_c t + a_1 A(t) \cos \omega_c t \cos \omega_0 t$$
$$+ a_2 A(t) \cos \omega_c t \cos 2\omega_0 t + \cdots, \qquad [170]$$

or, through trigonometric transformation,

$$i_p = a_0 A(t) \cos \omega_c t + \frac{a_1}{2} A(t) \cos (\omega_c - \omega_0)t + \frac{a_1}{2} A(t) \cos (\omega_c + \omega_0)t$$

$$+ \frac{a_2}{2} A(t) \cos (\omega_c - 2\omega_0)t + \frac{a_2}{2} A(t) \cos (\omega_c + 2\omega_0)t + \cdots. \quad [171]$$

Tuning the output transformer and load to the difference angular frequency $(\omega_c - \omega_0)$ in Eq. 171 suppresses the other components and gives an output that contains the original amplitude function $A(t)$ impressed on an angular frequency lower than ω_c. The second term on the right-hand side of Eq. 171 then becomes the intermediate-frequency amplitude-modulated output-current wave for the super-heterodyne receiver. Alternatively, the load may be tuned so that the n-th harmonic of the oscillator frequency is used, and the angular intermediate frequency becomes $(\omega_c - n\omega_0)$.

The *conversion transconductance* of a frequency converter is a quantity useful as a measure of its effectiveness. It is "the quotient of the magnitude of the desired output-frequency component of current by the magnitude of the input-frequency (signal) component of voltage when the impedance of the output external termination is negligible for all of the frequencies which may affect the result."[43] This definition involves the assumption that all the direct electrode voltages and the magnitude of the local-oscillator voltage are constant, and that the input-frequency voltage is effectively an infinitesimal. Thus, when the

---

[43] "Standards on Electron Tubes: Definitions of Terms, 1950," *I.R.E. Proc.*, *38* (1950), 431.

load is tuned so that the fundamental frequency of the oscillator voltage is used, the conversion transconductance is, in terms of the quantities in Eqs. 169 and 171,

$$\text{Conversion transconductance} = \frac{\frac{a_1}{2} A(t)}{A(t)} = \frac{a_1}{2}. \qquad [172]$$

If the second harmonic of the oscillator is used, the conversion transconductance becomes $a_2/2$, and so on. For any particular bias and waveform of oscillator voltage, the coefficient $a_1, a_2, \cdots$ may be evaluated by a graphical method such as that used in Art. 13, Ch. VIII, or Art. 4, Ch. X. If the oscillator voltage and bias are adjusted so that the waveform of the transconductance g_m in Fig. 32 approaches a square wave—that is, g_m is a constant, $g_{m_{max}}$, for half the cycle and zero the other half—the coefficient a_1 approaches $2g_{m_{max}}/\pi$. Correspondingly, the conversion transconductance approaches $g_{m_{max}}/\pi$.[44] Values of 25 to 30 per cent of $g_{m_{max}}$ are realized in practice.

The multigrid tube illustrated in Fig. 31 may also be used as a *pentagrid converter*. For such service the local oscillator function is commonly performed by the cathode and first two grids connected as a grounded-plate Hartley oscillator, the second grid serving as the plate.

18. FREQUENCY MODULATION

In Art. 15 of this chapter, a system of communication is described in which audio-frequency signals are transmitted by means of the amplitude variation of a high-frequency carrier wave. As the design of receiver circuits has progressed, a point has finally been reached where (technically speaking, at least) the sensitivity is no longer a problem. As more amplifier tubes are added to a receiver, the noise voltage caused by thermal agitation of electrons in the low-level input circuits is increased in the same ratio as the signal, so that no advantage is gained in making a receiver with greater sensitivity than that necessary to give appreciable response to the thermal noise level. If the efficiency of the receiving antenna system is increased beyond a certain point, a similar situation is reached in most localities—atmospheric and other electrical disturbances, including interference from other transmitters, are increased in the same proportion as the signal. The problem of improving reception quickly reduces itself to the problem of selecting the signal from the interference—the problem that has perhaps been the most important one for many years.

[44] C. F. Nesslage, E. W. Herold, and W. A. Harris, "A New Tube for Use in Super-heterodyne Frequency Conversion Systems," *I.R.E. Proc.*, **24** (1936), 207–218.

The history of this problem is exceptionally interesting because of the sharp controversy that has existed over fundamental ideas. It has long been known that a certain audio-frequency spectrum is necessary for the proper reproduction of speech and music. For 80 per cent intelligibility, telephone engineers have shown that a system capable of transmitting a band 2,500 cycles per second wide is necessary.[45] For adequate radio reception, a 5,000-cycle width is desirable, and improvement is noticeable up to the limits of audibility—about 17,000 cycles; perhaps a 10,000-cycle width is a fair compromise. When a carrier is amplitude-modulated by such a signal, the resultant high-frequency wave has a bandwidth either equal to or twice that of the audio modulation, depending on whether or not one of the side bands is suppressed. When static or other disturbance interferes with such a signal, that part of the disturbance that lies in the frequency range within the band of the desired signal cannot be separated from it. If the usual type of linear diode detector is used, the instantaneous-output voltage is proportional to the envelope of the carrier. If one broadcast signal interferes with another, the resultant is about the same as if the two audio signals were added in the same proportion as the carriers. Also an audible beat may be produced.[46] It was early concluded that the best to be hoped for was an interference proportional to the bandwidth of the receiver. For optimum results, the receiver band should cover the two side bands and nothing else.

18a. *Narrow-Band Frequency Modulation.* About 1920, engineers became interested in the possibility of improved radio communication by means of frequency modulation. Their proposals followed this reasoning: The frequency of a transmitter was to be varied linearly with the instantaneous value of an audio signal, but only by a small amount, say by ± 100 cycles at the peak amplitude of the signal. The receiver was then to include a crystal filter arranged in such a way as to make the output vary rapidly with frequency—by a circuit similar to a modern discriminator circuit but operating over a band only 200 cycles wide. It was believed that such a system would reduce static by narrowing the bandwidth from 20,000 cycles to 200 cycles, thereby reducing the interfering energy by 20 decibels. The fallacy in this argument was pointed out[47] by Carson in 1922.

[45] H. Fletcher, *Speech and Hearing* (New York: D. Van Nostrand Company, Inc., 1929), 280.

[46] This is an oversimplified discussion of a complicated process. The matter is treated in a lengthy paper by C. B. Aiken, "Theory of the Detection of Two Modulated Waves by a Linear Rectifier," *I.R.E. Proc.*, *21* (1933), 601–629, but the discussion there does not cover the case of a diode peak detector, which has become the most important type.

[47] J. R. Carson, "Notes on the Theory of Modulation," *I.R.E. Proc.*, *10* (1922), 57–64. This paper is very readable and has become a classic on the subject of modulation.

The difficulty may be seen from the following analysis. As is shown by Eq. 11, if the modulating wave is sinusoidal, a frequency-modulated wave is representable by an expression of the form

$$e(t) = A_c \cos (\omega_c t + \delta \sin \omega_m t), \qquad [173]$$

where

$$\delta = k_f A_m / \omega_m \qquad [174]$$

and is called the *modulation index*; that is, δ is the ratio of the maximum frequency deviation during a cycle to the modulating frequency. If Eq. 173 is expanded,

$$e(t) = A_c[\cos \omega_c t \cos (\delta \sin \omega_m t) - \sin \omega_c t \sin (\delta \sin \omega_m t)]. \qquad [175]$$

It is shown elsewhere[48] that

$$\cos (\delta \sin \omega_m t) = J_0(\delta) + 2J_2(\delta) \cos 2\omega_m t + 2J_4(\delta) \cos 4\omega_m t + \cdots \qquad [176]$$

and

$$\sin (\delta \sin \omega_m t) = 2J_1(\delta) \sin \omega_m t + 2J_3(\delta) \sin 3\omega_m t + \cdots , \qquad [177]$$

where the coefficients $J_n(\delta)$ are Bessel functions of δ and *do not involve* t. When Eqs. 176 and 177 are substituted in Eq. 175 and the products are put in terms of multiple angles, there results

$$e(t) = A_c[J_0(\delta) \cos \omega_c t + J_1(\delta) \cos (\omega_c + \omega_m)t - J_1(\delta) \cos (\omega_c - \omega_m)t$$
$$+ J_2(\delta) \cos (\omega_c + 2\omega_m)t + J_2(\delta) \cos (\omega_c - 2\omega_m)t + \cdots]. \qquad \blacktriangleright[178]$$

Note that regardless of how small the modulation index δ is, the spectrum of $e(t)$ consists solely of frequencies spaced by the *modulating frequency*—not the deviation frequency. Thus, if the frequency deviation is ± 100 cycles at a rate of 10 kilocycles per second,

$$e(t) = A_c[J_0(0.01) \cos \omega_c t + J_1(0.01) \cos (\omega_c + \omega_m)t$$
$$- J_1(0.01) \cos (\omega_c - \omega_m)t + J_2(0.01)\cos (\omega_c + 2\omega_m)t$$
$$+ J_2(0.01) \cos (\omega_c - 2\omega_m)t + \cdots] \qquad [179]$$

$$= A_c[0.999975 \cos \omega_c t + 0.005 \cos (\omega_c + \omega_m)t$$
$$- 0.005 \cos (\omega_c - \omega_m)t + 0.000012 \cos (\omega_c + 2\omega_m)t$$
$$+ 0.000012\cos (\omega_c - 2\omega_m)t + \cdots], \qquad [180]$$

which shows that such a signal consists essentially of a carrier with two side frequencies of 0.5 per cent amplitude displaced from it by 10 kilocycles. Clearly, a filter circuit that eliminates all but a band 200 cycles wide will pass an unmodulated carrier and nothing else—no information whatsoever will be conveyed, because the side frequencies will be eliminated.

[48] F. S. Woods, *Advanced Calculus* (new ed.; Boston: Ginn & Company, 1934), 281.

18b. *Wide-Band Frequency Modulation.* Armstrong, who was one of those attempting to eliminate static by the narrow-band system,[49] continued his efforts in a slightly different direction; he endeavored to find a man-made signal that was essentially different from a natural one. It occurred to him that, if static interference is largely of an amplitude-modulated type and has little frequency modulation, there might be a possibility of improving conditions by generating a signal with a very large amount of frequency modulation and passing it

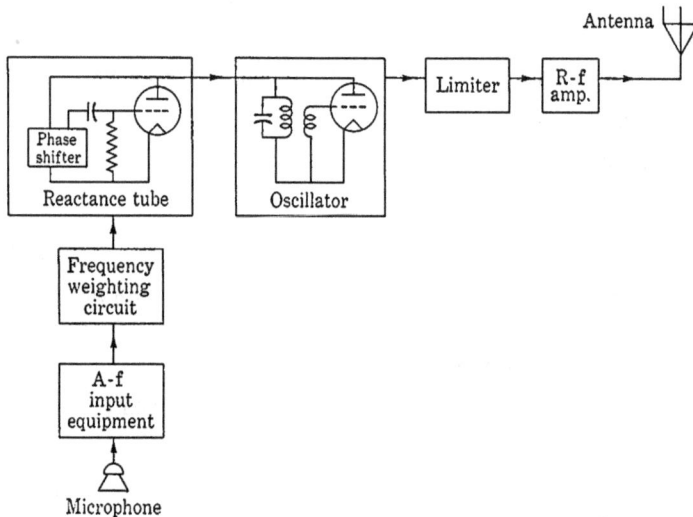

Fig. 33.   Frequency-modulated-transmitter diagram. The diagram is simplified by the omission of frequency-control equipment, frequency multipliers, direct-current power supplies, and so forth.

through a receiver unable to respond to amplitude modulation but sensitive to frequency changes. When first demonstrated, this system met with considerable skepticism, but was finally recognized as remarkably effective in suppressing interference and static.[50]

A typical frequency-modulation transmitter is shown in Fig. 33. The transmitter is an idealized one; a frequency-stabilization circuit, which is very important but is not essential for an understanding of the principles, is omitted. The reactance tube is arranged to place a reactance across the oscillator circuit that can be varied by the grid-bias

[49] E. H. Armstrong, "A Method of Reducing Disturbances in Radio Signalling by a System of Frequency Modulation," *I.R.E. Proc.*, *24* (1936), 689–740.

[50] I. R. Weir, "Field Tests on Frequency and Amplitude Modulation with Ultra-high-Frequency Waves," *G.E. Rev.*, *42* (1939), 188–191, 270–273; R. F. Guy and R. M. Morris, "NBC Frequency Modulation Field Test," *RCA Rev.*, *5* (1940), 190–225.

voltage at a modulation-frequency rate.* The net result is that the frequency is made to vary in direct proportion to the instantaneous value of an audio-frequency modulating signal. In present commercial broadcasting practice, the frequency is made to deviate by a peak of 75 kilocycles about a mean of approximately 100 megacycles. Although the reactance tube does not introduce much amplitude modulation, it is customary to include an amplitude limiter in the transmitter in order to reduce any that may be present. The purpose of the frequency-weighting circuit in the audio system is discussed subsequently. The frequency-modulated signal is finally amplified to the desired output power by a broad-band amplifier. The amplifier need not be linear

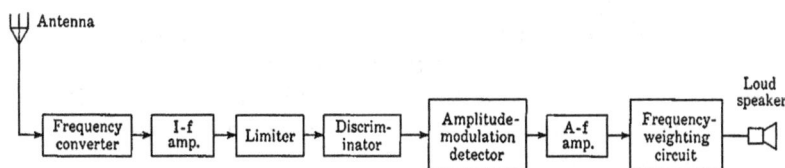

Fig. 34.   Frequency-modulation-receiver diagram.

with input voltage and can operate in a very efficient manner, since the output amplitude is not varied.

The transmitter shown in Fig. 33 is one of two types in common use at present. The other, developed by Armstrong,[51] makes use of phase modulation, in that the audio signal is first used to phase-modulate a 200-kilocycle wave. This phase-modulated wave is then put through frequency-multiplier circuits until the original frequency deviations have been amplified by several thousand, and the average frequency is brought to about 100 megacycles. The circuit has some interesting features that throw considerable light on the process of frequency modulation, and reference to the original description is well worth while.

A typical receiving circuit is indicated in Fig. 34. The incoming frequency-modulated signal is converted to a lower, or intermediate, frequency, which is then amplified and passed through an amplitude-

---

* The plate circuit of the reactance tube is a parallel current path across the tuned circuit. The grid voltage is maintained 90 degrees out of phase with the plate voltage by proper selection of the values of the resistance and the capacitance connecting the two. Consequently, the plate current contains a component 90 degrees out of phase with the plate voltage, and the tube therefore acts as a reactive load across the tuned circuit. This reactance can be changed at an audio-frequency rate by introduction of an audio signal at the control grid.

[51] E. H. Armstrong, "A Method of Reducing Disturbances in Radio Signalling by a System of Frequency Modulation." *I.R.E. Proc.*, *24* (1936), 692–696.

limiting circuit that removes any amplitude modulation introduced by interference from noise or other signals during transmission. After the removal of amplitude modulation by the limiter, the signal is passed through a frequency-sensitive network, the *discriminator*, which converts the frequency variations to amplitude variations. The resulting amplitude-modulated signal is then rectified and amplified in the same way that any other amplitude-modulated signal is, except that a frequency-weighting circuit described in Art. 18d is commonly used.

18c. *Frequency-Modulation Spectra.* As is indicated by Eq. 178, the frequency-modulated signal wave contains a carrier frequency and side frequencies, just as does the amplitude-modulated wave. However, the relationships between the number of components and the frequency interval between them in the modulating and modulated waves are quite different in the two types of modulation. In amplitude modulation, a sinusoidal modulating wave results in the production of only one side frequency on each side of the carrier; in frequency modulation it generally results in a large number.

The spectrum of a frequency-modulated signal may be obtained from Eq. 178, rewritten in the form

$$\frac{e(t)}{A_c} = J_0(\delta) \cos \omega_c t + J_1(\delta) \cos (\omega_c + \omega_m)t - J_1(\delta) \cos (\omega_c - \omega_m)t$$
$$+ J_2(\delta) \cos (\omega_c + 2\omega_m)t + J_2(\delta) \cos (\omega_c - 2\omega_m)t + \cdots, \qquad [181]$$

where

$$\delta = \frac{\text{Frequency deviation, } \Delta\omega_c/(2\pi)}{\text{Modulating frequency, } \omega_m/(2\pi)}, \qquad \blacktriangleright[182]$$

and the frequency deviation $\Delta\omega_c/(2\pi)$ is half the difference between the maximum limits between which the instantaneous frequency of the modulated wave varies.

For example, let

$$\delta = \frac{100 \text{ kilocycles}}{4 \text{ kilocycles}} = 25. \qquad [183]$$

Then the coefficients $J_n(\delta)$ have the values[52] as a function of n shown in Fig. 35a. The solid curve is a plot of $J_n(25)$ for continuous values of n, and the vertical lines are drawn at each integral value of n. The amplitudes of the frequency components in Eq. 181 are found by means of this curve and may be plotted as shown in Fig. 35b. The interval between the components in the spectrum is equal to the modulating frequency.

[52] E. Jahncke and F. Emde, *Tables of Functions with Formulae and Curves* (3rd ed.; Leipzig: B. G. Teubner, 1938), 171–179.

The spectra of frequency-modulated signals for several values of the modulation index are given in Figs. 36a and 36b. In the first of these series, Fig. 36a, the audio frequency $\omega_m/(2\pi)$ is held constant while the frequency deviation $\Delta\omega_c/(2\pi)$ takes on different discrete values. In the second series, Fig. 36b, the reverse is true. It is interesting to observe the amplitude of the carrier relative to that of the side frequencies as the modulation index is varied. For a certain value of the modulation index, the carrier amplitude disappears entirely from the resultant frequency-modulated wave.

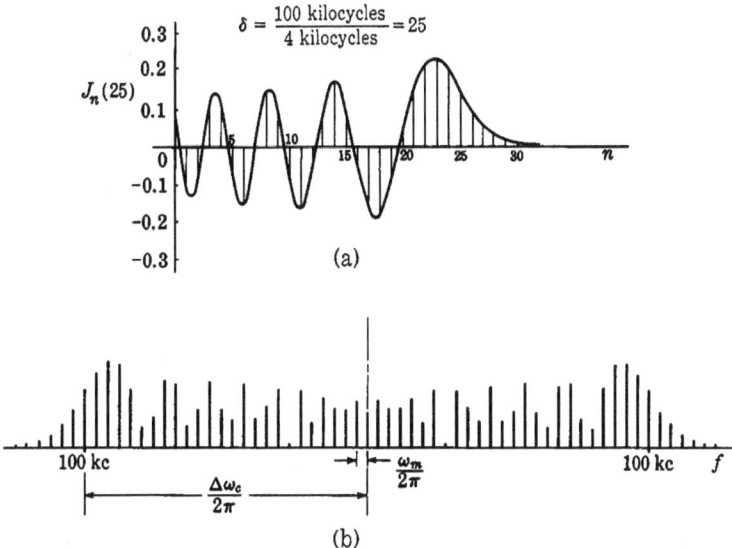

Fig. 35. Bessel functions and frequency-modulation spectrum for a modulation frequency of 4 kilocycles and a frequency deviation of 100 kilocycles.

18d. *Interference.* The primary advantage of frequency modulation as a means of communication is its relative freedom from interference.[53] In this connection, *interference* is to be interpreted as the presence of extraneous audio-frequency power in the output of the receiver as a result of a variety of causes, among which are atmospheric disturbances (static), noise from thermal agitation in tubes and

[53] In this article free use is made of the results of a doctoral dissertation, D. Pollack, *Interference between Stations in Frequency-Modulation Systems* (unpublished doctoral dissertation, Massachusetts Institute of Technology, Cambridge, Massachusetts, 1940), 42–54. See also O. E. Keall, "Interference in Relation to Amplitude-, Phase-, and Frequency-Modulated Systems," *Wireless Engr., 18* (1941), 6–17, 56–63; S. Goldman, "F-m Noise and Interference," *Electronics, 14* (August, 1941), 37–42; L. B. Arguimbau, *Vacuum-Tube Circuits* (New York: John Wiley & Sons, Inc., 1948), 523–533.

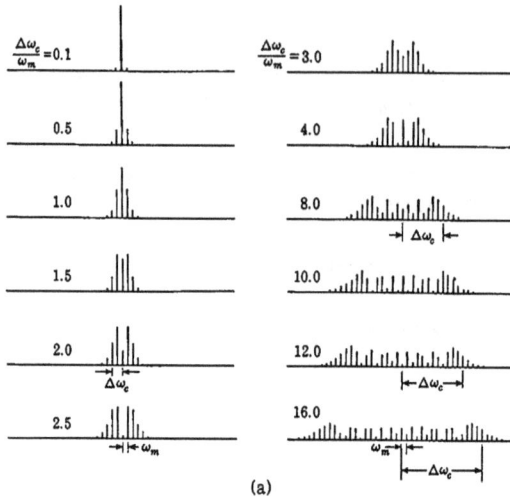

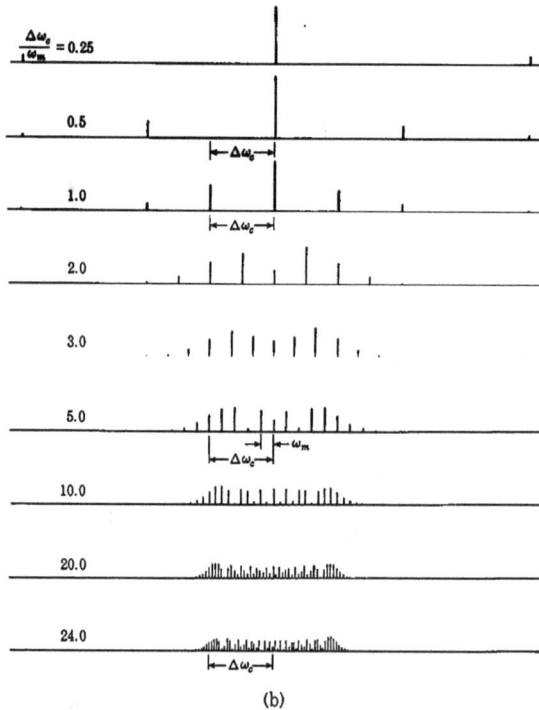

Fig. 36. Frequency-modulation spectra for different modulation indices.*

* This diagram is adapted from Balth van der Pol, "Frequency Modulation," *I.R.E. Proc.*, *18* (1930), Fig. 2, p. 1200, and Fig. 3, p. 1201, with permission.

conductors, disturbances from surrounding sources such as ignition systems and sparking at the commutators of electrical appliances, and radio-frequency signals other than the one desired. The statement is sometimes made that noise and static consist wholly of an amplitude-varying signal and that a receiver made insensitive to such variations (by an amplitude limiter) will not respond to these disturbances. This statement is too sweeping, but it does point in the right direction. The

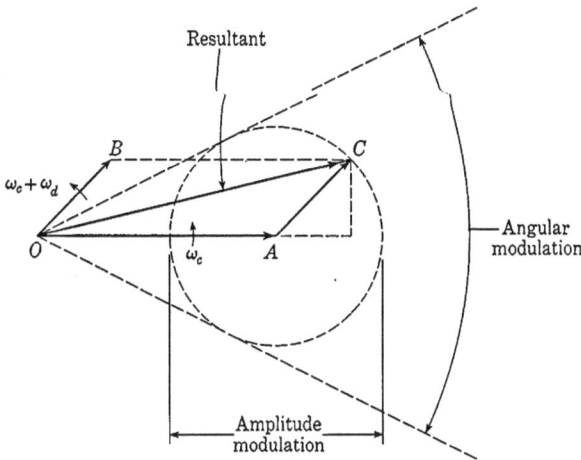

Fig. 37. Superposition of two sinusoidal signals. The projection of the rotating vector OA is cos $\omega_c t$; of OB is a cos $(\omega_c + \omega_d)t$.

reduction of interference is basically a problem that involves the transient behavior of the communication system, since most of the causes are nonperiodic. However, an approach based on steady-state behavior yields some results, and the discussion that follows is an analysis of one of the simpler causes of interference, namely, the simultaneous reception of two unmodulated carrier waves whose frequencies are so close together that neither is suppressed by the tuned circuits of the receiver. The amount of such interference is compared in the amplitude-modulation and frequency-modulation systems of communication.

The simultaneous reception of two unmodulated carrier waves at a receiver results in an input voltage that is the sum of the two waves. This sum may be shown vectorially, as in Fig. 37. For simplicity, it is assumed here that one signal wave—the desired one—has unit amplitude and that the other—the interfering wave—has an amplitude a. Thus a is the ratio of the amplitudes. The angular frequencies of the two waves are ω_c and $\omega_c + \omega_d$, respectively. The assumption is also made that the amplitude a of the interfering signal is smaller than unity and that the difference angular frequency ω_d is small compared with that of the desired signal ω_c.

The desired signal may be represented by the projection on a stationary axis of a unit vector OA rotating about the origin with an angular frequency ω_c, and similarly for the interfering signal. The instantaneous value of the result, which is merely the sum of the projections of the two vectors, is

$$e(t) = \cos \omega_c t + a \cos (\omega_c + \omega_d)t. \qquad [184]$$

Alternatively, however, the instantaneous value of the combination may be considered as the projection of the vector that results from the addition of the two vectors OA and OB, as indicated in Fig. 37—that is, as the projection of the vector OC. This vector may also be recognized as the resultant of a unit vector OA of angular velocity ω_c and a vector AC of length a rotating about the tip of OA at an angular velocity ω_d (with respect to the vector OA, not with respect to the stationary co-ordinates). Considered in this way, the resultant instantaneous value may be described by

$$e(t) = \sqrt{(1 + a \cos \omega_d t)^2 + (a \sin \omega_d t)^2} \cos (\omega_c t + \phi), \quad ▶[185]$$

where

$$\phi = \tan^{-1} \frac{a \sin \omega_d t}{1 + a \cos \omega_d t}. \qquad ▶[186]$$

The quantity $\sqrt{(1 + a \cos \omega_d t)^2 + (a \sin \omega_d t)^2}$ may be thought of as the "instantaneous amplitude," and, in accordance with Art. 2, the "instantaneous angular frequency" is given by

$$\frac{d}{dt} (\omega_c t + \phi) = \omega_c + \frac{d\phi}{dt}. \qquad ▶[187]$$

The representation in Eqs. 185 and 186 is exact, because it can be shown that the equations are mathematically identical with Eq. 184. Although, mathematically speaking, the wave is more complicated than a simple sinusoid, the concepts of instantaneous amplitude and instantaneous frequency are nevertheless useful, because the difference frequency is small compared with the frequency $\omega_c/(2\pi)$, from which fact it follows that neither ϕ nor the length of OC changes appreciably during any one revolution of the vectors about the origin. The instantaneous amplitude and frequency are functions of time, but they vary slowly in comparison with the rate at which the function as a whole varies. Consequently, the function is nearly sinusoidal during any one period of time equal to $2\pi/\omega_c$. Since both the instantaneous amplitude and the instantaneous frequency involve sinusoidal functions of time, the sum of the two sine waves has some of the characteristics that result from both amplitude and frequency modulation (see Art. 2). Subsequent discussion shows that a receiver for either amplitude modulation or frequency modulation will respond to it.

When the wave represented by Eqs. 185 and 186 is received by an amplitude-modulation receiver, the detector (assumed to be of the linear diode type) responds to the envelope of the wave. Inasmuch as the term $\cos(\omega_c t + \phi)$ varies rapidly in comparison with the variations in the instantaneous amplitude, the instantaneous amplitude is essentially the maximum value of $e(t)$ in the immediate neighborhood of the time t, and it therefore represents the envelope. If a is small compared with unity, the instantaneous amplitude (or the envelope) is given, to a close approximation, by $1 + a \cos \omega_d t$. Thus the output of the amplitude-modulation receiver is practically the same as if the original desired signal were amplitude-modulated by a modulation factor a at an angular frequency ω_d equal to the difference frequency.* The ratio of the amplitudes a is therefore the fractional utilization of the full range of the output that corresponds to the full permitted range of the modulating signal at the input of the system. A corresponding quantity, indicative of the interference for a frequency-modulation system, is developed subsequently.

When the wave represented by Eqs. 185 and 186 is received by a frequency-modulation receiver, the changes in frequency, not in amplitude, are responsible for the output amplitude. In the frequency-modulation receiver, the passage of the wave through the limiter shown in Fig. 34 produces essentially a square wave that crosses the axis at the same instants as the original wave and has the corresponding instantaneous polarity at all times. The limiter thus introduces extraneous high frequencies that are not essential, but these are eliminated by a fairly broad filter tuned to the intermediate frequency, and the output is then frequency-modulated with no amplitude modulation. In passing through the discriminator, the frequency variations are converted into amplitude variations; the relation between the amplitude of the output and the deviation of the instantaneous frequency from the mean value at the input is linear in the discriminator. When a is small compared with unity, Eq. 186 becomes

$$\phi = \tan^{-1}\frac{a \sin \omega_d t}{1 + a \cos \omega_d t} \approx \tan^{-1}(a \sin \omega_d t) \approx a \sin \omega_d t. \qquad [188]$$

---

* This point of view is useful in analyzing the distortion in single-sideband transmission and in beat-frequency oscillators.

The instantaneous frequency from Eq. 187 is then

$$f = \frac{1}{2\pi} \left(\omega_c + \frac{d\phi}{dt} \right) \approx \frac{1}{2\pi} (\omega_c + a\omega_d \cos \omega_d t). \qquad [189]$$

The instantaneous frequency therefore contains a sinusoidal term having a frequency equal to the difference frequency of the two original waves, and the output of the discriminator has an envelope whose variations are proportional to the variations in the instantaneous angular frequency $a\omega_d \cos \omega_d t$. The amplitude of the output of the linear diode detector corresponds to the envelope of the wave out of the discriminator. Consequently, it too is proportional to $a\omega_d \cos \omega_d t$. The audio output therefore has an amplitude proportional to $a\omega_d$ and a frequency $\omega_d/(2\pi)$.

In the frequency-modulation system, linearity between input and output is established by design for operation within the limits of a certain frequency deviation, which, for a typical system, is ± 75 kilocycles. Thus a frequency deviation of ± 75 kilocycles is the maximum permitted, just as 100 per cent modulation is the maximum permitted in an amplitude-modulation system. A deviation of ± 75 kilocycles corresponds to the full permitted range of the modulating signal at the input of the system. The fractional utilization of this full range by the interfering signal in the example considered here is then $a\omega_d/(2\pi \times 75)$, as far as the amplitude of the audio-frequency output is concerned. But the maximum frequency passed by the audio system, even in high-fidelity audio systems, is perhaps 15 kilocycles. Consequently, the fraction of the full output of the receiver utilized by the interference cannot be more than $(15/75)a$ or $(1/5)a$ for the frequency-modulation system, whereas it is a for the amplitude-modulation system. Evidently the use of a wider permitted frequency deviation would result in a reduction of the interference by an even larger factor.

In practice, interference is further reduced by the use of so-called *pre-emphasis circuits* that increase the high-frequency audio components before the modulation process and reduce them in the audio-frequency part of the receiver. This procedure is the same as that used in phonograph recording to reduce "scratch" and in the sound channel of television systems to reduce noise. The method is based on the fact that, for most audio programs, most of the power is contained in the low frequencies, so that multiplying all high-frequency components by a factor proportional to the frequency does not materially increase the peak amplitude. A weighting factor of $\sqrt{1 + (f/2{,}100)^2}$ has been found to be a good compromise. This factor corresponds, for example, to the fractional variation with frequency of the voltage across a series combination of inductance and resistance when it is fed with constant amplitude of current and its time constant is $1/(2\pi \times 2{,}100)$ second, or 75 microseconds. The use of such a factor does not appreciably change the peak amplitude of a signal wave; but, if noise or interference having power or energy uniformly distributed with respect to frequency is combined with the signal at some point in the system, the use of the reciprocal weighting factor at the receiver enormously reduces the interfering power at the output while merely restoring the signal to its original unweighted distribution of power.

Further reasoning of this sort may be used to determine the interference between two frequency-modulated signals, but the analysis is not simple. The results may be summarized by the statement that. in an ordinary amplitude-modulation broadcast radio system, to reduce the interference to one per cent, the amplitude of the interfering signal must be reduced to $\frac{1}{100}$ that of the desired signal, whereas in a frequency-modulation system a ratio of $\frac{1}{3}$ is sufficient.

In a similar way, it can be shown that static and thermal noise cause less interference in a frequency-modulation system. This reduction in noise has made the introduction of high-fidelity systems feasible, so that frequency-modulation systems can be made to operate with a 15-kilocycle audio band without appreciable noise in the output.

PROBLEMS

1. Figure 38 shows a Class C amplifier modulated by a Class B push-pull amplifier. The ratings of the tubes for the particular plate-supply voltages and conditions of operation are:

Type 1

As Plate-Modulated R-f Power Amplifier—Class C Telephony.

Carrier conditions (no speech input) for use with a maximum modulation factor of 1.0.

Direct plate voltage = 1,000 volts

Direct plate power input $(E_b I_b)$ = 175 watts

Plate dissipation = 67 watts.

Type 2

As A-f Power Amplifier and Modulator—Class B.

Values are for two tubes in push-pull.

Direct plate voltage = 750 volts

Maximum signal power output = 90 watts

Effective load resistance (plate-to-plate) = 6,400 ohms.

For analysis it may be assumed that (1) the Class C amplifier is so adjusted that the average component and the fundamental radio-frequency component of plate current in the Type 1 tube vary directly with the average plate voltage during a radio-frequency cycle, (2) the plate efficiency is constant during an audio-frequency cycle, and (3) the efficiency of the modulating transformer is 90 per cent.

(a) What maximum percentage modulation may be produced in the output current?

(b) What ratio of total primary turns to secondary turns should the modulating transformer have?

(c) What maximum power output in the carrier and side bands may be produced if the efficiency of power transfer through the tuned circuit to the coupled load is 90 per cent?

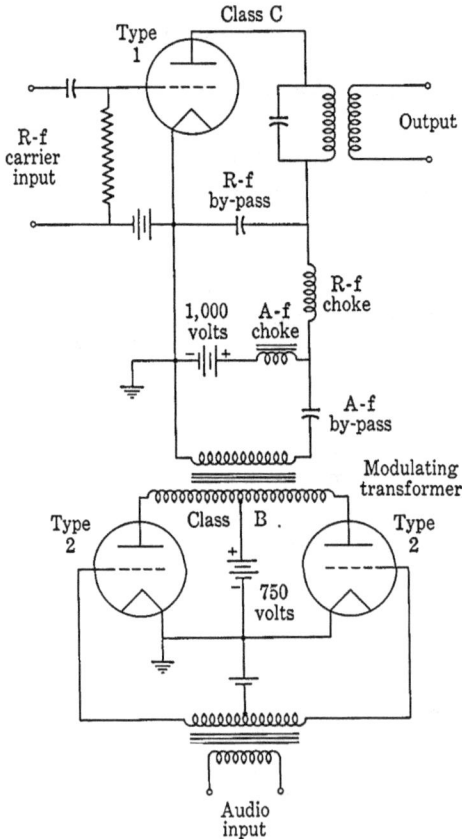

Fig. 38.   Class C plate-modulated r-f amplifier circuit for Prob. 1. The circuit has been simplified for clarity.

2. A Type 6H6 diode is used in the circuit of Fig. 16 as a linear detector. In the circuit, R_L is 100,000 ohms, R_g is 250,000 ohms, C_c is 0.1 μf, and C_s is 100 $\mu\mu$f. The carrier voltage is 10 volts rms at 10^6 cps and the modulating frequency is 10^3 cps.

(a) What is the limiting value of the percentage modulation above which the harmonic generation produced by the detector increases rapidly?

(b) What changes in the parameters would you recommend in order to increase this limiting value?

3. If the function $f(t)$ represents the voltage of an arbitrary audio-frequency signal, the equation

$$e = A_c[1 + Bf(t)] \cos \omega_c t,$$

where A_c is the carrier amplitude and B is a constant, expresses in analytic form the corresponding amplitude-modulated carrier of angular frequency ω_c radians per sec. Show analytically that, if this voltage is impressed upon an ideal linear diode rectifier, the audio-frequency content of the output is a perfect replica of the audio-frequency input $f(t)$ when the magnitude of $Bf(t)$ is never greater than unity.

4. The voltage impressed between the grid and cathode of the Van der Bijl modulation circuit of Fig. 25 is assumed to have the form

$$e_g = 0.5 \cos \omega_c t + 1.0 \cos \omega_m t \text{ volts,}$$

in which the carrier angular frequency ω_c is 10^7 radians per sec and the audio angular frequency ω_m is 10^3 radians per sec. The characteristics of the tube are given with sufficient accuracy by the relation

$$i_b = \tfrac{2}{3} \times 10^{-4} \left(e_c + \frac{e_b}{10}\right)^2 \text{ amp.}$$

In the circuit, the plate-voltage supply E_{bb} is 33 volts and the grid-bias voltage E_{cc} is -1.5 volts. Although the circuit parameters actually are chosen as discussed in Art. 12, the load for the tube may be considered to be a pure resistance of 20,000 ohms for purposes of analysis.

(a) Sketch the tube characteristic and show the operating conditions.
(b) Determine the numerical values of the coefficients in the relation

$$e_p = B_1 e_g + B_2 e_g{}^2,$$

which expresses the output voltage e_p in terms of e_g.
(c) Show that a portion of the output has essentially the form of an amplitude-modulated wave. Determine the magnitude of the carrier output in volts and the per cent modulation.
(d) If now the voltage impressed between the grid and the cathode is changed to the form

$$e_g = 0.5(1 + m \cos \omega_m t) \cos \omega_c t \text{ volts,}$$

show that the circuit operates as a square-law detector.
(e) Determine the magnitude of the audio-frequency component in the output voltage having the angular frequency ω_m and the per cent second harmonic of this frequency when the signal of (d) is 20 per cent and 100 per cent modulated.

5. An amplitude-modulated voltage wave

$$e = E(1 + m \cos \omega_m t) \cos \omega_c t$$

is transmitted to a receiver where it is demodulated by a square-law detector whose current-voltage characteristic is

$$i = ke^2.$$

What is the audio frequency in the detector output when

(a) one side frequency is eliminated from the transmitted wave?
(b) both side frequencies are eliminated from the transmitted wave?
(c) the carrier is eliminated from the transmitted wave?

6. In carrier-communication systems, which involve essentially the same principles as those for radio communication, the carrier is customarily suppressed at the sending end and supplied again at the receiving end before demodulation. Assuming a square-law detector,

(a) If both side bands are transmitted, show that for minimum audio-frequency distortion the carrier at the receiving end must be supplied in its original phase position relative to the side-band frequencies. Demonstrate this fact for an original amplitude-modulated carrier of the form

$$e = A_c(1 + m \cos \omega_m t) \cos \omega_c t,$$

and show specifically what would result after demodulation if the supplied carrier were given as

$$e_c = B \cos \left(\omega_c t + \frac{\pi}{2} \right).$$

(b) On the other hand, if one side band in addition to the carrier is suppressed before transmission, show that the phase of the supplied carrier may have any value without noticeably affecting the quality of the demodulated output in the steady state.

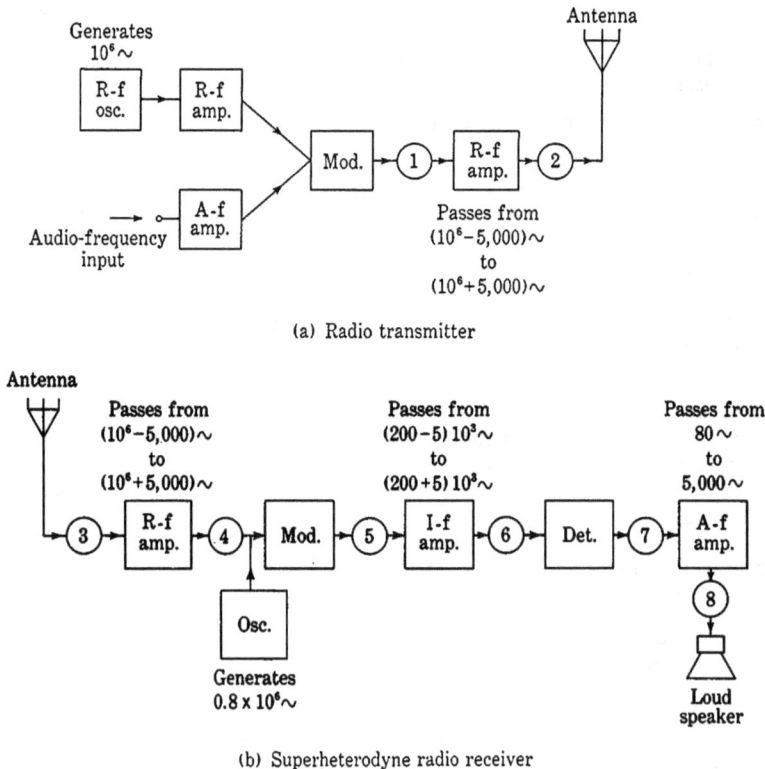

(a) Radio transmitter

(b) Superheterodyne radio receiver

Fig. 39.   Block-diagram representation of a typical radio-broadcast transmitter and receiver for use with Prob. 8.

7. A proposal is made to transmit information by modulation of a carrier followed by suppression of all frequencies except the upper side band before transmission. The missing carrier is to be supplied by a local oscillator at the receiver before detection is accomplished by a square-law detector.

(a) Discuss the effect upon the audio output of an oscillator angular frequency $\omega_c + \Delta\omega$ differing from the original carrier angular frequency by the amount $\Delta\omega$.

(b) Discuss the effect of small variations in $\Delta\omega$ upon the audio-frequency output.

8. Figure 39 shows in block-diagram form the general arrangement of the units in a radio-broadcast transmitter and a superheterodyne receiver. The modulators and the detector may be assumed to work as simple nonlinear resistive elements in series with the voltage sources and a resistive load. The nonlinear current-voltage characteristic of these devices may be assumed to be of the form

$$i = a_1 e + a_2 e^2,$$

where i and e are the current and terminal voltage, respectively, of the nonlinear device.

Determine all frequencies present at each of the eight points numbered in circles on the figure if

(a) there is no audio-frequency input at the transmitter,

(b) an audio-frequency signal of 500 cps is impressed,

(c) audio-frequency signals of 100 and 4,000 cps are simultaneously impressed.

9. A 100-megacycle-per-second carrier wave is frequency modulated by a 3-kilocycle-per-second sinusoidal audio modulating signal. The maximum frequency deviation is ±75 kilocycles per second.

(a) Determine the modulation index δ.

(b) Plot the coefficients $J_n(\delta)$ as a function of n.

(c) Plot the frequency spectrum of the frequency-modulated wave.

(d) Repeat (a), (b), and (c) for a 15-kilocycle-per-second signal, when the carrier and maximum deviation frequencies remain as before.

CHAPTER XIII

Semiconductor Rectifiers and Transistors

Semiconductors,[1] which are solids having a resistivity intermediate between that of an insulator and a metal, were used as detectors in radio receivers for years prior to the development of the electron tube. When that tube became available, however, the semiconductors were virtually abandoned because of the electron tube's ability to amplify and because of the difficulty of producing semiconductor devices with predetermined stable electrical characteristics. Recently, with the extension of radio and radar transmission into the microwave region, the unique properties of semiconductors have been again and further exploited by intensive research into the electrical properties of the solid state. This research has revealed many possibilities for replacement of electron tubes by rugged solid-state devices that may well have indefinitely long life. Thus, semiconductor rectifiers, photosensitive devices, and amplifiers are finding increased use as their development proceeds. In particular, the *transistor*,[2] which is a semiconductor amplifying device, has marked advantages over the electron tube for many applications. It is much smaller in size, requires no cathode heating power or warm-up time, has higher efficiency of power conversion, operates with much less direct input power, is free from microphonic troubles, and has potentially extremely long life. It does have the present disadvantages, however, of greater noise, lower frequency range, smaller power output, and lower permissible ambient temperature, which render it impractical for many purposes for which electron tubes are suited.

1. Conduction in semiconductors

For an explanation of the operation of semiconductor devices, at least a qualitative understanding of electronic conduction in solids is necessary. The discussion of conduction in metals in Art. 2, Ch. II, points out that when atoms are packed tightly together in a solid, the energy levels of the individual atoms expand into bands. Energy level diagrams showing such bands in various solids are given in Fig. 1.

[1] G. L. Pearson, "The Physics of Electronic Semiconductors," *A.I.E.E. Trans.*, 66 (1947), 209–214.

[2] J. Bardeen and W. H. Brattain, "The Transistor, A Semi-Conductor Triode," *Phys. Rev.*, 74 (1948), 230–231; J. A. Morton, "Present Status of Transistor Development," *B.S.T.J.*, 31 (1952), 411–442; R. F. Shea, Editor, *Principles of Transistor Circuits* (New York: John Wiley & Sons, Inc., 1953).

In an ideal *insulator*, the band of energy levels corresponding to the valence electrons is completely filled at a temperature of absolute zero, and a relatively wide gap of forbidden energy levels separates it from the next higher band of permitted energy levels, which is entirely empty. Despite any externally applied electric field, the net current associated with electron motion is zero for electrons occupying levels in the filled band because, when the band is full, for each electron moving in one direction there is another moving in exactly the opposite direction with exactly the same energy. Conduction can take place only if a means is provided for escape of an electron from the filled

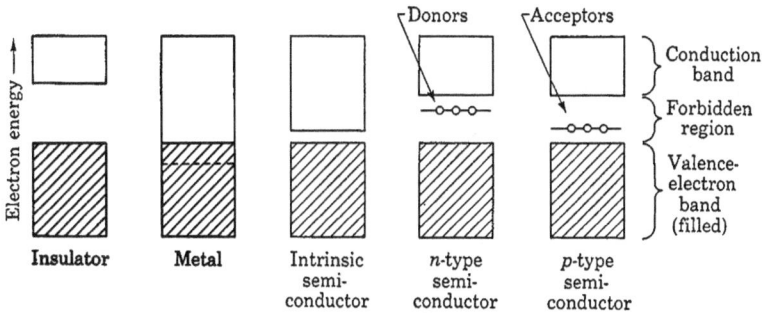

Fig. 1.   Energy-level bands in solids.

band into a permitted level in an unfilled band. When such escape does take place, the electron can drift in an externally applied electric field and contribute to electrical conduction through the solid. Furthermore, the vacancies left in the formerly filled band, called *holes*, can also drift in the electric field and contribute to the conduction just as though they were electrons with a positive instead of a negative charge. As the temperature of the insulator is increased from absolute zero, the added thermal energy tends to impel some of the valence electrons to jump the gap of forbidden energy levels and occupy unfilled levels above, leaving holes in the filled band. Drift of these electrons in one direction and of holes in the other could take place and constitute a conduction current, but the conductivity of such an insulator is small at ordinary temperatures because the forbidden region is so wide that only a negligible fraction of the thermally excited electrons can escape across it.

In a *metal*, the band of energy levels corresponding to the valence electrons is completely filled at a temperature of absolute zero, but a band of permitted and unoccupied levels overlaps it. Transfer of the valence electrons into these unoccupied levels takes place frequently when the temperature is raised. Electric conduction through motion of electrons occupying levels in this conduction band then occurs readily, and the conductivity is large.

In an electronic *semiconductor*, the energy-level distribution is similar to that of an insulator, as is shown in Fig. 1, but either the forbidden-energy gap is small compared with the thermal energy of electrons at room temperature, or additional allowed energy levels lie in the otherwise forbidden region. These additional energy levels are associated with the presence of impurities or of imperfections in the crystal lattice of a pure substance. Semiconductors generally fall into three classes. In an *intrinsic semiconductor*, the forbidden energy region is so narrow that even at room temperature appreciable numbers of thermally excited electrons can transfer across it. Electric conduction then takes place through simultaneous motion of holes and electrons. In an *n-type extrinsic semiconductor*, also called a *donor* and an *excess* type, localized impurity or imperfection levels that are filled at a temperature of absolute zero lie near the top of the otherwise forbidden region. Germanium having a small amount of antimony as an impurity is an example of an *n*-type semiconductor. When the temperature is elevated to room temperature, the thermal excitation of electrons from these levels into the unfilled band above is almost complete and can cause appreciable conductivity through drift of electrons alone. Finally, in a *p-type extrinsic semiconductor*, also called an *acceptor* and a *defect* type, impurity or imperfection levels that are unoccupied at a temperature of absolute zero lie near the bottom of the otherwise forbidden region. Germanium having a small amount of gallium or aluminum as an impurity is an example of a *p*-type semiconductor. At room temperature electrons are thermally excited from the filled band into almost all of these unoccupied levels and leave holes that can drift in an electric field and constitute a current. *P*-type and *n*-type semiconductors may be distinguished experimentally through the fact that the sign of the Hall[3] effect for one is opposite to that for the other. Generally, an impurity semiconductor contains some of both types of impurity, but one or the other predominates, and designation of the type is made accordingly. Since thermal excitation of electrons is an essential factor in the conductivity of semiconductors, the electrical characteristics of most practical semiconductor devices are markedly dependent on temperature.

2. SEMICONDUCTOR RECTIFIERS

The special properties of a junction between a metal and a semiconductor are employed in a large class of rectifying devices. One type

---

[3] Olof Lindberg, "Hall Effect," *I.R.E. Proc.*, *40* (1952), 1414–1419.

of semiconductor rectifier is the crystal diode,[4] a cross section of which is shown in Fig. 2b. It consists of a semiconductor crystal, such as germanium, silicon, or galena (lead sulphide), and two electrodes. One electrode is a wire point that makes contact with the crystal over a small area; the other is a metal surface that makes contact with the crystal over a large area. The wire point and crystal are usually encased in a wax-filled ceramic cartridge to hold them in contact.

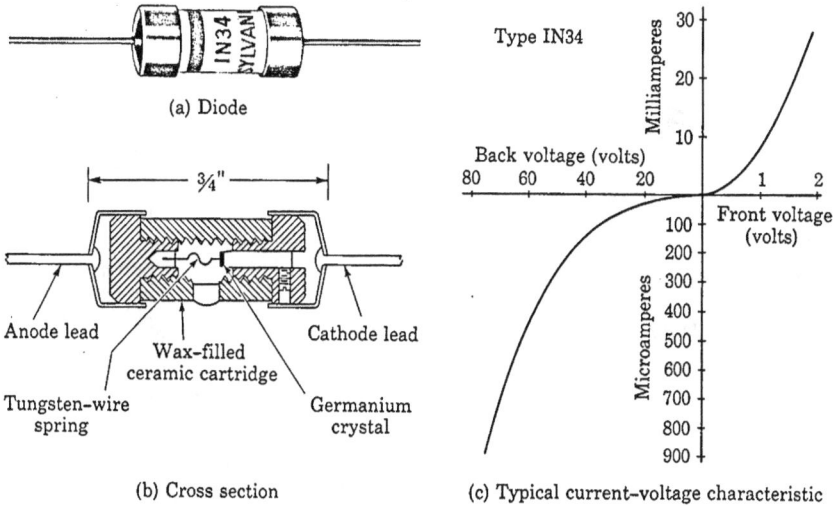

(a) Diode

(b) Cross section

(c) Typical current–voltage characteristic

Type IN34

Fig. 2. Germanium crystal diode. Rated maximum average anode current = 40 ma. Rated maximum peak inverse voltage = 60 volts. (*Adapted by courtesy of Sylvania Electric Products, Inc.*)

When the wire is positive in a typical diode, conduction of a particular value of current takes place at a lower voltage drop than when the semiconductor crystal is positive, and the over-all relation of current and voltage is correspondingly nonlinear, as is illustrated in Fig. 2c. In contrast to the vacuum diode, the point-contact crystal diode conducts an excessive reverse current upon application of an inverse voltage of a few tens of volts. This inability to withstand inverse voltage, the limitation on its forward-current rating imposed by the smallness of the contact area of the wire point, and the marked temperature dependence of its current-voltage characteristic are the chief disadvantages of the point-contact crystal diode as a power rectifier. Its major advantages over the vacuum diode for other uses

[4] E. C. Cornelius, "Germanium Crystal Diodes," *Electronics, 19* (February, 1946), 118–123; H. C. Torrey and C. A. Whitmer, *Crystal Rectifiers*, Massachusetts Institute of Technology Radiation Laboratory Series, Vol. 15 (New York: McGraw-Hill Book Company, Inc., 1948).

are its greater compactness, lack of need for an external source of cathode heating power, lower effective capacitance (a value of 1 micromicro-farad is typical), and much smaller transit time, all of which are particularly important features for high-frequency applications. Point-contact silicon diodes are effective rectifiers at frequencies as high as 25,000 megacycles per second.

A second type of semiconductor rectifier is the so-called *barrier-layer*, or *metallic, rectifier*. The selenium rectifier shown in Fig. 3 is an

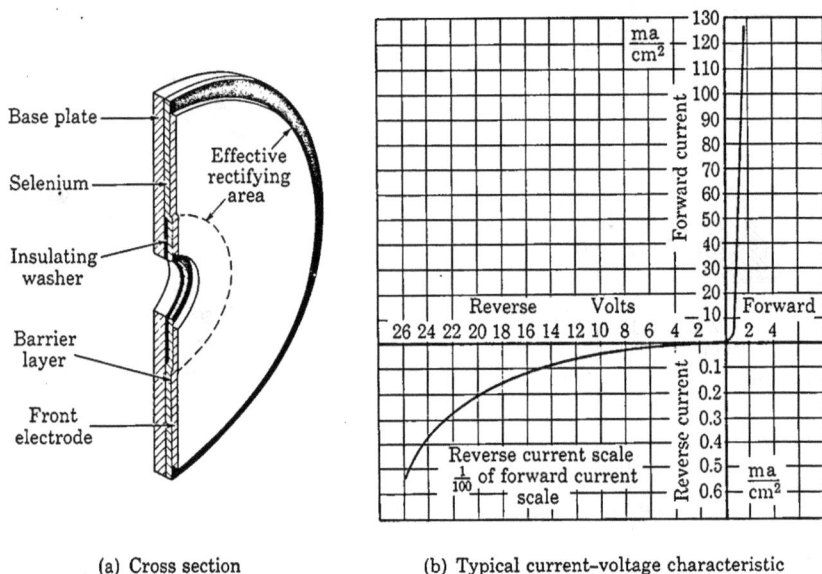

(a) Cross section   (b) Typical current–voltage characteristic

Fig. 3.   Selenium rectifier. (*Adapted by courtesy of Federal Telephone and Radio Corporation.*)

important example.[5] In it rectification takes place at the barrier layer between a front plate of sprayed metal and a thin selenium layer covering a back plate of steel. Conduction of a particular value of current occurs at a lower voltage drop when the semiconductor, selenium, is positive with respect to the front plate than for the opposite polarity. Thus the semiconductor is to be regarded as the anode and the front plate as the cathode. The direction of the current for the lower voltage drop in this barrier-layer rectifier is therefore

[5] E. A. Richards, "The Characteristics and Applications of the Selenium Rectifier," *I.E.E.J.* (*Part II*), *88* (1941), 423–442; C. A. Clarke, "Selenium Rectifier Characteristics, Applications, and Design Factors," *Electrical Communication*, No. 1, *20* (1941), 47–66; J. E. Yarnack, "Selenium Rectifiers for Closely Regulated Voltages," *Electronics*, *14* (September, 1941), 46–49.

opposite to that in the crystal rectifier of Fig. 2. Copper oxide,[6] copper sulphide, germanium,[7] and silicon[8] are other semiconductors commonly used. Current-voltage characteristics for typical germanium power rectifiers are shown in Fig. 4. Such diodes will withstand 100 to 400 volts in the inverse direction and conduct 500 amperes per square centimeter in the forward direction with a voltage drop of only 1 volt. Semiconductor rectifiers have a negative temperature coefficient of resistivity in both the forward and the inverse directions; hence the

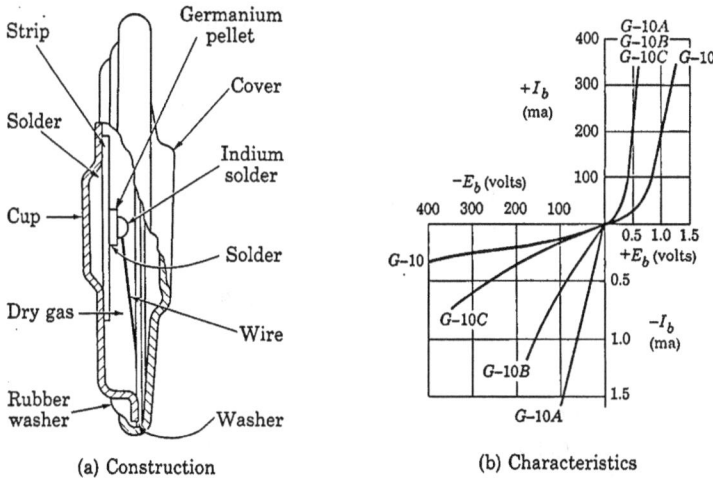

(a) Construction   (b) Characteristics

Fig. 4.   Germanium power rectifier and current-voltage characteristics.*

range of ambient temperature to which they are suited is limited. Since excessive current in either direction tends to overheat the rectifier through internal power loss, the alternating voltage and current that one disc will rectify are limited. For rectifying higher voltages, the

[6] L. O. Grondahl and P. H. Geiger, "A New Electronic Rectifier," *A.I.E.E.J.*, *46* (1927), 215–222; A. L. Williams and L. E. Thompson, "Metal Rectifiers," *I.E.E.J.* (*Part I*), *88* (1941), 353–383.

[7] R. N. Hall and W. C. Dunlap, "P-N Junctions Prepared by Impurity Diffusion," *Phys. Rev.*, *80* (1950), 467–468; F. J. Lingel, "Germanium Power-Rectifier Construction," *Electronics*, *25* (June, 1952), 210–212; L. D. Armstrong, "P-N Junctions by Impurity Introduction through an Intermediate Metal Layer," *I.R.E. Proc.*, *40* (1952), 1341–1342; R. N. Hall, "Power Rectifiers and Transistors," *I.R.E. Proc.*, *40* (1952), 1512–1518; C. L. Rouault and G. N. Hall, "A High-Voltage Medium-Power Rectifier," *I.R.E. Proc.*, *40* (1952), 1519–1521.

[8] G. L. Pearson and B. Sawyer, "Silicon P-N Junction Alloy Diodes," *I.R.E. Proc.*, *40* (1952), 1348–1351.

* The diagram of the construction is taken from F. J. Lingel, "Germanium Power-Rectifier Construction," *Electronics*, *25* (June, 1952), Fig. 2, p. 212, with permission. The characteristic curves are reproduced with the permission of the General Electric Company.

discs are stacked in series; for higher current, they are connected in parallel. Because of their simplicity and low cost, semiconductor rectifiers are widely used, particularly for heavy currents at low voltage.

3. TYPES OF TRANSISTORS[9]

Transistors fall generally into two classes, the *point-contact type* and the *junction type*. The schematic configuration of a representative point-contact transistor is shown in Fig. 5a. It comprises two closely

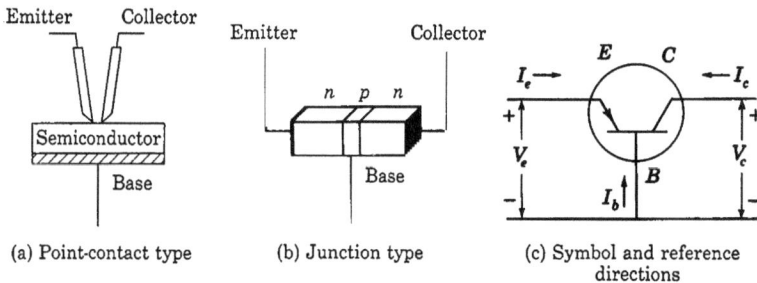

(a) Point-contact type   (b) Junction type   (c) Symbol and reference directions

Fig. 5. Schematic structures and symbols for transistors.

adjacent rectifying point contacts on the same small wafer of semiconductor, such as n-type germanium, and a third common connection to a larger area on the body of the semiconductor. Thus, it resembles two rectifying semiconductor diodes utilizing the same germanium crystal. Pointed wires as small as 0.002 inch in diameter are used as contacts and are spaced as close as 0.002 inch apart. The surface of the crystal is etched and oxidized, and an electrical forming treatment is given one of the contacts to improve the characteristics. The structure is ordinarily imbedded in wax or plastic to provide mechanical stability, and is sometimes enclosed in a metal cylinder connected to the semiconductor. One contact is designated the *emitter*, and the other the *collector*, for reasons that become apparent subsequently, and the germanium wafer is termed the *base*.

[9] The chief sources from which the facts on transistors in the remainder of this chapter are drawn are: J. A. Becker and J. N. Shive, "The Transistor—A New Semiconductor Amplifier," *E.E.*, *68* (1949), 215–221; R. M. Ryder and R. J. Kircher, "Some Circuit Aspects of the Transistor," *B.S.T.J.*, *28* (1949), 367–401; W. Shockley, *Electrons and Holes in Semiconductors* (New York: D. Van Nostrand Company, Inc., 1950); R. L. Wallace, Jr. and W. J. Pietenpol, "Some Circuit Properties and Applications of *n-p-n* Transistors," *I.R.E. Proc.*, *39* (1951), 753–767; W. Shockley, "Transistor Electronics: Imperfections, Unipolar and Analog Transistors," *I.R.E. Proc.*, *40* (1952), 1289–1313. An additional general reference is R. F. Shea, Editor, *Principles of Transistor Circuits* (New York: John Wiley & Sons, Inc., 1953). Other references are cited as used.

The structure of a typical junction-type transistor is shown schematically in Fig. 5b.[10] It consists of two regions of one conductivity type of semiconductor, separated by a region of the opposite type. The transistor shown is an *n-p-n* junction type; a *p-n-p* junction type is also possible. The center region of *p*-type germanium in Fig. 5b is the base, and may be less than 0.001 inch thick; one of the *n*-type regions is the emitter and the other the collector.

Amplification by a transistor depends upon the phenomenon of controlled injection of so-called minority carriers at the emitter contact or junction. To provide this injection, the emitter of all transistors is ordinarily biased in the forward-current direction, as on the characteristic in Fig. 2c, and the collector is ordinarily biased in the reverse-current direction so that a few milliamperes flow at each electrode. Thus the emitter-to-base resistance is small compared with the collector-to-base resistance. The emitter of a point-contact transistor containing an *n*-type semiconductor, for example, is biased positively for transistor action, and the current at the emitter contact is carried primarily by holes, instead of by electrons, which are the majority carriers in an *n*-type semiconductor. In other words, the current in the base material immediately adjacent to the emitter contact results primarily from the motion of holes, which are minority carriers in this instance. The negative collector attracts the positive holes thus injected by the emitter. Most of the holes travel to this near-by collector, and the corresponding current returns to the emitter through a load connected in series with this electrode, rather than through the more remote base connection. Thus, a large fraction of the emitter current flows through the collector circuit, and a change in the emitter current caused by a signal results in a change in the collector current. In fact, for reasons not fully understood, the change in collector current can be larger than the change in emitter current causing it, and some current amplification from the emitter circuit to the collector circuit can occur. Even when the change in collector current is smaller, however, considerable voltage and power amplification can take place because of the difference in internal impedances between the emitter and base and the collector and base. Since the emitter-to-base resistance is comparable to the low forward resistance of a crystal diode, but the collector-to-base resistance is comparable to the much larger reverse resistance, the collector and base terminal pair may be matched to a load resistance that is much larger than the emitter-to-base

[10] W. Shockley, M. Sparks, and G. K. Teal, "*p-n* Junction Transistors," *Phys. Rev.*, *83* (1951), 151–162; R. R. Law, C. W. Mueller, J. I. Pankove, and L. D. Armstrong, "A Developmental Germanium P-N-P Junction Transistor," *I.R.E. Proc.*, *40* (1952), 1352–1357.

resistance, and the power and voltage drop produced in this load by a particular increment of current is much larger than the power and voltage drop produced by approximately the same increment of current at the emitter and base terminal pair.

The behavior of the junction-type transistor is perhaps best understood if the emitter and collector voltages are first considered to be applied separately. For this discussion the transistor is assumed to be an *n-p-n* junction type, and the transistor materials are assumed to be such that electrons predominate as current carriers in the *n*-type emitter and collector regions and that the density of the holes present there is very much smaller than the density of the electrons. Also, it is assumed that holes predominate as carriers in the *p*-type base region and that the density of electrons available for conduction in this region is much smaller. If a positive voltage is applied to the collector while the emitter is disconnected, electrons that are prevalent as current carriers in the collector material are held there and cannot reach the base region, just as electrons do not leave a thermionic cathode in an electron tube when it is held at a positive voltage with respect to the adjacent electrodes. Likewise, holes, which are the majority carriers in the base region, cannot reach the positive collector. Hence, only a small current, which is caused by the thermal production of electron-hole pairs, as in an intrinsic semiconductor, flows in the collector circuit. In other words, a positive collector voltage biases the collector of an *n-p-n* type transistor in the reverse-current direction, as is required for amplification.

Similarly, if a positive voltage is applied to the emitter while the collector is disconnected, little emitter current results. But as the positive emitter voltage is reduced, some of the electrons in the emitter can travel against the retarding field and reach the base, whereupon the current corresponding to them returns through the base connection. At the same time, holes in the base region can flow into the emitter. In an actual *n-p-n* transistor, however, the density of the hole carriers in the base region is made much smaller than the density of the electron carriers in the emitter region—the conductivity of the emitter region is much larger than that of the base region. Consequently, the rate at which holes flow into the emitter is small, and the emitter current is essentially a current of electrons into the base. This current increases as the emitter bias is changed from positive to negative values. Thus a negative emitter voltage biases the emitter of an *n-p-n* type transistor in the forward-current direction, as is required for amplification. If the transistor were of the *p-n-p* type, a positive emitter-bias voltage and a negative collector-bias voltage would be required for amplification.

When the collector of the *n-p-n* type transistor is biased positively at the same time that the emitter is biased negatively, the electrons injected by the emitter, which are again minority carriers in the base, tend to diffuse through the thin base region without recombining and to pass into the closely adjacent collector and its external circuit, rather than to return through the remote base connection. Hence, the collector current almost equals the emitter current, and the base current is comparatively small. A relatively large current at a high impedance level in the collector circuit is thus controlled by an almost equal current in the low-impedance emitter circuit or by a much smaller current in the base circuit, and power and voltage gain result. The emitter may be likened to the cathode of an electron tube, the collector to the plate, and the base to the grid, for the base current is small, and the base-to-emitter voltage essentially controls the collector current.

The physical phenomena at the metal points of a point-contact transistor have many characteristics in common with those at the *p-n* junctions in a junction-type transistor, but additional factors enter into their explanation. In particular, the conditions conducive to effective electron or hole injection depend upon the surface energy states in which electrons can be tightly bound at the surface of the semiconductor,[11] and to some extent upon the work functions of the materials. Though the reasons for the existence of a current amplification greater than unity are not entirely clear, several theories have been advanced.[12] A forming process, which consists of the application of momentary excessive current at the collector and which increases the current amplification, is believed to result in formation of a *p-n* junction under the contact through localized heating.[13]

The point-contact and junction transistor triodes previously described are the basic types of transistors. Other types that have special properties have been developed or proposed. One such type is a *p-n-p-n* junction transistor tetrode, which is especially suited for

[11] J. Bardeen, "Surface States and Rectification at a Metal Semiconductor Contact," *Phys. Rev.*, *71* (1947), 717–727; W. H. Brattain and J. Bardeen, "Nature of the Forward Current in Germanium Point Contacts," *Phys. Rev.*, *74* (1948), 231–232; J. Bardeen and W. H. Brattain, "Physical Principles Involved in Transistor Action," *Phys. Rev.*, *75* (1949), 1208–1225.

[12] W. Shockley, "Theories of High Values of Alpha for Collector Contacts on Germanium," *Phys. Rev.*, *78* (1950), 294; W. R. Sittner, "Current Multiplication in a Type-A Transistor," *I.R.E. Proc.*, *40* (1952), 448–454; W. Shockley, "Transistor Electronics: Imperfections, Unipolar and Analog Transistors," *I.R.E. Proc.*, *40* (1952), 1309–1310.

[13] L. B. Valdes, "Transistor Forming Effects in *n*-Type Germanium," *I.R.E. Proc.*, *40* (1952), 445–448.

connection as a so-called hook-collector transistor.[14] Another special type is an *n-p-n* junction transistor tetrode involving two connections to the *p*-type base material.[15] These connections are located on opposite sides of the base bar, but are otherwise identical. Application of a transverse bias voltage of about 6 volts between the two base connections to provide a transverse field in the base gives high-frequency performance superior to that of triode transistors. Still another special type of transistor, for which a theory has been developed, is called a unipolar field-effect transistor.[16] This type is closely analogous to the vacuum tube in that the amplifying action involves currents carried predominantly by one type of carrier.

An actual emitter electrode is not essential for transistor action. Light or radiant energy impinging on a semiconductor diode in the vicinity of a point-contact or a *p-n* junction can cause a change in the conductivity of the diode.[17] Practical photosensitive devices are based on this principle.

4. CHARACTERISTIC CURVES OF TRANSISTORS

The conventional symbol for either basic type of transistor is shown in Fig. 5c; *E* is the emitter, *C* the collector, and *B* the base. The arrow at the electrode *E* is drawn in the direction of forward current at the emitter junction. As shown, the arrow designates a transistor utilizing an *n*-type semiconductor as a base. For a *p*-type base, the arrow is reversed. Symbols for the currents at the electrodes and the voltages between them, together with the conventional assigned positive reference directions, are also shown in Fig. 5c. Voltages for the transistor are conventionally referred to the base. Contrary to the standards for electron-tube circuits, however, capital letters are conventional to denote total instantaneous quantities, and increments from the quiescent values are denoted by the corresponding lower-case letters. The subscript *c* corresponds to the collector, *e* to the emitter, and *b* to the base. The symbol *v* is used for voltage to avoid confusion with the subscript *e*.

Relations among the two voltages and the two currents for the transistor may be shown by static characteristics, just as for the

[14] J. J. Ebers, "Four-Terminal P-N-P-N Transistors," *I.R.E. Proc.*, *40* (1952), 1361–1364.

[15] R. L. Wallace, Jr., L. G. Schimpf, and E. Dickten, "A Junction Transistor Tetrode for High-Frequency Use," *I.R.E. Proc.*, *40* (1952), 1395–1400.

[16] W. Shockley, "A Unipolar 'Field-Effect' Transistor," *I.R.E. Proc.*, *40* (1952), 1365–1376.

[17] J. N. Shive, "The Phototransistor," *Bell Lab. Rec.*, *28* (1950), 337–342; J. N. Shive, "Properties of the M-1740 P-N Junction Photocell," *I.R.E. Proc.*, *40* (1952), 1410–1413.

electron tube. A family of static characteristics may be constructed if any one current or voltage is chosen as being independent, a second as dependent, and a third as a parameter. Among the many such families possible, those of primary interest for circuit analysis are shown in Fig. 6. The currents I_e and I_c, rather than voltages, are chosen as the independent variables here, because as is discussed later, the curves sometimes exhibit more than one value of current for the same value of voltage, but always only one value of voltage for a particular value of current. That is, the voltage is always a single-valued function of the current, but the current may be a multi-valued function of the voltage. This multi-valued relationship may result in instability if the device is supplied from a constant-voltage source or one with a low internal resistance. The emitter and collector families in Fig. 6 relate the voltage on one side of the transistor to the current on the same side. The forward and feedback, or backward, families, on the other hand, relate a voltage on one side to the current on the other. Any two of these four families completely describe the behavior of the transistor, for the other two may be constructed from them. Furthermore, any pair of the two voltages and two currents determines the other pair because the transistor imposes a functional relationship among the four.

The curves in Fig. 6 are for an *n*-type point-contact transistor. Collector and forward transfer families for an *n-p-n* junction-type transistor are shown in Fig. 7.

5. GRAPHICAL DETERMINATION OF OPERATING POINTS

When the transistor is utilized as an amplifier, a source of input-signal voltage is connected to one pair of its terminals and a load to another pair. Six such connections are possible for this three-electrode device, just as for the vacuum triode discussed in Art. 10, Ch. VIII. One of these connections, which is called the *grounded-base connection* because the base terminal is common to the input and output circuits, is shown in Fig. 8a. A voltage V_{ee} to bias the emitter in the forward direction is connected in series with the source of input signal voltage, which is assumed to have an open-circuit voltage v_s and internal resistance R_s. The internal resistance of this source must be taken into account in a transistor circuit, because, unlike the negative-grid electron tube, the transistor cannot be considered to have infinite impedance. Conduction occurs in both the input and output circuits simultaneously, and conditions in one circuit depend on those in the other. The problem of determining the instantaneous operating

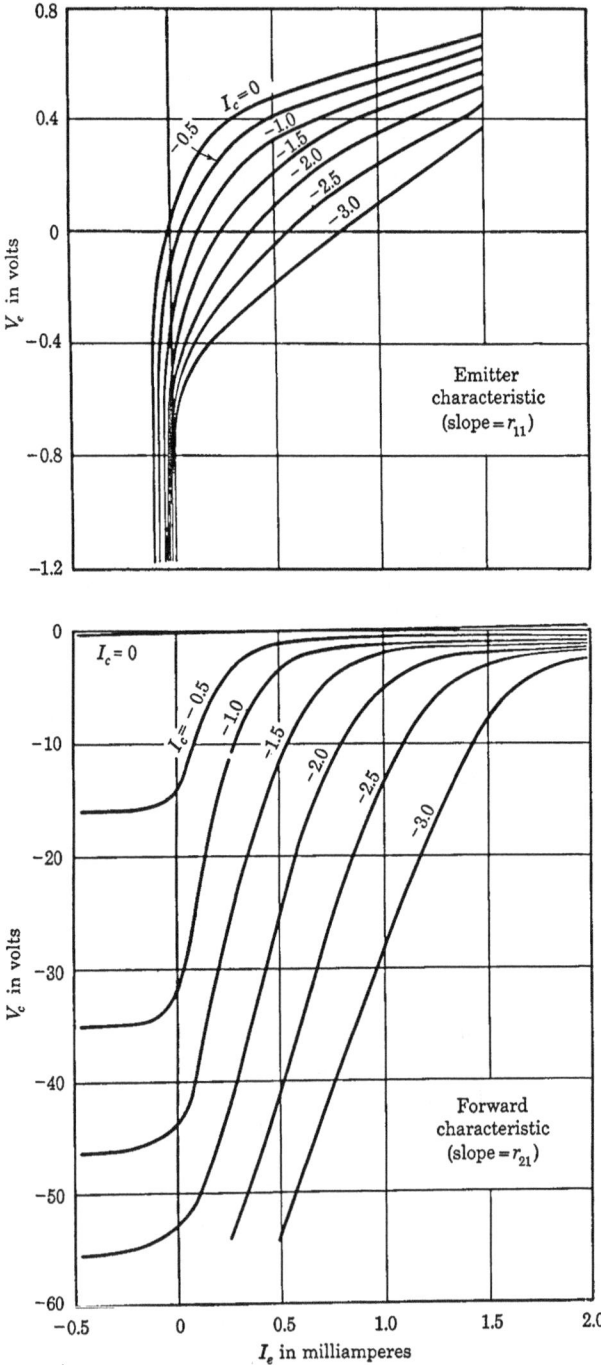

Fig. 6.   Static characteristics for a point-contact transistor.*

* This figure is adapted from R. M. Ryder and R. J. Kircher, "Some Circuit Aspects of the Transistor," *B.S.T.J.*, *28* (1949), Fig. 23, p. 390, with permission.

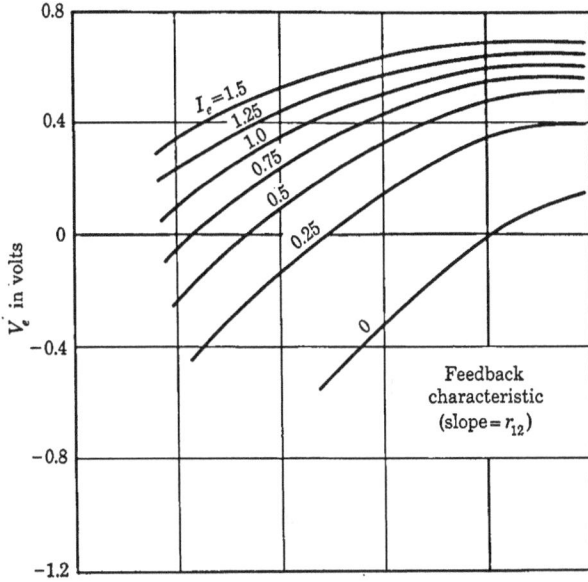

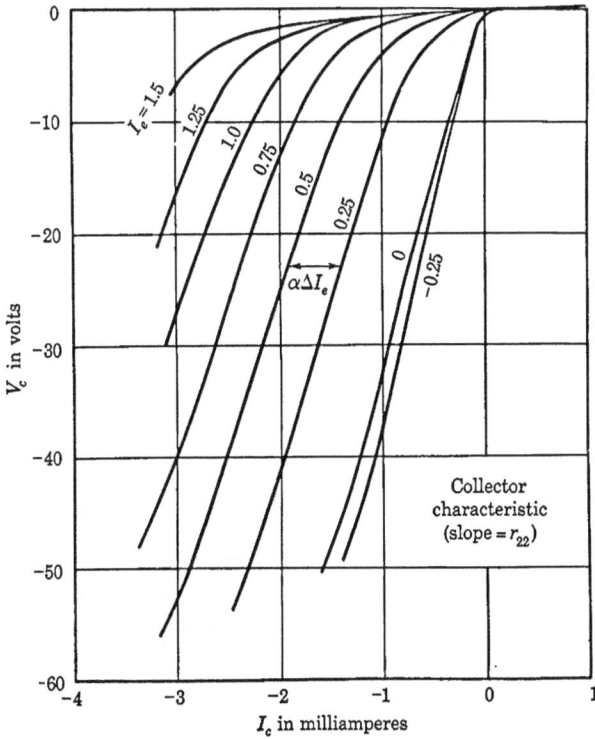

Fig. 6. (*continued*)

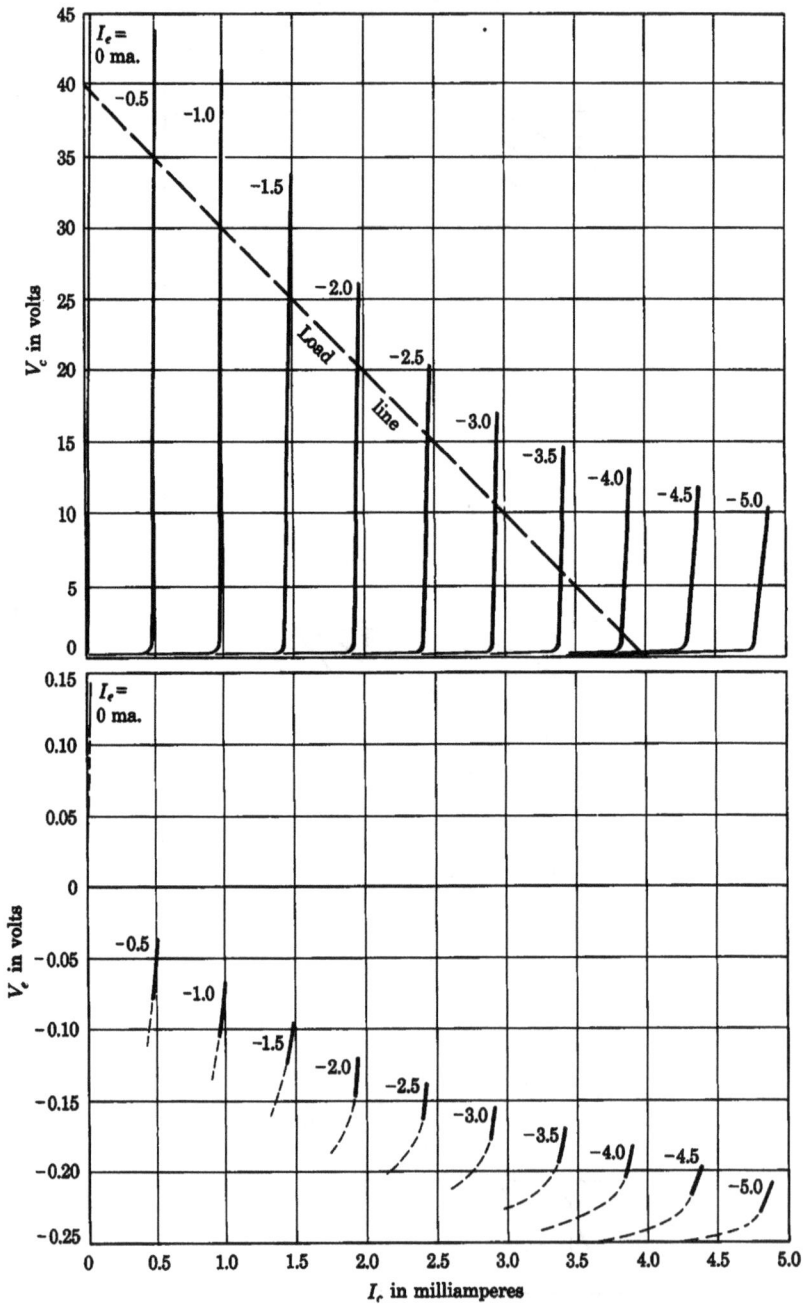

Fig. 7. Static characteristics of an *n-p-n* junction-type transistor.*

* This figure is taken from R. L. Wallace, Jr., and W. J. Pietenpol, "Some Circuit Properties and Applications of *n-p-n* Transistors," *B.S.T.J.*, *30* (1951), Fig. 4, p. 533, with permission.

conditions is exactly similar to that of an electron tube operating with a positive grid so that the grid current is appreciable.[18]

The instantaneous operating point for the amplifier of Fig. 8a may be found by the graphical method shown. Since in the circuit

$$V_c = V_{cc} - I_c R_L,$$ [1]

the collector current and voltage are constrained by the collector bias voltage V_{cc} and load R_L to lie on a straight load line on the collector characteristics, Fig. 8b, just as the operating point for an electron tube

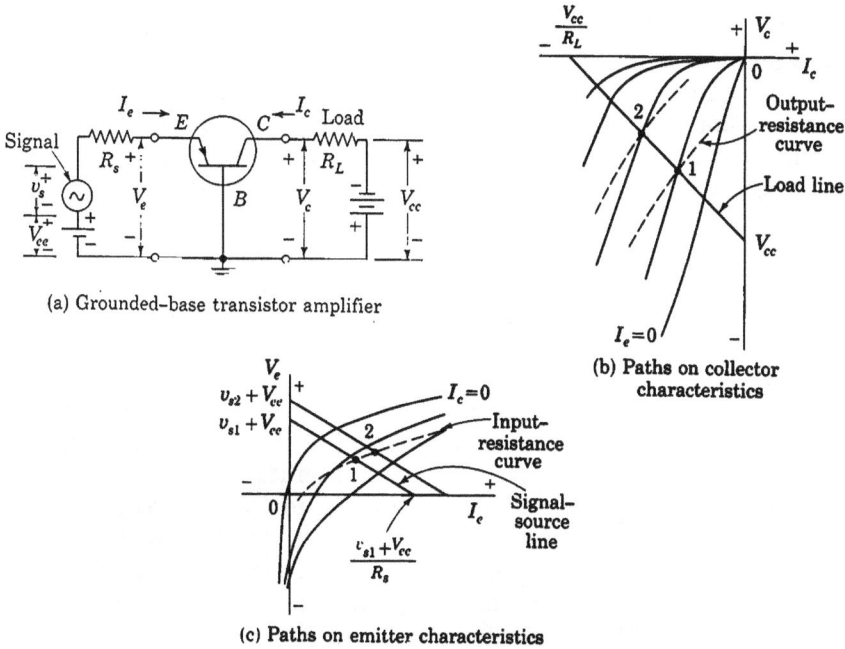

(a) Grounded-base transistor amplifier

(b) Paths on collector characteristics

(c) Paths on emitter characteristics

Fig. 8. Graphical construction for determination of operating point in a grounded-base transistor amplifier circuit.

lies on a straight load line on the plate characteristics. Intercepts of this straight resistance line are V_{cc} on the voltage axis, at a numerically negative voltage in accordance with the reference direction of V_{cc} and polarity of the battery in Fig. 8a, and V_{cc}/R_L on the current axis. The transistor behavior differs from that of an electron tube, however, in that for any particular value of input signal voltage v_{s1}, the emitter current I_e and voltage V_e are also constrained by the total input

[18] E. L. Chaffee, *Theory of Thermionic Vacuum Tubes* (New York: McGraw-Hill Book Company, Inc., 1933), 210–213.

voltage $v_{s1} + V_{ee}$ and internal source resistance R_s to lie on a straight line on the emitter characteristics in accordance with

$$V_e = v_{s1} + V_{ee} - I_e R_s.$$ [2]

This line is designated the *signal-source* line in Fig. 8c.

The two straight-line constraints may be combined on either family of characteristic curves to find the operating point that satisfies both. Thus, each pair of values of I_e and I_c along the load line on the collector family locates a point on a curve across the emitter family called the *input-resistance curve*. The intersection of this curve with the signal-source line at point 1 evidently satisfies simultaneously the voltage and current requirements set by the input circuit, output circuit, and transistor, and is thus the instantaneous operating point corresponding to v_{s1}. Similarly, each pair of current values along the signal-source line on the emitter family locates a point on a curve across the collector family called the *output-resistance curve*. The intersection of this curve with the load line at point 1 is the instantaneous operating point on the collector characteristics corresponding to v_{s1}. Construction of the input-resistance curve and output-resistance curve is similar to that for the dynamic transfer characteristic in Fig. 27, Ch. VIII.

When the input signal voltage changes to a new value v_{s2}, the signal-source line shifts to a new position parallel to the old, but the load line does not shift. Hence the input-resistance curve remains unchanged, and operation changes to point 2 on it. Correspondingly, the output-resistance curve shifts to a new position and intersects the load line at the new operating point 2 on the collector characteristics. The corresponding operating points 1 and 2 on the forward and feedback transfer characteristics, not shown in Fig. 8, are then readily located if desired. This graphical procedure thus determines the path of the instantaneous operating point on all four families of characteristic curves for any signal voltage source that varies so slowly that the transistor and circuit behavior is essentially resistive.

Feedback is frequently included in transistor circuits—either negative to stabilize the operation when linear amplification is desired, or positive to accentuate the tendency toward instability when oscillation or switching action is desired. One common method of applying feedback is the addition of a resistance R_b in series with the base, as is shown in Fig. 9. The graphical analysis in Fig. 8 indicates that a positive increment in I_e generally results in a negative increment in I_c. When the external base resistance R_b is present, and the resistance of the input signal source R_s is finite, a portion of the increment in I_c is forced to pass through the emitter and signal source, and this portion is in a direction such as to augment the original increment in I_e. Hence, the feedback caused by R_b is positive. Under the conditions described in Art. 10, this feedback may be large enough to cause instability even in the absence of an external base resistance. The internal resistance effectively in series with the base lead also produces positive feedback.

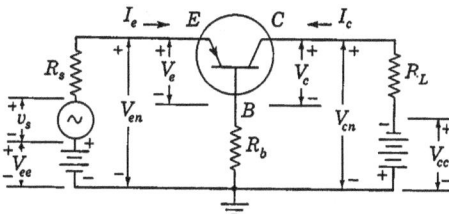

(a) Circuit connections

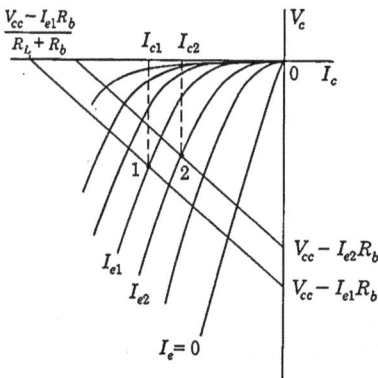

(b) Paths on collector characteristics

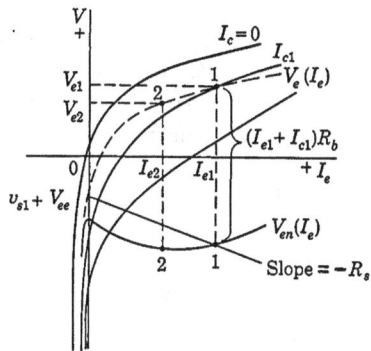

(c) Paths on emitter characteristics

Fig. 9.　Operating conditions for a grounded-base connection with external base resistance R_b.

The operating conditions when the external base resistance is included, and the effect of this resistance on the over-all input-resistance curve for the circuit, V_{en} as a function of I_e, may be found through the graphical construction in Fig. 9. An explicit solution for the output current resulting when an input signal voltage v_s is applied is not practicable, but an indirect solution is obtained if the voltage V_c is first expressed as

$$V_c = V_{cc} - I_c R_L - (I_c + I_e) R_b \qquad [3]$$

$$= (V_{cc} - I_e R_b) - I_c(R_L + R_b). \qquad [4]$$

This equation indicates that for any particular selected value I_{e1}, the voltage $(V_{cc} - I_{e1}R_b)$ is constant, and a straight line drawn through this voltage as an intercept on the voltage axis, with a slope corresponding to $R_L + R_b$, as is shown in Fig. 9b, represents the constraint imposed by the external collector circuit upon the collector current and voltage. The intersection of this line with the collector characteristic curve for the particular value I_{e1}, at I_{c1}, is the operating point, which may be transferred to the emitter characteristic to locate V_{e1} in Fig. 9c.

The emitter-to-ground voltage at the input, V_{en}, is given by

$$V_{en} = V_e + (I_e + I_c)R_b . \qquad [5]$$

Accordingly, algebraic addition of the voltage $(I_{e1} + I_{c1})R_b$, which ordinarily is numerically negative, to V_{e1} gives V_{en1}. This process determines point 1 on the curve of V_{en} as a function of I_e, indicated in Fig. 9c as $V_{en}(I_e)$. Since in the emitter circuit,

$$V_{en} = v_s + V_{ee} - I_e R_s , \qquad [6]$$

a line drawn through point 1 on the curve of $V_{en}(I_e)$ with a slope $-R_s$ intersects the voltage axis at $v_{s1} + V_{ee}$ and thereby determines the signal v_{s1} required to establish the current I_{e1} originally assumed.

Repetition of this procedure for a new value of emitter current establishes additional points on the operating paths so that loci may be drawn. For example, if a smaller value of emitter current I_{e2} is selected, a new load line parallel to the first must be drawn on the collector characteristics. This line intersects the collector characteristic for I_{e2} at point 2, which, when transferred to the emitter characteristics, gives V_{e2} at point 2, and algebraic addition of the corresponding voltage drop in R_b gives point 2 on the curve of $V_{en}(I_e)$.

Certain of the features of the graphical construction are exaggerated in Fig. 9 for clarity. For example, the load lines on the collector characteristics frequently lie much closer together than is illustrated. The shape of the over-all input-resistance curve $V_{en}(I_e)$ shown in Fig. 9c is typical for large values of R_b. Near the voltage axis in the lower right-hand quadrant it has a downward slope corresponding to a negative value of input resistance, a necessary feature for the switching and oscillation operations described in Art. 10.

The second basic connection for a transistor, called a grounded-emitter connection, is shown in Fig. 10. In this circuit, the source of signal voltage v_s having an internal resistance R_s is introduced in series with the base, a load resistance R_L is connected in series with the collector, and bias voltages V_{bb} and V_{cc} are included to bias the emitter positively and the collector negatively with respect to the base. Again, because of the interdependence of conditions in the base and collector circuits established by the transistor, an explicit graphical solution for the output current in terms of the signal voltage is not readily obtained. But the instantaneous values of all the currents and voltages for a particular value of v_s may be determined indirectly if the input volt-ampere characteristic between the base and emitter terminals is first determined and a resistance line corresponding to V_{bb}, v_s, and R_s is

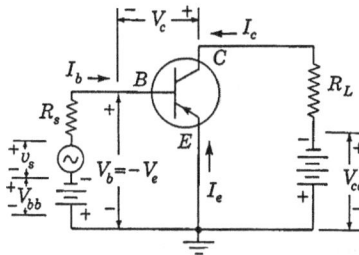

(a) Circuit connections

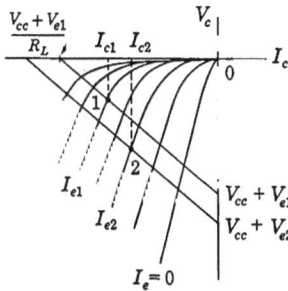

(b) Paths on collector characteristics   (c) Paths on emitter characteristics

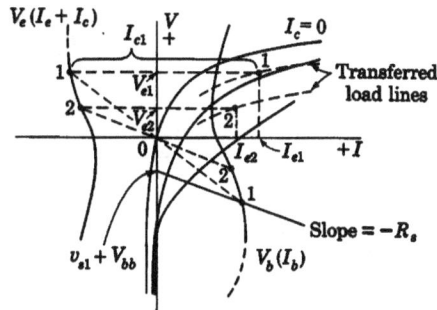

Fig. 10.   Operating conditions for a grounded-emitter connection.

then superposed on it; the intersection of the two determines both the base and collector currents.

The input characteristic needed for this solution is V_b as a function of I_b, or $V_b(I_b)$. But, since for this connection and the positive reference directions indicated,

$$V_b = -V_e,$$ [7]

and

$$I_b = -(I_e + I_c),$$ [8]

the required curve is also $-V_e$ as a **function of** $-(I_e + I_c)$. **Since the** load and the collector-supply voltage constrain the collector voltage and current to satisfy the relation

$$V_c = (V_e + V_{cc}) - I_c R_L,$$ [9]

for any selected value of emitter voltage V_{e1}, a load line on the collector characteristics having a voltage intercept $V_{e1} + V_{cc}$ and a slope $-R_L$, as in Fig. 10b, describes graphically the constraint imposed by the collector circuit external to the transistor. If this load line is transferred to the emitter characteristics as indicated in Fig. 10c, its abscissa at the ordinate V_{e1}, I_{e1} at point 1, is the emitter current corresponding to the selected value V_{e1}. The corresponding collector current I_{c1} is found at the intersection of the collector characteristic for I_{e1} and the load line as indicated on Fig. 10b. Adding this value I_{c1} algebraically to I_{e1} at the ordinate V_{e1} in Fig. 10c then locates point 1 on the curve $V_e(I_e + I_c)$ in the upper left-hand quadrant. Repeating this process for a new value V_{e2}, and so on, permits plotting a curve of $V_e(I_e + I_c)$. In accordance with Eqs. 7 and 8, the input characteristic $V_b(I_b)$ is this curve with the signs of both the voltage and the current reversed, that is, it is this curve rotated through 180 degrees so that it appears in the lower right-hand quadrant. Finally, since in the base circuit

$$V_b = (v_s + V_{bb}) - I_b R_s, \qquad [10]$$

a resistance line having a slope $-R_s$ through any point such as 1 on the curve $V_b(I_b)$ intersects the voltage axis at the voltage $v_{s1} + V_{bb}$ required to cause all the operating conditions designated as 1 on these diagrams.

Again, as in Fig. 9, certain of the graphical shapes and spacings are exaggerated in Fig. 10 for clarity. When the resistance in the collector circuit R_L is small, the input characteristic $V_b(I_b)$ for this connection may have a negative slope over its center portion, indicating a negative resistance, just like the center portion of the input characteristic for the grounded-base connection with external base resistance. This negative resistance may likewise be utilized for switching operation or oscillation, as is discussed in Art. 10.

6. Transistor coefficients and incremental equivalent circuits

Analysis of the relationships among the increments in the currents and voltages of a transistor is conveniently made in terms of an equivalent circuit for increments, which may be established in a manner similar to that used in Ch. IV for an electron tube. In functional notation, the emitter voltage may be written as

$$V_e = f_e(I_e, I_c), \qquad [11]$$

and the collector voltage as

$$V_c = f_c(I_e, I_c). \qquad [12]$$

Thus, if lower-case letters are used to denote incremental variations from a value corresponding to zero signal voltage,

$$v_e = \frac{\partial V_e}{\partial I_e} i_e + \frac{\partial V_e}{\partial I_c} i_c \qquad [13]$$

and

$$v_c = \frac{\partial V_c}{\partial I_e} i_e + \frac{\partial V_c}{\partial I_c} i_c \qquad [14]$$

are linear approximations to the incremental behavior of the transistor, just as Eq. 28, Ch. IV, is a linear approximation to the incremental behavior of an electron tube. In these equations, the partial derivatives have the dimensions of resistance; hence, symbols for them may be defined, and their incremental significance identified, as follows:

$$r_{11} = \frac{\partial V_e}{\partial I_e} = \text{incremental emitter resistance for an} \qquad [15]$$
$$\text{incrementally open-circuited collector,}$$

$$r_{22} = \frac{\partial V_c}{\partial I_c} = \text{incremental collector resistance for an} \qquad [16]$$
$$\text{incrementally open-circuited emitter,}$$

$$r_{21} = \frac{\partial V_c}{\partial I_e} = \text{incremental forward transfer resistance for} \qquad [17]$$
$$\text{an incrementally open-circuited collector,}$$

$$r_{12} = \frac{\partial V_e}{\partial I_c} = \text{incremental backward transfer resistance for} \qquad [18]$$
$$\text{an incrementally open-circuited emitter.}$$

Equations 13 and 14 then become

$$v_e = r_{11}i_e + r_{12}i_c, \qquad [19]$$

and

$$v_c = r_{21}i_e + r_{22}i_c. \qquad [20]$$

In accordance with Eqs. 15 through 18, each of the differential coefficients may be readily measured experimentally if one of the electrodes is supplied through an essentially infinite incremental resistance, such as the plate circuit of a vacuum pentode, so that the increment in one current is made zero. Graphically, each of the differential coefficients will be recognized as the slope at the operating point of a curve in one of the four families in Fig. 6.

A further significant differential coefficient is

$$\alpha = -\frac{\partial I_c}{\partial I_e}\bigg|_{V_c = \text{constant}} = \text{incremental current amplification for an} \qquad [21]$$
$$\text{incrementally short-circuited collector.}$$

It is approximately the ratio of the horizontal spacing between two collector characteristics to the difference between the values of emitter current corresponding to the individual curves, as is shown in Fig. 6. From Eq. 20, its value in terms of the slopes of the curves in the collector and forward families is

$$\alpha = \frac{r_{21}}{r_{22}}. \tag{22}$$

Equations 19 and 20 suggest the circuit in Fig. 11a as an equivalent circuit for the incremental behavior of the transistor. Other equivalent

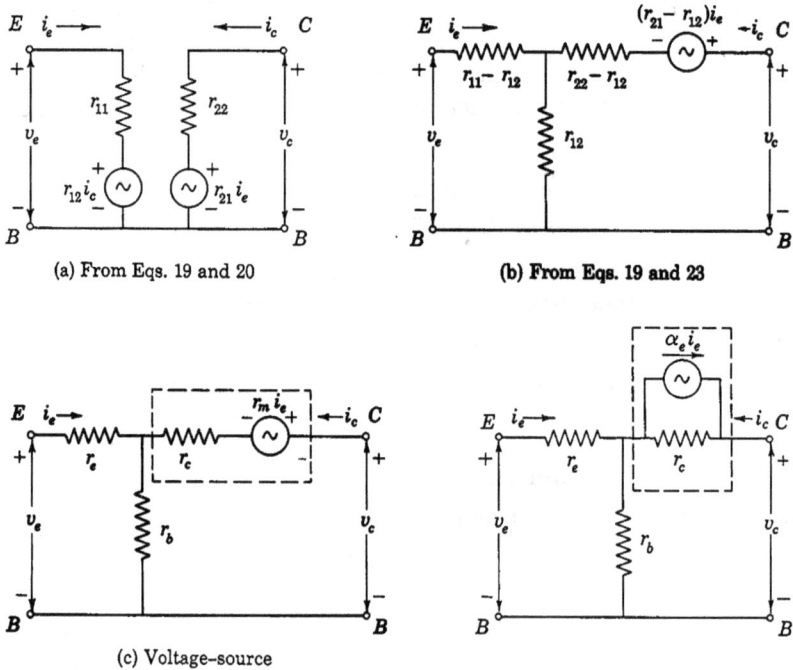

(a) From Eqs. 19 and 20

(b) From Eqs. 19 and 23

(c) Voltage-source

Fig. 11.

circuits also become evident if the circuit equations are rearranged.[19] One rearrangement is based upon the concept that if the transistor were a passive device so that the reciprocity theorem were applicable, r_{12} would equal r_{21}, but inspection of the slopes of the curves in Fig. 6 shows that these resistances are not equal—the transistor is an active element for which the reciprocity theorem is not applicable. Active elements—internal voltage or current sources—are required as

[19] L. C. Peterson, "Equivalent Circuits of Linear Active Four-Terminal Networks," *B.S.T.J.*, *27* (1948), 593–622; L. J. Giacoletto, "Junction Transistor Equivalent Circuits and Vacuum-Tube Analogy," *I.R.E. Proc.*, *40* (1952), 1490–1493.

part of the equivalent circuit, and four independent parameters are needed, not merely three as would be required for a passive circuit. Rearranging Eq. 20 in the form

$$v_c = r_{12}i_e + r_{22}i_c + (r_{21} - r_{12})i_e \qquad [23]$$

emphasizes the significance of the active nature of the transistor. Here the term $r_{12}i_e$ equals the only one involving i_e that would appear if the circuit were passive. The term $(r_{21} - r_{12})i_e$ is evidently associated with the fact that the transistor is an active rather than a passive device. Equations 19 and 23 suggest the equivalent circuit in Fig. 11b, rather than that in Fig. 11a. The tee network of resistances alone would represent the incremental behavior of the transistor if it were a passive element.

When new symbols for the combinations of incremental resistances associated with each terminal in Fig. 11b are adopted in accordance with the relations

Emitter incremental resistance $= r_e = r_{11} - r_{12}$, $\qquad [24]$

Base incremental resistance $= r_b = r_{12}$, $\qquad [25]$

Collector incremental resistance $= r_c = r_{22} - r_{12}$, $\qquad [26]$

Mutual incremental resistance $= r_m = r_{21} - r_{12}$, $\qquad [27]$

the incremental equivalent circuit becomes that shown in Fig. 11c. The resistances defined by Eqs. 24 through 27 may be determined through substitution of measured values of slopes of the curves in Fig. 6, or of the open-circuit resistances measured as described in Eqs. 15 through 18. Solving Eqs. 24 through 27 simultaneously gives for the resistances appearing in Eqs. 19 and 20

$$r_{11} = r_e + r_b; \qquad [28]$$

$$r_{22} = r_c + r_b; \qquad [29]$$

$$r_{12} = r_b; \qquad [30]$$

$$r_{21} = r_b + r_m. \qquad [31]$$

A fourth incremental equivalent circuit shown in Fig. 11d follows from that in Fig. 11c if the voltage source $r_m i_e$ and series resistance r_c are replaced by an equivalent current source $\alpha_e i_e$ and shunt resistance r_c. The value of the current ratio, α_e, is

$$\alpha_e = \frac{r_m}{r_c}. \qquad [32]$$

The role of r_m in Fig. 11c is analogous in some respects to that of g_m for an electron tube, and the role of the dimensionless ratio α_e is analogous to that of μ, because current in the transistor is chosen as the independent or controlling variable, whereas voltage is chosen as the independent or controlling variable for the electron tube. As a result of this choice, and of the inherent characteristics of the transistor, the principle of duality has been used as a guide in transistor-circuit design.[20]

TABLE I

TYPICAL OPERATING CONDITIONS AND COEFFICIENTS FOR TRANSISTORS*

	Point-Contact Type	Junction Type	
I_e	+0.6	−1.0	ma
I_c	−2.0	+1.0	ma
V_e	+0.7	−0.1	v
V_c	−40	+4.5	v
r_e	240	25.9	ohms
r_c	19,000	13.4×10^6	ohms
r_b	290	240	ohms
r_m	34,000	13.1×10^6	ohms
$r_c - r_m$	−15,000	$+0.288 \times 10^6$	ohms
α_e	1.79	0.9785	
r_{11}	530	266	ohms
r_{12}	290	240	ohms
r_{21}	34,300	13.1×10^6	ohms
r_{22}	19,200	13.4×10^6	ohms
α	1.78	0.9785	

A full understanding of the distinctive nature of the transistor is obtained only through appreciation of the magnitude of the resistances and current ratios effective in the equivalent circuits. Table I gives representative values for both the point-contact and the junction-type transistor. Three noteworthy facts are immediately apparent: First, the series resistance r_e at the emitter terminal is only a few tens or hundreds of ohms, whereas that at the collector terminal, r_c, is much larger—tens of thousands or millions of ohms. Second, the current ratio α_e is larger than unity for the point-contact type, but is

[20] R. L. Wallace, Jr., and G. Raisbeck, "Duality as a Guide in Transistor Circuit Design," *B.S.T.J.*, *30* (1951), 381–418.

* Most of the values in this table are taken from R. M. Ryder and R. J. Kircher, "Some Circuit Aspects of the Transistor," *B.S.T.J.*, *28* (1949), 373, and from R. L. Wallace, Jr., and W. J. Pietenpol, "Some Circuit Properties and Applications of *n-p-n* Transistors," *I.R.E. Proc.*, *39* (1951), 757, with permission.

slightly smaller than unity for the junction type. Third, the short-circuit current amplification, α, is almost equal to the current ratio α_e because, from Eqs. 22, 29, and 31,

$$\alpha = \frac{r_m + r_b}{r_c + r_b},$$ [33]

and r_b is small compared with both r_m and r_c. The ratio in Eq. 33 is thus approximately the same as that in Eq. 32. The significance of

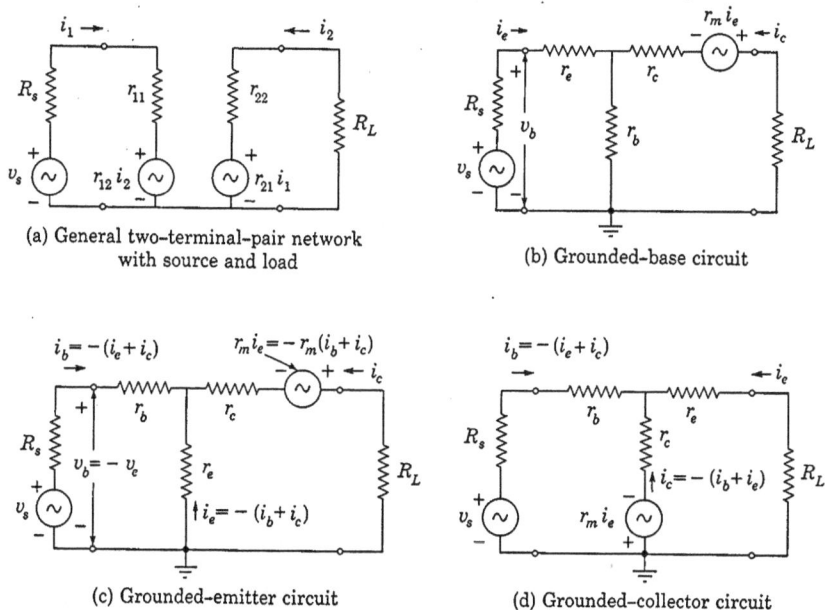

(a) General two–terminal–pair network
with source and load

(b) Grounded-base circuit

(c) Grounded-emitter circuit

(d) Grounded–collector circuit

Fig. 12.   Incremental equivalent circuits for three basic amplifier connections.

these magnitudes becomes apparent from analysis of the incremental operation of the transistor with the source and load connected, which follows in the next article.

7. SINGLE-STAGE TRANSISTOR AMPLIFIERS

When a transistor is connected as an amplifier with a signal source and a load, the incremental operation of the amplifier may be conveniently analyzed in terms of the equivalent circuit shown in Fig. 12a, regardless of the way the transistor is connected, because such an equivalent circuit can represent any linear two–terminal-pair network. For each of the three basic connections of the transistor mentioned in Art. 5, the incremental equivalent circuit may also be represented by one of the other three diagrams in Fig. 12. Thus, the currents i_1 and i_2 and the incremental resistances in Fig. 12a take on a new meaning for each different connection, but analysis of Fig. 12a serves for all. The resistances in Fig. 12a have a more general significance than those in Fig. 11a because the currents and voltages are not specified in terms of the electrode currents and voltages of the transistor. But once the transistor connection is stated, the self and transfer resistances in Fig. 12a are readily evaluated in terms of the resistances r_e, r_b, r_c, and r_m. For example, if the transistor is connected as a grounded-base amplifier, the incremental equivalent circuit in Fig. 12a must correspond with that in Fig. 12b, and Eqs. 28 through 31 then apply. If the transistor is connected as a grounded-emitter or a grounded-collector amplifier, however, Eqs. 28 through 31 are not applicable, and different relations among the resistances as developed subsequently must be used.

In terms of Fig. 12a, the simultaneous equations relating the incremental currents and voltages at the source and the load are

$$v_s = i_1(R_s + r_{11}) + i_2 r_{12} , \qquad [34]$$

$$0 = i_1 r_{21} \qquad\quad + i_2(r_{22} + R_L). \qquad [35]$$

From Eq. 35, the current amplification is

$$\text{Current amplification} = \frac{-i_2}{i_1} = \frac{r_{21}}{r_{22} + R_L} ; \qquad [36]$$

and the voltage amplification is, from Eqs. 34 and 35,

$$\begin{matrix}\text{Voltage}\\ \text{amplification}\end{matrix} = A = \frac{-i_2 R_L}{v_s} = \frac{r_{21} R_L}{(R_s + r_{11})(r_{22} + R_L) - r_{12} r_{21}}. \qquad [37]$$

As the load resistance approaches **infinity, this voltage amplification** approaches the open-circuit value

$$A_{oc} = \frac{r_{21}}{R_s + r_{11}}. \qquad [38]$$

The denominator of Eq. 37, which is the circuit determinant, may vanish for certain combinations of the transistor, source, and load resistances, because of the negative sign. But, as is mentioned in Ch. XI, such a condition corresponds to infinite voltage amplification so that an output can occur for zero input—the amplifier is unstable, and may become an oscillator. Thus, if the feedback resistance r_{12} is increased from zero, instability begins when the denominator in Eq. 37 decreases to zero. Further increase of r_{12} results in a stronger tendency toward instability and oscillation—if oscillations occur they build up until nonlinearity limits the amplitude, as is discussed in Art. 4, Ch. XI. Consequently, for freedom from such instability, the resistances in the circuit must satisfy the relation

$$r_{12}r_{21} < (R_s + r_{11})(r_{22} + R_L);$$ [39]

that is, the circuit determinant must remain positive.

For a determination of the power transfer from the signal source into the transistor at the input terminals and from the transistor into the load at the output terminals, the input and output resistances are of interest. The resistance faced by the signal-source voltage v_s is the total series resistance effective in the input circuit. From Eqs. 34 and 35, it is

$$\text{Total input-circuit resistance} = \frac{v_s}{i_1} = R_s + r_{11} - \frac{r_{12}r_{21}}{r_{22} + R_L}.$$ [40]

Hence, the input resistance faced at the terminals of the transistor by the whole driving source, which comprises the voltage v_s and internal resistance R_s is

$$\text{Input resistance} = R_{in} = r_{11} - \frac{r_{12}r_{21}}{r_{22} + R_L}.$$ [41]

Similarly, the output resistance faced by the load may be determined by considering first that an incremental voltage acts in series with the load while v_s is zero. The resistance this incremental voltage faces is then

$$\text{Total output-circuit resistance} = R_L + r_{22} - \frac{r_{12}r_{21}}{R_s + r_{11}};$$ [42]

hence, the output resistance faced by the load at the transistor terminals is

$$\text{Output resistance} = R_{out} = r_{22} - \frac{r_{12}r_{21}}{R_s + r_{11}}.$$ [43]

The stability criterion expressed in Eq. 39, which was established by the requirement that the voltage amplification remain finite, will also be recognized as a requirement that the total input-circuit resistance expressed by Eq. 40 be positive, and, simultaneously, that the total output-circuit resistance expressed by Eq. 42 likewise be positive.

As viewed from the load, the transistor behaves as its Thévenin's equivalent, a source having an open-circuit voltage $A_{oc}v_s$ and an internal resistance R_{out}. The power output of the transistor to the load resistance R_L in terms of the Thévenin equivalent is

$$P_{out} = \frac{(A_{oc}v_s)^2}{(R_L + R_{out})^2} R_L,$$ [44]

which may be evaluated in terms of the transistor parameters through substitution of Eqs. 38 and 43. The power sensitivity is the ratio of this power output to $v_s{}^2$.

For many purposes, the usefulness of a transistor as an amplifier is most effectively described in terms of the power gain obtainable under various circuit conditions. The ratio of the power output of the transistor amplifier to the power output that could be obtained if a matching transformer were substituted for the transistor is basic among these values of power gain. The power output when a matching transformer is substituted for the transistor is the *available power* from the input-signal source, which is discussed in Art. 15, Ch. VIII, and is given by

$$\text{Available input power} = v_s{}^2/(4R_s). \qquad [45]$$

The ratio of the actual power in the transistor load to this available input power is called the *operating power gain*. Thus,

$$\text{Operating power gain} = \frac{\text{Power to the load}}{\text{Available input power}}, \qquad [46]$$

which, from Eqs. 38, 43, 44, and 45, is

$$\text{Operating power gain} = 4R_sR_L \frac{r_{21}{}^2}{[(R_s + r_{11})(r_{22} + R_L) - r_{12}r_{21}]^2}. \qquad [47]$$

This operating power gain is the ratio that occurs in the general condition when neither the source nor the load is matched to the resistance it faces at the transistor terminals.

When the load is matched to the transistor, for example, by a matching output transformer, the power gain becomes the *available power gain*; that is,

$$\text{Available power gain} = \frac{\text{Power to a matched load}}{\text{Available input power}}. \qquad [48]$$

Here the numerator is the available power at the output; thus, the available power gain may also be expressed as the ratio of the available output power to the available input power for any input conditions—an impedance match at the input is not implied. Substitution of Eqs. 44 and 45 in this relation gives, for the condition that R_L equals R_{out},

$$\text{Available power gain} = A_{oc}{}^2 \frac{R_s}{R_{out}}, \qquad [49]$$

and use of Eqs. 38 and 43 gives, in terms of the transistor and circuit resistances,

$$\text{Available power gain} = \frac{r_{21}{}^2}{(R_s + r_{11})^2} \frac{R_s}{r_{22} - \dfrac{r_{12}r_{21}}{R_s + r_{11}}}. \qquad [50]$$

Finally, the *maximum available power gain* is of interest. It is the value obtained when the available power gain in Eq. 50 is maximized with respect to the internal source resistance R_s, and occurs when the input of the transistor is matched to the signal source while the output is maintained matched to the load. Thus, for the matched conditions, Eqs. 41 and 43 become

$$R_s = r_{11} - \frac{r_{12}r_{21}}{r_{22} + R_L}, \qquad [51]$$

$$R_L = r_{22} - \frac{r_{12}r_{21}}{R_s + r_{11}}. \qquad [52]$$

Solving these equations simultaneously gives for the required values of the source and load resistances,

$$\text{Matched } R_s = r_{11} \sqrt{1 - \frac{r_{12}r_{21}}{r_{11}r_{22}}} \qquad [53]$$

$$\text{Matched } R_L = r_{22} \sqrt{1 - \frac{r_{12}r_{21}}{r_{11}r_{22}}}. \qquad [54]$$

Substitution of the matched value from Eq. 53 in Eq. 50 gives the maximum available power gain,

$$\begin{array}{ll} \text{Maximum available} \\ \text{power gain} \end{array} = \frac{\begin{array}{c}\text{Power to a matched load} \\ \text{when input is matched}\end{array}}{\text{Available input power}} \qquad [55]$$

$$= \frac{r_{21}{}^2}{r_{11}r_{22}} \frac{1}{\left[1 + \sqrt{1 - \dfrac{r_{12}r_{21}}{r_{11}r_{22}}}\right]^2}. \qquad [56]$$

As the feedback resistance r_{12} approaches zero, the maximum available power gain approaches $r_{21}{}^2/(4r_{11}r_{22})$. It should be noted that the maximum available power gain sometimes cannot be realized because the values of the matched source and load resistance given by Eqs. 53 and 54 are so low as to cause violation of the stability criterion in Eq. 39.

When the transistor is connected as a grounded-base amplifier so that Fig. 12b applies, the values of the self resistances r_{11} and r_{22} and the transfer resistances r_{12} and r_{21} in Eqs. 34 through 56 are related to r_e, r_c, r_b, and r_m, by Eqs. 28 through 31. Substitution of these four relations in Eqs. 34 through 56 gives expressions for the performance of the grounded-base transistor amplifier in terms of the quantities indicated in Fig. 12b. Alternatively, each of these equations may be expressed in terms of α_e through substitution of $\alpha_e r_c$ for r_m in accordance with Eq. 32.

Since r_{11} is much smaller than r_{22} for the grounded-base connection, the input resistance in Eq. 41 is generally much smaller than the output resistance in Eq. 43. Hence, the circuit may be likened to the grounded-grid amplifier circuit, and A_{oc} is analogous to $(\mu + 1)$ for an electron tube.

An alternate form of the stability criterion in Eq. 39 obtainable by use of Eqs. 28 through 31, and applicable, therefore, to the grounded-base amplifier, is

$$\frac{r_m}{r_c + R_L} < 1 + \frac{R_s + r_e}{r_b} + \frac{R_s + r_e}{r_c + R_L}. \qquad [57]$$

This relation indicates that the most severe requirement on the transistor resistances occurs when both the source and load resistances are zero. Under these conditions, the criterion for stability becomes

$$\frac{r_m}{r_c} \text{ or } \alpha_e < 1 + \frac{r_e}{r_b} + \frac{r_e}{r_c}. \qquad [58]$$

For a junction-type transistor, this requirement is almost always satisfied, because α_e is generally smaller than unity. But, for a point-contact transistor, α_e may be much larger than unity, and r_b comparable with r_e, so that the criterion for stability is violated. Since such a transistor is then unstable while working between zero values of source and load resistances, it is said to be *short-circuit unstable*. Increase of the source resistance R_s tends to make such a transistor circuit stable because it increases the right-hand member in Eq. 57. Increase of the load resistance R_L also tends toward stability because it primarily decreases the left-hand member of the equation; its effect on the right-hand member is relatively small because r_c is generally large compared with r_b. Addition of an external resistance R_b in the base circuit, as in Fig. 9, effectively adds to r_b and may render the circuit unstable.

If the transistor is connected as a grounded-emitter amplifier, as in Fig. 10a, and the transistor is represented by the equivalent circuit in Fig. 11c, the incremental equivalent circuit for the amplifier is as shown in Fig. 12c. Relations among the currents are shown on the diagram, and the two loop-voltage equations in terms of the base and collector current at the input and output, respectively, are

$$v_s = i_b(R_s + r_e + r_b) + i_c r_e ,$$ [59]

$$0 = i_b(r_e - r_m) \qquad + i_c(r_c + r_e - r_m + R_L).$$ [60]

Comparison with Eqs. 34 and 35 shows that for this connection, the self and transfer resistances corresponding to Fig. 12a are

$$r_{11} = r_e + r_b ,$$ [61]

$$r_{22} = r_c + r_e - r_m ,$$ [62]

$$r_{12} = r_e ,$$ [63]

$$r_{21} = r_e - r_m .$$ [64]

Note that here the self and transfer resistances on the left-hand sides of these expressions have a significance that is different from that in Eqs. 28 through 31 because of the different connection of the transistor. Substitution of these four relations in Eqs. 36 through 56 gives expressions for the performance of the grounded-emitter transistor amplifier in terms of the quantities in Fig. 12c.

As the load resistance approaches zero, the current amplification approaches the value

$$\text{Short-circuit current amplification} = \frac{r_{21}}{r_{22}} = \frac{r_e - r_m}{r_c + r_e - r_m} \approx \frac{\alpha_e}{\alpha_e - 1},$$ [65]

since r_e is small compared with r_c. For a transistor having a value of α_e only slightly smaller than unity, such as a junction type, the short-circuit amplification is negative and its magnitude may be very large. For a transistor having α_e larger than unity, however, such as a point-contact type, this current amplification is positive and is generally not large. Also, since r_m is ordinarily large compared with r_e, r_{21} from Eq. 64 is negative in Eq. 37, whereas the denominator is positive in accordance with Eq. 39. Hence, the voltage amplification from Eq. 37 is negative. In other words, there is a polarity reversal between the incremental output and input voltages when they are referred to a common terminal. Furthermore, comparison of Eqs. 28 through 31 with Eqs. 61 through 64 indicates that for a particular quiescent operating point, the input resistance from Eq. 41 for the grounded-emitter connection is generally larger and the output resistance from Eq. 43 smaller than the corresponding values for the grounded-base connection because r_{22} is smaller, r_{21} is negative, and r_{11}

is unchanged. For these reasons, the grounded-emitter connection may be likened to the grounded-cathode connection for an electron tube.

When the grounded-emitter circuit is supplied from a high-resistance signal source that is essentially a current source, so that R_s is very large, the stability requirement expressed by Eq. 39 reduces to

$$\frac{r_m}{r_c} \text{ or } \alpha_e < 1 + \frac{R_L}{r_c} + \frac{r_e}{r_c}, \qquad [66]$$

when Eqs. 61 through 64 are substituted in it. Increase of R_L may therefore be necessary to maintain freedom from oscillation when a point-contact transistor connected as a grounded-emitter amplifier is supplied from a high-resistance signal source.

A grounded-collector transistor amplifier has the incremental equivalent circuit shown in Fig. 12d when the transistor is represented as in Fig. 11c. The corresponding circuit equations are

$$v_s = i_b(R_s + r_c + r_b) + i_e(r_c - r_m), \qquad [67]$$

$$0 = i_b r_c \qquad + i_e(r_c + r_e - r_m + R_L). \qquad [68]$$

Again, comparison of these relations with Eqs. 34 and 35 shows that for the grounded-collector connection the self and transfer resistances corresponding to Fig. 12a are

$$r_{11} = r_c + r_b, \qquad [69]$$

$$r_{22} = r_c + r_e - r_m, \qquad [70]$$

$$r_{12} = r_c - r_m, \qquad [71]$$

$$r_{21} = r_c, \qquad [72]$$

and substitution of these relations in Eqs. 36 through 56 gives expressions for its performance in terms of the quantities in Fig. 12d.

The stability requirement expressed by Eq. 39 becomes

$$(R_s + r_c + r_b)(r_c + r_e - r_m + R_L) > r_c(r_c - r_m), \qquad [73]$$

which reduces to the same relation as is expressed by Eq. 66 for the grounded-emitter connection when the circuit is supplied from a signal source having very large internal resistance R_s.

As α_e approaches unity, or r_m approaches r_c, r_{12} approaches zero; hence the matched source and load resistances in Eqs. 53 and 54 approach $r_b + r_c$ and r_e, respectively. Thus, the matched source resistance is in the order of megohms whereas the matched load resistance is a few hundred ohms or less for a junction-type transistor. Furthermore, the voltage amplification for small R_s and large R_L approaches $r_c/(r_b + r_c)$, which is nearly unity, since r_b is generally small, and there is no voltage-polarity inversion in this amplifier. Hence, this connection is the counterpart of the cathode-follower connection for an electron tube and has many of its characteristics.

The grounded-collector transistor amplifier can amplify in both directions—introduction of the signal voltage in series with R_L produces appreciable power in R_s. Analysis of the circuit shows that for α_e smaller than 2, the base-to-emitter power gain is larger than the emitter-to-base power gain; but, for α_e larger than 2, the opposite is true. A further feature is that the backward transmission involves a polarity reversal, whereas the forward transmission does not.

8. CASCADE TRANSISTOR AMPLIFIERS

Connecting transistor amplifier stages in cascade to obtain increased gain is practicable, but the design of such amplifiers is more complex than that of electron-tube amplifiers because of the bidirectional transmission, the relatively low input impedance, and, for the point-contact transistor, the tendency toward instability of the various types of transistor stages. The bidirectional transmission results from the presence of the forward and backward resistances, or of the base resistance r_b and mutual resistance r_m. These resistances cause feedback just as the grid-to-plate capacitances cause feedback in a resistance-capacitance-coupled amplifier. Analysis of a multistage transistor amplifier is thus comparable in difficulty with analysis of a multistage resistance-capacitance-coupled amplifier with the grid-to-plate capacitances taken into account—a problem whose solution is not ordinarily attempted rigorously because of its complexity. To insure stability in the complete transistor amplifier, the feedback must be considered. Methods of analysis developed for use in repetitive linear passive networks are applicable when extended to the active transistor network representation. In particular, matrix representation[21] provides a direct method of analyzing cascaded stages.

Transformers may sometimes be used to match the low input impedance of one stage to the high output impedance of the previous stage. But the transformer is often undesirable because of its comparative bulk and limited frequency response. Also, matching the impedances may lead to negative total input and output resistances and consequent instability when a point-contact transistor is used. Hence, a consideration of the possibility of cascading transistor amplifiers both with and without use of transformers is of interest.

[21] Jacob Shekel, "Matrix Representation of Transistor Circuits," *I.R.E. Proc.*, *40* (1952), 1493–1497; L. A. Zadeh, "A Note on the Analysis of Vacuum Tube and Transistor Circuits," *I.R.E. Proc.*, *41* (1953), 989–992.

When grounded-base stages are cascaded without transformers, each collector circuit faces the low input resistance of the next stage. Hence, the effective value for R_L in Eq. 36 is very small, and the current amplification per stage is approximately equal to the short-circuit current amplification α, or to α_e. Since α is generally smaller

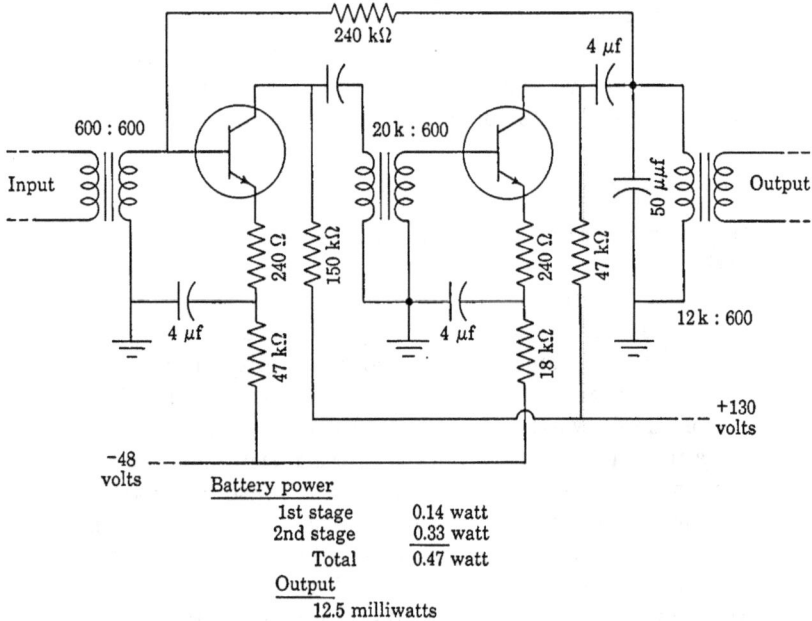

Fig. 13. Cascade transistor amplifier comprising two grounded-emitter stages with transformer coupling. (*Adapted by courtesy of Bell Telephone Laboratories, Inc.*)

than unity for junction-type transistors, increasing the number of cascaded grounded-base stages causes a reduction rather than an increase of the amplification, and a multistage amplifier gives less amplification than a single stage, when junction-type transistors are used.

Grounded-emitter stages cascaded without transformers can give substantial over-all power gain even for junction-type transistors having α_e smaller than unity, because the current amplification, as described by Eqs. 36 and 65, is large for this connection despite the mismatch between the output resistance of one stage and the input resistance of the next. Consequently, power gains up to about 1,000 per stage are obtained. Figure 13 shows an experimental two-stage transistor amplifier circuit that employs an interstage transformer and gives an over-all power gain of about 10,000 and a power output of

12.5 milliwatts without excessive distortion. Each stage is a grounded-emitter connection for the alternating components, but is effectively a grounded-base connection for the steady components because the transformer secondary windings have negligible resistance to direct current. In this amplifier, the stability and linearity are improved by 240-ohm resistances in series with each emitter to provide local negative feedback within each stage, and a 240,000-ohm resistance

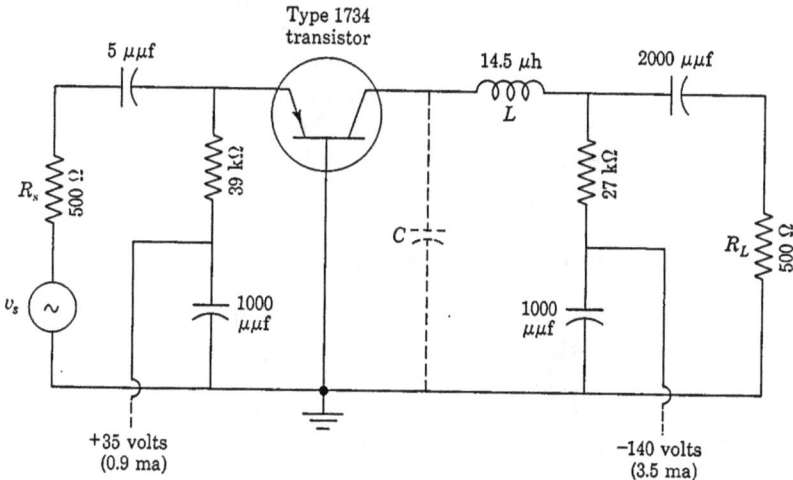

Fig. 14. Single-tuned intermediate-frequency transistor amplifier for use at 22 megacycles per second. (*Adapted by courtesy of Bell Telephone Laboratories, Inc.*)

from the output collector to the input base to provide negative feedback around both stages.

If grounded-emitter stages employing point-contact transistors are cascaded without interstage matching transformers, instability usually results because α_e is greater than unity and the effective load resistance R_L, which is the input resistance of the next stage, is so low that the requirement expressed in Eq. 66 is violated. Stabilization can be accomplished and the connection made practicable, however, through effectively increasing r_c by addition of external resistance in series with each collector terminal.

Unlike stages may also be connected in cascade to take advantage of the special properties of the different connections. For example, a combination of a grounded-base stage driving a grounded-collector stage in the direction from the emitter toward the base gives low resistance at both the input and the output, just as does a grounded-grid electron-tube stage driving a grounded-plate, or cathode-follower, stage.

Tuned transistor amplifiers may often have a configuration of circuit elements different from that for the corresponding electron-tube amplifier, because, as is mentioned previously, the transistor is in many respects the dual of the vacuum tube—current takes the place of voltage in the circuits, series resonance takes the place of parallel resonance, and so on. Figure 14 shows a single-tuned amplifier circuit for use as an intermediate-frequency amplifier at 22 megacycles per second in a superheterodyne system. The external collector inductance is resonated with the internal and stray collector-to-base capacitance. Design of cascaded tuned amplifiers often requires consideration of the performance at frequencies far outside as well as within the transmission band near the resonant frequency because some transistor stages tend to become unstable as the circuit impedances approach zero, and over-all instability may result.

9. TRANSISTOR PERFORMANCE LIMITATIONS

The performance of transistors is inferior to that of electron tubes in several respects. First, the noise inherent in the transistor limits the minimum signal that can be effectively amplified; second, the internal capacitance or transit-time effects limit the maximum frequency that can be amplified; third, the permissible internal heating and distortion limit the maximum power that can be delivered to a load; and, fourth, the large temperature coefficient imposes a limit on the ambient temperature permissible for satisfactory operation.

The noise inherent in the transistor is similar to that in a composition resistor or a metallic electrical contact,[22] and its characteristics differ from both those of the thermal noise in a resistor and those of the shot noise associated with a stream of particles. Nevertheless, its effect can be represented approximately by postulating the existence of noise voltages v_{ne} and v_{nc} in series with the emitter and collector terminals, respectively, as in Fig. 15. These two voltages account for all the noise developed in the transistor. In addition, when the transistor is connected to a source of signal voltage v_s having an internal resistance R_s, a voltage v_{ns} corresponding to the thermal noise in R_s as given by Eq. 7, Ch. IX, is effective in series with that resistance as shown.

For measurement of the separate noise voltages v_{ne} and v_{nc}, the transistor is commonly connected to its two bias voltages with large

[22] J. F. Blackburn, Editor, *Components Handbook*, Massachusetts Institute of Technology Radiation Laboratory Series, Vol. 17 (New York: McGraw-Hill Book Company, Inc., 1949), 49; R. L. Petritz, "On the Theory of Noise in *P-N* Junctions and Related Devices," *I.R.E. Proc., 40* (1952), 1440–1456.

inductances in series with the emitter and collector, and the voltages are adjusted so that operation takes place at the normal quiescent operating point.[23] Thus R_s and R_L in Fig. 15 are replaced by inductive reactances that are large compared with the resistances they face at all frequencies of interest, and the signal voltage v_s is zero. The voltages across the inductances are then v_{ne} and v_{nc}. The root-mean-square values of these voltages are usually measured as functions of frequency with a narrow-band tuned voltmeter having an effective bandwidth of

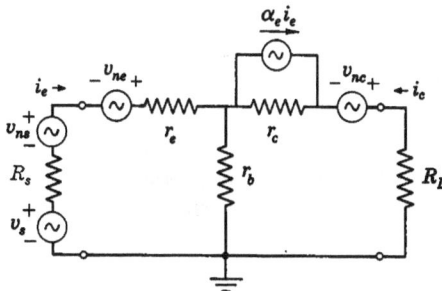

Fig. 15.   Noise voltages in a grounded-base transistor amplifier.

a few cycles per second, such as a commercial wave analyzer. It is to be emphasized that the noise voltages v_{ne} and v_{nc} are not single-frequency voltages. Rather, they correspond to the integrated heating effect over a narrow band in a spectrum of frequencies, for the distribution of noise power is a continuous spectrum just as is the distribution of power in a light beam from an incandescent source. If the transistor noise were simply thermal-agitation noise associated with equivalent resistances of the transistor, the indication of the voltmeter would be independent of the center frequency of the band over which the measurement is made. Actually, the two rms noise voltages are found to vary approximately inversely as the square root of the center frequency. More accurately, the mean-square voltages vary as $1/f^{1.1}$, or the noise power corresponding to the voltages decreases 11 db per decade as the frequency increases.

Specification of the noise voltages is commonly made in terms of the ratio of the rms voltage to the square root of the bandwidth at some selected center frequency. Typical values for a point-contact transistor are 100 microvolts in a one-cycle-per-second band at 1,000 cycles per second center frequency for v_{nc} and about one or two per cent of

---

[23] E. Keonjian and J. S. Schaffner, "An Experimental Investigation of Transistor Noise," *I.R.E. Proc.*, *40* (1952), 1456–1460; H. C. Montgomery, "Transistor Noise in Circuit Applications," *I.R.E. Proc.*, *40* (1952), 1461–1471.

that value for v_{ne} in the same band. Values for junction-type transistors are much smaller.

Noise in amplifiers is also frequently described in terms of the *noise figure*, mentioned in Art. 3, Ch. IX. This noise figure may be defined as the ratio of the noise power existing in the load in a specified frequency band as a result of all noise sources divided by the noise power existing in the load in the specified frequency band as a result only of the thermal noise in the internal resistance of the signal-voltage source. The noise figure is thus a measure of the noise that the transistor adds to the noise that would occur if all elements in the circuit were noise-free except the internal resistance of the signal-voltage source. Its significance is perhaps best understood from a consideration of methods by which it may be measured. One method involves, first, measuring the mean-square noise voltage across the load with a voltmeter such as the narrow-band wave analyzer mentioned above; second, measuring the voltage amplification A, in Eq. 37; and, third, computing the mean-square thermal noise voltage corresponding to R_s, which is $4kTR_s\,BW$, from known values of the temperature, resistance, and bandwidth of the voltmeter. Since the noise power in the load is the mean-square noise voltage divided by the load resistance, the noise figure is then given by

$$F = \frac{\text{Measured mean-square voltage at load in bandwidth } BW}{4kTR_s\,BW\,A^2}, \quad [74]$$

if A is uniform throughout the bandwidth.

Alternately, the noise figure may be computed from measured values of v_{ne} and v_{nc}. If all the noise voltages were statistically independent, the total noise power in the load would be the sum of the values computed for each source acting independently. Since the noise voltage v_{nc} in Fig. 15 is effectively in series with the load resistance R_L and the output resistance R_{out}, the noise power output due to this voltage alone is

$$\text{Noise power output due to } v_{nc} = \frac{v_{nc}{}^2 R_L}{(R_L + R_{out})^2}. \quad [75]$$

The noise power output due to v_{ns} alone may be computed from Eq. 46. It is

$$\text{Noise power output due to } v_{ns} = \frac{(A_{oc}v_{ns})^2 R_L}{(R_L + R_{out})^2}, \quad [76]$$

and the output caused by v_{ne} may be expressed similarly. The noise figure for the grounded-base connection when the noise voltages are statistically independent is then

$$F = \frac{(A_{oc}v_{ns})^2 + (A_{oc}v_{ne})^2 + v_{nc}{}^2}{(A_{oc}v_{ns})^2} = 1 + \frac{v_{ne}{}^2 + (v_{nc}/A_{oc})^2}{v_{ns}{}^2}. \quad [77]$$

The terms involving R_L and R_{out} cancel and vanish in this result. The noise figure is thus independent of the load resistance, but is dependent on the noise voltages and the open-circuit voltage amplification A_{oc}. It also depends on the source resistance R_s, for, by means of Eq. 7, Ch. IX, and Eqs. 28, 31, and 38, Eq. 77 may be put in the form

$$F = 1 + \frac{v_{ne}^2 + v_{nc}^2 \left(\dfrac{R_s + r_e + r_b}{r_m + r_b} \right)^2}{4kT R_s \, BW}, \qquad [78]$$

where it is understood that BW is the bandwidth in which v_{ne} and v_{nc} are effective. When v_{ne} and v_{nc} are not statistically independent, the correlation between them must be taken into account in the addition in the numerator of Eq. 78 and the noise figure is larger or smaller than Eq. 78 indicates.

Since the source resistance R_s enters into both the numerator and the denominator of the expression for the noise figure, equating to zero the derivative of the expression with respect to R_s gives the value of R_s at which F is a minimum when the correlation between v_{ne} and v_{nc} is zero, namely,

$$R_s \text{ for minimum noise figure} = \sqrt{(r_e + r_b)^2 + \left(\frac{v_{ne}}{v_{nc}} \right)^2 (r_m + r_b)^2}. \,[79]$$

Because the voltage ratio in Eq. 79 is small, the value for R_s is approximately $r_e + r_b$, which is the open-circuit input resistance of the circuit.

The minimum noise figure corresponding to the condition in Eq. 79 and a one-cycle-per-second bandwidth at one kilocycle per second for typical point-contact transistors is about 10^5, or 50 decibels, a value that is much larger than the representative noise figure for electron tubes, but junction types have shown values between 6 and 300, or 8 and 25 decibels.

The performance of transistors as amplifiers at high frequencies is limited by changes effective in some of the parameters shown in the equivalent circuits, Fig. 11, and by internal capacitance across the contacts or junctions. The resistances r_e, r_b, and r_c appear to be relatively constant with frequency up to tens of megacycles per second just as the forward and reverse resistances of a crystal diode rectifier are constant. But r_m decreases in magnitude and becomes a complex quantity with a phase angle as the frequency increases,[24] and α and α_e vary in a corresponding manner. In a typical point-contact transistor amplifier for high-frequency application, r_m, α, and α_e are

---

[24] R. L. Pritchard, "Frequency Variations of Current-Amplification Factor for Junction Transistors," *I.R.E. Proc.*, *40* (1952), 1476–1481.

reduced in magnitude to 0.707 times their low-frequency values when the frequency increases to a value lying between 10 and 20 megacycles per second. The output power from a grounded-base transistor amplifier circuit into a resistance load is approximately proportional to the square of the short-circuit current amplification α when the load resistance is small compared with the output resistance and the signal-source resistance is large compared with the input resistance. Hence the output power from such an amplifier is reduced by one-half when α is reduced by 0.707. The frequency at which this reduction occurs is therefore called the half-power frequency or the *cut-off frequency*. This cut-off frequency is roughly proportional to the collector voltage. The variation of r_m, α, and α_e is attributed to dispersion in the transit times of the separate holes or electrons injected by the emitter owing to the fact that they traverse paths of different lengths in their travel to the collector. Some transistors will operate at frequencies as high as 300 to 400 megacycles per second.[25]

Because of the larger area at the junction, capacitance across the junction is much larger in a junction-type transistor than in a point-contact type. A capacitance of 5 to 10 micromicrofarads is effective across r_c in the equivalent circuit of Fig. 11d in a typical transistor, and reduces the half-power frequency to a few thousands or tens of thousands of cycles per second. Transit-time dispersion also affects the frequency response as in the point-contact type, but its effect is usually outweighed by the collector capacitance for the grounded-base connection.

The alternating-current power output that may be obtained from a transistor amplifier is limited by the permissible internal power dissipation and the permissible harmonic distortion just as for an electron tube, and the power dissipation and distortion may be determined graphically from the characteristic curves. When the transistor is connected as a grounded-base amplifier with a resistance load, these quantities may be found from a load line drawn on the collector characteristics. For a junction-type transistor having collector characteristics such as those in Fig. 7, the curves are essentially straight and evenly spaced for equal increments of the emitter current over the full length of any load line confined to the quadrant shown; hence the full length can be utilized without incurring excessive distortion. If, to avoid excessive distortion, the operation is limited to this region of the characteristics so that the path of operation along the load line terminates practically on the two axes, conditions are essentially the same as for the ideal pentode discussed in Art. 16, Ch. VIII. The

[25] B. N. Slade, "The Control of Frequency Response and Stability of Point-Contact Transistors," *I.R.E. Proc.*, *40* (1952), 1382–1384.

operation of the transistor amplifier is said to be Class A. The efficiency of conversion of the direct-current power input, which is given by the product of the average values of the collector voltage and current, into alternating-current power output is 50 per cent, just as for the ideal pentode. As a result, the collector power dissipation equals the alternating-current power output. Consequently, the alternating-current power output of a junction-type transistor used as a Class A amplifier can be no larger than the permissible collector power dissipation, which ordinarily is a few hundred milliwatts. Class B operation offers the possibility of larger power output without excessive collector dissipation, as does Class C operation.

For a point-contact transistor having characteristics such as those shown in Fig. 6, the permissible excursion of the operating point is limited to the center region of the load line because crowding of the characteristics at each end leads to excessive distortion. The full length of the line cannot be utilized, and the efficiency of power conversion in the collector circuit is smaller than 50 per cent. Consequently, the alternating-current power output is smaller than the permissible collector dissipation when operation is restricted to Class A.

10. PULSE, OR SWITCHING, OPERATION[26]

Pulse, or switching, action may be obtained when transistors are operated so that the incremental resistance between a pair of terminals is negative. An example is the grounded-base circuit with added base resistance discussed in Art. 5, and shown in Fig. 16a. The graphical relation between V_{en} and I_e derived in Fig. 9 is shown in Fig. 16b as the N-shaped curve $V_{en}(I_e)$. When the incremental voltage ΔV is zero, and the emitter bias voltage V_{ee} and series resistance R_s are selected so that the corresponding resistance line intersects the N-shaped curve at three points, any one of the points represents a possible equilibrium condition at which the volt-ampere requirement of the transistor and that of the external circuit are satisfied simultaneously. Assume for discussion that operation starts at point 1. If a momentary positive incremental trigger voltage ΔV_1 is injected in series with V_{ee}, the resistance line shifts upward but remains parallel to its original position. As it passes the position of tangency to the N-shaped curve at 2, the only remaining equilibrium point is at 3. Hence, operation shifts to 3 and on up to 4 as the trigger voltage ΔV_1 continues to

[26] A. E. Anderson, "Transistors in Switching Circuits," *I.R.E. Proc.*, *40* (1952), 1541–1558; A. W. Lo, "Transistor Trigger Circuits," *I.R.E. Proc.*, *40* (1952), 1531–1541; J. H. Felker, "Regenerative Amplifier for Digital Computer Applications," *I.R.E. Proc.*, *40* (1952), 1584–1596.

increase. When the trigger voltage decreases to zero, the operating point drops back to 5 and remains there instead of returning to 1. Upon subsequent application of a negative incremental trigger voltage ΔV_2, a similar excursion occurs. Operation shifts from 5 to the tangent

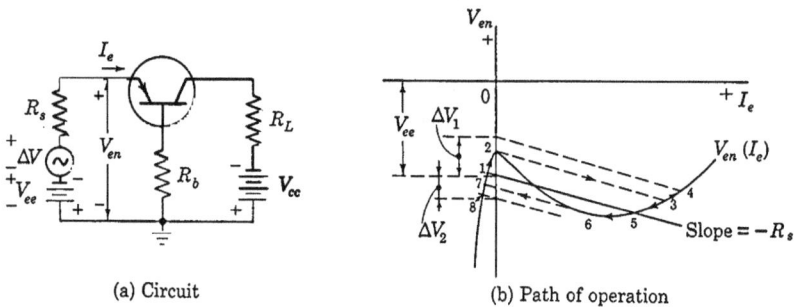

(a) Circuit   (b) Path of operation

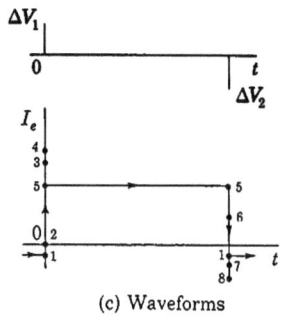

(c) Waveforms

Fig. 16.   Bistable switching operation in a grounded-base transistor circuit. When trigger voltage ΔV_1 occurs, operation shifts from 1 to 5; when ΔV_2 occurs, operation shifts from 5 back to 1.

point 6, jumps to 7, continues to 8 as the negative trigger voltage increases in magnitude, and finally returns to 1, the original stable operating point. Application of successive trigger voltages having opposite polarities thus causes the circuit conditions to change from one stable state to another. The circuit is said to be *bistable*, and the emitter current is either large or almost zero, as is shown in Fig. 16c.

If operation should begin at the middle intersection of the resistance line and the curve $V_{en}(I_e)$, which is on the negative-resistance, or downward-sloping portion, the first trigger voltage that occurred would shift the operating point to 1 or to 5. Subsequent trigger voltages would cause operation in the sequence described above, and the operating point would never again reach the middle intersection. The middle intersection is hence not of practical consequence.

In this and other transistor circuits that include an external base resistance R_b, some of which are discussed subsequently, the trigger signal is in practice often injected into the base circuit instead of into the emitter or the collector circuit.

Other types of switching operation involving the negative-resistance emitter characteristic are shown in Figs. 17 and 18. Figure 17 indicates

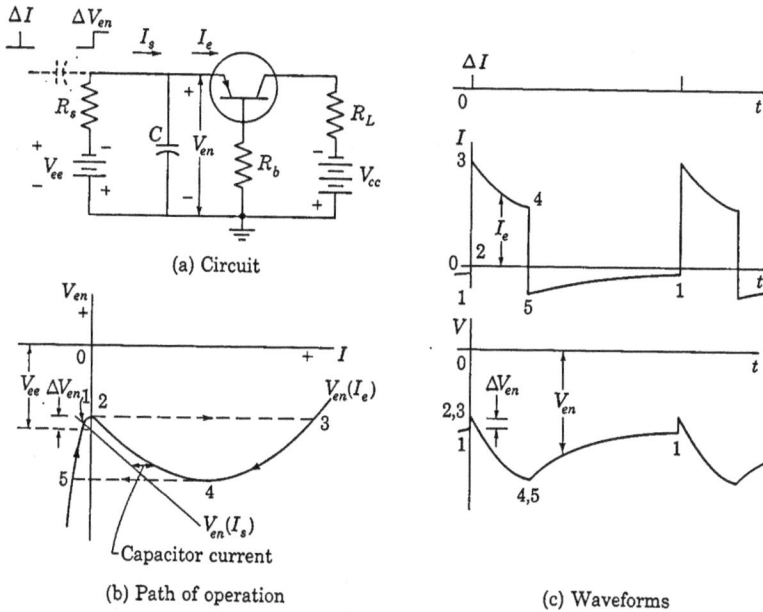

(a) Circuit

(b) Path of operation

(c) Waveforms

Fig. 17. Monostable, or "one-shot," switching operation in a grounded-base transistor circuit.

the operation when the resistance line intersects the $V_{en}(I_e)$ curve at only one point and when that point is on a positive-resistance, or upward-sloping, portion of the curve. If the capacitance C in Fig. 17a is zero, and a positive incremental trigger voltage is applied in series with R_s, or a trigger current is injected from a current source through the capacitor at the upper left-hand corner of Fig. 17a, operation shifts momentarily to the right-hand branch of the curve, as in Fig. 16b, but returns in a clockwise direction along the curve, jumping back to the left-hand branch and returning immediately to 1. For such operation, the circuit is frequently called a *regenerative pulse amplifier*. The operation is termed *monostable* because point 1 is the only stable operating point corresponding to the particular value of V_{ee}.

When the capacitance C is present in the circuit of Fig. 17a, the operation is considerably modified from that described above. In the first place, since C is connected at V_{en}, an increment of charge, or a corresponding *impulse* of input current I_s is required to change V_{en} suddenly by an amount ΔV_{en}. Thus, the source for triggering must be capable of supplying a large momentary current if it is to produce an almost suddenly occurring increment of voltage ΔV_{en}. Alternatively, C might be switched out of the circuit, charged to the voltage $V_{ee} + \Delta V_{en}$, and reconnected as shown. In the second place, the capacitor prevents sudden changes in the voltage V_{en} during subsequent operation of the circuit because only a finite current can reach it through the emitter circuit and the resistance R_s. Consequently, when a positive trigger voltage increment occurs, operation shifts from 1 to the tangent point

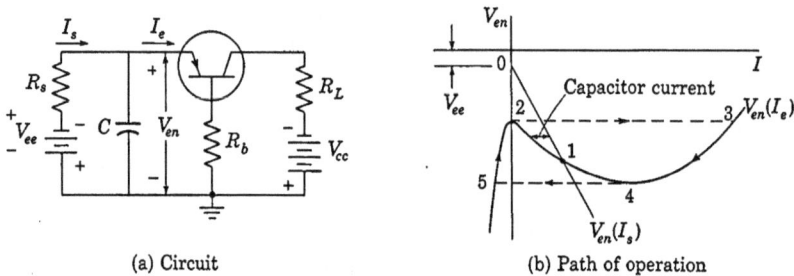

(a) Circuit   (b) Path of operation

Fig. 18.   Astable, or free-running, switching operation, called relaxation oscillation, in a grounded-base transistor circuit.

2, and then jumps very rapidly along the horizontal dotted line to 3. At any instant, the current in the capacitor is the horizontal distance between the straight resistance line and the operating point on the curve $V_{en}(I_e)$. When operation is at 3, this current is large and in the direction to discharge the capacitor—that is, to reduce V_{en}. Hence, V_{en} decreases, and the operating point corresponding to the current I_e and voltage V_{en} follows the curve from 3 to 4. The capacitor current meanwhile decreases to a relatively small value, so that the rate of change of voltage V_{en} across the capacitor decreases, but the current is still in a direction further to reduce V_{en}. Consequently, the operating point jumps horizontally from 4 to 5. The capacitor current simultaneously reverses; hence the capacitor recharges and its voltage returns to 1, the original and only stable operating point. A single trigger increment thus causes a complete excursion around the loop, and *"one-shot"* operation is said to occur. Waveforms are shown in Fig. 17c for V_{en} and I_e in time relation to the trigger current ΔI.

When a capacitance C is included in the circuit, and the resistance line corresponding to the bias voltage and resistance R_s intersects the curve $V_{en}(I_e)$ at only a single point and on the negative-resistance, or downward-sloping, portion of the curve, the operation is as shown in Fig. 18. If the operation should begin at intersection 1 on the negative-resistance part of the curve, any small variation, however slight, such as thermal agitation in the resistance, results in a capacitor current that tends to charge the capacitor and cause motion of the operating point along the $V_{en}(I_e)$ curve away from 1 toward 2 or 4. If the motion begins toward 2, say, it continues in that direction until 2 is reached. The capacitor current is then in a direction to cause further increase of V_{en}; hence the operating point jumps to 3 while the

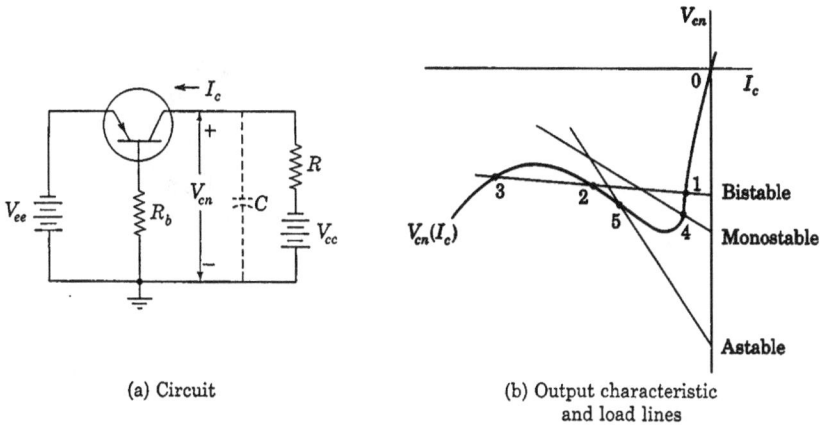

(a) Circuit

(b) Output characteristic
and load lines

Fig. 19. Switching operation in the collector circuit of a grounded-base transistor connection.

voltage V_{en} remains momentarily constant, just as is described previously for conditions in Fig. 17b. The capacitor current reverses during this jump and at 3 is in a direction to discharge C, so that the operating point moves from 3 to 4, jumps to 5, moves gradually to 2 while the capacitor recharges, jumps to 3, and so on, never again returning to 1, but effectively searching for a stable condition. The waveforms of the current and voltage resulting resemble those in Fig. 17c, with ΔV_{en} reduced to zero. The circuit conditions are said to be *astable*, or *free-running*, and the continuous oscillations that occur are called *relaxation oscillations*.

Switching action may also occur in the collector circuit because the volt-ampere relation at the collector and ground terminals is an N-shaped curve having a negative-resistance region, as is indicated in Fig. 19b. Bistable flip-flop behavior, monostable regenerative-amplifier or one-shot behavior, or astable relaxation oscillation may occur depending on the position and slope of the load line and the presence or absence of capacitance C between the collector and ground.

The volt-ampere input curve $V_b(I_b)$ between the base and emitter terminals for the grounded-emitter connection of the transistor developed in Fig. 10 also makes possible switching action. This curve

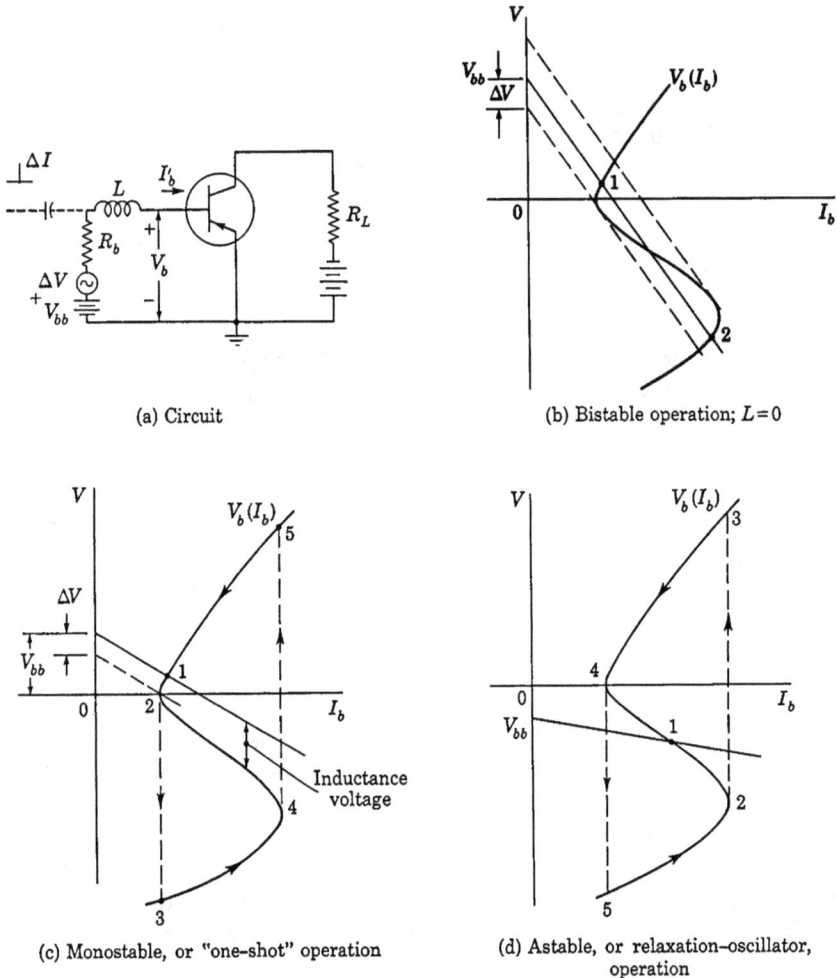

(a) Circuit

(b) Bistable operation; $L=0$

(c) Monostable, or "one-shot" operation

(d) Astable, or relaxation-oscillator, operation

Fig. 20. Switching operation in a grounded-emitter transistor circuit.

is S-shaped, as is shown in Fig. 10c, rather than N-shaped, as for the grounded-base connection. If the straight-line locus corresponding to V_{bb} and R_b in the grounded-emitter circuit of Fig. 20a intersects the curve $V_b(I_b)$ at three points, as in Fig. 20b, and the inductance is zero, application of a momentary negative trigger voltage in series with V_{bb}, which may be caused by a trigger current through R_b, causes a sudden shift of the operating point along the dotted line from stable point 1 to the lower branch of the curve and on to point 2 as the trigger voltage subsides to zero. A subsequent positive trigger voltage can cause the operating point to return to 1.

If the resistance line intersects the S-shaped curve at only a single point and on a positive-resistance portion, conditions are as shown in Fig. 20c. A negative trigger voltage ΔV in series with V_{bb} may shift the resistance line so that it intersects at the vertical tangent point 2. If a series inductance L is included in the circuit, further increase of the trigger voltage causes a jump to point 3 at constant current because the inductance prevents a sudden change of current. The voltage across the inductance for any condition of operation is the vertical displacement between the resistance line and the S-shaped curve as indicated. For operation at 3, this voltage is large and in a direction corresponding to an increasing current. Hence, the operating point travels gradually from 3 to 4 along the S-shaped curve as the current increases, then jumps to 5, and finally moves gradually back to 1. Thus, monostable, or one-shot, operation occurs.

If the resistance line intersects the S-shaped curve at only a single point and on the negative-resistance portion, as in Fig. 20d, relaxation oscillation occurs in a manner analogous to that illustrated in Fig. 18. Any slight statistical disturbance from the equilibrium conditions at point 1 results in development of a voltage across the inductance corresponding to an increase or decrease of current, so that the operating point moves to 2 or 4, respectively, and then follows counterclockwise around the loop continuously, never again returning to 1.

These various types of switching operation make possible a large number of functional operations by transistors. For example, transistors may serve for generation of essentially rectangular or sawtoothed waveforms, introduction of time delays, counting, gating, amplification of pulse amplitudes, and lengthening of pulses. Because of their small size, long life, low power consumption, and reliability, transistors are especially suitable for use in complex switching devices such as digital computers.

11. TRANSISTOR OSCILLATORS

Since transistors may amplify or have negative incremental resistance at pairs of terminals, they may be used as sinusoidal oscillators, and the principles outlined for vacuum-tube oscillators in Ch. XI are applicable to them. Inductive feedback from the collector to the emitter circuit, feedback through a parallel tuned circuit in series with the base terminal, and feedback from the collector to the emitter through a piezoelectric crystal acting as a tuned circuit or through an actual series-tuned circuit are but a few of the possible basic circuit configurations.

PROBLEMS

1. Plot on the collector characteristics for the point-contact transistor in Fig. 6 curves for the following constant emitter voltages: $+0.4$, $+0.2$, 0, -0.2, and -0.4 volt. Plot on the emitter characteristics curves for the following constant collector voltages: -5, -10, -20, -30, -40 volts. Is the incremental emitter-to-base resistance negative in any region of operation with constant collector voltage? Is the incremental collector-to-base resistance negative in any operating region with constant emitter voltage?

2. A point-contact transistor having the characteristics in Fig. 6 is connected as a grounded-base amplifier with a load resistance R_L of 10,000 ohms, a signal-source resistance of 500 ohms, and a collector bias voltage of -30 volts.

  (a) Plot the input-resistance curve.
  (b) Determine the operating conditions for input voltages $v_s + V_{ee}$ equal to 0, $+0.4$, and $+0.8$ volt.
  (c) Plot the output-resistance curve for each value of input voltage.

3. A point-contact transistor having the characteristics shown in Fig. 6 is connected as a grounded-base amplifier with added external base resistance R_b. The collector bias voltage V_{cc} is -20 volts, and the load resistance R_L is 10,000 ohms.

Plot on the emitter characteristics curves showing the emitter-to-ground voltage as a function of emitter current for R_b equal to 0, 500, 1,000, and 1,500 ohms. What is the minimum value of source resistance that will assure stable operation in this circuit?

4. For the operating conditions in Prob. 2 with $v_s + V_{ee}$ equal to $+0.4$ volt, determine r_e, r_b, r_c, r_m, and α_e by measurements from the curves.

Using these values, compute R_{in} and R_{out} and compare with the slopes of the input-resistance and output-resistance curves from Prob. 2, respectively, at the operating point.

Find the voltage gain and current gain graphically, and compare with the values computed from the measured values of r_e, r_b, r_c, and α_e.

5. Develop an expression for the backward operating power gain in each of the three basic transistor amplifier connections. Can all three amplify in the reverse direction?

6. For the *n-p-n* junction-type transistor described in Table I, and the grounded-base, grounded-emitter, and grounded-collector amplifier connections, find numerical values for the

  (a) incremental input resistances when the incremental load resistance is zero and when it is infinity,
  (b) incremental output resistances when the incremental resistance of the signal-voltage source is zero and when it is infinity,
  (c) matched input resistance,
  (d) matched output resistance,
  (e) maximum available power gain.

7. Show that for α or α_e equal to 2, the forward and backward operating-power gains of a grounded-collector amplifier are essentially equal.

8. Derive noise-figure relations corresponding to Eq. 78 but for the grounded-emitter and the grounded-collector amplifiers.

9. The rms noise voltage at the collector of a particular point-contact transistor is 100 microvolts in a 1 cycle-per-sec band at 1 kilocycle per sec and the rms noise voltage at the emitter is one per cent of that value. They may be assumed to be statistically independent. Each voltage varies approximately as the reciprocal of the square root of the center frequency of the band. The internal resistance of the source of signal voltage is 500 ohms, r_m is 30,000 ohms, r_e is 250 ohms, and r_b is 250 ohms. The temperature is 290 K, and Boltzmann's constant k is 1.38×10^{-23} watt-second per deg Kelvin.

Find the noise figure for the frequency band between 100 and 10,000 cps.

APPENDIX A

Name	Symbol	Value
Magnitude of charge of electron	Q_e	1.602×10^{-19} coulomb
Rest mass of electron	m_e	9.107×10^{-31} kilogram
$\dfrac{\text{Magnitude of charge of electron}}{\text{Rest mass of electron}}$	$\dfrac{Q_e}{m_e}$	1.759×10^{11} coulombs per kilogram
Electron volt	1 ev	1.602×10^{-19} joule
Speed of propagation of electromagnetic waves in free space	c	2.998×10^8 meters per second
Dielectric constant of free space	ε_v	8.854×10^{-12} farad per meter
Permeability of free space	μ_v	12.57×10^{-7} henry per meter
Planck constant	h	6.624×10^{-34} joule second
Boltzmann constant	k	1.380×10^{-23} joule per degree Kelvin
Stefan-Boltzmann constant	K	5.673×10^{-8} watt per square meter per degree Kelvin fourth
Gas constant	R	8.314 joules per degree per mole
Avogadro number	N	6.023×10^{23} per mole
$\dfrac{\text{Mass of atom of unit atomic weight}}{\text{Mass of electron}}$		1,823
Atomic weight of hydrogen		1.008
Atomic weight of helium		4.003
Atomic weight of neon		20.18
Atomic weight of argon		39.94
Atomic weight of krypton		83.7
Atomic weight of xenon		131.3
Atomic weight of mercury		200.6

* Except for the last six atomic weights, the constants in this table either are taken from R. T. Birge, "A New Table of Values of the General Physical Constants," *Rev. Mod. Phys.*, *13* (October, 1941), 233–239, with permission, or are calculated from the values given there. The atomic weights given are based on the atomic weight of oxygen being equal to 16.

APPENDIX B

UNITS AND CONVERSION FACTORS*

METER-KILOGRAM-SECOND (MKS) AND CENTIMETER-GRAM-SECOND (CGS) SYSTEMS

Amounts in any one row are equal; for example:

1 coulomb = 10^{-1} abcoulomb; or 0.004π oersted = 1 ampere-turn/meter.

Abbreviations *mksuru*, *esu*, and *emu* denote unrationalized units for which no name or combination of names has been approved.

Quantity	Symbol	MKS rationalized† units	MKS unrationalized units	CGS unrationalized electrostatic units (esu)	CGS unrationalized electromagnetic units (emu)
Length	l	1 meter	1 meter	10^2 centimeters	10^2 centimeters
Mass	m	1 kilogram	1 kilogram	10^3 grams	10^3 grams
Time	t	1 second	1 second	1 second	1 second
Force	F	1 newton	1 newton	10^5 dynes	10^5 dynes
Energy (work)	W	1 joule	1 joule	10^7 ergs	10^7 ergs
Power	P	1 watt	1 watt	$10^7 \dfrac{\text{ergs}}{\text{second}}$	$10^7 \dfrac{\text{ergs}}{\text{second}}$
Dielectric constant (permittivity)	ε	$1 \dfrac{\text{farad}}{\text{meter}}$	4π mksuru	$4\pi \times 10^{-11}c^2$ esu	$4\pi \times 10^{-11}$ emu
Charge	Q	1 coulomb	1 coulomb	$10^{-1}c$ statcoulombs	10^{-1} abcoulomb
Charge Density — Linear	q	$1 \dfrac{\text{coulomb}}{\text{meter}}$	$1 \dfrac{\text{coulomb}}{\text{meter}}$	$10^{-3}c \dfrac{\text{statcoulombs}}{\text{centimeter}}$	$10^{-3} \dfrac{\text{abcoulomb}}{\text{centimeter}}$
Charge Density — Surface	σ	$1 \dfrac{\text{coulomb}}{\text{meter}^2}$	$1 \dfrac{\text{coulomb}}{\text{meter}^2}$	$10^{-5}c \dfrac{\text{statcoulombs}}{\text{centimeter}^2}$	$10^{-5} \dfrac{\text{abcoulomb}}{\text{centimeter}^2}$
Charge Density — Volume	ρ	$1 \dfrac{\text{coulomb}}{\text{meter}^3}$	$1 \dfrac{\text{coulomb}}{\text{meter}^3}$	$10^{-7}c \dfrac{\text{statcoulombs}}{\text{centimeter}^3}$	$10^{-7} \dfrac{\text{abcoulomb}}{\text{centimeter}^3}$

* For further details, see E. E. Staff, M.I.T., *Electric Circuits* (Cambridge, Massachusetts: The Technology Press of M.I.T.; New York: John Wiley & Sons, Inc., 1940), Appendix C, 746–756; J. A. Stratton, *Electromagnetic Theory* (New York: McGraw-Hill Book Company, Inc., 1941), 16–23, 238–241: J. C. Slater and N. H. Frank, *Electromagnetism* (New York: McGraw-Hill Book Company, Inc., 1947), Appendix II, 205–216; N. H. Frank, *Introduction to Electricity and Optics* (2nd ed.; New York: McGraw-Hill Book Company, Inc., 1950), 39–41, 423–429; F. W. Sears, *Electricity and Magnetism* (Cambridge, Mass.: Addison-Wesley Press, Inc., 1951), 11–15, 226, 260–261; or other textbooks on electromagnetism.

† A rationalized system of units is one in which a factor 4π is absorbed in the permeability μ and in the dielectric constant ε so that this factor does not appear in Maxwell's equations nor in electromagnetic formulas for geometric configurations involving rectangular symmetry, such as formulas for the capacitance of a parallel-plate capacitor and the properties of a plane wave. But the factor 4π does appear in formulas for geometric configurations involving spherical or circular symmetry, such as in Coulomb's law and Ampère's law. See the footnote on page 17 for an explanation of the way rationalization affects the equations in this book.

Quantity	Symbol	MKS rationalized† units	MKS unrationalized units	CGS unrationalized electrostatic units (esu)	CGS unrationalized electromagnetic units (emu)
Electric Field Intensity	$\mathcal{E}$	$1\,\dfrac{\text{volt}}{\text{meter}}$	$1\,\dfrac{\text{volt}}{\text{meter}}$	$10^6 c^{-1}\,\dfrac{\text{statvolt}}{\text{centimeter}}$	$10^6\,\dfrac{\text{abvolts}}{\text{centimeter}}$
Potential Difference	E	1 volt	1 volt	$10^8 c^{-1}$ statvolt	10^8 abvolts
Electric Flux	ψ	1 coulomb	4π mksuru	$4\pi \times 10^{-1}c$ esu	$4\pi \times 10^{-1}$ emu
Electric Flux Displacement (electric flux density)	D	$1\,\dfrac{\text{coulomb}}{\text{meter}^2}$	4π mksuru	$4\pi \times 10^{-5}c$ esu	$4\pi \times 10^{-5}$ emu
Polarization	$\mathcal{P}$	$1\,\dfrac{\text{coulomb}}{\text{meter}^2}$	$1\,\dfrac{\text{coulomb}}{\text{meter}^2}$	$10^{-5}c\,\dfrac{\text{statcoulombs}}{\text{centimeter}^2}$	$10^{-5}\,\dfrac{\text{abcoulomb}}{\text{centimeter}^2}$
Capacitance	C	1 farad	1 farad	$10^{-9}c^2$ statfarads	10^{-9} abfarad
Elastance	S	1 daraf	1 daraf	$10^9 c^{-2}$ statdaraf	10^9 abdarafs
Current	I	1 ampere	1 ampere	$10^{-1}c$ statamperes	10^{-1} abampere
Current Density	J	$1\,\dfrac{\text{ampere}}{\text{meter}^2}$	$1\,\dfrac{\text{ampere}}{\text{meter}^2}$	$10^{-5}c\,\dfrac{\text{statamperes}}{\text{centimeter}^2}$	$10^{-5}\,\dfrac{\text{abampere}}{\text{centimeter}^2}$
Resistance	R	1 ohm	1 ohm	$10^9 c^{-2}$ statohm	10^9 abohms
Conductance	G	1 mho	1 mho	$10^{-9}c^2$ statmhos	10^{-9} abmho
Resistivity	ρ	1 ohm-meter	1 ohm-meter	$10^{11}c^{-2}$ statohm-cm	10^{11} abohm-cms
Conductivity	γ	$1\,\dfrac{\text{mho}}{\text{meter}}$	$1\,\dfrac{\text{mho}}{\text{meter}}$	$10^{-11}c^2\,\dfrac{\text{statmhos}}{\text{centimeter}}$	$10^{-11}\,\dfrac{\text{abmho}}{\text{centimeter}}$
Permeability	μ	$1\,\dfrac{\text{henry}}{\text{meter}}$	$\dfrac{1}{4\pi}\,\dfrac{\text{weber}}{\text{praoersted-m}^2}$	$\dfrac{10^7 c^{-2}}{4\pi}$ esu	$\dfrac{10^7}{4\pi}\,\dfrac{\text{gausses}}{\text{oersted}}$
Magnetic Field Intensity	H	$1\,\dfrac{\text{ampere-turn}}{\text{meter}}$	4π praoersteds	$4\pi \times 10^{-3}c$ esu	$4\pi \times 10^{-3}$ oersted
Magnetomotive Force	$\mathcal{F}$	1 ampere-turn	4π pragilberts	$4\pi \times 10^{-1}c$ esu	$4\pi \times 10^{-1}$ gilbert
Magnetic Flux	ϕ	1 weber	1 weber	$10^8 c^{-1}$ esu	10^8 maxwells
Magnetic Flux Density	B	$1\,\dfrac{\text{weber}}{\text{meter}^2}$	$1\,\dfrac{\text{weber}}{\text{meter}^2}$	$10^4 c^{-1}$ esu	10^4 gausses
Reluctance	$\mathcal{R}$	$1\,\dfrac{\text{ampere-turn}}{\text{weber}}$	$4\pi\,\dfrac{\text{pragilberts}}{\text{weber}}$	$4\pi \times 10^{-9}c^2$ esu	$4\pi \times 10^{-9}\,\dfrac{\text{gilbert}}{\text{maxwell}}$
Permeance	$\mathcal{P}$	$1\,\dfrac{\text{weber}}{\text{ampere-turn}}$	$\dfrac{1}{4\pi}\,\dfrac{\text{weber}}{\text{pragilbert}}$	$\dfrac{1}{4\pi}10^9 c^{-2}$ esu	$\dfrac{1}{4\pi}10^9\,\dfrac{\text{maxwells}}{\text{gilbert}}$
Inductance	L, M	1 henry	1 henry	$10^9 c^{-2}$ stathenry	10^9 abhenries

Numerical Values for Free Space

	ε_v	μ_v
MKS rationalized	$\dfrac{1}{4\pi}10^{11}c^{-2}\,\dfrac{\text{farad}}{\text{meter}}$	$4\pi \times 10^{-7}\,\dfrac{\text{henry}}{\text{meter}}$
CGS unrationalized emu	c^{-2} emu	$1\,\dfrac{\text{gauss}}{\text{oersted}}$
CGS unrationalized esu	1 esu	c^{-2} esu

$c = (2.997\ 76 \pm 0.000\ 04)\ 10^{10} \approx 3 \times 10^{10}$

$c^2 = (8.987\ 8 \pm 0.000\ 2)\ 10^{20} \approx 9 \times 10^{20}$

$c^{-1} = (0.333\ 560 \pm 0.000\ 004)10^{-10} \approx \tfrac{1}{3} \times 10^{-10}$

$c^{-2} = (0.111\ 262 \pm 0.000\ 003)10^{-20} \approx \tfrac{1}{9} \times 10^{-20}$

c is a numeric equal in magnitude to the velocity of electromagnetic wave propagation in free space, when expressed in centimeters per second.

Characteristic Curves of Representative Electron Tubes

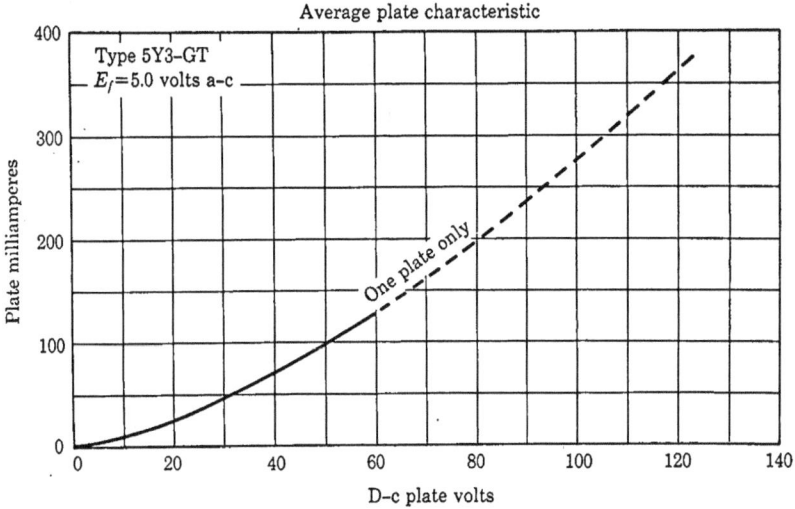

Average plate characteristic

Type 5Y3–GT
E_f=5.0 volts a–c

One plate only

Plate milliamperes

D–c plate volts

Fig. 1. 5Y3-G Full-wave vacuum rectifier. Peak inverse voltage = 1,400 volts max; peak plate current per plate = 400 ma max. (*Courtesy of Radio Corporation of America.*)

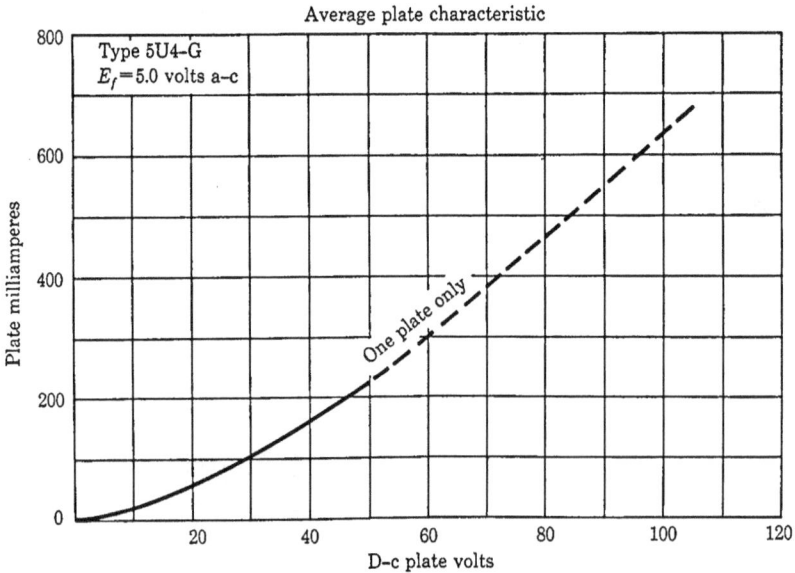

Average plate characteristic

Type 5U4–G
E_f=5.0 volts a–c

One plate only

Plate milliamperes

D–c plate volts

Fig. 2. 5U4-G Full-wave vacuum rectifier. Peak inverse voltage = 1,550 volts max; peak plate current per plate = 675 ma max. (*Courtesy of Radio Corporation of America.*)

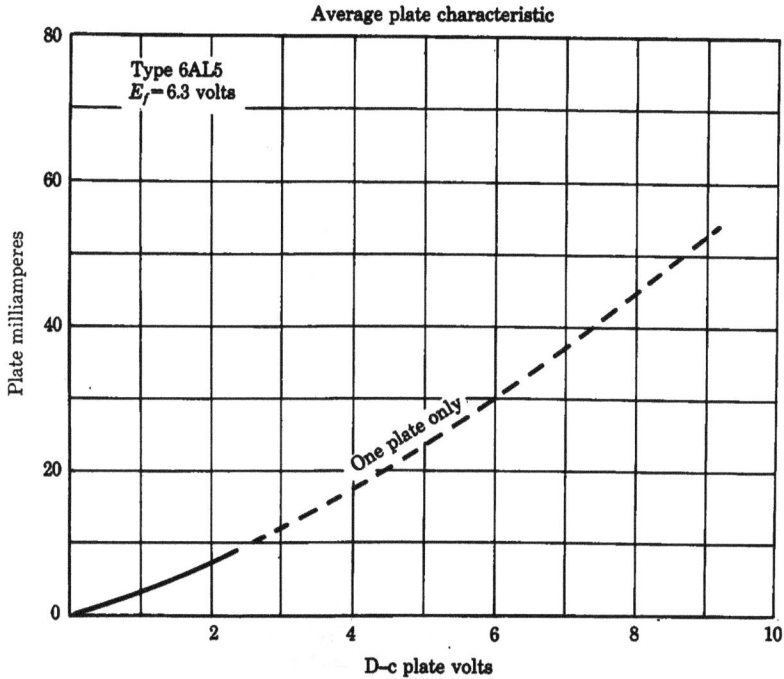

Fig. 3. 6AL5 Twin diode. Peak inverse voltage = 330 volts max; peak plate current per plate = 54 ma max. (*Courtesy of Radio Corporation of America.*)

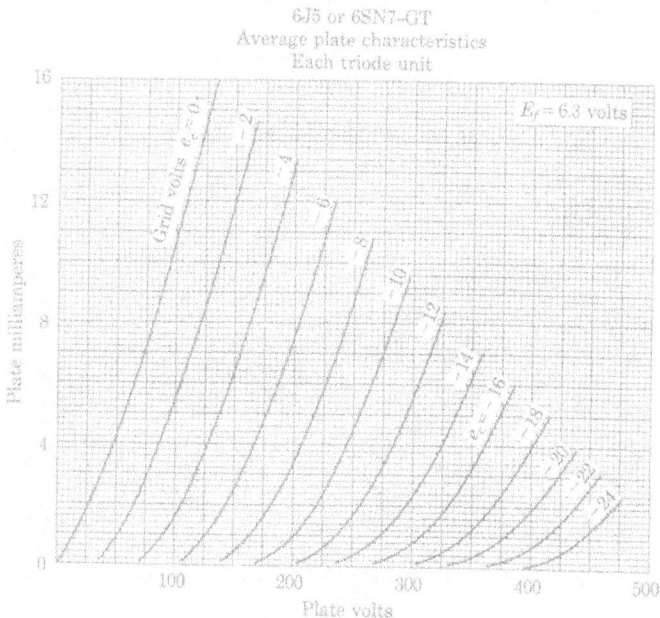

Fig. 4. 6J5 Medium-mu triode: $C_{gp} = 3.4\ \mu\mu f$; $C_{gk} = 3.4\ \mu\mu f$; $C_{pk} = 3.6\ \mu\mu f$. 6SN7-GT Twin-triode amplifier: $C_{gp} = 3.8$ and $4.0\ \mu\mu f$; $C_{gk} = 2.8$ and $3.0\ \mu\mu f$; $C_{pk} = 0.8$ and $1.2\ \mu\mu f$. Typical operation, Class A_1 amplifier: $E_b = 250$ volts; $E_c = -8$ volts; $I_b = 9$ ma; $\mu = 20$; $r_p = 7,700$ ohms; $g_m = 2,600\ \mu$mhos. (*Courtesy of Radio Corporation of America.*)

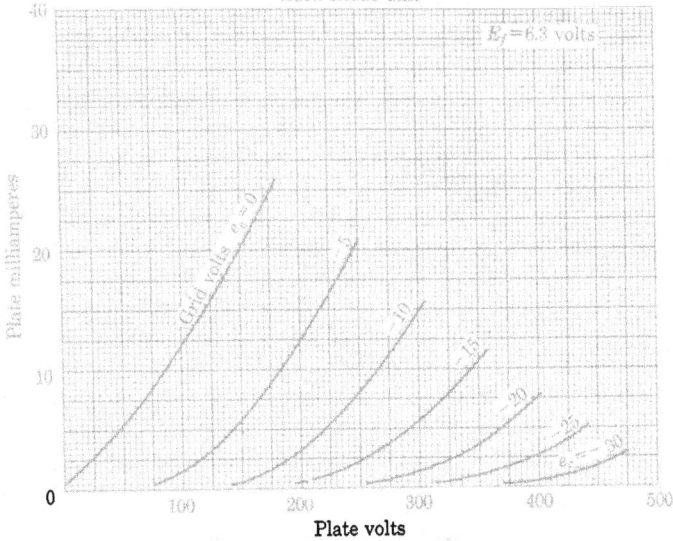

6C4 or 12AU7
Average plate characteristics
Each triode unit

Fig. 5.   6C4 H-f power triode: $C_{gp} = 1.6\ \mu\mu f$; $C_{gk} = 1.8\ \mu\mu f$; $C_{pk} = 1.3\ \mu\mu f$. 12AU7 Twin-triode amplifier: $C_{gp} = 1.5\ \mu\mu f$; $C_{gk} = 1.6\ \mu\mu f$; $C_{pk} = 0.50$ and $0.35\ \mu\mu f$. Typical operation, Class A_1 amplifier: $E_b = 250$ volts; $E_c = -8.5$ volts; $I_b = 10.5$ ma; $\mu = 17$; $r_p = 7,700$ ohms; $g_m = 2,200\ \mu$mhos. (*Courtesy of Radio Corporation of America.*)

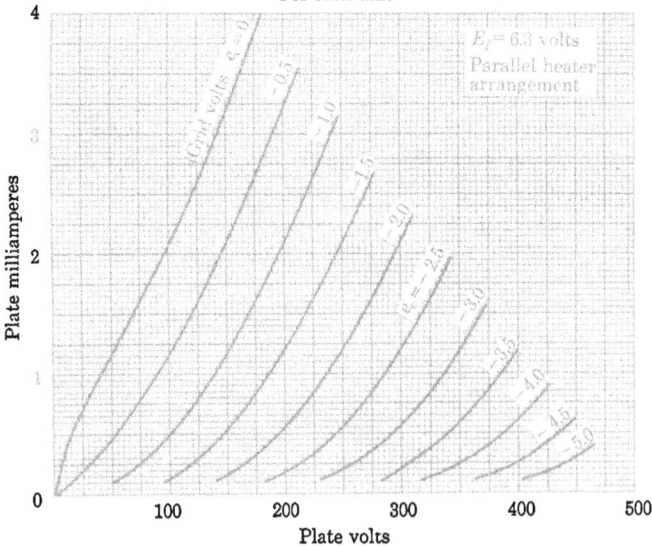

12AX7
Average plate characteristics
For each unit

Fig. 6.   12AX7 High-mu twin triode: $C_{gp} = 1.7\ \mu\mu f$; $C_{gk} = 1.6\ \mu\mu f$; $C_{pk} = 0.46$ and $0.34\ \mu\mu f$. Typical operation, Class A_1 amplifier: $E_b = 250$ volts; $E_c = -2$ volts; $I_b = 1.2$ ma; $\mu = 100$; $r_p = 62,500$ ohms; $g_m = 1,600\ \mu$mhos. (*Courtesy of Radio Corporation of America.*)

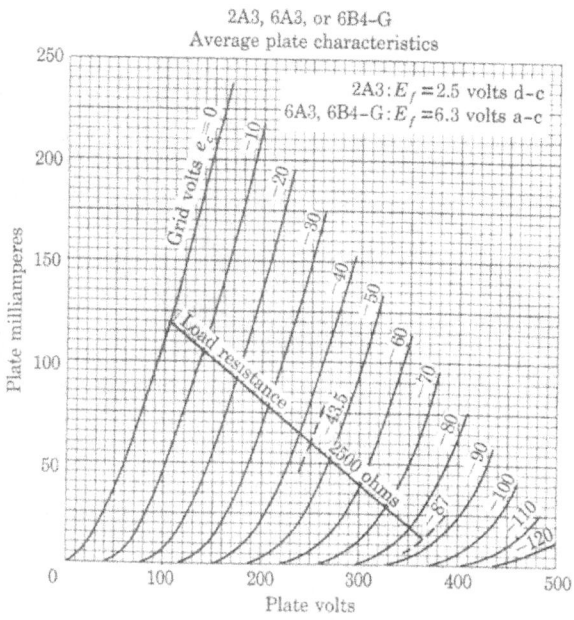

2A3, 6A3, or 6B4-G
Average plate characteristics

2A3: $E_f = 2.5$ volts d-c
6A3, 6B4-G: $E_f = 6.3$ volts a-c

Grid volts $e_c = 0$, -10, -20, -30, -40, -50, -60, -70, -80, -90, -100, -110, -120

Load resistance -43.5, -87, 2500 ohms

Plate milliamperes (vertical axis): 0, 50, 100, 150, 200, 250

Plate volts (horizontal axis): 0, 100, 200, 300, 400, 500

Fig. 7. 2A3, 6A3, or 6B4-G Power amplifier triode: $C_{gp} = 16.5$ $\mu\mu$f; $C_{gk} = 7.5$ $\mu\mu$f; $C_{pk} = 5.5$ $\mu\mu$f. Typical operation, Class A_1 amplifier: $E_b = 250$ volts; $E_c = -45$ volts; $I_b = 60$ ma; $\mu = 4.2$; $r_p = 800$ ohms; $g_m = 5,250$ μmhos. (*Courtesy of Radio Corporation of America.*)

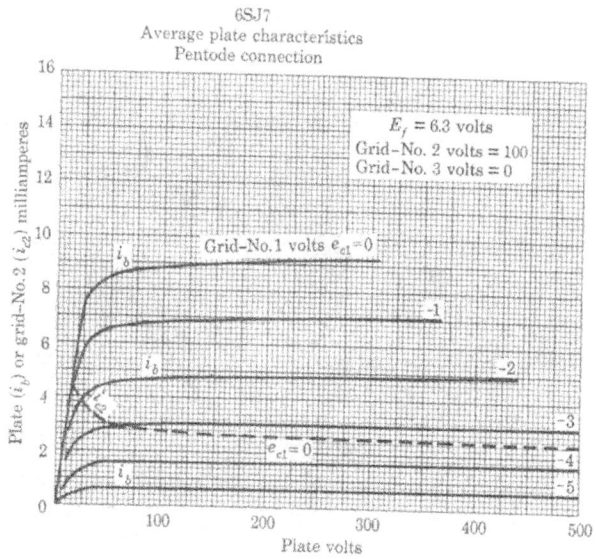

6SJ7
Average plate characteristics
Pentode connection

$E_f = 6.3$ volts
Grid-No. 2 volts $= 100$
Grid-No. 3 volts $= 0$

Grid-No.1 volts $e_{c1} = 0$, -1, -2, -3, -4, -5

i_b, i_{c2}, $e_{c1} = 0$

Plate (i_b) or grid-No.2 (i_{c2}) milliamperes (vertical axis): 0, 2, 4, 6, 8, 10, 12, 14, 16

Plate volts (horizontal axis): 100, 200, 300, 400, 500

Fig. 8. 6SJ7 Sharp-cutoff pentode: $C_{g1p} = 0.005$ $\mu\mu$f; $C_{in} = 6$ $\mu\mu$f; $C_{out} = 7$ $\mu\mu$f. Typical operation, Class A_1 amplifier: $E_b = 100$ volts; $E_{c3} = 0$ volt; $E_{c2} = 100$ volts; $E_{c1} = -3$ volts; $I_b = 2.9$ ma; $I_{c2} = 0.9$ ma; $g_m = 1,575$ μmhos; $r_p = 0.7$ megohm. (*Courtesy of Radio Corporation of America.*)

6K6-GT
Average plate characteristics
Pentode connection

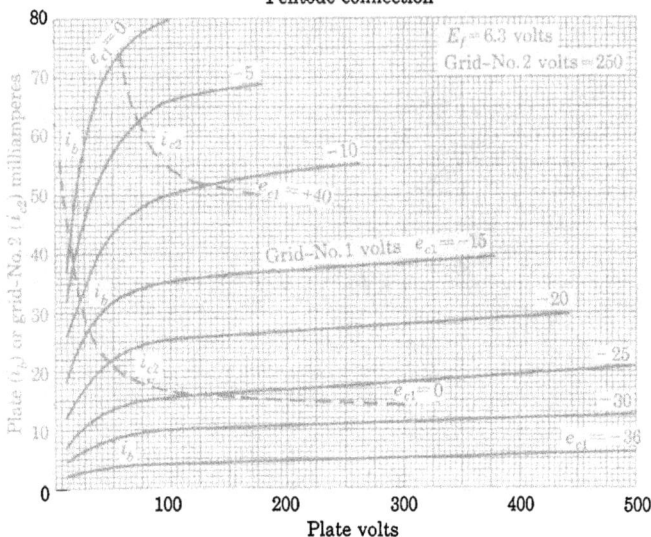

Fig. 9. 6K6-GT Power pentode: $C_{g1p} = 0.5\ \mu\mu f$; $C_{in} = 5.5\ \mu\mu f$; $C_{out} = 6.0\ \mu\mu f$. Typical operation, Class A_1 amplifier: $E_b = 250$ volts; $E_{c3} = 0$ volt; $E_{c2} = 250$ volts; $E_{c1} = -18$ volts; $I_b = 32$ ma; $I_{c2} = 5.5$ ma; $g_m = 2,300\ \mu$mhos; $r_p = 90,000$ ohms. (*Courtesy of Radio Corporation of America.*)

6CB6
Average plate characteristics

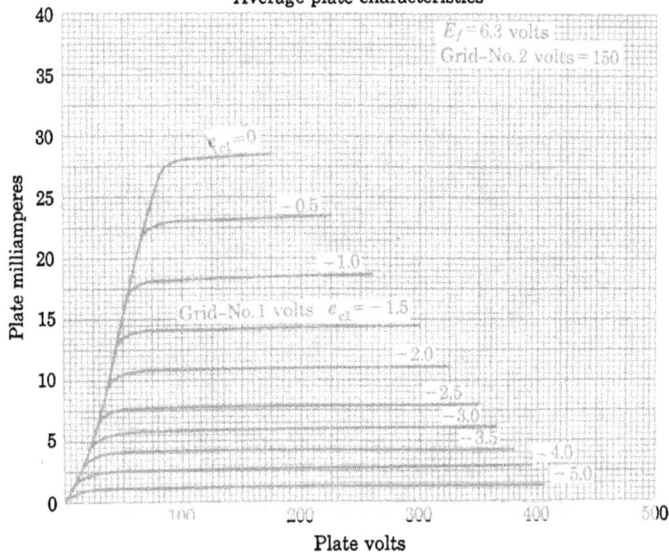

Fig. 10. 6CB6 Sharp-cutoff pentode: $C_{g1p} = 0.020\ \mu\mu f$; $C_{in} = 6.3\ \mu\mu f$; $C_{out} = 1.9\ \mu\mu f$. Typical operation, Class A_1 amplifier: $E_b = 200$ volts; $E_{c3} = 0$ volt; $E_{c2} = 150$ volts; $E_{c1} = -2.2$ volts; $I_b = 9.5$ ma; $I_{c2} = 2.8$ ma; $g_m = 6,200\ \mu$mhos; $r_p = 0.6$ megohm. (*Courtesy of Radio Corporation of America.*)

Answers to Representative Problems

CHAPTER I

1. 1,167 mi/sec.
3. (a) 0.0293 cm from k. (b) 0.121 μsec.
4. 0.176 from negative plate.
8. (a) 200 ev. (b) 3.37×10^{-9} sec. (c) 1.39×10^{-9} sec. (d) 1.24×10^{-9} sec.

13. 446 v.
16. 0.455 webers per m².
19. (a) 104 amp.
22. 1.99 amp.
26. (a) 0.398. (b) 1.0.

CHAPTER II

1. 0.0218 v.
5. 2,455 K.
8. 3,030 K.

12. (a) 165 ma. (b) 87.3 ma per w. (c) 0.0608. (d) No. (e) 1.3 amp; 346 ma per w; 0.0269.
15. 2.72; 3.76; 4.63.

CHAPTER III

1. 425 ma; 728 ma.
2. (a) 0.109 amp. (b) 54.7 v. (c) 9.53×10^{-3} weber per m².

6. (a) 4.25×10^{-6} mm Hg. (b) 0.0235; $1 - (6.3 \times 10^{-11})$.

CHAPTER IV

1. (a) 965 K. (b) 2.85×10^{-6} mm Hg.
3. (a) 3.5 cm; 8.82 cm. (b) 0.121 cm; 8.82 cm. (c) 64.2 amp; 3.42 v.

5. 2.53×10^{-9} sec; 2/3.
9. 10; 2,500 ohms; 4,000μmhos.

CHAPTER VI

7. 2.1%; 48%.
9. (a) 1, 1. (b) 2, 1. (c) 2, 1; 2, 2/3; 2, 1/2.
12. 4.52 h.
14. (a) 408 amp; 467 v. (b) 1,143 kw. (c) 0.551. (d) 28.2 kva.

16. (a) 4,080 v. (b) 9,560 v. (c) 30 amp. (d) 300 kva. (e) 0.955.
20. (a) 270 v. (b) 11,310 ohms. (c) 1,530 ohms. (d) 4.55 v.
25. 445 v; 150 ma.
28. 185 v; 2,470 ohms; 5,620 to 6,600 ohms.

CHAPTER VII

1. 48.3 v.
3. (a) 0.259 amp; 0.250 amp; 0.165 amp; 0.017 amp. (c) 21,900 ohms. (d) 7.52 w.

4. 1.63 amp.
12. (a) 23.7°. (b) 93%.

Chapter VIII

3. (a) 0.62 ma; 288 v; 80,600 ohms.
   (b) 1.1 ma; 262.5 v; −27.5 v.
6. (a) 13.2. (b) 385 v. (c) 1.35 h.
   (d) 250 v.
15. (a) 120 v. (b) 6.2%. (c) 5.8%.

17. (a) 3.18. (b) 45 v. (c) 21.7 w.
    (d) 12.1 w. (e) 0.81 w. (f) 11.3 w.
    (g) 6.7%.
21. (a) 24.4 v. (b) 0. (c) 3.13 ma. (d) 0.
    (e) 76.3 mw. (f) 4.54 w.
24. (a) 1.09. (b) 13.2 v. (c) 1.15 w.
    (d) 74 v. (e) 15.5 ma. (f) 0.

Chapter IX

2. (a) 148,000 ohms; 106,000 ohms.
   (b) 0.938 v. (c) 0.00642 v.
3. 30,900 ohms; 0.00527 μf.
12. (a) 445 ohms. (b) 10.85. (c) 2.25 w.
    (d) 76.4 cps.

15. 18.2; 1.68 and 1.50 megacycles per
    sec.
18. 1 : 50.4.
19. (a) 0.00298. (b) 0.0172.

Chapter X

5. (a) Class B. (b) 12.7 ma. (c) 141 v;
   0.8 v. (d) 2 w. (e) 1.81 w.
6. 1770 ohms; 237 v; 165 w.
7. (a) 5180 ohms. (b) −80 v. (c) 0.225
   amp. (d) 0.25 amp. (e) 1295 v.
   (f) 177.5 v. (g) 2.5 amp. (h) 324 w.
   (i) 450 w. (j) 126 w. (k) 72%.
   (l) 165 μh. (m) 615 $\mu\mu$f.

9. (a) 35 w. (b) 50 w. (c) 70%. (d) 15 w.
   (e) 0.65 w. (f) 38 μh; 19.8 ohms;
   667 $\mu\mu$f.

Chapter XI

1. (a) $g_m = L(R_t + r_p)/(MR_i r_p)$.
3. (a) $R = r_p(R_t + r_p)/[R_t(\mu^2 - 1)$
   $-r_p]$.
   (b) $f = 1/(2\pi\sqrt{LC})$.

8. (a) 38 μh; 667 $\mu\mu$f. (b) 10.7 μh; 5,550
   ohms; 0.61 w; 33.75 w; 300 $\mu\mu$f.

Chapter XII

1. (a) 96.3%. (b) 1.06. (c) 97.2 w; 22.5 w.

5. (a) ω_m. (b) Zero. (c) $2\omega_m$.

Chapter XIII

6. (a) 31.1 and 266 ohms; 1,440 and
   266 ohms; 1,445 and 13.4 × 10⁶
   ohms.
   (b) 1.56 × 10⁶ and 13.4 × 10⁶ ohms;
   1.56 × 10⁶ and 0.288 × 10⁶
   ohms; 31.1 and 0.288 × 10⁶
   ohms.

   (c) 91 ohms; 619 ohms; 139,000
   ohms.
   (d) 4.58 × 10⁶ ohms; 0.671 × 10⁶
   ohms; 2,990 ohms. (e) 2.7 × 10⁴;
   2.02 × 10⁵; 46.5.
9. 7 × 10⁵.

Bibliography

Since electronics is a rapidly growing field of electrical engineering, much of the literature on the subject is scattered in the form of articles and reports in technical periodicals. Study of the field therefore involves reading in a large number of different publications.

Footnotes throughout this book include references to both books and periodicals, which are selected because of their bearing on the particular subject under discussion. The majority of these references in turn contain additional references to earlier literature on the same subject. The footnote references therefore serve as a guide to further study in the more important branches of electronics.

As an aid to acquiring a knowledge of electronics on a broader scope than that covered by the footnotes, a list of bibliographies in the field is given below. These bibliographies have appeared from time to time in various forms, such as textbook documentation by means of footnote references, chapter references, or complete bibliographies, as bibliographies in survey articles in technical journals, and as exhaustive bibliographies giving a critical estimate of available literature and published separately. Some of them are continued and brought up to date annually. These authoritative bibliographies will be found to be valuable tools for independent study of the diverse applications of electronics.

The following abbreviations for periodicals and circulars are used in footnotes and in this bibliography of bibliographies.

A.I.E.E. J.	American Institute of Electrical Engineers *Journal*
A.I.E.E. Technical Paper	American Institute of Electrical Engineers *Technical Paper*
A.I.E.E. Trans.	American Institute of Electrical Engineers *Transactions*
A.S.A.	American Standards Association
Ann. d. Phys.	*Annalen der Physik*
Arch. f. Elek.	*Archiv für Elektrotechnik*
Astrophys. J.	*Astrophysical Journal*
Bell Lab. Rec.	*Bell Laboratories Record*
B.S.T.J.	*Bell System Technical Journal*
Camb. Phil. Soc. Proc.	Cambridge Philosophical Society *Proceedings*
E.E.	*Electrical Engineering*
E.N.T.	*Elektrische Nachrichten-Technik*
E.T.Z.	*Elektrotechnische Zeitschrift*
Exp. Wireless and Wireless Eng.	*Experimental Wireless and Wireless Engineering*
G.E. Rev.	*General Electric Review*
Gen. Rad. Exp.	*General Radio Experimenter*
I.R.E. Proc.	Institute of Radio Engineers *Proceedings*
I.E.E. J.	Institution of Electrical Engineers *Journal*
I.E.E. Proc.	Institution of Electrical Engineers *Proceedings*
J. App. Phys.	*Journal of Applied Physics*
J.F.I.	*Journal* of the Franklin Institute
J.O.S.A.	*Journal* of the Optical Society of America

J.S.I.	*Journal of Scientific Instruments*
K. Akad. Amsterdam Proc.	*Proceedings* of the Koninklijke Akademie van Wetenschappen te Amsterdam
Nat. Acad. Sci. Proc.	National Academy of Sciences *Proceedings*
Philips Tech. Rev.	*Philips Technical Review*
Phil. Mag.	*Philosophical Magazine*
Phys. Rev.	*Physical Review*
Phys. Soc. Proc.	Physical Society *Proceedings*
Phys. Zeits.	*Physikalische Zeitschrift*
RCA Rev.	*RCA Review*
Rev. Mod. Phys.	*Reviews of Modern Physics*
R.S.I.	*Review of Scientific Instruments*
Roy. Soc. Proc. (London)	Royal Society of London *Proceedings*
Nat. Bur. Stand. Sci. Paper	National Bureau of Standards *Scientific Paper*
Verh. d. D. Phys. Ges.	*Verhandlungen der Deutschen Physikalischen Gesellschaft*
Wireless Engr.	*Wireless Engineer*
Zeits. f. Phys.	*Zeitschrift für Physik*
Zeits. f. tech. Phys.	*Zeitschrift für technische Physik*

PHYSICAL PHENOMENA

L. Marton, Editor, *Advances in Electronics* (New York: Academic Press, Inc., 1948 to date). This volume, which is published annually, is an authoritative review of current progress in various fields and contains extensive bibliographies.

A. C. Strickland, Editor, *Reports on Progress in Physics* (London: The Physical Society, 1934 to date). This also is an annual report comprising articles having extensive bibliographies.

S. Dushman, "Thermionic Emission," *Rev. Mod. Phys.*, 3 (1930), 381–476. Bibliography: 222 footnote references. These references form a comprehensive list of existing (1930) publications on thermionic emission, including electron-emission theory and technical data on the various types of cathodes.

A. L. Reimann, *Thermionic Emission* (New York: John Wiley & Sons, Inc., 1934). Bibliography: extensive lists at the ends of the chapters.

C. Herring and M. H. Nichols, "Thermionic Emission," *Rev. Mod. Phys.*, 21 (1949), 185–270. This review article contains an extensive bibliography.

A. M. Glover, "A Review of the Development of Sensitive Phototubes," *I.R.E. Proc.*, 29 (1941), 413–423. Bibliography: 73 footnotes and 10 references, p. 423. The footnote references, which are briefly discussed in the text, and the general bibliography constitute a very comprehensive review of the literature of photoelectricity and photoemissive surfaces.

V. K. Zworykin and E. G. Ramberg, *Photoelectricity and Its Application* (New York: John Wiley & Sons, Inc., 1949). This revision of an earlier book contains numerous references and an index of authors.

K. T. Compton and I. Langmuir, "Electrical Discharges in Gases; Part I, Survey of Fundamental Processes," *Rev. Mod. Phys.*, 2 (1930), 123–242; "Part II, Fundamental Phenomena in Electrical Discharges," *Rev. Mod. Phys.*, 3 (1931), 191–257. Bibliography: 301 footnote references. These references give a very complete documentation of the subject. Titles are omitted, but the subject matter of each reference is clearly indicated in the text, which is a summary of existing (1930) knowledge of gaseous-discharge phenomena.

L. B. Loeb, *Fundamental Processes of Electrical Discharges in Gases* (New York: John Wiley & Sons, Inc., 1939). This authoritative reference book is fully documented and contains an index of authors.

J. D. Cobine, *Gaseous Conductors* (New York: McGraw-Hill Book Company, Inc., 1941). Frequent footnote references and 21 general references are included in this textbook.

F. A. Maxfield and R. R. Benedict, *Theory of Gaseous Conduction and Electronics* (New York: McGraw-Hill Book Company, Inc., 1941). This textbook has footnote references and a bibliography for each chapter.

L. O. Grondahl, "Copper-Cuprous-Oxide Rectifier and Photoelectric Cell," *Rev. Mod. Phys.*, *5* (1933), 141–168. Bibliography: 141 references, pp. 165–168. This article deals mainly with physical phenomena, but the bibliography covers applications also. The references are arranged chronologically in three groups: rectifiers, photoelectric cells, and properties of cuprous oxide.

W. Shockley, *Electrons and Holes in Semiconductors* (New York: D. Van Nostrand Company, Inc., 1950). References are collected at the beginning of each of the three parts of this classic reference book.

R. W. Ditchburn and N. F. Mott, Organizers, and H. K. Henisch, Editor, *Semi-Conducting Materials* (New York: Academic Press, Inc., Publishers, 1951). Twenty-five modern articles with a few references each are contained in this proceedings of a conference.

N. G. Parke, III, *Guide to the Literature of Mathematics and Physics* (New York: McGraw-Hill Book Company, Inc., 1947). This guide offers suggestions for reference, reading, and study in the field of physics and contains 2,300 bibliographical references under 150 subject headings.

ELECTRON TUBES AND GENERAL APPLICATIONS

G. G. Blake, *History of Radio Telegraphy and Telephony* (London: Radio Press Ltd., 1926). Bibliography: 1,125 references, pp. 353–403. This very complete bibliography of early radio communication includes references to original sources of basic data and to patents in the field of electronics.

J. W. Horton, "Use of Vacuum Tubes in Measurements," *A.I.E.E. Trans.*, *54* (1935), 93–102. Bibliography: 604 references, pp. 94–102. References selected from some 1,500 papers published between 1902 and 1934 are classified under specific subjects and arranged chronologically under each subdivision. The first part of the bibliography refers to publications concerning characteristics of vacuum tubes and theoretical aspects of their behavior; the second part refers to publications concerning their application to measuring methods and apparatus.

I. E. Mouromtseff, "Electronics and Development of Electron Tubes," *J.F.I.*, *240* (1945), 171–192. This survey contains 74 references.

W. C. White, "Trends in Electron Tube Design," *Elec. Eng.*, *67* (1948), 517–530. Thirty-six references are included in this review.

W. H. Kohl, *Materials Technology for Electron Tubes* (New York: Reinhold Publishing Corporation, 1951). This reference book contains extensive bibliographies on many of the practical aspects of electron tube construction.

Institute of Radio Engineers Annual Review Committee, "Radio Progress During——," *I.R.E. Proc.* This report, issued annually, is divided into parts, each of which includes a brief summary of developments in the specific field and a list of references to outstanding articles in current literature. Similar annual reports of progress have been published in the *I.R.E. Proceedings* since 1935; they often contain more than 1,000 references. The summaries are based on material prepared by the I.R.E. Annual Review Committee. Individual reports are prepared by the Chairmen of I.R.E. Technical Committees.

"Abstracts and References." *Wireless Engineer*, monthly, and *I.R.E. Proceedings*, monthly. Compiled by the Radio Research Organization and reproduced by arrangement with the Department of Scientific and Industrial Research [Great Britain], this monthly abstract section gives a very complete survey of current technical literature in the fields of radio communication and electronics. Articles in less accessible foreign journals are abstracted in considerable detail.

"Bibliography," *Electronics* (Buyer's Guide), *19* (June 15, 1946), 239–243. Approximately 500 books about the scientific, engineering, and technological aspects of electronics and related fields are listed in this bibliography.

F. Rockett, "Books about Electronics," *Electronics* (Buyer's Guide), *20* (June, 1947), 142–144. This list includes about 125 books relating to electronics and electronic applications issued during the previous year.

H. J. Reich, *Theory and Applications of Electron Tubes* (2nd ed.; New York: McGraw-Hill Book Company, Inc., 1944). The bibliography in this textbook comprises footnotes and references at the end of each chapter, and is particularly complete. An index of authors is included.

K. R. Spangenberg, *Vacuum Tubes* (New York: McGraw-Hill Book Company, Inc., 1948). This extensive treatment includes numerous footnote references and an index of authors.

P. Parker, *Electronics* (London: Edward Arnold and Co., 1950). Numerous book and periodical references are contained in the notes grouped at the end of each chapter in this textbook, and an index of authors is included.

W. C. White, "Electronic Uses in Industry," *Electronic Industries*, *1* (June, 1943), 72–76, 140, 142, 144, 146, 148, 150; *3* (February, 1944), 96–98, 176, 178, 180, 182, 184, 186; *4* (February, 1945), 102–105, 172, 174, 176, 178, 180, 182, 184, 186, 188, 190, 192, 194; *5* (June, 1946), 66–69, 100, 102, 104, 106, 108, 110, 111. This sequence of annual bibliographies on electron tube applications contains 1,078 references grouped according to the application, and an index of subjects.

J. G. Brainerd, G. Koehler, H. J. Reich, and L. F. Woodruff, *Ultra-High-Frequency Techniques* (New York: D. Van Nostrand Company, Inc., 1942). Bibliography: Chapter 16, *Guide to the Literature of Ultra-High-Frequency Techniques*, by Ruth McG. Lane, 439 references. In tracing the development of ultra-high-frequency technique from background theory to current trends, this selective bibliography includes many basic references in the field of electronics. Under three main divisions (Introduction, Fundamentals for UHF, UHF Techniques) subdivisions include references to theory and operation of electronic devices in general, and to ultra-high-frequency techniques in particular, the latter including ultra-high-frequency generators, transmission, radiation, propagation, and wave guides. Arrangement of references is chronologic in each section, basic textbook references are included, and a subject index is added.

F. Vilbig, *Schrifttumsverzeichnis zum Lehrbuch der Hochfrequenztechnik* (*Bibliography for the Textbook on High-Frequency Technique*) (2nd ed.; Leipzig: Akademische Verlagsgesellschaft M.B.H., 1939). Bibliography: approximately 5,000 references. This extremely comprehensive bibliography, published as a supplement to the textbook on high-frequency technique by the same author and publisher, follows the arrangement of the textbook, and includes chapters on tubes, amplifiers, rectifiers, and other applications. References for each chapter are arranged first chronologically, then alphabetically by author under each year. Titles are in German, but English translations are given for all references to other than German publications.

M. J. O. Strutt, *Moderne Mehrgitter-Elektronenröhren* (*Modern Multigrid Electron Tubes*) (2nd ed.; Berlin: Julius Springer, 1940). Bibliography: 323 references, pp. 265–278. This is a comprehensive list of references on multigrid tubes. The references are unclassified, but are arranged alphabetically by author.

ELECTRON DEVICES; SPECIAL APPLICATIONS

J. T. MacGregor-Morris and J. A. Henley, *Cathode-Ray Oscillography* (London: Chapman & Hall, Ltd., 1936). Bibliography: 102 references, pp. 240–243. Brief references concerning hot- and cold-cathode oscillographs, their operation, and applications, are listed in the order of their citation in chapters of the text. More important references are indicated by an asterisk, and most informative by a double asterisk.

G. Parr, *Cathode-Ray Tube and Its Applications* (2nd ed.; London: Chapman & Hall, Ltd., 1941). Bibliography: 737 references, pp. 161–176. References to general cathode-ray-tube theory, including beam properties, magnetic and gas focusing, electron optics, and so on, are followed by references on time bases, radio and industrial applications, and television. Brief bibliographic data, with titles omitted, are listed in a tabular form which expedites reference use. The arrangement is chronologic under each subject division.

J. F. Rider and S. D. Uslan, *Encyclopedia on Cathode-Ray Oscilloscopes and Their Uses* (New York: J. F. Rider Publisher, Inc., 1950), 969–974. An extensive bibliography arranged by chapters is included in this book.

E. Brüche and O. Scherzer, *Geometrische Elektronenoptik: Grundlagen und Anwendungen* (*Geometric Electron Optics: Fundamentals and Applications*) (Berlin: Julius Springer, 1934). Bibliography: 452 references, pp. 321–328. This is the first book to present experimental and theoretical data concerning electron behavior in relation to pure geometric optics. Numbered references throughout the text refer to the comprehensive bibliography, which is arranged alphabetically by author.

H. Busch and E. Brüche, *Beiträge zur Elektronenoptik* (*Contributions to Electron Optics*) (Leipzig: Johann Ambrosius Barth, 1937). Bibliography: 245 references, pp. 149–152. This bibliography, which supplements the one in the book by Brüche and Scherzer cited above, includes references to television development, and gives many patent references.

L. M. Meyers, *Electron Optics* (New York: D. Van Nostrand Company, Inc., 1939). Bibliography: 801 references, pp. 585–608. A very complete list of references, arranged alphabetically by author and numbered, but without titles, is given. Bibliographic numbers in the text guide the reader to the references pertaining to specific subjects such as electron lenses, the electron microscope, and the electron multiplier.

V. K. Zworykin, G. A. Morton, E. G. Ramberg, J. Hillier, and A. W. Vance, *Electron Optics and the Electron Microscope* (New York: John Wiley & Sons, Inc., 1945). This reference book contains 404 references arranged at the ends of the chapters, and an index of authors.

G. A. Morton, "Electron Guns for Television Applications," *Rev. Mod. Phys.*, *18* (1946), 362–378. This is a review article containing 10 references.

H. A. Thomas, *Theory and Design of Valve Oscillators* (London: Chapman & Hall, Ltd., 1951). Bibliography: 172 references, pp. 304–310. Many of the more important references to the literature on oscillator theory and development are included in this comprehensive bibliography, which is arranged alphabetically by author.

J. F. Rider, *Oscillator at Work* (New York: John F. Rider Publisher, Inc., 1940). Bibliography: 128 references, pp. 237–243. This classified list of references covers various types of oscillators: ultra-high-frequency, magnetron, magnetostriction, tuning-fork, crystal-controlled, negative-resistance, relaxation, beat-frequency, and multivibrators. References are arranged alphabetically under each type.

August Hund, *Frequency Modulation* (New York: McGraw-Hill Book Company, Inc., 1942). Bibliography: 179 references, pp. 361–368. These references are grouped according to topics, among which are frequency distribution of energy of modulated waves, wave propagation, antenna systems, noise and interference, and fundamentals and apparatus of frequency modulation.

V. K. Zworykin and G. A. Morton, *Television: The Electronics of Image Transmission* (New York: John Wiley & Sons, Inc., 1940). Bibliography: references at the end of each chapter. This text gives a comprehensive survey of television, and to each chapter is added a list of selected references for further reading.

G. E. Anner, *Elements of Television Systems* (New York: Prentice-Hall, Inc., 1951). This book contains many footnote references and an index of authors.

D. G. Fink, *Television Engineering* (2nd ed.; New York: McGraw-Hill Book Company, Inc., 1952). Approximately 425 references are included in footnotes and chapter bibliographies in this revised book, and an index of authors is appended.

R. Latham, A. H. King, and L. Rushforth, *The Magnetron* (London: Chapman & Hall, Ltd., 1952). This monograph contains numerous references at the ends of the chapters, and an index of authors.

F. E. Terman, *Radio Engineers Handbook* (New York: McGraw-Hill Book Company, Inc., 1943). This authoritative reference book contains many footnote references and an index of authors.

R. F. Shea, Editor, *Principles of Transistor Circuits* (New York: John Wiley & Sons, Inc., 1953). This reference book contains a list of 205 references grouped according to the year of publication and arranged alphabetically within each group.

Author Index

Subject Index